EXCITING ACTIVE MODEL EXERCISES

Active Models are interactive Excel spreadsheets that correspond to examples in the textbook and allow students to graphically explore and better understand important quantitative concepts. Students or instructors can adjust inputs to the mo and, in effect, answer a whole series of "what if" scenarios. These Active Models are great for classroom demonstration and/or homework. **Over 25 of these models are included on the Free Student CD-ROM and many are featured in the text** (see below for a full listing).

Chapter	Title	Active Model
3	Project Management	• Gantt Chart
4	Forecasting	• Moving Averages
		• Exponential Smoothing
		• Exponential Smoothing with Trend Adjustment
		• Trend Projections
5	Design of Goods and Services	• Decision Tree
6	Managing Quality	• Pareto Chart
6 Supplement	Statistical Process Control	• p-Chart
		• Process Capability
		• OC Curve
7	Process Strategy	• Crossover Chart
7 Supplement	Capacity Planning	• Capacity
		• Breakeven Analysis
8	Location Strategies	• Center of Gravity
9	Layout Strategy	• Process Layout
10 Supplement	Work Measurement	• Work Sampling
12	Inventory Management	• Economic Order Quantity Model
		• Production Order Quantity Model

Chapter	Title	Active Model
13	Aggregate Planning	• Leveling Strategies
14	Material Requirements Planning (MRP) and ERP	• Order Releases
15	Short-Term Scheduling	• Job Shop Sequencing
17	Maintenance and Reliability	• Series Reliability
		• Redundancy
Quantitative Modules		
B	Linear Programming	• LP Graph
D	Waiting-Line Models	• Single-Server Model
		• Multiple Server System with Costs
		• Constant Service Times
E	Learning Curves	• Unit Curve, Cumulative Curve, and Costs

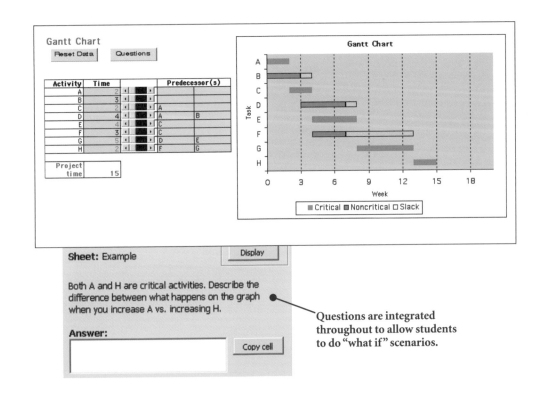

Questions are integrated throughout to allow students to do "what if" scenarios.

OPERATIONS MANAGEMENT *Rocks* WITH ITS THOROUGH SERVICE INTEGRATION THROUGHOUT!

In this edition, we illustrate how operations management is put into practice at Hard Rock Cafe, one of the most widely recognized company names in the world. Hard Rock Cafe invited us to come in and shoot the "behind the scenes" operations functions of their organization, giving students a real inside look at such issues as global strategy, project management, forecasting, location, scheduling, human resources, and more.

This modern and exciting corporation, emphasizing operations in a service environment, is featured throughout the text in Examples, Photos, Video Cases, and a Global Company Profile. The seven new video cases are listed below:

Hard Rock Cafe: Operations Management in Services (Ch. 1)
Hard Rock Cafe's Global Strategy (Ch. 2)
Managing Hard Rock's Rockfest (Ch. 3)
Forecasting at Hard Rock Cafe (Ch. 4)
Where to Place Hard Rock's Next Cafe (Ch. 8)
Hard Rock's Human Resource Strategy (Ch. 10)
Scheduling at Hard Rock Cafe (Ch. 15)

In addition to Hard Rock Cafe, the new editions contain an extensive amount of service applications to make the course even more relevant for students.

Chapter	Topic/Company Illustration
1	Global Company Profile: **Operations Management at Hard Rock Cafe** Operations in the Service Sector OM in Action: Increasing Productivity in the L.A. Motor Pool Productivity and the Service Sector OM in Action: Taco Bell Uses Productivity to Lower Costs Case Study: National Air Express Case Study: **Hard Rock Cafe: Operations Management in Services**
2	Mission Statements for Federal Express, Merck, and **Hard Rock Cafe** Goods vs. Services Applied to the 10 OM Decisions Case Study: Minit-Lube, Inc. Case Study: **Hard Rock Cafe's Global Strategy**
3	Global Company Profile: Bechtel Group OM in Action: Delta's Ground Crew Orchestrates a Smooth Takeoff Milwaukee General Hospital Project Management Example OM in Action: Project Management and Software Development Case Study: Southwestern University: (A) Case Study: Bay Community Hospital Case Study: **Managing Hard Rock's Rockfest**
4	OM in Action: Forecasting at Disney World San Diego Hospital Seasonality and Trend Forecasting in the Service Sector Case Study: Southwestern University: (B) Case Study: **Forecasting at Hard Rock Cafe**
5	Goods and Services Selection Service Design Case Study: De Mar's Product Strategy
6	OM in Action: L.L. Bean's Reputation Makes It a Benchmark Favorite OM in Action: TQM Improves Copier Service Service Industry Inspection TQM in Services

Chapter	Topic/Company Illustration
6	OM in Action: Richey International's Spies Case Study: Southwestern University: (C) Case Study: Quality at the Ritz-Carlton Hotel Company
6 Supplement	Insurance Records *p*-chart example OM in Action: Unisys Corp.'s Costly Experiment in Health Care Services Case Study: Alabama Airlines's On-Time Schedule
7	OM in Action: Mass Customization at Borders Books and at Smooth FM Radio Service Blueprinting Service Process Design OM in Action: Technology Changes in the Hotel Industry
7 Supplement	Concert Hall Capacity Motel Capacity Case Study: Capacity Planning at Shouldice Hospital
8	Global Company Profile: Federal Express Center-of-Gravity Method Service Location Strategy OM in Action: Hospitals Think Location, Location, Location How Hotel Chains Select Sites The Telemarketing Industry Case Study: The Ambrose Distribution Center Case Study: **Where to Place Hard Rock's Next Cafe**
9	Global Company Profile: McDonald's Layout in Emergency Rooms Example Office Layout OM in Action: Shopping Mall Layout Meets the Internet Retail Layout Servicescapes Case Study: State Auto License Renewals
10	Global Company Profile: Southwest Airlines OM in Action: Empowerment at the Ritz-Carlton Ergonomics and Work Methods Case Study: Karstadt versus J.C. Penney Case Study: **Hard Rock's Human Resource Strategy**

Table continued on next page.

Chapter	Topic/Company Illustration
10 Supplement	OM in Action: UPS: The Tightest Ship in the Shipping Business Service Labor Standards with MTM-HC Work Sampling
11	OM in Action: A Rose Is a Rose, But Only if It Is Fresh OM in Action: RF Tags: Keeping the Shelves Stocked Distribution Systems OM in Action: DHL's Role in the Supply Chain
11 Supplement	Internet Trading Exchanges Health Care Exchange Scheduling and Logistics Improvements Case Study: E-Commerce at Amazon.com
12	Global Company Profile: Amazon.com OM in Action: What the Marines Learned about Inventory from Wal-Mart Control of Service Inventories Case Study: Southwestern University: (D) Case Study: Mayo Medical Center
13	OM in Action: A Tale of Two Delivery Services Aggregate Planning in Services Yield Management OM in Action: Yield Management at Hertz Case Study: Southwestern University: (E) Case Study: Andrew-Carter, Inc.
14	MRP in Services Distribution Resource Planning (DRP) OM in Action: Managing Benetton with ERP Software ERP in the Service Sector Case Study: Ikon's Attempt at ERP
15	Global Company Profile: Delta Airlines Scheduling Printing with Gantt Charts and the Assignment Methods Priority Rules for Sequencing Architectural Jobs Scheduling for Services

Chapter	Topic/Company Illustration
15	OM in Action: Scheduling Workers Who Fall Asleep On the Job Is Not Easy OM in Action: Banking and the Theory of Constraints (TOC) OM in Action: Scheduling Aircraft Turnaround Cyclical Scheduling Case Study: Payroll Planning, Inc. Case Study: **Scheduling at Hard Rock Cafe**
16	JIT in Services
17	Global Company Profile: NASA OM in Action: Tomcat F-14 Pilots Love Redundancy
Quantitative Modules	
A	Case Study: Tom Tucker's Liver Transplant
B	Using LP at American Airlines OM in Action: Using LP to Select Tenants in a Mall OM in Action: Scheduling Planes at Delta with LP Labor Scheduling in Banking Case Study: Golding Landscaping and Plants, Inc. Rental Car Movements
C	Case Study: Custom Vans, Inc.
D	Queues at Disney Common Waiting-Line Situations OM in Action: L.L. Bean Turns to Queuing Theory OM in Action: Shortening Arraignment Times at NYC Police Department Case Study: The Winter Park Hotel
E	Heart Surgery Learning Curves
F	Bay Medical Center OM in Action: Simulating Jackson Memorial Hospital's Operating Rooms Case Study: Alabama Airlines's Reservation System
CD-ROM Tutorial	
5	Vehicle Routing and Scheduling Case Study: Routing and Scheduling Phlebotomists

OPERATIONS MANAGEMENT FLEXIBLE VERSION

Seventh Edition

Jay Heizer

Jesse H. Jones Professor of Business Administration
Texas Lutheran University

Barry Render

Charles Harwood Professor of Operations Management
Crummer Graduate School of Business
Rollins College

Upper Saddle River, New Jersey 07458

Library of Congress Cataloging-in-Publication Data

Heizer, Jay H.

  Operations management / Jay Heizer, Barry Render.—7th ed.

    p. cm.

  ISBN 0-13-142272-3 (paperback)

   1. Production management.   I. Render, Barry. II. Title.

TS155.H3725 2003

658.5—dc22

                             2003062339

Executive Editor: Tom Tucker
Editor-in-Chief: P.J. Boardman
Assistant Editor: Erika Rusnak
Editorial Assistant: Dawn Stapleton
Senior Media Project Manager: Nancy Welcher
Executive Marketing Manager: Debbie Clare
Marketing Assistant: Amanda Fisher
Managing Editor (Production): Cynthia Regan
Permissions Supervisor: Suzanne Grappi
Production Manager: Arnold Vila
Design Manager: Maria Lange
Designer: Blair Brown
Interior/Cover Design: Blair Brown
Cover Photo: Courtesy of Hard Rock Cafe International, Inc.
Photo Researcher: Mary Ann Price
Image Permission Coordinator: Carolyn Gauntt
Manager, Print Production: Christy Mahon
Composition/Illustration (Interior): UG / GGS Information Services, Inc.
Full-Service Project Management: UG / GGS Information Services, Inc.
Printer/Binder: R.R. Donnelley & Sons Company

Credits and acknowledgments borrowed from other sources and reproduced, with permission, in this textbook appear on appropriate page within text and on page xvii.

Microsoft Excel, Solver, and Windows are registered trademarks of Microsoft Corporation in the U.S.A. and other countries. Screen shots and icons reprinted with permission from the Microsoft Corporation. This book is not sponsored or endorsed by or affiliated with Microsoft Corporation.

Pearson Education LTD.
Pearson Education Australia PTY, Limited
Pearson Education Singapore, Pte. Ltd
Pearson Education North Asia Ltd
Pearson Education, Canada, Ltd
Pearson Educación de Mexico, S.A. de C.V.
Pearson Education—Japan
Pearson Education Malaysia, Pte. Ltd

10 9 8 7 6 5 4 3 2
ISBN 0-13-142272-3

To our families:

Kay, Donna, Kira, and Janée

Donna, Charlie, and Jesse

And to the following practicing managers who have enriched our lives and indirectly this text, reminding us that the theory, concepts, and techniques of operations management have real and meaningful applications that bring us all a higher standard of living.

Pete Beaudrault
President and CEO, Hard Rock Cafe International

Bob Buerlein
President, American Historical Foundation

Harold Cole
Chairman and CEO, Truxmore Industries

Bob Collins
President, Wheeled Coach

Don Collins
Chairman, Collins Industries

Phil Crosby
President, Crosby Associates

Horace Faber
Chairman and CEO, Halifax Paper Co.

Ron Herrmann
Chairman, Columbia Industries, Inc.

Charles Hicks
Chief Financial Officer, Dixie Container Corp.

Emil Holzwart
President, Holzwart-Heizer, Inc.

Tim Kuch
Vice President, Regal Marine

Alan Nagle
CEO, (Retired) Tupperware International

Cliff Schroeder
Chairman and CEO, Dixie Container Corp.

Michael Shader
President, First Printing

Jay Heizer holds the Jesse H. Jones Chair of Business Administration at Texas Lutheran University in Seguin, Texas. He received his B.B.A. and M.B.A. from the University of North Texas and his Ph.D. in Management and Statistics from Arizona State University. He was previously a member of the faculty at the University of Memphis, the University of Oklahoma, Virginia Commonwealth University, and the University of Richmond. He has also held visiting positions at Boston University, George Mason University, the Czech Management Center, and the Otto-Von-Guericke University Magdeburg.

Dr. Heizer's industrial experience is extensive. He learned the practical side of operations management as a machinist apprentice at Foringer and Company, production planner for Westinghouse Airbrake, and at General Dynamics, where he worked in engineering administration. Additionally, he has been actively involved in consulting in the OM and MIS areas for a variety of organizations including Philip Morris, Firestone, Dixie Container Corporation, Columbia Industries, and Truxmore Industries. He holds the CPIM certification from the American Production and Inventory Control Society.

Professor Heizer has co-authored five books and has published over thirty articles on a variety of management topics. His papers have appeared in the *Academy of Management Journal*, *Journal of Purchasing*, *Personnel Psychology*, *Production & Inventory Control Management*, *APICS-The Performance Advantage*, *Journal of Management History*, *IIE Solutions*, and *Engineering Management*, among others. He has taught operations management courses in undergraduate, graduate, and executive programs.

Barry Render holds the Charles Harwood Endowed Professorship in Operations Management at the Crummer Graduate School of Business at Rollins College, in Winter Park, Florida. He received his B.S. in Mathematics and Physics at Roosevelt University, and his M.S. in Operations Research and Ph.D. in Quantitative Analysis at the University of Cincinnati. He previously taught at George Washington University, University of New Orleans, Boston University, and George Mason University, where he held the GM Foundation Professorship in Decision Sciences and was Chair of the Decision Science Department. Dr. Render has also worked in the aerospace industry for General Electric, McDonnell Douglas, and NASA.

Professor Render has co-authored ten textbooks with Prentice Hall, including *Managerial Decision Modeling with Spreadsheets*, *Quantitative Analysis for Management*, *Service Management*, *Introduction to Management Science*, and *Cases and Readings in Management Science*. *Quantitative Analysis for Management* is now in its eighth edition and is a leading text in that discipline in the U.S. and globally. His more than one hundred articles on a variety of management topics have appeared in *Decision Sciences*, *Production and Operations Management*, *Interfaces*, *Information and Management*, *Journal of Management Information Systems*, *Socio-Economic Planning Sciences*, *IIE Solutions*, and *Operations Management Review*, among others.

Dr. Render has also been honored as an AACSB Fellow and was twice named as a Senior Fullbright Scholar. He was vice-president of the Decision Science Institute Southeast Region and served as Software Review Editor for *Decision Line* for six years. He has also served as Editor of the *New York Times* Operations Management special issues from 1996 to 2001. Finally, Professor Render has been actively involved in consulting for government agencies and for many corporations, including NASA, FBI, U.S. Navy, Fairfax County, Virginia, and C&P Telephone.

He teaches operations management courses in Rollins College's MBA and Executive MBA programs. In 1995 he was named as that school's Professor of the Year, and in 1996 was selected by Roosevelt University to receive the St. Claire Drake Award for Outstanding Scholarship.

When the first Hard Rock Cafe opened on June 14, 1971, no one could have guessed the worldwide impact that it would have. The achievement of 92 percent brand awareness and the celebration we experience in being the ninth most recognizable logo in the world remains exciting to all of the Hard Rock family.

But here we are more than three decades later and I am endlessly asked, "What is the secret to your success?" To be perfectly honest with you, I am unsure. However, I am sure our mantra "Love All, Serve All" isn't just a cool slogan, but rather our way of life. I am also sure we have the greatest people working for us at all levels of the organization. As an inter-dependent group, we are passionately committed to subordinating ourselves to the greater cause: That being, creating 'Raving Fans' of our over 35 million guests annually throughout our businesses around the globe, in 38 countries.

The Hard Rock is now comprised of many business extensions beyond our core cafe business. They include Hard Rock Hotel & Casinos, Hard Rock Live, Hard Rock Academy, Hard Rock Vault, Hard Rock Records, Hard Rock Radio Show and Hard Rock Getaways. However, while all these are important brand extensions, none is more mission critical than the Hard Rock Cafes themselves, as they truly are the engine that drives this wonderful brand. To this point, we have as an organization, adopted an operations management driven culture, supported by the illustration of an inverted triangle where at the bottom resides myself, the CEO, and at the top our guests, then our staff, and then our operations team. While all our key disciplines (i.e. marketing, finance, IT, human resources, etc.), are extremely important resources for our business to experience sustainable growth, no discipline is as mission critical to our short- and long-term objectives than that of our day-to-day operations. Without question, operations management delivers on the moment of truth with our guests, day in and day out.

Throughout this great book you will enjoy a look backstage at Hard Rock via several insightful case studies that I am sure you will find entertaining and meaningful to you both personally and educationally. I have the distinction and honor of being the first Hard Rock Cafe President and CEO who possesses a purely operations management background. I strongly believe this is a true advantage I enjoy when dealing with our people, product, and physical plants of our massive engine . . . the cafes themselves.

Let it be known that there are no cash registers in our corporate offices, just support, guidance and unparallel passion for our operations team and to our loyal guests' dining, retail and entertainment experience worldwide.

As you move through this textbook and your OM course, we invite you to visit us many times, both online and offline. Your book's Web site has a direct link to Hard Rock Cafe's site. There you will find contests, discounts, chances for free hotel nights and other specials just for you.

All is One . . . , Love All . . . , Serve All . . . , Save the Planet . . . , Take Time To Be Kind . . . , Humanity is Instrumental . . . , Music for Life . . . , Rock & Roll Rides Again.

Peace without the quiet.

PETE BEAUDRAULT
President & CEO
Hard Rock Cafe International, Inc.
Orlando, Florida, USA

Brief Contents

Contents

CD-ROM TUTORIALS

Photo Credits

Preface

Welcome to your Operations Management (OM) course. In this book, we present a state-of-the-art view of the activities of the operations function. Operations is an exciting area of management that has a profound effect on the productivity of both manufacturing and services. Indeed, few activities have as much impact on the quality of our lives. The goal of this text is to present a broad introduction to the field of operations in a realistic, practical manner. Operations management includes a blend of topics from accounting, industrial engineering, management, management science, and statistics. Even if you are not planning on a career in the operations area, you will likely be interfacing with people who are. Therefore, having a solid understanding of the role of operations in an organization is of substantial benefit to you. This book will also help you understand how OM affects society and your life. Certainly, you will better understand what goes on behind the scenes when you buy a meal at Hard Rock Cafe, place an order through Amazon.com, or buy a customized Dell Computer over the Internet.

Although many of our readers are not OM majors, we know that marketing, finance, accounting, and MIS students will find the material both interesting and useful because we develop a fundamental working knowledge of the firm. Over 350,000 readers of our earlier editions seem to have endorsed this premise.

FEATURES OF THE FLEXIBLE VERSION

The primary goal of the *Flexible Version* is to provide the content and pedagogy in a flexible, easy-to-use environment that will meet the needs of students and professors. The focus is on helping the student move from a passive learner to an active participant in the learning process. The *Flexible Version* includes three powerful student resources: a paperback text, a *Student Lecture Guide*, and a *Student CD-ROM*.

THE PAPERBACK TEXT

This text contains the same content as *Operations Management*, Seventh Edition, but without any of the end-of-chapter homework material, which has been moved to the *Student Lecture Guide*.

THE STUDENT LECTURE GUIDE

The *Student Lecture Guide* is designed for portability so the student can take it to class. The *Student Lecture Guide* bridges the gap between the text and the lecture, and by taking notes directly in the lecture guide, the student will take an active part in learning. Each chapter begins with a set of study questions to alert the student to important concepts in the chapter. The *Student Lecture Guide* also outlines each major section of the chapter, often with key words or formulas. Additionally, most techniques are demonstrated via Practice Problems. There is ample space for the student to take notes and work out the Practice Problems. Discussion Questions and homework problems (from the text) are also in the lecture guide, so the instructor can assign or discuss these in class.

The Practice Problems are also available on a separate set of PowerPoints should the instructor want to present the material this way. If the instructor wishes to use his or her *own* examples, there is Additional Practice Problem Space to accommodate this approach.

THE STUDENT CD-ROM

The *Student CD-ROM* includes ExcelOM, Active Model Exercises, Example Data Files, 29 Video Clips, the Practice Problems, and PowerPoint Lecture Notes.

The *Flexible Version* is also supported by an extensive Web site (**www.prenhall.com/heizer**) that contains the following resources:

- Quizzes
- Internet Exercises
- Current Articles and Research

- Virtual Company Tours
- Supplemental Internet Homework Problems
- Internet Cases
- Videos

Blackboard, Web CT, and Course Compass are also available.

OPERATIONS MANAGEMENT BY HEIZER AND RENDER IS AVAILABLE IN THREE VERSIONS

The three versions are:

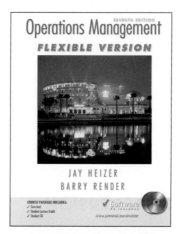

Operations Management, Seventh Edition, *Flexible Version*, consists of a paperback text containing traditional text material minus the end-of-chapter material, a unique *Student Lecture Guide*, and a *Student CD-ROM*.

Copyright 2005/ISBN: 0-13-105845-2

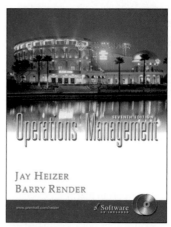

Operations Management, Seventh Edition, which is hard cover.

Copyright 2004/ISBN: 0-13-104638-8

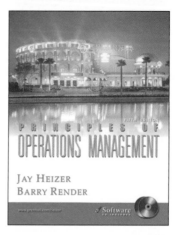

Principles of Operations Management, Fifth Edition, a paperback.

Copyright 2004/ISBN: 0-13-140639-6

All three versions include identical core chapters 1–17. However, *Operations Management: Flexible Version* and *Operations Management*, Seventh Edition, also include six quantitative modules in Part IV.

OPERATIONS MANAGEMENT, SEVENTH EDITION

PART I INTRODUCTION

1. Operations and Productivity
2. Operations Strategy in a Global Environment
3. Project Management
4. Forecasting

PART II DESIGNING OPERATIONS

5. Design of Goods and Services
6. Managing Quality
6S. Statistical Process Control
7. Process Strategy
7S. Capacity Planning
8. Location Strategies
9. Layout Strategy
10. Human Resources and Job Design
10S. Work Measurement

PART III MANAGING OPERATIONS

11. Supply-Chain Management
11S. E-Commerce and Operations Management
12. Inventory Management
13. Aggregate Planning
14. Material Requirements Planning (MRP) and ERP
15. Short-Term Scheduling
16. Just-in-Time and Lean Production Systems
17. Maintenance and Reliability

PART IV QUANTITATIVE MODULES

A. Decision-Making Tools
B. Linear Programming
C. Transportation Models
D. Waiting-Line Models
E. Learning Curves
F. Simulation

PRINCIPLES OF OPERATIONS MANAGEMENT, FIFTH EDITION

PART I INTRODUCTION

1. Operations and Productivity
2. Operations Strategy in a Global Environment
3. Project Management
4. Forecasting

PART II DESIGNING OPERATIONS

5. Design of Goods and Services
6. Managing Quality
6S. Statistical Process Control
7. Process Strategy
7S. Capacity Planning
8. Location Strategies
9. Layout Strategy
10. Human Resources and Job Design
10S. Work Measurement

PART III MANAGING OPERATIONS

11. Supply-Chain Management
11S. E-Commerce and Operations Management
12. Inventory Management
13. Aggregate Planning
14. Material Requirements Planning (MRP) and ERP
15. Short-Term Scheduling
16. Just-in-Time and Lean Production Systems
17. Maintenance and Reliability

FOCUS OF THIS EDITION

This new edition continues to place a special focus on important aspects of Operations Management including:

- **Strategy**—as our unifying theme in every chapter.
- **Global Operations**—and how this impacts product and process design, location, human resources, and other issues.
- **Service Operations**—recognizing the dominant proportion of jobs and operations decisions in services.
- **Software for OM**—our free Excel OM add-in and Microsoft Project 2002 software are included on the student CD-ROM packaged with the text.
- **Modern topical coverage**—with coverage of the Internet, Microsoft Project 2002, E-Commerce, ERP, yield management, and mass customization.
- **Real world examples of operations management**—to maximize student interest and excitement.

NEW TO THIS EDITION

Hard Rock Cafe Integration In this edition, we illustrate how operations management is put into practice at Hard Rock Cafe, one of the most widely recognized company names in the world. Hard Rock invited us to come in and shoot the "behind the scenes" operations functions of their organization, giving students a real inside look at such issues as global strategy, project management, forecasting, location, scheduling, human resources, and more. This modern and exciting corporation, emphasizing operations in a service environment, is featured throughout the text in examples, photos, video cases, and a Global Company Profile. A VHS tape is available to adopters which includes seven 5–7 minute segments of each topic. The student CD-ROM also contains 2-minute, abbreviated clips of these overall topics. This is the perfect way to integrate service application into the OM course.

VIDEO CASE STUDY

Where to Place Hard Rock's Next Cafe

Some people would say that Oliver Munday, Hard Rock's vice president for cafe development, has the best job in the world. Travel the world to pick a country for Hard Rock's next cafe, select a city, and find the ideal site. It's true that selecting a site involves lots of incognito walking around, visiting nice restaurants, and drinking in bars. But that is not where Mr. Munday's work begins, nor where it ends. At the front end, selecting the country and city first involves a great deal of research. At the back end, Munday not only picks the final site and negotiates the deal, but then works with architects and planners, and stays with the project through the opening and first year's sales.

Munday is currently looking heavily into global expansion in Europe, Latin America, and Asia. "We've got to look at political risk, currency, and social norms—how does our brand fit into the country," he says. Once the country is selected, Munday focuses on the region and city. His research checklist is extensive.

Site location now tends to focus on the tremendous resurgence of "city centers," where nightlife tends to concentrate. That's what Munday selected in Moscow and Bogota, although in both locations he chose to find a local partner and franchise the operation. In these two political environments, "Hard Rock wouldn't dream of operating by ourselves," says Munday. The location decision also is at least a 10-to-15-year commitment by Hard Rock, which employs tools such as break-even analysis to help decide whether to purchase land and build, or to remodel an existing facility.

Discussion Questions*

1. From Munday's checklist, select any other four categories, such as population (A1), hotels (B2), or nightclubs (E), and provide three subcategories that should be evaluated. (See item C1 (airport) for a guide.)
2. Why is site selection more than just evaluating the best nightclubs and restaurants in a city?

Active Model Exercises Active Model Exercises are interactive Excel spreadsheets of examples in the textbook that allow the student to explore and better understand these important quantitative concepts. Students and instructors can adjust inputs to the model and, in effect, answer a whole series of "what if" questions that is provided (e.g., What if one activity in a PERT network takes 3 days longer? Chapter 3. What if holding cost or demand in an inventory model doubles? Chapter 12. What if the exponential smoothing constant is 0.3 instead of 0.5? Chapter 4). These Active Models are great for classroom presentation and/or homework. Over 25 of these models are included on the student CD-ROM and many are featured in the text.

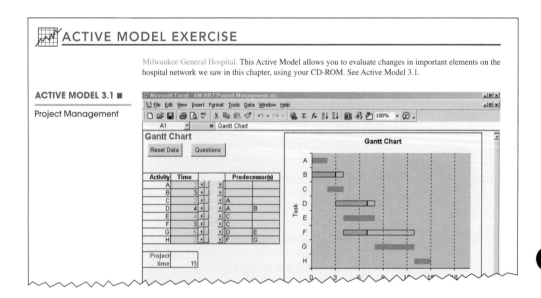

ACTIVE MODEL EXERCISE

Milwaukee General Hospital. This Active Model allows you to evaluate changes in important elements on the hospital network we saw in this chapter, using your CD-ROM. See Active Model 3.1.

ACTIVE MODEL 3.1 ■

Project Management

Expanded Homework Problem Sets One of the trademarks of our text has always been a large selection of examples, solved problems, and homework problems. For this edition, each chapter's homework set has been expanded, with scores of new problems. The result is 587 new, existing, and rewritten problems, giving us the largest, clearest, most diverse problem sets of any OM text. These problems focus on problem formulation and interpretation as well as calculation. Each problem is identified as belonging to one of three levels: introductory (one dot), moderate (two dots), and challenging (three dots). To make the transition to this edition seamless, all of the even-numbered problems from the previous edition are unchanged (and their answers appear in Appendix VI.) Plus, solutions to all of the problems appear in the Instructor's Solutions Manual, written by the authors themselves to ensure the quality and accuracy you've come to expect!

New Case Studies There are 55 cases studies, with 18 being new to this edition. The cases are generally one to two pages in length, making them short enough to cover in weekly assignments but detailed enough to add depth to each topic they represent. Forty of the cases focus on the service sector, such as Shouldice Hospital, IKON, Alabama Airlines, Ritz-Carlton, Mayo Clinic, Mutual Insurance of Iowa, and Hard Rock Cafe. In addition, our Web site, at www.prenhall.com/heizer, includes over 53 additional case studies. All are solved in the Instructor's Solutions Manual.

Harvard Case Study Links to Our Book Harvard Business School has selected this text as the model for its matching case study program, and has selected two to five cases to "match" to each of our 17 chapters. At the end of each chapter, we briefly describe each Harvard case and provide a link to the Harvard Web site for further details.

Internet Homework Problems In addition to the problems appearing in the text itself, there are now 165 homework problems on the text's Web site, all of which are included in the Instructor's Solutions Manual. These "new" problems are derived from the odd-numbered problems replaced in the previous edition, meaning that instructors who have favorite problems can still find them as Internet Homework Problems.

CHAPTER-BY-CHAPTER CHANGES

To highlight the extent of the revision of our previous edition, here are a few of the changes on a chapter-by-chapter basis. Active Model exercises appear in most chapters.

Chapter 1: Operations and Productivity Hard Rock Cafe is introduced in this chapter with a *Global Company Profile* and a video case study. There is another new case study, "Zychol Chemicals Corp.," and five new homework problems.

Chapter 2: Operations Strategy in a Global Environment Chapter 3 from the previous edition, *Operations in a Global Environment*, is now merged into the *Operations Strategy* chapter. Global issues and critical success factors dealing with location decisions are now in Chapter 8. A new video case study on Hard Rock's global strategy and that firm's strategy of experience differentiation are included.

Chapter 3: Project Management There are major changes to this chapter, which was Chapter 16 in the previous edition but has been moved forward to give the pervasive nature of project management added emphasis. Activity-on-node (AON) is now emphasized more than activity-on-arrow (AOA) and a hospital example of its use is followed throughout the chapter. A major section on Microsoft's Project 2002 software is new, with numerous MS Project screen captures and a version of the program loaded on the student CD-ROM. There is a new *Critical Thinking Exercise* that focuses on the Colorado Rockies baseball team stadium, nine new homework problems, including two on work breakdown structure (WBS), and two new case studies, including one dealing with how Hard Rock manages its annual Rockfest.

Chapter 4: Forecasting The chapter now includes coverage of Mean Absolute Percent Error (MAPE), has two new *OM in Action* boxes (including one on Disney's forecasting methods), 15 new homework problems, and two new case studies: "Analog Cell Phone, Inc." and "Forecasting at Hard Rock Cafe."

Chapter 5: Design of Goods and Services Computer-Aided Design (CAD), Computer-Aided Manufacturing (CAM), and Virtual Reality Technology (all formerly in Chapter 7 supplement on Operations Technology) have been moved to Chapter 5. There are four new homework problems.

Chapter 6: Managing Quality ISO 9000 has been expanded, a new section on six sigma is included, and there are nine additional homework problems.

Supplement to Chapter 6: Statistical Process Control Process Capability (C_p) is a new topic, a useful table that helps decide which control chart to use (Figure S6.2) is added, and there are 19 new homework problems, including ones dealing with C_p, C_{pk}, and acceptance sampling. Alabama Airlines is a new case study using p-charts.

Chapter 7: Process Strategy Capacity Planning has been moved to a new supplement to Chapter 7 and material on Production Technologies (such as Process Control, Vision Systems, Robots, ASRS, AGV, FMS, and CIM) now appears in this chapter instead of the Operations Technology Supplement in the prior edition. There is a new *OM in Action* box on technology in the hotel industry and two new homework problems. The Rochester Manufacturing case, formerly in the Chapter 7 Supplement, now appears in Chapter 7.

Supplement to Chapter 7: Capacity Planning This new supplement places greater emphasis on the topic of capacity planning, which was covered as part of Chapter 7 in the previous edition. The treatment is expanded, with new examples, text, and numerous additional homework problems. Break-even analysis, decision trees, and strategy-driven investments are included in the supplement.

Chapter 8: Location Strategies There is increased emphasis on global issues affecting location decisions and on graphical information systems (GIS). A new *OM in Action* box looks at Alabama 10 years after it lured Mercedes to open a plant there. There are two new case studies: "Ambrose Distribution Center" and "Where to Place Hard Rock's Next Cafe," as well as eight new homework problems. The global factor rating homework problems that were in Chapter 3 in the previous edition are now in Chapter 8.

Chapter 9: Layout Strategy This chapter now includes the retail layout issue of slotting fees, has a new *OM in Action* box, "Shopping Mall Layout Meets the Internet," and four new homework problems.

Chapter 10: Human Resources and Job Design This chapter has four additional homework problems and two new case studies: "Karstadt vs. J.C. Penney" and "Hard Rock's Human Resource Strategy."

Supplement to Chapter 10: Work Measurement This supplement contains eight new homework problems and a new case study: "Jackson Manufacturing Company."

Chapter 11: Supply-Chain Management Major revisions in this chapter include topics on managing the supply chain, integrating the supply chain, the bullwhip effect, and the cost of shipping alternatives. There are three new homework problems (that deal with shipping alternatives) and a new case study, "Dell's Supply Chain and the Impact of E-Commerce." New *OM in Action* boxes deal with speeding roses to the U.S. market from Latin America, using radio frequency (RF) tags to keep the shelves stocked at Wal-Mart, and DHL's role in the global supply chain.

Supplement to Chapter 11: E-Commerce and Operations Management This supplement is updated with expanded material on the Internet, Internet trading exchanges, and Internet outsourcing. New *OM in Action* boxes are "The Face of Covisint's Online Exchange" and "Internet Keeps Burger King Manager in the Know." There are three new homework problems and a new case study, "E-Commerce at Amazon.com."

Chapter 12: Inventory Management A new *Global Company Profile* on Amazon.com opens the chapter and a new *OM in Action* box, "What the Marines Learned About Inventory from Wal-Mart," appears. The topic of fixed period inventory systems is expanded and there are 14 additional homework problems.

Chapter 13: Aggregate Planning The topic of yield management is now treated in detail, including a graphical presentation and examples. There are seven new homework problems, including several on aggregate planning in services and on yield management.

Chapter 14: Material Requirements Planning and ERP Enterprise Resource Planning (ERP) is now a major part of the MRP chapter. Some of this ERP material had been in the supplement to Chapter 7 in the previous edition, but the topic is treated in more depth in this edition. A new *OM in Action* box, "There is Nothing Easy About ERP," appears, as well as nine new homework problems, and the case study, "Ikon's Attempt at ERP."

Chapter 15: Short-Term Scheduling A new *OM in Action* box, "Banking and the Theory of Constraints," appears in this edition, along with six new homework problems and a video case study, "Scheduling at Hard Rock Cafe."

Chapter 16: Just-In-Time and Lean Production Systems This new chapter, expanded from the previous edition (where it was the supplement to Chapter 12), includes a *Global Company Profile* on Green Gear Cycling, new material on Kanban and reducing lot sizes, and two new *OM in Action* boxes: "Lean Production at Cessna Aircraft" and "Dell's Lean Production." There are also three new homework problems and a new case study of Toyota's JIT practices.

Chapter 17: Maintenance and Reliability We have added a new *OM in Action* box, "Tomcat F-14 Pilots Love Redundancy," and eight new homework problems.

Quantitative Module A: Decision-Making Tools There are six new homework problems and a new case study, "Ski Right Corp."

Quantitative Module B: Linear Programming There is a new focus on sensitivity analysis, by both the graphical approach and using Excel's Solver. There are also 12 new homework problems.

Quantitative Module C: Transportation Models We added five new homework problems to this module.

Quantitative Module D: Waiting-Line Models There are 10 new homework problems.

Quantitative Module E: Learning Curves We have added nine new homework problems.

Quantitative Module F: Simulation We have added a new *OM in Action* box, "Simulating Jackson Memorial Hospital's Operating Rooms," and nine new homework problems.

CD-ROM Tutorials Four mini chapters from the previous edition are unchanged, but a fifth topic is added. The tutorials are: Tutorial 1, Statistical Tools for Managers; Tutorial 2, Acceptance Sampling; Tutorial 3, The Simplex Method of Linear Programming; Tutorial 4, The MODI and VAM Methods of Solving Transportation Problems. The new CD-ROM Tutorial 5, Vehicle Routing and Scheduling, provides a complete treatment of this valuable topic, including algorithms, discussion questions, and homework problems.

TRADEMARK FEATURES

Our goal is to provide students with the finest pedagogical devices to help enhance learning and teaching.

- **Global Company Profiles** Each chapter opens with a two-page, full-color analysis of a leading global organization. These include Amazon, Volkswagen, Dell, NASA, Delta Airlines, McDonald's, Boeing, and many more.

GLOBAL COMPANY PROFILE:

Inventory Management Provides Competitive Advantage at Amazon.com

When Jeff Bezos opened his revolutionary business in 1995, Amazon.com was intended to be a "virtual" retailer—no inventory, no warehouses, no overhead—just a bunch of computers taking orders and authorizing others to fill them. Things clearly didn't work out that way. Now Amazon stocks millions of items of inventory, amid hundreds of thousands of bins on metal shelves, in warehouses around the country that have twice the floor space of the Empire State Building.

Precisely managing this massive inventory has forced Amazon into becoming a world-class leader in warehouse management and automation. This profile shows what goes on behind the scenes. When you place an order at Amazon.com, you are not only doing business with an Internet company, you are doing business with a company that obtains competitive advantage through inventory management.

Sources: New York Times (January 21, 2002): C-3; Time (December 27, 1999): 68–73; and the Wall Street Journal (November 22, 2002): A1, A6.

*1. **You order three items, and a computer in Seattle takes charge.** A computer assigns your order—a book, a game, and a digital camera—to one of Amazon's massive U.S. distribution centers, such as the 800,000 square foot facility in McDonough, Georgia.*
*2. **The "flow meister" in McDonough receives your order** (right). She determines which workers go where to fill your order.*

*3. **Rows of red lights show which products are ordered** (left). Workers move from bulb to bulb, retrieving an item from the shelf above and pressing a button that resets the light. This is known as a "pick-to-light" system. This system doubles the picking speed of manual operators and drops the error rate to nearly zero.*
*4. **Your items are put into crates on moving belts** (below). Each item goes into a large green crate that contains many customers' orders. When full, the crates ride a series of conveyor belts that winds more than 10 miles through the plant at a constant speed of 2.9 feet per second. The bar code on each item is scanned 15 times, by machines and by many of the 600 workers. The goal is to reduce errors to zero—returns are very expensive.*

■ **OM in Action Boxes** Fifty-three half-page examples of recent OM practices are drawn from a wide variety of sources, including the *Wall Street Journal*, *New York Times*, *Fortune*, *Forbes*, and *Harvard Business Review*. These boxes bring OM to life.

OM IN ACTION

What the Marines Learned about Inventory from Wal-Mart

The U.S. Marine Corps knew it had inventory problems. A few years ago, when a soldier at Camp Pendleton, near San Diego, put in an order for a spare part, it took him a week to get it—from the other side of the base. Worse, the force had 207 computer systems worldwide. Called the "Rats' Nest" by marine techies, most systems didn't even talk to each other.

To execute a victory over uncontrolled supplies, the corps studied Wal-Mart, Caterpillar, Inc., and UPS. "We're in the middle of a revolution," says General Gary McKissock. McKissock aims to reduce inventory for the corps by half, saving $200 million, and to shift 2,000 marines from inventory detail to the battlefield.

By replacing inventory with information, the corps won't have to stockpile tons of supplies near the battlefield, like it did during the Gulf War, only to find it couldn't keep track of what was in containers. Then there was the marine policy requiring a 60-day supply of everything. McKissock figured out there was no need to overstock commodity items, like office supplies, that can be obtained anywhere. And with advice from the private sector, the marines have been upgrading warehouses, adding wireless scanners for real-time inventory placement and tracking. Now, if containers need to be sent into a war zone, they will have radio frequency transponders which, when scanned, will link to a database detailing what's inside.

Sources: Business Week (December 24, 2001): 24; and Federal Computer Week (December 11, 2000): 9.

Other student resources include Critical Thinking Exercises, Solved Problems, and Marginal Notes and Definitions. Solutions to Even-Numbered Problems appear in the *Student Lecture Guide*.

FREE STUDENT CD-ROM WITH EVERY NEW TEXT

Packaged free with every new copy of the text is a student CD-ROM that contains exciting resources to liven up the course and help students learn the content material.

- **Fifteen Exciting Video Cases** These video cases feature real companies and allow students to watch short video clips, read about the key topics, answer questions, and then e-mail their answers to their instructors. These case studies can also be assigned without using class time to show the videos.

- **CD-ROM Video Clips** Another expanded feature on the student CD-ROM is twenty-nine 1- to 2-minute videos, which appear throughout the book and are noted in the margins. These video clips illustrate chapter-related topics with videos at Harley-Davidson, Ritz-Carlton, Hard Rock Cafe, and other firms.

- **Active Models** The 29 Active Models, described earlier, appear in files on the student CD-ROM. Samples of the Models appear in most text chapters.

- **PowerPoint Lecture Notes** Based on an extensive set of over 1,000 PowerPoint slides, these lecture notes provide reinforcement to the main points of each chapter and allow students to review chapter material.

- **Additional Practice Problems** Provide problem-solving experience. They supplement the examples and solved problems found in each chapter.

- **Self-Review Quizzes** For each chapter, a link is provided to our text Web site, where these quizzes allow students to test their understanding of each topic.

- **Problem-Solving Software** Excel OM is our exclusive user-friendly Excel add-in. Excel OM automatically creates worksheets to model and solve problems. Users select a topic from the pull-down menu, fill in the data, and then Excel will display and graph (where appropriate) the results. This software is great for student homework, "what if" analysis, or classroom demonstrations. Examples in the text that can be solved with Excel OM appear on the data file on the CD-ROM and are identified by an icon in the margin of the text.

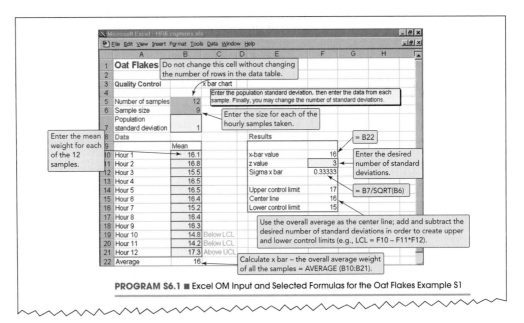

PROGRAM S6.1 ■ Excel OM Input and Selected Formulas for the Oat Flakes Example S1

- **Microsoft Project 2002** MS Project, the most popular and powerful project management package, is now on the student CD-ROM. This version is documented in Chapter 3 and can be activated to work for 120 days.

- **Five Bonus Chapters** "Statistical Tools for Managers," "Acceptance Sampling," "The Simplex Method of Linear Programming," "The MODI and VAM Methods of Solving Transportation Problems," and "Vehicle Routing and Scheduling" are provided as additional material.

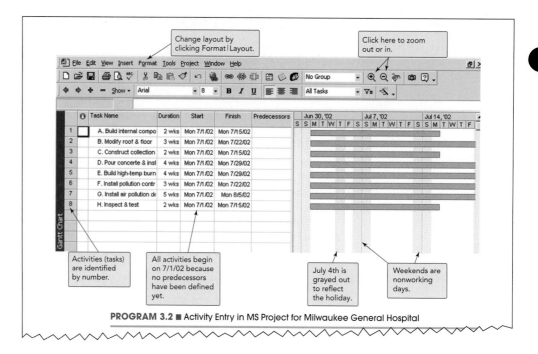

PROGRAM 3.2 ■ Activity Entry in MS Project for Milwaukee General Hospital

INSTRUCTOR'S RESOURCES

Test Item File The test item file, updated by Ross Fink of Bradley University, contains a variety of true/false, multiple choice, fill-in-the-blank, short answer, and problem-solving questions for each chapter. The test item file can also be downloaded from Prentice Hall's Web site at http://www.prenhall.com/heizer.

New TestGen-EQ Software The print Test Banks are designed for use with the TestGen-EQ test-generating software. This computerized package allows instructors to custom design, save, and generate classroom tests. The test program permits instructors to edit, add, or delete questions from the test banks; edit existing graphics and create new graphics; analyze test results; and organize a database of tests and student results. This new software allows for greater flexibility and ease of use. It provides many options for organizing and displaying tests, along with a search and sort feature.

Instructor's Solutions Manual The Instructor's Solutions Manual, written by the authors, contains the answers to all of the discussion questions, critical thinking exercises, active models, and cases in the text, as well as worked-out solutions to all of the end-of-chapter problems, internet problems, and internet cases. The Instructor's Solutions Manual can also be downloaded from Prentice Hall's Web site at http://www.prenhall.com/heizer.

PowerPoint Presentations An extensive set of PowerPoint presentations, created by John Swearingen of Bryant College, is available for each chapter. These slides can also be downloaded from Prentice Hall's Web site at http://www.prenhall.com/heizer.

Instructor's Resource Manual The Instructor's Resource Manual, updated by John Swearingen of Bryant College, contains many useful resources for the instructor—course outlines, video notes, Internet exercises, additional teaching resources, and faculty notes. The Instructor's Resource Manual can also be downloaded from Prentice Hall's Web site at http://www.prenhall.com/heizer.

Instructor's Resource CD-ROM The Instructor's Resource CD-ROM provides the electronic files for the entire Instructor's Solutions Manual (in MS Word), PowerPoint presentations (in PowerPoint), Test Item File (in MS Word), and computerized test bank (TestGen-EQ).

Video Package Designed specifically for the Heizer/Render texts, the video package contains the following videos:

Regal Marine: Operations Strategy
Ritz-Carlton: Quality
Regal Marine: Product Design
Wheeled Coach: Process Strategy
Wheeled Coach: Facility Layout
Regal Marine: Supply Chain Management
Wheeled Coach: Inventory Control
Wheeled Coach: Materials Requirements
   Planning
Operations Management at Hard Rock
Hard Rock Cafe's Global Strategy
Managing Hard Rock's Rockfest
Forecasting at Hard Rock Cafe
Where to Place Hard Rock's Next Cafe
Hard Rock Cafe's Human Resource Strategy
Scheduling at Hard Rock Cafe

Competitiveness and Continuous
   Improvement at Xerox
Teams and Employee Involvement
   at Hewlett Packard
Statistical Process Control
   at Kurt Manufacturing
Process Strategy and Selection
Technology and Manufacturing:
   Flexible Manufacturing Systems
Overview of OM and Strategy
   at Whirlpool
Product Design and Supplier Partnerships
   at Motorola
Service Quality and Design at Marriott
A Plant Tour of Winnebago Industries
E-Commerce and Teva Sports Sandals

INTERNET SUPPORT AND RESOURCE

By logging on to www.prenhall.com/heizer, you will find the most advanced, text-specific, and *personalized* site available on the Web. Students can log on and have a dialogue with their peers, talk to a tutor, take a quiz and get immediate feedback, and read articles about current events—all with the click of a mouse. Our goal is to build an online community dedicated to excellence in teaching and optimal learning. Teaching and learning extends far beyond the classroom, and this partnership is dedicated to providing readers with the very best possible support and service. We encourage professors to integrate the Internet into the operations management course. Some of the resources you will find include:

FOR STUDENTS:

Online Quizzes These extensive quizzes contain a broad assortment of questions, 20–25 per chapter, which include multiple choice, true or false, and Internet essay questions. The quiz questions are graded and can be transmitted to the instructor for extra credit or serve as practice exams.

Virtual Tours These company tours provide direct links to companies ranging from a hospital to an auto manufacturer, that practice key concepts. After touring each Web site, students are asked questions directly related to the concepts discussed in the chapter.

In the News New current events articles are added throughout the year. Each article is summarized by our teams of expert professors, and is fully supported by exercises, activities, and instructor materials.

Internet Resources Provide discipline-specific sites, including preview information that allows you to review site information before you view the site, ensuring you visit the best available business resources found by our learning community.

Notes Allows you to add personal notes to our resources for personal reminders and references.

FOR FACULTY:

This password-protected area provides faculty with the most current and advanced support materials available: The Instructor's Resource Manual, Instructor's Solutions Manual, PowerPoint slides, and Test Questions. For more information on the full list of support material, please look inside the front cover of the text.

ACKNOWLEDGMENTS

We thank the many individuals who were kind enough to assist us in this endeavor. The following professors provided insights that guided us in this revision:

Suad Alwan
Chicago State University

Henry S. Maddux III
Sam Houston State University

Mark McKay
University of Washington

Philip F. Musa
University of Alabama at Birmingham

Susan K. Norman
Northern Arizona University

V. Udayabhanu
San Francisco State University

We also wish to acknowledge the help of the reviewers of the earlier editions of this text. Without the help of these fellow professors, we would never have received the feedback needed to put together a teachable text. The reviewers are listed in alphabetical order.

Sema Alptekin
University of Missouri-Rolla

Jean-Pierre Amor
University of San Diego

Moshen Attaran
California State University-Bakersfield

John H. Blackstone
University of Georgia

Ali Behnezhad
California State University-Northridge

Theodore Boreki
Hofstra University

Rick Carlson
Metropolitan State University

Wen-Chyuan Chiang
University of Tulsa

Mark Coffin
Eastern California University

Henry Crouch
Pittsburgh State University

Warren W. Fisher
Stephen F. Austin State University

Larry A. Flick
Norwalk Community Technical College

Barbara Flynn
Wake Forest University

Damodar Golhar
Western Michigan University

Jim Goodwin
University of Richmond

James R. Gross
University of Wisconsin-Oshkosh

Donald Hammond
University of South Florida

John Harpell
West Virginia University

Marilyn K. Hart
University of Wisconsin-Oshkosh

James S. Hawkes
University of Charleston

George Heinrich
Wichita State University

Sue Helms
Wichita State University

Johnny Ho
Columbus State University

Zialu Hug
University of Nebraska-Omaha

Peter Ittig
University of Massachussetts

Paul Jordan
University of Alaska

Larry LaForge
Clemson University

Hugh Leach
Washburn University

B.P. Lingeraj
Indiana University

Andy Litteral
University of Richmond

Laurie E. Macdonald
Bryant College

Mike Maggard
Northeastern University

Arthur C. Meiners, Jr.
Marymount University

Zafar Malik
Governors State University

Doug Moodie
Michigan Tech University

Joao Neves
Trenton State College

John Nicolay
University of Minnesota

Niranjan Pati
University of Wisconsin-LaCrosse

Michael Pesch
St. Cloud State University

David W. Pentico
Duquesne University

Leonard Presby
William Patterson State College-NJ

Zinovy Radovilsky
California State University-Hayward

Ranga V. Ramasesh
Texas Christian University

Emma Jane Riddle
Winthrop University

M.J. Riley
Kansas State University

Narendrea K. Rustagi
Howard University

Teresita S. Salinas
Washburn University

Chris Sandvig
Western Washington University

Ronald K. Satterfield
University of South Florida

Robert J. Schlesinger
San Diego State University

Shane J. Schvaneveldt
Weber State University

Avanti P. Sethi
Wichita State University

Girish Shambu
Canisius College

L.W. Shell
Nicholls State University

Susan Sherer
Lehigh University

Vicki L. Smith-Daniels
Arizona State University

Vic Sower
Sam Houston State University

Stan Stockton
Indiana University

John Swearingen
Bryant College

Susan Sweeney
Providence College

Kambiz Tabibzadeh
Eastern Kentucky University

Rao J. Taikonda
University of Wisconsin-Oshkosh

Cecelia Temponi
Southwestern Texas State University

Madeline Thimmes
Utah State University

Doug Turner
Auburn University

John Visich-Disc
University of Houston

Rick Wing
San Francisco State University

Bruce M. Woodworth
University of Texas-El Paso

In addition, we appreciate the wonderful people at Prentice Hall who provided both help and advice: Tom Tucker, our decision sciences executive editor, Debbie Clare, our executive marketing manager; Erika Rusnak, our assistant editor; Nancy Welcher, our senior media project manager; Cynthia Regan, our managing editor; and Dawn Stapleton, our editorial assistant. Reva Shader developed the exemplary subject indexes for this text. Karina Mayer provided research assistance. Donna Render and Kay Heizer provided the accurate typing and proofing so critical in a rigorous textbook. We are truly blessed to have such a fantastic team of experts directing, guiding, and assisting us.

We also appreciate the efforts of colleagues who have helped to shape the entire learning package that accompanies this text. Wayne Shell (Nicholls State University) helped create our new problem set and edited/checked the old one; Professor Howard Weiss (Temple University) developed the Active Models, Excel OM, and POM for Windows microcomputer software; Professor John Swearingen (Bryant College) created the PowerPoints. Professor Swearingen also wrote the Instructor's Resource Manual; Dr. Vijay Gupta developed the Excel OM and POM for Windows Data Disks; Dr. Ross Fink (Bradley University) prepared the Test Bank; and Beverly Amer (Northern Arizona University) produced and directed our video and CD-ROM case series. We have been fortunate to have been able to work with all these people.

We wish you a pleasant and productive introduction to operations management.

BARRY RENDER
GRADUATE SCHOOL OF BUSINESS
ROLLINS COLLEGE
WINTER PARK, FL 32789
PHONE: (407) 646-2657
FAX: (407) 646-1550
EMAIL: BARRY.RENDER@ROLLINS.EDU

JAY HEIZER
TEXAS LUTHERAN UNIVERSITY
1000 W. COURT STREET
SEGUIN, TX 78155
PHONE: (830) 372-6056
FAX: (830) 372-6065
EMAIL: JHEIZER@TLU.EDU

PART I

INTRODUCTION TO OPERATIONS MANAGEMENT

Chapter 1

Operations and Productivity

Chapter Outline

LEARNING OBJECTIVES

When you complete this chapter you should be able to

IDENTIFY OR DEFINE:

- Production and productivity
- Operations management (OM)
- What operations managers do
- Services

DESCRIBE OR EXPLAIN:

- A brief history of operations management
- Career opportunities in operations management
- The future of the discipline
- Measuring productivity

Operations Management at Hard Rock Cafe

Operations managers throughout the world are producing products every day to provide for the well-being of society. These products take on a multitude of forms. They may be washing machines at Maytag, motion pictures at Dreamworks, rides at Disney World, or food at Hard Rock Cafe. These firms produce thousands of complex products every day—to be delivered as the customer ordered them, when the customers wants them, and where the customer wants them. Hard Rock does this for over 35 million guests worldwide every year. This is a challenging task and the operations manager's job, whether at Maytag, Dreamworks, Disney, or Hard Rock, is demanding.

Orlando-based Hard Rock Cafe opened its first restaurant in London in 1971, making it over 30 years old and the granddaddy of theme restaurants. Although other theme restaurants have come and gone, Hard Rock is still going strong with 110 restaurants in 38 countries—with new restaurants opening each year. Hard Rock made its name with rock music memorabilia, having started when Eric Clapton, a

Efficient kitchen layouts, motivated personnel, tight schedules, and the right ingredients at the right place at the right time are required to delight the customer.

regular customer, marked his favorite bar stool by hanging his guitar on the wall in the London cafe. Now Hard Rock has millions of dollars invested in memorabilia. To keep customers com-

ing back time and again, Hard Rock creates value in the form of good food and entertainment.

The operations managers at Hard Rock Cafe at Universal Studios in

Hard Rock Cafe in Orlando, Florida, prepares over 3,500 meals each day. Seating over 1,500 people, it is one of the largest restaurants in the world. But Hard Rock's operations managers serve the hot food hot and the cold food cold when and where the customer wants it.

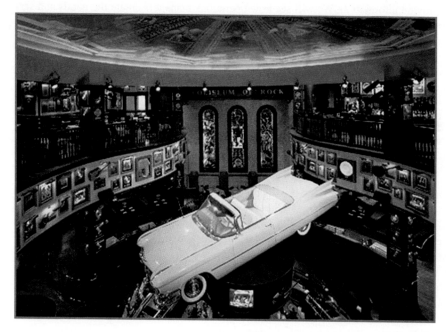

Operations managers are interested in the attractiveness of the layout, but they must be sure that the facility contributes to the efficient movement of people and material with the necessary controls to ensure that proper portions are served.

Orlando provide over 3,500 custom products, in this case meals, every day. These products must be designed and tested, then analyzed for cost of ingredients and labor requirements, as well as for customer satisfaction. On approval, a menu item is put into production using ingredients from qualified suppliers. The production process, from receiving, to cold storage, to grilling or baking or frying, and a dozen other steps, is designed and maintained to yield a quality meal. Operations managers must also prepare efficient employee schedules and design efficient layouts, using the best people they can recruit and train. Managers who successfully design and deliver goods and services throughout the world understand operations management. In this text, we look not only at how Hard Rock's managers create value, but also how a variety of managers in other industries do so. Operations management is demanding and challenging, but it is also exciting. And it impacts our lives every day.

Lots of work goes into designing, testing, and costing meals. Then suppliers deliver quality products on time, every time for well-trained cooks to prepare quality meals. But none of that matters unless an enthusiastic wait staff, such as the one shown here, is doing its job.

Video 1.1

Operations Management
at Hard Rock

Operations management (OM) is a discipline that applies to restaurants like Hard Rock Cafe as well as factories like Sony, Ford, and Maytag. The techniques of OM apply throughout the world to virtually all productive enterprises. It doesn't matter if the application is in an office, a warehouse, a restaurant, a department store, or a factory—the production of goods and services requires operations management. And the *efficient* production of goods and services requires effective applications of the concepts, tools, and techniques of OM that we introduce in this book.

As we progress through this text, we will discover how to manage operations in a changing global economy. An array of informative examples, charts, text discussions, and pictures illustrate concepts and provide information. We will see how operations managers create the goods and services that enrich our lives.

In this chapter, we first define *operations management*, explaining its heritage and exploring the exciting role operations managers play in a huge variety of businesses. Then we discuss production and productivity in both goods- and service-producing firms. This is followed by a discussion of operations in the service sector and the challenge of managing an effective production system.

WHAT IS OPERATIONS MANAGEMENT?

Production
The creation of goods and services.

Operations management (OM)
Activities that relate to the creation of goods and services through the transformation of inputs to outputs.

Production is the creation of goods and services. **Operations management (OM)** is the set of activities that creates value in the form of goods and services by transforming inputs into outputs. Activities creating goods and services take place in all organizations. In manufacturing firms, the production activities that create goods are usually quite obvious. In them, we can see the creation of a tangible product such as a Sony TV or a Harley Davidson motorcycle.

In organizations that do not create physical products, the production function may be less obvious. It may be "hidden" from the public and even from the customer. Examples are the transformations that take place at a bank, hospital, airline office, or college.

Often when services are performed, no tangible goods are produced. Instead, the product may take such forms as the transfer of funds from a savings account to a checking account, the transplant of a liver, the filling of an empty seat on an airline, or the education of a student. Regardless of whether the end product is a good or service, the production activities that go on in the organization are often referred to as operations or *operations management*.

ORGANIZING TO PRODUCE GOODS AND SERVICES

To create goods and services, all organizations perform three functions (see Figure 1.1). These functions are the necessary ingredients not only for production but also for an organization's survival. They are:

1. *Marketing*, which generates the demand, or at least takes the order for a product or service (nothing happens until there is a sale).
2. *Production/operations*, which creates the product.
3. *Finance/accounting*, which tracks how well the organization is doing, pays the bills, and collects the money.

Universities, churches or synagogues, and businesses all perform these functions. Even a volunteer group such as the Boy Scouts of America is organized to perform these three basic functions. Figure 1.1 shows how a bank, an airline, and a manufacturing firm organize themselves to perform these functions. The blue-shaded areas of Figure 1.1 show the operations functions in these firms.

WHY STUDY OM?

We study OM for four reasons:

1. OM is one of the three major functions of any organization, and it is integrally related to all the other business functions. All organizations market (sell), finance (account), and produce (operate), and it is important to know how the OM activity functions. Therefore, we study *how people organize themselves for productive enterprise*.
2. We study OM because we want to know *how goods and services are produced*. The production function is the segment of our society that creates the products we use.
3. We study OM to *understand what operations managers do*. By understanding what these managers do, you can develop the skills necessary to become such a manager. This will help you explore the numerous and lucrative career opportunities in OM.

4. We study OM *because it is such a costly part of an organization*. A large percentage of the revenue of most firms is spent in the OM function. Indeed, OM provides a major opportunity for an organization to improve its profitability and enhance its service to society. Example 1 considers how a firm might increase its profitability via the production function.

FIGURE 1.1 ■

Organization Charts for Two Service and One Manufacturing Organization

(A) A bank, (B) an airline, and (C) a manufacturing organization. The blue areas are OM activities.

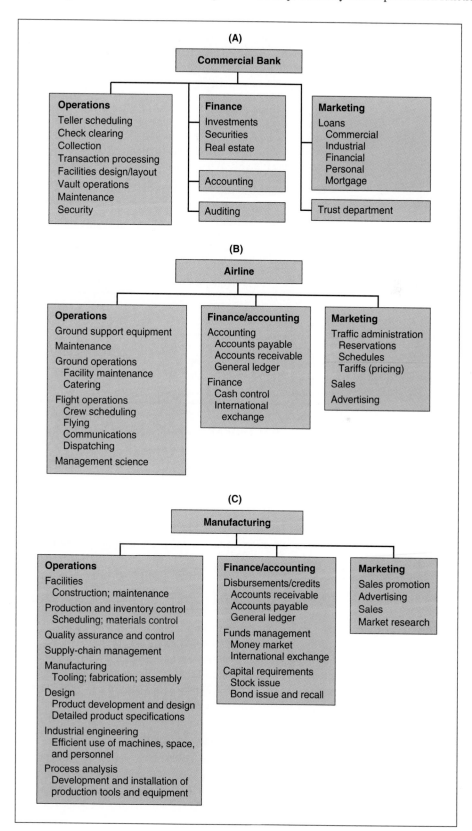

Example 1

Fisher Technologies is a small firm that must double its dollar contribution to fixed cost and profit in order to be profitable enough to purchase the next generation of production equipment. Management has determined that if the firm fails to increase contribution, its bank will not make the loan and the equipment cannot be purchased. If the firm cannot purchase the equipment, the limitations of the old equipment will force Fisher to go out of business and, in doing so, put its employees out of work and discontinue producing goods and services for its customers.

Table 1.1 shows a simple profit-and-loss statement and three strategic options for the firm. The first option is a *marketing option* where good marketing management may increase sales by 50%. By increasing sales by 50%, contribution will in turn increase 71%, but increasing sales 50% may be difficult; it may even be impossible.

The second option is a *finance/accounting option* where finance costs are cut in half through good financial management. But even a reduction of 50% is still inadequate for generating the necessary increase in contribution. Contribution is increased by only 21%.

The third option is an *OM option* where management reduces production costs by 20% and increases contribution by 114%. Given the conditions of our brief example, Fisher Technologies has increased contribution from $10,500 to $22,500 and will now have a bank willing to lend it additional funds.

TABLE 1.1 ■ Options for Increasing Contribution

	CURRENT	MARKETING OPTION[a] INCREASE SALES REVENUE 50%	FINANCE/ ACCOUNTING OPTION[b] REDUCE FINANCE COSTS 50%	OM OPTION[c] REDUCE PRODUCTION COSTS 20%
Sales	$100,000	$150,000	$100,000	$100,000
Costs of goods	−80,000	−120,000	−80,000	−64,000
Gross margin	20,000	30,000	20,000	36,000
Finance costs	− 6,000	− 6,000	− 3,000	− 6,000
Subtotal	14,000	24,000	17,000	30,000
Taxes at 25%	− 3,500	− 6,000	− 4,250	− 7,500
Contribution[d]	$ 10,500	$ 18,000	$ 12,750	$ 22,500

[a]Increasing sales 50% increases contribution by $7,500 or 71% (7,500/10,500).

[b]Reducing finance costs 50% increases contribution by $2,250 or 21% (2,250/10,500).

[c]Reducing production costs 20% increases contribution by $12,000 or 114% (12,000/10,500).

[d]Contribution to fixed cost (excluding finance costs) and profit.

Example 1 underscores the importance of an effective operations activity of a firm. Development of increasingly effective operations is the approach taken by many companies as they face growing global competition.

WHAT OPERATIONS MANAGERS DO

Management process
The application of planning, organizing, staffing, leading, and controlling to the achievement of objectives.

All good managers perform the basic functions of the management process. The **management process** consists of *planning*, *organizing*, *staffing*, *leading*, and *controlling*. Operations managers apply this management process to the decisions they make in the OM function. The 10 major decisions of OM are shown in Table 1.2. Successfully addressing each of these decisions requires planning, organizing, staffing, leading, and controlling. Typical issues relevant to these decisions and the chapter where each is discussed are also shown.

How This Book Is Organized

The 10 decisions shown in Table 1.2 are activities required of operations managers. The ability to make good decisions in these areas and allocate resources to ensure their effective execution goes a long way toward an efficient operations function. The text is structured around these 10 decisions. Throughout the book, we will discuss the issues and tools that help managers make these

TABLE 1.2 ■

Ten Critical Decisions of Operations Management

TEN DECISION AREAS	ISSUES	CHAPTER(S)
Service and product design	What good or service should we offer? How should we design these products?	5
Quality management	Who is responsible for quality? How do we define the quality?	6, 6 Supplement
Process and capacity design	What process and what capacity will these products require? What equipment and technology is necessary for these processes?	7, 7 Supplement
Location	Where should we put the facility? On what criteria should we base the location decision?	8
Layout design	How should we arrange the facility? How large must the facility be to meet our plan?	9
Human resources and job design	How do we provide a reasonable work environment? How much can we expect our employees to produce?	10, 10 Supplement
Supply-chain management	Should we make or buy this component? Who are our suppliers and who can integrate into our e-commerce program?	11, 11 Supplement
Inventory, material requirements planning, and JIT (just-in-time)	How much inventory of each item should we have? When do we reorder?	12, 14, 16
Intermediate and short-term scheduling	Are we better off keeping people on the payroll during slowdowns? Which job do we perform next?	13, 15
Maintenance	Who is responsible for maintenance? When do we do maintenance?	17

10 decisions. We will also consider the impact that these decisions can have on the firm's strategy and productivity.

Where Are the OM Jobs?   How does one get started on a career in operations? The 10 OM decisions identified in Table 1.2 are made by individuals who work in the disciplines shown in the blue areas of Figure 1.1. Competent business students who know their accounting, statistics, finance, and OM have an opportunity to enter the entry-level positions in all of these areas. As you read this text, identify disciplines that can assist you in making these decisions. Then take courses in those areas. The more background an OM student has in accounting, statistics, information systems, and mathematics, the more job opportunities will be available. About 40% of *all* jobs are in OM. Figure 1.2 shows some recent job opportunities.

THE HERITAGE OF OPERATIONS MANAGEMENT

The field of OM is relatively young, but its history is rich and interesting. Our lives and the OM discipline have been enhanced by the innovations and contributions of numerous individuals. We now introduce a few of these people, and we provide a summary of significant events in operations management in Figure 1.3.

Eli Whitney (1800) is credited for the early popularization of interchangeable parts, which was achieved through standardization and quality control. Through a contract he signed with the U.S. government for 10,000 muskets, he was able to command a premium price because of their interchangeable parts.

FIGURE 1.2 ■

Many Opportunities Exist for Operations Managers

PLANT MANAGER

Division of Fortune 1000 company seeks plant manager for plant located in the upper Hudson Valley area. This plant manufactures loading dock equipment for commercial markets.

The Candidate must be experienced in plant management including expertise in production planning, purchasing and inventory management. Good written and oral communication skills are a must along with excellent understanding and application skills in managing people.

Director of Purchasing

Well-established full line food distributor is seeking an experienced purchasing agent to support rapidly expanding food service sales. Must have thorough knowledge of day to day purchasing functions, ability to review vendor programs, establish operating par levels, and coordinate activities with operations. The candidate must be prepared to work with vendors to develop Internet catalogues. Must be well versed in all food categories, a team worker, and bottom line oriented. Salary commensurate with experience.

Quality Manager

Several openings exist in our small package processing facilities in the Northeast, Florida, and Southern California for quality managers. These highly visible positions require extensive use of statistical tools to monitor all aspects of service timeliness and workload measurement. The work involves (1) a combination of hands-on applications and detailed analysis using databases and spreadsheets, (2) process audits to identify areas for improvement, and (3) manage implementation of changes. Positions involve night hours and weekends. Send resume.

Process Improvement Consultants

An expanding consulting firm is seeking consultants to design and implement lean production and cycle time reduction plans in both service and manufacturing processes. Our firm is currently working with an international bank to improve its back office operations as well as several manufacturing firms. A business degree required, APICS certification a plus.

Supply Chain Manager and Planner

Responsibilities entail negotiating contracts and establishing long-term relationships with suppliers. We will rely on the selected candidate to maintain accuracy in the purchasing system, invoices, and product returns. A bachelor's degree and up to 2 years related experience are required. Working knowledge of MRP, ability to use feedback to master scheduling and suppliers and consolidate orders for best price and delivery are necessary. Proficiency in all PC Windows applications, particularly Excel and Word, is essential. Knowledge of Oracle business system I is a plus. Effective verbal and written communication skills are essential.

Customization Focus

**Mass Customization Era
1995–2010**
Globalization
Internet
Enterprise Resource Planning
Learning Organization
International Quality Standards
Finite Scheduling
Supply Chain Management
Agile Manufacturing
E-Commerce

Quality Focus

**Lean Production Era
1980–1995**
Just-in-Time
Computer-Aided Design
Electronic Data Interchange
Total Quality Management
Baldrige Award
Empowerment
Kanbans

Cost Focus

**Early Concepts
1776–1880**
Labor Specialization
    (Smith, Babbage)
Standardized Parts (Whitney)

**Scientific Management Era
1880–1910**
Gantt Charts (Gantt)
Motion & Time Studies
    (Gilbreth)
Process Analysis (Taylor)
Queuing Theory (Erlang)

**Mass Production Era
1910–1980**
Moving Assembly Line
    (Ford/Sorensen)
Statistical Sampling
    (Shewhart)
Economic Order
    Quantity (Harris)
Linear Programming
    PERT/CPM (DuPont)
Material Requirements
    Planning

FIGURE 1.3 ■ Significant Events in Operations Management

Frederick W. Taylor's *Principles of Scientific Management* revolutionized manufacturing. A scientific approach to the analysis of daily work and the tools of industry frequently increased productivity 400%.

Frederick W. Taylor (1881), known as the father of scientific management, contributed to personnel selection, planning and scheduling, motion study, and the now popular field of ergonomics. One of his major contributions was his belief that management should be much more resourceful and aggressive in the improvement of work methods. Taylor and his colleagues, Henry L. Gantt and Frank and Lillian Gilbreth, were among the first to systematically seek the best way to produce.

Another of Taylor's contributions was the belief that management should assume more responsibility for:

1. Matching employees to the right job.
2. Providing the proper training.
3. Providing proper work methods and tools.
4. Establishing legitimate incentives for work to be accomplished.

Charles Sorensen towed an automobile chassis on a rope over his shoulders through the Ford plant while others added parts.

By 1913, Henry Ford and Charles Sorensen combined what they knew about standardized parts with the quasi-assembly lines of the meatpacking and mail-order industries and added the revolutionary concept of the assembly line where men stood still and material moved.[1]

Quality control is another historically significant contribution to the field of OM. Walter Shewhart (1924) combined his knowledge of statistics with the need for quality control and provided the foundations for statistical sampling in quality control. W. Edwards Deming (1950) believed, as did Frederick Taylor, that management must do more to improve the work environment and processes so that quality can be improved.

Operations management will continue to progress with contributions from other disciplines, including *industrial engineering* and *management science*. These disciplines, along with statistics, management, and economics, have contributed substantially to greater productivity.

Innovations from the *physical sciences* (biology, anatomy, chemistry, physics) have also contributed to advances in OM. These innovations include new adhesives, chemical processes for printed circuit boards, gamma rays to sanitize food products, and molten tin tables on which to float higher-quality molten glass as it cools. The design of products and processes often depends on the biological and physical sciences.

Especially important contributions to OM have come from the *information sciences*, which we define as the systematic processing of data to yield information. The information sciences, the Internet, and e-commerce are contributing in a major way toward improved productivity while providing society with a greater diversity of goods and services.

Decisions in operations management require individuals who are well versed in management science, in information science, and often in one of the biological or physical sciences. In this textbook, we take a look at the diverse ways a student can prepare for a career in operations management.

OPERATIONS IN THE SERVICE SECTOR

Manufacturers produce a tangible product, and service products are often intangible. But many products are a combination of a good and a service, which complicates the definition of a service. Even the U.S. government has trouble generating a consistent definition. Because definitions vary, much of the data and statistics generated about the service sector are inconsistent. However, we will define **services** as including repair and maintenance, government, food and lodging, transportation, insurance, trade, financial, real estate, education, legal, medical, entertainment, and other professional occupations.[2]

Services
Those economic activities that typically produce an intangible product (such as education, entertainment, lodging, government, financial and health services).

Differences between Goods and Services

Let's examine some of the differences between goods and services:

- Services are usually *intangible* (for example, your purchase of a ride in an empty airline seat between two cities) as opposed to a tangible good.
- Services are often *produced and consumed simultaneously*; there is no stored inventory. For instance, the beauty salon produces a haircut that is "consumed" simultaneously, or the doctor produces an operation that is "consumed" as it is produced. We have not yet figured out how to inventory haircuts or appendectomies.

---

[1]Jay Heizer, "Determining Responsibility for the Development of the Moving Assembly Line," *Journal of Management History* 4, no. 2 (1998): 94–103.

[2]This definition is similar to the categories used by the U.S. Bureau of Labor Statistics.

- Services are often *unique*. Your mix of financial coverage, such as investments and insurance policies, may not be the same as anyone else's, just as the medical procedure or a haircut produced for you is not exactly like anyone else's.
- Services have *high customer interaction*. Services are often difficult to standardize, automate, and make as efficient as we would like because customer interaction demands uniqueness. In fact, in many cases this uniqueness is what the customer is paying for; therefore, the operations manager must ensure that the product is designed so that it can be delivered in the required unique manner.
- Services have *inconsistent product definition*. Product definition may be rigorous, as in the case of an auto insurance policy, but inconsistent because policyholders change cars and mature.
- Services are often *knowledge-based*, as in the case of educational, medical, and legal services, and therefore hard to automate.
- Services are frequently *dispersed*. Dispersion occurs because services are frequently brought to the client/customer via a local office, a retail outlet, or even a house call.

Table 1.3 indicates some additional differences between goods and services that impact OM decisions. Although service products are different from goods, the operations function continues to transform resources into products. Indeed, the activities of the operations function are often very similar for both goods and services. For instance, both goods and services must have quality standards established, and both must be designed and processed on a schedule, in a facility where human resources are employed.

Having made the distinction between goods and services, we should point out that in many cases, the distinction is not clear-cut. In reality, almost all services are a mixture of a service and a tangible product; similarly, the sale of most goods includes or requires a service. For instance, many products have the service components of financing and transportation (e.g., automobile sales). Many also require after-sale training and maintenance (e.g., office copiers). Still other services, such as consulting or counseling, may require a tangible report.

Moreover, many "service" activities take place within goods-producing operations. Human resource management, logistics, accounting, training, field service, and repair are all service activities, but they take place within a manufacturing organization.

Pure service
A service that does not include a tangible product.

When a tangible product is *not* included in the service, we may call it a **pure service**. Although there are not very many pure services, one example is counseling. Figure 1.4 shows the range of *services* in a product. The range is extensive and shows the pervasiveness of service activities.

Growth of Services

Services now constitute the largest economic sector in advanced societies. For example, service-sector employment in the U.S. is shown in Figure 1.5(a). Until about 1900, most Americans were employed in agriculture. Increased agricultural productivity allowed people to leave the farm and seek employment in the city. The manufacturing and service sectors began to grow, with services becoming the dominant employer in the early 1920s and manufacturing employment peaking at about 32% in 1950. Similarly, as Figures 1.5(a) and (b) show, productivity increases in manufacturing have allowed more of our economic resources to be devoted to ser-

TABLE 1.3 ■

Differences between Goods and Services

ATTRIBUTES OF GOODS (TANGIBLE PRODUCT)	ATTRIBUTES OF SERVICES (INTANGIBLE PRODUCT)
Product can be resold.	Reselling a service is unusual.
Product can be inventoried.	Many services cannot be inventoried.
Some aspects of quality are measurable.	Many aspects of quality are difficult to measure.
Selling is distinct from production.	Selling is often a part of the service.
Product is transportable.	Provider, not product, is often transportable.
Site of facility is important for cost.	Site of facility is important for customer contact.
Often easy to automate.	Service is often difficult to automate.
Revenue is generated primarily from the tangible product.	Revenue is generated primarily from the intangible services.

FIGURE 1.4 ■

Most Goods Contain
a Service and Most
Services Contain
a Good

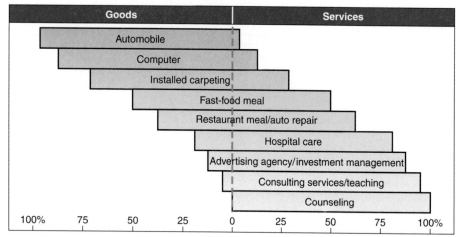

vices. Consequently, much of the world can now enjoy the pleasures of education, health services, entertainment, and myriad other things that we call services. Examples of firms and percentage of employment in the **service sector** are shown in Table 1.4. Table 1.4 also provides employment percentages for the nonservice sectors of manufacturing, construction, mining, and agriculture on the bottom four lines.

Service sector
That segment of the economy that includes trade, financial, lodging, education, legal, medical, and other professional occupations.

Service Pay

Although there is a common perception that service industries are low-paying, in fact, many service jobs pay very well. Operations managers in the maintenance facility of an airline are very well paid, as are the operations managers who supervise computer services to the financial community. About 42% of all service workers receive wages above the national average. However, the service-sector average is driven down because 14 of the Commerce Department categories of the 33 service industries do indeed pay below the all-private industry average. Of these, retail trade, which pays only 61% of the national private industry average, is large. But even considering the retail sector, the average wage of all service workers is about 96% of the average of all private industries.[3]

FIGURE 1.5 ■

Development of the
Service Economy

Sources: Adapted from U.S.
Labor Statistics Bureau; OECD;
national statistics; U.S.
Commerce Department U.S.
Foreign Trade Highlights; *The
Economist* (May 29, 1999): 100;
*Statistical Abstract of the United
States,* 2001.

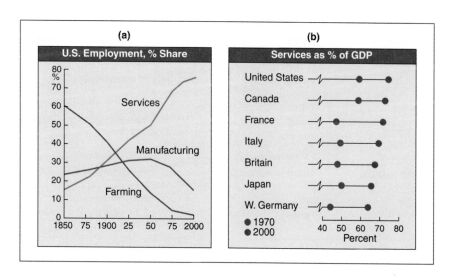

[3]Herbert Stein and Murray Foss, *The New Illustrated Guide to the American Economy* (Washington, DC: The AIE Press, 1995): 30.

TABLE 1.4 ■

Examples of
Organizations
in Each Sector

Source: *Statistical Abstract
of the United States* (2001),
Table 596.

SECTOR	EXAMPLE	PERCENTAGE OF ALL JOBS
Service Sector		
Professional Services, Education, Legal, Medical	New York City P.S. 108, Notre Dame University, San Diego Zoo	24.3
Trade (retail, wholesale)	Walgreen's, Wal-Mart, Nordstrom's	20.6
Utilities, Transportation	Pacific Gas & Electric, American Airlines, Santa Fe R.R., Roadway Express	7.2
Business and Repair Services	Snelling and Snelling, Waste Management, Inc., Pitney-Bowes	7.1
Finance, Insurance, Real Estate	Citicorp, American Express, Prudential, Aetna, Trammell Crow	6.5
Food, Lodging, Entertainment	McDonald's, Hard Rock Cafe, Motel 6, Hilton Hotels, Walt Disney, Paramount Pictures	5.2
Public Administration	U.S., State of Alabama, Cook County	4.5
Manufacturing Sector	General Electric, Ford, U.S. Steel, Intel	14.8
Construction Sector	Bechtel, McDermott	7.0
Agriculture	King Ranch	2.4
Mining Sector	Homestake Mining	.4
Grand Total		100.0

Service Sector subtotal: 75.4

EXCITING NEW TRENDS IN OPERATIONS MANAGEMENT

One of the reasons OM is such an exciting discipline is that the operations manager is confronted with an ever-changing world. Both the approach to and the results of the 10 OM decisions in Table 1.2 are subject to change. These dynamics are the result of a variety of forces, from globalization of world trade to the transfer of ideas, products, and money at electronic speeds. The direction now being taken by OM—where it has been and where it is going—is shown in Figure 1.6. We now introduce some of the challenges shown in Figure 1.6.

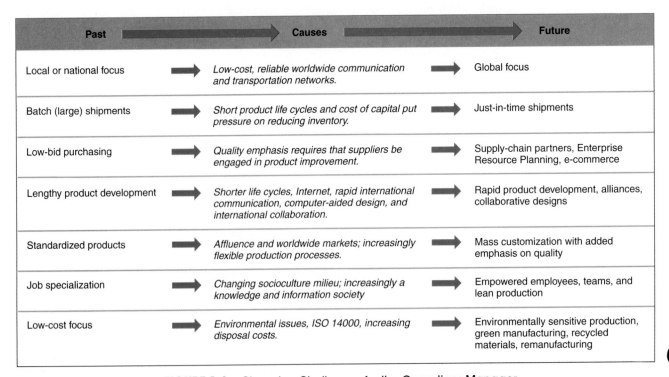

Past	Causes	Future
Local or national focus	Low-cost, reliable worldwide communication and transportation networks.	Global focus
Batch (large) shipments	Short product life cycles and cost of capital put pressure on reducing inventory.	Just-in-time shipments
Low-bid purchasing	Quality emphasis requires that suppliers be engaged in product improvement.	Supply-chain partners, Enterprise Resource Planning, e-commerce
Lengthy product development	Shorter life cycles, Internet, rapid international communication, computer-aided design, and international collaboration.	Rapid product development, alliances, collaborative designs
Standardized products	Affluence and worldwide markets; increasingly flexible production processes.	Mass customization with added emphasis on quality
Job specialization	Changing socioculture milieu; increasingly a knowledge and information society	Empowered employees, teams, and lean production
Low-cost focus	Environmental issues, ISO 14000, increasing disposal costs.	Environmentally sensitive production, green manufacturing, recycled materials, remanufacturing

FIGURE 1.6 ■ Changing Challenges for the Operations Manager

- *Global focus:* The rapid decline in communication and transportation costs has made markets global. But at the same time, resources in the form of materials, talent, and labor have also become global. Contributing to this rapid globalization are countries throughout the world that are vying for economic growth and industrialization. Operations managers are responding with innovations that generate and move ideas, parts, and finished goods rapidly, wherever and whenever needed.

- *Just-in-time performance:* Vast financial resources are committed to inventory, making it costly. Inventory also impedes response to rapid changes in the marketplace. Operations managers are viciously cutting inventories at every level, from raw materials to finished goods.

- *Supply-chain partnering:* Shorter product life cycles, as well as rapid changes in material and process technology, require more participation by suppliers. Suppliers usually supply over half of the value of products. Consequently, operations managers are building long-term partnerships with critical players in the supply chain.

- *Rapid product development:* Rapid international communication of news, entertainment, and lifestyles is dramatically chopping away at the life span of products. Operations managers are responding with technology and alliances (partners) that are faster and management that is more effective.

- *Mass customization:* Once we begin to consider the world as the marketplace, then the individual differences become quite obvious. Cultural differences, compounded by individual differences, in a world where consumers are increasingly aware of options, places substantial pressure on firms to respond. Operations managers are responding with production processes that are flexible enough to cater to individual whims of consumers. The goal is to produce individual products, whenever and wherever needed.

- *Empowered employees:* The knowledge explosion and a more technical workplace have combined to require more competence at the workplace. Operations managers are responding by moving more decision making to the individual worker.

- *Environmentally sensitive production:* The operation manager's continuing battle to improve productivity is increasingly concerned with designing products and processes that are environmentally friendly. That means designing products that are biodegradable, or automobile components that can be reused or recycled, or more efficient packaging.

These and many more topics that are part of the exciting challenges to operations managers are discussed in this text.

THE PRODUCTIVITY CHALLENGE

Productivity
The ratio of outputs (goods and services) divided by one or more inputs (such as labor, capital, or management).

Video 1.2

The Transformation Process at Regal Marine

The creation of goods and services requires changing resources into goods and services. The more efficiently we make this change, the more productive we are and the more value is added to the good or service provided. **Productivity** is the ratio of outputs (goods and services) divided by the inputs (resources, such as labor and capital) (see Figure 1.7). The operations manager's job is to enhance (improve) this ratio of outputs to inputs. Improving productivity means improving efficiency.[4]

This improvement can be achieved in two ways: a reduction in inputs while output remains constant, or an increase in output while inputs remain constant. Both represent an improvement in productivity. In an economic sense, inputs are labor, capital, and management, which are integrated into a production system. Management creates this production system, which provides the conversion of inputs to outputs. Outputs are goods and services, including such diverse items as guns, butter, education, improved judicial systems, and ski resorts. *Production* is the making of goods and services. High production may imply only that more people are working and that employment levels are high (low unemployment), but it does not imply high *productivity*.

---

[4]*Efficiency* means doing the job well—with a minimum of resources and waste. Note the distinction between being *efficient*, which implies doing the job well, and *effective*, which means doing the right thing. A job well done—say, by applying the 10 decisions of operations management—helps us be *efficient*; developing and using the correct strategy helps us be *effective*.

FIGURE 1.7 ■

The Economic System Adds Value by Transforming Inputs to Outputs

An effective feedback loop evaluates process performance against a plan or standard. It also evaluates customer satisfaction and sends signals to those controlling the inputs and process.

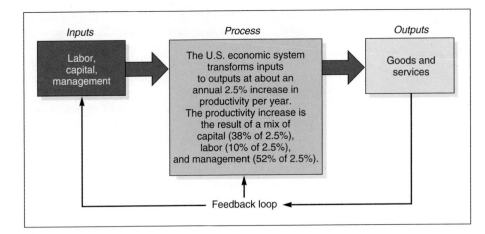

Measurement of productivity is an excellent way to evaluate a country's ability to provide an improving standard of living for its people. *Only through increases in productivity can the standard of living improve.* Moreover, only through increases in productivity can labor, capital, and management receive additional payments. If returns to labor, capital, or management are increased without increased productivity, prices rise. On the other hand, downward pressure is placed on prices when productivity increases, because more is being produced with the same resources.

The benefits of increased productivity are illustrated in the *OM in Action* box, "Increasing Productivity in the L.A. Motor Pool."

For over 100 years (from about 1869 to 1973), the U.S. was able to increase productivity at an average rate of almost 2.5% per year. Such growth doubled U.S. wealth every 30 years. However, from 1973 until the mid-1990s the U.S. was unable to sustain that increase and productivity dropped to about 1% per year. This was disastrous and explains much of the national concern with economic growth. But the U.S. productivity picture is improving. The manufacturing sector, although a decreasing portion of the U.S. economy, has in recent years seen productivity increases exceeding 4% per year, and the service sector, with increases of almost 1%, has also shown some improvement. The combination has moved U.S. annual productivity growth as we move into the twenty-first century back to the 2.5% range for the economy as a whole.[5]

In this text we examine how to improve productivity through the operations function. Productivity is a significant issue for the world and one that the operations manager is uniquely qualified to address.

Productivity Measurement

The measurement of productivity can be quite direct. Such is the case when productivity can be measured as labor-hours per ton of a specific type of steel or as the energy necessary to generate a kilowatt of electricity.[6] An example of this can be summarized in the following equation:

Single-factor productivity
Indicates the ratio of one resource (input) to the goods and services produced (outputs).

$$\text{Productivity} = \frac{\text{Units produced}}{\text{Input used}} \qquad (1\text{-}1)$$

For example, if units produced = 1,000 and labor-hours used is 250, then:

$$\text{Productivity} = \frac{\text{Units produced}}{\text{Labor-hours used}} = \frac{1,000}{250} = 4 \text{ units per labor-hour}$$

Multifactor productivity
Indicates the ratio of many or all resources (inputs) to the goods and services produced (outputs).

The use of just one resource input to measure productivity, as shown above, is known as **single-factor productivity**. However, a broader view of productivity is **multifactor productivity**, which includes all inputs (e.g., labor, material, energy, capital). Multifactor productivity is also known as

[5]According to the *Statistical Abstract of the United States*, non-farm business sector productivity increase for 1995 was 0.9%; 1996, 2.5%; 1997, 2.0%; 1998, 2.7%; 1999, 2.6%; 2000, 4.3% (see Table 613). Productivity increase for the recession year of 2001 was 1.1%. (See the *Wall Street Journal* (August 12, 2002): A2.

[6]The quality and time period are assumed to remain constant.

OM IN ACTION

Increasing Productivity in the L.A. Motor Pool

The newly elected mayor of Los Angeles faced many problems. One of them was a 21,000-vehicle motor pool with bloated expenses and poor vehicle availability. On any given day, as many as 30% of the city's 900 trash trucks and 11% of the police department's cars were in the repair shop. The problems included too many vehicles in some agencies, vehicle sabotage and abuse, missed repairs, and vehicles never serviced. The L.A. motor pool and its $120-million-a-year maintenance operation needed improved productivity.

The mayor implemented seven simple operations management innovations: (1) Individual drivers were turned into team players who helped complete each other's routes; (2) trucks were assigned specific parking places so they could be easily located each morning; (3) tire pressure was checked on every truck every night to avoid flat tires during working hours; (4) all trucks were emptied every night to avoid such dangers as leftover cinders igniting a fire; (5) standard customer pickups were established (this alone saved the city $12 million per year); (6) the utility department installed a computerized fleet management system (to track vehicle use and to charge departments); and (7) mechanics were moved to night shifts so vehicles were not in the shop during the day.

As a result of these management changes, the department cut its total fleet by 500 vehicles; its inventory of parts dropped 20%, freeing up $5.4 million dollars a year; and out-of-service garbage trucks dropped from that embarrassing 30% to 18%.

Sources: The Wall Street Journal (July 6, 1995): A1, A10; and American City & County (July 1997): FM1-FM4.

total factor productivity. Multifactor productivity is calculated by combining the input units, as shown below:

$$\text{Productivity} = \frac{\text{Output}}{\text{Labor + Material + Energy + Capital + Miscellaneous}} \quad (1\text{-}2)$$

To aid in the computation of multifactor productivity, the individual inputs (the denominator) can be expressed in dollars and summed as shown in Example 2.

Example 2

Collins Title Company has a staff of 4, each working 8 hours per day (for a payroll cost of $640/day) and overhead expenses of $400 per day. Collins processes and closes on 8 titles each day. The company recently purchased a computerized title-search system that will allow the processing of 14 titles per day. Although the staff, their work hours, and pay will be the same, the overhead expenses are now $800 per day.

Labor productivity with the old system: $\frac{8 \text{ titles per day}}{32 \text{ labor-hours}} = .25$ titles per labor-hour

Labor productivity with the new system: $\frac{14 \text{ titles per day}}{32 \text{ labor-hours}} = .4375$ titles per labor-hour

Multifactor productivity with the old system: $\frac{8 \text{ titles per day}}{\$640 + 400} = .0077$ titles per dollar

Multifactor productivity with the new system: $\frac{14 \text{ titles per day}}{\$640 + 800} = .0097$ titles per dollar

Labor productivity has increased from .25 to .4375. The change is .4375/.25 = 1.75, or a 75% increase in labor productivity. Multifactor productivity has increased from .0077 to .0097. This change is .0097/.0077 = 1.259, or a 25.9% increase in multifactor productivity.

Use of productivity measures aids managers in determining how well they are doing. The multifactor-productivity measures provide better information about the trade-offs among factors, but substantial measurement problems remain. Some of these measurement problems are:

1. *Quality* may change while the quantity of inputs and outputs remains constant. Compare a radio of this decade with one of the 1940s. Both are radios, but few people would deny that the quality has improved. The unit of measure—a radio—is the same, but the quality has changed.

2. *External elements*[7] may cause an increase or decrease in productivity for which the system under study may not be directly responsible. A more reliable electric power service may greatly improve production, thereby improving the firm's productivity because of this support system rather than because of managerial decisions made within the firm.

3. *Precise units of measure* may be lacking. Not all automobiles require the same inputs: Some cars are subcompacts, others 911 Turbo Porsches.

Productivity measurement is particularly difficult in the service sector, where the end product can be hard to define. For example, the quality of your haircut, the outcome of a court case, or service at a retail store are all ignored in the economic data. In some cases, adjustments are made for the quality of the product sold, but *not* the quality of the sales performance or a broader product selection. Productivity measurements are in specific inputs and outputs, while a free economy is producing worth—what people want. People may want customized products along with convenience, speed, and safety. Traditional measures of inputs and outputs may be a very poor measure of these other factors. Note the quality-measurement problems in a law office, where each case is different, altering the accuracy of the measure "cases per labor-hour" or "cases per employee."

Video 1.3

Productivity at Whirlpool

Productivity Variables

As we saw in Figure 1.6, productivity increases are dependent upon three **productivity variables**:

Productivity variables
The three factors critical to productivity improvement—labor, capital, and the arts and science of management.

1. *Labor*, which contributes about 10% of the annual increase.
2. *Capital*, which contributes about 38% of the annual increase.
3. *Management*, which contributes about 52% of the annual increase.

These three factors are critical to improved productivity. They represent the broad areas in which managers can take action to improve productivity.[8]

Labor   Improvement in the contribution of labor to productivity is the result of a healthier, better-educated, and better-nourished labor force. Some increase may also be attributed to a shorter workweek. Historically, about 10% of the annual improvement in productivity is attributed to improvement in the quality of labor. Three key variables for improved labor productivity are:

1. Basic education appropriate for an effective labor force.
2. Diet of the labor force.
3. Social overhead that makes labor available, such as transportation and sanitation.

Many American high schools exceed a 50% dropout rate in spite of offering a wide variety of programs.

In developed nations, a fourth challenge to management is *maintaining and enhancing the skills of labor* in the midst of rapidly expanding technology and knowledge. Recent data suggest that the average American 17-year-old knows significantly less mathematics than the average Japanese at the same age, and about half cannot answer the questions in Figure 1.8. Moreover, more than 38% of American job applicants tested for basic skills were deficient in reading, writing, or math.[9]

FIGURE 1.8 ■

About Half of the 17-Year-Olds in the U.S. Cannot Correctly Answer Questions of This Type

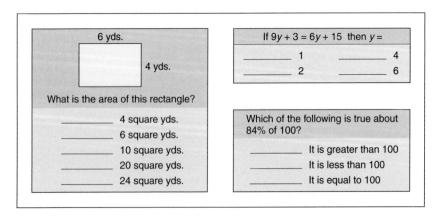

---

[7]These are exogenous variables—that is, variables outside of the system under study that influence it.

[8]The percentages are from Herbert Stein and Murray Foss, *The New Illustrated Guide to the American Economy* (Washington, DC: the AIE Press, 1995): 67.

[9]Rodger Doyle, "Can't Read, Can't Count," *Scientific American* (October 2001): 24.

Between 20% and 30% of U.S. workers lack the basic skills they need for their current jobs.
(*Source:* Nan Stone, *Harvard Business Review.*)

Overcoming shortcomings in the quality of labor while other countries have a better labor force is a major challenge. Perhaps improvements can be found not only through increasing competence of labor but also via a fifth item, *better utilized labor with a stronger commitment.* Training, motivation, team building, and the human resource strategies discussed in Chapter 10, as well as improved education, may be among the many techniques that will contribute to increased labor productivity. Improvements in labor productivity are possible; however, they can be expected to be increasingly difficult and expensive.

Capital　Human beings are tool-using animals. Capital investment provides those tools. Capital investment has increased in the U.S. every year except during a few very severe recession periods. Annual capital investment in the U.S. has increased at an annual rate of 1.5% after allowances for depreciation.

Inflation and taxes increase the cost of capital, making capital investment increasingly expensive. When the capital invested per employee drops, we can expect a drop in productivity. Using labor rather than capital may reduce unemployment in the short run, but it also makes economies less productive and therefore lowers wages in the long run. Capital investment is often a necessary, but seldom a sufficient ingredient in the battle for increased productivity.

The trade-off between capital and labor is continually in flux. The higher the interest rate, the more projects requiring capital are "squeezed out": They are not pursued because the potential return on investment for a given risk has been reduced. Managers adjust their investment plans to changes in capital cost.

Management　Management is a factor of production and an economic resource. Management is responsible for ensuring that labor and capital are effectively used to increase productivity. Management accounts for over half of the annual increase in productivity. It includes improvements made through the application of technology and the utilization of knowledge.

This application of technology and utilization of new knowledge requires training and education. Education will remain an important high-cost item in postindustrial societies. Postindustrial societies are technological societies requiring training, education, and knowledge. Consequently, they are also known as knowledge societies. **Knowledge societies** are those in which much of the labor force has migrated from manual work to technical and information-processing tasks requiring education and knowledge. Effective operations managers build workforces and organizations that recognize the continuing need for education and knowledge. *They ensure that technology, education, and knowledge are used effectively.*

More effective utilization of capital, as opposed to the investment of additional capital, is also important. The manager, as a productivity catalyst, is charged with the task of making improvements in capital productivity within existing constraints. Productivity gains in knowledge societies require managers who are comfortable with technology and management science.

Knowledge society
A society in which much of the labor force has migrated from manual work to work based on knowledge.

Siemens, the multi-billion-dollar German conglomerate, has long been known for its apprentice programs in its home country. Because education is often the key to efficient operations in a technological society, Siemens has spread its apprentice-training programs to its U.S. plants. These programs are laying the foundation for the highly skilled workforce that is essential for global competitiveness.

The productivity challenge is difficult. A country cannot be a world-class competitor with second-class inputs. Poorly educated labor, inadequate capital, and dated technology are second-class inputs. High productivity and high-quality outputs require high-quality inputs.

Productivity and the Service Sector

The service sector provides a special challenge to the accurate measurement of productivity and productivity improvement. The traditional analytical framework of economic theory is based primarily on goods-producing activities. Consequently, most published economic data relate to goods production. But the data do indicate that, as our contemporary service economy has increased in size, we have had slower growth in productivity.

Productivity of the service sector has proven difficult to improve because service-sector work is:

1. Typically labor-intensive (for example, counseling, teaching).
2. Frequently individually processed (for example, investment counseling).
3. Often an intellectual task performed by professionals (for example, medical diagnosis).
4. Often difficult to mechanize and automate (for example, a haircut).
5. Often difficult to evaluate for quality (for example, performance of a law firm).

Playing a Mozart string quartet still takes four musicians the same length of time.

The more intellectual and personal the task, the more difficult it is to achieve increases in productivity. Low-productivity improvement in the service sector is also attributable to the growth of low-productivity activities in the service sector. These include activities not previously a part of the measured economy, such as child care, food preparation, house cleaning, and laundry service. These activities have moved out of the home and into the measured economy as more and more women have joined the workforce. Inclusion of these activities has probably resulted in lower measured productivity for the service sector, although, in fact, actual productivity has probably increased because these activities are now more efficiently produced than previously.[10]

However, in spite of the difficulty of improving productivity in the service sector, improvements are being made. And this text presents a multitude of ways to do it. Indeed, a recent article in the *Harvard Business Review* reinforces the concept that managers can improve service productivity. The authors argue that "the primary reason why the productivity growth rate has stagnated in the service sector is management,"[11] and they find astonishing what can be done when management pays attention to how work actually gets done.

Although the evidence indicates that all industrialized countries have the same problem with service productivity, the U.S. remains the world leader in overall productivity *and* service productivity. Retailing is twice as productive in the U.S. as in Japan, where laws protect shopkeepers from discount chains. The U.S. telephone industry is at least twice as productive as Germany's. The U.S. banking system is also 33% more efficient than Germany's banking oligopolies. However, because productivity is central to the operations manager's job and because the service sector is so large, we take special note in this text of how to improve productivity in the service sector. (See, for instance, the *OM in Action* box, "Taco Bell Uses Productivity to Lower Costs.")

THE CHALLENGE OF SOCIAL RESPONSIBILITY

Operations managers function in a system where they are subjected to constant changes and challenges: These come from stakeholders such as customers, suppliers, owners, lenders, and employees. These stakeholders and government agencies require that managers respond in a socially responsible way in maintaining a clean environment, a safe workplace, and ethical behavior. If operations managers focus on increasing productivity in an open system in which all stakeholders have a voice, then many of these challenges are mitigated. The company will use fewer resources, the employees will be committed, and the ethical climate will be enhanced. Throughout this text, we emphasize a variety of ways in which operations managers can take responsible actions to address these challenges successfully.

[10]Allen Sinai and Zaharo Sofianou, "The Service Economy–Productivity Growth Issues" (CSI Washington, DC), *The Service Economy* (January 1992): 11–16.

[11]Michael van Biema and Bruce Greenwald, "Managing Our Way to Higher Service-Sector Productivity," *Harvard Business Review* 75, no. 4 (July–August 1997): 89. Their conclusions are not unique. Management *does* make a difference:

OM IN ACTION

Taco Bell Uses Productivity to Lower Costs

Founded in 1962 by Glenn Bell, Taco Bell is seeking competitive advantage via low cost. Like many services, Taco Bell increasingly relies on its operations function to improve productivity and reduce cost.

First, it revised the menu and designed meals that were easy to prepare. Taco Bell then shifted a substantial portion of food preparation to suppliers who could perform food processing more efficiently than a stand-alone restaurant. Ground beef was precooked prior to arrival and then reheated, as were many dishes that arrived in plastic boil bags for easy sanitary reheating. Similarly, tortillas arrived already fried and onions prediced. Efficient layout and automation cut to 8 seconds the time needed to prepare tacos and burritos. These advances have been combined with training and empowerment to increase the span of management from one supervisor for 5 restaurants to one supervisor for 30 or more.

Operations managers at Taco Bell believe they have cut in-store labor by 15 hours per day and reduced floor space by more than 50%. The result is a store that can handle twice the volume with half the labor. Effective operations management has resulted in productivity increases that support Taco Bell's low-cost strategy. Taco Bell is now the fast-food low-cost leader and has a 73% share of the Mexican fast-food market.

Sources: Nation's Restaurant News (January 15, 2001): 57; and *Interfaces* (January–February 1998): 75–91.

SUMMARY

Operations, marketing, and finance/accounting are the three functions basic to all organizations. The operations function creates goods and services. Much of the progress of operations management has been made in the twentieth century, but since the beginning of time, humankind has been attempting to improve its material well-being. Operations managers are key players in the battle for improved productivity.

However, as societies have become increasingly affluent, more of their resources are devoted to services. In the U.S., three quarters of the workforce is employed in the service sector. Although productivity improvements are difficult to achieve in the service sector, operations management is the primary vehicle for making improvements.

KEY TERMS

Production *(p. 4)*
Operations management (OM) *(p. 4)*
Management process *(p. 6)*
Services *(p. 9)*
Pure service *(p. 10)*
Service sector *(p. 11)*

Productivity *(p. 13)*
Single-factor productivity *(p. 14)*
Multifactor productivity *(p. 14)*
Productivity variables *(p. 16)*
Knowledge society *(p. 17)*

SOLVED PROBLEMS

Solved Problem 1.1

Productivity can be measured in a variety of ways, such as by labor, capital, energy, material usage, and so on. At Modern Lumber, Inc., Art Binley, president and producer of apple crates sold to growers, has been able, with his current equipment, to produce 240 crates per 100 logs. He currently purchases 100 logs per day, and each log requires 3 labor-hours to process. He believes that he can hire a professional buyer who can buy a better-quality log at the same cost. If this is the case, he can increase his production to 260 crates per 100 logs. His labor-hours will increase by 8 hours per day.

What will be the impact on productivity (measured in crates per labor-hour) if the buyer is hired?

SOLUTION

(a) Current labor productivity $= \dfrac{240 \text{ crates}}{100 \text{ logs} \times 3 \text{ hours/log}}$

$$= \frac{240}{300}$$

$$= .8 \text{ crates per labor-hour}$$

(b)

Labor productivity with buyer $= \dfrac{260 \text{ crates}}{(100 \text{ logs} \times 3 \text{ hours/log}) + 8 \text{ hours}}$

$$= \frac{260}{308}$$

$$= .844 \text{ crates per labor-hour}$$

Using current productivity (.80 from [a]) as a base, the increase will be 5.5% (.844/.8 = 1.055, or a 5.5% increase).

Solved Problem 1.2

Art Binley has decided to look at his productivity from a multifactor (total factor productivity) perspective (refer to Solved Problem 1.1). To do so, he has determined his labor, capital, energy, and material usage and decided to use dollars as the common denominator. His total labor-hours are now 300 per day and will increase to 308 per day. His capital and energy costs will remain constant at $350 and $150 per day, respectively. Material costs for the 100 logs per day are $1,000 and will remain the same. Because he pays an average of $10 per hour (with fringes), Binley determines his productivity increase as follows:

SOLUTION

CURRENT SYSTEM			SYSTEM WITH PROFESSIONAL BUYER	
Labor:	300 hrs. @ $10 =	$3,000	308 hrs. @ $10 =	$3,080
Material:	100 logs/day	1,000		1,000
Capital:		350		350
Energy:		150		150
Total Cost		$4,500		$4,580

Multifactor productivity of current system:
= 240 crates/4,500 = .0533 crates/dollar

Multifactor productivity of proposed system:
= 260 crates/4,580 = .0568 crates/dollar

Using current productivity (.0533) as a base, the increase will be .0656. That is, .0568/.0533 = 1.0656, or a 6.56% increase.

INTERNET AND STUDENT CD-ROM EXERCISES

Visit our home page or use your student CD-ROM to help with material in this chapter.

 On Our Home Page, www.prenhall.com/heizer

- Self-Tests
- Practice Problems
- Internet Exercises
- Current Articles and Research
- Virtual Company Tour
- Internet Homework Problems

 On Your Student CD-ROM

- Power Point Lecture
- Video Clips and Video Case
- Practice Problems

ADDITIONAL CASE STUDY

Harvard has selected this Harvard Business School case to accompany this chapter (textbookcasematch.hbsp.harvard.edu**):**

- **Taco Bell Corp.** (#692-058): Illustrates the power of breakthrough thinking in a service industry.

Chapter 2

Operations Strategy in a Global Environment

Chapter Outline

LEARNING OBJECTIVES

When you complete this chapter you should be able to

IDENTIFY OR DEFINE:

Mission

Strategy

Ten decisions of OM

Multinational Corporation

DESCRIBE OR EXPLAIN:

Specific approaches used by OM to achieve strategies

Differentiation

Low Cost

Response

Four global operations strategies

Why global issues are important

GLOBAL COMPANY PROFILE:

Boeing's Global Strategy Yields Competitive Advantage

When building its newest product, the 777, Boeing took a huge financial risk of $4 billion. Global competition meant finding not only exceptional suppliers, wherever they might be, but also suppliers willing to step up to the risk associated with new products. Boeing found its 777 partners in over a dozen countries; a few of them are shown in the table below. These "partners," investing $1.5 billion, not only spread the risk, but also added the advantage to all parties of including local content in the plane. Countries that have a manufacturing stake in the 777, says Boeing, are more likely to buy from Boeing than from European competitor Airbus Industries.

Boeing's worldwide purchasing effort results in components coming together on the 777 assembly line in Everett, Washington. While components come from throughout the world, about 20% of the 777's structure comes from a Japanese consortium of Fuji, Kawasaki, and Mitsubishi.

Some International Suppliers of Boeing 777 Components

FIRM	COUNTRY	PARTS
Alenia	Italy	Wing flaps
AeroSpace Technologies	Australia	Rudder
CASA	Spain	Ailerons
Fuji	Japan	Landing gear doors, wing section
GEC Avionics	United Kingdom	Flight computers
Hawker de Havilland	Australia	Elevators
Kawasaki	Japan	Fuselage sections 43, 44, cargo doors
Korean Air	Korea	Flap supports
Menasco Aerospace	Canada	Landing gears
Messier-Bugatti	France	Landing gears
Mitsubishi	Japan	Fuselage sections 46, 47, 48, passenger doors
Short Brothers	Ireland	Landing gear doors
Singapore Aerospace	Singapore	Landing gear doors
Smiths Industries	United Kingdom	Electronic systems

BOEING

Tight supplier contracts ensure that quality components are delivered on time and within cost to Boeing's final assembly line. It all comes together in Everett. For a discussion of assembly lines, see Chapter 9, "Layout Strategy."

Aircraft fuselage sections are manufactured in Japan. These 14,000-pound pieces (shown in the center of the photo) house the aft lower cargo hold and the economy section for the interior passenger cabin.

One such nation is Japan, whose consortium of Fuji, Kawasaki, and Mitsubishi complete one-fifth of the plane's body, including most of the fuselage, parts of the wing, and landing gear doors.

Boeing has contracted more than half its work outside the United States, a savings of $600 million per year. But this strategy is not without its detractors. To Boeing machinists, who earn an average of $25 per hour, this practice means pickets with signs such as "Export Planes—Not Our Jobs." But Boeing claims it is impossible to export planes without sending a significant number of jobs overseas. When a foreign nation now agrees to buy planes from Boeing, it typically does so on condition that some work will be done in that country. Union workers disagree. Our biggest competitor is not Airbus, they say; it is workers in Mexico, Korea, and Poland.

Today's operations manager must have a global view of operations strategy. Rapid growth in world trade and emerging markets like China and Eastern Europe means that many organizations must extend their operations globally. Making a product only in the U.S. and then exporting it no longer guarantees success or even survival. There are new standards of global competitiveness that include quality, variety, customization, convenience, timeliness, and cost. This globalization of strategy contributes efficiency and adds value to products and services offered the world, but it also complicates the operations manager's job.

Companies today respond to the global environment with strategies and speeds unheard of in the past. For instance:

- Boeing is competitive because both its sales and production are worldwide.
- Italy's Benetton moves inventory to stores around the world faster than its competition by building flexibility into design, production, and distribution.
- Sony purchases components from suppliers in Thailand, Malaysia, and around the world for assembly in its electronic products.
- General Motors is simultaneously building four similar plants in Argentina, Poland, China, and Thailand so they can learn from each other and drive down cost while increasing quality.

The opportunities of the global environment are often enticing, but the operations manager must realize that barriers are also created. Complexity, risk, and competition are intensified; companies must carefully account for them.[1]

A GLOBAL VIEW OF OPERATIONS[2]

There are many reasons why a domestic business operation will decide to change to some form of international operation. These can be viewed as a continuum ranging from tangible reasons to intangible reasons (see Figure 2.1). Let us examine, in turn, each of the six reasons listed in Figure 2.1.

German apparel maker Hugo Boss AG is being forced by high German labor costs to shift over half its production to central Europe and North America.

Reduce Costs   Many international operations seek to take advantage of the tangible opportunities to reduce their costs. Foreign locations with lower wages can help lower both direct and indirect costs. (See the *OM in Action* box, "U.S. Cartoon Production at Home in Manila.") Less stringent government regulations on a wide variety of operation practices (e.g., environmental control, health and safety, etc.) reduce costs. Opportunities to cut the cost of taxes and tariffs also encourage foreign operations. In Mexico, the creation of **maquiladoras** (free trade zones) allows manufacturers to cut their costs of taxation by paying only on the value added by Mexican workers.[3] If a U.S. manufacturer, such as IBM, brings a $500 computer to a maquiladora operation for assembly work costing $25, tariff duties will be charged only on the $25 of work performed in Mexico.

Maquiladoras
Mexican factories located along the U.S.–Mexico border that receive preferential tariff treatment.

Shifting low-skilled jobs to another country has several potential advantages. First, and most obvious, the firm may reduce costs. Second, moving the lower skilled jobs to a lower cost location frees higher cost workers for more valuable tasks. Third, reducing wage costs allows the savings to be invested in improved products and facilities (and the retraining of existing workers if necessary) at the home location. The impact of this approach is shown in the *OM in Action* box, "A Global Perspective Provides Competitive Advantage."

FIGURE 2.1 ■

Reasons to Globalize Operations

Source: Adapted from M. J. Schniederjans, *Operations Management: A Global Context* (New York: Quorum Books, 1998).

	Reasons to Globalize
Tangible Reasons	• Reduce costs (labor, taxes, tariffs, etc.)
	• Improve supply chain
	• Provide better goods and services
	• Attract new markets
Intangible Reasons	• Learn to improve operations
	• Attract and retain global talent

---

[1] See related discussion in Pankaj Ghemawat, "Distance Still Matters," *Harvard Business Review* 79, no. 8 (September 2001): 137–147.

[2] The authors wish to thank Professor Marc J. Schniederjans, University of Nebraska, for his extensive input and assistance in developing this section of this chapter. Professor Schniederjans is author of *Operations Management: A Global Context* and *International Facility Location and Acquisition Analysis*, both published by Quorum Books, 1998.

[3] L. E. Koslow, *Business Abroad* (Houston: Gulf Publishing, 1996): 55–56.

OM IN ACTION

U.S. Cartoon Production at Home in Manila

Fred Flintstone is not from Bedrock. He is actually from Manila, capital of the Philippines. So are Tom and Jerry, Aladdin, and Donald Duck. More than 90% of American television cartoons are produced in Asia, with the Philippines leading the way. With their natural advantage of English as an official language and a strong familiarity with U.S. culture, animation companies in Manila now employ more than 1,700 people. Filipinos think Western, and "you need to have a group of artists that can understand the humor that goes with it," says Bill Dennis, a Hanna-Barbera executive.

Major studios like Disney, Marvel, Warner Brothers, and Hanna-Barbera send *storyboards*—cartoon action outlines—and voice tracks to the Philippines. Artists there draw, paint, and film about 20,000 sketches for a 30-minute episode. The cost of $130,000 to produce an episode in the Philippines compares to $160,000 in Korea and $500,000 in the U.S.

Sources: Variety (January 7–13, 2002): 43; and the *New York Times* (June 25, 1999): 60.

World Trade Organization (WTO)
An international organization that helps promote world trade by lowering barriers to the free flow of goods across borders.

NAFTA
A free trade agreement between Canada, Mexico, and the U.S.

European Union (EU)
The EU has 15 member states (Belgium, Germany, France, Italy, Luxembourg, Netherlands, Denmark, Ireland, United Kingdom, Greece, Spain, Portugal, Austria, Finland, Sweden) and is preparing for 13 eastern and southern European countries to join.

Trade agreements have also helped reduce tariffs and thereby reduce the cost of operating facilities in foreign countries. The **World Trade Organization (WTO)** has helped to reduce tariffs from 40% in 1940 to 3% in 1995. Another important trade agreement is the **North American Free Trade Agreement (NAFTA)**. NAFTA seeks to phase out all trade and tariff barriers among Canada, Mexico, and the U.S. Other trade agreements that are accelerating global trade include APEC (the Pacific rim countries), SEATO (Australia, New Zealand, Japan, Hong Kong, South Korea, New Guinea, and Chile), and MERCOSUR (Argentina, Brazil, Paraguay, Uruguay).

Another trading group is the **European Union (EU)**. The European Union has reduced trade barriers between the participating European nations through standardization and a common currency, the euro. However, this major U.S. trading partner, with 375 million people, is also placing some of the world's most restrictive conditions on products sold in the EU. Everything from recycling standards to automobile bumpers to hormone-free farm products must meet EU standards, complicating international trade of these items.

Improve the Supply Chain   The supply chain can often be improved by locating facilities in countries where unique resources are available. These resources may be expertise, labor, or raw material. For example, auto-styling studios from throughout the world are migrating to the auto mecca of southern California, to ensure the necessary expertise in contemporary auto design. Similarly, world athletic shoe production has migrated from South Korea to Guangzhou, China: This location takes advantage of the low-cost labor and production competence in a city where 40,000 people work making athletic shoes for the world. And a perfume essence manufacturer wants a presence in Grasse, France, where much of the world's perfume essences are prepared from the flowers of the Mediterranean.

Provide Better Goods and Services   While the characteristics of goods and services can be objective and measurable (e.g., number of on-time deliveries), they can also be subjective and less measurable (e.g., sensitivity to culture). We need an ever better understanding of differences in culture and of the way business is handled in different countries. Improved understanding as the result of a local presence permits firms to customize products and services to meet unique cultural needs in foreign markets.

Another reason for international operations is to reduce response time to meet customers' changing product and service requirements. Customers who purchase goods and services from U.S. firms are increasingly located in foreign countries. Providing them with quick and adequate service is often improved by locating facilities in their home countries.

Attract New Markets   Because international operations require interaction with foreign customers, suppliers, and other competitive businesses, international firms inevitably learn about opportunities for new products and services. Knowledge of these markets not only helps to increase

OM IN ACTION

A Global Perspective Provides Competitive Advantage

Aerovox Inc., based in New Bedford, Massachusetts, was on the verge of bankruptcy when it shifted 300 of its 700 jobs to Juarez, Mexico. The $2 capacitors that Aerovox made in New Bedford store an electrical charge for appliances such as refrigerators. Aerovox flipped this money-losing low-margin capacitor business into a profit by moving it to Juarez. Earnings from this venture allowed the firm to complete two acquisitions, one in Huntsville, Alabama, and the other in the United Kingdom. These acquisitions in turn let Aerovox develop more sophisticated products and retool to produce them. Aerovox employees in the U.S. now make custom-designed parts for portable defibrillators. These components sell for $30 to $5,000 a piece. Today, Aerovox has 1,600 employees in its worldwide workforce and sales have more than doubled.

With a slightly different global approach, Dana Corp., based in Toledo, Ohio, established a joint venture with Cardanes S.A. to produce truck transmissions in Queretaro, Mexico. Then Dana switched 288 U.S. employees in its Jonesboro, Arkansas, plant from producing truck transmissions at breakeven to axle production at a profit. Productivity is up in Jonesboro, and the Mexican joint venture is making money. Employees in both Jonesboro and Queretaro, as well as stockholders, came out ahead on the move.

Resourceful organizations like Aerovox and Dana use a global perspective to become more efficient, which allows them to develop new products, retrain employees, and invest in new plant and equipment.

Sources: The *Wall Street Journal* (November 15, 1999): A28; *Quality Progress* (September 2001): 51–61; and www.dana.com/news/.

sales, but also permits organizations to diversify their customer base and smooth the business cycle. Global operations may add production flexibility so products and services can be switched between economies that are booming and those that are not.

Another reason to go into foreign markets is the opportunity to expand the *life cycle* (i.e., stages a product goes through; see Chapter 5) of an existing product. While some products in the U.S. are in a "mature" stage of their product life cycle, they may represent state-of-the-art products in less developed countries. For example, the U.S. market for personal computers could be characterized as "mature," but as in the "introductory" stage in many developing countries such as Albania, China, and Burma (Myanmar).

A worldwide strategy places added burdens on operations management. Because of economic and lifestyle differences, designers must target products to each market. For instance, clothes washers sold in northern countries must spin-dry clothes much better than those in warmer climates, where consumers are likely to line-dry them. Similarly, as shown here, Whirlpool refrigerators sold in Bangkok are manufactured in bright colors because they are often put in living rooms.

Learn to Improve Operations

Learning does not take place in isolation. Firms serve themselves and their customers well when they remain open to the free flow of ideas. For example, General Motors found that it could improve operations by jointly building and running, with the Japanese, an auto assembly plant in San Jose, California. This strategy allows GM to contribute its capital and knowledge of U.S. labor and environmental laws while the Japanese contribute production and inventory ideas. GM also used its employees and experts from Japan to help design its U.S. Saturn plant around production ideas from Japan. Similarly, operations managers have improved equipment and layout by learning from the ergonomic competence of the Scandinavians.

Attract and Retain Global Talent

Global organizations can attract and retain better employees by offering more employment opportunities. They need people in all functional areas and areas of expertise worldwide. Global firms can recruit and retain good employees because they provide both greater growth opportunities and insulation against unemployment during times of economic downturn. During economic downturns in one country or continent, a global firm has the means to relocate unneeded personnel to more prosperous locations. Global organizations also provide incentives for people who like to travel or take vacations in foreign countries.

So, to recap Figure 2.1, successfully achieving a competitive advantage in our shrinking world means maximizing all of the possible opportunities, from tangible to intangible, that international operations can offer.

Cultural and Ethical Issues

One of the great challenges as operations go global is reconciling differences in social and cultural behavior. With issues ranging from bribery, to child labor, to the environment, managers sometimes do not know how to respond when operating in a different culture. What one country's culture deems acceptable may be considered unacceptable or illegal in another.

In the last decade, changes in international laws, agreements, and codes of conduct have been applied to define ethical behavior among managers around the world.[4] The World Trade Organization, for example, helps to make uniform the protection of both governments and industries from foreign firms that engage in unethical conduct. Even on issues where significant differences between cultures exist, as in the area of bribery or the protection of intellectual property, global uniformity is slowly being accepted by most nations.

In spite of cultural and ethical differences, we live in a period of extraordinary mobility of capital, information, goods, and even people. We can expect this to continue. The financial sector, the telecommunications sector, and the logistics infrastructure of the world are healthy institutions that foster efficient and effective use of capital, information, and goods. Globalization, with all of its opportunities and risks, is here and will continue. It must be considered as managers develop their missions and strategies.

DEVELOPING MISSIONS AND STRATEGIES

An effective operations management effort must have a *mission* so it knows where it is going and a *strategy* so it knows how to get there. This is the case for a small or domestic organization, as well as a large international organization.

Mission

Economic success, indeed survival, is the result of identifying missions to satisfy a customer's needs and wants. We define the organization's **mission** as its purpose—what it will contribute to society. Mission statements provide boundaries and focus for organizations and the concept around which the firm can rally. The mission states the rationale for the organization's existence. Developing a good strategy is difficult, but it is much easier if the mission has been well defined. Figure 2.2 provides examples of mission statements.

Globalization may take us to the floating factory: A six-person crew will take a factory from port to port in order to obtain the best market, material, labor, and tax advantages.

The service industry, by way of the floating resort (the cruise ship), already provides such an example.

"The ethics of the world market are very clear. Manufacturers will move wherever it is cheapest or most convenient to their interests."
Carlos Arias Macelli, owner of a Guatemala plant that supplies J.C. Penney

Mission
The purpose or rationale for an organization's existence.

---

[4] S. J. Carroll and M. J. Gannon, *Ethical Dimensions of International Management* (Thousand Oaks, CA: Sage Publications, 1997).

FIGURE 2.2 ■

Mission Statements for
Three Organizations

Source: Annual reports:
courtesy of Merck and FedEx.
Hard Rock Cafe: *Employee
Handbook,* 2001, p. 3.

FedEx
FedEx is committed to our People-Service-Profit philosophy. We will produce outstanding financial returns by providing totally reliable, competitively superior, global air-ground transportation of high-priority goods and documents that require rapid, time-certain delivery. Equally important, positive control of each package will be maintained utilizing real time electronic tracking and tracing systems. A complete record of each shipment and delivery will be presented with our request for payment. We will be helpful, courteous, and professional to each other and the public. We will strive to have a completely satisfied customer at the end of each transaction.
Merck
The mission of Merck is to provide society with superior products and services—innovations and solutions that improve the quality of life and satisfy customer needs—to provide employees with meaningful work and advancement opportunities and investors with a superior rate of return.
Hard Rock Cafe
Our Mission: To spread the spirit of Rock 'n' Roll by delivering an exceptional entertainment and dining experience. We are committed to being an important, contributing member of our community and offering the Hard Rock family a fun, healthy, and nurturing work environment while ensuring our long-term success.

Once an organization's mission has been decided, each functional area within the firm determines its supporting mission. By "functional area" we mean the major disciplines required by the firm, such as marketing, finance/accounting, and production/operations. Missions for each function are developed to support the firm's overall mission. Then within that function lower-level supporting missions are established for the OM functions. Figure 2.3 provides such a hierarchy of sample missions.

Strategy

Strategy
How an organization
expects to achieve its
missions and goals.

With the mission established, strategy and its implementation can begin. **Strategy** is an organization's action plan to achieve the mission. Each functional area has a strategy for achieving its mission and for helping the organization reach the overall mission. These strategies exploit opportunities and strengths, neutralize threats, and avoid weaknesses. In the following sections we will describe how strategies are developed and implemented.

Firms achieve missions in three conceptual ways: (1) differentiation, (2) cost leadership, and (3) quick response.[5] This means operations managers are called on to deliver goods and services that are (1) *better*, or at least different, (2) *cheaper*, and (3) more *responsive*. Operations managers translate these *strategic concepts* into tangible tasks to be accomplished. Any one or combination of these three strategic concepts can generate a system that has a unique advantage over competitors. For example, Hunter Fan has differentiated itself as a premier maker of quality ceiling fans that lower heating and cooling costs for its customers. Nucor Steel, on the other hand, satisfies customers by being the lowest-cost steel producer in the world. And Dell achieves rapid response by building personal computers with each customer's requested software in a matter of hours.

Clearly strategies differ. And each strategy puts different demands on operations management. Hunter Fan's strategy is one of *differentiating* itself via quality from others in the industry. Nucor focuses on value at *low cost*, while Dell's dominant strategy is quick, reliable *response*.

Video 2.1

Operations Strategy at
Regal Marine

---

[5] See related discussion in Michael E. Porter, *Competitive Strategy: Techniques for Analyzing Industries and Competitors* (New York: The Free Press, 1980). Also see Donald C. Hambrick and James W. Fredrickson, "Are You Sure You Have a Strategy?" *Academy of Management Executive,* 15, no. 4 (November 2001): 48–59.

FIGURE 2.3 ■

Sample Missions for a Company, the Operations Function, and Major Departments in an Operations Function

Sample Company Mission	
To manufacture and service a growing and profitable worldwide microwave communications business that exceeds our customers' expectations.	
Sample Operations Management Mission	
To produce products consistent with the company's mission as the worldwide low-cost manufacturer.	
Sample OM Department Missions	
Product design	To lead in research and engineering competencies in all areas of our primary business, designing and producing products and services with outstanding quality and inherent customer value.
Quality management	To attain the exceptional value that is consistent with our company mission and marketing objectives by close attention to design, procurement, production, and field service opportunities.
Process design	To determine and design or produce the production process and equipment that will be compatible with low-cost product, high quality, and a good quality of work life at economical cost.
Location selection	To locate, design, and build efficient and economical facilities that will yield high value to the company, its employees, and the community.
Layout design	To achieve, through skill, imagination, and resourcefulness in layout and work methods, production effectiveness and efficiency while supporting a high quality of work life.
Human resources	To provide a good quality of work life, with well-designed, safe, rewarding jobs, stable employment, and equitable pay, in exchange for outstanding individual contribution from employees at all levels.
Supply-chain management	To cooperate with suppliers to develop innovative products from stable, effective, and efficient sources of supply.
Inventory	To achieve low investment in inventory consistent with high customer service levels and high facility utilization.
Scheduling	To achieve high levels of throughput and timely customer delivery through effective scheduling.
Maintenance	To achieve high utilization of facilities and equipment by effective preventive maintenance and prompt repair of facilities and equipment.

ACHIEVING COMPETITIVE ADVANTAGE THROUGH OPERATIONS

Competitive advantage

The creation of a unique advantage over competitors.

Each of the three strategies provides an opportunity for operations managers to achieve competitive advantage. **Competitive advantage** implies the creation of a system that has a unique advantage over competitors. The idea is to create customer value in an efficient and sustainable way. Pure forms of these strategies may exist, but operations managers will more likely be called on to implement some combination of them. Let us briefly look at how managers achieve competitive advantage via *differentiation*, *low cost*, and *response*.

Competing on Differentiation

Safeskin Corporation is number one in latex exam gloves because it has differentiated itself and its products. It did so by producing gloves that were designed to prevent allergic reactions about which doctors were complaining. When other glove makers caught up, Safeskin developed hypoallergenic gloves. Then it added texture to its gloves. Then it developed a synthetic disposable glove for those allergic to latex—always staying ahead of the competition. Safeskin's strategy is to develop a reputation for designing and producing reliable state-of-the-art gloves, thereby differentiating itself.

Differentiation
To distinguish the offerings of the organization in any way that the customer perceives as adding value.

Differentiation is concerned with providing *uniqueness*. A firm's opportunities for creating uniqueness are not located within a particular function or activity, but can arise in virtually everything that the firm does. Moreover, because most products include some service and most services include some product, the opportunities for creating this uniqueness are limited only by imagination. Indeed, **differentiation** should be thought of as going beyond both physical characteristics and service attributes to encompass everything about the product or service that influences the value that the customers derive from it. Therefore, effective operations managers assist in defining everything about a product or service that will influence the potential value to the customer. This may be the convenience of a broad product line, product features, or a service related to the product. Such services can manifest themselves through convenience (location of distribution centers or stores), training, product delivery and installation, or repair and maintenance services.

Experience differentiation
Engages the customer with the product through imaginative use of the five senses, so the customer "experiences" the product.

In the service sector, one option for extending product differentiation is through an *experience*. Differentiation by experience in services is a manifestation of the growing "experience economy."[6] The idea of **experience differentiation** is to engage the customer—to use people's five senses so they become immersed and perhaps even an active participant in the product. Disney does this with the Magic Kingdom. People no longer just go on a ride; they are immersed in the Magic Kingdom— surrounded by a dynamic visual and sound experience that complements the physical ride. Some rides further engage the customer by having them steer the ride or shoot targets or villains.

Theme restaurants, such as Hard Rock Cafe, likewise differentiate themselves by providing an "experience." Hard Rock engages the customer with classic rock music, big-screen rock videos, memorabilia, and staff who can tell stories. In many instances, a full-time guide is available to explain the displays, and there is always a convenient retail store so the guest can take home a tangible part of the experience. The result is a "dining experience" rather than just a meal. In a less dramatic way, your local supermarket delivers an experience when it provides music and the aroma of freshly baked bread, and when it has samples for you to taste.

Video 2.2

Hard Rock's Global Strategy

Competing on Cost

Southwest Airlines has been a consistent moneymaker while other U.S. airlines have lost billions. Southwest has done this by fulfilling a need for low-cost and short-hop flights. Its operations strategy has included use of secondary airports and terminals, first-come, first-served seating, few fare options, smaller crews flying more hours, snacks-only or no-meal flights, and no downtown ticket offices.

Additionally, and less obviously, Southwest has very effectively matched capacity to demand and effectively utilized this capacity. It has done this by designing a route structure that matches the capacity of its Boeing 737, the only plane in its fleet. Second, it achieves more air miles than other airlines by faster turnarounds—its planes are on the ground less.

One driver of a low-cost strategy is a facility that is effectively utilized. Southwest and others with low-cost strategies understand this and utilize resources effectively. Identifying the optimum size can allow firms to spread overhead costs over enough units to drive down costs and provide a cost advantage. For instance, Wal-Mart continues to pursue its low-cost strategy with superstores, open 24 hours a day. For 20 years, it has successfully grabbed market share. Wal-Mart has driven down store overhead costs, shrinkage, and distribution costs. Its rapid transportation of goods, reduced warehousing costs, and direct shipment from manufacturers have resulted in high inventory turnover and made it a low-cost leader.

[6]For an engaging book on the experience economy, see Joseph Pine II and James H. Gilmore, *The Experience Economy*, (Boston: Harvard Business School Press, 1999). Also see Leonard L. Berry, Lewis P. Carbone, and Stephan H. Haeckel, "Managing the Total Customer Experience," *MIT Sloan Management Review* (spring 2002): 85–90.

Low-cost leadership
Achieving maximum value as perceived by the customer.

Low-cost leadership entails achieving maximum *value* as defined by your customer. It requires examining each of the 10 OM decisions in a relentless effort to drive down costs while meeting customer expectations of value. A low-cost strategy does *not* imply low value or low quality.

Competing on Response

The third strategy option is response. Response is often thought of as *flexible* response, but it also refers to *reliable* and *quick* response. Indeed, we define **response** as including the entire range of values related to timely product development and delivery, as well as reliable scheduling and flexible performance.

Response
That set of values related to rapid, flexible, and reliable performance.

Flexible response may be thought of as the ability to match changes in a marketplace where design innovations and volumes fluctuate substantially.

Hewlett-Packard is an exceptional example of a firm that has demonstrated flexibility in both design and volume changes in the volatile world of personal computers. HP's products often have a life cycle of months, and volume and cost changes during that brief life cycle are dramatic. However, HP has been successful at institutionalizing the ability to change products and volume to respond to dramatic changes in product design and costs—thus building a *sustainable competitive advantage*.

The second aspect of response is the *reliability* of scheduling. One way the German machine industry has maintained its competitiveness despite having the world's highest labor costs is through reliable response. This response manifests itself in reliable scheduling. German machine firms have meaningful schedules—and they perform to these schedules. Moreover, the results of these schedules are communicated to the customer and the customer can, in turn, rely on them. Consequently, the competitive advantage generated through reliable response has value to the end customer.

"In the future, there will be just two kinds of firms: those who disrupt their markets and those who don't survive the assault."
Professor Richard D'Aveni, author of Hypercompetition

The third aspect of response is *quickness*. Johnson Electric, discussed in the *OM in Action* box, competes on speed—speed in design, production, and delivery. Whether it is a production system at Johnson Electric, a lunch delivered in 15 minutes at Bennigan's, or customized pagers delivered in three days from Motorola, the operations manager who develops systems that respond quickly can have a competitive advantage.

In practice, these three *concepts*—differentiation, low cost, and response—are often implemented via the six *specific strategies* shown in Figure 2.4: (1) flexibility in design and volume, (2) low price, (3) delivery, (4) quality, (5) after-sale service, and (6) a broad product line. Through these

OM IN ACTION

Global Strategy at Hong Kong's Johnson Electric

Patrick Wang, managing director of Johnson Electric Holdings, Ltd., walks through his Hong Kong headquarters with a micromotor in his hand. This tiny motor, about twice the size of his thumb, powers a Dodge Viper power door lock. Although most people have never heard of Johnson Electric, we all have several of its micromotors nearby. This is because Johnson is the world's leading producer of micromotors for cordless tools, household appliances (such as coffee grinders and food processors), personal care items (such as hair dryers and electric shavers), and cars. A luxury Mercedes, with its headlight wipers, power windows, power seat adjustments, and power side mirrors, may use 50 Johnson micromotors.

Like all truly global businesses, Johnson spends liberally on communications to tie together its global network of factories, R&D facilities, and design centers. For example, Johnson Electric installed a $20-million videoconferencing system that allows engineers in Cleveland, Ohio, and Stuttgart, Germany, to monitor trial production of their micromotors in China.

Johnson's first strength is speed in product development, speed in production, and speed in delivering—13 million motors a month, mostly assembled in China but delivered throughout the world. Its second strength is the ability to stay close to its customers. Johnson has design and technical centers scattered across the U.S., Europe, and Japan. "The physical limitations of the past are gone" when it comes to deciding where to locate a new center, says Patrick Wang. "Customers talk to us where they feel most comfortable, but products are made where they are most competitive."

Sources: Far Eastern Economic Review (May 16, 2002): 44–45; and South China Morning Post (December 8, 2000): 10.

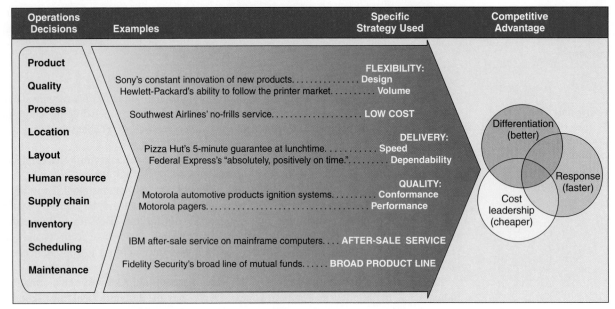

FIGURE 2.4 ■ Operations Management's Contribution to Strategy

Source: For related presentation, see Jeffrey G. Miller and Aleda Roth, "A Taxonomy of Manufacturing Strategies," *Management Science* 40, no. 3 (March 1994): 285–304.

six specific strategies, OM can increase productivity and generate a sustainable competitive advantage. Proper implementation of the following decisions by operations managers will allow these strategies to be achieved.

TEN STRATEGIC OM DECISIONS

Operations decisions
The strategic decisions of OM are product design, quality, process design, location selection, layout design, human resources and job design, supply-chain management, inventory, scheduling, and maintenance.

Differentiation, low cost, and response can be achieved when managers make effective decisions in 10 areas of OM. These are collectively known as **operations decisions**. The 10 decisions of OM that support missions and implement strategies follow:

1. *Goods and service design.* Designing goods and services defines much of the transformation process. Costs, quality, and human resource decisions are often determined by design decisions. Designs usually determine the lower limits of cost and the upper limits of quality.
2. *Quality.* The customer's quality expectations must be determined and policies and procedures established to identify and achieve that quality.
3. *Process and capacity design.* Process options are available for products and services. Process decisions commit management to specific technology, quality, human resource use, and maintenance. These expenses and capital commitments will determine much of the firm's basic cost structure.
4. *Location selection.* Facility location decisions for both manufacturing and service organizations may determine the firm's ultimate success. Errors made at this juncture may overwhelm other efficiencies.
5. *Layout design.* Material flows, capacity needs, personnel levels, technology decisions, and inventory requirements influence layout.
6. *Human resources and job design.* People are an integral and expensive part of the total system design. Therefore, the quality of work life provided, the talent and skills required, and their costs must be determined.

7. *Supply-chain management.* These decisions determine what is to be made and what is to be purchased. Consideration is also given to quality, delivery, and innovation, all at a satisfactory price. Mutual trust between buyer and supplier is necessary for effective purchasing.

8. *Inventory.* Inventory decisions can be optimized only when customer satisfaction, suppliers, production schedules, and human resource planning are considered.

9. *Scheduling.* Feasible and efficient schedules of production must be developed; the demands on human resources and facilities must be determined and controlled.

10. *Maintenance.* Decisions must be made regarding desired levels of reliability and stability, and systems must be established to maintain that reliability and stability.

Operations managers implement these 10 decisions by identifying key tasks and the staffing needed to achieve them. However, the implementation of decisions is influenced by a variety of issues, including a product's proportion of goods and services (see Table 2.1). Few products are either all goods or all services. While the 10 decisions remain the same for both goods and services, their relative importance and method of implementation depend on this ratio of goods and services. Throughout this text, we discuss how strategy is selected and implemented for both goods and services through these 10 operations management decisions.

TABLE 2.1 ■

The Differences Between Goods and Services Influence How the 10 Operations Management Decisions Are Applied

OPERATIONS DECISIONS	GOODS	SERVICES
Goods and service design	Product is usually tangible.	Product is not tangible. A new range of product attributes—a smile.
Quality	Many objective quality standards.	Many subjective quality standards—nice color.
Process and capacity design	Customer is not involved in most of the process.	Customer may be directly involved in the process—a haircut.
		Capacity must match demand to avoid lost sales—customers often avoid waiting.
Location selection	May need to be near raw materials or labor force.	May need to be near customer—car rental.
Layout design	Layout can enhance production efficiency.	Can enhance product as well as production—layout of a fine-dining restaurant.
Human resources and job design	Workforce focused on technical skills. Labor standards can be consistent. Output-based wage system possible.	Direct workforce usually needs to be able to interact well with customer—bank teller. Labor standards vary depending on customer requirements—legal cases.
Supply-chain management	Supply-chain relationships critical to final product.	Supply-chain relationships important but may not be critical.
Inventory	Raw materials, work-in-process, and finished goods may be inventoried.	Most services cannot be stored, so other ways must be found to accommodate changes in demand—can't store haircuts.
Scheduling	Ability to inventory may allow leveling of production rates.	Often concerned with meeting the customer's immediate schedule with human resources.
Maintenance	Maintenance is often preventive and takes place at the production site.	Maintenance is often "repair" and takes place at the customer's site.

Let's look at an example of strategy development through one of the 10 decisions.

Example 1

Pierre Alexander has just completed chef school and is ready to open his own restaurant. After examining both the external environment and his prospective strengths and weaknesses, he makes a decision on the mission for his restaurant, which he defines as "To provide outstanding French fine dining for the people of Chicago." His supporting operations strategy is to ignore the options of *cost leadership* and *quick response* and focus on *differentiation*. Consequently, his operations strategy requires him to evaluate product designs (menus and meals) and selection of process, layout, and location. He must also evaluate the human resources, suppliers, inventory, scheduling, and maintenance that will support his mission and a differentiation strategy.

Examining just one of these 10 decisions, *process design*, requires that Pierre consider the issues presented in the following figure.

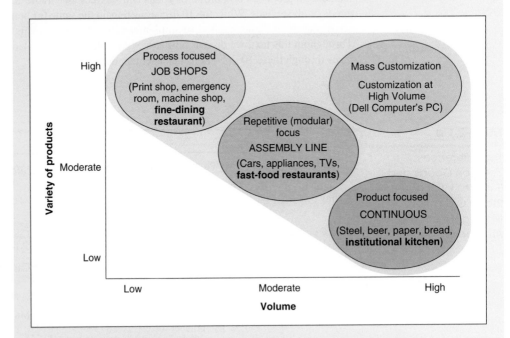

The first option is to operate in the lower right corner of the preceding figure, where he could produce high volumes of food with a limited variety, much as in an institutional kitchen. Such a process could produce large volumes of standard items such as baked goods and mashed potatoes prepared with state-of-the-art automated equipment. Alexander concludes that this is not an acceptable process option.

Alternatively, he can move to the middle of the figure, where he could produce more variety and lower volumes. Here he would have less automation and use prepared modular components for meals, much as a fast-food restaurant does. Again, he deems such process designs inappropriate for his mission.

Another option is to move to the upper right corner and produce a high volume of customized meals, but neither Pierre nor anyone else knows how to do this with gourmet meals.

Finally, Alexander can design a process that operates in the upper left corner of the figure, which requires little automation but lends itself to high variety. This process option suggests that he build an extremely flexible kitchen suitable for a wide variety of custom meals catering to the whims of each customer. With little automation, such a process would be suitable for a huge variety. This process strategy will support his mission and desired product differentiation. Only with a process such as this can he provide the fine French-style gourmet dining that he has in mind.

The 10 decisions of operations management are implemented in ways that provide competitive advantage, not just for fine-dining restaurants, but for all the goods and services that enrich our lives. How this might be done for two drug companies, one seeking a competitive advantage via differentiation, and the other via low cost, is shown in Table 2.2.

TABLE 2.2 ■ Operations Strategies of Two Drug Companies

COMPETITIVE ADVANTAGE	BRAND NAME DRUGS, INC. — PRODUCT DIFFERENTIATION	GENERIC DRUG CORP. — LOW COST
PRODUCT SELECTION AND DESIGN	Heavy R&D investment; extensive labs; focus on development in a broad range of drug categories	Low R&D investment; focus on development of generic drugs
QUALITY	Quality is major priority, standards exceed regulatory requirements	Meets regulatory requirements on a country by country basis as necessary
PROCESS	Product and modular production process; tries to have long product runs in specialized facilities; builds capacity ahead of demand	Process focused; general production processes; "job shop" approach, short-run production; focus on high utilization
LOCATION	Still located in city where it was founded	Recently moved to low-tax, low-labor-cost environment
SCHEDULING	Centralized production planning	Many short-run products complicate scheduling
LAYOUT	Layout supports automated product-focused production	Layout supports process-focused "job shop" practices
HUMAN RESOURCES	Hire the best; nationwide searches	Very experienced top executives provide direction; other personnel paid below industry average
SUPPLY CHAIN	Long-term supplier relationships	Tends to purchase competitively to find bargains
INVENTORY	Maintains high finished goods inventory primarily to ensure all demands are met	Process focus drives up work-in-process inventory; finished goods inventory tends to be low
MAINTENANCE	Highly trained staff; extensive parts inventory	Highly trained staff to meet changing demands

ISSUES IN OPERATIONS STRATEGY

Once a firm has formed a mission, developing and implementing a specific strategy requires that the operations manager consider a number of issues. We will examine these issues in three ways. First, we look at what *research* tells us about effective operations management strategies. Second, we identify some of the *preconditions* to developing effective OM strategy. Third, we look at the *dynamics* of OM strategy development.

Research

PIMS

A program established in cooperation with GE to identify characteristics of high-return-on-investment firms.

Strategic insight has been provided by the findings of the Strategic Planning Institute.[7] Its **PIMS** program (profit impact of market strategy) was established in cooperation with the General Electric Corporation. PIMS has collected nearly 100 data items from about 3,000 cooperating organizations. Using the data collected and high *return on investment* (ROI)[8] as a measure of success, PIMS has been able to identify some characteristics of high-ROI firms. Among those characteristics that impact strategic OM decisions are:

1. High product quality (relative to the competition).
2. High capacity utilization.
3. High operating efficiency (the ratio of expected to actual employee productivity).
4. Low investment intensity (the amount of capital required to produce a dollar of sales).
5. Low direct cost per unit (relative to the competition).

These five findings support a high return on investment and should therefore be considered as an organization develops a strategy. In the analysis of a firm's relative strengths and weaknesses, these characteristics can be measured and evaluated. The specific strategic approaches suggested earlier in Figure 2.3 indicate where an operations manager may want to go, but without achieving the five characteristics of firms with a high return on investment, that journey may not be successful.

Another research study indicates the significant role that OM can play in competitive strategy. When a wide mix of 248 businesses were asked to evaluate the importance of 32 categories in obtaining a sustainable competitive advantage, 28% of the categories selected fell under operations

[7]R. D. Buzzel and B. T. Gale, *The PIMS Principles* (New York: The Free Press, 1987).

[8]Like other performance measures, *return on investment* (ROI) has limitations, including sensitivity to the business cycle, depreciation policies and schedules, book value (goodwill), and transfer pricing.

management. When quality/service is added, the total goes to 44%. The study supports the major role OM strategy plays in developing a competitive advantage.[9]

Preconditions

Before establishing and attempting to implement a strategy, the operations manager needs to understand that the firm is operating in an open system in which a multitude of factors exists. These factors influence strategy development and execution. The more thorough the analysis and understanding of both the external and internal factors, the more the likelihood of success. Although the list of factors to be considered is extensive, at a minimum it entails an understanding of:

"To the Japanese, strategy is so dynamic as to be thought of as 'accommodation' or 'adaptive persistence.'"
Richard Pascale,
Sloan Management Review

1. Strengths and weaknesses of competitors, as well as possible new entrants into the market, substitute products, and commitment of suppliers and distributors.
2. Current and prospective environmental, technological, legal, and economic issues.
3. Product life cycle, which may dictate the limitations of operations strategy.
4. Resources available within the firm and within the OM function.
5. Integration of the OM strategy with the company's strategy and other functional areas.

Dynamics

Strategies change for two reasons. First, strategy is dynamic because of *changes within the organization*. All areas of the firm are subject to change. Changes may occur in a variety of areas, including personnel, finance, technology, and product life. All may make a difference in an organization's strengths and weaknesses and therefore its strategy. Figure 2.5 shows possible change

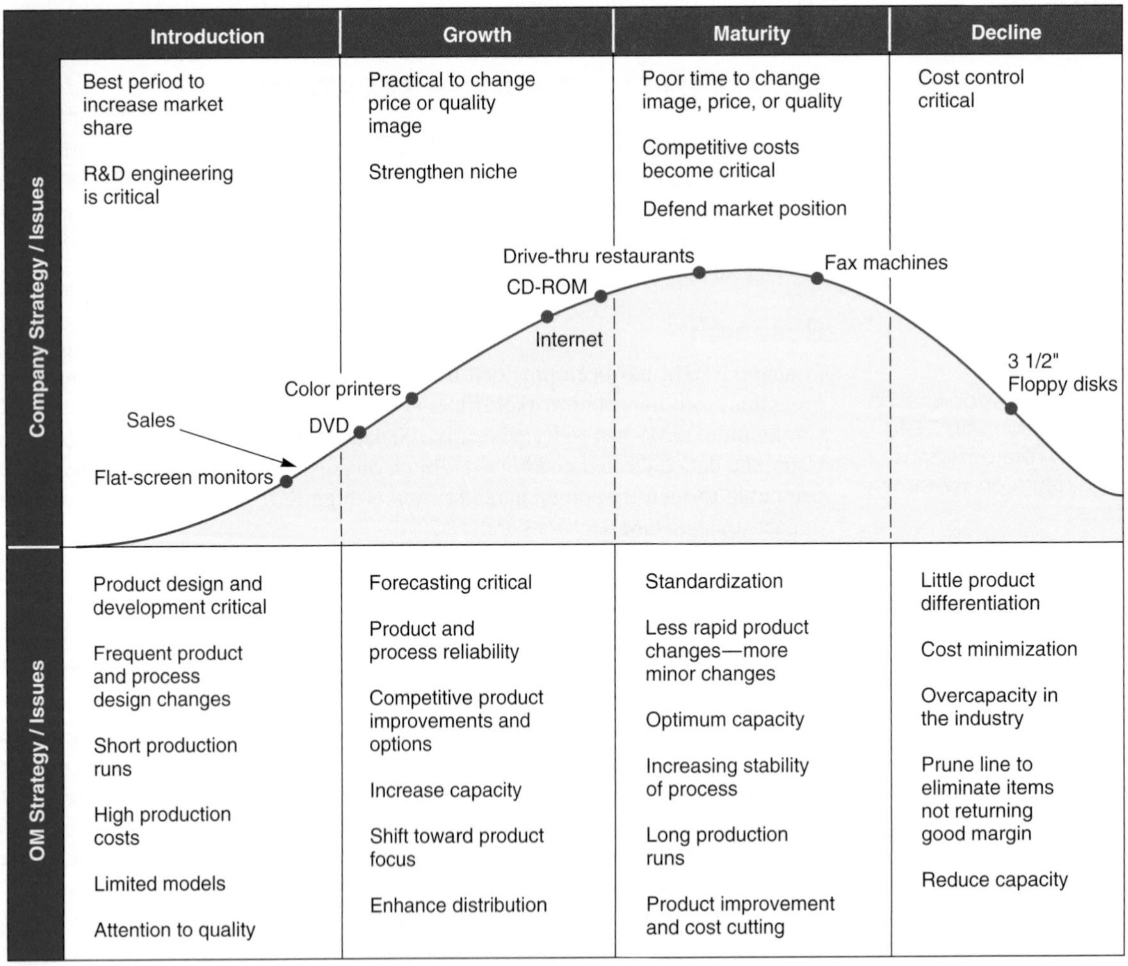

	Introduction	Growth	Maturity	Decline
Company Strategy / Issues	Best period to increase market share R&D engineering is critical	Practical to change price or quality image Strengthen niche	Poor time to change image, price, or quality Competitive costs become critical Defend market position	Cost control critical
OM Strategy / Issues	Product design and development critical Frequent product and process design changes Short production runs High production costs Limited models Attention to quality	Forecasting critical Product and process reliability Competitive product improvements and options Increase capacity Shift toward product focus Enhance distribution	Standardization Less rapid product changes—more minor changes Optimum capacity Increasing stability of process Long production runs Product improvement and cost cutting	Little product differentiation Cost minimization Overcapacity in the industry Prune line to eliminate items not returning good margin Reduce capacity

Curve labels: Sales, Flat-screen monitors, DVD, Color printers, Internet, CD-ROM, Drive-thru restaurants, Fax machines, 3 1/2" Floppy disks

FIGURE 2.5 ■ Strategy and Issues During a Product's Life

[9]See David A. Aaker, "Creating a Sustainable Competitive Advantage," *California Management Review* (winter 1989): 91–106.

in both overall strategy and OM strategy during the product's life. For instance, as a product moves from introduction to growth, product and process design typically move from development to stability. As the product moves to the growth stage, forecasting and capacity planning become issues.

Strategy is also dynamic because of *changes in the environment*. Boeing provides an example, in the opening *Global Company Profile* of this chapter, of how strategy must change as the environment changes. Its strategies, like many OM strategies, are increasingly global. Microsoft also had to adapt quickly to a changing environment. Microsoft's shift in strategy was caused by changing customer demand and the Internet. Microsoft moved from operating systems to office products, to an Internet service provider.

STRATEGY DEVELOPMENT AND IMPLEMENTATION

SWOT analysis
Determining internal strengths and weaknesses and external opportunities and threats.

Once firms understand the issues involved in developing an effective strategy, they evaluate their internal strengths and weaknesses as well as the opportunities and threats of the environment. This is known as **SWOT analysis** (for *S*trength, *W*eakness, *O*pportunities, and *T*hreats). Beginning with SWOT analyses, firms position themselves, through their strategy, to have a competitive advantage. The firm may have excellent design skills or great talent at identifying outstanding locations. However, the firm may recognize limitations of its manufacturing process or in finding good suppliers. The idea is to maximize opportunities and minimize threats in the environment while maximizing the advantages of the organization's strengths and minimizing the weaknesses. Any preconceived ideas about mission are then reevaluated to ensure they are consistent with the SWOT analysis. Subsequently, a strategy for achieving the mission is developed. This strategy is continually evaluated against the value provided customers and competitive realities. The process is shown in Figure 2.6. From this process critical success factors are identified.

Identify Critical Success Factors

Critical success factors
Those activities or factors that are *key* to achieving competitive advantage.

Because no firm does everything exceptionally well, a successful strategy implementation requires identifying those tasks that are critical to success. The operations manager asks, "What tasks must be done particularly well for a given operations strategy to succeed? Which elements contain the highest likelihood of failure, and which will require additional commitment of managerial, monetary, technological, and human resources? Which activities will help the OM function provide a competitive advantage?"

Critical success factors are selected in light of achieving the mission, as well as the organization's internal strengths. **Critical success factors** (CSFs) are those relatively few activities that make a difference between having and not having a competitive advantage. Ultimately the CSFs

FIGURE 2.6 ■

Strategy Development Process

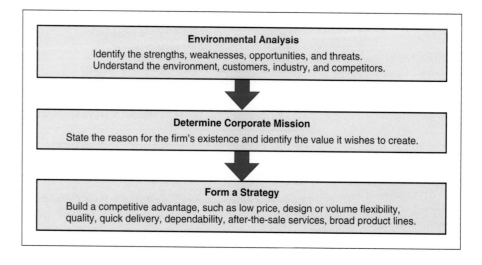

make a difference between an organization's success and failure. Successful organizations identify and use critical success factors to develop a unique and distinct competence that allows them to achieve a competitive advantage.[10]

The critical success factors can overlap functional areas of the firm such as marketing or finance, or they may be within one functional area. In this text, we are, of course, going to focus primarily on the 10 decisions within the operations management function that often are critical success factors. Potential CSFs for marketing, finance, and operations are shown in Figure 2.7.

The 10 operations management decisions developed in this text provide an excellent checklist for determining the critical success factors within the operations function. For instance, the 10 decisions and related CSFs can manifest themselves in a firm's ability to differentiate. That differentiation may be via innovation and new products, where the CSF is product design, as is the case for 3M and Rubbermaid. Similarly, differentiation may be via quality, where the CSF is institutionalizing that quality, as at McDonald's. And differentiation may be via maintenance, where the CFSs are providing reliability and after-sale service, as is the case at IBM.

Activity map

A graphical link of competitive advantage, CSFs, and supporting activities.

Whatever the CSFs, they must be supported by the related activities. One approach to identifying the activities is an **activity map**, which links competitive advantage, CSFs, and supporting activities.[11] For example, Figure 2.8 shows how Southwest Airlines has built a set of integrated activities to support its low-cost competitive advantage. Notice how the critical success factors are supported by other activities. The activities fit together and reinforce each other. And the better they fit and reinforce each other, the more sustainable the competitive advantage. By identifying a competitive advantage and focusing on the critical success factors and the supporting set of activities, Southwest Airlines has become one of the great airline success stories.

FIGURE 2.7 ■

Implement the Strategy by Identifying the Critical Success Factors

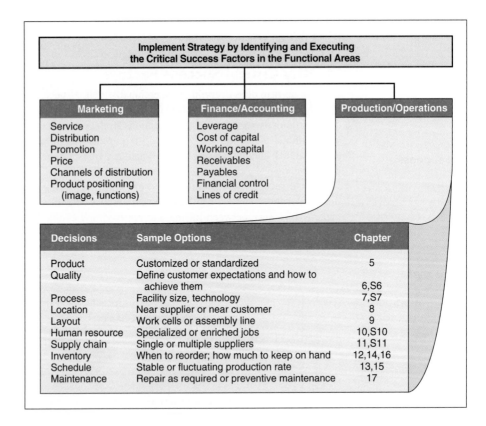

---

[10]For a discussion of distinct competence, see Michael E. Porter, *Competitive Advantage* (New York: The Free Press, 1985).

[11]Michael E. Porter and C. Roland Christensen, "What Is Strategy?" *Harvard Business Review* (November–December 1996): 61–75.

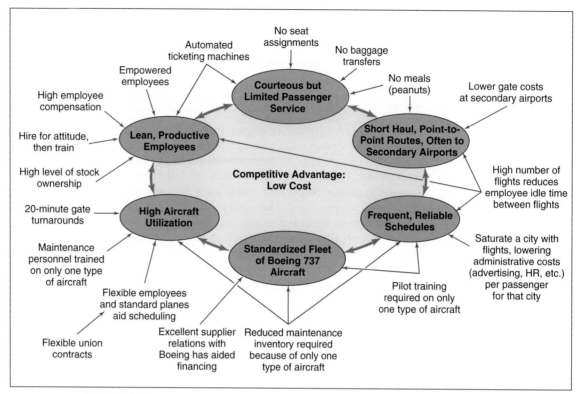

FIGURE 2.8 ■ Activity Mapping of Southwest Airlines' Low-Cost Competitive Advantage

To achieve a low-cost competitive advantage, Southwest has identified a number of critical success factors (connected by red arrows) and support activities (shown by blue arrows). As this figure indicates, a low-cost advantage is highly dependent upon a very well-run operations function.

Build and Staff the Organization

The operations manager's job is a three-step process. Once a strategy and critical success factors have been identified, the second step is to group the necessary activities into an organizational structure. The third step is to staff it with personnel who will get the job done. The manager works with subordinate managers to build plans, budgets, and programs that will successfully implement strategies that achieve missions. Firms tackle this organization of the operations function in a variety of ways. The organization charts shown in Chapter 1 (Figure 1.2) indicate the way some firms have organized to perform the required activities.

Integrate OM With Other Activities

The organization of the operations function and its relationship to other parts of the organization vary with the OM mission. Moreover, the operations function is most likely to be successful when the operations strategy is integrated with other functional areas of the firm, such as marketing, finance, MIS, and human resources. In this way all of the areas support the company's objectives. For example, short-term scheduling in the airline industry is dominated by volatile customer travel patterns. Day-of-week preference, holidays, seasonality, college schedules, and so on, all play a role in changing flight schedules. Consequently, airline scheduling, although an OM activity, can be a part of marketing. Effective scheduling in the trucking industry is reflected in the amount of time trucks travel loaded. However, scheduling of trucks requires information from delivery and pickup points, drivers, and other parts of the organization. When the organization of the OM function results in effective scheduling in the air passenger and commercial trucking industries, a competitive advantage can exist.

"The manufacturing business of tomorrow will not be run by financial executives, marketers, or lawyers inexperienced in manufacturing, as so many U.S. companies are today."

Peter Drucker

The operations manager provides a means of transforming inputs into outputs. The transformations may be in terms of storage, transportation, manufacturing, dissemination of information, and utility of the product or service. *The operations manager's job is to implement an OM strategy, provide competitive advantage, and increase productivity.*

GLOBAL OPERATIONS STRATEGY OPTIONS

As we suggested early in this chapter, many operations strategies now require an international dimension. We tend to call a firm with an international dimension an international business or a multinational corporation. An **international business** is any firm that engages in international trade or investment. This is a very broad category and is the opposite of a domestic, or local, firm.

A **multinational corporation (MNC)** is a firm with *extensive* international business involvement. MNCs buy resources, create goods or services, and sell goods or services in a variety of countries. The term *multinational corporation* applies to most of the world's large, well-known businesses. Certainly IBM is a good example of an MNC. It imports electronics components to the U.S. from over 50 countries, exports computers to over 130 countries, has facilities in 45 countries, and earns more than half of its sales and profits abroad.

Operations managers of international and multinational firms approach global opportunities with one of four operations strategies. They are: *International*, *Multidomestic*, *Global*, and *Transnational* (see Figure 2.9). The matrix of Figure 2.9 has a vertical axis of cost reduction and a horizontal axis of local responsiveness. Local responsiveness implies quick response and/or the differentiation necessary for the local market. The operations manager must know how to position the firm in this matrix. Let us briefly examine each of the four strategies.

International business
A firm that engages in cross-border transactions.

Multinational corporation (MNC)
A firm that has extensive involvement in international business, owning or controlling facilities in more than one country.

FIGURE 2.9 ■

Four International Operations Strategies

Sources: See similar presentations in C. Hill and G. Jones, *Strategic Management,* 5th ed. (New York: Houghton-Mifflin, 2002); and M. Hitt, R. D. Ireland, and R. E. Hoskisson, *Strategic Management, Competitiveness and Globalization,* 5th ed (Cincinnati: Southwestern College Publishing, 2003).

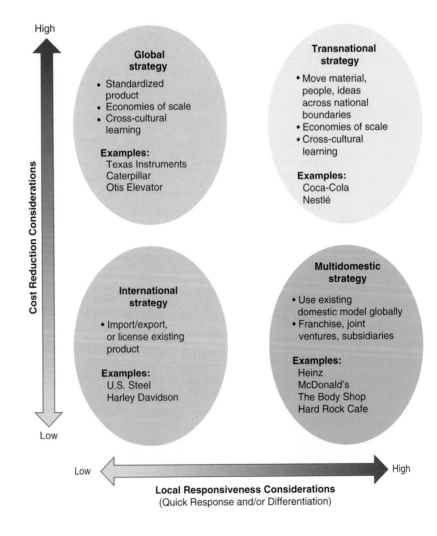

Some international businesses provide the same levels of technology, compensation, safety, and environmental awareness in every country. This IBM plant in Brazil, for example, uses the same protective "bunnysuits" and "cleanrooms" as IBM plants in the U.S. and Japan.

International Strategy

International strategy
Global markets are penetrated using exports and licenses.

An **international strategy** uses exports and licenses to penetrate the global arena. As Figure 2.9 suggests, the international strategy is the least advantageous, with little local responsiveness and little cost advantage. There is little responsiveness because we are exporting or licensing a good from the home country. And the cost advantages may be few because we are using the existing production process at some distance from the new market. However, an international strategy is often the easiest, as exports can require little change in existing operations, and licensing agreements often leave much of the risk to the licensee.

Multidomestic Strategy

Multidomestic strategy
Operating decisions are decentralized to each country to enhance local responsiveness.

With 2,000 restaurants in Japan, and a presence of more than a generation, the average Japanese family thinks Japan invented McDonald's.
                        CEO Jack Greenberg

The **multidomestic strategy** has decentralized authority with substantial autonomy at each business. Organizationally these are typically subsidiaries, franchises, or joint ventures with substantial independence. The advantage of this strategy is maximizing a competitive response for the local market. However, the strategy has little or no cost advantage. Many food producers such as Heinz use a multidomestic strategy to accommodate local tastes because global integration of the production process is not critical. The concept is one of "we were successful in the home market, let's export the management talent and processes, not necessarily the product, to accommodate another market." McDonald's is operating primarily as a multidomestic, which gives it the local responsiveness needed to modify its menu country by country. McDonald's can then serve beer in Germany, wine in France, McHuevo (poached egg hamburger) in Uruguay, and hamburgers without beef in India. Interestingly, McDonald's prefers to call itself *multilocal*.[12]

Global Strategy

Global strategy
Operating decisions are centralized and headquarters coordinates the standardization and learning between facilities.

A **global strategy** has a high degree of centralization, with headquarters coordinating the organization to seek out standardization and learning between plants, thus generating economies of scale. This strategy is appropriate when the strategic focus is cost reduction, but has little to recommend it when the demand for local responsiveness is high. Caterpillar, the world leader in earth-moving equipment,

[12]James L. Watson, ed., *Golden Arches East: McDonald's in East Asia* (Stanford University Press, 1997): 12. *Note:* McDonald's also operates with some of the advantages of a global organization. By using very similar product lines throughout the world, McDonald's obtains some of the standardization advantages of a global strategy. However, it manages to retain the advantages of a multidomestic.

In its continuing fierce worldwide battle with Caterpillar for the global heavy equipment customer, Komatsu is building equipment throughout the world as cost and logistics dictate. This worldwide strategy allows Komatsu to move production as markets and exchange rates change.

and Texas Instruments, a world leader in semiconductors, pursue global strategies. Caterpillar and Texas Instruments find this strategy advantageous because the end products are similar throughout the world. Earth-moving equipment is the same in Nigeria as in Iowa, which allows Caterpillar to have individual factories focus on a limited line of products to be shipped worldwide. This results in economies of scale and learning within each facility. A global strategy also allows Texas Instruments to build optimum-size plants with similar process and to then maximize learning by aggressive communication between plants. The result is an effective cost reduction advantage for Texas Instruments.

Transnational Strategy

Transnational strategy
Combines the benefits of global-scale efficiencies with the benefits of local responsiveness.

A **transnational strategy** exploits the economies of scale and learning, as well as pressure for responsiveness, by recognizing that core competence does not reside in just the "home" country, but can exist anywhere in the organization. **Transnational** describes a condition in which material, people, and ideas cross—or *transgress*—national boundaries. These firms have the potential to pursue all three operations strategies (i.e., differentiation, low cost, and response). Such firms can be thought of as "world companies" whose country identity is not as important as its interdependent network of worldwide operations. Key activities in a transnational company are neither centralized in the parent company nor decentralized so that each subsidiary can carry out its own tasks on a local basis. Instead, the resources and activities are dispersed, but specialized, so as to be both efficient and flexible in an interdependent network.[13] Nestlé is a good example of such a company. Although it is legally Swiss, 95% of its assets are held and 98% of its sales are made outside of Switzerland. Less than 10% of its workers are Swiss. Similarly, service firms such as Asea Brown Boveri (an engineering firm that is Swedish but headquartered in Switzerland), Reuters (a news agency), Bertelsmann (a publisher), and Citicorp (a banking corporation) can be viewed as transnationals. As the national identities of these transnationals fade, Alvin Toffler has suggested that they may become stateless.[14]

SUMMARY

Global operations provide an increase in both the challenges and opportunities for operations managers. Although it is a challenging task, operations managers can improve productivity in a competitive, dynamic global economy. They can build and manage OM functions that contribute in a significant way to competitiveness. Organizations identify their strengths and weaknesses. They then develop effective missions and strategies that account for these strengths and weaknesses and complement the opportunities and threats in the environment. If this procedure is performed well, the organization can have competitive advantage through some combination of product differentiation, low cost, and response. This competitive advantage is often achieved via a move to international, multidomestic, global, or transnational strategies.

[13]Christopher Bartlett and Sumantra Ghoshal, *Transnational Management* (Homewood, IL: Richard B. Irwin, 1992): 14.

[14]Alvin Toffler quoted in "Recipe for Intelligence," *Information Today* (March 1994): 61–63.

Effective use of resources, whether domestic or international, is the responsibility of the professional manager, and professional managers are among the few in our society who *can* achieve this performance. The challenge is great, and the rewards to the manager and to society substantial.

KEY TERMS

Maquiladoras *(p. 24)*
World Trade Organization (WTO) *(p. 25)*
North American Free Trade Agreement (NAFTA) *(p. 25)*
European Union (EU) *(p. 25)*
Mission *(p. 27)*
Strategy *(p. 28)*
Competitive advantage *(p. 29)*
Differentiation *(p. 30)*
Experience differentiation *(p. 30)*
Low-cost leadership *(p. 31)*
Response *(p. 31)*

Operations decisions *(p. 32)*
PIMS *(p. 35)*
SWOT analysis *(p. 37)*
Critical success factors *(p. 37)*
Activity map *(p. 38)*
International business *(p. 40)*
Multinational corporation (MNC) *(p. 40)*
International strategy *(p. 41)*
Multidomestic strategy *(p. 41)*
Global Strategy *(p. 41)*
Transnational strategy *(p. 42)*

SOLVED PROBLEM

Strategy at Pirelli SpA

The global tire industry continues to consolidate. Michelin buys Goodrich and Uniroyal and builds plants throughout the world. Bridgestone buys Firestone, expands its research budget, and focuses on world markets. Goodyear spends almost 4% of its sales revenue on research. These three aggressive firms have come to dominate the world tire market with a 15% to 20% market share each. Against this formidable array the old-line Italian tire company Pirelli SpA responded, but with two mistakes: the purchase of Armstrong Tire and a disastrous bid to take over the German tire maker Continental AG. Pirelli still had only 5% of the market and by 1991 was losing $500 million a year while the competition was getting stronger. Tires are a tough, competitive business that rewards companies with strong market shares and long production runs.

Use a SWOT analysis to establish a feasible strategy for Pirelli.

SOLUTION

1. Find an opportunity in the world market that avoids the mass-market onslaught by the big three tire makers.

2. Maximize the internal strength represented by Pirelli tires winning World Rally Championships in 1995 and 1996 and having one of the world's strongest brand names.

Pirelli established exclusive deals with Jaguar's XJ-8 and Lotus Elise and takes a large share of tire sales on Porsches, S-Class Mercedes, BMWs, and Saabs. People are willing to pay a premium for Pirellis. Pirelli also switched out of low-margin standard tires and into higher-margin performance tires. The operations function responded by focusing its design efforts on performance tires and developing a system of modular tire manufacture that allows much faster switching between models. This modular system, combined with investments in new manufacturing flexibility, has driven batch sizes down to as small as 150 to 200, making small-lot performance tires economically feasible. A threat from the big three going after the performance market remains, but Pirelli has bypassed its weakness of having a small market share. And the firm has returned to profitability.

Sources: Forbes (May 19, 1997): 106–113; and the *Wall Street Journal* (August 1, 1997): B3.

INTERNET AND STUDENT CD-ROM EXERCISES

Visit our home page or use your student CD-ROM to help with material in this chapter.

 On Our Home Page, www.prenhall.com/heizer

- Self-Tests
- Practice Problems
- Internet Exercises
- Current Articles and Research
- Virtual Company Tour
- Internet Cases

 On Your Student CD-ROM

- Power Point Lecture
- Practice Problems
- Video Clips and Video Cases

ADDITIONAL CASE STUDIES

Internet Case Studies: Visit our Web site at www.prenhall.com/heizer **for these free case studies:**

- **Johannsen Steel Company:** Discusses a specialty steel company and its difficulty in making strategy adjustments.

- **International Operations at General Motors:** Deals with GM's global expansion strategic plans.

- **Motorola's Global Strategy:** Focuses on Motorola's international strategy.

Harvard has selected these Harvard Business School cases to accompany this chapter (textbookcasematch.hbsp.harvard.edu**):**

- **Fresh Connections** (#600-022): Investigates how to structure operations to take advantage of the continued growth in the home meal replacement market.

- **Komatsu Ltd.** (#398-016): Describes strategic and organizational transformations at Komatsu, a major Japan-based producer of construction equipment.

- **Toys "Я" Us Japan** (#796-077): Documents Toys "Я" Us difficulties as it enters the Japanese toy market.

- **Lenzing AG: Expanding in Indonesia** (#796-099): Presents the issues surrounding expansion in a foreign country.

Project Management

Chapter Outline

LEARNING OBJECTIVES

When you complete this chapter you should be able to

IDENTIFY OR DEFINE:

Work breakdown structure

Critical path

AOA and AON networks

Forward and backward passes

Variability in activity times

DESCRIBE OR EXPLAIN:

The role of the project manager

Program evaluation and review technique (PERT)

Critical path method (CPM)

Crashing a project

The use of MS Project

GLOBAL COMPANY PROFILE:

Project Management Provides a Competitive Advantage for Bechtel

Now in its 105th year, the San Francisco-based Bechtel Group is the world's premier manager of massive construction and engineering projects. Known for billion-dollar projects, Bechtel is famous for its construction feats on the Hoover Dam and the Boston Central Artery/Tunnel project, and more recently the rebuilding of Kuwait's oil and gas infrastructure after the invasion by Iraq's Saddam Hussein.

Even for Bechtel, whose competitive advantage is project management, restoring the 650 blazing oil wells lit by Iraqi sabotage in 1990 was a logistical nightmare. The panorama of destruction in Kuwait was breathtaking, with

Workers wrestle with a 1,500-ton boring machine, measuring 25 feet in diameter, that was used to dig the Eurotunnel in the early 1990's. With overruns that boosted the cost of the project to $13 billion, a Bechtel Group VP was brought in to head operations.

In Kuwait, Bechtel's fire-fighting crews relied on explosives and heavy machinery to put out well fires started by retreating Iraqi troops. More than 200 lagoons were built and filled with seawater so pumps could also hose down the flames.

fire roaring out of the ground from virtually every compass point. Kuwait had no water, electricity, food, or facilities. The country was also littered with unexploded mines, bombs, grenades, and shells, while lakes of oil covered its roads.

With a major global procurement program, Bechtel specialists tapped the company's network of suppliers and buyers worldwide. At the port of Dubai, 550 miles southeast of Kuwait, the firm established a central transshipment point, deploying 520,000 tons of equipment and supplies. Creating a workforce of 16,000, Bechtel mobilized 742 airplanes and ships and more than 5,800 bulldozers, ambulances, and other pieces of operating equipment from 40 countries on five continents.

BECHTEL GROUP

Managing massive construction projects such as this is the strength of Bechtel. With large penalties for late completion and incentive for early completion, a good project manager is worth his or her weight in gold.

Now, over a decade later, the fires are long out and Kuwait is again shipping oil. Bechtel's more recent projects include:

- Building 26 massive distribution centers, in just 2 years, for the Internet company Webvan Group.
- Constructing 30 high-security data centers worldwide for Equinix, Inc.
- Building and running a rail line between London and the Channel Tunnel ($4.6 billion).

- Developing an oil pipeline from the Caspian Sea region to Russia ($850 million).
- Expanding the Dubai Airport in the United Arab Emirates ($600 million) and the Miami International Airport ($2 billion).
- Building liquefied natural gas plants in Trinidad, West Indies ($1 billion).
- Building a new subway for Athens, Greece ($2.6 billion).
- Constructing a natural gas pipeline in Thailand ($700 million).

- Building a highway to link the north and south of Croatia ($303 million).

When companies or countries seek out firms to manage these massive projects, they go to Bechtel, which, again and again, through outstanding project management, has demonstrated its competitive advantage.

Source: Courtesy of Bechtel.

47

THE IMPORTANCE OF PROJECT MANAGEMENT

- When the Bechtel project management team entered Kuwait, it quickly had to mobilize an international force of nearly 8,000 manual workers, 1,000 construction professionals, 100 medical personnel, and 2 helicopter evacuation teams. It also had to set up 6 full-service dining halls to provide 27,000 meals a day and build a 40-bed field hospital.
- When Microsoft Corporation set out to develop Windows XP—its biggest, most complex, and most important program to date—time was the critical thing for the project manager. With hundreds of programmers working on millions of lines of code in a program costing hundreds of millions to develop, immense stakes rode on the project being delivered on time.
- When Hard Rock Cafe sponsors Rockfest, hosting 100,000 plus fans at its annual concert, the project manager begins his planning some 9 months earlier. Using the software package MS Project, described in this chapter, each of the hundreds of details can be monitored and controlled. When a band can't reach the Rockfest site by bus because of massive traffic jams, Hard Rock's project manager is ready with a helicopter backup.

Video 3.1

Project Management at
Hard Rock's Rockfest

Bechtel, Microsoft, and Hard Rock are just three examples of firms that face a modern phenomenon: growing project complexity and collapsing product/service life cycles. This change stems from awareness of the strategic value of time-based competition and a quality mandate for continuous improvement. Each new product/service introduction is a unique event—a project. In addition, projects are a common part of our everyday life. We may be planning a wedding or a surprise birthday party, remodeling a house, or preparing a semester-long class project.

Scheduling projects is a difficult challenge to operations managers. The stakes in project management are high. Cost overruns and unnecessary delays occur due to poor scheduling and poor controls.

Projects that take months or years to complete are usually developed outside the normal production system. Project organizations within the firm may be set up to handle such jobs and are often disbanded when the project is complete. On other occasions, managers find projects just a part of their job. The management of projects involves three phases (see Figure 3.1):

1. *Planning.* This phase includes goal setting, defining the project, and team organization.
2. *Scheduling.* This phase relates people, money, and supplies to specific activities and relates activities to each other.
3. *Controlling.* Here the firm monitors resources, costs, quality, and budgets. It also revises or changes plans and shifts resources to meet time and cost demands.

We will begin this chapter with a brief overview of these functions. Three popular techniques to allow managers to plan, schedule, and control—Gantt charts, PERT, and CPM—are also described.

PROJECT PLANNING

Project organization
An organization formed to ensure that programs (projects) receive the proper management and attention.

Projects can be defined as a series of related tasks directed toward a major output. In some firms a **project organization** is developed to make sure existing programs continue to run smoothly on a day-to-day basis while new projects are successfully completed.

For companies with multiple large projects, such as a construction firm, a project organization is an effective way of assigning the people and physical resources needed. It is a temporary organization structure designed to achieve results by using specialists from throughout the firm. NASA and many other organizations use the project approach. You may recall Project Gemini and Project Apollo. These terms were used to describe teams that NASA organized to reach space exploration objectives.

The project organization works best when:

1. Work can be defined with a specific goal and deadline.
2. The job is unique or somewhat unfamiliar to the existing organization.
3. The work contains complex interrelated tasks requiring specialized skills.
4. The project is temporary but critical to the organization.
5. The project cuts across organizational lines.

Project Planning,
Scheduling, and
Controlling

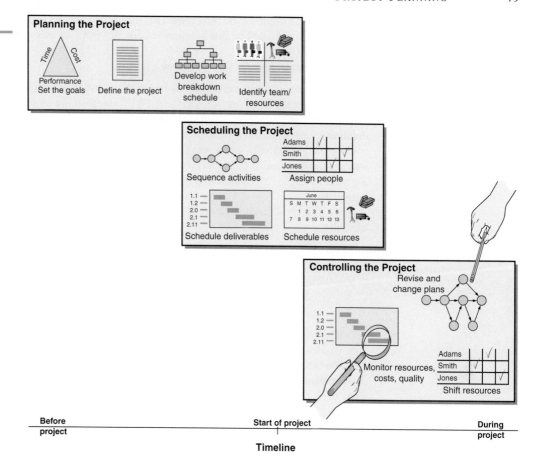

Planning the Project

Time Cost
Performance
Set the goals   Define the project   Develop work
breakdown
schedule   Identify team/
resources

Scheduling the Project

Sequence activities   Assign people

Adams √
Smith   √
Jones

Schedule deliverables   Schedule resources

June
S M T W T F S
      1 2 3 4 5 6
7 8 9 10 11 12 13

1.1
1.2
2.0
2.1
2.11

Controlling the Project

Revise and
change plans

1.1
1.2
2.0
2.1
2.11

Monitor resources,
costs, quality

Adams   √
Smith √
Jones   √

Shift resources

Before
project   Start of project   During
project

Timeline

The Project Manager

When a project
organization is made
permanent it is usually
called a "matrix
organization."

An example of a project organization is shown in Figure 3.2. Project team members are temporarily assigned to a project and report to the project manager. The manager heading the project coordinates activities with other departments and reports directly to top management. Project managers receive high visibility in a firm and are responsible for making sure that (1) all necessary activities are finished in proper sequence and on time; (2) the project comes in within budget; (3) the project meets its quality goals; and (4) the people assigned to the project receive the motivation, direction,

A Sample Project
Organization

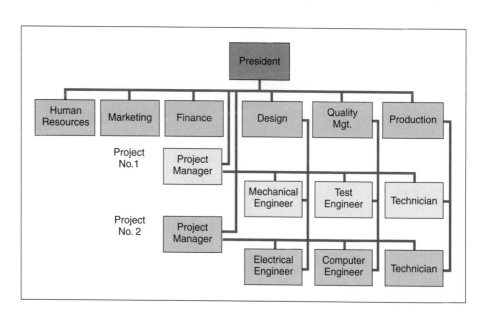

President

Human Resources   Marketing   Finance   Design   Quality Mgt.   Production

Project No.1   Project Manager

Mechanical Engineer   Test Engineer   Technician

Project No. 2   Project Manager

Electrical Engineer   Computer Engineer   Technician

and information needed to do their jobs. This means that project managers should be good coaches and communicators, and be able to organize activities from a variety of disciplines.

Work Breakdown Structure

Work breakdown structure (WBS)
Dividing a project into more and more detailed components.

The project management team begins its task well in advance of project execution so that a plan can be developed. One of its first steps is to carefully establish the project's objectives, then break the project down into manageable parts. This **work breakdown structure (WBS)** defines the project by dividing it into its major subcomponents (or tasks), which are then subdivided into more detailed components, and finally into a set of activities and their related costs. The division of the project into smaller and smaller tasks can be difficult, but is critical to managing the project and to scheduling success. Gross requirements for people, supplies, and equipment are also estimated in this planning phase.

The work breakdown structure typically decreases in size from top to bottom and is indented like this:

```
Level
1    Project
2       Major tasks in the project
3          Subtasks in major tasks
4             Activities (or "work packages") to be completed
```

This hierarchical framework can be illustrated with the development of Microsoft's operating system, Windows XP. As we see in Figure 3.3, the project, creating a new operating system, is labeled 1.0. The first step is to identify the major tasks in the project (level 2). Two examples would be development of graphic user interfaces or GUIs (1.1), and creating compatibility with previous versions of Windows (1.2). The major subtasks for 1.2 are creating a team to handle compatibility with Windows 98 (1.21), a compatibility team for Windows NT (1.22), and compatibility with Windows 2000 (1.23). Then each major subtask is broken down into level-4 activities that need to be done, such as "importing files" created in Windows 2000 (1.231). There are usually many level-4 activities.

PROJECT SCHEDULING

Project scheduling involves sequencing and allotting time to all project activities. At this stage, managers decide how long each activity will take and compute how many people and materials will be needed at each stage of production. Managers also chart separate schedules for personnel needs by type of skill (management, engineering, or pouring concrete, for example). Charts also can be developed for scheduling materials.

Gantt charts
Planning charts used to schedule resources and allocate time.

One popular project scheduling approach is the Gantt chart. **Gantt charts** are low-cost means of helping managers make sure that (1) all activities are planned for, (2) their order of performance is accounted for, (3) the activity time estimates are recorded, and (4) the overall project time is developed. As Figure 3.4 shows, Gantt charts are easy to understand. Horizontal bars are drawn for each project activity along a time line. This illustration of a routine servicing of a Delta jetliner during a 60-minute layover shows that Gantt charts also can be used for scheduling repetitive operations. In this case, the chart helps point out potential delays. The *OM in Action* box on Delta provides additional insights. (A second illustration of a Gantt chart is also provided in Chapter 15, Figure 15.4.)

FIGURE 3.3 ■

Work Breakdown Structure

Level	Level ID Number	Activity
1	1.0	Develop/launch Windows XP Operating System
2	1.1	Development of GUIs
2	1.2	Ensure compatibility with earlier Windows versions
3	1.21	Compatibility with Windows 98
3	1.22	Compatibility with Windows NT
3	1.23	Compatibility with Windows 2000
4	1.231	Ability to import files

OM IN ACTION

Delta's Ground Crew Orchestrates a Smooth Takeoff

Flight 199's three engines screech its arrival as the wide-bodied jet lumbers down Orlando's taxiway with 200 passengers arriving from San Juan. In an hour, the plane is to be airborne again.

However, before this jet can depart, there is business to attend to: hundreds of passengers plus tons of luggage and cargo to unload and load; hundreds of meals, thousands of gallons of jet fuel, countless soft drinks and bottles of liquor to restock; cabin and restrooms to clean; toilet holding tanks to drain; and engines, wings, and landing gear to inspect.

The 12-person ground crew knows that a miscue anywhere—a broken cargo loader, lost baggage, misdirected passengers—can mean a late departure and trigger a chain reaction of headaches from Orlando to Dallas to every destination of a connecting flight.

Dennis Dettro, the operations manager for Delta's Orlando International Airport, likes to call the turnaround operation "a well-orchestrated symphony." Like a pit crew awaiting a race car, trained crews are in place for Flight 199 with baggage carts and tractors, hydraulic cargo loaders, a truck to load food and drinks, another to lift the cleanup crew, another to put fuel on, and a fourth to take water off. The "orchestra" usually performs so smoothly that most passengers never suspect the proportions of the effort. Gantt charts, such as the one in Figure 3.4, aid Delta and other airlines with the staffing and scheduling that are necessary for this symphony to perform.

Sources: New York Times (January 21, 1997): C1, C20; and *USA Today* (March 17, 1998): 10E.

Gantt charts are an example of a widely used, nonmathematical technique that is very popular with managers because it is simple and visual.

On simple projects, scheduling charts such as these can be used alone. They permit managers to observe the progress of each activity and to spot and tackle problem areas. Gantt charts, though, do not adequately illustrate the interrelationships between the activities and the resources.

PERT and CPM, the two widely used network techniques that we shall discuss shortly, *do* have the ability to consider precedence relationships and interdependency of activities. On complex projects, the scheduling of which is almost always computerized, PERT and CPM thus have an edge over the simpler Gantt charts. Even on huge projects, though, Gantt charts can be used as summaries of project status and may complement the other network approaches.

To summarize, whatever the approach taken by a project manager, project scheduling serves several purposes:

1. It shows the relationship of each activity to others and to the whole project.
2. It identifies the precedence relationships among activities.
3. It encourages the setting of realistic time and cost estimates for each activity.
4. It helps make better use of people, money, and material resources by identifying critical bottlenecks in the project.

FIGURE 3.4 ■

Gantt Chart of Service Activities for a Delta Jet during a 60-Minute Layover

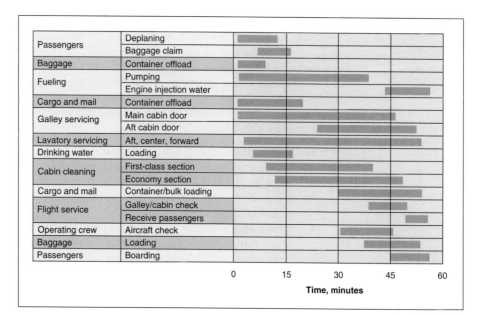

PROJECT CONTROLLING

The control of large projects, like the control of any management system, involves close monitoring of resources, costs, quality, and budgets. Control also means using a feedback loop to revise the project plan and having the ability to shift resources to where they are needed most. Computerized PERT/CPM reports and charts are widely available today on personal computers. Some of the more popular of these programs are Primavera (by Primavera Systems, Inc.), MacProject (by Apple Computer Corp.), Pertmaster (by Westminster Software, Inc.), VisiSchedule (by Paladin Software Corp.), Time Line (by Symantec Corp.), and MS Project (by Microsoft Corp.), which we illustrate in this chapter.

These programs produce a broad variety of reports, including (1) detailed cost breakdowns for each task, (2) total program labor curves, (3) cost distribution tables, (4) functional cost and hour summaries, (5) raw material and expenditure forecasts, (6) variance reports, (7) time analysis reports, and (8) work status reports.

PROJECT MANAGEMENT TECHNIQUES: PERT AND CPM

Program evaluation and review technique (PERT)
A project management technique that employs three time estimates for each activity.

Critical path method (CPM)
A project management technique that uses only one time factor per activity.

Critical path
The computed longest time path(s) through a network.

Program evaluation and review technique (PERT) and the **critical path method (CPM)** were both developed in the 1950s to help managers schedule, monitor, and control large and complex projects. CPM arrived first, in 1957, as a tool developed by J. E. Kelly of Remington Rand and M. R. Walker of duPont to assist in the building and maintenance of chemical plants at duPont. Independently, PERT was developed in 1958 by Booz, Allen, and Hamilton for the U.S. Navy.

The Framework of PERT and CPM

PERT and CPM both follow six basic steps:

1. Define the project and prepare the work breakdown structure.
2. Develop the relationships among the activities. Decide which activities must precede and which must follow others.
3. Draw the network connecting all of the activities.
4. Assign time and/or cost estimates to each activity.
5. Compute the longest time path through the network. This is called the **critical path**.
6. Use the network to help plan, schedule, monitor, and control the project.

Step 5, finding the critical path, is a major part of controlling a project. The activities on the critical path represent tasks that will delay the entire project unless they are completed on time. Managers can gain the flexibility needed to complete critical tasks by identifying noncritical activities and replanning, rescheduling, and reallocating labor and financial resources.

Although PERT and CPM differ to some extent in terminology and in the construction of the network, their objectives are the same. Furthermore, the analysis used in both techniques is very similar. The major difference is that PERT employs three time estimates for each activity. These time estimates are used to compute expected values and standard deviations for the activity. CPM makes the assumption that activity times are known with certainty, and hence requires only one time factor for each activity.

For purposes of illustration, the rest of this section concentrates on a discussion of PERT. Most of the comments and procedures described, however, apply just as well to CPM.

PERT and CPM are important because they can help answer questions such as the following about projects with thousands of activities:

1. When will the entire project be completed?
2. What are the critical activities or tasks in the project—that is, which activities will delay the entire project if they are late?
3. Which are the noncritical activities—the ones that can run late without delaying the whole project's completion?
4. What is the probability that the project will be completed by a specific date?
5. At any particular date, is the project on schedule, behind schedule, or ahead of schedule?
6. On any given date, is the money spent equal to, less than, or greater than the budgeted amount?
7. Are there enough resources available to finish the project on time?
8. If the project is to be finished in a shorter amount of time, what is the best way to accomplish this goal at the least cost?

The Navy, under the direction of Admiral Rickover, successfully used PERT to build the first Polaris submarine ahead of schedule.

Network Diagrams and Approaches

Activity-on-Node (AON)
A network diagram in which nodes designate activities.

Activity-on-Arrow (AOA)
A network diagram in which arrows designate activities.

The first step in a PERT or CPM network is to divide the entire project into significant activities in accordance with the work breakdown structure. There are two approaches for drawing a project network: **activity on node (AON)** and **activity on arrow (AOA)**. Under the AON convention, *nodes* designate activities. Under AOA, *arrows* represent activities. Activities consume time and resources. The basic difference between AON and AOA is that the nodes in an AON diagram represent activities. In an AOA network, the nodes represent the starting and finishing times of an activity and are also called events. So nodes in AOA consume neither time nor resources.

Figure 3.5 illustrates both conventions for a small portion of the airline turnaround Gantt chart (in Figure 3.4). The examples provide some background for understanding six common activity relationships in networks. In Figure 3.5(a), activity A must be finished before activity B is started, and B must, in turn, be completed before C begins. Activity A might represent "deplaning passengers," while B is "cabin cleaning," and C is "boarding new passengers."

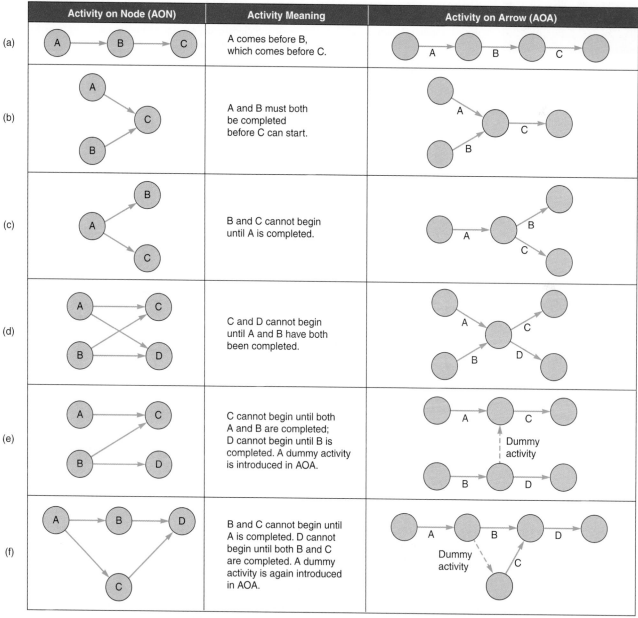

Figure 3.5 ■ A Comparison of AON and AOA Network Conventions

Dummy activities
An activity having no time, inserted into the network to maintain the logic of the network.

Figures 3.5(e) and 3.5(f) illustrate that the AOA approach sometimes needs the addition of a **dummy activity** to clarify relationships. A dummy activity consumes no time or resources, but is required when a network has two activities with identical starting and ending events, or when two or more follow some, but not all, "preceding" activities. The use of dummy activities is also important when computer software is employed to determine project completion time. A dummy activity has a completion time of zero.

Although both AON and AOA are popular in practice, many of the project management software packages, including Microsoft Project 2002, use AON networks. For this reason, although we illustrate both types of networks in the next example, we focus on AON networks in subsequent discussions in this chapter.

Activity-on-Node Example

Example 1

**Excel OM Data File
Ch03Ex1.xla**

Given the following information, develop a table showing activity precedence relationships.

Milwaukee General Hospital, located in downtown Milwaukee, has long been trying to avoid the expense of installing air pollution control equipment in its extensive laundry/cleaning operations facility. The environmental protection agency has recently given the hospital 16 weeks to install a complex air filter system. Milwaukee General has been warned that it may be forced to close the laundry facility unless the device is installed in the allotted period. Dr. Joni Steinberg, the hospital administrator, wants to make sure that installation of the filtering system progresses smoothly and on time.

Milwaukee General has identified the eight activities that need to be performed in order for the project to be completed. When the project begins, two activities can be simultaneously started: building the internal components for the device (activity A) and the modifications necessary for the floor and roof (activity B). The construction of the collection stack (activity C) can begin when the internal components are completed. Pouring the concrete floor and installation of the frame (activity D) can be started as soon as the internal components are completed and the roof and floor have been modified.

After the collection stack has been constructed, two activities can begin: building the high-temperature burner (activity E) and installing the pollution control system (activity F). The air pollution device can be installed (activity G) after the concrete floor has been poured, the frame has been installed, and the high-temperature burner has been built. Finally, after the control system and pollution device have been installed, the system can be inspected and tested (activity H).

Activities and precedence relationships may seem rather confusing when they are presented in this descriptive form. It is therefore convenient to list all the activity information in a table, as shown in Table 3.1. We see in the table that activity A is listed as an *immediate predecessor* of activity C. Likewise, both activities D and E must be performed prior to starting activity G.

TABLE 3.1 ■ Milwaukee General Hospital's Activities and Predecessors

ACTIVITY	DESCRIPTION	IMMEDIATE PREDECESSORS
A	Build internal components	—
B	Modify roof and floor	—
C	Construct collection stack	A
D	Pour concrete and install frame	A, B
E	Build high-temperature burner	C
F	Install pollution control system	C
G	Install air pollution device	D, E
H	Inspect and test	F, G

It is enough to list only the immediate predecessors for each activity.

Note that in Example 1 it is enough to list just the *immediate predecessors* for each activity. For instance, in Table 3.1, since activity A precedes activity C and activity C precedes activity E, the fact that activity A precedes activity E is *implicit*. This relationship need not be explicitly shown in the activity precedence relationships.

Networks consist of nodes that are connected by lines (or arcs).

When there are many activities in a project with fairly complicated precedence relationships, it is difficult for an individual to comprehend the complexity of the project from just the tabular information. In such cases, a visual representation of the project, using a *project network*, is convenient and useful. A project network is a diagram of all the activities and the precedence relationships that exist between these activities in a project. We now illustrate how to construct a project network for Milwaukee General Hospital.

Example 2

Draw the AON network for Milwaukee General Hospital, using the data in Example 1.

Recall that in the AON approach, we denote each activity by a node. The lines, or arcs, represent the precedence relationships between the activities.

In this example, there are two activities (A and B) that do not have any predecessors. We draw separate nodes for each of these activities, as shown in Figure 3.6. Although not required, it is usually convenient to have a unique starting activity for a project. We have therefore included a *dummy activity* called Start in Figure 3.6. This dummy activity does not really exist and takes up zero time and resources. Activity Start is an immediate predecessor for both activities A and B, and serves as the unique starting activity for the entire project.

FIGURE 3.6 ■

Beginning AON Network for Milwaukee General Hospital

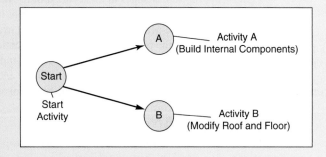

We now show the precedence relationships using lines with arrow symbols. For example, an arrow from activity Start to activity A indicates that Start is a predecessor for activity A. In a similar fashion, we draw an arrow from Start to B.

Next, we add a new node for activity C. Since activity A precedes activity C, we draw an arc from node A to node C (see Figure 3.7). Likewise, we first draw a node to represent activity D. Then, since activities A and B both precede activity D, we draw arrows from A to D, and B to D (see Figure 3.7).

FIGURE 3.7 ■

Intermediate AON Network for Milwaukee General Hospital

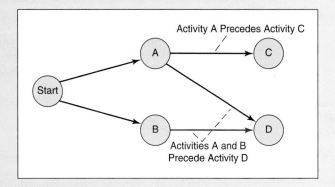

We proceed in this fashion, adding a separate node for each activity and a separate line for each precedence relationship that exists. The complete AON project network for the Milwaukee General Hospital project is shown in Figure 3.8.

FIGURE 3.8 ■

Complete AON Network for Milwaukee General Hospital

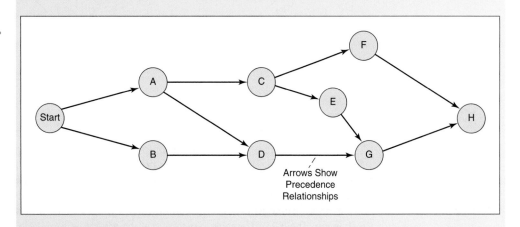

Drawing a project network properly takes some time and experience. When we first draw a project network, it is not unusual that we place our nodes (activities) in the network in such a fashion that the arrows (precedence relationships) are not straight lines. That is, the lines could be intersecting each other, and even facing in opposite directions. For example, if we had switched the location of the nodes for activities E and F in Figure 3.8, the lines from F to H and E to G would have intersected. Although such a project network is perfectly valid, it is good practice to have a well-drawn network. One rule that we especially recommend is to place the nodes in such a fashion that all arrows point in the same direction. To achieve this, we suggest that you first get a rough draft version of the network, making sure all the relationships are shown. Then you can redraw the network to make appropriate changes in the location of the nodes.

As with the unique starting node, it is convenient to have the project network finish with a unique ending node. In the Milwaukee General Hospital example, it turns out that a unique activity, H, is the last activity in the project. We therefore automatically have a unique ending node.

> It is convenient, but not required, to have unique starting and ending activities in a project.

In situations in which a project has multiple ending activities, we include a "dummy" ending activity. This dummy activity has all the multiple ending activities in the project as immediate predecessors. We illustrate this type of situation in Solved Problem 3.5 at the end of this chapter.

Activity-on-Arrow Example

We saw earlier, that in an AOA project network, we can represent activities by arrows. A node represents an *event*, which marks the start or completion time of an activity. We usually identify an event (node) by a number.

Example 3

Draw the complete AOA project network for Milwaukee General Hospital's problem.

Using the data from the table in Example 1, we see that activity A starts at event 1 and ends at event 2. Likewise, activity B starts at event 1 and ends at event 3. Activity C, whose only immediate predecessor is activity A, starts at node 2 and ends at node 4. Activity D, however, has two predecessors (i.e., A and B). Hence, we need both activities A and B to end at event 3, so that activity D can start at that event. However, we cannot have multiple activities with common starting and ending nodes in an AOA network. To overcome this difficulty, in such cases, we may need to add a dummy line (activity) to enforce the precedence relationship. The dummy activity, shown in Figure 3.9 as a dashed line, is inserted between events 2 and 3 to make the diagram reflect the precedence between A and D. Recall that the dummy activity does not really exist in the project and takes up zero time. The remainder of the AOA project network for Milwaukee General Hospital's example is also shown.

FIGURE 3.9 ■

Complete AOA Network (with Dummy Activity) for Milwaukee General Hospital

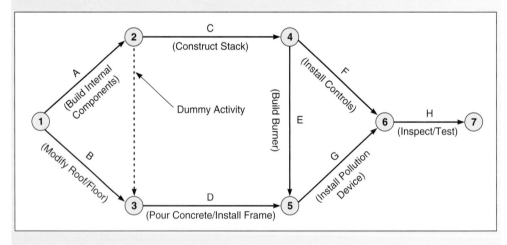

DETERMINING THE PROJECT SCHEDULE

Look back to Figure 3.8 (in Example 2) for a moment to see Milwaukee General Hospital's completed AON project network. Once this project network has been drawn to show all the activities and their precedence relationships, the next step is to determine the project schedule. That is, we need to identify the planned starting and ending time for each activity.

Let us assume Milwaukee General estimates the time required for each activity, in weeks, as shown in Table 3.2. The table indicates that the total time for all eight of the hospital's activities is

TABLE 3.2 ■

TABLE 3.2 ■

Time Estimates for
Milwaukee General
Hospital

ACTIVITY	DESCRIPTION	TIME (WEEKS)
A	Build internal components	2
B	Modify roof and floor	3
C	Construct collection stack	2
D	Pour concrete and install frame	4
E	Build high-temperature burner	4
F	Install pollution control system	3
G	Install air pollution device	5
H	Inspect and test	2
	Total time (weeks)	25

Critical path analysis
Helps determine the
project schedule.

25 weeks. However, since several activities can take place simultaneously, it is clear that the total project completion time may be less than 25 weeks. To find out just how long the project will take, we perform the **critical path analysis** for the network.

As mentioned earlier, the critical path is the *longest* time path through the network. To find the critical path, we calculate two distinct starting and ending times for each activity. These are defined as follows:

Earliest start (ES) = earliest time at which an activity can start, assuming all predecessors have been completed

Earliest finish (EF) = earliest time at which an activity can be finished

Latest start (LS) = latest time at which an activity can start so as to not delay the completion time of the entire project

Latest finish (LF) = latest time by which an activity has to finish so as to not delay the completion time of the entire project

We use a two-pass
procedure to find the
project schedule.

Forward pass
Identifies all the earliest
times.

We use a two-pass process, consisting of a forward pass and a backward pass, to determine these time schedules for each activity. The early start and finish times (ES and EF) are determined during the **forward pass**. The late start and finish times (LS and LF) are determined during the backward pass.

Forward Pass

To clearly show the activity schedules on the project network, we use the notation shown in Figure 3.10. The ES of an activity is shown in the top left corner of the node denoting that activity. The EF is shown in the top right corner. The latest times, LS and LF, are shown in the bottom left and bottom right corners, respectively.

FIGURE 3.10 ■

Notation Used in Nodes
for Forward and
Backward Pass

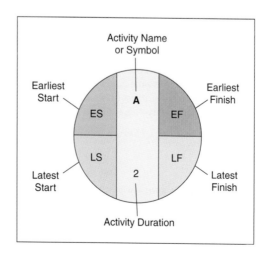

Earliest Start Time Rule   Before an activity can start, *all* its immediate predecessors must be finished.

All predecessor activities must be completed before an activity can begin.

- If an activity has only a single immediate predecessor, its ES equals the EF of the predecessor.
- If an activity has multiple immediate predecessors, its ES is the maximum of all EF values of its predecessors. That is,

$$ES = Max\{EF \text{ of all immediate predecessors}\} \qquad (3\text{-}1)$$

Earliest Finish Rule   The earliest finish time (EF) of an activity is the sum of its earliest start time (ES) and its activity time. That is,

$$EF = ES + \text{Activity time} \qquad (3\text{-}2)$$

Example 4

Calculate the earliest start and finish times for the activities in the Milwaukee General Hospital project. Table 3.2 on page 57 contains the activity times.

Figure 3.11 shows the complete project network for the hospital's project, along with the ES and EF values for all activities. We now describe how these values are calculated.

FIGURE 3.11 ■

Earliest Start and Earliest Finish Times for Milwaukee General Hospital

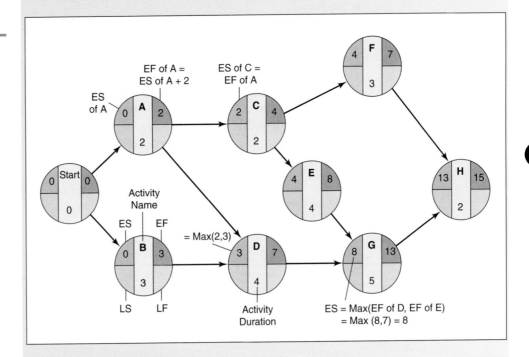

Since activity Start has no predecessors, we begin by setting its ES to 0. That is, activity Start can begin at the *end* of week 0, which is the same as the beginning of week 1.[1] If activity Start has an ES of 0, its EF is also 0, since its activity time is 0.

Next, we consider activities A and B, both of which have only Start as an immediate predecessor. Using the earliest start time rule, the ES for both activities A and B equals zero, which is the EF of activity Start. Now, using the earliest finish time rule, the EF for A is 2 (= 0 + 2), and the EF for B is 3 (= 0 + 3).

Since activity A precedes activity C, the ES of C equals the EF of A (= 2). The EF of C is therefore 4 (= 2 + 2).

---

[1]In writing all earliest and latest times, we need to be consistent. For example, if we specify that the ES value of activity *i* is week 4, do we mean the *beginning* of week 4 or the *end* of week 4? Note that if the value refers to the *beginning* of week 4, it means that week 4 is also available for performing activity *i*. In our discussions, *all* earliest and latest time values correspond to the *end* of a period. That is, if we specify that the ES of activity *i* is week 4, it means that activity *i* starts work only at the beginning of week 5.

We now come to activity D. Both activities A and B are immediate predecessors for B. Whereas A has an EF of 2, activity B has an EF of 3. Using the earliest finish time rule, we compute the ES of activity D as follows:

$$ES \text{ of } D = Max(EF \text{ of } A, EF \text{ of } B) = Max(2, 3) = 3$$

The EF of D equals 7 (= 3 + 4). Next, both activities E and F have activity C as their only immediate predecessor. Therefore, the ES for both E and F equals 4 (= EF of C). The EF of E is 8 (= 4 + 4), and the EF of F is 7 (= 4 + 3).

Activity G has both activities D and E as predecessors. Using the earliest start time rule, its ES is therefore the maximum of the EF of D and the EF of E. Hence, the ES of activity G equals 8 (= maximum of 7 and 8), and its EF equals 13 (= 8 + 5).

Finally, we come to activity H. Since it also has two predecessors, F and G, the ES of H is the maximum EF of these two activities. That is, the ES of H equals 13 (= maximum of 13 and 7). This implies that the EF of H is 15 (= 13 + 2). Since H is the last activity in the project, this also implies that the earliest time in which the entire project can be completed is 15 weeks.

Although the forward pass allows us to determine the earliest project completion time, it does not identify the critical path. In order to identify this path, we need to now conduct the backward pass to determine the LS and LF values for all activities.

Backward Pass

Backward pass
Finds all latest times.

Just as the forward pass began with the first activity in the project, the **backward pass** begins with the last activity in the project. For each activity, we first determine its LF value, followed by its LS value. The following two rules are used in this process.

Latest Finish Time Rule   This rule is again based on the fact that before an activity can start, all its immediate predecessors must be finished.

- If an activity is an immediate predecessor for just a single activity, its LF equals the LS of the activity that immediately follows it.
- If an activity is an immediate predecessor to more than one activity, its LF is the minimum of all LS values of all activities that immediately follow it. That is,

LF of an activity = minimum LS of all activities that follow.

$$LF = Min\{LS \text{ of all immediate following activities}\} \qquad (3\text{-}3)$$

Latest Start Time Rule   The latest start time (LS) of an activity is the difference of its latest finish time (LF) and its activity time. That is,

$$LS = LF - \text{Activity time} \qquad (3\text{-}4)$$

Example 5

Calculate the latest start and finish times for each activity in Milwaukee General's pollution project. Use Figure 3.11 as a beginning point.

Figure 3.12 shows the complete project network for Milwaukee General, along with LS and LF values for all activities. In what follows, we see how these values were calculated.

We begin by assigning an LF value of 15 weeks for activity H. That is, we specify that the latest finish time for the entire project is the same as its earliest finish time. Using the latest start time rule, the LS of activity H is equal to 13 (= 15 − 2).

Since activity H is the lone succeeding activity for both activities F and G, the LF for both F and G equals 13. This implies that the LS of G is 8 (= 13 − 5), and the LS of F is 10 (= 13 − 3).

Proceeding in this fashion, the LF of E is 8 (= LS of G), and its LS is 4 (= 8 − 4). Likewise, the LF of D is 8 (= LS of G), and its LS is 4 (= 8 − 4).

We now consider activity C, which is an immediate predecessor to two activities: E and F. Using the latest finish time rule, we compute the LF of activity C as follows:

$$LF \text{ of } C = Min(LS \text{ of } E, LS \text{ of } F) = Min(4, 10) = 4$$

The LS of C is computed as 2 (= 4 − 2). Next, we compute the LF of B as 4 (= LS of D), and its LS as 1 (= 4 − 3).

Latest Start and Latest Finish Times for Milwaukee General Hospital

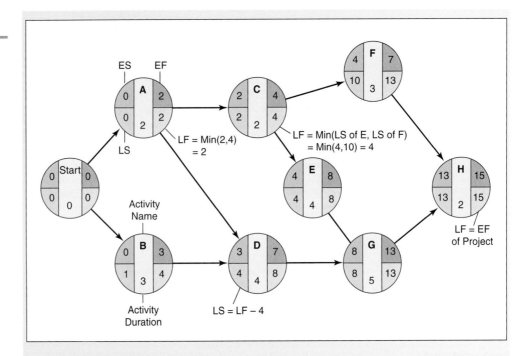

We now consider activity A. We compute its LF as 2 (= minimum of LS of C and LS of D). Hence, the LS of activity A is 0 (= 2 − 2). Finally, both the LF and LS of activity Start are equal to 0.

Calculating Slack Time and Identifying the Critical Path(s)

After we have computed the earliest and latest times for all activities, it is a simple matter to find the amount of **slack time**, or free time, that each activity has. Slack is the length of time an activity can be delayed without delaying the entire project. Mathematically,

Slack time
Free time for an activity.

$$\text{Slack} = \text{LS} - \text{ES} \qquad \text{or} \qquad \text{Slack} = \text{LF} - \text{EF} \qquad (3\text{-}5)$$

Example 6

Active Model 3.1

This example is further illustrated in Active Model 3.1 on the Student CD-ROM and in the Exercise located in your Student Lecture Guide.

Calculate the slack for the activities in the Milwaukee General Hospital project, starting with the data in Figure 3.12 in the previous example.

Table 3.3 summarizes the ES, EF, LS, LF, and slack time for all of the hospital's activities. Activity B, for example, has 1 week of slack time since its LS is 1 and its ES is 0 (alternatively, its LF is 4 and its EF is 3). This means that activity B can be delayed by up to 1 week, and the whole project can still finish in 15 weeks.

On the other hand, activities A, C, E, G and H have *no* slack time. This means that none of them can be delayed without delaying the entire project. Conversely, if hospital administrator Dr. Joni Steinberg wants to reduce the total project times, she will have to reduce the length of one of these activities.

Milwaukee General Hospital's Schedule and Slack Times

ACTIVITY	EARLIEST START ES	EARLIEST FINISH EF	LATEST START LS	LATEST FINISH LF	SLACK LS − ES	ON CRITICAL PATH
A	0	2	0	2	0	Yes
B	0	3	1	4	1	No
C	2	4	2	4	0	Yes
D	3	7	4	8	1	No
E	4	8	4	8	0	Yes
F	4	7	10	13	6	No
G	8	13	8	13	0	Yes
H	13	15	13	15	0	Yes

Critical activities have no slack time.

The activities with zero slack are called *critical activities* and are said to be on the critical path. The critical path is a continuous path through the project network that:

- starts at the first activity in the project (Start in our example),
- terminates at the last activity in the project (H in our example), and
- includes only critical activities (i.e., activities with no slack time).

Example 7

FIGURE 3.13 ■

Critical Path and Slack Times for Milwaukee General Hospital

Critical path is the longest path through the network.

Show Milwaukee General's critical path, Start-A-C-E-G-H, in network form.

Figure 3.13 indicates that the total project completion time of 15 weeks corresponds to the longest path in the network.

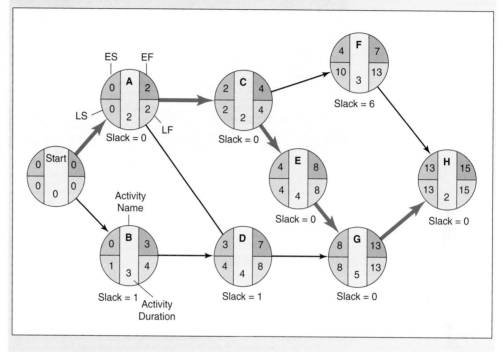

Total Slack Time versus Free Slack Time Look again at the project network in Figure 3.13. Consider activities B and D, which have slack of 1 week each. Does it mean that we can delay *each* activity by 1 week, and still complete the project in 15 weeks? The answer is no.

Let's assume that activity B is delayed by 1 week. It has used up its slack of 1 week and now has an EF of 4. This implies that activity D now has an ES of 4 and an EF of 8. Note that these are also its LS and LF values, respectively. That is, activity D also has no slack time now. Essentially, the slack of 1 week that activities B and D had is, for that path, *shared* between them. Delaying either activity by 1 week causes not only that activity, but also the other activity, to lose its slack. This type of a slack time is referred to as **total slack**. Typically, when two or more noncritical activities appear successively in a path, they share total slack.

In contrast, consider the slack time of 6 weeks in activity F. Delaying this activity decreases only its slack time and does not impact the slack time of any other activity. This type of a slack time is referred to as **free slack**. Typically, if a noncritical activity has critical activities on either side of it in a path, its slack time is free slack.

Total slack
Time shared among more than one activity.

Free slack
Time associated with a single activity.

VARIABILITY IN ACTIVITY TIMES

In identifying all earliest and latest times so far, and the associated critical path(s), we have adopted the CPM approach of assuming that all activity times are known and fixed constants. That is, there is no variability in activity times. However, in practice, it is likely that activity completion times vary depending on various factors.

For example, building internal components (activity A) for Milwaukee General Hospital is estimated to finish in 2 weeks. Clearly, factors such as late arrival of raw materials, absence of key personnel, and so on, could delay this activity. Suppose activity A actually ends up taking 3 weeks.

Activity times are subject to variability.

To plan, monitor, and control the huge number of details involved in sponsoring a rock festival attended by over 100,000 fans, Hard Rock Cafe uses MS Project and the tools you see in this chapter. For more details, read the Video Case Study, "Managing Hard Rock's Rockfest," located in your Student Lecture Guide.

Since A is on the critical path, the entire project will now be delayed by 1 week to 16 weeks. If we had anticipated completion of this project in 15 weeks, we would obviously miss our deadline.

Although some activities may be relatively less prone to delays, others could be extremely susceptible to delays. For example, activity B (modify roof and floor) could be heavily dependent on weather conditions. A spell of bad weather could significantly impact its completion time.

This means that we cannot ignore the impact of variability in activity times when deciding the schedule for a project. PERT addresses this issue.

Three Time Estimates in PERT

In PERT, we employ a probability distribution based on three time estimates for each activity, as follows:

Optimistic time (a) = time an activity will take if everything goes as planned. In estimating this value, there should be only a small probability (say, 1/100) that the activity time will be $< a$.

Pessimistic time (b) = time an activity will take assuming very unfavorable conditions. In estimating this value, there should also be only a small probability (also, 1/100) that the activity time will be $> b$.

Most likely time (m) = most realistic estimate of the time required to complete an activity.

When using PERT, we often assume that activity time estimates follow the **beta probability distribution** (see Figure 3.14). This continuous distribution is often appropriate for determining the expected value and variance for activity completion times.

To find the *expected activity time*, t, the beta distribution weights the three time estimates as follows

$$t = (a + 4m + b)/6 \qquad (3\text{-}6)$$

That is, the most likely time (m) is given four times the weight as the optimistic time (a) and pessimistic time (b). The time estimate t computed using Equation 3-6 for each activity is used in the project network to compute all earliest and latest times.

Optimistic time
The "best" activity completion time that could be obtained in a PERT network.

Pessimistic time
The "worst" activity time that could be expected in a PERT network.

Most likely time
The most probable time to complete an activity in a PERT network.

Beta probability distribution
A mathematical distribution that may describe the activity time estimate distributions in a PERT network.

FIGURE 3.14 ■

Beta Probability
Distribution with Three
Time Estimates

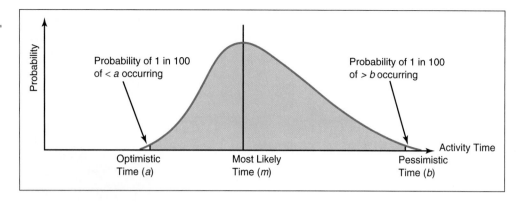

To compute the *dispersion* or *variance of activity completion time*, we use the formula:[2]

$$\text{Variance} = [(b - a)/6]^2 \qquad (3-7)$$

Example 8

**Excel OM Data File
Ch03Ex8.xla**

Suppose Dr. Steinberg and the project management team at Milwaukee General Hospital developed the following time estimates for Activity F (Installing the Pollution Control System):

$$a = 1 \text{ week}, m = 2 \text{ weeks}, b = 9 \text{ weeks}$$

a. Find the expected time and variance for Activity F.
b. Then compute the expected time and variance for all of the other activities in the pollution control project. Use the time estimates in Table 3.4.

SOLUTION

a. The expected time for Activity F is:

$$t = \frac{a + 4m + b}{6} = \frac{1 + 4(2) + 9}{6} = \frac{18}{6} = 3 \text{ weeks}$$

The variance for Activity F is:

$$\text{Variance} = \left[\frac{(b - a)}{6}\right]^2 = \left[\frac{(9 - 1)}{6}\right]^2 = \left(\frac{8}{6}\right)^2 = \frac{64}{36} = 1.78$$

b. The rest of the calculations follow in Table 3.4.

TABLE 3.4 ■ Time Estimates (in weeks) for Milwaukee General's Project

ACTIVITY	OPTIMISTIC a	MOST LIKELY m	PESSIMISTIC b	EXPECTED TIME $t = (a + 4m + b)/6$	VARIANCE $[(b - a)/6]^2$
A	1	2	3	2	$[(3 - 1)/6]^2 = 4/36 = .11$
B	2	3	4	3	$[(4 - 2)/6]^2 = 4/36 = .11$
C	1	2	3	2	$[(3 - 1)/6]^2 = 4/36 = .11$
D	2	4	6	4	$[(6 - 2)/6]^2 = 16/36 = .44$
E	1	4	7	4	$[(7 - 1)/6]^2 = 36/36 = 1.00$
F	1	2	9	3	$[(9 - 1)/6]^2 = 64/36 = 1.78$
G	3	4	11	5	$[(11 - 3)/6]^2 = 64/36 = 1.78$
H	1	2	3	2	$[(3 - 1)/6]^2 = 4/36 = .11$

The expected times in this table are, in fact, the activity times we used in our earlier computation and identification of the critical path.

---

[2]This formula is based on the statistical concept that from one end of the beta distribution to the other is 6 standard deviations (± 3 standard deviations from the mean). Since $(b - a)$ is 6 standard deviations, the variance is $[(b - a/6]^2$.

We see here a ship being built at the Hyundi shipyard, Asia's largest shipbuilder, in Korea. Managing this project uses the same techniques as managing the remodeling of a store or installing a new production line.

Probability of Project Completion

The critical path analysis helped us determine that Milwaukee General Hospital's expected project completion time is 15 weeks. Dr. Joni Steinberg knows, however, that there is significant variation in the time estimates for several activities. Variation in activities that are on the critical path can affect the overall project completion time—possibly delaying it. This is one occurrence that worries the hospital administrator considerably.

We compute the project variance by summing variances of only those activities on the critical path.

PERT uses the variance of critical path activities to help determine the variance of the overall project. Project variance is computed by summing variances of critical activities:

$$\sigma_p^2 = \text{Project variance} = \Sigma \ (\text{variances of activities on critical path}) \qquad (3\text{-}8)$$

Example 9

From Example 8 (see Table 3.4), we know that the variance of activity A is 0.11, variance of activity C is 0.11, variance of activity E is 1.00, variance of activity G is 1.78, and variance of activity H is 0.11.
   Compute the total project variance and project standard deviation.

$$\text{Project variance } (\sigma_p^2) = 0.11 + 0.11 + 1.00 + 1.78 + 0.11 = 3.11$$

which implies

$$\text{Project standard deviation } (\sigma_p) = \sqrt{\text{Project variance}} = \sqrt{3.11} = 1.76 \text{ weeks}$$

How can this information be used to help answer questions regarding the probability of finishing the project on time? PERT makes two more assumptions: (1) total project completion times follow a normal probability distribution and (2) activity times are statistically independent. With these assumptions, the bell-shaped normal curve shown in Figure 3.15 can be used to represent project

FIGURE 3.15 ■

Probability Distribution for Project Completion Times at Milwaukee General

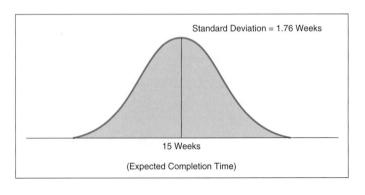

completion dates. This normal curve implies that there is a 50% chance that the hospital's project completion time will be less than 15 weeks and a 50% chance that it will exceed 15 weeks.

Example 10

Dr. Joni Steinberg would like to find the probability that her project will be finished on or before the 16 week deadline.

To do so, she needs to determine the appropriate area under the normal curve. The standard normal equation can be applied as follows:

$$Z = \text{(due date} - \text{expected date of completion)}/\sigma_p \qquad (3\text{-}9)$$
$$= (16 \text{ weeks} - 15 \text{ weeks})/1.76 \text{ weeks} = 0.57$$

where Z is the number of standard deviations the due date or target date lies from the mean or expected date.

Referring to the Normal Table in Appendix I, we find a probability of 0.7157. Thus, there is a 71.57% chance that the pollution control equipment can be put in place in 16 weeks or less. This is shown in Figure 3.16.

FIGURE 3.16 ■

Probability of Milwaukee General Hospital Meeting the 16-Week Deadline

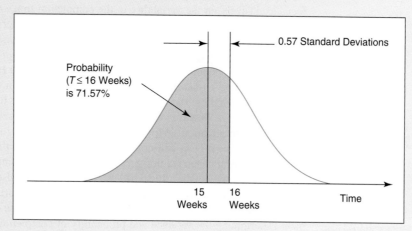

Determining Project Completion Time for a Given Confidence Level   Let's say Dr. Steinberg is worried that there is only a 71.57% chance that the pollution control equipment can be put in place in 16 weeks or less. She thinks that it may be possible to plead with the environmental group for more time. However, before she approaches the group, she wants to arm herself with sufficient information about the project. Specifically, she wants to find the deadline by which she has a 99% chance of completing the project. She hopes to use her analysis to convince the group to agree to this extended deadline.

Clearly, this due date would be greater than 16 weeks. However, what is the exact value of this new due date? To answer this question, we again use the assumption that Milwaukee General's project completion time follows a normal probability distribution with a mean of 15 weeks and a standard deviation of 1.76 weeks.

Example 11

FIGURE 3.17 ■

Z-Value for 99% Probability of Project Completion at Hospital

Dr. Steinberg wants to find the due date under which her hospital's project has a 99% chance of completion. She first needs to compute the Z-value corresponding to 99%, as shown in Figure 3.17.

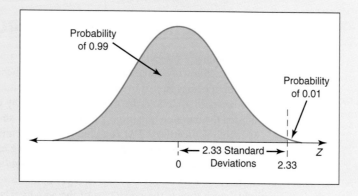

Referring again to the Normal Table in Appendix I, we identify a Z value of 2.33 as being closest to the probability of 0.99. That is, Dr. Steinberg's due date should be 2.33 standard deviations above the mean project completion time. Starting with the standard normal equation (see Equation 3-9), we can solve for the due date and rewrite the equation as

$$\text{Due date} = \text{Expected completion time} + Z \times \sigma_p \tag{3-10}$$

$$= 15 + 2.33 \times 1.76 = 19.1 \text{ weeks}$$

Hence, if Steinberg can get the environmental group to agree to give her a new deadline of 19.1 weeks (or more), she can be 99% sure of finishing the project on time.

Variability in Completion Time of Noncritical Paths   In our discussion so far, we focus exclusively on the variability in the completion times of activities on the critical path. This seems logical since these activities are, by definition, the more important activities in a project network. However, when there is variability in activity times, it is important that we also investigate the variability in the completion times of activities on *noncritical* paths.

Consider, for example, activity D in Milwaukee General Hospital's project. Recall from Figure 3.13 (in Example 7) that this is a noncritical activity, with a slack time of 1 week. We have therefore not considered the variability in D's time in computing the probabilities of project completion times. We observe, however, that D has a variance of 0.44 (see Table 3.4 in Example 8). In fact, the pessimistic completion time for D is 6 weeks. This means that if D ends up taking its pessimistic time to finish, the project will not finish in 15 weeks, even though D is not a critical activity.

For this reason, when we find probabilities of project completion times, it may be necessary for us to not focus only on the critical path(s). We may need to also compute these probabilities for noncritical paths, especially those that have relatively large variances. It is possible for a noncritical path to have a smaller probability of completion within a due date, when compared with the critical path. Determining the variance and probability of completion for a noncritical path is done in the same manner as Examples 9 and 10.

Noncritical paths with large variances should also be closely monitored.

What Project Management Has Provided So Far   Project management techniques have thus far been able to provide Dr. Joni Steinberg with several valuable pieces of management information:

1. The project's expected completion date is 15 weeks.
2. There is a 71.57% chance that the equipment will be in place within the 16-week deadline. PERT analysis can easily find the probability of finishing by any date Steinberg is interested in.
3. Five activities (A, C, E, G, and H) are on the critical path. If any one of these is delayed for any reason, the entire project will be delayed.
4. Three activities (B, D, F) are not critical but have some slack time built in. This means that Steinberg can borrow from their resources, and, if necessary, she may be able to speed up the whole project.
5. A detailed schedule of activity starting and ending dates has been made available (see Table 3.3 in Example 6).

COST-TIME TRADE-OFFS AND PROJECT CRASHING

While managing a project, it is not uncommon for a project manager to be faced with either (or both) of the following situations: (1) the project is behind schedule and (2) the scheduled project completion time has been moved forward. In either situation, some or all of the remaining activities need to be speeded up in order to finish the project by the desired due date. The process by which we shorten the duration of a project in the cheapest manner possible is called project **crashing**.

As mentioned earlier, CPM is a deterministic technique in which each activity has two sets of times. The first is the *normal* or *standard* time that we used in our computation of earliest and latest times. Associated with this normal time is the *normal* cost of the activity. The second time is the *crash time*, which is defined as the shortest duration required to complete an activity. Associated with this crash time is the *crash cost* of the activity. Usually, we can shorten an activity by adding

Crashing
Shortening activity time in a network to reduce time on the critical path so total completion time is reduced.

Crash time is the shortest duration of an activity.

extra resources (e.g., equipment, people) to it. Hence, it is logical for the crash cost of an activity to be higher than its normal cost.

The amount by which an activity can be shortened (i.e., the difference between its normal time and crash time) depends on the activity in question. We may not be able to shorten some activities at all. For example, if a casting needs to be heat-treated in the furnace for 48 hours, adding more resources does not help shorten the time. In contrast, we may be able to shorten some activities significantly (e.g., frame a house in 3 days instead of 10 days by using three times as many workers.)

> We want to find the cheapest way of crashing a project to the desired due date.

Likewise, the cost of crashing (or shortening) an activity depends on the nature of the activity. Managers are usually interested in speeding up a project at the least additional cost. Hence, when choosing which activities to crash, and by how much, we need to ensure the following:

- the amount by which an activity is crashed is, in fact, permissible;
- taken together, the shortened activity durations will enable us to finish the project by the due date;
- the total cost of crashing is as small as possible.

Crashing a project involves four steps, as follows:

> This assumes crash costs are linear over time.

Step 1: Compute the crash cost per week (or other time period) for each activity in the network. If crash costs are linear over time, the following formula can be used:

$$\text{Crash cost per period} = \frac{(\text{Crash cost} - \text{Normal cost})}{(\text{Normal time} - \text{Crash time})} \qquad (3\text{-}11)$$

Step 2: Using the current activity times, find the critical path(s) in the project network. Identify the critical activities.

Step 3: If there is only one critical path, then select the activity on this critical path that (a) can still be crashed and (b) has the smallest crash cost per period. Crash this activity by one period.

If there is more than one critical path, then select one activity from each critical path such that (a) each selected activity can still be crashed and (b) the total crash cost per period of *all* selected activities is the smallest. Crash each activity by one period. Note that the same activity may be common to more than one critical path.

Step 4: Update all activity times. If the desired due date has been reached, stop. If not, return to Step 2.

We illustrate project crashing in Example 12.

Example 12

Excel OM Data File Ch03Ex12.xla

Suppose that Milwaukee General Hospital has been given only 13 weeks (instead of 16 weeks) to install the new pollution control equipment or face a court-ordered shutdown. As you recall, the length of Joni Steinberg's critical path was 15 weeks. Which activities should Steinberg crash, and by how much, in order to meet this 13-week due date? Naturally, Steinberg is interested in speeding up the project by 2 weeks, at the least additional cost.

The hospital's normal and crash times, and normal and crash costs, are shown in Table 3.5. Note, for example, that activity B's normal time is 3 weeks (the estimate used in computing the critical path), and its crash time is 1 week. This means that activity B can be shortened by up to 2 weeks if extra resources are provided. The cost of these additional resources is $4,000 (= difference between the crash cost of $34,000 and the normal cost of $30,000). If we assume that the crashing cost is linear over time (i.e., the cost is the same each week), activity B's crash cost per week is $2,000 (= $4,000/2).

TABLE 3.5 ■ Normal and Crash Data for Milwaukee General Hospital

ACTIVITY	TIME (WEEKS) NORMAL	TIME (WEEKS) CRASH	COST ($) NORMAL	COST ($) CRASH	CRASH COST PER WEEK ($)	CRITICAL PATH?
A	2	1	22,000	22,750	750	Yes
B	3	1	30,000	34,000	2,000	No
C	2	1	26,000	27,000	1,000	Yes
D	4	3	48,000	49,000	1,000	No
E	4	2	56,000	58,000	1,000	Yes
F	3	2	30,000	30,500	500	No
G	5	2	80,000	84,500	1,500	Yes
H	2	1	16,000	19,000	3,000	Yes

This calculation for Activity B is shown in Figure 3.18. Crash costs for all other activities can be computed in a similar fashion.

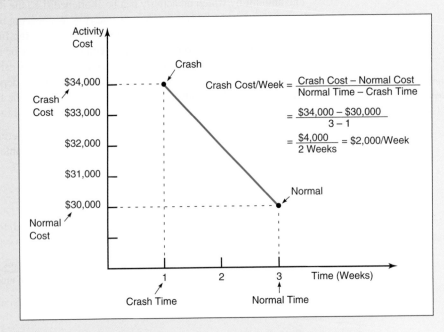

Steps 2, 3, and 4 can now be applied to reduce Milwaukee General's project completion time at a minimum cost. We show the project network for Milwaukee General again in Figure 3.19.

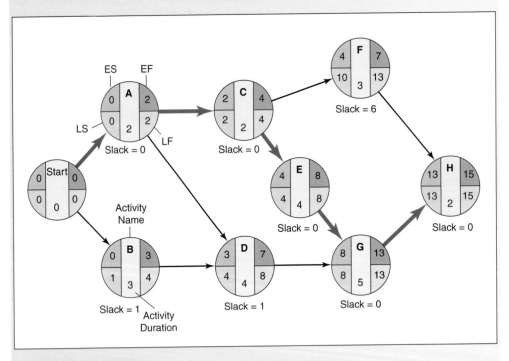

The current critical path (using normal times) is Start-A-C-E-G-H, in which Start is just a dummy starting activity. Of these critical activities, activity A has the lowest crash cost per week of $750. Joni Steinberg should therefore crash activity A by 1 week to reduce the project completion time to 14 weeks. The cost is an additional $750. Note that activity A cannot be crashed any further since it has reached its crash limit of 1 week.

At this stage, the original path Start-A-C-E-G-H remains critical with a completion time of 14 weeks. However, a new path Start-B-D-G-H is also critical now, with a completion time of 14 weeks. Hence, any further crashing must be done to both critical paths.

On each of these critical paths, we need to identify one activity that can still be crashed. We also want the total cost of crashing an activity on each path to be the smallest. We might be tempted to simply pick the activities with the smallest crash cost per period in each path. If we do this, we would select activity C from the first path and activity D from the second path. The total crash cost would then be $2,000 (= $1,000 + $1,000).

But we spot that activity G is common to both paths. That is, by crashing activity G, we will simultaneously reduce the completion time of both paths. Even though the $1,500 crash cost for activity G is higher than that for activities C and D, we would still prefer crashing G since the total cost is now only $1,500 (compared with the $2,000 if we crash C and D).

Hence, to crash the project down to 13 weeks, Dr. Steinberg should crash activity A by 1 week, and activity G by 1 week. The total additional cost is $2,250 (= $750 + $1,500).

Crashing is especially important when contracts for projects include bonuses or penalties for early or late finishes.

A CRITIQUE OF PERT AND CPM

As a critique of our discussions of PERT, here are some of its features about which operations managers need to be aware:

Advantages
1. Especially useful when scheduling and controlling large projects.
2. Straightforward concept and not mathematically complex.
3. Graphical networks help to perceive relationships among project activities quickly.
4. Critical path and slack time analyses help pinpoint activities that need to be closely watched.
5. Project documentation and graphs point out who is responsible for various activities.
6. Applicable to a wide variety of projects.
7. Useful in monitoring not only schedules, but costs as well.

Limitations

In large networks there are too many activities to monitor closely, but managers can concentrate on the critical activities.

1. Project activities have to be clearly defined, independent, and stable in their relationships.
2. Precedence relationships must be specified and networked together.
3. Time estimates tend to be subjective and are subject to fudging by managers who fear the dangers of being overly optimistic or not pessimistic enough.
4. There is the inherent danger of too much emphasis being placed on the longest, or critical, path. Near-critical paths need to be monitored closely as well.

OM IN ACTION

Project Management and Software Development

Although computers have revolutionized how companies conduct business and allowed some organizations to achieve a long-term competitive advantage in the marketplace, the software that controls these computers is often more expensive than intended and takes longer to develop than expected. In some cases, large software projects are never fully completed.

The London Stock Exchange, for example, had an ambitious software project called TAURUS that was intended to improve computer operations at the exchange. After numerous delays and cost overruns, however, the project, which cost hundreds of millions of dollars, was finally halted. The FLORIDA system, an ambitious software development project for the department of Health and Rehabilitative Services (HRS) for the State of Florida, which was also delayed, cost $100 million more than expected, and didn't operate as everyone had hoped. While not all software development projects are delayed or over budget, it has been estimated that more than half of all software projects cost over double their original projections.

To control large software projects, many companies are now using project management techniques. Ryder Systems, Inc., American Express Financial Advisors, and United Airlines have all created project management departments for their software and information systems projects. These departments have the authority to monitor large software projects and make changes to deadlines, budgets, and resources used to complete software development efforts.

Sources: Computerworld (April 22, 2002): 43; and Information Week (January 25, 1999): 140.

USING MICROSOFT PROJECT TO MANAGE PROJECTS

The approaches discussed so far are effective for managing small projects. However, for large or complex projects, specialized project management software is much preferred. In this section, we provide a brief introduction to the most popular example of such specialized software, Microsoft Project 2002.

We should note that at this introductory level, our intent here is not to describe the full capabilities of this program. Rather, we illustrate how it can be used to perform some of the basic calculations in managing projects. We leave it to you to explore the advanced capabilities and functions of Microsoft Project 2002 (or any other project management software) in greater detail. A time-limited version of MS Project appears on the CD that comes with this text.

MS Project is useful for project scheduling and control.

Microsoft Project is extremely useful in drawing project networks, identifying the project schedule, and managing project costs and other resources. It does not, however, perform PERT probability calculations.

Creating a Project Schedule Using MS Project

Let us again consider the Milwaukee General Hospital project. Recall that this project has eight activities (repeated in the margin). The first step is to define the activities and their precedence relationships. To do so, we start Microsoft Project and click FileⅠNew to open a blank project. We can now enter the project start date in the summary information that is first presented (see Program 3.1). Note that dates are referred to by actual calendar dates rather than as day 0, day 1, and so on. For example, we have used July 1, 2002 as our project starting date in Program 3.1. Microsoft Project will automatically update the project finish date once we have entered all the project information. In Program 3.1, we have specified the current date as August 12, 2002.

First, we define a new project.

Entering Activity Information   After entering the summary information, we now use the window shown in Program 3.2 to enter all activity information. For each activity (or task, as Microsoft Project calls it), we enter its name and duration. Microsoft Project identifies tasks by numbers (e.g., 1, 2) rather than letters. Hence, for convenience, we have shown both the letter (e.g., A, B) and the description of the activity in the *Task Name* column in Program 3.2. By default, the duration is measured in days. To specify weeks, we include the letter "*w*" after the duration of each activity. For example, we enter the duration of activity A as 2*w*.

Next, we enter the activity information.

DURATIONS	
ACTIVITY	TIME IN WEEKS
A	2
B	3
C	2
D	4
E	4
F	3
G	5
H	2

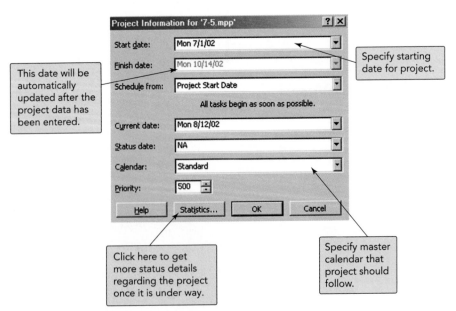

PROGRAM 3.1 ■ Project Summary Information in MS Project

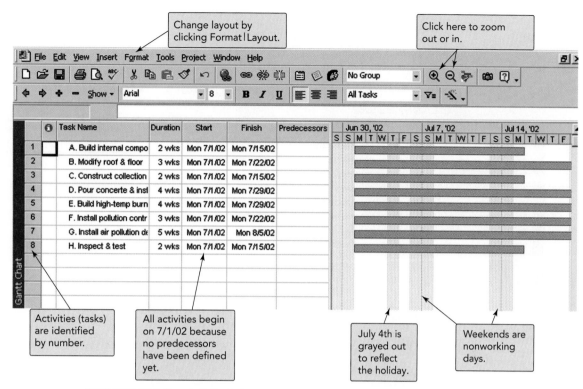

PROGRAM 3.2 ■ Activity Entry in MS Project for Milwaukee General Hospital

As we enter the activities and durations, the software automatically inserts start and finish dates. Note that all activities have the same start date (i.e., 7/1/02) since we have not yet defined the precedence relationships. Also, as shown in Program 3.2, if the Gantt Chart option is selected in the View menu, a horizontal bar corresponding to the duration of each activity appears on the right pane of the window.

Observe that Saturdays and Sundays are automatically grayed out in the Gantt chart to reflect the fact that these are nonworking days. In most project management software, the entire project is linked to a master calendar (or alternatively, each activity is linked to its own specific calendar). Additional nonworking days can be defined using these calendars. For example, we have used Tools|Change Working Time to specify July 4, 2002, as a nonworking day in Program 3.2. This automatically extends all activity completion times by one day. Since activity A starts on Monday, July 1, 2002, and takes 2 weeks (i.e., 10 working days), its finish time is now Monday, July 15, 2002 (rather than Friday, July 12, 2002).

Defining Precedence Relationships The next step is to define precedence relationships (or links) between these activities. There are two ways of specifying these links. The first is to enter the relevant activity numbers (e.g., 1, 2) in the *Predecessor* column, as shown in Program 3.3 for activities C and D. The other approach uses the Link icon. For example, to specify the precedence relationship between activities C and E, we click activity C first, hold the Ctrl key down, and then click activity E. We then click the Link icon, as shown in Program 3.3. As soon as we define a link, the bars in the Gantt chart are automatically repositioned to reflect the new start and finish times for the linked activities. Further, the link itself is shown as an arrow extending from the predecessor activity.

The schedule automatically takes nonworking days into account.

PRECEDENCES	
ACTIVITY	PREDECESSORS
A	—
B	—
C	A
D	A, B
E	C
F	C
G	D, E
H	F, G

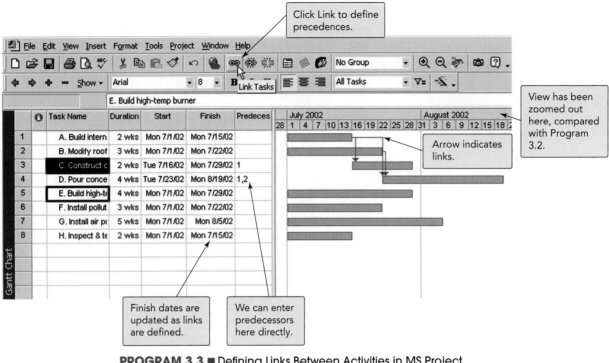

PROGRAM 3.3 ■ Defining Links Between Activities in MS Project

The project can be viewed either as a Gantt chart or as a network.

Viewing the Project Schedule   When all links have been defined, the complete project schedule can be viewed as a Gantt chart, as shown in Program 3.4. We can also select View|Network Diagram to view the schedule as a project network (shown in Program 3.5). The critical path is shown in red on the screen (bold in Program 3.5) in the network diagram. We can click on any of the activities in the project network to view details of the activities. Likewise, we can

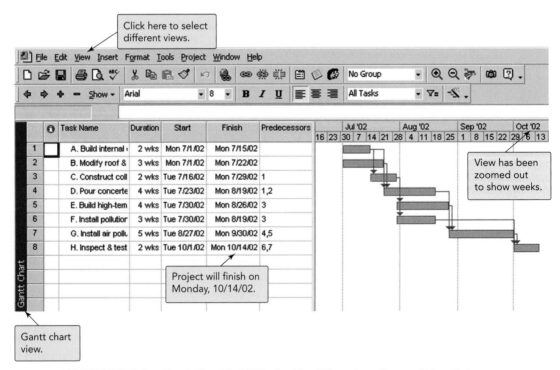

PROGRAM 3.4 ■ Gantt Chart in MS Project for Milwaukee General Hospital

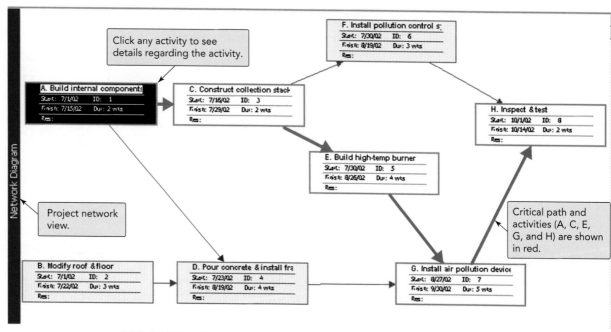

PROGRAM 3.5 ■ Project Network in MS Project for Milwaukee General Hospital

easily add or remove activities and/or links from the project network. Each time we do so, MS Project automatically updates all start dates, finish dates, and the critical path(s). If desired, we can manually change the layout of the network (e.g., reposition activities) by changing the options in Format|Layout.

Programs 3.4 and 3.5 show that if Milwaukee General's project starts on July 1, 2002, it can be finished on October 14, 2002. The start and finish dates for all activities are also clearly identified. This schedule takes into account the nonworking days on all weekends, and on July 4. These programs illustrate how the use of specialized project management software can greatly simplify the scheduling procedures discussed earlier in this chapter.

PERT Analysis   As mentioned, MS Project does not perform the PERT probability calculations discussed in Examples 10 and 11. However, by clicking View|Toolbars|PERT Analysis, we can get Microsoft Project to allow us to enter optimistic, most likely, and pessimistic times for each activity. We can then choose to view Gantt charts based on any of these three times for each activity.

Tracking Progress and Managing Costs Using MS Project

Perhaps the biggest advantage of using specialized software to manage projects is that they can track the progress of the project. In this regard, Microsoft Project has many features available to track individual activities in terms of time, cost, resource usage, and so on. In this section, we illustrate how we can track the progress of a project in terms of time.

Tracking the Time Status of a Project   An easy way to track the time progress of tasks is to enter the percent of work completed for each task. One way to do so is to double-click on any activity in the Task Name column in Program 3.4. A window, like the one shown in Program 3.6 is displayed. Let us now enter the percent of work completed for each task.

The table in the margin provides data regarding the percent of each of Milwaukee General's activities as of today. (Assume today is Monday, August 12, 2002, i.e., the end of the sixth week of the project schedule.)[3] Program 3.6 shows that activity A is 100% complete. We enter the percent completed for all other activities in a similar fashion.

The biggest benefit of using software is to track a project.

HOSPITAL PROJECT PERCENT COMPLETED ON AUG. 12, 2002	
ACTIVITY	COMPLETED
A	100
B	100
C	100
D	10
E	20
F	20
G	0
H	0

[3]Remember that the nonworking day on July 4 has moved all schedules by one day. Therefore, activities end on Mondays rather than on Fridays.

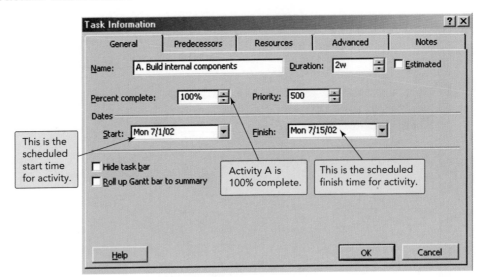

PROGRAM 3.6 ■ Updating Activity Progress in MS Project

"Poorly managed projects are costly, not only financially, but also in wasted time and demoralized personnel. But failure is almost never the result of poor software."

C. Fujinami and A. Marshall, consultants at Kepner Tregoe, Inc.

As shown in Program 3.7, the Gantt chart immediately reflects this updated information by drawing a thick line within each activity's bar. The length of this line is proportional to the percent of that activity's work that has been completed.

How do we know if we are on schedule? Notice that there is a vertical line shown on the Gantt chart corresponding to today's date. Microsoft Project will automatically move this line to correspond with the current date. If the project is on schedule, we should see all bars to the *left* of today's line indicate that they have been completed. For example, Program 3.7 shows that activities A, B, and C are on schedule. In contrast, activities D, E, and F appear to be behind schedule. These activities need to be investigated further to determine the reason for the delay. This type of easy *visual* information is what makes such software so useful in practice for project management.

In addition to reading this section on MS project, we encourage you to load the software from your Student CD-ROM and try these procedures.

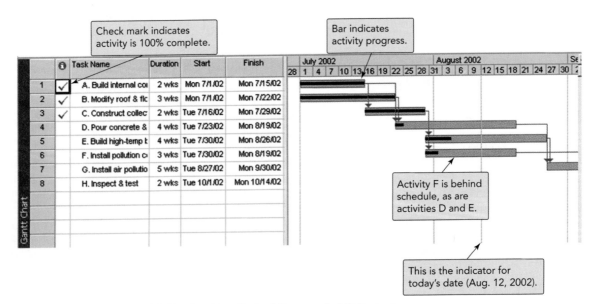

PROGRAM 3.7 ■ Tracking Project Progress in MS Project

SUMMARY

PERT, CPM, and other scheduling techniques have proven to be valuable tools in controlling large and complex projects. With these tools, managers understand the status of each activity and know which activities are critical and which have slack; in addition, they know where crashing makes the most sense. Projects are segmented into discrete activities, and specific resources are identified. This allows project managers to respond aggressively to global competition. Effective project management also allows firms to create products and services for global markets. As with MS Project provided on your Student CD-ROM and illustrated in this chapter, a wide variety of software packages are available to help managers handle network modeling problems.

PERT and CPM do not, however, solve all the project scheduling and management problems. Good management practices, clear responsibilities for tasks, and straightforward and timely reporting systems are also needed. It is important to remember that the models we described in this chapter are only tools to help managers make better decisions.

KEY TERMS

Project organization (p. 48)
Work breakdown structure (WBS) (p. 50)
Gantt charts (p. 50)
Program evaluation and review technique (PERT) (p. 52)
Critical path method (CPM) (p. 52)
Critical path (p. 52)
Activity-on-node (AON) (p. 53)
Activity-on-arrow (AOA) (p. 53)
Dummy activities (p. 54)
Critical path analysis (p. 57)

Forward pass (p. 57)
Backward pass (p. 59)
Slack time (p. 60)
Total slack (p. 61)
Free slack (p. 61)
Optimistic time (p. 62)
Pessimistic time (p. 62)
Most likely time (p. 62)
Beta probability distribution (p. 62)
Crashing (p. 66)

USING EXCEL OM

In addition to the Microsoft Project software just illustrated, Excel OM has a Project Scheduling module. Program 3.8 uses the data from the hospital example in this chapter (see Examples 4 and 5). The PERT/CPM analysis also handles activities with 3 time estimates.

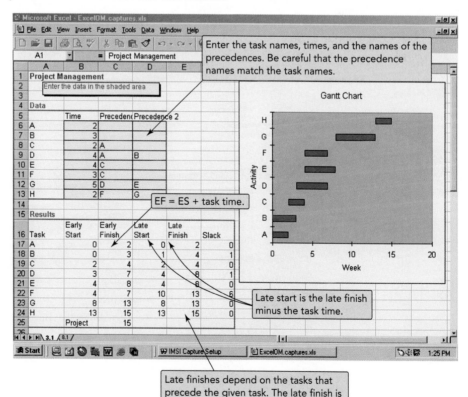

PROGRAM 3.8 ■

Excel OM's Use of Milwaukee General Hospital's Data from Examples 4 and 5

[P] USING POM FOR WINDOWS

POM for Window's Project Scheduling module can also find the expected project completion time for a CPM and PERT network with either one or three time estimates. POM for Windows also performs project crashing. For further details refer to Appendix V.

SOLVED PROBLEMS

Solved Problem 3.1

Construct an AON network based on the following:

ACTIVITY	IMMEDIATE PREDECESSOR(S)
A	—
B	—
C	—
D	A, B
E	C

SOLUTION

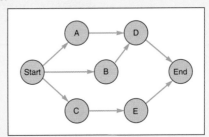

Solved Problem 3.2

Insert a dummy activity and event to correct the following AOA network:

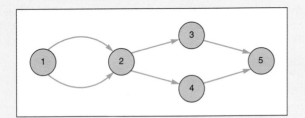

SOLUTION

We add the following dummy activity and dummy event to obtain the correct AOA network:

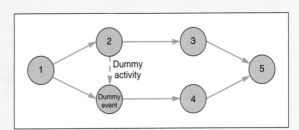

Solved Problem 3.3

Calculate the critical path, project completion time T, and variance σ_p^2, based on the following AON network information:

ACTIVITY	TIME	VARIANCE	ES	EF	LS	LF	SLACK
A	2	$^2/_6$	0	2	0	2	0
B	3	$^2/_6$	0	3	1	4	1
C	2	$^4/_6$	2	4	2	4	0
D	4	$^4/_6$	3	7	4	8	1
E	4	$^2/_6$	4	8	4	8	0
F	3	$^1/_6$	4	7	10	13	6
G	5	$^1/_6$	8	13	8	13	0

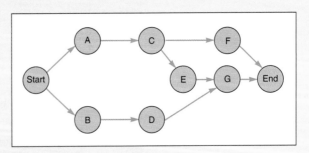

SOLUTION

We conclude that the critical path is Start, A, C, E, G, End.

$$\text{Total project time} = T = 2 + 2 + 4 + 5 = 13$$

and

$$\sigma_p^2 = \Sigma \text{ Variances on the critical path} = \frac{2}{6} + \frac{4}{6} + \frac{2}{6} + \frac{1}{6} = \frac{9}{6} = 1.5$$

Solved Problem 3.4

To complete the wing assembly for an experimental aircraft, Jim Gilbert has laid out the seven major activities involved. These activities have been labeled A through G in the following table, which also shows their estimated completion times (in weeks) and immediate predecessors. Determine the expected time and variance for each activity.

ACTIVITY	a	m	b	IMMEDIATE PREDECESSORS
A	1	2	3	—
B	2	3	4	—
C	4	5	6	A
D	8	9	10	B
E	2	5	8	C, D
F	4	5	6	D
G	1	2	3	E

SOLUTION

Expected times and variances can be computed using Formulas 3-6 and 3-7 presented on pages 62–63 in this chapter. The results are summarized in the following table:

ACTIVITY	EXPECTED TIME (IN WEEKS)	VARIANCE
A	2	$\frac{1}{9}$
B	3	$\frac{1}{9}$
C	5	$\frac{1}{9}$
D	9	$\frac{1}{9}$
E	5	1
F	5	$\frac{1}{9}$
G	2	$\frac{1}{9}$

Solved Problem 3.5

Referring to Solved Problem 3.4, now Jim Gilbert would like to determine the critical path for the entire wing assembly project as well as the expected completion time for the total project. In addition, he would like to determine the earliest and latest start and finish times for all activities.

SOLUTION

The AON network for Gilbert's project is shown in Figure 3.20. Note that this project has multiple activities (A and B) with no immediate predecessors, and multiple activities (F and G) with no successors. Hence, in addition to a unique starting activity (Start), we have included a unique finishing activity (End) for the project.

Figure 3.20 shows the earliest and latest times for all activities. The results are also summarized in the following table:

ACTIVITY	ACTIVITY TIME ES	EF	LS	LF	SLACK
A	0	2	5	7	5
B	0	3	0	3	0
C	2	7	7	12	5
D	3	12	3	12	0
E	12	17	12	17	0
F	12	17	14	19	2
G	17	19	17	19	0

Expected project length = 19 weeks

Variance of the critical path = 1.333

Standard deviation of the critical path = 1.155 weeks

The activities along the critical path are B, D, E, and G. These activities have zero slack as shown in the table.

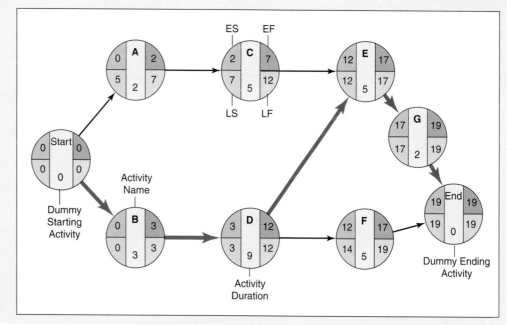

FIGURE 3.20 ■

Critical Path for Solved Problem 3.5

Solved Problem 3.6

The following information has been computed from a project:

$$\text{Expected total project time} = T = 62 \text{ weeks}$$
$$\text{Project variance} = \sigma_p^2 = 81$$

What is the probability that the project will be completed 18 weeks *before* its expected completion date?

SOLUTION

The desired completion date is 18 weeks before the expected completion date, 62 weeks. The desired completion date is 44 (or 62 − 18) weeks.

$$Z = \frac{\text{Due date} - \text{expected completion date}}{\sigma_p} = \frac{44-62}{9} = \frac{-18}{9} = -2.0$$

The normal curve appears as follows:

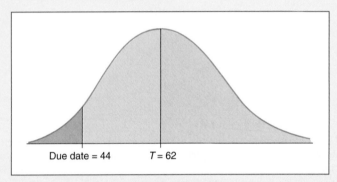

Due date = 44   T = 62

Because the normal curve is symmetrical and table values are calculated for positive values of Z, the area desired is equal to 1 − (table value). For Z = +2.0, the area from the table is .97725. Thus, the area corresponding to a Z value of −2.0 is .02275 (or 1 − 0.97725). Hence, the probability of completing the project 18 weeks before the expected completion date is approximately .023, or 2.3%.

Solved Problem 3.7

Determine the least cost of reducing the project completion date by 3 months based on the following information:

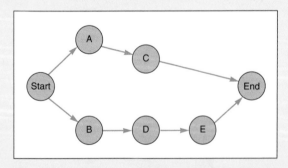

ACTIVITY	NORMAL TIME (MONTHS)	CRASH TIME (MONTHS)	NORMAL COST	CRASH COST
A	6	4	$2,000	$2,400
B	7	5	3,000	3,500
C	7	6	1,000	1,300
D	6	4	2,000	2,600
E	9	8	8,800	9,000

SOLUTION

The first step in this problem is to compute ES, EF, LS, LF, and slack for each activity.

ACTIVITY	ES	EF	LS	LF	SLACK
A	0	6	9	15	9
B	0	7	0	7	0
C	6	13	15	22	9
D	7	13	7	13	0
E	13	22	13	22	0

The critical path consists of activities B, D, and E.

Next, crash cost/month must be computed for each activity.

ACTIVITY	NORMAL TIME − CRASH TIME	CRASH COST − NORMAL COST	CRASH COST/MONTH	CRITICAL PATH?
A	2	$400	$200/month	No
B	2	500	250/month	Yes
C	1	300	300/month	No
D	2	600	300/month	Yes
E	1	200	200/month	Yes

Finally, we will select that activity on the critical path with the smallest crash cost/month. This is activity E. Thus, we can reduce the total project completion date by 1 month for an additional cost of $200. We still need to reduce the project completion date by 2 more months. This reduction can be achieved at least cost along the critical path by reducing activity B by 2 months for an additional cost of $500. This solution is summarized in the following table:

ACTIVITY	MONTHS REDUCED	COST
E	1	$200
B	2	500
		Total: $700

INTERNET AND STUDENT CD-ROM EXERCISES

Visit our home page or use your student CD-ROM to help you with the material in this chapter.

 On Our Home Page, www.prenhall.com/heizer

- Self-Tests
- Practice Problems
- Internet Exercises
- Current Articles and Research
- Virtual Company Tour
- Internet Homework Problems
- Internet Cases

 On Your Student CD-ROM

- PowerPoint Lecture
- Video Clip and Video Case
- Practice Problems
- ExcelOM
- Excel OM Data Files
- MS Project
- Active Model Exercise

ADDITIONAL CASE STUDIES

Internet Case Studies: Visit our Web site at www.prenhall.com/heizer **for these free case studies:**

- **Haywood Brothers Construction Company:** Involves finding the likelihood a project will be completed as scheduled.

- **Family Planning Research Center of Nigeria:** Deals with critical path scheduling, crashing, and personnel smoothing at an African clinic.

- **Shale Oil Company:** This oil refinery must shut down for maintenance of a major piece of equipment.

Harvard has selected these Harvard Business School cases to accompany this chapter
(textbookcasematch.hbsp.harvard.edu):

- **Microsoft Office 2000** (#600-097): An analysis of the evolution of the Office 2000 project.

- **Chrysler and BMW: Tritec Engine Joint Venture** (#600-004): A gifted project leader defines a new product strategy.

- **BAE Automated Systems (A): Denver International Baggage-Handling System** (#396-311): The project management of the construction of Denver's baggage-handling system.

- **Turner Construction Co.** (#190-128): Deals with the project management control system at a construction company.

Forecasting

Chapter Outline

LEARNING OBJECTIVES

When you complete this chapter you should be able to

IDENTIFY OR DEFINE:

Forecasting

Types of forecasts

Time horizons

Approaches to forecasts

DESCRIBE OR EXPLAIN:

Moving averages

Exponential smoothing

Trend projections

Regression and correlation analysis

Measures of forecast accuracy

Forecasting Provides Tupperware's Competitive Advantage

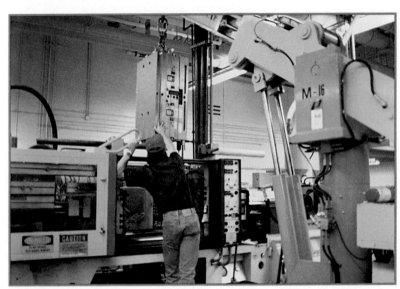

At Tupperware, stainless steel alloy molds, each requiring over 1,000 hours of skilled handcrafting, are the heart of the manufacturing process. Each mold creates the exact shape of a new product: Molds cost an average of $100,000 and can weigh up to 5 tons. When a specific product is scheduled for a production run, its mold is carefully placed, as we see in the photo, into an injection molding machine.

tered "consultants" or sales representatives, (2) the percentage of currently "active" dealers (this number changes each week and month), and (3) sales per active dealer, on a weekly basis. Forecasts incorporate historical data, recent events, and promotional events.

Tupperware maintains its edge over strong competitors like Rubbermaid by using a *group process* to refine its statistical forecasts. Although inputs come from sales, marketing, finance, and production, final forecasts are the consensus of all participating managers. This final step is Tupperware's version of the "jury of executive opinion" described in this chapter.

When most people think of Tupperware, they envision plastic food-storage containers sold through home parties. However, Tupperware happens to be a successful global manufacturer, with more than 85% of its $1.1 billion in sales outside the U.S. A household name in nearly 100 countries, the firm has 13 plants located around the world: one in South Carolina, three in Latin America, one in Africa, four in Europe, and four in Asia. Throughout the world, Tupperware stands for quality, providing a lifetime warranty that each of its 400 plastic products will not chip, crack, break, or peel.

Forecasting demand at Tupperware is a critical, never-ending process. Each of its 50 profit centers around the world is responsible for computerized

monthly, quarterly, and 12-month sales projections. These are aggregated by region and then globally at Tupperware's world headquarters in Orlando, Florida. These forecasts drive production at each plant.

The variety of statistical forecasting models used at Tupperware includes every technique discussed in this chapter, including moving averages, exponential smoothing, and regression analysis. At world headquarters, huge databases are maintained to map the sales of each product, the test-market results of each *new* product (20% of the firm's sales come from products less than two years old), and the stage of each product in its own life cycle.

Three factors are key in Tupperware's sales forecasts: (1) the number of regis-

The plastic pellets that are melted at 500 degrees into Tupperware products are dropped through pipes from second-floor bins into the machine holding a mold. After being injected into water-cooled molds at a pressure up to 20,000 lbs per square inch, the product cools and is removed and inspected.

TUPPERWARE CORPORATION

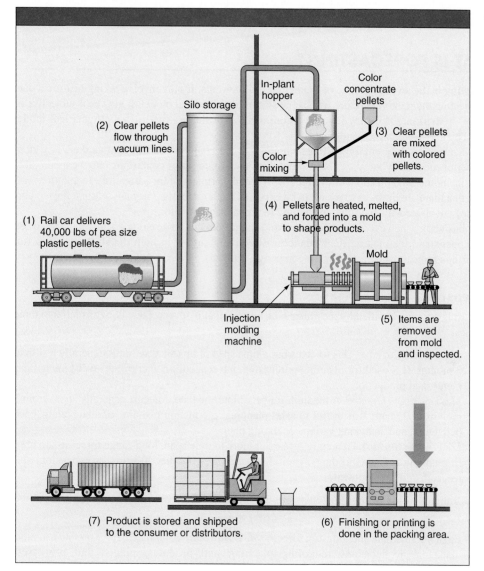

Tupperware's
Manufacturing Process

In-plant hopper

Color concentrate pellets

Silo storage

(2) Clear pellets flow through vacuum lines.

(3) Clear pellets are mixed with colored pellets.

Color mixing

(4) Pellets are heated, melted, and forced into a mold to shape products.

(1) Rail car delivers 40,000 lbs of pea size plastic pellets.

Mold

Injection molding machine

(5) Items are removed from mold and inspected.

(7) Product is stored and shipped to the consumer or distributors.

(6) Finishing or printing is done in the packing area.

Every day managers like those at Tupperware make decisions without knowing what will happen in the future. They order inventory without knowing what sales will be, purchase new equipment despite uncertainty about demand for products, and make investments without knowing what profits will be. Managers are always trying to make better estimates of what will happen in the future in the face of uncertainty. Making good estimates is the main purpose of forecasting.

In this chapter, we examine different types of forecasts and present a variety of forecasting models. Our purpose is to show that there are many ways for managers to forecast. We also provide an overview of business sales forecasting and describe how to prepare, monitor, and judge the accuracy of a forecast. Good forecasts are an *essential* part of efficient service and manufacturing operations.

WHAT IS FORECASTING?

Forecasting

The art and science of predicting future events.

Forecasting is the art and science of predicting future events. It may involve taking historical data and projecting them into the future with some sort of mathematical model. It may be a subjective or intuitive prediction. Or it may involve a combination of these—that is, a mathematical model adjusted by a manager's good judgment.

As we introduce different forecasting techniques in this chapter, you will see that there is seldom one superior method. What works best in one firm under one set of conditions may be a complete disaster in another organization, or even in a different department of the same firm. In addition, you will see that there are limits as to what can be expected from forecasts. They are seldom, if ever, perfect. They are also costly and time-consuming to prepare and monitor.

Few businesses, however, can afford to avoid the process of forecasting by just waiting to see what happens and then taking their chances. Effective planning in both the short and long run depends on a forecast of demand for the company's products.

Forecasting Time Horizons

A forecast is usually classified by the *future time horizon* that it covers. Time horizons fall into three categories:

1. *Short-range forecast.* This forecast has a time span of up to 1 year but is generally less than 3 months. It is used for planning purchasing, job scheduling, workforce levels, job assignments, and production levels.
2. *Medium-range forecast.* A medium-range, or intermediate, forecast generally spans from 3 months to 3 years. It is useful in sales planning, production planning and budgeting, cash budgeting, and analyzing various operating plans.
3. *Long-range forecast.* Generally 3 years or more in time span, long-range forecasts are used in planning for new products, capital expenditures, facility location or expansion, and research and development.

Medium-range and long-range forecasts are distinguished from short-range forecasts by three features:

1. First, intermediate and long-run forecasts *deal with more comprehensive issues* and support management decisions regarding planning and products, plants, and processes. Implementing some facility decisions, such as GM's decision to open a new Brazilian manufacturing plant, can take 5 to 8 years from inception to completion.
2. Second, short-term forecasting usually *employs different methodologies* than longer-term forecasting. Mathematical techniques, such as moving averages, exponential smoothing, and trend extrapolation (all of which we shall examine shortly), are common to short-run projections. Broader, *less* quantitative methods are useful in predicting such issues as whether a new product, like the optical disk recorder, should be introduced into a company's product line.
3. Finally, as you would expect, short-range forecasts *tend to be more accurate* than longer-range forecasts. Factors that influence demand change every day. Thus, as the time horizon lengthens, it is likely that one's forecast accuracy will diminish. It almost goes without saying, then, that sales forecasts must be updated regularly in order to maintain their value and integrity. After each sales period, forecasts should be reviewed and revised.

Our forecasting ability has improved, but it has been outpaced by an increasingly complex world economy.

The Influence of Product Life Cycle

Another factor to consider when developing sales forecasts, especially longer ones, is product life cycle. Products, and even services, do not sell at a constant level throughout their lives. Most successful products pass through four stages: (1) introduction, (2) growth, (3) maturity, and (4) decline.

Products in the first two stages of the life cycle (such as virtual reality and high-definition TVs) need longer forecasts than those (such as $3\frac{1}{2}''$ floppy disks and station wagons) in the maturity and decline stages. Forecasts that reflect life cycle are useful in projecting different staffing levels, inventory levels, and factory capacity as the product passes from the first to the last stage. The challenge of introducing new products is treated in more detail in Chapter 5.

TYPES OF FORECASTS

Economic forecasts
Planning indicators valuable in helping organizations prepare medium- to long-range forecasts.

Technological forecasts
Long-term forecasts concerned with the rates of technological progress.

Demand forecasts
Projections of a company's sales for each time period in the planning horizon.

Organizations use three major types of forecasts in planning future operations:

1. **Economic forecasts** address the business cycle by predicting inflation rates, money supplies, housing starts, and other planning indicators.
2. **Technological forecasts** are concerned with rates of technological progress, which can result in the birth of exciting new products, requiring new plants and equipment.
3. **Demand forecasts** are projections of demand for a company's products or services. These forecasts, also called *sales forecasts*, drive a company's production, capacity, and scheduling systems and serve as inputs to financial, marketing, and personnel planning.

Economic and technological forecasting are specialized techniques that may fall outside the role of the operations manager. The emphasis in this book will therefore be on demand forecasting.

THE STRATEGIC IMPORTANCE OF FORECASTING

Good forecasts are of critical importance in all aspects of a business: *The forecast is the only estimate of demand until actual demand becomes known.* Forecasts of demand therefore drive decisions in many areas. Let's look at the impact of product forecast on three activities: (1) human resources, (2) capacity, and (3) supply-chain management.

Video 4.1

Forecasting at Hard Rock Cafe

Human Resources

Hiring, training, and laying off workers all depend on anticipated demand. If the human resources department must hire additional workers without warning, the amount of training declines and the quality of the workforce suffers. A large Louisiana chemical firm almost lost its biggest customer when a quick expansion to round-the-clock shifts led to a total breakdown in quality control on the second and third shifts.

Capacity

When capacity is inadequate, the resulting shortages can mean undependable delivery, loss of customers, and loss of market share. This is exactly what happened to Nabisco when it underestimated the huge demand for its new low-fat Snackwell Devil's Food Cookies. Even with production lines working overtime, Nabisco could not keep up with demand, and it lost customers. When excess capacity is built, on the other hand, costs can skyrocket.

Supply-Chain Management

Good supplier relations and the ensuing price advantages for materials and parts depend on accurate forecasts. For example, auto manufacturers who want TRW Corp. to guarantee sufficient airbag capacity must provide accurate forecasts to justify TRW plant expansions. In the global marketplace, where expensive components for Boeing 777 jets are manufactured in dozens of countries, coordination driven by forecasts is critical. Scheduling transportation to Seattle for final assembly at the lowest possible cost means no last-minute surprises that can harm already low profit margins.

SEVEN STEPS IN THE FORECASTING SYSTEM

Forecasting follows seven basic steps. We use Tupperware Corporation, the focus of this chapter's *Global Company Profile*, as an example of each step.

1. *Determine the use of the forecast.* Tupperware uses demand forecasts to drive production at each of its 13 plants.
2. *Select the items to be forecasted.* For Tupperware, there are over 400 products, each with its own SKU (stock-keeping unit). Tupperware, like other firms of this type, does demand forecasts by families (or groups) of SKUs.
3. *Determine the time horizon of the forecast.* Is it short-, medium-, or long-term? Tupperware develops forecasts monthly, quarterly, and for 12-month sales projections.
4. *Select the forecasting model(s).* Tupperware uses a variety of statistical models that we shall discuss, including moving averages, exponential smoothing, and regression analysis. It also employs judgmental, or nonquantitative, models.
5. *Gather the data needed to make the forecast.* Tupperware's world headquarters maintains huge databases to monitor the sale of each product.
6. *Make the forecast.*
7. *Validate and implement the results.* At Tupperware, forecasts are reviewed in sales, marketing, finance, and production departments to make sure that the model, assumptions, and data are valid. Error measures are applied; then the forecasts are used to schedule material, equipment, and personnel at each plant.

These seven steps present a systematic way of initiating, designing, and implementing a forecasting system. When the system is to be used to generate forecasts regularly over time, data must be routinely collected. Then actual computations are usually made by computer.

Regardless of the system that firms like Tupperware use, each company faces several realities:

1. Forecasts are seldom perfect. This means that outside factors that we cannot predict or control often impact the forecast. Companies need to allow for this reality.
2. Most forecasting techniques assume that there is some underlying stability in the system. Consequently, some firms automate their predictions using computerized forecasting software, then closely monitor only the product items whose demand is erratic.
3. Both product family and aggregated forecasts are more accurate than individual product forecasts. Tupperware, for example, aggregates product forecasts by both family (e.g., mixing bowls versus cups versus storage containers) and region. This approach helps balance the over- and underpredictions of each product and country.

FORECASTING APPROACHES

There are two general approaches to forecasting, just as there are two ways to tackle all decision modeling. One is quantitative analysis; the other is a qualitative approach. **Quantitative forecasts** use a variety of mathematical models that rely on historical data and/or causal variables to forecast demand. Subjective or **qualitative forecasts** incorporate such factors as the decision maker's intuition, emotions, personal experiences, and value system in reaching a forecast. Some firms use one approach and some use the other. In practice, a combination of the two is usually most effective.

Overview of Qualitative Methods

In this section, we consider four different *qualitative* forecasting techniques:

1. **Jury of executive opinion.** Under this method, the opinions of a group of high-level experts or managers, often in combination with statistical models, are pooled to arrive at a group estimate of demand. Bristol-Meyers Squibb Company, for example, uses 220 well-known research scientists as its jury of executive opinion to get a grasp on future trends in the world of medical research.

"Those who can predict the future have never been appreciated in their own times."

Quality expert Philip Crosby

Quantitative forecasts
Forecasts that employ one or more mathematical models that rely on historical data and/or causal variables to forecast demand.

Qualitative forecasts
Forecasts that incorporate such factors as the decision maker's intuition, emotions, personal experiences, and value system.

Jury of executive opinion
A forecasting technique that takes the opinion of a small group of high-level managers and results in a group estimate of demand.

Delphi method
A forecasting technique using a group process that allows experts to make forecasts.

2. **Delphi method.** There are three different types of participants in the Delphi method: decision makers, staff personnel, and respondents. Decision makers usually consist of a group of 5 to 10 experts who will be making the actual forecast. Staff personnel assist decision makers by preparing, distributing, collecting, and summarizing a series of questionnaires and survey results. The respondents are a group of people, often located in different places, whose judgments are valued. This group provides inputs to the decision makers before the forecast is made.

    The state of Alaska, for example, has used the Delphi method to develop its long-range economic forecast. An amazing 90% of the state's budget is derived from 1.5 million barrels of oil pumped daily through a pipeline at Prudhoe Bay. The large Delphi panel of experts had to represent all groups and opinions in the state and all geographic areas. Delphi was the perfect forecasting tool because panelist travel could be avoided. It also meant that leading Alaskans could participate because their schedules were not impacted by meetings and distances.

Sales force composite
A forecasting technique based upon salespersons' estimates of expected sales.

3. **Sales force composite.** In this approach, each salesperson estimates what sales will be in his or her region. These forecasts are then reviewed to ensure that they are realistic. Then they are combined at the district and national levels to reach an overall forecast.

Consumer market survey
A forecasting method that solicits input from customers or potential customers regarding future purchasing plans.

4. **Consumer market survey.** This method solicits input from customers or potential customers regarding future purchasing plans. It can help not only in preparing a forecast but also in improving product design and planning for new products. The consumer market survey and sales force composite methods can, however, suffer from overly optimistic forecasts that arise from customer input. The 2001 crash of the telecommunication industry was the result of overexpansion to meet "explosive customer demand." Where did this data come from? Oplink Communications, a Nortel Networks supplier, says its "company forecasts over the last few years were based mainly on informal conversations with customers."[1]

Overview of Quantitative Methods

Five quantitative forecasting methods, all of which use historical data, are described in this chapter. They fall into two categories:

1. Naive approach
2. Moving averages
3. Exponential smoothing — **time-series models**
4. Trend projection
5. Linear regression — **associative model**

Time-Series Models   **Time-series** models predict on the assumption that the future is a function of the past. In other words, they look at what has happened over a period of time and use a series of past data to make a forecast. If we are predicting weekly sales of lawn mowers, we use the past weekly sales for lawn mowers when making the forecasts.

Time series
A forecasting technique that uses a series of past data points to make a forecast.

Associative Models   Associative (or causal) models, such as linear regression, incorporate the variables or factors that might influence the quantity being forecast. For example, an associative model for lawn mower sales might include such factors as new housing starts, advertising budget, and competitors' prices.

Two famous quotes:
"You can never plan the future from the past."
Sir Edmund Burke
"I know of no way of judging the future but by the past."
Patrick Henry

TIME-SERIES FORECASTING

A time series is based on a sequence of evenly spaced (weekly, monthly, quarterly, and so on) data points. Examples include weekly sales of Nike Air Jordans, quarterly earnings reports of Microsoft stock, daily shipments of Coors beer, and annual consumer price indices. Forecasting time-series data implies that future values are predicted *only* from past values and that other variables, no matter how potentially valuable, may be ignored.

[1]"Lousy Sales Forecasts Helped Fuel the Telecom Mess," the *Wall Street Journal* (July 9, 2001): B1–B4.

Decomposition of a Time Series

Analyzing time series means breaking down past data into components and then projecting them forward. A time series has four components: trend, seasonality, cycles, and random variation.

1. *Trend* is the gradual upward or downward movement of the data over time. Changes in income, population, age distribution, or cultural views may account for movement in trend.
2. *Seasonality* is a data pattern that repeats itself after a period of days, weeks, months, or quarters. There are six common seasonality patterns:

PERIOD OF PATTERN	"SEASON" LENGTH	NUMBER OF "SEASONS" IN PATTERN
Week	Day	7
Month	Week	4–4 ½
Month	Day	28–31
Year	Quarter	4
Year	Month	12
Year	Week	52

Restaurants and barber shops, for example, experience weekly seasons, with Saturday being the peak of business. Beer distributors forecast yearly patterns, with monthly seasons. Three "seasons"—May, July, and September—each contain a big beer-drinking holiday.

3. *Cycles* are patterns in the data that occur every several years. They are usually tied into the business cycle and are of major importance in short-term business analysis and planning. Predicting business cycles is difficult because they may be affected by political events or by international turmoil.
4. *Random variations* are "blips" in the data caused by chance and unusual situations. They follow no discernible pattern, so they cannot be predicted.

Figure 4.1 illustrates a demand over a 4-year period. It shows the average, trend, seasonal components, and random variations around the demand curve. The average demand is the sum of the demand for each period divided by the number of data periods.

During stable times, forecasting is easy; it is just this year's performance plus or minus a few percentage points.

Naive Approach

The simplest way to forecast is to assume that demand in the next period will be equal to demand in the most recent period. In other words, if sales of a product—say, Motorola cellular phones—were 68 units in January, we can forecast that February's sales will also be 68 phones. Does this make any

FIGURE 4.1 ■

Product Demand Charted over 4 Years with a Growth Trend and Seasonality Indicated

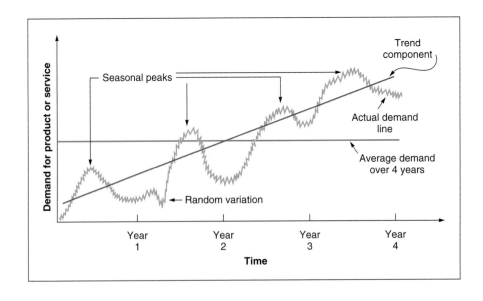

Naive approach
A forecasting technique that assumes demand in the next period is equal to demand in the most recent period.

Moving averages
A forecasting method that uses an average of the *n* most recent periods of data to forecast the next period.

sense? It turns out that for some product lines, this **naive approach** is the most cost-effective and efficient objective forecasting model. At least it provides a starting point against which more sophisticated models that follow can be compared.

Moving Averages

A **moving-average** forecast uses a number of historical actual data values to generate a forecast. Moving averages are useful *if we can assume that market demands will stay fairly steady over time.* A 4-month moving average is found by simply summing the demand during the past 4 months and dividing by 4. With each passing month, the most recent month's data are added to the sum of the previous 3 months' data, and the earliest month is dropped. This practice tends to smooth out short-term irregularities in the data series.

Mathematically, the simple moving average (which serves as an estimate of the next period's demand) is expressed as

$$\text{Moving average} = \frac{\Sigma \text{ demand in previous } n \text{ periods}}{n} \tag{4-1}$$

where *n* is the number of periods in the moving average—for example, 4, 5, or 6 months, respectively, for a 4-, 5-, or 6-period moving average.

Example 1 shows how moving averages are calculated.

Example 1

**Excel OM Data File
Ch04Ex1.xla**

Active Model 4.1

Example 1 is further illustrated as Active Model 4.1 on your CD-ROM.

Storage shed sales at Donna's Garden Supply are shown in the middle column of the table below. A 3-month moving average appears on the right.

MONTH	ACTUAL SHED SALES	3-MONTH MOVING AVERAGE
January	10	
February	12	
March	13	
April	16	$(10 + 12 + 13)/3 = 11\frac{2}{3}$
May	19	$(12 + 13 + 16)/3 = 13\frac{2}{3}$
June	23	$(13 + 16 + 19)/3 = 16$
July	26	$(16 + 19 + 23)/3 = 19\frac{1}{3}$
August	30	$(19 + 23 + 26)/3 = 22\frac{2}{3}$
September	28	$(23 + 26 + 30)/3 = 26\frac{1}{3}$
October	18	$(26 + 30 + 28)/3 = 28$
November	16	$(30 + 28 + 18)/3 = 25\frac{1}{3}$
December	14	$(28 + 18 + 16)/3 = 20\frac{2}{3}$

Thus we see that the forecast for December is $20\frac{2}{3}$. To project the demand for sheds in the coming January, we sum the October, November, and December sales and divide by 3: January forecast = (18 + 16 + 14)/3 = 16.

When a detectable trend or pattern is present, *weights* can be used to place more emphasis on recent values. This practice makes forecasting techniques more responsive to changes because more recent periods may be more heavily weighted. Choice of weights is somewhat arbitrary because there is no set formula to determine them. Therefore, deciding which weights to use requires some experience. For example, if the latest month or period is weighted too heavily, the forecast might reflect a large unusual change in the demand or sales pattern too quickly.

A weighted moving average may be expressed mathematically as

$$\begin{array}{c}\text{Weighted} \\ \text{moving average}\end{array} = \frac{\Sigma \text{ (weight for period } n\text{)(demand in period } n\text{)}}{\Sigma \text{ weights}} \tag{4-2}$$

Example 2 shows how to calculate a weighted moving average.

Example 2

**Excel OM Data File
Ch04Ex2.xla**

Donna's Garden Supply (see Example 1) decides to forecast storage shed sales by weighting the past 3 months as follows:

WEIGHTS APPLIED	PERIOD
3	Last month
2	Two months ago
1	Three months ago
6	Sum of weights

Forecast for this month =

$$\frac{3 \times \text{sales last mo.} + 2 \times \text{sales 2 mos. ago} + 1 \times \text{sales 3 mos. ago}}{6 \quad \text{sum of the weights}}$$

The results of this weighted-average forecast are as follows:

MONTH	ACTUAL SHED SALES	THREE-MONTH WEIGHTED MOVING AVERAGE
January	10	
February	12	
March	13	
April	16	$[(3 \times 13) + (2 \times 12) + (10)]/6 = 12\,\frac{1}{6}$
May	19	$[(3 \times 16) + (2 \times 13) + (12)]/6 = 14\,\frac{1}{3}$
June	23	$[(3 \times 19) + (2 \times 16) + (13)]/6 = 17$
July	26	$[(3 \times 23) + (2 \times 19) + (16)]/6 = 20\,\frac{1}{2}$
August	30	$[(3 \times 26) + (2 \times 23) + (19)]/6 = 23\,\frac{5}{6}$
September	28	$[(3 \times 30) + (2 \times 26) + (23)]/6 = 27\,\frac{1}{2}$
October	18	$[(3 \times 28) + (2 \times 30) + (26)]/6 = 28\,\frac{1}{3}$
November	16	$[(3 \times 18) + (2 \times 28) + (30)]/6 = 23\,\frac{1}{3}$
December	14	$[(3 \times 16) + (2 \times 18) + (28)]/6 = 18\,\frac{2}{3}$

In this particular forecasting situation, you can see that more heavily weighting the latest month provides a much more accurate projection.

Using data that are 20 years old (e.g., from housing prices or tuition rates) may not be so useful that all the past data are needed to forecast next year's values. It is not always necessary to use *all* data.

Both simple and weighted moving averages are effective in smoothing out sudden fluctuations in the demand pattern in order to provide stable estimates. Moving averages do, however, present three problems:

1. Increasing the size of *n* (the number of periods averaged) does smooth out fluctuations better, but it makes the method less sensitive to *real* changes in the data.
2. Moving averages cannot pick up trends very well. Because they are averages, they will always stay within past levels and will not predict changes to either higher or lower levels. That is, they *lag* the actual values.
3. Moving averages require extensive records of past data.

Figure 4.2, a plot of the data in Examples 1 and 2, illustrates the lag effect of the moving-average models. Note that both the moving average and weighted moving-average lines lag the actual demand from April on. The weighted moving average, however, usually reacts more quickly to demand changes. Even in periods of downturn (see November and December), it more closely tracks the demand.

Exponential Smoothing

Exponential smoothing
A weighted moving-average forecasting technique in which data points are weighted by an exponential function.

Exponential smoothing is a sophisticated weighted moving-average forecasting method that is still fairly easy to use. It involves very *little* record keeping of past data. The basic exponential smoothing formula can be shown as follows:

New forecast = last period's forecast
          + α (last period's actual demand − last period's forecast)   (4-3)

FIGURE 4.2 ■

Actual Demand vs. Moving-Average and Weighted Moving-Average Methods for Donna's Garden Supply

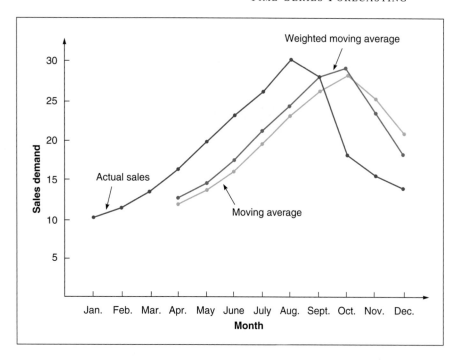

Smoothing constant
The weighting factor used in an exponential smoothing forecast, a number between 0 and 1.

where α is a weight, or **smoothing constant**, chosen by the forecaster, that has a value between 0 and 1. Equation (4-3) can also be written mathematically as

$$F_t = F_{t-1} + \alpha(A_{t-1} - F_{t-1}) \qquad (4\text{-}4)$$

where
F_t = new forecast
F_{t-1} = previous forecast
α = smoothing (or weighting) constant $(0 \le \alpha \le 1)$
A_{t-1} = previous period's actual demand

The concept is not complex. The latest estimate of demand is equal to our old estimate adjusted by a fraction of the difference between the last period's actual demand and the old estimate. Example 3 shows how to use exponential smoothing to derive a forecast.

Example 3

In January, a car dealer predicted February demand for 142 Ford Mustangs. Actual February demand was 153 autos. Using a smoothing constant chosen by management of $\alpha = .20$, we can forecast March demand using the exponential smoothing model. Substituting our sample data into the formula, we obtain

New forecast (for March demand) = 142 + .2(153 − 142) = 142 + 2.2

= 144.2

Thus, the March demand forecast for Ford Mustangs is rounded to 144.

Exponential smoothing may have an obscure-sounding name, but it is actually widely used in business and is an important part of many computerized inventory control systems.

The *smoothing constant*, α, is generally in the range from .05 to .50 for business applications. It can be changed to give more weight to recent data (when α is high) or more weight to past data (when α is low). When α reaches the extreme of 1.0, then in Equation (4-4), $F_t = 1.0A_{t-1}$. All the older values drop out, and the forecast becomes identical to the naive model mentioned earlier in this chapter. That is, the forecast for the next period is just the same as this period's demand.

The following table helps illustrate this concept. For example, when $\alpha = .5$, we can see that the new forecast is based almost entirely on demand in the last three or four periods. When $\alpha = .1$, the forecast places little weight on recent demand and takes many periods (about 19) of historic values into account.

Just 3 weeks after IBM announced a new home computer line a few years ago, the firm sold out its supply through year's end and was unable to fill holiday orders. Why? IBM attributes the shortage to conservative forecasting—a chronic problem in miscalculating demand for PCs. The potential revenue loss of $100 million repeats similar forecasting problems for IBM's popular Think Pad portable PC two years earlier.

	WEIGHT ASSIGNED TO				
SMOOTHING CONSTANT	MOST RECENT PERIOD (α)	2ND MOST RECENT PERIOD $\alpha(1 - \alpha)$	3RD MOST RECENT PERIOD $\alpha(1 - \alpha)^2$	4TH MOST RECENT PERIOD $\alpha(1 - \alpha)^3$	5TH MOST RECENT PERIOD $\alpha(1 - \alpha)^4$
$\alpha = .1$	.1	.09	.081	.073	.066
$\alpha = .5$	.5	.25	.125	.063	.031

Selecting the Smoothing Constant   The exponential smoothing approach is easy to use, and it has been successfully applied in virtually every type of business. However, the appropriate value of the smoothing constant, α, can make the difference between an accurate forecast and an inaccurate forecast. High values of α are chosen when the underlying average is likely to change. Low values of α are used when the underlying average is fairly stable. In picking a value for the smoothing constant, the objective is to obtain the most accurate forecast.

Measuring Forecast Error

The forecast error tells us how well the model performed against itself using past data.

The overall accuracy of any forecasting model—moving average, exponential smoothing, or other—can be determined by comparing the forecasted values with the actual or observed values. If F_t denotes the forecast in period t, and A_t denotes the actual demand in period t, the *forecast error* (or deviation) is defined as

$$\text{Forecast error} = \text{Actual demand} - \text{Forecast value}$$
$$= A_t - F_t$$

There are several measures commonly used in practice to calculate the overall forecast error. These measures can be used to compare different forecasting models, as well as to monitor forecasts to ensure they are performing well. Three of the most popular measures are mean absolute deviation (MAD), mean squared error (MSE), and mean absolute percent error (MAPE). We now describe and give an example of each.

Mean Absolute Deviation   The first measure of the overall forecast error for a model is the **mean absolute deviation (MAD)**. This value is computed by taking the sum of the absolute values of the individual forecast errors and dividing by the number of periods of data (*n*):

Mean absolute deviation (MAD)
A measure of the overall forecast error for a model.

$$\text{MAD} = \frac{\sum|\text{actual} - \text{forecast}|}{n} \qquad (4\text{-}5)$$

Example 4 applies this concept with a trial-and-error testing of two values of α.

Example 4

During the past 8 quarters, the Port of Baltimore has unloaded large quantities of grain from ships. The port's operations manager wants to test the use of exponential smoothing to see how well the technique works in predicting tonnage unloaded. He guesses that the forecast of grain unloaded in the first quarter was 175 tons. Two values of α are examined: $\alpha = .10$ and $\alpha = .50$. The following table shows the *detailed* calculations for $\alpha = .10$ only:

Excel OM Data Files
Ch04Ex4a.xla,
Ch04Ex4b.xla

Active Model 4.2

Example 4 is further illustrated in Active Model 4.2 on the CD-ROM and in the Exercise located in your Student Lecture Guide.

Quarter	Actual Tonnage Unloaded	Rounded Forecast with $\alpha = .10$[a]	Rounded Forecast with $\alpha = .50$[a]
1	180	175	175
2	168	176 = 175.00 + .10(180 − 175)	178
3	159	175 = 175.50 + .10(168 − 175.50)	173
4	175	173 = 174.75 + .10(159 − 174.75)	166
5	190	173 = 173.18 + .10(175 − 173.18)	170
6	205	175 = 173.36 + .10(190 − 173.36)	180
7	180	178 = 175.02 + .10(205 − 175.02)	193
8	182	178 = 178.02 + .10(180 − 178.02)	186
9	?	179 = 178.22 + .10(182 − 178.22)	184

[a]Forecasts rounded to the nearest ton.

To evaluate the accuracy of each smoothing constant, we can compute forecast errors in terms of absolute deviations and MADs.

Quarter	Actual Tonnage Unloaded	Rounded Forecast with $\alpha = .10$	Absolute Deviation for $\alpha = .10$	Rounded Forecast with $\alpha = .50$	Absolute Deviation for $\alpha = .50$
1	180	175	5	175	5
2	168	176	8	178	10
3	159	175	16	173	14
4	175	173	2	166	9
5	190	173	17	170	20
6	205	175	30	180	25
7	180	178	2	193	13
8	182	178	4	186	4
	Sum of absolute deviations		84		100
	$\text{MAD} = \dfrac{\sum|\text{deviations}|}{n}$		10.50		12.50

On the basis of this analysis, a smoothing constant of $\alpha = .10$ is preferred to $\alpha = .50$ because its MAD is smaller.

Most computerized forecasting software includes a feature that automatically finds the smoothing constant with the lowest forecast error. Some software modifies the α value if errors become larger than acceptable.

Mean squared error (MSE)
The average of the squared differences between the forecasted and observed values.

Mean Squared Error   The **mean squared error (MSE)** is a second way of measuring overall forecast error. MSE is the average of the squared differences between the forecasted and observed values. Its formula is

$$MSE = \frac{\Sigma(\text{forecast errors})^2}{n} \qquad (4\text{-}6)$$

Example 5 finds the MSE for the Port of Baltimore introduced in Example 4.

Example 5

QUARTER	ACTUAL TONNAGE UNLOADED	FORECAST FOR $\alpha = .10$	(ERROR)^2
1	180	175	$5^2 = 25$
2	168	176	$(-8)^2 = 64$
3	159	175	$(-16)^2 = 256$
4	175	173	$2^2 = 4$
5	190	173	$17^2 = 289$
6	205	175	$30^2 = 900$
7	180	178	$2^2 = 4$
8	182	178	$4^2 = 16$
			Sum of errors squared = 1,558

$$MSE = \frac{\Sigma\ (\text{forecast errors})^2}{n} = 1,558/8 = 194.75$$

Is this MSE good or bad? It all depends on the MSEs for other values of α. As a practice exercise, find the MSE for $\alpha = .50$. (You should get MSE = 201.5.) The result indicates that $\alpha = .10$ is a better choice because we want to minimize MSE. Coincidentally, this confirms the conclusion we reached using MAD in Example 4.

MSE accentuates large deviations.

A drawback of using the MSE is that it tends to accentuate large deviations due to the squared term. For example, if the forecast error for period 1 is twice as large as the error for period 2, the squared error in period 1 is four times as large as that for period 2. Hence, using MSE as the measure of forecast error typically indicates that we prefer to have several smaller deviations rather than even one large deviation.

Mean absolute percent error (MAPE)
The average of the absolute differences between the forecast and actual values, expressed as a percent of actual values.

Mean Absolute Percent Error   A problem with both the MAD and MSE is that their values depend on the magnitude of the item being forecast. If the forecast item is measured in thousands, the MAD and MSE values can be very large. To avoid this problem, we can use the **mean absolute percent error (MAPE)**. This is computed as the average of the absolute difference between the forecasted and actual values, expressed as a percentage of the actual values. That is, if we have forecasted and actual values for n periods, the MAPE is calculated as:

$$MAPE = \frac{100 \sum_{i=1}^{n} |\text{actual}_i - \text{forecast}_i| / \text{actual}_i}{n} \qquad (4\text{-}7)$$

Example 6 illustrates the calculations using the data from Examples 4 and 5.

Example 6

QUARTER	ACTUAL TONNAGE UNLOADED	FORECAST FOR $\alpha = .10$	ABSOLUTE PERCENT ERROR 100(|ERROR|/ACTUAL)
1	180	175	100(5/180) = 2.77%
2	168	176	100(8/168) = 4.76%
3	159	175	100(16/159) = 10.06%
4	175	173	100(2/175) = 1.14%
5	190	173	100(17/190) = 8.95%
6	205	175	100(30/205) = 14.63%
7	180	178	100(2/180) = 1.11%
8	182	178	100(4/182) = 2.20%
			Sum of % errors = 45.62%

$$MAPE = \frac{\Sigma\ \text{absolute percent errors}}{n} = \frac{45.62\%}{8} = 5.70\%$$

OM IN ACTION

Forecasting at Disney World

When Disney chairman Michael Eisner receives a daily report from his main theme parks in Orlando, Florida, the report contains only two numbers: the *forecast* of yesterday's attendance at the parks (Magic Kingdom, Epcot, Fort Wilderness, MGM Studios, and Blizzard Beach) and the *actual* attendance. An error close to zero (using MAPE as the measure) is expected. Eisner takes his forecasts very seriously.

The forecasting team at Disney World doesn't just do a daily prediction, however, and Eisner is not its only customer. The team also provides daily, weekly, monthly, annual, and 5-year forecasts to the labor management, maintenance, operations, finance, and park scheduling departments. Forecasters use judgmental models, econometric models, moving-average models, and regression analysis. The team's annual forecast of total

volume, conducted in 1999 for the year 2000, resulted in a MAPE of 0.

With 20% of Disney World's customers coming from outside the United States, its economic model includes such variables as consumer confidence and the gross domestic product of seven countries. Disney also surveys one million people each year to examine their future travel plans and their experiences at the parks. This helps forecast not only attendance, but behavior at each ride (how long people will wait and how many times they will ride). Inputs to the monthly forecasting model include airline specials, speeches by the Chair of the Federal Reserve, and Wall Street trends. Disney even monitors 3,000 school districts inside and outside the U.S. for holiday/vacation schedules.

Source: J. Newkirk and M. Haskell. "Forecasting in the Service Sector," presentation at the 12th Annual Meeting of the Production and Operations Management Society. April 1, 2001, Orlando, FL.

MAPE expresses the error as a percentage of the actual values.

The MAPE is perhaps the easiest measure to interpret. For example, a result that the MAPE is 6% is a clear statement that is not dependent on issues such as the magnitude of the input data.

Exponential Smoothing with Trend Adjustment

Simple exponential smoothing, the technique we just illustrated in Examples 3 to 6, is like *any* moving-average technique: It fails to respond to trends. Other forecasting techniques that can deal with trends are certainly available. However, because exponential smoothing is such a popular modeling approach in business, let us look at it in more detail.

Here is why exponential smoothing must be modified when a trend is present. Assume that demand for our product or service has been increasing by 100 units per month and that we have been forecasting with $\alpha = 0.4$ in our exponential smoothing model. The following table shows a severe lag in the 2nd, 3rd, 4th, and 5th months, even when our initial estimate for month 1 is perfect.

MONTH	ACTUAL DEMAND	FORECAST FOR MONTH $T(F_T)$
1	100	$F_1 = 100$ (given)
2	200	$F_2 = F_1 + \alpha(A_1 - F_1) = 100 + .4(100 - 100) = 100$
3	300	$F_3 = F_2 + \alpha(A_2 - F_2) = 100 + .4(200 - 100) = 140$
4	400	$F_4 = F_3 + \alpha(A_3 - F_3) = 140 + .4(300 - 140) = 204$
5	500	$F_5 = F_4 + \alpha(A_4 - F_4) = 204 + .4(400 - 204) = 282$

To improve our forecast, let us illustrate a more complex exponential smoothing model, one that adjusts for trend. The idea is to compute an exponentially smoothed average of the data and then adjust for positive or negative lag in trend. The new formula is

$$\text{Forecast including trend } (FIT_t) = \text{exponentially smoothed forecast } (F_t) \\ + \text{exponentially smoothed trend } (T_t) \tag{4-8}$$

With trend-adjusted exponential smoothing, estimates for both the average and the trend are smoothed. This procedure requires two smoothing constants, α for the average and β for the trend. We then compute the average and trend each period:

$$F_t = \alpha(\text{Actual demand last period}) + (1 - \alpha)(\text{Forecast last period} + \text{Trend estimate last period})$$

or

$$F_t = \alpha(A_{t-1}) + (1 - \alpha)(F_{t-1} + T_{t-1})$$ (4-9)

$$T_t = \beta(\text{Forecast this period} - \text{Forecast last period})$$
$$+ (1 - \beta)(\text{Trend estimate last period})$$

or

$$T_t = \beta(F_t - F_{t-1}) + (1 - \beta)T_{t-1}$$ (4-10)

where F_t = exponentially smoothed forecast of the data series in period t
T_t = exponentially smoothed trend in period t
A_t = actual demand in period t
α = smoothing constant for the average $(0 \le \alpha \le 1)$
β = smoothing constant for the trend $(0 \le \beta \le 1)$

So the three steps to compute a trend-adjusted forecast are:

Step 1: Compute F_t, the exponentially smoothed forecast for period t, using Equation (4-9).

Step 2: Compute the smoothed trend, T_t, using Equation (4-10).

Step 3: Calculate the forecast including trend, FIT_t, by the formula $FIT_t = F_t + T_t$.

Example 7 shows how to use trend-adjusted exponential smoothing.

Example 7

**Excel OM Data File
Ch04Ex7.xla**

Active Model 4.3

Example 7 is further
illustrated in Active Model
4.3 on the CD-ROM.

A large Portland manufacturer uses exponential smoothing to forecast demand for a piece of pollution-control equipment. It appears that an increasing trend is present.

MONTH (t)	ACTUAL DEMAND (A_t)	MONTH (t)	ACTUAL DEMAND (A_t)
1	12	6	21
2	17	7	31
3	20	8	28
4	19	9	36
5	24	10	?

Smoothing constants are assigned the values of $\alpha = .2$ and $\beta = .4$. Assume the initial forecast for month 1 (F_1) was 11 units and the trend over that period (T_1) was 2 units.

Step 1: Forecast for month 2:

$$F_2 = \alpha A_1 + (1 - \alpha)(F_1 + T_1)$$
$$F_2 = (.2)(12) + (1 - .2)(11 + 2)$$
$$= 2.4 + (.8)(13) = 2.4 + 10.4 = 12.8 \text{ units}$$

Step 2: Compute the trend in period 2:

$$T_2 = \beta(F_2 - F_1) + (1 - \beta)T_1$$
$$= .4(12.8 - 11) + (1 - .4)(2)$$
$$= (.4)(1.8) + (.6)(2) = .72 + 1.2 = 1.92$$

Step 3: Compute the forecast including trend (FIT_t):

$$FIT_2 = F_2 + T_2$$
$$= 12.8 + 1.92$$
$$= 14.72 \text{ units}$$

We will also do the same calculations for the third month.

Step 1: F_3 = $\alpha A_2 + (1 - \alpha)(F_2 + T_2) = (.2)(17) + (1 - .2)(12.8 + 1.92)$
= $3.4 + (.8)(14.72) = 3.4 + 11.78 = 15.18$

Step 2: T_3 = $\beta(F_3 - F_2) + (1 - \beta)T_2 = (.4)(15.18 - 12.8) + (1 - .4)(1.92)$
= $(.4)(2.38) + (.6)(1.92) = .952 + 1.152 = 2.10$

Step 3: FIT_3 = $F_3 + T_3$
= $15.18 + 2.10 = 17.28.$

Table 4.1 completes the forecasts for the 10-month period. Figure 4.3 compares actual demand to forecast including trend (FIT_t).

TABLE 4.1 ■ Forecast with $\alpha = .2$ and $\beta = .4$

MONTH	ACTUAL DEMAND	SMOOTHED FORECAST, F_t	SMOOTHED TREND, T_t	FORECAST INCLUDING TREND FIT_t
1	12	11	2	13.00
2	17	12.80	1.92	14.72
3	20	15.18	2.10	17.28
4	19	17.82	2.32	20.14
5	24	19.91	2.23	22.14
6	21	22.51	2.38	24.89
7	31	24.11	2.07	26.18
8	28	27.14	2.45	29.59

FIGURE 4.3 ■

Exponential Smoothing with Trend-Adjustment Forecasts Compared to Actual Demand Data

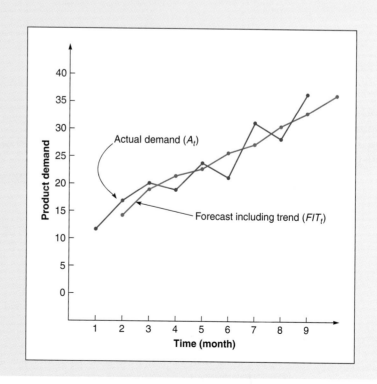

The value of the trend-smoothing constant, β, resembles the α constant because a high β is more responsive to recent changes in trend. A low β gives less weight to the most recent trends and tends to smooth out the present trend. Values of β can be found by the trial-and-error approach or by using sophisticated commercial forecasting software, with the MAD used as a measure of comparison.

Simple exponential smoothing is often referred to as *first-order smoothing*, and trend-adjusted smoothing is called *second-order*, or *double*, *smoothing*. Other advanced exponential-smoothing models are also used, including seasonal-adjusted and triple smoothing, but these are beyond the scope of this book.[2]

Trend Projections

Trend projection
A time-series forecasting method that fits a trend line to a series of historical data points and then projects the line into the future for forecasts.

The last time-series forecasting method we will discuss is **trend projection**. This technique fits a trend line to a series of historical data points and then projects the line into the future for medium-to-long-range forecasts. Several mathematical trend equations can be developed (for example, exponential and quadratic), but in this section, we will look at *linear* (straight-line) trends only.

If we decide to develop a linear trend line by a precise statistical method, we can apply the *least squares method*. This approach results in a straight line that minimizes the sum of the squares of the vertical differences or deviations from the line to each of the actual observations. Figure 4.4 illustrates the least squares approach.

A least squares line is described in terms of its *y*-intercept (the height at which it intercepts the *y*-axis) and its slope (the angle of the line). If we can compute the *y*-intercept and slope, we can express the line with the following equation:

$$\hat{y} = a + bx \tag{4-11}$$

where $\hat{y}$ (called "y hat") = computed value of the variable to be predicted (called the dependent variable)

a = y-axis intercept

b = slope of the regression line (or the rate of change in y for given changes in x)

x = the independent variable (which in this case is *time*)

FIGURE 4.4 ■

The Least Squares Method for Finding the Best-Fitting Straight Line, Where the Asterisks Are the Locations of the Seven Actual Observations or Data Points

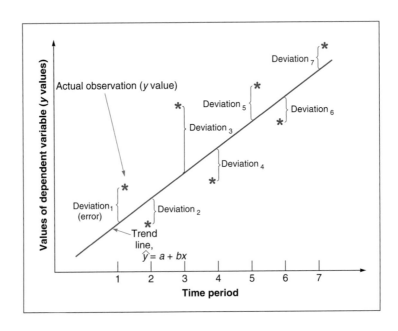

[2]For more details, see D. Groebner, P. Shannon, P. Fry, and K. Smith, *Business Statistics*, 5th ed. (Upper Saddle River, NJ: Prentice Hall, 2001).

Statisticians have developed equations that we can use to find the values of a and b for any regression line. The slope b is found by

$$b = \frac{\sum xy - n\bar{x}\bar{y}}{\sum x^2 - n\bar{x}^2} \tag{4-12}$$

where
b = slope of the regression line
$\sum$ = summation sign
x = known values of the independent variable
y = known values of the dependent variable
$\bar{x}$ = average of the value of the x's
$\bar{y}$ = average of the value of the y's
n = number of data points or observations

We can compute the y-intercept a as follows:

$$a = \bar{y} - b\bar{x} \tag{4-13}$$

Example 8 shows how to apply these concepts.

Example 8

Excel OM Data File Ch04Ex8.xla

Active Model 4.4

Example 8 is further illustrated in Active Model 4.4 on the CD-ROM.

The demand for electrical power at N.Y. Edison over the period 1997 to 2003 is shown below, in megawatts. Let's forecast 2004 demand by fitting a straight-line trend to these data.

YEAR	ELECTRICAL POWER DEMAND	YEAR	ELECTRICAL POWER DEMAND
1997	74	2001	105
1998	79	2002	142
1999	80	2003	122
2000	90		

With a series of data over time, we can minimize the computations by transforming the values of x (time) to simpler numbers. Thus, in this case, we can designate 1997 as year 1, 1998 as year 2, and so on.

YEAR	TIME PERIOD (x)	ELECTRIC POWER DEMAND (Y)	x^2	XY
1997	1	74	1	74
1998	2	79	4	158
1999	3	80	9	240
2000	4	90	16	360
2001	5	105	25	525
2002	6	142	36	852
2003	7	122	49	854
	$\sum x = 28$	$\sum y = 692$	$\sum x^2 = 140$	$\sum xy = 3,063$

$$\bar{x} = \frac{\sum x}{n} = \frac{28}{7} = 4 \qquad \bar{y} = \frac{\sum y}{n} = \frac{692}{7} = 98.86$$

$$b = \frac{\sum xy - n\bar{x}\bar{y}}{\sum x^2 - n\bar{x}^2} = \frac{3,063 - (7)(4)(98.86)}{140 - (7)(4^2)} = \frac{295}{28} = 10.54$$

$$a = \bar{y} - b\bar{x} = 98.86 - 10.54(4) = 56.70$$

Thus, the least squares trend equation is $\hat{y} = 56.70 + 10.54x$. To project demand in 2004, we first denote the year 2004 in our new coding system as $x = 8$:

$$\text{Demand in 2004} = 56.70 + 10.54(8)$$

$$= 141.02, \text{ or } 141 \text{ megawatts}$$

We can estimate demand for 2005 by inserting $x = 9$ in the same equation:

$$\text{Demand in 2005} = 56.70 + 10.54(9)$$
$$= 151.56, \text{ or } 152 \text{ megawatts}$$

To check the validity of the model, we plot historical demand and the trend line in Figure 4.5. In this case, we may wish to be cautious and try to understand the 2002 to 2003 swing in demand.

FIGURE 4.5 ■

Electrical Power and the
Computed Trend Line

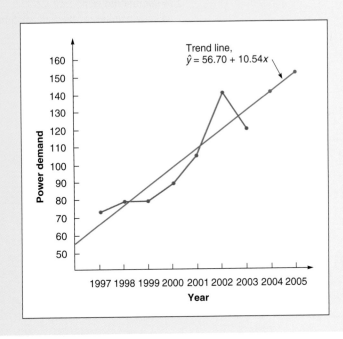

Notes on the Use of the Least Squares Method   Using the least squares method implies that we have met three requirements:

1. We always plot the data, because least squares data assume a linear relationship. If a curve appears to be present, curvilinear analysis is probably needed.
2. We do not predict time periods far beyond our given database. For example, if we have 20 months' worth of average prices of Microsoft stock, we could forecast only 3 or 4 months into the future. Forecasts beyond that have little statistical validity. Thus, you cannot take 5 years' worth of sales data and project 10 years into the future. The world is too uncertain.
3. Deviations around the least squares line (see Figure 4.4) are assumed to be random. They are normally distributed with most observations close to the line and only a smaller number farther out.

Seasonal Variations in Data

Seasonal variations
Regular upward or downward movements in a time series that tie to recurring events.

Seasonal variations in data are regular up and down movements in a time series that relate to recurring events such as weather or holidays. Demand for coal and fuel oil, for example, peaks during cold winter months. Demand for golf clubs or suntan lotion may be highest in summer.

Seasonality may be applied to hourly, daily, weekly, monthly, or other recurring patterns. Fast-food restaurants experience *daily* surges at noon and again at 5 P.M. Movie theaters see higher demand on Friday and Saturday evenings. The post office, Toys "Я" Us, The Christmas Store, and Hallmark Card Shops also exhibit seasonal variation in customer traffic and sales.

Similarly, understanding seasonal variations is important for capacity planning in organizations that handle peak loads. These include electric power companies during extreme cold and warm periods, banks on Friday afternoons, and buses and subways during the morning and evening rush hours.

Time-series forecasts like that in Example 8 involve reviewing the trend of data over a series of time. The presence of seasonality makes adjustments in trend-line forecasts necessary. Seasonality is expressed in terms of the amount that actual values differ from average values in the time series. Analyzing data in monthly or quarterly terms usually makes it easy for a statistician to spot seasonal patterns. Seasonal indices can then be developed by several common methods.

In what is called a *multiplicative seasonal model*, seasonal factors are multiplied by an estimate of average demand to produce a seasonal forecast. Our assumption in this section is that trend has been removed from the data. Otherwise, the magnitude of the seasonal data will be distorted by the trend.

Here are the steps we will follow for a company that has "seasons" of 1 month:

1. Find the *average historical demand each season* (or month in this case) by summing the demand for that month in each year and dividing by the number of years of data available. For example, if, in January, we have seen sales of 8, 6, and 10 over the past 3 years, average January demand equals $(8 + 6 + 10)/3 = 8$ units.
2. Compute the *average demand over all months* by dividing the total average annual demand by the number of seasons. For example, if the total average demand for a year is 120 units and there are 12 seasons (each month), the average monthly demand is $120/12 = 10$ units.
3. Compute a *seasonal index* for each season by dividing that month's actual historical demand (from step 1) by the average demand over all months (from step 2). For example, if the average historical January demand over the past 3 years is 8 units and the average demand over all months is 10 units, the seasonal index for January is $8/10 = .80$. Likewise, a seasonal index of 1.20 for February would mean that February's demand is 20% larger than the average demand over all months.
4. Estimate next year's total annual demand.
5. Divide this estimate of total annual demand by the number of seasons, then multiply it by the seasonal index for that month. This provides the *seasonal forecast*.

Example 9 illustrates this procedure as it computes seasonal factors from historical data.

Because John Deere understands seasonal variations in sales, it has been able to obtain 70% of its orders in advance of seasonal use (through price reductions and incentives such as free interest) so it can smooth production.

Demand for many products is seasonal. Kawasaki, the manufacturer of these jet skis and snowmobiles, produces products with complementary demands to address seasonal fluctuations.

Example 9

Monthly demand for IBM laptop computers at a Des Moines distributor for 2000 to 2002 is shown in the following table:

	DEMAND			AVERAGE 2000–2002 DEMAND	AVERAGE MONTHLY DEMAND[a]	SEASONAL INDEX[b]
MONTH	2000	2001	2002			
Jan.	80	85	105	90	94	.957 (= 90/94)
Feb.	70	85	85	80	94	.851 (= 80/94)
Mar.	80	93	82	85	94	.904 (= 85/94)
Apr.	90	95	115	100	94	1.064 (= 100/94)
May	113	125	131	123	94	1.309 (= 123/94)
June	110	115	120	115	94	1.223 (= 115/94)
July	100	102	113	105	94	1.117 (= 105/94)
Aug.	88	102	110	100	94	1.064 (= 100/94)
Sept.	85	90	95	90	94	.957 (= 90/94)
Oct.	77	78	85	80	94	.851 (= 80/94)
Nov.	75	82	83	80	94	.851 (= 80/94)
Dec.	82	78	80	80	94	.851 (= 80/94)

Total average annual demand = 1,128

[a]Average monthly demand = $\dfrac{1,128}{12 \text{ months}} = 94$

[b]Seasonal index = $\dfrac{\text{average 2000–2002 monthly demand}}{\text{average monthly demand}}$

If we expected the 2003 annual demand for computers to be 1,200 units, we would use these seasonal indices to forecast the monthly demand as follows:

MONTH	DEMAND	MONTH	DEMAND
Jan.	$\dfrac{1,200}{12} \times .957 = 96$	July	$\dfrac{1,200}{12} \times 1.117 = 112$
Feb.	$\dfrac{1,200}{12} \times .851 = 85$	Aug.	$\dfrac{1,200}{12} \times 1.064 = 106$
Mar.	$\dfrac{1,200}{12} \times .904 = 90$	Sept.	$\dfrac{1,200}{12} \times .957 = 96$
Apr.	$\dfrac{1,200}{12} \times 1.064 = 106$	Oct.	$\dfrac{1,200}{12} \times .851 = 85$
May	$\dfrac{1,200}{12} \times 1.309 = 131$	Nov.	$\dfrac{1,200}{12} \times .851 = 85$
June	$\dfrac{1,200}{12} \times 1.223 = 122$	Dec.	$\dfrac{1,200}{12} \times .851 = 85$

For simplicity, only 3 periods are used for each monthly index in the above example. Example 10 illustrates how indices that have already been prepared can be applied to adjust trend-line forecasts for seasonality.

Example 10

A San Diego hospital used 66 months of adult inpatient hospital days to reach the following equation:

$$\hat{y} = 8,090 + 21.5x$$

where

$$\hat{y} = \text{patient days}$$
$$x = \text{time, in months}$$

Based on this model, which reflects only trend data, the hospital forecasts patient days for the next month (period 67) to be

Patient days = 8,090 + (21.5)(67) = 9,530 (trend only)

While this model, as plotted in Figure 4.6, recognized the upward trend line in the demand for inpatient services, it ignored the seasonality that the administration knew to be present.

FIGURE 4.6 ■

Trend Data for San Diego Hospital

Source: From "Modern Methods Improve Hospital Forecasting" by W. E. Sterk and E. G. Shryock from *Healthcare Financial Management,* Vol. 41, no. 3, p. 97. Reprinted by permission of Healthcare Financial Management Association.

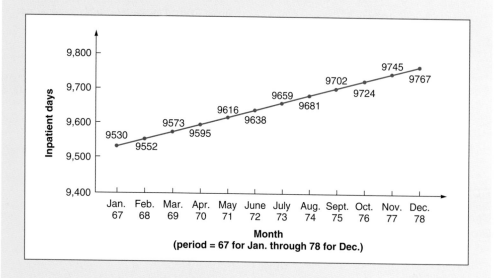

The table below provides seasonal indices based on the same 66 months. Such seasonal data, by the way, were found to be typical of hospitals nationwide.

SEASONALITY INDICES FOR ADULT INPATIENT DAYS AT SAN DIEGO HOSPITAL

MONTH	SEASONALITY INDEX	MONTH	SEASONALITY INDEX
January	1.04	July	1.03
February	0.97	August	1.04
March	1.02	September	0.97
April	1.01	October	1.00
May	0.99	November	0.96
June	0.99	December	0.98

These seasonal indices are graphed in Figure 4.7. Note that January, March, July, and August seem to exhibit significantly higher patient days on average, while February, September, November, and December experience lower patient days.

FIGURE 4.7 ■

Seasonal Index for San Diego Hospital

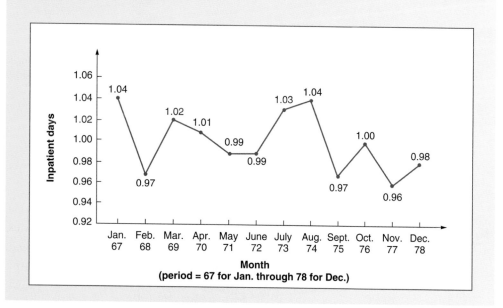

However, neither the trend data nor the seasonal data alone provide a reasonable forecast for the hospital. Only when the hospital multiplied the trend-adjusted data times the appropriate seasonal index did it obtain good forecasts. Thus, for period 67 (January):

$$\text{Patient Days} = (\text{trend adjusted forecast})(\text{monthly seasonal index}) = (9{,}530)(1.04) = 9{,}911$$

The patient days for each month are:

PERIOD	67	68	69	70	71	72	73	74	75	76	77	78
MONTH	Jan.	Feb.	March	April	May	June	July	Aug.	Sept.	Oct.	Nov.	Dec.
FORECAST WITH TREND & SEASONAL	9,911	9,265	9,764	9,691	9,520	9,542	9,949	10,068	9,411	9,724	9,355	9,572

A graph showing the forecast using both trend and seasonality appears in Figure 4.8.

FIGURE 4.8 ■

Combined Trend and Seasonal Forecast

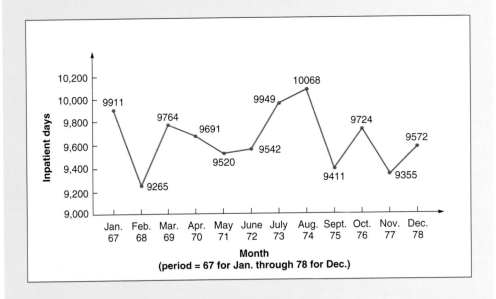

Notice that with trend only, the September forecast is 9,702, but with both trend and seasonal adjustments the forecast is 9,411. By combining trend and seasonal data the hospital was better able to forecast inpatient days and the related staffing and budgeting vital to effective operations.

Example 11 further illustrates seasonality for quarterly data at a department store.

Example 11

Management at Davis's Department Store has used time-series regression to forecast retail sales for the next 4 quarters. Sales estimates are $100,000, $120,000, $140,000, and $160,000 for the respective quarters. Seasonal indices for the 4 quarters have been found to be 1.30, .90, .70, and 1.15, respectively.

To compute a seasonalized or adjusted sales forecast, we just multiply each seasonal index by the appropriate trend forecast:

$$\hat{y}_{\text{seasonal}} = \text{Index} \times \hat{y}_{\text{trend forecast}}$$

Thus for

Quarter I: $\hat{y}_{\text{I}} = (1.30)(\$100{,}000) = \$130{,}000$

Quarter II: $\hat{y}_{\text{II}} = (.90)(\$120{,}000) = \$108{,}000$

Quarter III: $\hat{y}_{\text{III}} = (.70)(\$140{,}000) = \$98{,}000$

Quarter IV: $\hat{y}_{\text{IV}} = (1.15)(\$160{,}000) = \$184{,}000$

Cyclical Variations in Data

Cycles
Patterns in the data that occur every several years.

Cycles are like seasonal variations in data, but occur every several *years*, not weeks, months, or quarters. Forecasting them from a time series of data is difficult because it is very hard to predict the turning points that indicate a new cycle is beginning.

The best way to predict business cycles is by finding a *leading variable* with which the data series seems to correlate. For example, birthrates "lead" college enrollments by about 18 years. When the Ohio Board of Regents looks at long-term cycles in attendance at the 70 public colleges in that state, changes in births 18 years earlier is a good predictor of swings in enrollment. Likewise, housing construction permits are an excellent leading variable for such related items as sales of refrigerators, lawn services, and school enrollments.

Developing causal or associative techniques of variables that impact one another is our next topic.

ASSOCIATIVE FORECASTING METHODS: REGRESSION AND CORRELATION ANALYSIS

Unlike time-series forecasting, *associative forecasting* models usually consider *several* variables that are related to the quantity being predicted. Once these related variables have been found, a statistical model is built and used to forecast the item of interest. This approach is more powerful than the time-series methods that use only the historic values for the forecasted variable.

Linear-regression analysis
A straight-line mathematical model to describe the functional relationships between independent and dependent variables.

Many factors can be considered in an associative analysis. For example, the sales of IBM PCs might be related to IBM's advertising budget, the company's prices, competitors' prices and promotional strategies, and even the nation's economy and unemployment rates. In this case, PC sales would be called the *dependent variable* and the other variables would be called *independent variables*. The manager's job is to develop *the best statistical relationship between PC sales and the independent variables*. The most common quantitative associative forecasting model is **linear-regression analysis**.

Using Regression Analysis to Forecast

We can use the same mathematical model that we employed in the least squares method of trend projection to perform a linear-regression analysis. The dependent variables that we want to forecast will still be $\hat{y}$. But now the independent variable, x, need no longer be time. We use the equation

$$\hat{y} = a + bx$$

where $\hat{y}$ = value of the dependent variable (in our example, sales)
     a = y-axis intercept
     b = slope of the regression line
     x = independent variable

Example 12 shows how to use linear regression.

Example 12

**Excel OM Data File
Ch04Ex12.xla**

Nodel Construction Company renovates old homes in West Bloomfield, Michigan. Over time, the company has found that its dollar volume of renovation work is dependent on the West Bloomfield area payroll. The following table lists Nodel's revenues and the amount of money earned by wage earners in West Bloomfield during the past 6 years.

NODEL'S SALES ($000,000), y	LOCAL PAYROLL ($000,000,000), x	NODEL'S SALES ($000,000), y	LOCAL PAYROLL ($000,000,000), x
2.0	1	2.0	2
3.0	3	2.0	1
2.5	4	3.5	7

Nodel management wants to establish a mathematical relationship to help predict sales. First, it needs to determine whether there is a straight-line (linear) relationship between area payroll and sales, so it plots the known data on a scatter diagram.

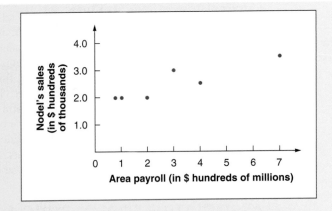

It appears from the six data points that there is a slight positive relationship between the independent variable (payroll) and the dependent variable (sales): As payroll increases, Nodel's sales tend to be higher. We can find a mathematical equation by using the least squares regression approach.

Sales, y	Payroll, x	x^2	xy
2.0	1	1	2.0
3.0	3	9	9.0
2.5	4	16	10.0
2.0	2	4	4.0
2.0	1	1	2.0
3.5	7	49	24.5
$\Sigma y = 15.0$	$\Sigma x = 18$	$\Sigma x^2 = 80$	$\Sigma xy = 51.5$

$$\bar{x} = \frac{\Sigma x}{6} = \frac{18}{6} = 3$$

$$\bar{y} = \frac{\Sigma y}{6} = \frac{15}{6} = 2.5$$

$$b = \frac{\Sigma xy - n\bar{x}\bar{y}}{\Sigma x^2 - n\bar{x}^2} = \frac{51.5 - (6)(3)(2.5)}{80 - (6)(3^2)} = .25$$

$$a = \bar{y} - b\bar{x} = 2.5 - (.25)(3) = 1.75$$

The estimated regression equation, therefore, is

$$\hat{y} = 1.75 + .25x$$

or

$$\text{Sales} = 1.75 + .25 \, (\text{payroll})$$

If the local chamber of commerce predicts that the West Bloomfield area payroll will be $600 million next year, we can estimate sales for Nodel with the regression equation:

$$\text{Sales (in hundred thousands)} = 1.75 + .25(6)$$
$$= 1.75 + 1.50 = 3.25$$

or

$$\text{Sales} = \$325,000$$

The final part of Example 12 shows a central weakness of associative forecasting methods like regression. Even when we have computed a regression equation, we must provide a forecast of the independent variable x—in this case, payroll—before estimating the dependent variable y for the next time period. Although this is not a problem for all forecasts, you can imagine the difficulty of determining future values of *some* common independent variables (such as unemployment rates, gross national product, price indices, and so on).

Standard Error of the Estimate

Standard error of the estimate
A measure of variability around the regression line—its standard deviation.

The forecast of $325,000 for Nodel's sales in Example 12 is called a *point estimate* of *y*. The point estimate is really the *mean*, or *expected value*, of a distribution of possible values of sales. Figure 4.9 illustrates this concept.

To measure the accuracy of the regression estimates, we must compute the **standard error of the estimate**, $S_{y,x}$. This computation is called the *standard deviation of the regression:* It measures the error from the dependent variable, *y*, to the regression line, rather than to the mean. Equation (4-14) is a similar expression to that found in most statistics books for computing the standard deviation of an arithmetic mean:

$$S_{y,x} = \sqrt{\frac{\Sigma(y - y_c)^2}{n - 2}}$$

(4-14)

where y = *y*-value of each data point
y_c = computed value of the dependent variable, from the regression equation
n = number of data points

Equation (4-15) may look more complex, but it is actually an easier-to-use version of Equation (4-14). Both formulas provide the same answer and can be used in setting up prediction intervals around the point estimate.[3]

$$S_{y,x} = \sqrt{\frac{\Sigma y^2 - a\,\Sigma y - b\,\Sigma xy}{n - 2}}$$

(4-15)

Example 13 shows how we would calculate the standard error of the estimate in Example 12.

FIGURE 4.9 ■

Distribution about the Point Estimate of $600 Million Payroll

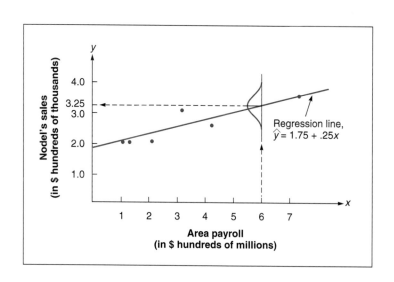

---

[3]When the sample size is large ($n > 30$), the prediction interval value of *y* can be computed using normal tables. When the number of observations is small, the *t*-distribution is appropriate. See D. Groebner et al., *Business Statistics*, 5th ed. (Upper Saddle River, NJ: Prentice Hall, 2001).

Glidden Paints assembly lines fill thousands of cans per hour. To predict demand, the firm uses associative forecasting methods such as linear regression, with independent variables such as disposable personal income and GNP. Although housing starts would be a natural variable, Glidden found that it correlated poorly with past sales. It turns out that most Glidden paint is sold through retailers to customers who already own homes or businesses.

Example 13

To compute the standard error of the estimate for Nodel's data in Example 12, the only number we need that is not available to solve for $S_{y,x}$ is Σy^2. Some quick addition reveals $\Sigma y^2 = 39.5$. Therefore:

$$S_{y,x} = \sqrt{\frac{\Sigma y^2 - a\,\Sigma y - b\,\Sigma xy}{n-2}}$$

$$= \sqrt{\frac{39.5 - 1.75(15.0) - .25(51.5)}{6-2}}$$

$$= \sqrt{.09375} = .306 \text{ (in \$ hundred thousands)}$$

The standard error of the estimate is then \$30,600 in sales.

Correlation Coefficients for Regression Lines

The regression equation is one way of expressing the nature of the relationship between two variables. Regression lines are not "cause-and-effect" relationships. They merely describe the relationships among variables. The regression equation shows how one variable relates to the value and changes in another variable.

Coefficient of correlation

A measure of the strength of the relationship between two variables.

Another way to evaluate the relationship between two variables is to compute the **coefficient of correlation**. This measure expresses the degree or strength of the linear relationship. Usually identified as r, the coefficient of correlation can be any number between $+1$ and -1. Figure 4.10 illustrates what different values of r might look like.

To compute r, we use much of the same data needed earlier to calculate a and b for the regression line. The rather lengthy equation for r is

$$r = \frac{n\Sigma xy - \Sigma x\,\Sigma y}{\sqrt{\left[n\,\Sigma x^2 - (\Sigma x)^2\right]\left[n\,\Sigma y^2 - (\Sigma y)^2\right]}} \qquad (4\text{-}16)$$

FIGURE 4.10 ■

Four Values of the Correlation Coefficient

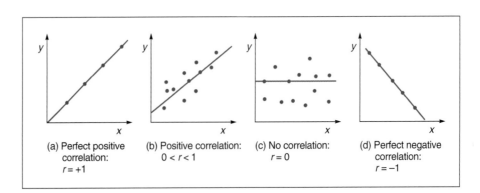

(a) Perfect positive correlation: $r = +1$

(b) Positive correlation: $0 < r < 1$

(c) No correlation: $r = 0$

(d) Perfect negative correlation: $r = -1$

Example 14 shows how to calculate the coefficient of correlation for the data given in Examples 12 and 13.

Example 14

A high *r* doesn't always mean one variable will be a good predictor of the other. Skirt lengths and stock market prices may be correlated, but raising one doesn't mean the other will also go up.

In Example 12, we looked at the relationship between Nodel Construction Company's renovation sales and payroll in its hometown of West Bloomfield. To compute the coefficient of correlation for the data shown, we need add only one more column of calculations (for y^2) and then apply the equation for r:

y	x	x^2	xy	y^2
2.0	1	1	2.0	4.0
3.0	3	9	9.0	9.0
2.5	4	16	10.0	6.25
2.0	2	4	4.0	4.0
2.0	1	1	2.0	4.0
3.5	7	49	24.5	12.25
$\Sigma y = 15.0$	$\Sigma x = 18$	$\Sigma x^2 = 80$	$\Sigma xy = 51.5$	$\Sigma y^2 = 39.5$

$$r = \frac{(6)(51.5) - (18)(15.0)}{\sqrt{[(6)(80) - (18)^2][(6)(39.5) - (15.0)^2]}}$$

$$= \frac{309 - 270}{\sqrt{(156)(12)}} = \frac{39}{\sqrt{1,872}}$$

$$= \frac{39}{43.3} = .901$$

This *r* of .901 appears to be a significant correlation and helps to confirm the closeness of the relationship between the two variables.

Coefficient of determination

A measure of the amount of variation in the dependent variable about its mean that is explained by the regression equation.

Although the coefficient of correlation is the measure most commonly used to describe the relationship between two variables, another measure does exist. It is called the **coefficient of determination** and is simply the square of the coefficient of correlation—namely, r^2. The value of r^2 will always be a positive number in the range of $0 \leq r^2 \leq 1$. The coefficient of determination is the percent of variation in the dependent variable (y) that is explained by the regression equation. In Nodel's case, the value of r^2 is .81, indicating that 81% of the total variation is explained by the regression equation.

Multiple-Regression Analysis

Multiple regression

A causal forecasting method with more than one independent variable.

Multiple regression is a practical extension of the simple regression model we just explored. It allows us to build a model with several independent variables instead of just one variable. For example, if Nodel Construction wants to include average annual interest rates in its model for forecasting renovation sales, the proper equation would be

$$\hat{y} = a + b_1 x_1 + b_2 x_2 \tag{4-17}$$

where
- $\hat{y}$ = dependent variable, sales
- a = a constant
- x_1 and x_2 = values of the two independent variables, area payroll and interest rates, respectively
- b_1 and b_2 = coefficients for the two independent variables

The mathematics of multiple regression becomes quite complex (and is usually tackled by computer), so we leave the formulas for a, b_1, and b_2 to statistics textbooks. However, Example 15 shows how to interpret Equation (4-17) in forecasting Nodel's sales.

OM IN ACTION

Forecasting Manpower with Multiple Regression at TransAlta Utilities

TransAlta Utilities (TAU) is a $1.6 billion energy company operating in Canada, New Zealand, Australia, Argentina, and the United States. Headquartered in Alberta, Canada, TAU is Canada's largest publicly owned utility. It serves 340,000 customers in Alberta through 57 customer-service facilities, each of which is staffed by 5–20 customer service linemen. The 270 linemen's jobs are to handle new connections, repairs, patrol power lines, and check substations. This existing system was not the result of some optimal central planning but was put in place incrementally as the company grew.

With help from the University of Alberta, TAU developed a causal model to decide how many linemen should be assigned to each facility. The research team decided to build a multiple regression model with three independent variables. The hardest part of the task was to select variables that were easy to quantify with available data. In the end, the explanatory variables were number of urban customers, number of rural customers, and the geographic size of a service area. The model assumes that the time spent on customers is proportional to the number of customers and the time spent on facilities (line patrol and substation checks) and travel is proportional to the size of the service region. By definition, the unexplained time is time that is not explained by the three variables (e.g., meetings, breaks, unproductive time).

Not only did the results of the model please TAU managers, but the cost savings of the project (which included optimizing the number of facilities and their locations) is $4 million per year.

Source: E. Erkut, T. Myroon, and K. Strangway. "TransAlta Redesigns its Service-Delivery Network." *Interfaces* (March–April, 2000: 54–69).

Example 15

The new multiple-regression line for Nodel Construction, calculated by computer software, is

$$\hat{y} = 1.80 + .30x_1 - 5.0x_2$$

We also find that the new coefficient of correlation is .96; implying the inclusion of the variable x_2, interest rates, adds even more strength to the linear relationship.

We can now estimate Nodel's sales if we substitute values for next year's payroll and interest rate. If West Bloomfield's payroll will be $600 million and the interest rate will be .12 (12%), sales will be forecast as

$$\text{Sales (\$ hundred thousands)} = 1.80 + .30(6) - 5.0(.12)$$
$$= 1.8 + 1.8 - .6$$
$$= 3.00$$

or

$$\text{Sales} = \$300,000$$

MONITORING AND CONTROLLING FORECASTS

Once a forecast has been completed, it should not be forgotten. No manager wants to be reminded that his or her forecast is horribly inaccurate, but a firm needs to determine why actual demand (or whatever variable is being examined) differed significantly from that projected. If the forecaster is accurate, that individual usually makes sure that everyone is aware of his or her talents. Very seldom does one read articles in *Fortune*, *Forbes*, or the *Wall Street Journal*, however, about money managers who are consistently off by 25% in their stock market forecasts.

One way to monitor forecasts to ensure that they are performing well is to use a tracking signal. A **tracking signal** is a measurement of how well the forecast is predicting actual values. As forecasts are updated every week, month, or quarter, the newly available demand data are compared to the forecast values.

Tracking signal
A measurement of how well the forecast is predicting actual values.

Bias
A forecast that is consistently higher or consistently lower than actual values of a time series.

The tracking signal is computed as the *running sum of the forecast errors (RSFE)* divided by the *mean absolute deviation (MAD)*:

$$\left(\begin{array}{c}\text{Tracking}\\\text{signal}\end{array}\right) = \frac{\text{RSFE}}{\text{MAD}}$$

$$= \frac{\sum(\text{actual demand in period } i - \text{forecast demand in period } i)}{\text{MAD}}$$

(4-18)

where

$$\text{MAD} = \frac{\sum|\text{actual} - \text{forecast}|}{n}$$

as seen earlier in Equation (4-5).

Positive tracking signals indicate that demand is *greater* than forecast. *Negative* signals mean that demand is *less* than forecast. A good tracking signal—that is, one with a low RSFE—has about as much positive error as it has negative error. In other words, small deviations are okay, but positive and negative errors should balance one another so that the tracking signal centers closely around zero. A consistent tendency for forecasts to be greater or less than the actual values (that is, for a high RSFE) is called a **bias** error. Bias could occur if, for example, the wrong variables or trend line are used or if a seasonal index is misapplied.

Once tracking signals are calculated, they are compared to predetermined control limits. When a tracking signal exceeds an upper or lower limit, there is a problem with the forecasting method, and management may want to reevaluate the way it forecasts demand. Figure 4.11 shows the graph of a tracking signal that is exceeding the range of acceptable variation. If the model being used is exponential smoothing, perhaps the smoothing constant needs to be readjusted.

How do firms decide what the upper and lower tracking limits should be? There is no single answer, but they try to find reasonable values—in other words, limits not so low as to be triggered with every small forecast error, and not so high as to allow bad forecasts to be regularly overlooked. George Plossl and Oliver Wight, two inventory control experts, have suggested using maximums of ±4 MADs for high-volume stock items and ±8 MADs for lower-volume items.[4] Other forecasters suggest slightly lower ranges. Because one MAD is equivalent to approximately .8 standard deviations, ±2 MADs = ±1.6 standard deviations, ±3 MADs = ±2.4 standard deviations, and ±4 MADs = ±3.2 standard deviations. This fact suggests that for a forecast to be "in control," 89% of the errors are expected to fall within ±2 MADs, 98% within ±3 MADs, or 99.9% within ±4 MADs.[5]

Example 16 shows how the tracking signal and RSFE can be computed.

FIGURE 4.11 ■

A Plot of Tracking Signals

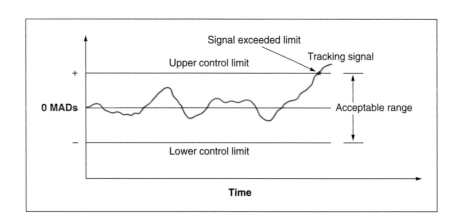

[4]See G. W. Plossl and O. W. Wight, *Production and Inventory Control* (Upper Saddle River, NJ: Prentice Hall, 1967).

[5]To prove these three percentages to yourself, just set up a normal curve for ±1.6 standard deviations (z values). Using the normal table in Appendix I, you find that the area under the curve is .89. This represents ±2 MADs. Likewise, ±3 MADs = ±2.4 standard deviations encompass 98% of the area, and so on for ±4 MADs.

Example 16

Rick Carlson Bakery's quarterly sales of croissants (in thousands), as well as forecast demand and error computations, are shown below. The objective is to compute the tracking signal and determine whether forecasts are performing adequately.

QUARTER	ACTUAL DEMAND	FORECAST DEMAND	ERROR	RSFE	ABSOLUTE FORECAST ERROR	CUMULATIVE ABSOLUTE FORECAST ERROR	CUMULATIVE MAD	TRACKING SIGNAL (RSFE/MAD)
1	90	100	−10	−10	10	10	10.0	−10/10 = −1
2	95	100	−5	−15	5	15	7.5	−15/7.5 = −2
3	115	100	+15	0	15	30	10.0	0/10 = 0
4	100	110	−10	−10	10	40	10.0	−10/10 = −1
5	125	110	+15	+5	15	55	11.0	+5/11 = +0.5
6	140	110	+30	+35	30	85	14.2	+35/14.2 = +2.5

$$\text{At the end of quarter 6, } \text{MAD} = \frac{\sum |\text{forecast errors}|}{n} = \frac{85}{6} = 14.2$$

$$\text{and Tracking signal} = \frac{\text{RSFE}}{\text{MAD}} = \frac{35}{14.2} = 2.5 \text{ MADs}$$

This tracking signal is within acceptable limits. We see that it drifted from −2.0 MADs to +2.5 MADs.

Adaptive smoothing
An approach to exponential smoothing forecasting in which the smoothing constant is automatically changed to keep errors to a minimum.

Focus forecasting
Forecasting that tries a variety of computer models and selects the best one for a particular application.

Adaptive Smoothing

Adaptive forecasting refers to computer monitoring of tracking signals and self-adjustment if a signal passes a preset limit. For example, when applied to exponential smoothing, the α and β coefficients are first selected on the basis of values that minimize error forecasts, and then adjusted accordingly whenever the computer notes an errant tracking signal. This process is called **adaptive smoothing**.

Focus Forecasting

Rather than adapt by choosing a smoothing constant, computers allow us to try a variety of forecasting models. Such an approach is called focus forecasting. **Focus forecasting** is based on two principles:

1. Sophisticated forecasting models are not always better than simple ones.
2. There is no single technique that should be used for all products or services.

Bernard Smith, inventory manager for American Hardware Supply, coined the term *focus forecasting*. Smith's job was to forecast quantities for 100,000 hardware products purchased by American's 21 buyers.[6] He found that buyers neither trusted nor understood the exponential smoothing model then in use. Instead, they used very simple approaches of their own. So Smith developed his new computerized system for selecting forecasting methods.

Smith chose seven forecasting methods to test. They ranged from the simple ones that buyers used (such as the naive approach) to statistical models. Every month, Smith applied the forecasts of all seven models to each item in stock. In these simulated trials, the forecast values were subtracted from the most recent actual demands, giving a simulated forecast error. The forecast method yielding the least error is selected by the computer, which then uses it to make next month's forecast. Although buyers still have an override capability, American Hardware finds that focus forecasting provides excellent results.

FORECASTING IN THE SERVICE SECTOR

Forecasting in the service sector presents some unusual challenges. A major technique in the retail sector is tracking demand by maintaining good short-term records. For instance, a barbershop catering to men expects peak flows on Fridays and Saturdays. Indeed, most barbershops will be closed on

---

[6]Bernard T. Smith, *Focus Forecasting: Computer Techniques for Inventory Control* (Boston: CBI Publishing, 1978).

FIGURE 4.12 ■

Forecast of Sales by
Hour for a Fast-Food
Restaurant

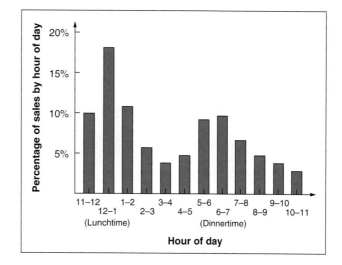

Sunday and Monday, and many call in extra help on Friday and Saturday. A downtown restaurant, on the other hand, may need to track conventions and holidays for effective short-term forecasting.

Specialty Retail Shops   Specialty retail facilities, such as flower shops, may have other unusual demand patterns, and those patterns will differ depending on the holiday. When Valentine's Day falls on a weekend, for example, flowers can't be delivered to offices and those romantically inclined are likely to celebrate with outings rather than flowers. If a holiday falls on a Monday, some of the celebration may also take place on the weekend, reducing flower sales. However, when Valentine's Day falls in midweek, busy midweek schedules often make flowers the optimal way to celebrate. Because flowers for Mother's Day are to be delivered on Saturday or Sunday, this holiday forecast varies less. Due to special demand patterns, many service firms maintain records of sales, noting not only the day of the week but also unusual events, including the weather, so that patterns and correlations that influence demand can be developed.

Fast-Food Restaurants   Fast-food restaurants are well aware not only of weekly, daily, and hourly but even 15 minute variations in demands that influence sales. Therefore, detailed forecasts of demand are needed. Figure 4.12 shows the hourly forecast for a typical fast-food restaurant. Note the lunchtime and dinnertime peaks.

Firms like Taco Bell now use point-of-sale computers that track sales every quarter hour. Taco Bell found that a 6-week moving average was the forecasting technique that minimized its mean squared error (MSE) of these quarter hour forecasts. Building this forecasting methodology into each of Taco Bell's 6,500 stores' computers, the model makes weekly projections of customer transactions. These in turn are used by store managers to schedule staff, who begin in 15-minute increments, not 1-hour blocks as in other industries. The forecasting model has been so successful that Taco Bell has increased customer service while documenting more than $50 million in labor cost savings in 4 years of use.[7]

SUMMARY

Forecasts are a critical part of the operations manager's function. Demand forecasts drive a firm's production, capacity, and scheduling systems and affect the financial, marketing, and personnel planning functions.

There are a variety of qualitative and quantitative forecasting techniques. Qualitative approaches employ judgment, experience, intuition, and a host of other factors that are difficult to quantify. Quantitative forecasting uses historical data and causal, or associative, relations to project future demands. Table 4.2 summarizes the formulas we introduced in quantitative forecasting. Forecast calculations are seldom performed by hand. Most operations managers turn to software packages such as Forecast PRO, SAP, tsMetrix, AFS, SAS, SPSS, or Excel.

[7]J. Hueter and W. Swart. "An Integrated Labor Management System for Taco Bell." *Interfaces* 28, no. 1 (January–February 1998): 75–91.

TABLE 4.2

Summary of Forecasting
Formulas

Moving averages—forecasts based on an average of recent values

$$\text{Moving average} = \frac{\Sigma \text{ demand in previous } n \text{ periods}}{n} \tag{4-1}$$

Weighted moving averages—a moving average with weights that vary

$$\text{Weighted moving average} = \frac{\Sigma \text{ (weight for period } n)(\text{demand in period } n)}{\Sigma \text{ weights}} \tag{4-2}$$

Exponential smoothing—a moving average with weights following an exponential distribution

$$\text{New forecast} = \text{last period's forecast} + \alpha \text{ (last period's actual} \atop \text{demand} - \text{last period's forecast)} \tag{4-3}$$

$$F_t = F_{t-1} + \alpha(A_{t-1} - F_{t-1}) \tag{4-4}$$

Mean absolute deviation—a measure of overall forecast error

$$\text{MAD} = \frac{\Sigma |\text{forecast errors}|}{n} \tag{4-5}$$

Mean squared error—a second measure of forecast error

$$\text{MSE} = \frac{\Sigma \text{ (forecast errors)}^2}{n} \tag{4-6}$$

Mean absolute percent error—a third measure of forecast error

$$\text{MAPE} = \frac{100 \sum_{i=1}^{n} |\text{actual}_i - \text{forecast}_i| / \text{actual}_i}{n} \tag{4-7}$$

Exponential smoothing with trend adjustment—an exponential smoothing model that can accommodate trend

$$\text{Forecast including trend } (FIT_t) = \text{exponentially smoothed forecast } (F_t) \atop + \text{ exponentially smoothed trend } (T_t) \tag{4-8}$$

$$F_t = \alpha(A_{t-1}) + (1 - \alpha)(F_{t-1} + T_{t-1}) \tag{4-9}$$

$$T_t = \beta(F_t - F_{t-1}) + (1 - \beta)T_{t-1} \tag{4-10}$$

Trend projection and regression analysis—fitting a trend line to historical data or a regression line to an independent variable

$$\hat{y} = a + bx \tag{4-11}$$

$$b = \frac{\Sigma xy - n\bar{x}\bar{y}}{\Sigma x^2 - n\bar{x}^2} \tag{4-12}$$

$$a = \bar{y} - b\bar{x} \tag{4-13}$$

Multiple regression analysis—a regression model with more than one independent (predicting) variable

$$\hat{y} = a + b_1 x_1 + b_2 x_2 + \cdots + b_n x_n \tag{4-17}$$

Tracking signal—a measurement of how well the forecast is predicting actual values

$$\text{Tracking signal} = \frac{\text{RSFE}}{\text{MAD}} = \frac{\Sigma \text{ (actual demand in period } i - \text{forecast demand in period } i)}{\text{MAD}} \tag{4-18}$$

No forecasting method is perfect under all conditions. And even once management has found a satisfactory approach, it must still monitor and control forecasts to make sure errors do not get out of hand. Forecasting can often be a very challenging, but rewarding, part of managing.

KEY TERMS

Forecasting *(p. 84)*
Economic forecasts *(p. 85)*
Technological forecasts *(p. 85)*
Demand forecasts *(p. 85)*
Quantitative forecasts *(p. 86)*
Qualitative forecasts *(p. 86)*
Jury of executive opinion *(p. 86)*
Delphi method *(p. 87)*
Sales force composite *(p. 87)*
Consumer market survey *(p. 87)*
Time series *(p. 87)*
Naive approach *(p. 89)*
Moving averages *(p. 89)*
Exponential smoothing *(p. 90)*
Smoothing constant *(p. 91)*

Mean absolute deviation (MAD) *(p. 93)*
Mean squared error (MSE) *(p. 94)*
Mean absolute percent error (MAPE) *(p. 94)*
Trend projection *(p. 98)*
Seasonal variations *(p. 100)*
Cycles *(p. 105)*
Linear-regression analysis *(p. 105)*
Standard error of the estimate *(p. 107)*
Coefficient of correlation *(p. 108)*
Coefficient of determination *(p. 109)*
Multiple regression *(p. 109)*
Tracking signal *(p. 110)*
Bias *(p. 111)*
Adaptive smoothing *(p. 112)*
Focus forecasting *(p. 112)*

USING EXCEL SPREADSHEETS IN FORECASTING

Excel and spreadsheets in general are frequently used in forecasting. Both exponential smoothing and regression analysis (simple and multiple) are supported by built-in Excel functions. You may also use Excel OM's forecasting module, which has five components: (1) moving averages, (2) weighted moving averages, (3) exponential smoothing, (4) regression (with one variable only), and (5) decomposition. Excel OM's error analysis is much more complete than that available with the Excel add-in.

Program 4.1 illustrates Excel OM's input and output, using Example 2's weighted moving average data.

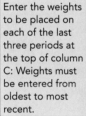

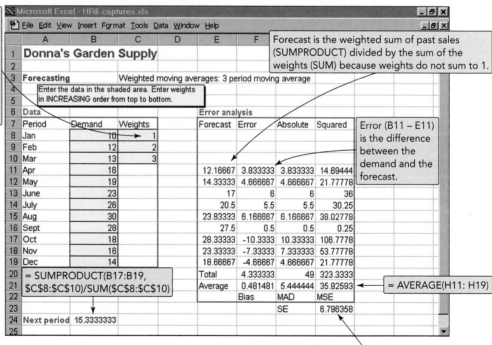

PROGRAM 4.1 ■ Analysis of Excel OM's Weighted Moving-Average Program, Using Data from Example 2 as Input

Program 4.2 provides an Excel OM regression analysis, using the electrical power data from Example 8.

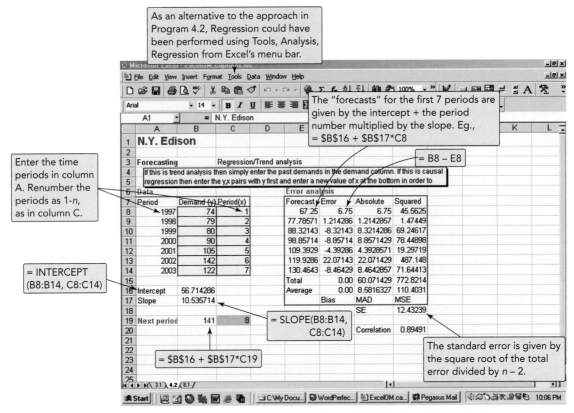

As an alternative to the approach in Program 4.2, Regression could have been performed using Tools, Analysis, Regression from Excel's menu bar.

PROGRAM 4.2 ■ Excel OM's Regression Analysis, with Data from Example 8

As an alternative, you may want to experiment with Excel's built-in regression analysis. To do so, under the *Tools* menu bar selection choose *Analysis,* then *Regression.* Enter your *Y* and *X* data into two columns (say C and D). When the regression window appears, enter the *Y* and *X* ranges, then select *OK.* Excel offers several plots and tables to those interested in more rigorous analysis of regression problems.

USING POM FOR WINDOWS IN FORECASTING

POM for Windows can project moving averages (both simple and weighted), handle exponential smoothing (both simple and trend adjusted), forecast with least squares trend projection, and solve linear-regression (causal) models. A summary screen of error analysis and a graph of the data can also be generated. As a special example of exponential smoothing adaptive forecasting, when using an alpha of 0, POM for Windows will find the alpha value that yields the minimum MAD.

Appendix V provides further details.

SOLVED PROBLEMS

Solved Problem 4.1

Sales of Volkswagen's popular Beetle have grown steadily at auto dealerships in Nevada during the past 5 years (see table to the right). The sales manager had predicted in 1997 that 1998 sales would be 410 VWs. Using exponential smoothing with a weight of $\alpha = .30$, develop forecasts for 1999 through 2003.

YEAR	SALES	FORECAST
1998	450	410
1999	495	
2000	518	
2001	563	
2002	584	
2003	?	

SOLUTION

YEAR	FORECAST
1998	410.0
1999	$422.0 = 410 + .3\,(450 - 410)$
2000	$443.9 = 422 + .3\,(495 - 422)$
2001	$466.1 = 443.9 + .3\,(518 - 443.9)$
2002	$495.2 = 466.1 + .3\,(563 - 466.1)$
2003	$521.8 = 495.2 + .3\,(584 - 495.2)$

Solved Problem 4.2

In Example 7, we applied trend-adjusted exponential smoothing to forecast demand for a piece of pollution-control equipment for months 2 and 3 (out of 9 months of data provided). Let us now continue this process for month 4. We want to confirm the forecast for month 4 shown in Table 4.1 (p. 97) and Figure 4.3 (p. 97).

For month 4, $A_4 = 19$, with $\alpha = .2$, and $\beta = .4$

SOLUTION

$$F_4 = \alpha A_3 + (1 - \alpha)(F_3 + T_3)$$
$$= (.2)(20) + (1 - .2)(15.18 + 2.10)$$
$$= 4.0 + (.8)(17.28)$$
$$= 4.0 + 13.82$$
$$= 17.82$$
$$T_4 = \beta(F_4 - F_3) + (1 - \beta)T_3$$
$$= (.4)(17.82 - 15.18) + (1 - .4)(2.10)$$
$$= (.4)(2.64) + (.6)(2.10)$$
$$= 1.056 + 1.26$$
$$= 2.32$$
$$FIT_4 = 17.82 + 2.32$$
$$= 20.14$$

Solved Problem 4.3

Room registrations in the Toronto Towers Plaza Hotel have been recorded for the past 9 years. In order to project future occupancy, management would like to determine the mathematical trend of guest registration. This estimate would help the hotel determine whether future expansion will be needed. Given the following time-series data, develop a regression equation relating registrations to time (e.g., a trend equation). Then forecast 2005 registrations. Room registrations are in the thousands:

1995: 17   1996: 16   1997: 16   1998: 21   1999: 20
2000: 20   2001: 23   2002: 25   2003: 24

SOLUTION

YEAR	TRANSFORMED YEAR, x	REGISTRANTS, y (IN THOUSANDS)	x^2	xy
1995	1	17	1	17
1996	2	16	4	32
1997	3	16	9	48
1998	4	21	16	84
1999	5	20	25	100
2000	6	20	36	120
2001	7	23	49	161
2002	8	25	64	200
2003	9	24	81	216
	$\Sigma x = 45$	$\Sigma y = y\,182$	$\Sigma x^2 = 285$	$\Sigma xy = 978$

$$b = \frac{\Sigma xy - n\bar{x}\bar{y}}{\Sigma x^2 - n\bar{x}^2} = \frac{978 - (9)(5)(20.22)}{285 - (9)(25)} = \frac{978 - 909.9}{285 - 225} = \frac{68.1}{60} = 1.135$$
$$a = \bar{y} - b\bar{x} = 20.22 - (1.135)(5) = 20.22 - 5.675 = 14.545$$
$$\hat{y}(\text{registrations}) = 14.545 + 1.135x$$

The projection of registrations in the year 2005 (which is $x = 11$ in the coding system used) is

$$\hat{y} = 14.545 + (1.135)(11) = 27.03$$
or 27,030 guests in 2005

Solved Problem 4.4

Quarterly demand for Jaguar XJ8s at a New York auto dealer is forecast with the equation

$$\hat{y} = 10 + 3x$$

where x = quarters, and

Quarter I of 2002 = 0
Quarter II of 2002 = 1
Quarter III of 2002 = 2
Quarter IV of 2002 = 3
Quarter I of 2003 = 4
and so on

and

$\hat{y}$ = quarterly demand

The demand for sports sedans is seasonal, and the indices for Quarters I, II, III, and IV are 0.80, 1.00, 1.30, and 0.90, respectively. Forecast demand for each quarter of 2004. Then seasonalize each forecast to adjust for quarterly variations.

SOLUTION

Quarter II of 2003 is coded $x = 5$; Quarter III of 2003, $x = 6$; and Quarter IV of 2003, $x = 7$. Hence, Quarter I of 2004 is coded $x = 8$; Quarter II, $x = 9$; and so on.

$\hat{y}(2004 \text{ Quarter I}) = 10 + 3(8) = 34$
$\hat{y}(2004 \text{ Quarter II}) = 10 + 3(9) = 37$
$\hat{y}(2004 \text{ Quarter III}) = 10 + 3(10) = 40$
$\hat{y}(2004 \text{ Quarter IV}) = 10 + 3(11) = 43$

Adjusted forecast = (.80)(34) = 27.2
Adjusted forecast = (1.00)(37) = 37
Adjusted forecast = (1.30)(40) = 52
Adjusted forecast = (.90)(43) = 38.7

INTERNET AND STUDENT CD-ROM EXERCISES

Visit our home page or use your student CD-ROM to help with material in this chapter.

 On Our Home Page, www.prenhall.com/heizer

- Self-Tests
- Practice Problems
- Internet Exercises
- Current Articles and Research
- Virtual Company Tour
- Internet Homework Problems
- Internet Cases

 On Your Student CD-ROM

- PowerPoint Lecture
- Video Clip and Video Case
- Practice Problems
- Active Model Exercises
- Excel OM
- Excel OM Data Files

ADDITIONAL CASE STUDIES

Internet Case Studies: Visit our Web site at www.prenhall.com/heizer **for these free case studies:**

- **Akron Zoological Park:** Involves forecasting attendance at Akron's zoo.

- **Human Resources, Inc.:** Requires developing a forecasting model best suited to a small company that conducts management seminars.

- **North–South Airline:** Reflects the merger of two airlines and addresses their maintenance costs.

Harvard has selected these Harvard Business School case studies to accompany this chapter (textbookcasematch.hbsp.harvard.edu)**:**

- **Merchandising at Nine West Retail Stores** (# 698-098): This large retail shoe store chain faces a merchandising decision.

- **New Technologies, New Markets: The Launch of Hong Kong Telecom's Video-on-Demand** (# HKU-011): Asks students to examine the forecasting behind a new technology.

- **Sport Obermeyer Ltd.** (# 695-022): This skiwear company has short life-cycle products with uncertain demand and a globally dispersed supply chain.

- **L.L. Bean, Inc.** (# 893-003): L.L. Bean must forecast and manage thousands of inventory items sold through its catalogs.

Design of Goods and Services

Chapter Outline

LEARNING OBJECTIVES

When you complete this chapter you should be able to

IDENTIFY OR DEFINE:

Product life cycle

Product development team

Manufacturability and value engineering

Robust design

Time-based competition

Modular design

Computer-aided design

Value analysis

Group technology

Configuration management

EXPLAIN:

Alliances

Concurrent engineering

Product-by-value analysis

Product documentation

GLOBAL COMPANY PROFILE:

Product Strategy Provides Competitive Advantage at Regal Marine

Twenty-eight years after its founding by potato farmer Paul Kuck, Regal Marine has become a powerful force on the waters of the world. The world's third largest boat manufacturer (by global sales), Regal exports to 30 countries, including Russia and China. Almost one-third of its sales are overseas.

Product design is critical in the highly competitive pleasure boat business: "We keep in touch with our customers and we respond to the marketplace," says Kuck. "We're introducing six new models this year alone. I'd say we're definitely on the aggressive end of the spectrum."

With changing consumer tastes, compounded by material changes and ever improving marine engineering, the design function is under constant pressure. Added to these pressures is the constant issue of cost competitiveness combined with the need to provide good value for customers.

Consequently, Regal Marine is a frequent user of computer-aided design (CAD). New designs come to life via Regal's three-dimensional CAD system,

CAD/CAM is used to design the hull of a new product. This process results in faster and more efficient design and production.

Once a hull has been pulled from the mold, it travels down a monorail assembly path. JIT inventory delivers engines, wiring, seats, flooring, and interiors when needed.

borrowed from automotive technology. Regal's naval architects' goal is to continue to reduce the time from concept to prototype to production. The sophisticated CAD system not only has reduced product development time but also has reduced problems with tooling and production, resulting in a superior product.

All of Regal's products, from its $14,000 18-foot boat to the $500,000 42-foot Commodore yacht, follow a similar production process. Hulls and bows are separately hand-produced by spraying preformed molds with three to five layers of a fiberglass laminate. The hulls and bows harden and are removed to become the lower and upper structure of the boat. As they move to the assembly line, they are joined and components added at each workstation.

Wooden decks, precut in-house by computer-driven routers, are delivered

REGAL MARINE

on a just-in-time basis for installation at one station. Engines—one of the few purchased components—are installed at another. Racks of electrical wiring harnesses, engineered and rigged in-house, are then installed. An in-house upholstery department delivers customized seats, beds, dashboards, or other cushioned components. Finally, chrome fixtures are put in place, and the boat is sent to Regal's test tank for watertight, gauge, and system inspection.

At the final stage, boats are placed in this test tank, where a rain machine ensures watertight fits.

Product decision
The selection, definition,
and design of products.

Global firms like Regal Marine know that the basis for an organization's existence is the good or service it provides society. Great products are the keys to success. Anything less than an excellent product strategy can be devastating to a firm. To maximize the potential for success, top companies focus on only a few products and then concentrate on those products. For instance, Honda's focus is engines. Virtually all of Honda's sales (autos, motorcycles, generators, lawn mowers) are based on its outstanding engine technology. Likewise, Intel's focus is on computer chips, and Microsoft's is PC software. However, because most products have a limited and even predictable life cycle, companies must constantly be looking for new products to design, develop, and take to market. Good operations managers insist on strong communication between customer, product, processes, and suppliers that results in a high success rate for their new products. Benchmarks, of course, vary by industry, but Regal introduces six new boats a year, and Rubbermaid introduces a new product each day!

One product strategy is to build particular competence in customizing an established family of goods or services. This approach allows the customer to choose product variations while reinforcing the organization's strength. Dell Computers, for example, has built a huge market by delivering computers with the exact hardware and software desired by end users. And Dell does it fast—it understands that speed to market is imperative to gain a competitive edge.

Note that many service firms also refer to their offerings as products. For instance, when Allstate Insurance offers a new homeowner's policy, it is referred to as a new "product." Similarly, when Citicorp opens a mortgage department, it offers a number of new mortgage "products." Although the term *products* may often refer to tangible goods, it also refers to offerings by service organizations.

An effective product strategy links product decisions with investment, market share, and product life cycle, and defines the breadth of the product line. The objective of the **product decision** is to develop and implement a product strategy that meets the demands of the marketplace with a competitive advantage. As one of the ten decisions of OM, product strategy may focus on developing a competitive advantage via differentiation, low cost, rapid response, or a combination of these.

GOODS AND SERVICES SELECTION

Product Strategy Options Support Competitive Advantage

A world of options exists in the selection, definition, and design of products. Product selection is choosing the good or service to provide customers or clients. For instance, hospitals specialize in various types of patients and various types of medical procedures. A hospital's management may decide to operate a general-purpose hospital or a maternity hospital or, as in the case of the Canadian hospital Shouldice, specialize in hernias. Hospitals select their products when they decide what kind of hospital to be. Numerous other options exist for hospitals, just as they exist for McDonald's or General Motors.

Organizations like Shouldice Hospital *differentiate* themselves through their product. Shouldice differentiates itself by offering a distinctly unique and high-quality product. Its hernia-repair service is so effective it allows patients to return to normal living in 8 days as opposed to the average of 2 weeks—and with very few complications. Shouldice customers come from throughout the world, and the hospital is so popular that it can't handle all those desiring service.

Taco Bell has developed and executed a *low-cost* strategy through product design. By designing a product (its menu) that can be produced with a minimum of labor in small kitchens, Taco Bell has developed a product line that is both low cost and high value. Successful product design has allowed Taco Bell to increase the food content of its products from 27¢ to 45¢ of each sales dollar.

Toyota's strategy is *rapid response* to changing consumer demand. By executing the fastest automobile design in the industry, Toyota has driven the speed of product development down to well under 2 years in an industry whose standard is still close to 3 years. Although competitors often operate in a 3-year design cycle, shorter design time allows Toyota to get a car to market before consumer tastes change.

Video 5.1

Product Strategy at
Regal Marine

Product selection occurs in services as well as manufacturing. Shown here is a lounge at Shouldice Hospital. Shouldice is renowned for its world-class specialization in hernia repair—no emergency room, no maternity ward, no open heart surgery, just hernias. The entire production system is designed for this one product. Local anesthetics are used; patients enter and leave the operating room on their own; rooms are spartan and meals are served in a common dining room, encouraging patients to get out of bed for meals and join fellow patients in the lounge. As Shouldice has demonstrated, product selection affects the entire production system.

Product decisions are fundamental to an organization's strategy and have major implications throughout the operations function. For instance, GM's steering columns are a good example of the strong role product design plays in both quality and efficiency. The new steering column has a simpler design, with about 30% fewer parts than its predecessor. The result: Assembly time is one-third of the older column, and the new column's quality is about seven times higher. As an added bonus, machinery on the new line costs a third less than that in the old line.

Product Life Cycles

Products are born. They live and they die. They are cast aside by a changing society. It may be helpful to think of a product's life as divided into four phases. Those phases are introduction, growth, maturity, and decline.

Product life cycles may be a matter of a few hours (a newspaper), months (seasonal fashions and personal computers), years (phonograph records), or decades (Volkswagen Beetle). Regardless of the length of the cycle, the task for the operations manager is the same: to design a system that helps introduce new products successfully. If the operations function cannot perform effectively at this stage, the firm may be saddled with losers—products that cannot be produced efficiently and perhaps not at all.

Figure 5.1 shows the four life cycle stages and the relationship of product sales, cash flow, and profit over the life cycle of a product. Note that typically a firm has a negative cash flow while it develops a product. When the product is successful, those losses may be recovered. Eventually, the successful product may yield a profit prior to its decline. However, the profit is fleeting. Hence, the constant demand for new products.

FIGURE 5.1 ■

Product Life Cycle, Sales, Cost, and Profit

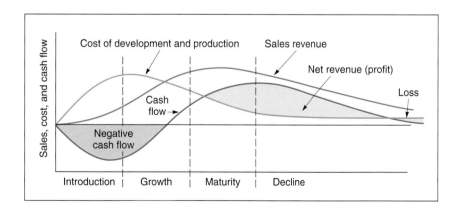

Life Cycle and Strategy

Just as operations managers must be prepared to develop new products, they must also be prepared to develop *strategies* for new and existing products. Periodic examination of products is appropriate because *strategies change as products move through their life cycle*. Successful product strategies require determining the best strategy for each product based on its position in its life cycle. A firm, therefore, identifies products or families of products and their position in the life cycle. Let us review some strategy options as products move through their life cycles.

Introductory Phase   Because products in the introductory phase are still being "fine-tuned" for the market, as are their production techniques, they may warrant unusual expenditures for (1) research, (2) product development, (3) process modification and enhancement, and (4) supplier development. For example, when cellular phones were first introduced, the features desired by the public were still being determined. At the same time, operations managers were still groping for the best manufacturing techniques.

Growth Phase   In the growth phase, product design has begun to stabilize, and effective forecasting of capacity requirements is necessary. Adding capacity or enhancing existing capacity to accommodate the increase in product demand may be necessary.

Maturity Phase   By the time a product is mature, competitors are established. So high-volume, innovative production may be appropriate. Improved cost control, reduction in options, and a paring down of the product line may be effective or necessary for profitability and market share.

Decline Phase   Management may need to be ruthless with those products whose life cycle is at an end. Dying products are typically poor products in which to invest resources and managerial talent. Unless dying products make some unique contribution to the firm's reputation or its product line or can be sold with an unusually high contribution, their production should be terminated.[1]

Product-by-Value Analysis

The effective operations manager selects items that show the greatest promise. This is the Pareto principle (i.e., focus on the critical few, not the trivial many) applied to product mix: Resources are to be invested in the critical few and not the trivial many. **Product-by-value analysis** lists products in descending order of their *individual dollar contribution* to the firm. It also lists the *total annual dollar contribution* of the product. Low contribution on a per-unit basis by a particular product may look substantially different if it represents a large portion of the company's sales.

A product-by-value report allows management to evaluate possible strategies for each product. These might include increasing cash flow (for example, increasing contribution by raising selling price or lowering cost), increasing market penetration (improving quality and/or reducing cost or price), or reducing costs (improving the production process). The report may also tell management which product offerings should be eliminated and which fail to justify further investment in research and development or capital equipment. The report focuses management's attention on the strategic direction for each product.

Product-by-value analysis
A listing of products in descending order of their individual dollar contribution to the firm, as well as the *total annual* dollar contribution of the product.

GENERATING NEW PRODUCTS

Because products die; because products must be weeded out and replaced; because firms generate most of their revenue and profit from new products— product selection, definition, and design take place on a continuing basis. Knowing how to successfully find and develop new products is a requirement.

New Product Opportunities

One technique to generate new product ideas is brainstorming. **Brainstorming** is a technique in which a diverse group of people share, without criticism, ideas on a particular topic. The goal in this application is to generate an open discussion that will yield creative ideas about possible products

Brainstorming
A team technique to generate creative ideas on a particular subject. Ideas are not reviewed until after the brainstorming session.

---

[1]*Contribution* is defined as the difference between direct cost and selling price. Direct costs are labor and material that go into the product.

OM IN ACTION

Stryker's Product Ideas Come from Its Customers

Homer Stryker's hospital products firm has made *Forbes*'s list of the Best Small Companies in America for 10 years straight. From its humble start 50 years ago by its clever orthopedist founder, Stryker Corporation now offers an array of niche products, including bone drills and saws, hospital beds, hip implants, and video cameras for internal surgery.

Churning out new products has been Stryker's strength. By operating with autonomous divisions, each with its own highly trained, specialized sales staff, Stryker knows how to listen to its customers. Stryker's salespeople act as a de facto research-and-development team. Most of the company's new product ideas come from salespeople working directly with surgeons—and often standing in the operating room next to the physician. There they can observe the doctor in action, write down

their comments, and come up with ways to improve a saw, a hip implant, or a bed.

As a case in point, eye surgeons kept complaining about a bed's lack of flexibility at the head level. It was hard, they said, to position a patient's head. Stryker people took note and the firm rolled out a profitable bed with a moveable headrest.

Another hot new product is a tiny $18,000 video camera used inside a long tube inserted through the abdomen for gallbladder surgery. Aided by the camera, the surgeon can swiftly remove the gallbladder with only a minute incision. Instead of a 1-week hospital stay, the patient is out the next day. These products and similar products from Stryker's 12 divisions reflect the continuing focus on holding down health-care costs while producing brisk sales. Effective product development remains a key ingredient of Stryker's profitability.

Sources: Forbes (May 13, 2002): 104–108; *Barron's* (May 7, 2001): 43; and the *Wall Street Journal* (July 7, 1997): B10.

and product improvements. Although firms may include brainstorming in various stages of new-product development, directly and energetically focusing on the specific opportunities, as noted below, is often rewarding.

> "Companies that are successful today . . . are those that can get closest to their customers' needs."
>
> Federal Reserve Board

1. *Understanding the customer* is the premier issue in new-product development. Many commercially important products are initially thought of and even prototyped by users rather than producers. Such products tend to be developed by "lead users"—companies, organizations, or individuals that are well ahead of market trends and have needs that go far beyond those of average users.[2] The operations manager must be "tuned in" to the market and particularly these lead users. The *OM in Action* box, "Stryker's Product Ideas Come from Its Customers," discusses how Stryker stays tuned in and maintains its flow of new ideas.
2. *Economic change* brings increasing levels of affluence in the long run but economic cycles and price changes in the short run. In the long run, for instance, more and more people can afford automobiles, but in the short run, a recession may weaken the demand for automobiles.
3. *Sociological and demographic change* may appear in such factors as decreasing family size. This trend alters the size preference for homes, apartments, and automobiles.
4. *Technological change* makes possible everything from hand-held computers to cellular phones to artificial hearts.
5. *Political/legal change* brings about new trade agreements, tariffs, and government contract requirements.
6. Other changes may be brought about through *market practice*, *professional standards*, *suppliers*, and *distributors*.

Operations managers must be aware of these factors and be able to anticipate changes in product opportunities, the products themselves, product volume, and product mix.

Importance of New Products

More than 30% of Rubbermaid sales each year come from products less than 5 years old.

The importance of new products cannot be overestimated. As Figure 5.2 shows, leading companies generate a substantial portion of their sales from products less than 5 years old. This is why Gillette developed its new three-blade razor, in spite of continuing high sales of its phenomenally successful Sensor razor.

[2]Eric von Hipple, Stefan Thomke, and Mary Sonnack, "Creating Breakthroughs at 3M," *Harvard Business Review* 71, no. 5 (September–October 1999): 47–57.

FIGURE 5.2 ■

Percent of Sales from Products Introduced in the Last 5 Years

The higher the percent of sales from products introduced in the last 5 years, the more likely the firm is to be a leader.

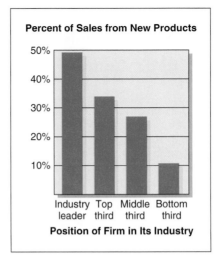

Percent of Sales from New Products

Position of Firm in Its Industry

Despite constant efforts to introduce viable new products, many new products do not succeed. Indeed, for General Mills to come up with a winner in the breakfast cereal market—defined as a cereal that gets a scant one-half of 1% of the market—isn't easy. Among the top 10 brands of cereal, the youngest, Honey Nut Cheerios, was created in 1979.[3] DuPont estimates that it takes 250 ideas to yield one *marketable* product.[4]

As one can see, product selection, definition, and design occur frequently—perhaps hundreds of times for each financially successful product. Operations managers and their organizations must be able to accept risk and tolerate failure. They must accommodate a high volume of new product ideas while maintaining the activities to which they are already committed.

Motorola went through 3,000 working models to develop its first pocket-size cellular telephone.

PRODUCT DEVELOPMENT

Product Development System

An effective product strategy links product decisions with cash flow, market dynamics, product life cycle, and the organization's capabilities. A firm must have the cash for product development, understand the changes constantly taking place in the marketplace, and have the necessary talents and resources available.[5] The product development system may well determine not only product success but also the firm's future. Figure 5.3 shows the stages of product development. In this system, product options go through a series of steps, each having its own screening and evaluation criteria and providing feedback to prior steps.[6]

The screening process extends to the operations function. Optimum product development depends not only on support from other parts of the firm, but also the successful integration of all 10 of the OM decisions, from product design to maintenance. Identifying products that appear likely to capture market share, be cost effective, and profitable, but are in fact very difficult to produce, may lead to failure rather than success.[7]

---

[3]Richard Gibson, "A Cereal Maker's Quest for the Next Grape-Nuts," the *Wall Street Journal* (January 23, 1997): B1.

[4]Rosabeth Kanter, John Kao, and Fred Wiersema, *Innovation Breakthrough Thinking at 3M, DuPont, GE, Pfizer, and Rubbermaid* (New York: Harper-Business, 1997).

[5]Ming Ding and Jehosua Eliashberg, "Structuring the New Product Development Pipeline," *Management Science* 48, no. 3 (March 2002): 343–363.

[6]For insight into optimal testing strategies, see Stefan Thomke and David E. Bell, "Sequential Testing in Product Development," *Management Science* 47, no. 2 (February 2001): 308–323.

[7]Rohit Verma, Gary M. Thompson, William L. Moore, and Jordan J. Louviere, "Effective Design of Products/Services: An Approach Based on Integration of Marketing and Operations Management Decisions," *Decision Sciences* 32, no. 1 (winter 2001): 165–193.

FIGURE 5.3 ■

Product Development Stages

Product concepts are developed from a variety of sources, both external and internal to the firm. Concepts that survive the product idea stage progress through various stages, with nearly constant review, feedback, and evaluation in a highly participative environment to minimize failure.

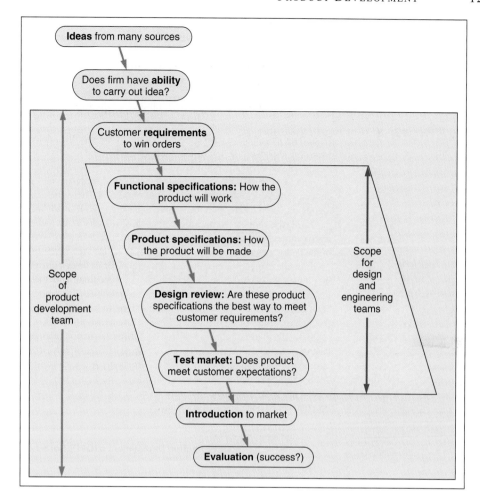

Quality Function Deployment (QFD)

Quality function deployment (QFD)
A process for determining customer requirements (customer "wants") and translating them into the attributes (the "hows") that each functional area can understand and act on.

Quality function deployment (QFD) refers to both (1) determining what will satisfy the customer and (2) translating those customer desires into the target design.[8] The idea is to capture a rich understanding of customer wants and to identify alternative process solutions. This information is then integrated into the evolving product design. QFD is used early in the design process to help determine *what will satisfy the customer* and *where to deploy quality efforts.*

One of the tools of QFD is the house of quality. The **house of quality** is a graphic technique for defining the relationship between customer desires and product (or service). Only by defining this relationship in a rigorous way can operations managers build products and processes with features desired by customers. Defining this relationship is the first step in building a world-class production system. To build the house of quality, we perform six basic steps:

House of quality
A part of the quality function deployment process that utilizes a planning matrix to relate customer "wants" to "how" the firm is going to meet those "wants."

1. Identify customer *wants.* (What do prospective customers want in this product?)
2. Identify *how* the good/service will satisfy customer wants. (Identify specific product characteristics, features, or attributes and show how they will satisfy customer *wants.*)
3. Relate customer *wants* to product *hows.* (Build a matrix, as in Example 1, that shows this relationship.)
4. Identify relationships between the firm's *hows.* (How do our *hows* tie together? For instance, in the following example, there is a high relationship between low electricity requirements and auto focus, auto exposure, and auto film advance because they all require electricity. This relationship is shown in the "roof" of the house in Example 1.)

[8]See Yoji Akao, ed., *Quality Function Deployment: Integrating Customer Requirements into Product Design* (Cambridge, MA: Productivity Press, 1990).

5. Develop importance ratings. (Using the *customer's* importance ratings and weights for the relationships shown in the matrix, compute *our* importance ratings, as in Example 1.)

6. Evaluate competing products. (How well do competing products meet customer wants? Such an evaluation, as shown in the two columns on the right of the figure in Example 1, would be based on market research.)

Example 1 shows how to construct a house of quality.

Example 1

First, through extensive market research, Great Cameras, Inc., determined what the customer *wants*. Those *wants* are shown on the left of the house of quality and are: lightweight, easy to use, reliable, easy to hold steady, and no double exposures. Second, the product development team determined *how* the organization is going to translate those customer *wants* into product design and process attribute targets. These *hows* are entered across the top portion of the house of quality. These characteristics are low electricity requirements, aluminum components, auto focus, auto exposure, auto film advance, and ergonomic design.

Third, the product team evaluated each of the customer *wants* against the *hows*. In the relationship matrix of the house, the team evaluated how well its design will meet customer needs. Fourth, in the "roof" of the house, the product development team developed the relationship between the attributes.

Fifth, the team developed importance ratings for its design attributes on the bottom row of the table. This was done by assigning values (5 for high, 3 for medium, and 1 for low) to each entry in the relationship matrix, and then multiplying each of these values by the customer's importance rating. These values in the "Our importance ratings" row provide a ranking of how to proceed with product and process design, with the highest values being the most critical to a successful product.

Sixth, the house of quality is also used for the evaluation of competitors. How well do *competitors* meet customer demand? The two columns on the right indicate how market research thinks competitors satisfy customer wants (**G**ood, **F**air, or **P**oor). So company A does a good job on "lightweight," "easy to use," and "easy to hold steady," a fair job on "reliability," and a poor job on "no double exposures." Company B does a good job with "reliability" but poor on other attributes. Products from other firms and even the proposed product can be added next to company B.

QFD Capture Software is a management aid to prioritize choices for better products and services. A free evaluation version is available at www.qfdcapture.com.

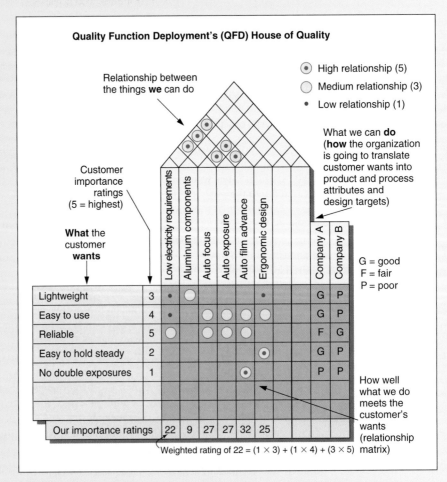

Quality Function Deployment's (QFD) House of Quality

○ High relationship (5)
○ Medium relationship (3)
• Low relationship (1)

Relationship between the things **we** can do

What we can **do** (**how** the organization is going to translate customer wants into product and process attributes and design targets)

Customer importance ratings (5 = highest)

What the customer **wants**

G = good
F = fair
P = poor

		Low electricity requirements	Aluminum components	Auto focus	Auto exposure	Auto film advance	Ergonomic design		Company A	Company B
Lightweight	3	•	○				•		G	P
Easy to use	4	•		○	○	○	○		G	P
Reliable	5	○		○	○	○			F	G
Easy to hold steady	2						○		G	P
No double exposures	1					○			P	P
Our importance ratings		22	9	27	27	32	25			

Weighted rating of 22 = (1 × 3) + (1 × 4) + (3 × 5)

How well what we do meets the customer's wants (relationship matrix)

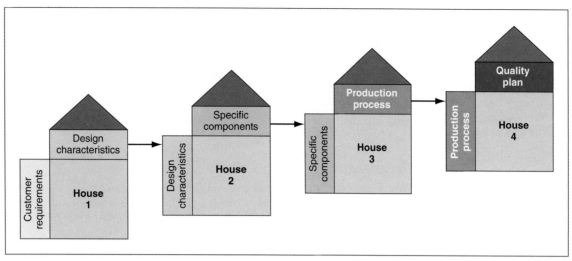

FIGURE 5.4 ■ House of Quality Sequence Indicates How to Deploy Resources to Achieve Customer Requirements

Another use of quality function deployment (QFD) is to show how the quality effort will be *deployed*. As Figure 5.4 shows, *design characteristics* of House 1 become the inputs to House 2, which are satisfied by *specific components* of the product. Similarly, the concept is carried to House 3, where the specific components are to be satisfied through particular *production processes*. Once those production processes are defined, they become requirements of House 4 to be satisfied by a *quality plan* that will ensure conformance of those processes. The quality plan is a set of specific tolerances, procedures, methods, and sampling techniques that will ensure that the production process meets the customer requirements.

> **Product excellence means determining what the customer wants and providing it.**

Much of the QFD literature and effort is devoted to meeting customer requirements with design characteristics (House 1 in Figure 5.4), and its importance is not to be underestimated. However, the *sequence* of houses is a very effective way of identifying, communicating, and allocating resources throughout the system. The series of houses helps operations managers determine where to *deploy* quality resources. In this way we meet customer requirements, produce quality products, and win orders.

Organizing for Product Development

The traditional U.S. approach to product development is an organization with distinct departments. These departments are: first, a research and development department to do the necessary research; then an engineering department to design the product; then a manufacturing engineering department to design a product that can be produced; and finally, a production department that produces the product. The distinct advantage of this approach is that fixed duties and responsibilities exist. The distinct disadvantage is lack of forward thinking: How will downstream departments in the process deal with the concepts, ideas, and designs presented to them, and ultimately what will the customer think of the product? A second and popular approach is to assign a product manager to "champion" the product through the product development system and related organizations. However, a third, and perhaps the best, product development approach used in the U.S. seems to be the use of teams. Such teams are known variously as *product development teams, design for manufacturability teams*, and *value engineering teams*.

The Japanese bypass the team issue by not subdividing organizations into research and development, engineering, production, and so forth. Consistent with the Japanese style of group effort and teamwork, these activities are all in one organization. Japanese culture and management style are more collegial and the organization less structured than in most Western countries. Therefore, the Japanese find it unnecessary to have "teams" provide the necessary communication and coordination. However, the typical Western style and the conventional wisdom is to use teams.

> **Product development teams**
>
> Teams charged with moving from market requirements for a product to achieving product success.

Product development teams are charged with the responsibility of moving from market requirements for a product to achieving a product success (refer back to Figure 5.3 on page 129). Such teams often include representatives from marketing, manufacturing, purchasing, quality assurance, and field service personnel. Many teams also include representatives from vendors.

Regardless of the formal nature of the product development effort, research suggests that success is more likely in an open, highly participative environment where those with potential contributions are allowed to make them. The objective of a product development team is to make the good or service a success. This includes marketability, manufacturability, and serviceability.

Concurrent engineering
Use of participating teams in design and engineering activities.

Use of such teams is also called **concurrent engineering** and implies a team representing all affected areas (known as a *cross-functional* team). Concurrent engineering also implies speedier product development through simultaneous performance of various aspects of product development.[9] The team approach is the dominant structure for product development by leading organizations in the U.S.[10]

Manufacturability and Value Engineering

Manufacturability and value engineering
Activities that help improve a product's design, production, maintainability, and use.

Manufacturability and value engineering activities are concerned with improvement of design and specifications at the research, development, design, and production stages of product development. In addition to immediate, obvious cost reduction, design for manufacturability and value engineering may produce other benefits. These include:

1. Reduced complexity of the product.
2. Additional standardization of components.
3. Improvement of functional aspects of the product.
4. Improved job design and job safety.
5. Improved maintainability (serviceability) of the product.
6. Robust design.

Manufacturability and value engineering activities may be the best cost-avoidance technique available to operations management. They yield value improvement by focusing on achieving the functional specifications necessary to meet customer requirements in an optimal way. Value engineering programs, when effectively managed, typically reduce costs between 15% and 70% without reducing quality. Some studies have indicated that for every dollar spent on value engineering, $10 to $25 in savings can be realized.

Product design affects virtually all aspects of operating expense. Consequently, the development process needs to ensure a thorough evaluation of design prior to a commitment to produce. The cost reduction achieved for a specific bracket via value engineering is shown in Figure 5.5.

ISSUES FOR PRODUCT DESIGN

In addition to developing an effective system and organization structure for product development, several *techniques* are important to the design of a product. We will now review seven of these: (1) robust design, (2) modular design, (3) computer-aided design (CAD), (4) computer-aided manufacturing (CAM), (5) virtual reality technology, (6) value analysis, and (7) environmentally friendly designs.

FIGURE 5.5 ■

Cost Reduction of a Bracket via Value Engineering

Each time the bracket is redesigned and simplified, we are able to produce it for less.

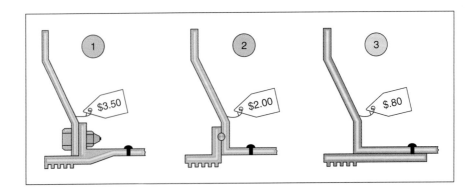

[9]Firms that have high technological or product change in their competitive environment tend to use more concurrent engineering practices. See Xenophon Koufteros, Mark Vonderembse, and William Doll, "Concurrent Engineering and its Consequences," *Journal of Operations Management* 19, no. 1 (January 2001): 97–115.

[10]"Best Practices Survey 1994: Product Definition," *Target* 11, no. 3 (May–June 1995): 22–24.

Robust Design

Robust design means that the product is designed so that small variations in production or assembly do not adversely affect the product. For instance, AT&T developed an integrated circuit that could be used in many products to amplify voice signals. As originally designed, the circuit had to be manufactured very precisely to avoid variations in the strength of the signal. Such a circuit would have been costly to make because of stringent quality controls needed during the manufacturing process. However, after testing and analyzing the design, AT&T engineers realized that if the resistance of the circuit were reduced—a minor change with no associated costs—the circuit would be far less sensitive to manufacturing variations. The result was a 40% improvement in quality.

Modular Design

Products designed in easily segmented components are known as **modular designs**. Modular designs offer flexibility to both production and marketing. The production department typically finds modularity helpful because it makes product development, production, and subsequent changes easier. Moreover, marketing may like modularity because it adds flexibility to the ways customers can be satisfied. For instance, virtually all premium high-fidelity stereos are produced and sold this way. The customization provided by modularity allows customers to mix and match to their own taste. This is also the approach taken by Harley-Davidson, where relatively few different engines, chassis, gas tanks, and suspension systems are mixed to produce a huge variety of motorcycles. It has been estimated that many automobile manufacturers can, by mixing the available modules, never make two cars alike. This same concept of modularity is carried over to many industries, from airframe manufacturers to fast-food restaurants. Airbus uses the same wing modules on several planes, just as McDonald's and Burger King use relatively few modules (cheese, lettuce, buns, sauces, pickles, meat patties, French fries, etc.) to make a variety of meals.

Video 5.2

Modular Assembly at Harley-Davidson

Computer-Aided Design (CAD)

Computer-aided design (CAD) is the use of computers to interactively design products and prepare engineering documentation. Although the use and variety of CAD software is extensive, most of it is still used for drafting and three-dimensional drawings. However, its use is rapidly expanding. CAD software allows designers to save time and money by shortening development cycles for virtually all products. The speed and ease with which sophisticated designs can be manipulated, analyzed, and modified with CAD makes review of numerous options possible before final commitments are made.

Computer-aided design: B.F. Goodrich engineers and managers use software from Structural Dynamics Research Corporation (SDRC) to model wheel and brake assemblies. By analyzing stress and heat, they can often avoid costly design and production mistakes. Whereas errors that are found at the design stage on a CRT screen can often be fixed for a nominal cost, the cost is substantial once production has begun.

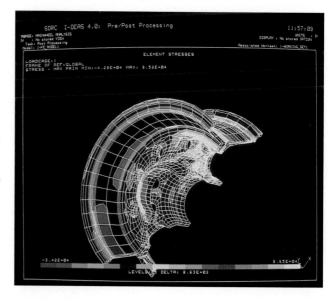

Faster development, better products, accurate flow of information to other departments—all contribute to a tremendous payoff for CAD. The payoff is particularly significant because most product costs are determined at the design stage.

One extension of CAD is **Design for Manufacture and Assembly (DFMA)** software, which focuses on the effect of design upon assembly. It allows designers to examine the integration of product designs before the product is manufactured. For instance, DFMA allows automobile designers to examine how a transmission will be placed in a car on the production line, even while both the transmission and the car are still in the design stage.

A second CAD extension is **3-D object modeling**. The technology is particularly useful for small prototype development (as shown in the photo below). 3-D object modeling rapidly builds up a model in very thin layers of synthetic materials for evaluation. This technology speeds development by avoiding a more lengthy and formal manufacturing process.

Some CAD systems have moved to the Internet through e-commerce, where they link computerized design with purchasing, outsourcing, manufacturing, and long-term maintenance. This move supports rapid product change and the growing trend toward "mass customization." With CAD on the Internet, customers can enter a supplier's design libraries and make design changes. The supplier's software can then automatically generate the drawings, update the bill of material, and prepare instructions for the supplier's production process. The result is customized products produced faster and cheaper.[11]

CAD technologies are based on electronic product-design information in digital form. This digital information has proven so important that a standard for its exchange has been developed and is known as **Standard for the Exchange of Product Data (STEP)**. STEP permits manufacturers to express 3-D product information in a standard format so it can be exchanged internationally, allowing geographically dispersed manufacturers to integrate design, manufacture, and support processes.[12]

Computer-Aided Manufacturing (CAM)

Computer-aided manufacturing (CAM) refers to the use of specialized computer programs to direct and control manufacturing equipment. When computer-aided design (CAD) information is translated into instructions for computer-aided manufacturing (CAM), the result of these two technologies is CAD/CAM.

Design for Manufacture and Assembly (DFMA)
Software that allows designers to look at the effect of design on manufacturing of the product.

3-D object modeling
An extension of CAD that builds small prototypes.

Standard for the Exchange of Product Data (STEP)
Provides a format allowing the electronic transmittal of three-dimensional data.

Computer-aided manufacturing (CAM)
The use of information technology to control machinery.

This prototype wheel for an auto tire (at the left of the photo) is being built using 3-D System's Sterolithography technology, a 3-D object modeling system. This technology uses laser to build structures layer by layer in .001 of an inch increments. The technique reduces the time it takes to create a sample from weeks to hours while also reducing costs. The technique is also known as rapid prototyping.

---

[11]Christopher M. Wright, "Collaborative Manufacturing Technology Ushers in a New Era," *APICS—The Performance Advantage* (March 2002): 33–36.

[12]The STEP format is documented in the European Community's standard called ISO 10303.

OM IN ACTION

Toyota Is Revving Up

Toyota is putting the pedal to the metal. With a billion-dollar purchase, the Japanese giant is buying hardware, software, and services that will further reduce its new-car design time. For a company that already designs cars faster than anyone in the industry, the move is significant. Already holding the reputation as the best automobile company in the world, Toyota has just raised the bar.

Toyota intends to model every aspect of car production, from styling to parts, to assembly sequence, to design of the factory itself. The CAD/CAM capabilities of the software will let Toyota's designers collaborate with each other and with their worldwide design suppliers. The new software allows not only the testing of designs of parts and assemblies for "manufacturability," but the digital testing of component installation as the car rolls down the assembly line. Ultimately the system will be used to digitally model the entire factory, specifying each step in the production process: which tools, supplies, and parts are used and where; how many people are needed at each assembly point; and exactly what they will do.

The software is Dassault Systems S.A.'s 3D Product Life Cycle Management (PLM) suite. In addition to PLM, the package includes design collaboration and production support applications. The product will link Toyota's 56 plants in 25 countries and its 1,000 plus suppliers.

Sources: Information Week (April 1, 2002): 16–18, the *Wall Street Journal* (March 26, 2002): B7; and *Asia Computer Weekly* (April 8, 2002): 1.

Proctor & Gamble used CAD when designing its Crest toothpaste pump dispenser.

The benefits of CAD and CAM include:

1. *Product quality.* CAD permits the designer to investigate more alternatives, potential problems, and dangers.
2. *Shorter design time.* A shorter design phase lowers cost and allows a more rapid response to the market.
3. *Production cost reductions.* Reduced inventory, more efficient use of personnel through improved scheduling, and faster implementation of design changes lower costs.
4. *Database availability.* Consolidating accurate product data so everyone is operating from the same information results in dramatic cost reductions.
5. *New range of capabilities.* For instance, the ability to rotate and depict objects in three-dimensional form, to check clearances, to relate parts and attachments, to improve the use of numerically controlled machine tools—all provide new capability for manufacturing. CAD/CAM removes substantial detail work, allowing designers to concentrate on the conceptual and imaginative aspects of their task.

Many designers, such as Toyota Motor Corp., are moving ahead with the next wave of digital tools, as shown in the *OM in Action* box, "Toyota Is Revving Up."

Virtual Reality Technology

Virtual reality
A visual form of communication in which images substitute for reality and typically allow the user to respond interactively.

Virtual reality is a visual form of communication in which images substitute for the real thing, but still allow the user to respond interactively. The roots of virtual reality technology in operations are in computer-aided design. Once design information is in a CAD system, it is also in electronic digital form for other uses. For instance, General Motors creates its version of a "virtual car" using ceiling-mounted video projectors to project stereoscopic images on the floor of a small, stark room. After donning a special pair of glasses, both designers and customers see a three-dimensional model of what the inside of a new design looks like. Virtual reality is also being used to develop 3-D layouts of everything from restaurants to amusement parks. Changes to the car, restaurant, or ride are made much less expensively at this design stage than they can be later.

Like Toyota and GM, many firms throughout the world are now using these design technologies to speed up product development, drive down costs, and improve products.

Value analysis
A review of successful products that takes place during the production process.

Value Analysis

Although value engineering (discussed on page 132) focuses on *preproduction* design improvement, value analysis, a related technique, takes place *during* the production process, when it is clear that a new product is a success. **Value analysis** seeks improvements that lead to either a better product or a product made more economically. The techniques and advantages for value analysis are the same as for value engineering, although minor changes in implementation may be necessary because value analysis is taking place while the product is being produced.

Environmentally Friendly Designs

One of the operations manager's most environmentally sound activities is to enhance productivity. Planet Earth is finite; managers who squeeze more out of its resources are its heroes. Good operations managers can drive down costs while preserving those resources. DuPont, for example, designs its polyester film stronger and thinner so it uses less material and costs less to make. Also, because the film performs better, customers are willing to pay more for it.[13]

Bristol-Meyers Squibb has responded to environmental issues with a pollution prevention program called Environment 2000. This program addresses environmental, health, and safety issues at all stages of the product life cycle. Ban Roll-On was one of the first products studied. Repackaging Ban in smaller cartons resulted in a reduction of 600 tons of recycled paperboard. The product then required 55% less shelf space for display. As a result, not only is pollution prevented but store operating costs are also reduced.

Environmental Teams   One way to accomplish programs like these is to add an environmental charge to the value engineering and value analysis teams. With employees from different functional areas working together, a wider range of environmental issues can be addressed. These teams should consider two issues. First, they should view the impact of product design from a "systems" perspective, that is, view the product in terms of its impact on the entire economy. For example, between styrofoam or paper containers, which one is really "best," and by what criteria? We may know which is most economical for the firm, but is that one also most economical for society? Second, the teams should consider the life cycle of the product, from raw material, to installation, to use, to disposal. The goal is to reduce the environmental impact of a product throughout its life, but the task is challenging.[14] The goals of such a strategy include:

1. Developing safe and more environmentally sound products.
2. Minimizing waste of raw materials and energy.
3. Differentiating products from the competition.
4. Reducing environmental liabilities.
5. Increasing cost-effectiveness of complying with environmental regulations.
6. Being recognized as a good corporate citizen.

The German auto firm BMW has successfully addressed the decline stage of the life cycle by being environmentally friendly at the design stage; its designs now include recyclable plastic components, as shown in the photo on the next page. This effort is consistent with the environmental issues raised by the ISO 14000 standard, a topic we address in Chapter 6.

Green manufacturing
Sensitivity to a wide variety of environmental issues in production processes.

Green Manufacturing   The concept of **green manufacturing**—that is, making environmentally sound products through efficient processes—can be good business.[15] Companies can show their sensitivity to green manufacturing in product and process design in several ways:

1. *Make products recyclable.* Germany, a leader in the "green movement," has passed a packaging ordinance requiring beer brewers to use refillable bottles.
2. *Use recycled materials.* Scotch-Brite soap pads at 3M are designed to use recycled plastics, as are the park benches and other products at Plastic Recycling Corporation.

---

[13]A. B. Lovins, L. H. Lovins, and P. Hawken, "A Road Map for Natural Capitalism," *Harvard Business Review* 77, no. 3 (May–June 1999): 153.

[14]Chialen Chen, in "Design for the Environment: A Quality-Based Model for Green Product Development" (*Management Science* 47, no. 2 [February 2001]: 250–263) suggests that " . . . green product development and stricter environmental standards might not necessarily benefit the environment."

[15]For an article on management's perception of "environmentally responsible manufacturing," see Steven A. Melnyk, Robert Sroufe, and Frank Montabon, "How Does Management View Environmentally Responsible Manufacturing?" *Production and Inventory Management Journal* 42, nos. 3 and 4 (third and fourth quarters 2001): 55–63.

BMW uses parts made of recycled plastics (blue) and parts that can be recycled (green). "Green manufacturing" means companies can reuse, refurbish, or dispose of a product's components safely and reduce total life cycle product costs.

3. *Use less harmful ingredients.* Standard Register, like most of the printing industry, has replaced environmentally dangerous inks with soybean-based inks that reduce air and water pollution.

4. *Use lighter components.* The auto industry continues to expand the use of aluminum and plastic components to reduce weight. This change in material, while expensive, makes autos more environmentally friendly by improving mileage.

5. *Use less energy.* While the auto industry is redesigning autos to improve mileage, General Electric is redesigning a new generation of refrigerators that require substantially less electricity during their lifetime. DuPont is so good at energy efficiency that it has turned its expertise into a consulting business.

6. *Use less material.* Most companies waste material—in the plant and in the packaging. An employee team at a Sony semiconductor plant achieved a 50% reduction in the amount of chemicals used in the silicon wafer etching process. This and similar successes reduce both production costs and environmental concerns. To conserve packaging, Boston's Park Plaza Hotel eliminated bars of soap and bottles of shampoo by installing pump dispensers in its bathrooms. This saved the need for 1 million plastic containers a year.

Green manufacturing is appreciated by the public, and it can save money, material, and the environment we live in. These are the kind of win–win situations that operations managers seek.

TIME-BASED COMPETITION

As product life cycles shorten, the need for faster product development increases. Additionally, as technological sophistication of new products increases, so do the expense and risk. For instance, drug firms invest an average of 12 to 15 years and $400 million before receiving regulatory approval of each new drug. And even then, only 1 of 5 will actually be a success.[16] Those operations managers who master this art of product development continually gain on slower product developers. To the swift goes the competitive advantage. This concept is called **time-based competition**.

Often, the first company into production may have its product adopted for use in a variety of applications that will generate sales for years. It may become the "standard." Consequently, there is often more concern with getting the product to market than with optimum product design or process efficiency. Even so, rapid introduction to the market may be good management because until competition begins to introduce copies or improved versions, the product can sometimes be priced high enough to justify somewhat inefficient production design and methods. For example, when Kodak first introduced its Ektar film, it sold for 10% to 15% more than conventional film. Motorola's innovative pocket-sized cellular telephone was 50% smaller than any competitor's and sold for twice the price.

Time-based competition

Competition based on time; rapidly developing products and moving them to market.

[16]*The Emerging BioEconomy* (Washington, DC: New Economy Strategies, Inc., April 2002): 9.

FIGURE 5.6 ■

Product Development
Continuum

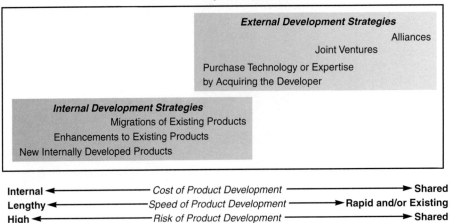

Product Development Continuum

Internal ◄───	Cost of Product Development ───►	Shared
Lengthy ◄───	Speed of Product Development ───►	Rapid and/or Existing
High ◄───	Risk of Product Development ───►	Shared

Much of the current competitive battlefield is focused around the speed of product to market. If an organization loses here, catching up in other areas is very difficult. The president of one huge U.S. firm says: "If I miss one product cycle, I'm dead."

Because time-based competition is so important, instead of developing new products from scratch (which has been the focus thus far in this chapter) a number of other strategies can be used. Figure 5.6 shows a continuum that goes from new, internally developed products (on the lower left) to "alliances." *Enhancements* and *migrations* use the organization's existing product strengths for innovation and therefore are typically faster, while at the same time being less risky than developing entirely new products. Enhancements may be modest changes in color, size, weight, or features, such as are taking place in cellular phones, or even changes in commercial aircraft. Boeing's approach to enhancements of the 737 is described in the *OM in Action* box. However, Boeing also uses its engineering prowess to *migrate* from one model to the next, as it has done when moving from the 757 to the 767 to the 777. These approaches allow Boeing to speed development while reducing both cost and risk for new designs.

The product development strategies on the lower left of Figure 5.6 are internal development strategies, while the three approaches we now introduce can be thought of as external development strategies. Firms use both. The external strategies are: (1) purchasing technology, (2) establishing joint ventures, and (3) developing alliances.

OM IN ACTION

Enhancing the Boeing 737 Life Cycle

Many firms have found that they can extend the life cycles of their products by enhancing them. In this way, they prolong earnings streams and generate additional profits. One multinational successfully doing this is Boeing. Its key products include the 737, the 747, the 757, the 767, and the 777.

Boeing delivered its first 737 in 1967. The plane sold well for several years; many carriers initially bought 20 or 30 at a time. However, in the mid-1970s, it began to lose ground. Boeing decided that the 737's life cycle was ending and was preparing to end its production. It decided to try one last measure, however, by marketing the plane to fledgling airlines in developing countries. Instead of trying to sell 20 planes at a time to KLM or United, Boeing intended to sell 1 or 2 at a time to small airlines in countries in Africa, South America, and other developing regions.

Boeing first realized that it had to make a few modifications to the basic 737 so that it would better fit local conditions. For example, pilots in developing countries were not as skilled as their Western counterparts and tended to "bounce" more during landing. So Boeing redesigned the landing system to be better able to handle extreme landing conditions.

The enhancement plan was a big success. Boeing sold enough 737s, even in small quantities, to justify keeping the plane in production. And as those small airlines began to grow, they continued to buy 737s as well as upgrading to newer, larger Boeing aircraft. Surprisingly, domestic orders also continued to come in on a regular basis from airlines such as Southwest, which uses 737s exclusively. As a result, Boeing continues to make the 737, which recently became the largest-selling commercial aircraft in aviation history.

Sources: Forbes (May 13, 2002): 82–86; and *Aircraft Value News* (January 18, 1999): 1.

Purchase of Technology by Acquiring a Firm

Microsoft and Cisco Systems are examples of companies on the cutting edge of technology that often speed development by *acquiring entrepreneurial firms* that have already developed the technology that fits their mission. The issue then becomes fitting the purchased organization, its technology, and its product lines into the buying firm, rather than a product development issue.

Joint Ventures

Joint ventures are combined ownership, usually between just two firms, to form a new entity. Ownership can be fifty-fifty or one owner can assume a larger portion to ensure tighter control. Joint ventures are often appropriate to exploit specific product opportunities that may not be central to the firm's mission. Such ventures are more likely to work when the risks are known and can be equitably shared. For instance, GM and Toyota formed a joint venture with their NUMMI plant in northern California to produce the GM Prism and the Toyota Corolla. Both companies saw a learning opportunity, as well as a product they both needed in the North American market. Toyota wanted to learn about building and managing a plant in North America, and GM wanted to learn about manufacturing a small car with Toyota's manufacturing techniques. The risks are well understood, as were the respective commitments. Similarly, Fuji-Xerox, a manufacturer and marketer of photocopiers, is a joint venture of Xerox, the U.S. maker of photocopiers, and Fuji, Japan's largest manufacturer of film.

Alliances

Alliances are cooperative agreements that allow firms to remain independent, but use complementing strengths to pursue strategies consistent with their individual missions. When new products are central to the mission, but substantial resources are required and sizeable risk is present, then alliances may be a good strategy for product development. Alliances are particularly beneficial when the products to be developed also have technologies that are in ferment. Additionally, if the boundaries between firms will be difficult to specify, alliances may be the best strategy. For example, Microsoft is pursuing a number of alliances with a variety of companies to deal with the convergence of computing, the Internet, and television broadcasting. Alliances in this case are appropriate because the technological unknowns, capital demands, and risks are significant. Similarly, three firms, DaimlerChrysler, Ford Motor, and Ballard Power Systems, have formed an alliance to develop "green" cars powered by fuel cells by the year 2004.[17] However, alliances are much more difficult to achieve and maintain than joint ventures because of the ambiguities associated with them. It may be helpful to think of an alliance as an incomplete contract between the firms. The firms remain separate.

Enhancements, migration, acquisitions, joint ventures, and alliances are all strategies for speeding product development. Moreover, they typically reduce the risk associated with product development while enhancing the human and capital resources available.

DEFINING THE PRODUCT

Once new goods or services are selected for introduction, they must be defined. First, a good or service is defined in terms of its *functions*—that is, what it is to *do*. The product is then designed and the firm determines how the functions are to be achieved. Management typically has a variety of options as to how a product should achieve its functional purpose. For instance, when an alarm clock is produced, aspects of design such as the color, size, or location of buttons may make substantial differences in ease of manufacture, quality, and market acceptance.

Rigorous specifications of a product are necessary to assure efficient production. Equipment, layout, and human resources cannot be determined until the product is defined, designed, and documented. Therefore, every organization needs documents to define its products. This is true of everything from meat patties, to cheese, to computers, to a medical procedure. In the case of cheese, a written specification is typical. Indeed, written specifications or standard grades exist and provide the definition for many products. For instance, Monterey Jack cheese has a written description that specifies the characteristics necessary for each Department of Agriculture grade. A portion

---

Joint ventures
Firms establishing joint ownership to pursue new products or markets.

Alliances
Cooperative agreements that allow firms to remain independent, but that pursue strategies consistent with their individual missions.

Enhancements, migration, purchasing, and alliances are strategies for speeding product development and reducing risk.

---

[17]Jeffrey Ball, "Auto Makers are Racing to Market 'Green' Cars Powered by Fuel Cells," the *Wall Street Journal* (March 15, 1999): A1–A8.

FIGURE 5.7 ■

Monterey Jack

A portion of the general requirements for the U.S. grades of Monterey cheese is shown here.

---

§ 58.2469 Specifications for U.S. grades of Monterey (Monterey Jack) cheese

(a) *U.S. grade AA.* Monterey Cheese shall conform to the following requirements:
(1) *Flavor.* Is fine and highly pleasing, free from undesirable flavors and odors. May possess a very slight acid or feed flavor.
(2) *Body and texture.* A plug drawn from the cheese shall be reasonably firm. It shall have numerous small mechanical openings evenly distributed throughout the plug. It shall not possess sweet holes, yeast holes, or other gas holes.

(3) *Color.* Shall have a natural, uniform, bright, attractive appearance.
(4) *Finish and appearance—bandaged and paraffin-dipped.* The rind shall be sound, firm, and smooth, providing a good protection to the cheese.

Code of Federal Regulation, Parts 53 to 109, Revised as of Jan. 1, 1985, General Service Administration.

---

of the Department of Agriculture grade for Monterey Jack Grade AA is shown in Figure 5.7. Similarly, McDonald's Corp. has 60 specifications for potatoes that are to be made into French fries.

Most manufactured items as well as their components are defined by a drawing, usually referred to as an engineering drawing. An **engineering drawing** shows the dimensions, tolerances, materials, and finishes of a component. The engineering drawing will be an item on a bill of material. An engineering drawing is shown in Figure 5.8. The **bill of material (BOM)** lists the components, their description, and the quantity of each required to make one unit of a product. A bill of material for a manufactured item is shown in Figure 5.9(a). Note that subassemblies and components (lower level items) are indented at each level to indicate their subordinate position. An engineering drawing shows how to make one item on the bill of material.

In the food service industry, bills of material manifest themselves in *portion-control standards*. The portion-control standard for Hard Rock Cafe's hickory BBQ bacon cheeseburger is shown in Figure 5.9(b). In a more complex product, a bill of material is referenced on other bills of material of which they are a part. In this manner, subunits (subassemblies) are part of the next higher unit (their parent bill of material) that ultimately makes a final product. In addition to being defined by written specifications, portion-control documents, or bills of material, products can be defined in other ways. For example, products such as chemicals, paints, and petroleums may be defined by formulas or proportions that describe how they are to be made. Movies are defined by scripts, and insurance coverage by legal documents known as policies.

Engineering drawing
A drawing that shows the dimensions, tolerances, materials, and finishes of a component.

Bill of material (BOM)
A listing of the components, their description, and the quantity of each required to make one unit of a product.

Make-or-Buy Decisions

For many components of products, firms have the option of producing the components themselves or purchasing them from outside sources. Choosing between these options is known as the make-or-buy decision. The **make-or-buy decision** distinguishes between what the firm wants to *produce* and what it wants to *purchase*. Because of variations in quality, cost, and delivery schedules, the make-or-buy decision is critical to product definition. Many items can be purchased as a "standard item"

Make-or-buy decision
The choosing between producing a component or a service and purchasing it from an outside source.

FIGURE 5.8 ■

Engineering Drawings Such as This One Show Dimensions, Tolerances, Materials, and Finishes

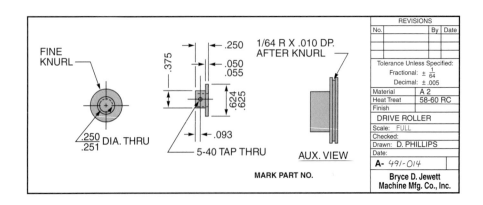

FIGURE 5.9 ■

Bills of Material Take
Different Forms in a
Manufacturing Plant (a)
and a Restaurant (b),
but in Both Cases, the
Product Must Be Defined

(a) Bill of Material for a Panel Weldment		
NUMBER	DESCRIPTION	QTY
A 60-71	PANEL WELDM'T	1
A 60-7	LOWER ROLLER ASSM.	1
R 60-17	ROLLER	1
R 60-428	PIN	1
P 60-2	LOCKNUT	1
A 60-72	GUIDE ASSM. REAR	1
R 60-57-1	SUPPORT ANGLE	1
A 60-4	ROLLER ASSEM.	1
02-50-1150	BOLT	1
A 60-73	GUIDE ASSM. FRONT	1
A 60-74	SUPPORT WELDM'T	1
R 60-99	WEAR PLATE	1
02-50-1150	BOLT	1

(b) Hard Rock Cafe's Hickory BBQ Bacon Cheeseburger	
DESCRIPTION	QTY
Bun	1
Hamburger patty	8 oz.
Cheddar cheese	2 slices
Bacon	2 strips
BBQ onions	1/2 cup
Hickory BBQ sauce	1 oz.
Burger set	
Lettuce	1 leaf
Tomato	1 slice
Red onion	4 rings
Pickle	1 slice
French fries	5 oz.
Seasoned salt	1 tsp.
11-inch plate	1
HRC flag	1

produced by someone else. Such a standard item does not require its own bill of material or engineering drawing because its specification as a standard item is adequate. Examples are the standard bolts listed on the bill of material shown in Figure 5.9(a), for which there will be SAE (Society of Automotive Engineers) specifications. Therefore, there typically is no need for the firm to duplicate this specification in another document. We discuss what is known as the make-or-buy decision in more detail in Chapter 11.

Group Technology

Group technology

A product and component coding system that specifies the type of processing and the parameters of the processing; it allows similar products to be grouped.

Engineering drawings may also include codes to facilitate group technology. **Group technology** requires that components be identified by a coding scheme that specifies the type of processing (such as drilling) and the parameters of the processing (such as size). This facilitates standardization of materials, components, and processes as well as the identification of families of parts. As families of parts are identified, activities and machines can be grouped to minimize setups, routings, and material handling. An example of how families of parts may be grouped is shown in Figure 5.10. Group technology provides a systematic way to review a family of components to see if an existing component might suffice on a new project. Using existing or standard components eliminates all the costs connected with the design and development of the new part, which is a major cost

FIGURE 5.10 ■

A Variety of Group
Technology Coding
Schemes Move
Manufactured
Components from
(a) Ungrouped to
(b) Grouped (families
of parts)

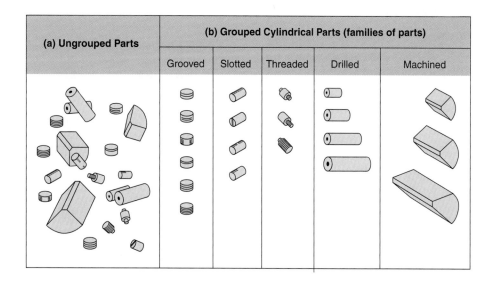

Each year the JR Simplot potato-processing facility in Caldwell, Idaho, produces billions of French fries for McDonald's. Sixty specifications define how these potatoes become French fries. The specifications, for instance, require a special blend of frying oil, a unique steaming process, and exact time and temperature for prefrying and drying. The product is further defined by requiring that 40% of all French fries be between 2 and 3 inches long. Another 40% must be over 3 inches long. A few stubby ones can constitute the final 20%.

reduction. For these reasons, successful implementation of group technology leads to the following advantages:

1. Improved design (because more design time can be devoted to fewer components).
2. Reduced raw material and purchases.
3. Simplified production planning and control.
4. Improved layout, routing, and machine loading.
5. Reduced tooling setup time, and work-in-process and production time.

The application of group technology helps the entire organization, as many costs are reduced.

DOCUMENTS FOR PRODUCTION

Once a product is selected and designed, its production is assisted by a variety of documents. We will briefly review some of these.

An **assembly drawing** simply shows an exploded view of the product. An assembly drawing is usually a three-dimensional drawing, known as an *isometric drawing*; the relative locations of components are drawn in relation to each other to show how to assemble the unit (see Figure 5.11[a]).

The **assembly chart** shows in schematic form how a product is assembled. Manufactured components, purchased components, or a combination of both may be shown on an assembly chart. The assembly chart identifies the point of production at which components flow into subassemblies and ultimately into a final product. An example of an assembly chart is shown in Figure 5.11(b).

Assembly drawing
An exploded view of the product, usually via a three-dimensional or isometric drawing.

Assembly chart
A graphic means of identifying how components flow into subassemblies and ultimately into a final product.

FIGURE 5.11 ■

Assembly Drawing and Assembly Chart

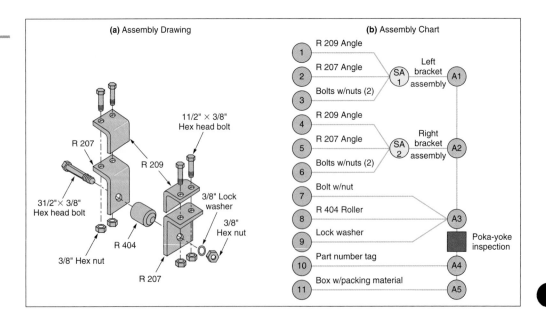

Route sheet
A listing of the operations necessary to produce the component with the material specified in the bill of material.

Work order
An instruction to make a given quantity of a particular item, usually to a given schedule.

Engineering change notice (ECN)
A correction or modification of an engineering drawing or bill of material.

Configuration management
A system by which a product's planned and changing components are accurately identified and for which control and accountability of change are maintained.

The **route sheet** lists the operations (including assembly and inspection) necessary to produce the component with the material specified in the bill of material. The route sheet for an item will have one entry for each operation to be performed on the item. When route sheets include specific methods of operation and labor standards, they are often known as *process sheets*.

The **work order** is an instruction to make a given quantity of a particular item, usually to a given schedule. The ticket that a waiter writes in your favorite restaurant is a work order. In a hospital or factory, the work order is a more formal document that provides authorization to draw various pharmaceuticals or items from inventory, to perform various functions, and to assign personnel to perform those functions.

Engineering change notices (ECNs) change some aspect of the product's definition or documentation, such as an engineering drawing or a bill of material. For a complex product that has a long manufacturing cycle, such as a Boeing 777, the changes may be so numerous that no two 777s are built exactly alike—which is indeed the case. Such dynamic design change has fostered the development of a discipline known as configuration management, which is concerned with product identification, control, and documentation. **Configuration management** is the system by which a product's planned and changing configurations are accurately identified and for which control and accountability of change are maintained.

SERVICE DESIGN

Much of our discussion so far has focused on what we can call tangible products, that is, goods. On the other side of the product coin are, of course, services. Service industries include banking, finance, insurance, transportation, and communications. The products offered by service firms range from a medical procedure that leaves only the tiniest scar after an appendectomy, to a shampoo and cut at a hair salon, to a great movie.

Designing services is challenging because they often have unique characteristics. One reason productivity improvements in services are so low is because both the design and delivery of service products include customer interaction. When the customer participates in the design process, the service supplier may have a menu of services from which the customer selects options (see Figure 5.12[a]). At this point, the customer may even participate in the *design* of the service. Design specifications may take the form of a contract or a narrative description with photos (such as for cosmetic surgery or a hairstyle). Similarly, the customer may be involved in the *delivery* of a service (see Figure 5.12[b]) or in both design and delivery, a situation that maximizes the product design challenge (see Figure 5.12[c]).

However, like goods, a large part of cost and quality of a service is defined at the design stage. Also like goods, a number of techniques can both reduce costs and enhance the product. One technique is to design the product so that *customization is delayed* as late in the process as possible. This is the way a hair salon operates: Although shampoo and rinse are done in a standard way with lower-cost labor, the tint and styling (customizing) are done last. It is also the way most restaurants operate: "How would you like that cooked?" "Which dressing would you prefer with your salad?"

The second approach is to *modularize* the product so that customization takes the form of changing modules. This strategy allows modules to be designed as "fixed," standard entities. The modular approach to product design has applications in both manufacturing and service. Just as modular design allows you to buy a Harley-Davidson motorcycle or a high-fidelity stereo with just the features you want, modular flexibility also lets you buy meals, clothes, and insurance on a mix-and-match (modular) basis. Similarly, investment portfolios are put together on a modular basis. Certainly college curricula are another example of how the modular approach can be used to customize a service (in this case, education).

A third approach to the design of services is to divide the service into small parts and identify those parts that lend themselves to *automation* or *reduced customer interaction*. For instance, by isolating check-cashing activity via ATM machines, banks have been very effective at designing a product that both increases customer service and reduces costs. Similarly, airlines are moving to ticketless service. Because airlines spend $15 to $30 to produce a single ticket (including labor, printing, and travel agent's commission), ticketless systems save the industry a billion dollars a year.

FIGURE 5.12 ■

Customer Participation
in the Design of Services

Source: *Robert Murdick, Barry
Render, and Roberta Russell,*
Service Operations
Management *(Boston: Allyn &
Bacon, 1990).*

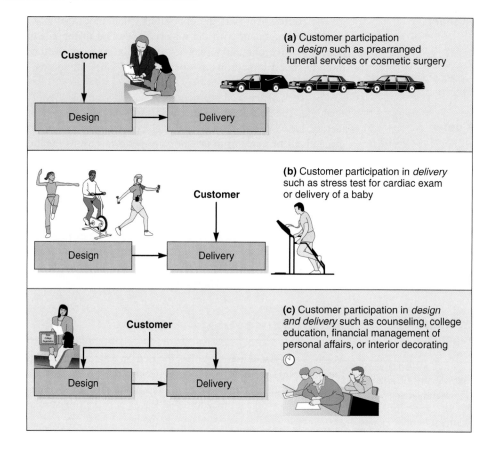

Reducing both costs and lines at airports—and thereby increasing customer satisfaction—provides a win–win "product" design.

Because of the high customer interaction in many service industries, a fourth technique is to focus design on the so-called *moment-of-truth*. Jan Carlzon, former president of Scandinavian Airways, believes that in the service industry there is a moment-of-truth when the relationship between the provider and the customer is crucial.[18] At that moment, the customer's satisfaction with the service is defined. The **moment-of-truth** is the moment that exemplifies, enhances, or detracts from the customer's expectations. That moment may be as simple as a smile or having the checkout clerk focus on you rather than talking over his shoulder to the clerk at the next counter. Moments-of-truth can occur when you order at McDonald's, get a haircut, or register for college courses. Figure 5.13 shows a moment-of-truth analysis for a computer company's customer service hotline. The operations manager's task is to identify moments-of-truth and design operations that meet or exceed the customer's expectations.

Moment-of-truth
In the service industry,
that crucial moment
between the service
provider and the
customer that exemplifies,
enhances, or detracts
from the customer's
expectations.

Documents for Services

Because of the high customer interaction of most services, the documents for moving the product to production are different from those used in goods-producing operations. The documentation for a service will often take the form of explicit job instructions that specify what is to happen at the moment-of-truth. For instance, regardless of how good a bank's products may be in terms of checking, savings, trusts, loans, mortgages, and so forth, if the moment-of-truth is not done well, the product may be poorly received. Example 2 shows the kind of documentation a bank may use to move a product (drive-up window banking) to "production." In a telemarketing service, the product design and its related transmittal to production may take the form of telephone script, and a storyboard (see the photo on the next page) is frequently used for a motion picture.

[18]Jan Carlzon, *Moments of Truth* (Cambridge: Ballinger Publishing, 1987).

Experience Detractors

- I had to call more than once to get through.
- A recording spoke to me rather than a person.
- While on hold, I get silence, and I wonder if I am disconnected.
- The technician sounded like he was reading a form of routine questions.
- The technician sounded uninterested.
- The technician rushed me.

Standard Expectations

- Only one local number needs to be dialed.
- I never get a busy signal.
- I get a human being to answer my call quickly and he or she is pleasant and responsive to my problem.
- A timely resolution to my problem is offered.
- The technician is able to explain to me what I can expect to happen next.

Experience Enhancers

- The technician was sincerely concerned and apologetic about my problem.
- He asked intelligent questions that allowed me to feel confident in his abilities.
- The technician offered various times to have work done to suit my schedule.
- Ways to avoid future problems were suggested.

FIGURE 5.13 ■ Moment-of-Truth: The Customer Contacts the Service Hotline at a Computer Company

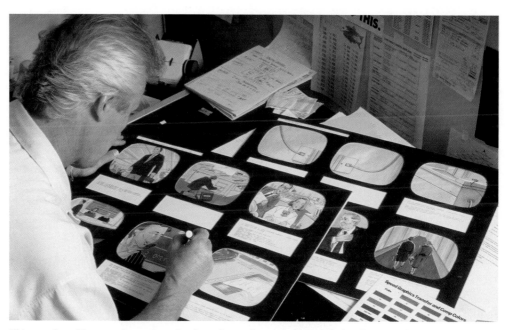

This storyboard lays out the product clearly so that each activity is identified and its contribution to the process known.

Example 2

> **Documentation for Moving a Service Product to Production**
>
> Customers who use drive-up teller stations rather than walk-in lobbies require different customer relations techniques. The distance and machinery between you and the customer raises communication barriers. Communication tips to improve customer relations at a drive-up window are:
>
> - Be especially discreet when talking to the customer through the microphone.
> - Provide written instructions for customers who must fill out forms you provide.
> - Mark lines to be completed or attach a note with instructions.
> - Always say "please" and "thank you" when speaking through the microphone.
> - Establish eye contact with the customer if the distance allows it.
> - If a transaction requires that the customer park the car and come into the lobby, apologize for the inconvenience.
>
> *Source:* Adapted with permission from *Teller Operations* (Chicago, IL: The Institute of Financial Education, 1999): 32.

APPLICATION OF DECISION TREES TO PRODUCT DESIGN

Decision trees can be used for new-product decisions as well as for a wide variety of other management problems. They are particularly helpful when there are a series of decisions and various outcomes that lead to *subsequent* decisions followed by other outcomes. To form a decision tree, we use the following procedure:

1. Be sure that all possible alternatives and states of nature are included in the tree. This includes an alternative of "doing nothing."
2. Payoffs are entered at the end of the appropriate branch. This is the place to develop the payoff of achieving this branch.
3. The objective is to determine the expected value of each course of action. We accomplish this by starting at the end of the tree (the right-hand side) and working toward the beginning of the tree (the left), calculating values at each step and "pruning" alternatives that are not as good as others from the same node.

Example 3 shows the use of a decision tree applied to product design.

Example 3

Active Model 5.1

Example 3 is further illustrated in Active Model 5.1 on the CD-ROM and in the Exercise located in your Student Lecture Guide.

> Silicon, Inc., a semiconductor manufacturer, is investigating the possibility of producing and marketing a microprocessor. Undertaking this project will require either purchasing a sophisticated CAD system or hiring and training several additional engineers. The market for the product could be either favorable or unfavorable. Silicon, Inc., of course, has the option of not developing the new product at all.
>
> With favorable acceptance by the market, sales would be 25,000 processors selling for $100 each. With unfavorable acceptance, sales would be only 8,000 processors selling for $100 each. The cost of CAD equipment is $500,000, but that of hiring and training three new engineers is only $375,000. However, manufacturing costs should drop from $50 each when manufacturing without CAD to $40 each when manufacturing with CAD.
>
> The probability of favorable acceptance of the new microprocessor is .40; the probability of unfavorable acceptance is .60. See Figure 5.14.
>
> The expected monetary values (EMVs) have been circled at each step of the decision tree. For the top branch:
>
> $$\text{EMV (purchase CAD system)} = (.4)(\$1,000,000) + (.6)(-\$20,000)$$
> $$= \$388,000$$
>
> This figure represents the results that will occur if Silicon, Inc., purchases CAD.
>
> The expected value of hiring and training engineers is the second series of branches:
>
> $$\text{EMV (hire/train engineers)} = (.4)(\$875,000) + (.6)(\$25,000)$$
> $$= \$365,000$$
>
> The EMV of doing nothing is $0.
>
> Because the top branch has the highest expected monetary value (an EMV of $388,000 vs. $365,000 vs. $0), it represents the best decision. Management should purchase the CAD system.

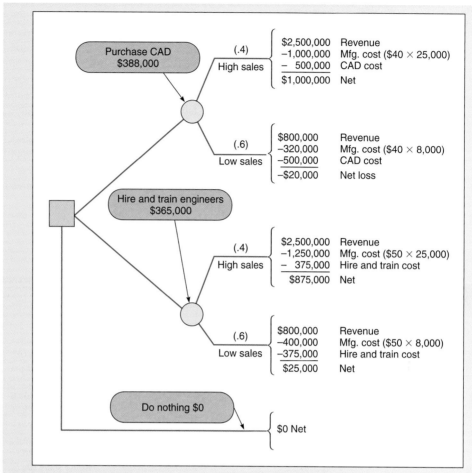

FIGURE 5.14 ■ Decision Tree for Development of a New Product

TRANSITION TO PRODUCTION

Eventually, our product, whether a good or service, has been selected, designed, and defined. It has progressed from an idea to a functional definition, and then perhaps to a design. Now, management must make a decision as to further development and production or termination of the product idea. One of the arts of modern management is knowing when to move a product from development to production; this move is known as *transition to production*. The product development staff is always interested in making improvements in a product. Because this staff tends to see product development as evolutionary, they may never have a completed product, but as we noted earlier, the cost of late product introduction is high. Although these conflicting pressures exist, management must make a decision—more development or production.

Once this decision is made, there is usually a period of trial production to ensure that the design is indeed producible. This is the manufacturability test. This trial also gives the operations staff the opportunity to develop proper tooling, quality control procedures, and training of personnel to ensure that production can be initiated successfully. Finally, when the product is deemed both marketable and producible, line management will assume responsibility.

Some companies appoint a *project manager*, while others use *product development teams* to ensure that the transition from development to production is successful. Both approaches allow a wide range of resources and talents to be brought to bear to ensure satisfactory production of a product that is still in flux. A third approach is *integration of the product development and manufacturing organizations*. This approach allows for easy shifting of resources between the two organizations as needs change. The operations manager's job is to make the transition from R&D to production seamless or as smooth as possible.

SUMMARY

Effective product strategy requires selecting, designing, and defining a product and then transitioning that product to production. Only when this strategy is carried out effectively can the production function contribute its maximum to the organization. The operations manager must build a product development system that has the ability to conceive, design, and produce products that will yield a competitive advantage for the firm. As products move through their life cycle (introduction, growth, maturity, and decline), the options that the operations manager should pursue change. Both manufactured and service products have a variety of techniques available to aid in performing this activity efficiently.

Written specifications, bills of material, and engineering drawings aid in defining products. Similarly, assembly drawings, assembly charts, route sheets, and work orders are often used to assist in the actual production of the product. Once a product is in production, value analysis is appropriate to ensure maximum product value. Engineering change notices and configuration management provide product documentation.

KEY TERMS

Product decision *(p. 124)*
Product-by-value analysis *(p. 126)*
Brainstorming *(p. 126)*
Quality function deployment (QFD) *(p. 129)*
House of quality *(p. 129)*
Product development teams *(p. 131)*
Concurrent engineering *(p. 132)*
Manufacturability and value engineering *(p. 132)*
Robust design *(p. 133)*
Modular design *(p. 133)*
Computer-aided design (CAD) *(p. 133)*
Design for manufacture and assembly (DFMA) *(p. 134)*
3-D object modeling *(p. 134)*
Standard for the Exchange of Product Data (STEP) *(p. 134)*
Computer-aided manufacturing (CAM) *(p. 134)*

Virtual reality *(p. 135)*
Value analysis *(p. 135)*
Green manufacturing *(p. 136)*
Time-based competition *(p. 137)*
Joint ventures *(p. 139)*
Alliances *(p. 139)*
Engineering drawing *(p. 140)*
Bill of material (BOM) *(p. 140)*
Make-or-buy decision *(p. 140)*
Group technology *(p. 141)*
Assembly drawing *(p. 142)*
Assembly chart *(p. 142)*
Route sheet *(p. 143)*
Work order *(p. 143)*
Engineering change notice (ECN) *(p. 143)*
Configuration management *(p. 143)*
Moment-of-truth *(p. 144)*

SOLVED PROBLEM

Solved Problem 5.1

Sarah King, president of King Electronics, Inc., has two design options for her new line of high-resolution cathode-ray tubes (CRTs) for computer-aided design workstations. The life cycle sales forecast for the CRT is 100,000 units.

Design option A has a .90 probability of yielding 59 good CRTs per 100 and a .10 probability of yielding 64 good CRTs per 100. This design will cost $1,000,000.

Design option B has a .80 probability of yielding 64 good units per 100 and a .20 probability of yielding 59 good units per 100. This design will cost $1,350,000.

Good or bad, each CRT will cost $75. Each good CRT will sell for $150. Bad CRTs are destroyed and have no salvage value. Because units break up when thrown in the trash, there is little disposal cost. Therefore, we ignore any disposal costs in this problem.

SOLUTION

We draw the decision tree to reflect the two decisions and the probabilities associated with each decision. We then determine the payoff associated with each branch. The resulting tree is shown in Figure 5.15.

For design A,

$$\text{EMV (design A)} = (.9)(\$350,000) + (.1)(\$1,100,000)$$
$$= \$425,000$$

For design B,

$$\text{EMV (design B)} = (.8)(\$750,000) + (.2)(\$0)$$
$$= \$600,000$$

The highest payoff is design option B at $600,000.

FIGURE 5.15 ■

Decision Tree for Solved
Problem 5.1

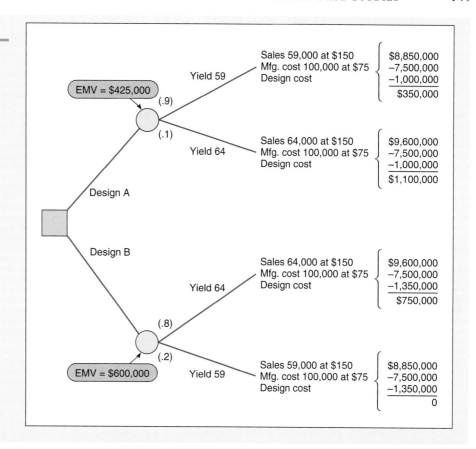

ADDITIONAL CASE STUDIES

Harvard has selected these Harvard Business School cases to accompany this chapter
(textbookcasematch.hbsp.harvard.edu)**:**

- **The Ritz-Carlton** (#601–163): Allows students to examine innovation and service improvement in the hospitality industry.

- **Product Development at Dell Computer Corp.** (#699–010): Focuses on how Dell redesigned its new-product development process.

- **Innovation at 3M Corp. (A)** (#699–012): Describes how 3M Corp.'s new-product development process obtains customer input.

- **CIBA Vision: The Daily Disposable Lens Project (A)** (#696–100): Examines CIBA Vision's evaluation of a new low-cost disposable contact lens.

Managing Quality

Chapter Outline

LEARNING OBJECTIVES

When you complete this chapter you should be able to

IDENTIFY OR DEFINE:

Quality

Malcolm Baldrige National Quality Award

ISO international quality standards

Taguchi concepts

DESCRIBE OR EXPLAIN:

Why quality is important

Total quality management (TQM)

Pareto charts

Process charts

Quality robust products

Inspection

Deming, Juran, and Crosby's ideas

Managing Quality Provides a Competitive Advantage at Motorola

Motorola decided some years ago to be a world leader in quality. Indeed, Motorola is so good that it became the first winner of the Malcolm Baldrige National Quality Award. Motorola believes in total quality management and practices it from the top, specifically from Honorary Chairman Robert Galvin. It achieves outstanding quality through a demonstrated top-management commitment that permeates this entire global organization.

To make the quality focus work, Motorola did a number of things:

- Aggressively began a worldwide education program to be sure that employees understood quality and statistical process control.
- Established goals—namely, its Six Sigma program. Motorola's Six Sigma program means that it can expect to have a defect rate of no more than a few parts per million.
- Established extensive employee participation and employee teams. More than 4,000 Total Customer Satisfaction teams from throughout the world vie for awards based on team performance.

Motorola's divisions can expect a quality service review every 2 years. Five-member teams are selected from various parts of the company to perform the review. After the review, the general manager and staff have a session with the teams and go over the review. The strengths and weaknesses are discussed, and recommendations are given to the local management about improvements that must be made.

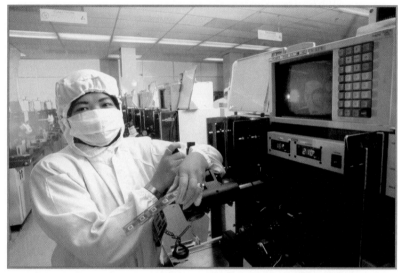

While Motorola's Accelerated Life Testing (ALT) facility tests its pagers for extreme conditions of temperature shock, dust, water, and vibrations, the electronic capability must also be tested, as is the case here.

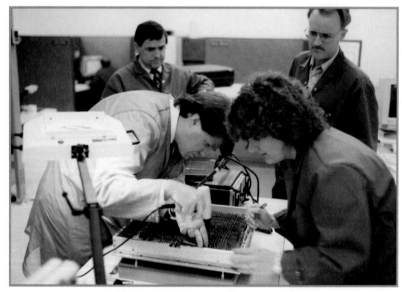

Motorola's aggressive commitment to quality requires an equally aggressive move toward employee training, making Motorola one of the leaders in employee training. Here, employees in the Tempe, Arizona, plant work on a piece of equipment during a training session.

Motorola's strong emphasis on employee participation and total quality management often puts employees in teams that are responsible for evaluating and improving their own processes. This team is from Penang, Malaysia.

The system is working; it gives Motorola uniformity and consistency. Corporate goals receive commitment throughout the organization, and that is a powerful quality tool. The quality effort has allowed Motorola to move from 6,000 rejects per million just 5 years ago to only 40 defects per million now. Motorola believes it has saved $700 million in manufacturing costs over those 5 years.

QUALITY AND STRATEGY

As Motorola and many other firms have found, quality is a wonderful tonic for improving operations. Managing quality helps build successful strategies of *differentiation, low cost,* and *response.* For instance, defining customer quality expectations has helped Bose Corp. successfully *differentiate* its stereo speakers as among the best in the world. Nucor has learned to produce quality steel at *low cost* by developing efficient processes that produce consistent quality. And Dell Computers rapidly *responds* to customer orders because quality systems, with little rework, have allowed it to achieve rapid throughput in its plants. Indeed, quality may be the critical success factor for these firms just as it is at Motorola.

As Figure 6.1 suggests, improvements in quality help firms increase sales and reduce costs, both of which can increase profitability. Increases in sales often occur as firms speed response, lower selling prices as a result of economies of scale, and improve their reputation for quality products. Similarly, improved quality allows costs to drop as firms increase productivity and lower rework, scrap, and warranty costs.

One analysis of air conditioner manufacturers has documented that quality and productivity are positively related. In that study, companies with the highest quality were five times as productive (as measured by units produced per labor-hour) as companies with the poorest quality. Indeed, when the implications of an organization's long-term costs and the potential for increased sales are considered, total costs may well be at a minimum when 100% of the goods or services are perfect and defect-free.

Quality, or the lack of quality, impacts the entire organization from supplier to customer and from product design to maintenance. However, perhaps more importantly, *building* an organization that can achieve quality also affects the entire organization—and it is a demanding task. Figure 6.2 lays out the flow of activities for an organization to use to achieve total quality management (TQM). A successful set of activities begins with an organizational environment that fosters quality, followed by an understanding of the principles of quality, and then an effort to engage employees in the necessary activities to implement quality. When these things are done well, the organization typically satisfies its customers and obtains a competitive advantage. The ultimate goal is to win customers. Because quality causes so many other good things to happen, it is a great place to start.

DEFINING QUALITY

Quality

The ability of a product or service to meet customer needs.

Total quality management systems are driven by identifying and satisfying customer needs. Total quality management takes care of the customer. Consequently, we accept the definition of **quality** as adopted by the American Society for Quality: "The totality of features and characteristics of a product or service that bears on its ability to satisfy stated or implied needs."[1]

However, others believe that definitions of quality fall into several categories. Some definitions are *user-based.* They propose that quality "lies in the eyes of the beholder." Marketing people like this approach and so do customers. To them, higher quality means better performance, nicer features, and other (sometimes costly) improvements. To production managers, quality is *manufacturing-based.* They believe that quality means conforming to standards and "making it right the first time."

FIGURE 6.1 ■

Ways Quality Improves
Profitability

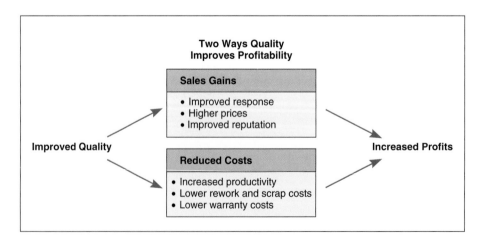

[1]See the American Society for Quality Web site at www.asq.org.

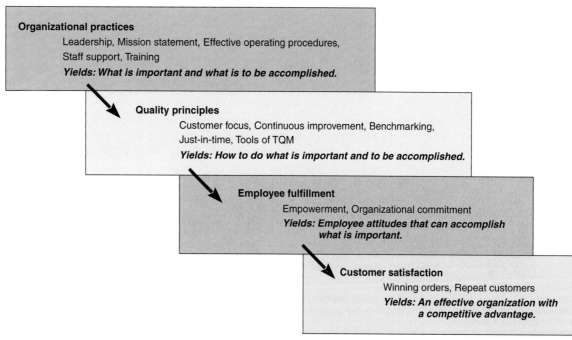

FIGURE 6.2 ■ The Flow of Activities that Are Necessary to Achieve Total Quality Management

Quality may be in the eyes of the beholder, but to create a good or a service, operations managers must define what the beholder (the consumer) expects.

Yet a third approach is *product-based*, which views quality as a precise and measurable variable. In this view, for example, really good ice cream has high butterfat levels.

This text develops approaches and techniques to address all three categories of quality. The characteristics that connote quality must first be identified through research (a user-based approach to quality). These characteristics are then translated into specific product attributes (a product-based approach to quality). Then the manufacturing process is organized to ensure that products are made precisely to specifications (a manufacturing-based approach to quality). A process that ignores any one of these steps will not result in a quality product.

Implications of Quality

In addition to being a critical element in operations, quality has other implications. Here are three other reasons why quality is important:

Video 6.1

TQM at Ritz-Carlton Hotels

1. *Company reputation.* An organization can expect its reputation for quality—be it good or bad—to follow it. Quality will show up in perceptions about the firm's new products, employment practices, and supplier relations. Self-promotion is not a substitute for quality products.
2. *Product liability.* The courts increasingly hold organizations that design, produce, or distribute faulty products or services liable for damages or injuries resulting from their use. Legislation such as the Consumer Product Safety Act sets and enforces product standards by banning products that do not reach those standards. Impure foods that cause illness, nightgowns that burn, tires that fall apart, or auto fuel tanks that explode on impact can all lead to huge legal expenses, large settlements or losses, and terrible publicity.
3. *Global implications.* In this technological age, quality is an international, as well as OM, concern. For both a company and a country to compete effectively in the global economy, products must meet global quality, design, and price expectations. Inferior products harm a firm's profitability and a nation's balance of payments.

Malcolm Baldrige National Quality Award

For further information regarding the Baldrige Award and its 1,000-point scoring system, visit www.quality.nist.gov.

The global implications of quality are so important that the U.S. has established the *Malcolm Baldrige National Quality Award* for quality achievement. The award is named for former Secretary of Commerce Malcolm Baldrige. Winners include such firms as Motorola, Milliken, Xerox, Federal Express, Ritz-Carlton Hotels, AT&T, Cadillac, and Texas Instruments.

The Japanese have a similar award, the Deming Prize, named after an American, Dr. W. Edwards Deming.

Cost of Quality (COQ)

Four major categories of costs are associated with quality. Called the **cost of quality (COQ)**, they are:

- *Prevention costs*—costs associated with reducing the potential for defective parts or services (e.g., training, quality improvement programs).
- *Appraisal costs*—costs related to evaluating products, processes, parts, and services (e.g., testing, labs, inspectors).
- *Internal failure*—costs that result from production of defective parts or services before delivery to customers (e.g., rework, scrap, downtime).
- *External costs*—costs that occur after delivery of defective parts or services (e.g., rework, returned goods, liabilities, lost goodwill, costs to society).

The first three costs above can be reasonably estimated, but external costs are very hard to quantify. When GE had to recall 3.1 million dishwashers in 1999 (because of a defective switch alleged to have started seven fires), the cost of repairs exceeded the value of all the machines. This leads to the belief by many experts that the cost of poor quality is consistently underestimated.

Observers of quality management, including Philip Crosby and Genichi Taguchi, believe that, on balance, the cost of quality products is only a fraction of the benefits. They think the real losers are organizations that fail to work aggressively at quality. For instance, Philip Crosby stated that quality is free. "It is not a gift, but it is free. What costs money are the unquality things—all the actions that involve not doing it right the first time."[2]

INTERNATIONAL QUALITY STANDARDS

ISO 9000

Quality is so important globally that the world is uniting around a single quality standard, **ISO 9000**. ISO 9000 is the only quality standard with international recognition. In 1987, 91 member nations (including the U.S.) published a series of quality assurance standards, known collectively as ISO 9000. The U.S., through the American National Standards Institute, has adopted the ISO 9000 series as the ANSI/ASQ Q9000 series.[3] The focus of the standards is to establish quality management procedures, through leadership, detailed documentation, work instructions, and recordkeeping. These procedures, we should note, say nothing about the actual quality of the product—they deal entirely with standards to be followed.

To become ISO 9000 certified, organizations go through a 9-to-18-month process that involves documenting quality procedures, an on-site assessment, and an ongoing series of audits of their products or services. To do business globally—and especially in Europe—being listed in the ISO directory is critical. As of 2003, there were well over 400,000 certifications awarded to firms in 158 countries. About 40,000 U.S. firms are ISO 9000 certified.

ISO revised its standards in December 2000 into more of a quality management system, which is detailed in its ISO 9001: 2000 component. Leadership by top management and customer requirements and satisfaction play a much larger role, while documented procedures receive less emphasis under ISO 9001: 2000.[4]

---

[2]Philip B. Crosby, *Quality Is Free* (New York: McGraw-Hill, 1979). Further, J. M. Juran states, in his book *Juran on Quality by Design* (The Free Press 1992, p. 119) that costs of poor quality "are huge, but the amounts are not known with precision. In most companies the accounting system provides only a minority of the information needed to quantify this cost of poor quality. It takes a great deal of time and effort to extend the accounting system so as to provide full coverage."

[3]ASQ is the American Society for Quality.

[4]Craig Cochran, "The ISO to Know," *IIE Solutions* (December 2001): 30–34.

The ISO 9000 Certified sign is up, but this Bridgestone/Firestone plant in Decatur, Illinois, produced thousands of defective tires. The basics of making tires have remained the same for over 100 years at this plant, which produces more than 25,000 largely handmade tires a day.

ISO 14000

ISO 14000
An environmental management standard established by the International Standards Organization (ISO).

The continuing internationalization of quality is evident with the development of **ISO 14000**. ISO 14000 is an environmental management standard that contains five core elements: (1) environmental management, (2) auditing, (3) performance evaluation, (4) labeling, and (5) life-cycle assessment. The new standard could have several advantages:

- Positive public image and reduced exposure to liability.
- Good systematic approach to pollution prevention through the minimization of ecological impact of products and activities.
- Compliance with regulatory requirements and opportunities for competitive advantage.
- Reduction in need for multiple audits.

This standard is being accepted worldwide.

TOTAL QUALITY MANAGEMENT

Total quality management (TQM)
Management of an entire organization so that it excels in all aspects of products and services that are important to the customer.

Total quality management (TQM) refers to a quality emphasis that encompasses the entire organization, from supplier to customer. TQM stresses a commitment by management to have a continuing companywide drive toward excellence in all aspects of products and services that are important to the customer.

TQM is important because quality decisions influence each of the 10 decisions made by operations managers. Each of those 10 decisions deals with some aspect of identifying and meeting customer expectations. Meeting those expectations requires an emphasis on TQM if a firm is to compete as a leader in world markets.

Quality expert W. Edwards Deming used 14 points (see Table 6.1) to indicate how he implemented TQM.[5] We develop these into six concepts for an effective TQM program: (1) continuous improvement, (2) employee empowerment, (3) benchmarking, (4) just-in-time (JIT), (5) Taguchi concepts, and (6) knowledge of TQM tools.

Critical to improving quality is management leadership. The reason ISO revamped ISO 9000 was because earlier versions were weak in what they required of top management.

Continuous Improvement

Total quality management requires a never-ending process of continuous improvement that covers people, equipment, suppliers, materials, and procedures. The basis of the philosophy is that every aspect of an operation can be improved. The end goal is perfection, which is never achieved but always sought.

---

[5]John C. Anderson, Manus Rungtusanatham, and Roger G. Schroeder, "A Theory of Quality Management Underlying the Deming Management Method," *Academy of Management Review* 19, no. 3 (1994): 472–509.

TABLE 6.1 ■

Deming's 14 Points for Implementing Quality Improvement

1. Create consistency of purpose.
2. Lead to promote change.
3. Build quality into the product; stop depending on inspections to catch problems.
4. Build long-term relationships based on performance instead of awarding business on the basis of price.
5. Continuously improve product, quality, and service.
6. Start training.
7. Emphasize leadership.
8. Drive out fear.
9. Break down barriers between departments.
10. Stop haranguing workers.
11. Support, help, and improve.
12. Remove barriers to pride in work.
13. Institute a vigorous program of education and self-improvement.
14. Put everybody in the company to work on the transformation.

Source: Deming revised his 14 points a number of times over the years. See W. Edwards Deming, "Philosophy Continues to Flourish," *APICS—The Performance Advantage* 1, no. 4 (October 1991): 20.

PDCA
A continuous improvement model of plan, do, check, act.

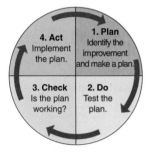

FIGURE 6.3 ■
PDCA Cycle

Kaizen
The Japanese word for the ongoing process of incremental improvement.

Six sigma
A quality program that yields 99.9997% accurate products or services.

Employee empowerment
Enlarging employee jobs so that the added responsibility and authority is moved to the lowest level possible in the organization.

Plan-Do-Check-Act   Walter Shewhart, a pioneer in quality management, developed a circular model known as **PDCA** (plan, do, check, act) as his version of continuous improvement. Deming later took this concept to Japan during his work there after World War II. The PDCA cycle is shown in Figure 6.3 as a circle to stress the continuous nature of the improvement process.

Six Sigma   The Japanese use the word **kaizen** to describe this ongoing process of unending improvement—the setting and achieving of ever-higher goals. In the U.S., *TQM* and *zero defects* are also used to describe continuous improvement efforts.

The term **six sigma**, popularized by Motorola, Honeywell, and General Electric, refers to a TQM program with extremely high process capability (99.9997% accuracy). For example, if 20,000,000 passengers pass through London's Heathrow Airport with checked baggage each year, a six sigma program for baggage handling will result in only 72 passengers with misplaced luggage. The more common 3-sigma program (which we will address in the supplement to this chapter) would result in 3,660 passengers with misplaced bags *every day*. General Electric's approach is to certify employees as "black belts" in six sigma after they complete thorough statistical training—and then to send them out to train their colleagues.

Whether it's PDCA, *kaizen*, zero defects, or six sigma, the operations manager is a key player in building a work culture that endorses continuous improvement.

Employee Empowerment

Employee empowerment means involving employees in every step of the production process. Consistently, business literature suggests that some 85% of quality problems have to do with materials and processes, not with employee performance. Therefore, the task is to design equipment and processes that produce the desired quality. This is best done with a high degree of involvement by those who understand the shortcomings of the system. Those dealing with the system on a daily basis understand it better than anyone else. One study indicated that TQM programs that delegate responsibility for quality to shop-floor employees tend to be twice as likely to succeed as those implemented with "top-down" directives.[6]

When nonconformance occurs, the worker is seldom wrong. Either the product was designed wrong, the system that makes the product was designed wrong, or the employee was improperly trained. Although the employee may be able to help solve the problem, the employee rarely causes it.

Techniques for building employee empowerment include (1) building communication networks that include employees; (2) developing open, supportive supervisors; (3) moving responsibility from both managers and staff to production employees; (4) building high-morale organizations; (5) and creating such formal organization structures as teams and quality circles.

---

[6]"The Straining of Quality," *The Economist* (January 14, 1995): 55. We also see that this is one of the strengths of Southwest Airlines, which offers bare-boned domestic service, but whose friendly and humorous employees help it obtain number 1 ranking for quality. (See the *Wall Street Journal* [April 27, 2000].)

Teams can be built to address a variety of issues. One popular focus of teams is quality. Such teams are often known as quality circles. A **quality circle** is a group of employees who meet regularly to solve work-related problems. The members receive training in group planning, problem solving, and statistical quality control. They generally meet once a week (usually after work, but sometimes on company time). Although the members are not rewarded financially, they do receive recognition from the firm. A specially trained team member, called the facilitator, usually helps train the members and keeps the meetings running smoothly. Teams with a quality focus have proven to be a cost-effective way to increase productivity as well as quality.

Benchmarking

Benchmarking is another ingredient in an organization's TQM program. **Benchmarking** involves selecting a demonstrated standard of products, services, costs, or practices that represent the very best performance for processes or activities very similar to your own. The idea is to develop a target at which to shoot and then to develop a standard or benchmark against which to compare your performance. The steps for developing benchmarks are:[7]

- Determine what to benchmark.
- Form a benchmark team.
- Identify benchmarking partners.
- Collect and analyze benchmarking information.
- Take action to match or exceed the benchmark.

In the ideal situation, you find one or more similar organizations that are leaders in the particular areas you want to study. Then you compare yourself (benchmark yourself) against them. The company need not be in your industry. Indeed, to establish world-class standards, it may be best to look outside of your industry. If one industry has learned how to compete via rapid product development while yours has not, it does no good to study your industry. As discussed in the *OM in Action* box, "L.L. Bean's Reputation Makes It a Benchmark Favorite," this is exactly what Xerox and DaimlerChrysler did when they went to L.L. Bean for order-filling and warehousing benchmarks.

Quality circle
A group of employees meeting regularly with a facilitator to solve work-related problems in their work area.

Benchmarking
Selecting a demonstrated standard of performance that represents the very best performance for a process or activity.

Video 6.2

Xerox's Benchmarking Strategy

W. Edwards Deming (left). In his quality crusade, Deming insisted that management accept responsibility for building good systems. The employee, he believed, cannot produce products that on the average exceed the quality of what the process is capable of producing. Dr. Deming died in 1993.

J. M. Juran (middle). A pioneer in teaching the Japanese how to improve quality, Juran believes strongly in top-management commitment, support, and involvement in the quality effort. He is also a believer in teams that continually seek to raise quality standards. Juran varies from Deming somewhat in focusing on the customer and defining quality as fitness for use, not necessarily the written specifications.

Philip B. Crosby (right). Quality Is Free was Crosby's attention-getting book published in 1979. Crosby believed that in the traditional trade-off between the cost of improving quality and the cost of poor quality, the cost of poor quality is understated. The cost of poor quality should include all of the things that are involved in not doing the job right the first time. Mr. Crosby died in 2001.

"There is absolutely no reason for having errors or defects in any product or service."
Philip Crosby

[7]Adapted from Michael J. Spendolini, *The Benchmarking Book* (New York: AMACOM, 1992).

OM IN ACTION

L.L. Bean's Reputation Makes It a Benchmark Favorite

When Xerox set out to improve its order filling, it went to L.L. Bean. What did copier parts have in common with Bean's outdoor paraphernalia? Nothing. But Xerox managers felt that their order-filling processes were similar: They both involve handling products so varied in size and shape that the work must be done by hand. Bean, it turns out, was able to "pick" orders three times as fast as Xerox. The lesson learned, Xerox pared its warehouse costs by 10%. "Too many companies suffer because they refuse to believe others can do things better," says Robert Camp, Xerox's benchmarking manager.

Then DaimlerChrysler came to study Bean's warehousing methods. Bean employees use flowcharts to spot wasted motions. This practice resulted in an employee suggestion to stock high-volume items close to packing stations. So impressed was DaimlerChrysler that it decided to follow suit and rely more on problem solving at the worker level.

L.L. Bean now receives up to five requests a week for benchmark visits—too many to handle. The company schedules only those with a "genuine interest in quality, not the merely curious," says Bean plant manager Robert Olive.

Sources: Catalog Age (April 2002): 35; and *Business Week* (September 18, 1995): 122–132.

Benchmarks often take the form of "best practices" found in other firms. Table 6.2 illustrates best practices for resolving customer complaints.

TABLE 6.2 ■

Best Practices for Resolving Customer Complaints

- *Make it easy for clients to complain:* It is free market research.
- *Respond quickly to complaints:* It adds customers and loyalty.
- *Resolve complaints on the first contact:* It reduces cost.
- *Use computers to manage complaints:* Discover trends, share them, and align your services.
- *Recruit the best for customer service jobs:* It should be part of formal training and career advancement.

Source: Canadian Government Guide on Complaint Mechanism.

Benchmarks can and should be established in a variety of areas. Total quality management requires no less.[8]

Just-in-Time (JIT)

The philosophy behind just-in-time (JIT) is one of continuing improvement and enforced problem solving. JIT systems are designed to produce or deliver goods just as they are needed. JIT is related to quality in three ways.

- *JIT cuts the cost of quality.* This occurs because scrap, rework, inventory investment, and damage costs are directly related to inventory on hand. Because there is less inventory on hand with JIT, costs are lower. Additionally, inventory hides bad quality whereas JIT immediately *exposes* bad quality.
- *JIT improves quality.* As JIT shrinks lead time, it keeps evidence of errors fresh and limits the number of potential sources of error. JIT creates, in effect, an early warning system for quality problems, both within the firm and with vendors.
- *Better quality means less inventory and a better, easier-to-employ JIT system.* Often the purpose of keeping inventory is to protect against poor production performance resulting from unreliable quality. If consistent quality exists, JIT allows firms to reduce all the costs associated with inventory.

Taguchi Concepts

Most quality problems are the result of poor product and process design.[9] Genichi Taguchi has provided us with three concepts aimed at improving both product and process quality. They are: *quality robustness*, *quality loss function*, and *target-oriented quality*.

---

[8]Note that benchmarking is good for evaluating how well you are doing the thing you are doing compared to the industry, but the more imaginative approach to process improvement is to ask, "Should we be doing this at all?" Comparing your warehousing operations to the marvelous job that L.L. Bean does is fine, but maybe you should have a "pass-through" warehouse (see Chapter 11 supplement) or be outsourcing the warehousing function.

[9]Glen Stuart Peace, *Taguchi Methods: A Hands-On Approach* (Reading, MA: Addison-Wesley, 1993).

FIGURE 6.4 ■

(a) Quality Loss
Function; (b) Distribution
of Products Produced

*Taguchi aims for the target
because products
produced near the upper
and lower acceptable
specifications result in
higher quality loss function.*

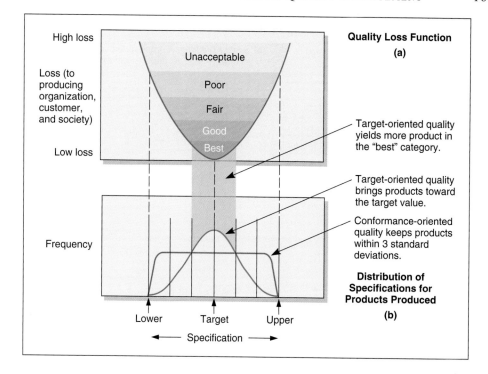

Quality robust
Products that are
consistently built to meet
customer needs in spite of
adverse conditions in the
production process.

**Quality loss
function (QLF)**
A mathematical function
that identifies all costs
connected with poor
quality and shows how
these costs increase as
product quality moves
from what the customer
wants.

Quality robust products are products that can be produced uniformly and consistently in adverse manufacturing and environmental conditions. Taguchi's idea is to remove the *effects* of adverse conditions instead of removing the causes. Taguchi suggests that removing the effects is often cheaper than removing the causes and more effective in producing a robust product. In this way, small variations in materials and process do not destroy product quality.

A **quality loss function (QLF)** identifies all costs connected with poor quality and shows how these costs increase as the product moves away from being exactly what the customer wants. These costs include not only customer dissatisfaction but also warranty and service costs; internal inspection, repair, and scrap costs; and costs that can best be described as costs to society. Notice that Figure 6.4(a) shows the quality loss function as a curve that increases at an increasing rate. It takes the general form of a simple quadratic formula:

$$L = D^2 C$$

where L = loss to society
D^2 = square of the distance from the target value
C = cost of the deviation at the specification limit

All the losses to society due to poor performance are included in the loss function. The smaller the loss, the more desirable the product. The farther the product is from the target value, the more severe the loss.

Taguchi observed that traditional conformance-oriented specifications (that is, the product is good as long as it falls within the tolerance limits) are too simplistic. As shown in Figure 6.4(b), conformance-oriented quality accepts all products that fall within the tolerance limits, producing more units farther from the target. Therefore, the loss (cost) is higher in terms of customer satisfaction and benefits to society. Target-oriented quality, on the other hand, strives to keep the product at the desired specification, producing more (and better) units near the target. **Target-oriented quality** is a philosophy of continuous improvement to bring the product exactly on target.

Target-oriented quality
A philosophy of continuous
improvement to bring the
product exactly on target.

Knowledge of TQM Tools

To empower employees and implement TQM as a continuing effort, everyone in the organization must be trained in the techniques of TQM. In the following section, we focus on some of the diverse and expanding tools that are used in the TQM crusade.

Tools for Generating Ideas

(a) *Check Sheet:* An organized method of recording data.

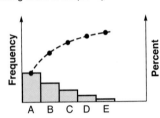

Defect	Hour							
	1	2	3	4	5	6	7	8
A	///	/		/	/	/	///	/
B	//	/	/	/			//	///
C	/	//					//	////

(b) *Scatter Diagram:* A graph of the value of one variable vs. another variable.

Productivity · **Absenteeism**

(c) *Cause and Effect Diagram:* A tool that identifies process elements (causes) that might effect an outcome.

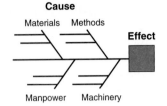
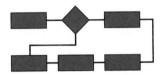

Cause

Materials   Methods

Effect

Manpower   Machinery

Tools to Organize the Data

(d) *Pareto Charts:* A graph to identify and plot problems or defects in descending order of frequency.

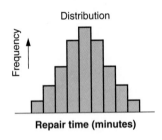

Frequency · Percent · A B C D E

(e) *Flow Charts (Process Diagrams):* A chart that describes the steps in a process.

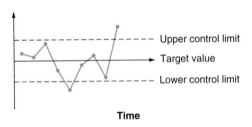

Tools for Identifying Problems

(f) *Histogram:* A distribution showing the frequency of occurrences of a variable.

Distribution

Frequency

Repair time (minutes)

(g) *Statistical Process Control Chart:* A chart with time on the horizontal axis to plot values of a statistic.

———— Upper control limit
——→ Target value
———— Lower control limit

Time

FIGURE 6.5 ■ Seven Tools of TQM

TOOLS OF TQM

Seven tools that are particularly helpful in the TQM effort are shown in Figure 6.5. We will now introduce these tools.

Check Sheets

A check sheet is any kind of a form that is designed for recording data. In many cases, the recording is done so the patterns are easily seen while the data are being taken (see Figure 6.5[a]). Check sheets help analysts find the facts or patterns that may aid subsequent analysis. An example might be a drawing that shows a tally of the areas where defects are occurring or a check sheet showing the type of customer complaints.

Scatter Diagrams

Scatter diagrams show the relationship between two measurements. An example is the positive relationship between length of a service call and the number of trips the repairperson makes back to the truck for parts (as discussed in the *OM in Action* box, "TQM Improves Copier Service"). Another

OM IN ACTION

TQM Improves Copier Service

In the copier industry, technology in copier design has blurred the distinction between most companies' products. Savin, a copier manufacturer owned by Japan's Ricoh Corp., believes that competitive advantage is to be found in service and is stressing customer service rather than product specifications. Says Savin VP Robert Williams: "A company's fortunes ride on the quality of its service."

Here are two ways in which Savin reduced expenses while improving service quality:

- Using the tools of TQM, Savin found that significant time on service calls was being wasted when engineers had to go back to their trucks for spare parts. The firm assembled a "call kit," which allows engineers to carry onto customer premises all parts with highest probability for use. Now service calls are faster and cost less, and more can be made per day.
- The Pareto principle, that 20% of your staff causes 80% of your errors, was used to tackle the "call-back" problem. Callbacks meant the job was not done right the first time and that a second visit, at Savin's expense, was needed. Retraining only the 11% of customer engineers with the most call-backs resulted in a 19% drop in return visits.

"Total quality management," according to Williams, "is an approach to doing business that should permeate every job in the service industry."

Sources: The Wall Street Journal (May 19, 1998): B8; and Office Systems (December 1998): 40–44.

---

example might be a plot of productivity and absenteeism as shown in Figure 6.5(b). If the two items are closely related, the data points will form a tight band. If a random pattern results, the items are unrelated.

Cause-and-effect diagram

A schematic technique used to discover possible locations of quality problems.

Cause-and-Effect Diagrams

Another tool for identifying quality issues and inspection points is the **cause-and-effect diagram**, also known as an **Ishikawa diagram** or a **fish-bone chart**. Figure 6.6 illustrates a chart (note the shape resembling the bones of a fish) for an everyday quality control problem—a dissatisfied airline customer. Each "bone" represents a possible source of error.

FIGURE 6.6 ■

Fish-Bone Chart (or Cause-and-Effect Diagram) for Problems with Airline Customer Service.

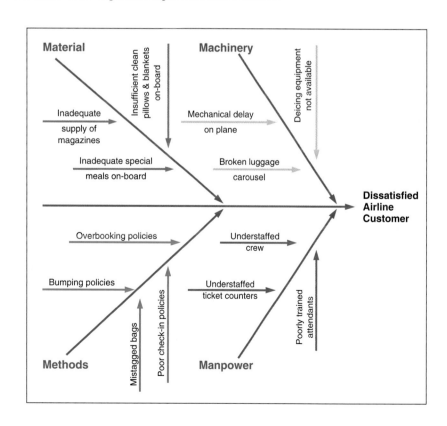

The operations manager starts with four categories: material, machinery/equipment, manpower, and methods. These four *M*s are the "causes." They provide a good checklist for initial analysis. Individual causes associated with each category are tied in as separate bones along that branch, often through a brainstorming process. For example, the machinery branch in Figure 6.6 has problems caused by deicing equipment, mechanical delays, and broken carousels. When a fish-bone chart is systematically developed, possible quality problems and inspection points are highlighted.

Pareto Charts

Pareto charts
A graphic way of identifying the few critical items as opposed to many less important ones.

Pareto charts are a method of organizing errors, problems, or defects to help focus on problem-solving efforts. They are based on the work of Vilfredo Pareto, a nineteenth-century economist. Joseph M. Juran popularized Pareto's work when he suggested that 80% of a firm's problems are a result of only 20% of the causes.

Example 1 indicates that of the five types of complaints identified, the vast majority were of one type, poor room service.

Example 1

The Hard Rock Hotel in Bali has just collected the data from 75 complaint calls to the general manager during the month of October. The manager decides to prepare a Pareto analysis of the complaints. The data provided are room service, 54; check-in delays, 12; hours the pool is open, 4; minibar prices, 3; and miscellaneous, 2.

The Pareto chart shown indicates that 72% of the calls were the result of one cause, room service. The majority of complaints will be eliminated when this one cause is corrected.

Active Model 6.1

Example 1 is further illustrated in Active Model 6.1 in the CD-ROM and in the Exercise located in your Student Lecture Guide.

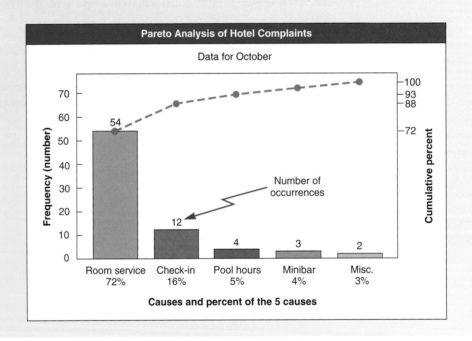

Pareto analysis indicates which problems may yield the greatest payoff. Pacific Bell discovered this when it tried to find a way to reduce damage to buried phone cable, the number 1 cause of phone outages. Pareto analysis showed that 41% of cable damage was caused by construction work. Armed with this information, Pacific Bell was able to devise a plan to reduce cable cuts by 24% in one year, saving $6 million.

Flow Charts

Flow charts
Block diagrams that graphically describe a process or system.

Flow charts graphically present a process or system using annotated boxes and interconnected lines (see Figure 6.5[e]). They are a simple, but great tool for trying to make sense of a process or explain a process. Example 2 uses a flow chart to show the process in the packing and shipping department of a chicken processing plant.

Example 2

The WJC Chicken Processing Plant in Little Rock, Arkansas, would like its new employees to understand more about the packing and shipping process. They have prepared the following chart to aid the new-employee training program.

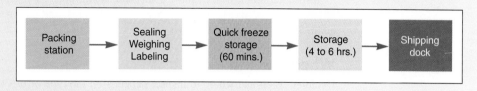

Histograms

Histograms show the range of values of a measurement and the frequency with which each value occurs (see Figure 6.5[f]). They show the most frequently occurring readings as well as the variations in the measurements. Descriptive statistics, such as the average and standard deviation, may be calculated to describe the distribution. However, the data should always be plotted so the shape of the distribution can be "seen." A visual presentation of the distribution may also provide insight into the cause of the variation.

Statistical Process Control (SPC)

Statistical process control monitors standards, makes measurements, and takes corrective action as a product or service is being produced. Samples of process outputs are examined; if they are within acceptable limits, the process is permitted to continue. If they fall outside certain specific ranges, the process is stopped and, typically, the assignable cause located and removed.

Control charts are graphic presentations of data over time that show upper and lower limits for the process we want to control (see Figure 6.5[g]). Control charts are constructed in such a way that new data can be quickly compared to past performance data. We take samples of the process output and plot the average of these samples on a chart that has the limits on it. The upper and lower limits in a control chart can be in units of temperature, pressure, weight, length, and so on.

Figure 6.7 shows the plot of percentages of a sample in a control chart. When the average of the samples falls within the upper and lower control limits and no discernible pattern is present, the process is said to be in control with only natural variation present. Otherwise, the process is out of control or out of adjustment.

The supplement to this chapter details how control charts of different types are developed. It also deals with the statistical foundation underlying the use of this important tool.

THE ROLE OF INSPECTION

To make sure a system is producing at the expected quality level, control of the process is needed. The best processes have little variation from the standard expected. The operations manager's task is to build such systems and to verify, often by inspection, that they are performing to standard. This **inspection** can involve measurement, tasting, touching, weighing, or testing of the product (sometimes even

Statistical process control (SPC)
A process used to monitor standards, making measurements and taking corrective action as a product or service is being produced.

Control charts
Graphic presentations of process data over time with predetermined control limits.

Inspection
A means of ensuring that an operation is producing at the quality level expected.

FIGURE 6.7 ■

Control Chart for Percentage of Free Throws Missed by the Chicago Bulls in Their First Nine Games of the New Season

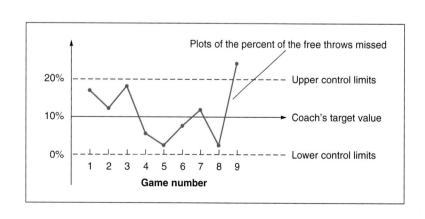

destroying it when doing so). Its goal is to detect a bad process immediately. Inspection does not correct deficiencies in the system or defects in the products; nor does it change a product or increase its value. Inspection only finds deficiencies and defects, and it is expensive.

Inspection should be thought of as an audit. Audits do not add value to the product. However, operations managers, like financial managers, need audits, and they need to know when and where to audit. So there are two basic issues relating to inspection: (1) *when to inspect* and (2) *where to inspect*.

When and Where to Inspect

Deciding when and where to inspect depends on the type of process and the value added at each stage. Inspections (audits) can take place at any of the following points:

1. At your supplier's plant while the supplier is producing.
2. At your facility upon receipt of goods from your supplier.
3. Before costly or irreversible processes.
4. During the step-by-step production process.
5. When production or service is complete.
6. Before delivery from your facility.
7. At the point of customer contact.

One of the themes of our treatment of quality is that "quality cannot be inspected into a product."

The seven tools of TQM discussed in the previous section aid in this "when and where to inspect" decision. However, inspection is not a substitute for a robust product produced by well-trained employees in a good process. In one well-known experiment conducted by an independent research firm, 100 defective pieces were added to a "perfect" lot of items and then subjected to 100% inspection.[10] The inspectors found only 68 of the defective pieces in their first inspection. It took another three passes by the inspectors to find the next 30 defects. The last two defects were never found. So the bottom line is that there is variability in the inspection process. Additionally, inspectors are only human: They become bored, they become tired, and the inspection equipment itself has variability. Even with 100% inspection, inspectors cannot guarantee perfection. Therefore, good processes and employee empowerment are usually a better solution than trying to find defects by inspection.

For example, at Velcro Industries, as in many organizations, quality was viewed by machine operators as the job of "those quality people." Inspections were based on random sampling, and if a part showed up bad, it was thrown out. The company decided to pay more attention to operators, machine repair and design, measurement methods, communications, and responsibilities, and to invest more money in training. Over time as defects declined, Velcro was able to pull half its quality control people out of the process.

Source Inspection

Source inspection
Controlling or monitoring at the point of production or purchase—at the source.

The best inspection can be thought of as no inspection at all; this "inspection" is always done at the source—it is just doing the job properly with the operator ensuring that this is so. This may be called **source inspection** (or source control) and is consistent with the concept of employee empowerment, where individual employees self-check their own work. The idea is that each supplier, process, and employee *treats the next step in the process as the customer*, ensuring perfect product to the next "customer." This inspection may be assisted by the use of checklists and controls such as a fail-safe device called a *poka-yoke*, a name borrowed from the Japanese.

Poka-yoke
Literally translated, "foolproof"; it has come to mean a device or technique that ensures the production of a good unit every time.

A **poka-yoke** is a foolproof device or technique that ensures production of good units every time.[11] These special devices avoid errors and provide quick feedback of problems. A simple example of a poka-yoke device is the diesel or leaded gas pump nozzle that will not fit into the "unleaded" gas tank opening on your car. In McDonald's, the French-fry scoop and standard-size bag used to measure the correct quantity are poka-yokes. Similarly, in a hospital, the prepackaged surgical coverings that contain exactly the items needed for a medical procedure are poka-yokes. Checklists are another type of poka-yoke. The idea of source inspection and poka-yokes is to ensure that 100% good product or service is provided at each step in the process.

---

[10]*Statistical Quality Control* (Springfield, MA: Monsanto Chemical Company, n.d.): 19.

[11]For further discussion, see Alan Robinson, *Modern Approaches to Management Improvement: The Shingo System* (Cambridge, MA: Productivity Press, 1990).

Good methods analysis and the proper tools can result in poka-yokes that improve both quality and speed. Here, two poka-yokes are demonstrated. First, the aluminum scoop automatically positions the French fries vertically, and second, the properly sized container ensures the portion served is correct. This combination also speeds delivery, ensuring French fries delivered just as the customer requests them.

Service Industry Inspection

In *service*-oriented organizations, inspection points can be assigned at a wide range of locations, as illustrated in Table 6.3. Again, the operations manager must decide where inspections are justified and may find the seven tools of TQM useful when making these judgments.

Inspection of Attributes versus Variables

Attribute inspection
An inspection that classifies items as being either good or defective.

Variable inspection
Classifications of inspected items as falling on a continuum scale such as dimension, size, or strength.

When inspections take place, quality characteristics may be measured as either *attributes* or *variables*. **Attribute inspection** classifies items as being either good or defective. It does not address the *degree* of failure. For example, the lightbulb burns or it does not. **Variable inspection** measures such dimensions as weight, speed, height, or strength to see if an item falls within an acceptable range. If a piece of electrical wire is supposed to be 0.01 inch in diameter, a micrometer can be used to see if the product is close enough to pass inspection.

Knowing whether attributes or variables are being inspected helps us decide which statistical quality control approach to take, as we will see in the supplement to this chapter.

TABLE 6.3 ■

Examples of Inspection in Services

Organization	What Is Inspected	Standard
Jones Law Offices	Receptionist performance	Phone answered by the second ring
	Billing	Accurate, timely, and correct format
	Attorney	Promptness in returning calls
Hard Rock Hotel	Reception desk	Use customer's name
	Doorman	Greet guest in less than 30 seconds
	Room	All lights working, spotless bathroom
	Minibar	Restocked and charges accurately posted to bill
Bayfield Community Hospital	Billing	Accurate, timely, and correct format
	Pharmacy	Prescription accuracy, inventory accuracy
	Lab	Audit for lab-test accuracy
	Nurses	Charts immediately updated
	Admissions	Data entered correctly and completely
Hard Rock Cafe	Busboy	Serves water and bread within 1 minute
	Busboy	Clears all entrée items and crumbs prior to dessert
	Waiter	Knows and suggests specials, desserts
Nordstrom's Department Store	Display areas	Attractive, well organized, stocked, good lighting
	Stockrooms	Rotation of goods, organized, clean
	Salesclerks	Neat, courteous, very knowledgeable

TQM IN SERVICES

The personal component of services is more difficult to measure than the quality of the tangible component. Generally, the user of a service, like the user of a good, has features in mind that form a basis for comparison among alternatives. Lack of any one feature may eliminate the service from further consideration. Quality also may be perceived as a bundle of attributes in which many lesser characteristics are superior to those of competitors. This approach to product comparison differs little between goods and services. However, what is very different about the selection of services is the poor definition of the (1) *intangible differences between products* and (2) *the intangible expectations customers have of those products.*[12] Indeed, the intangible attributes may not be defined at all. They are often unspoken images in the purchaser's mind. This is why all of those marketing issues such as advertising, image, and promotion can make a difference.

The operations manager plays a significant role in addressing several major aspects of service quality. First, the *tangible component of many services is important.* How well the service is designed and produced does make a difference. This might be how accurate, clear, and complete your checkout bill at the hotel is, how warm the food is at Taco Bell, or how well your car runs after you pick it up at the repair shop.

Second, another aspect of service and service quality is the process. Notice in Table 6.4 that 9 out of 10 of the determinates of service quality are related to *the service process.* Such things as reliability and courtesy are part of the process. The operations manager can *design processes (service products) that have these attributes* and can ensure their quality through the TQM techniques discussed in this chapter.

Third, the operations manager should realize that the customer's expectations are the standard against which the service is judged. Customers' perceptions of service quality result from a comparison of their before-service expectations with their actual-service experience. In other words, service quality is judged on the basis of whether it meets expectations. The *manager may be able to influence both the quality of the service and the expectation.* Don't promise more than you can deliver.

Fourth, the manager must expect exceptions. There is a standard quality level at which the regular service is delivered, such as the bank teller's handling of a transaction. However, there are "exceptions" or "problems" initiated by the customer or by less than optimal operating conditions

TABLE 6.4 ■

Determinants of Service Quality

Reliability involves consistency of performance and dependability. It means that the firm performs the service right the first time and that the firm honors its promises.

Responsiveness concerns the willingness or readiness of employees to provide service. It involves timeliness of service.

Competence means possession of the required skills and knowledge to perform the service.

Access involves approachability and ease of contact.

Courtesy involves politeness, respect, consideration, and friendliness of contact personnel (including receptionists, telephone operators, etc.).

Communication means keeping customers informed in language they can understand and listening to them. It may mean that the company has to adjust its language for different consumers—increasing the level of sophistication with a well-educated customer and speaking simply and plainly with a novice.

Credibility involves trustworthiness, believability, and honesty. It involves having the customer's best interests at heart.

Security is the freedom from danger, risk, or doubt.

Understanding/knowing the customer involves making the effort to understand the customer's needs.

Tangibles include the physical evidence of the service.

Source: Excerpted with permission from *Journal of Marketing,* published by the American Marketing Association, from A. Parasuranam, Valarie A. Zeithaml, and Leonard L. Berry (fall 1985): 44.

[12]L. Berry, V. Zeithaml, and A. Parasuraman, "Quality Counts in Services, Too," *Business Horizons* (May–June 1985): 45–46.

Designing a high-quality process that fills these pharmaceutical bottles in sterile conditions is much more fruitful than having an inspector evaluate the bacteria count on bottles filled in a poor system. Good quality systems focus on quality processes, not after-the-fact inspections.

(for example, the computer "crashed"). This implies that the quality control system must recognize and *have a set of alternative plans for less than optimal operating conditions.*

Designing the product, managing the service process, matching customer expectations to the product, and preparing for the exceptions are keys to quality services. The *OM in Action* box, "Richey International's Spies," provides another glimpse of how OM managers improve quality in services.

OM IN ACTION

Richey International's Spies

How do luxury hotels maintain quality? They inspect. But when the product is one-on-one service, largely dependent on personal behavior, how do you inspect? You hire spies!

Richey International is the spy. Preferred Hotels and Resorts Worldwide and Intercontinental Hotels have both hired Richey to do quality evaluations via spying. Richey employees posing as customers perform the inspections. However, even then management must have established what the customer expects and specific services that yield customer satisfaction. Only then do managers know where and how to inspect. Aggressive training and objective inspections reinforce behavior that will meet those customer expectations.

The hotels use Richey's undercover inspectors to ensure performance to exacting standards. The hotels do not know when the evaluators will arrive or what aliases they will use. Over 50 different standards are eval-

uated before the inspectors even check in at a luxury hotel. Over the next 24 hours, using checklists, tape recordings, and photos, written reports are prepared and include evaluation of standards such as:

- Does the doorman greet each guest in less than 30 seconds?
- Does the front-desk clerk use the guest's name during check-in?
- Is the bathroom tub and shower spotlessly clean?
- How many minutes does it take to get coffee after the guest sits down for breakfast?
- Did the waiter make eye contact?
- Were minibar charges posted correctly on the bill?

Established standards, aggressive training, and inspections are part of the TQM effort at these hotels. Quality does not happen by accident.

Sources: The Wall Street Journal (May 12, 1999): B1, B12; and *Forbes* (October 5, 1998): 88–89.

SUMMARY

Quality is a term that means different things to different people. It is defined in this chapter as "the totality of features and characteristics of a product or service that bears on its ability to satisfy stated or implied needs." Defining quality expectations is critical to effective and efficient operations.

Quality requires building a total quality management (TQM) environment because quality cannot be inspected into a product. The chapter also addresses six TQM concepts: continuous improvement, employee empowerment, benchmarking, just-in-time, Taguchi concepts, and knowledge of TQM tools. The seven TQM tools introduced in this chapter are check sheets, scatter diagrams, cause-and-effect diagrams, Pareto charts, flow charts, histograms, and statistical process control (SPC).

KEY TERMS

Quality *(p. 154)*
Cost of Quality (COQ) *(p. 156)*
ISO 9000 *(p. 156)*
ISO 14000 *(p. 157)*
Total quality management (TQM) *(p. 157)*
PDCA *(p. 158)*
Kaizen *(p. 158)*
Six sigma *(p. 158)*
Employee empowerment *(p. 158)*
Quality circle *(p. 159)*
Benchmarking *(p. 159)*
Quality robust *(p. 161)*
Quality loss function (QLF) *(p. 161)*

Target-oriented quality *(p. 161)*
Cause-and-effect diagram, Ishikawa diagram, or fish-bone chart *(p. 163)*
Pareto charts *(p. 164)*
Flow charts *(p. 164)*
Statistical process control (SPC) *(p. 165)*
Control charts *(p. 165)*
Inspection *(p. 165)*
Source inspection *(p. 166)*
Poka-yoke *(p. 166)*
Attribute inspection *(p. 167)*
Variable inspection *(p. 167)*

INTERNET AND STUDENT CD-ROM EXERCISES

Visit our home page or your student CD-ROM to help with material in this chapter.

 On Our Home Page, www.prenhall.com/heizer

- Self-Tests
- Practice Problems
- Internet Exercises
- Current Articles and Research
- Virtual Company Tour
- Internet Homework Problems
- Internet Cases

 On Your Student CD-ROM

- PowerPoint Lecture
- Practice Problems
- Video Clips and Video Case
- Active Model Exercise

ADDITIONAL CASE STUDIES

Internet Case Studies: Visit our Web site at www.prenhall.com/heizer **for these free case studies:**

- **Westover Electrical, Inc.:** This electric motor manufacturer has a large log of defects in its wiring process.
- **Falls Church General Hospital:** Establishing quality standards in a 615-bed hospital.
- **Quality Cleaners:** This small cleaners needs a quality management system.
- **Belair Casino Hotel, Zimbabwe:** A resort hotel in Africa needs to analyze its customer comment cards.

Harvard has selected these Harvard Business School cases to accompany this chapter
(textbookcasematch.hbsp.harvard.edu)**:**

- **Wainwright Industries (A): Beyond the Baldrige** (#396-219): Traces the growth of an auto supply company and its culture of quality.
- **Romeo Engine Plant** (#197-100): The employees at this auto engine plant must solve problems and ensure quality, not watch parts being made.
- **Motorola-Penang** (#494-135): The female manager of this Malaysia factory is skeptical of empowerment efforts at other Motorola sites.
- **Measure of Delight: The Pursuit of Quality at AT&T Universal Card Service (A)** (#694-047): Links performance measurement and compensation policies to precepts of quality management.

Statistical Process Control

Supplement Outline

LEARNING OBJECTIVES

When you complete this supplement you should be able to

IDENTIFY OR DEFINE:

Natural and assignable causes of variations

Central limit theorem

Attribute and variable inspection

Process control

$\bar{x}$-charts and *R*-charts

LCL and UCL

P-charts and *c*-charts

C_p and C_{pk}

Acceptance sampling

OC curve

AQL and LTPD

AOQ

Producer's and consumer's risk

DESCRIBE OR EXPLAIN:

The role of statistical quality control

BetzDearborn, A Division of Hercules Incorporated, is headquartered in Trevose, Pennsylvania. It is a global supplier of specialty chemicals for the treatment of industrial water, wastewater, and process systems. The company uses statistical process control to monitor the performance of treatment programs in a wide variety of industries throughout the world. BetzDearborn's quality assurance laboratory (shown here) also uses statistical sampling techniques to monitor manufacturing processes at all of the company's production plants.

Statistical process control (SPC)
A process used to monitor standards, making measurements and taking corrective action as a product or service is being produced.

In this supplement, we address statistical process control—the same techniques used at BetzDearborn, at IBM, at GE, and at Motorola to achieve quality standards. We also introduce acceptance sampling. **Statistical process control** is the application of statistical techniques to the control of processes. *Acceptance sampling* is used to determine acceptance or rejection of material evaluated by a sample.

STATISTICAL PROCESS CONTROL (SPC)

Statistical process control (SPC) is a statistical technique that is widely used to ensure that processes meet standards. All processes are subject to a certain degree of variability. While studying process data in the 1920s, Walter Shewhart of Bell Laboratories made the distinction between the common and special causes of variation. Many people now refer to these variations as *natural* and *assignable* causes. He developed a simple but powerful tool to separate the two—the control chart.

Control chart
A graphic presentation of process data over time.

We use statistical process control to measure performance of a process. A process is said to be operating *in statistical control* when the only source of variation is common (natural) causes. The process must first be brought into statistical control by detecting and eliminating special (assignable) causes of variation.[1] Then its performance is predictable, and its ability to meet customer expectations can be assessed. The *objective* of a process control system is to *provide a statistical signal when assignable causes of variation are present*. Such a signal can quicken appropriate action to eliminate assignable causes.

---

[1]Removing assignable causes is work. As quality guru W. Edwards Deming observed, "A state of statistical control is not a natural state for a manufacturing process. It is instead an achievement, arrived at by elimination, one by one, by determined effort, of special causes of excessive variation." See W. Edwards Deming, "On Some Statistical Aids toward Economic Production," *Interfaces* 5, no. 4 (1975): 5.

Natural Variations Natural variations affect almost every production process and are to be expected. **Natural variations** are the many sources of variation that occur within a process that is in statistical control. Natural variations behave like a constant system of chance causes. Although individual values are all different, as a group they form a pattern that can be described as a *distribution*. When these distributions are *normal*, they are characterized by two parameters. These parameters are:

- mean, μ (the measure of central tendency—in this case, the average value)
- standard deviation, σ (the measure of dispersion)

As long as the distribution (output measurements) remains within specified limits, the process is said to be "in control," and natural variations are tolerated.

Assignable Variations **Assignable variation** in a process can be traced to a specific reason. Factors such as machine wear, misadjusted equipment, fatigued or untrained workers, or new batches of raw material are all potential sources of assignable variations.

Natural and assignable variations distinguish two tasks for the operations manager. The first is to *ensure that the process is capable* of operating under control with only natural variation. The second is, of course, to *identify and eliminate assignable variations* so that the processes will remain under control.

Samples Because of natural and assignable variation, statistical process control uses averages of small samples (often of four to eight items) as opposed to data on individual parts. Individual pieces tend to be too erratic to make trends quickly visible.

Figure S6.1 provides a detailed look at the important steps in determining process variation. The horizontal scale can be weight (as in the number of ounces in boxes of cereal) or length (as in fence

Natural variations
Variabilities that affect every production process to some degree and are to be expected; also known as common causes.

Assignable variation
Variation in a production process that can be traced to specific causes.

FIGURE S6.1 ■

Natural and Assignable Variation

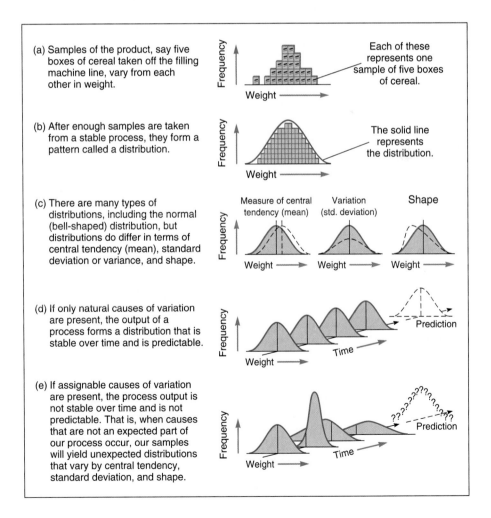

posts) or any physical measure. The vertical scale is frequency. The samples of five boxes of cereal in Figure S6.1 (a) are weighed; (b) they form a distribution, (c) that can vary. The distributions formed in (b) and (c) will fall in a predictable pattern, (d) if only natural variation is present. If assignable causes of variation are present, then we can expect either the mean to vary or the dispersion to vary, as is the case in (e).

Control Charts The process of building control charts is based on the concepts presented in Figure S6.2. This figure shows three distributions that are the result of outputs from three types of processes. We plot small samples and then examine characteristics of the resulting data to see if the process is within "control limits." The purpose of control charts is to help distinguish between natural variations and variations due to assignable causes. As seen in Figure S6.2, a process is (a) in control *and the process is capable of producing within established control limits*, (b) in control *but the process is not capable of producing within established limits*, or (c) out of control. We now look at ways to build control charts that help the operations manager keep a process under control.

Control Charts for Variables

Variables are characteristics that have continuous dimensions. They have an infinite number of possibilities. Examples are weight, speed, length, or strength. Control charts for the mean, $\bar{x}$ or x-bar, and the range, R, are used to monitor processes that have continuous dimensions. The $\bar{x}$-chart tells us whether changes have occurred in the central tendency (the mean, in this case) of a process. These changes might be due to such factors as tool wear, a gradual increase in temperature, a different method used on the second shift, or new and stronger materials. The **R-chart** values indicate that a gain or loss in dispersion has occurred. Such a change might be due to worn bearings, a loose tool, an erratic flow of lubricants to a machine, or to sloppiness on the part of a machine operator. The two types of charts go hand in hand when monitoring variables because they measure the two critical parameters, central tendency and dispersion.

$\bar{x}$-chart
A quality control chart for variables that indicates when changes occur in the central tendency of a production process.

R-chart
A control chart that tracks the "range" within a sample; indicates that a gain or loss in uniformity has occurred in dispersion of a production process.

FIGURE S6.2 ■

Process Control: Three Types of Process Outputs

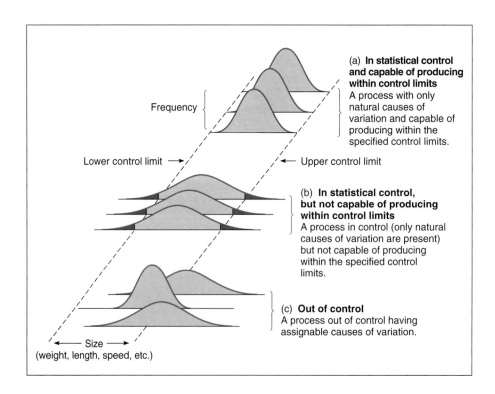

Frequency

Lower control limit →   ← Upper control limit

(a) In statistical control and capable of producing within control limits
A process with only natural causes of variation and capable of producing within the specified control limits.

(b) In statistical control, but not capable of producing within control limits
A process in control (only natural causes of variation are present) but not capable of producing within the specified control limits.

(c) Out of control
A process out of control having assignable causes of variation.

Size →
(weight, length, speed, etc.)

The Central Limit Theorem

Central limit theorem
The theoretical foundation for $\bar{x}$-charts that states that regardless of the distribution of the population of all parts or services, the distribution of $\bar{x}$s will tend to follow a normal curve as the number of samples increases.

The theoretical foundation for $\bar{x}$-charts is the **central limit theorem**. This theorem states that regardless of the distribution of the population, the distribution of $\bar{x}$s (each of which is a mean of a sample drawn from the population) will tend to follow a normal curve as the number of samples increases. Fortunately, even if the sample (n) is fairly small (say, 4 or 5), the distributions of the averages will still roughly follow a normal curve. The theorem also states that: (1) the mean of the distribution of the $\bar{x}$s (called $\bar{\bar{x}}$) will equal the mean of the overall population (called μ); and (2) the standard deviation of the *sampling distribution*, $\sigma_{\bar{x}}$, will be the *population standard deviation*,[2] σ, divided by the square root of the sample size, n. In other words:

$$\bar{\bar{x}} = \mu \qquad \text{(S6-1)}$$

The two parameters are:
Mean → measure of central tendency.
Range → measure of dispersion.

and

$$\sigma_{\bar{x}} = \frac{\sigma}{\sqrt{n}} \qquad \text{(S6-2)}$$

Figure S6.3 shows three possible population distributions, each with its own mean, μ, and standard deviation, σ. If a series of random samples ($\bar{x}_1, \bar{x}_2, \bar{x}_3, \bar{x}_4$, and so on), each of size n, is drawn from any population distribution (which could be normal, beta, uniform, and so on), the resulting distribution of $\bar{x}_i$s will appear as they do in Figure S6.3.

Moreover, the sampling distribution, as is shown in Figure S6.4, will have less variability than the process distribution. Because the sampling distribution is normal, we can state that:

- 95.45% of the time, the sample averages will fall within $\pm 2\sigma_{\bar{x}}$ if the process has only natural variations.
- 99.73% of the time, the sample averages will fall within $\pm 3\sigma_{\bar{x}}$ if the process has only natural variations.

If a point on the control chart falls outside of the $\pm 3\sigma_{\bar{x}}$ control limits, then we are 99.73% sure the process has changed. This is the theory behind control charts.

FIGURE S6.3 ■

The Relationship between Population and Sampling Distributions

Regardless of the population distribution (e.g., normal, beta, uniform), each with its own mean (μ) and standard deviation (σ), the distribution of sample means is normal.

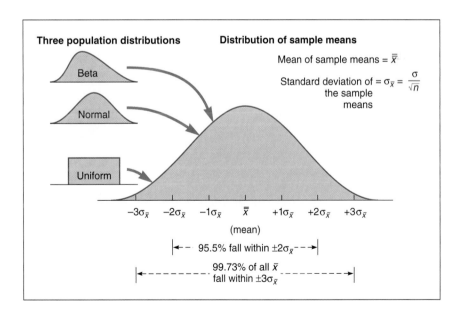

[2]*Note:* The standard deviation is easily calculated as: $\sigma = \sqrt{\dfrac{\sum_{i=1}^{n}(x_i - \bar{x})^2}{n-1}}$.

FIGURE S6.4 ■

The Sampling Distribution of Means Is Normal and Has Less Variability Than the Process Distribution

In this figure, the process distribution from which the sample was drawn was also normal, but it could have been any distribution.

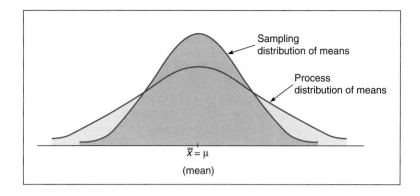

Setting Mean Chart Limits (x̄-Charts)

If we know, through past data, the standard deviation of the process population, σ, we can set upper and lower control limits by these formulas:

$$\text{Upper control limit (UCL)} = \bar{\bar{x}} + z\sigma_{\bar{x}} \tag{S6-3}$$

$$\text{Lower control limit (LCL)} = \bar{\bar{x}} - z\sigma_{\bar{x}} \tag{S6-4}$$

where $\bar{\bar{x}}$ = mean of the sample means or a target value set for the process

z = number of normal standard deviations (2 for 95.45% confidence, 3 for 99.73%)

$\sigma_{\bar{x}}$ = standard deviation of the sample means = $\sigma/\sqrt{n}$

σ = population (process) standard deviation

n = sample size

"All quality control does is find our mistakes. I want to start avoiding them."

Example S1 shows how to set control limits for sample means using standard deviations.

Example S1

Excel OM Data File Ch06SExS1.xla

The weights of boxes of Oat Flakes within a large production lot are sampled each hour. To set control limits that include 99.73% of the sample means, samples of nine boxes are randomly selected and weighed. Here are the nine boxes chosen for Hour 1:

The average weight in the first sample = $\dfrac{17 + 13 + 16 + 18 + 17 + 16 + 15 + 17 + 16}{9}$

= 16.1 oz.

Also, the *population* standard deviation (σ) is known to be 1 ounce. We do not show each of the boxes sampled in Hours 2 through 12, but here are the rest of the results:

	WEIGHT OF SAMPLE		WEIGHT OF SAMPLE		WEIGHT OF SAMPLE
HOUR	(AVG. OF 9 BOXES)	HOUR	(AVG. OF 9 BOXES)	HOUR	(AVG. OF 9 BOXES)
1	16.1	5	16.5	9	16.3
2	16.8	6	16.4	10	14.8
3	15.5	7	15.2	11	14.2
4	16.5	8	16.4	12	17.3

The average mean of the 12 samples is calculated to be exactly 16 ounces. We therefore have $\bar{\bar{x}} = 16$ ounces, $\sigma = 1$ ounce, $n = 9$, and $z = 3$. The control limits are:

$$\text{UCL}_{\bar{x}} = \bar{\bar{x}} + z\sigma_{\bar{x}} = 16 + 3\left(\frac{1}{\sqrt{9}}\right) = 16 + 3\left(\frac{1}{3}\right) = 17 \text{ ounces}$$

$$\text{LCL}_{\bar{x}} = \bar{\bar{x}} - z\sigma_{\bar{x}} = 16 - 3\left(\frac{1}{\sqrt{9}}\right) = 16 - 3\left(\frac{1}{3}\right) = 15 \text{ ounces}$$

The 12 samples are then plotted on the control chart shown below. Because the means of recent sample averages fall outside the upper and lower control limits of 17 and 15, we can conclude that the process is becoming erratic and *not* in control.

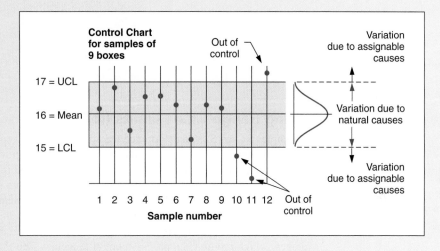

The range is the difference between the largest and the smallest items in a sample.

Because process standard deviations are either not available or difficult to compute, we usually calculate control limits based on the average *range* values rather than on standard deviations. Table S6.1 provides the necessary conversion for us to do so. The *range* is defined as the difference between the largest and smallest items in one sample. For example, the heaviest box of Oat Flakes in Hour 1 of Example S1 was 18 ounces and the lightest was 13 ounces, so the range for that hour is 5 ounces. We use Table S6.1 and the equations

$$\text{UCL}_{\bar{x}} = \bar{\bar{x}} + A_2\bar{R} \tag{S6-5}$$

and

$$\text{LCL}_{\bar{x}} = \bar{\bar{x}} - A_2\bar{R} \tag{S6-6}$$

where $\bar{R}$ = average range of the samples
A_2 = value found in Table S6.1
$\bar{\bar{x}}$ = mean of the sample means

TABLE S6.1 ■

Factors for Computing Control Chart Limits (3 sigma)

SAMPLE SIZE, n	MEAN FACTOR, A_2	UPPER RANGE, D_4	LOWER RANGE, D_3
2	1.880	3.268	0
3	1.023	2.574	0
4	.729	2.282	0
5	.577	2.115	0
6	.483	2.004	0
7	.419	1.924	0.076
8	.373	1.864	0.136
9	.337	1.816	0.184
10	.308	1.777	0.223
12	.266	1.716	0.284

Source: Reprinted by permission of American Society for Testing Materials. Copyright 1951. Taken from Special Technical Publication 15-C, "Quality Control of Materials," pp. 63 and 72.

Example S2 shows how to set control limits for sample means using Table S6.1 and the average range.

Example S2

Excel OM Data File
Ch06SExS2.xla

Super Cola bottles soft drinks labeled "net weight 16 ounces." An overall process average of 16.01 ounces has been found by taking several batches of samples, in which each sample contained 5 bottles. The average range of the process is .25 ounce. Determine the upper and lower control limits for averages in this process.

Looking in Table S6.1 for a sample size of 5 in the mean factor A_2 column, we find the number .577. Thus, the upper and lower control chart limits are:

$$UCL_{\bar{x}} = \bar{\bar{x}} + A_2\bar{R}$$
$$= 16.01 + (.577)(.25)$$
$$= 16.01 + .144$$
$$= 16.154 \text{ ounces}$$
$$LCL_{\bar{x}} = \bar{\bar{x}} - A_2\bar{R}$$
$$= 16.01 - .144$$
$$= 15.866 \text{ ounces}$$

Setting Range Chart Limits (R-Charts)

In Examples S1 and S2, we determined the upper and lower control limits for the process *average*. In addition to being concerned with the process average, operations managers are interested in the process *dispersion*, or *range*. Even though the process average is under control, the dispersion of the process may not be. For example, something may have worked itself loose in a piece of equipment that fills boxes of Oat Flakes. As a result, the average of the samples may remain the same, but the variation within the samples could be entirely too large. For this reason, operations managers use control charts for ranges in order to monitor the process variability, as well as control charts for averages, which monitor the process central tendency. The theory behind the control charts for ranges is the same as that for process average control charts. Limits are established that contain ±3 standard deviations of the distribution for the average range $\bar{R}$. We can use the following equations to set the upper and lower control limits for ranges:

$$UCL_R = D_4\bar{R} \tag{S6-7}$$

$$LCL_R = D_3\bar{R} \tag{S6-8}$$

It's important to note that when building the UCL_R and LCL_R, we use the *average* range, $\bar{R}$. But when plotting points once the R-chart is developed, we use the *individual* range values for each sample.

where UCL_R = upper control chart limit for the range
LCL_R = lower control chart limit for the range
D_4 and D_3 = values from Table S6.1

Example S3 shows how to set control limits for sample ranges using Table S6.1 and the average range.

Example S3

The average *range* of a process for loading trucks is 5.3 pounds. If the sample size is 5, determine the upper and lower control chart limits.

Looking in Table S6.1 for a sample size of 5, we find that $D_4 = 2.115$ and $D_3 = 0$. The range control limits are:

$$UCL_R = D_4\bar{R} = (2.115)(5.3 \text{ pounds}) = 11.2 \text{ pounds}$$
$$LCL_R = D_3\bar{R} = (0)(5.3 \text{ pounds}) = 0$$

Using Mean and Range Charts

The normal distribution is defined by two parameters, the *mean* and *standard deviation*. The $\bar{x}$ (mean)-chart and the R-chart mimic these two parameters. The $\bar{x}$-chart is sensitive to shifts in the process mean, whereas the R-chart is sensitive to shifts in the process standard deviation. Consequently, by using both charts we can track changes in the process distribution.

FIGURE S6.5 ■

Mean and Range Charts Complement Each Other by Showing the Mean and Dispersion of the Normal Distribution

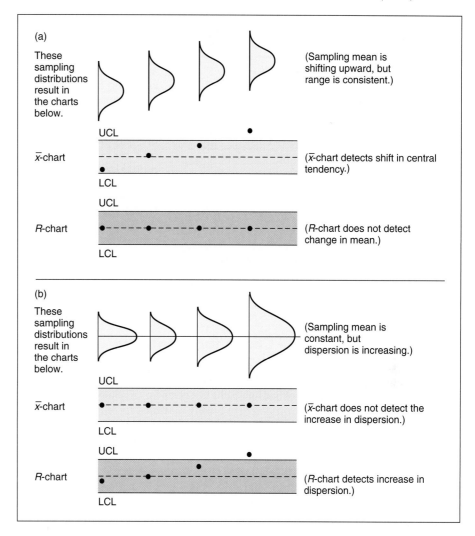

Video S6.1

SPC at Harley-Davidson

For instance, the samples and the resulting $\bar{x}$-chart in Figure S6.5(a) show the shift in the process mean, but because the dispersion is constant, no change is detected by the R-chart. Conversely, the samples and the $\bar{x}$-chart in Figure S6.5(b) detect no shift (because none is present), but the R-chart does detect the shift in the dispersion. Both charts are required to track the process accurately.

Steps to Follow When Using Control Charts   There are five steps that are generally followed in using $\bar{x}$- and R-charts:

1. Collect 20 to 25 samples of $n = 4$ or $n = 5$ each from a stable process and compute the mean and range of each.
2. Compute the overall means ($\bar{\bar{x}}$ and $\bar{R}$), set appropriate control limits, usually at the 99.73% level, and calculate the preliminary upper and lower control limits. *If the process is not currently stable*, use the desired mean, μ, instead of $\bar{\bar{x}}$ to calculate limits.
3. Graph the sample means and ranges on their respective control charts and determine whether they fall outside the acceptable limits.
4. Investigate points or patterns that indicate the process is out of control. Try to assign causes for the variation and then resume the process.
5. Collect additional samples and, if necessary, revalidate the control limits using the new data.

Applications of control charts appear in examples in this supplement, as well as in the *OM in Action* box, "Green Is the Color of Money for DuPont and the Environment."

OM IN ACTION

Green Is the Color of Money for DuPont and the Environment

DuPont has found that statistical process control (SPC) is an excellent approach to solving environmental problems. With a goal of slashing manufacturing waste and hazardous waste disposals by 35%, DuPont brought together information from its quality control systems and its material management databases.

Cause-and-effect diagrams and Pareto charts revealed where major problems occurred. Then the company began reducing waste materials through improved SPC standards for production. Tying together shop-floor information-based monitoring systems with air-quality standards, DuPont identified ways to reduce emissions. Using a vendor evaluation system linked to JIT purchasing

requirements, the company initiated controls over incoming hazardous materials.

DuPont now saves more than 15 million pounds of plastics annually by recycling them into products rather than dumping them into landfills. Through electronic purchasing, the firm has reduced wastepaper to a trickle, and by using new packaging designs, it has cut in-process material wastes by nearly 40%.

By integrating SPC with environmental-compliance activities, DuPont has made major quality improvements that far exceed regulatory guidelines. DuPont's innovations in solving environmental problems have, at the same time, realized huge cost savings.

Sources: P. E. Barnes, *Business and Economic Review* (January–March 1998): 21–24; *Environmental Quality Management* (summer 1998): 97–110; and *Purchasing* (November 6, 1997): 114.

Control Charts for Attributes

Control charts for $\bar{x}$ and R do not apply when we are sampling *attributes*, which are typically classified as *defective* or *nondefective*. Measuring defectives involves counting them (for example, number of bad lightbulbs in a given lot, or number of letters or data entry records typed with errors), whereas *variables* are usually measured for length or weight. There are two kinds of attribute control charts: (1) those that measure the *percent* defective in a sample—called *p*-charts—and (2) those that count the *number* of defects—called *c*-charts.

p-charts
A quality control chart that is used to control attributes.

p-Charts   Using **p-charts** is the chief way to control attributes. Although attributes that are either good or bad follow the binomial distribution, the normal distribution can be used to calculate *p*-chart limits when sample sizes are large. The procedure resembles the $\bar{x}$-chart approach, which was also based on the central limit theorem.

A Mitutoyo Corporation variable chart display using SPC software that is directly linked to measuring devices.

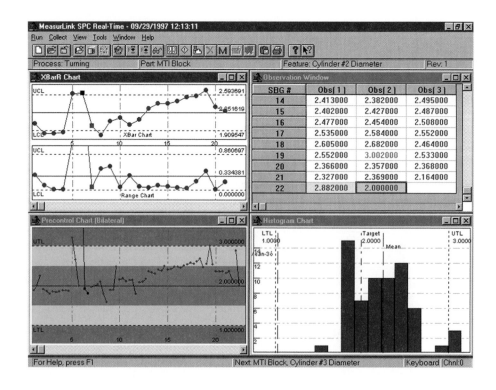

Although SPC charts can be generated by computer, this one is being prepared by hand. This chart is updated each hour and reflects a week of workshifts.

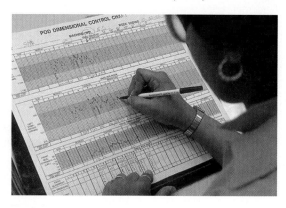

The formulas for *p*-chart upper and lower control limits follow:

$$\text{UCL}_p = \bar{p} + z\sigma_{\hat{p}} \qquad \text{(S6-9)}$$

$$\text{LCL}_p = \bar{p} - z\sigma_{\hat{p}} \qquad \text{(S6-10)}$$

where $\bar{p}$ = mean fraction defective in the sample

           z = number of standard deviations ($z = 2$ for 95.45% limits; $z = 3$ for 99.73% limits)

           $\sigma_{\hat{p}}$ = standard deviation of the sampling distribution

$\sigma_{\hat{p}}$ is estimated by the formula:

$$\sigma_{\hat{p}} = \sqrt{\frac{\bar{p}(1 - \bar{p})}{n}} \qquad \text{(S6-11)}$$

where n = size of *each* sample.

Example S4 shows how to set control limits for *p*-charts for these standard deviations.

Example S4

Excel OM Data File
Ch06SExS4.xla

Data-entry clerks at ARCO key in thousands of insurance records each day. Samples of the work of 20 clerks are shown in the table. One hundred records entered by each clerk were carefully examined and the number of errors counted. The fraction defective in each sample was then computed.

Set the control limits to include 99.73% of the random variation in the entry process when it is in control.

SAMPLE NUMBER	NUMBER OF ERRORS	FRACTION DEFECTIVE	SAMPLE NUMBER	NUMBER OF ERRORS	FRACTION DEFECTIVE
1	6	.06	11	6	.06
2	5	.05	12	1	.01
3	0	.00	13	8	.08
4	1	.01	14	7	.07
5	4	.04	15	5	.05
6	2	.02	16	4	.04
7	5	.05	17	11	.11
8	3	.03	18	3	.03
9	3	.03	19	0	.00
10	2	.02	20	4	.04
				80	

$$\bar{p} = \frac{\text{total number of errors}}{\text{total number of records examined}} = \frac{80}{(100)(20)} = .04$$

$$\sigma_{\hat{p}} = \sqrt{\frac{(.04)(1 - .04)}{100}} = .02 \text{ (rounded up from 0.196)}$$

(*Note:* 100 is the size of *each* sample = n)

Active Model S6.1

Example S4 is further illustrated in Active Model S6.1 on the CD-ROM and in the Exercise located in your Student Lecture Guide.

$$UCL_p = \bar{p} + z\sigma_{\hat{p}} = .04 + 3(.02) = .10$$
$$LCL_p = \bar{p} - z\sigma_{\hat{p}} = .04 - 3(.02) = 0$$

(because we cannot have a negative percent defective)

When we plot the control limits and the sample fraction defectives below, we find that only one data-entry clerk (number 17) is out of control. The firm may wish to examine that individual's work a bit more closely to see if a serious problem exists (see Figure S6.6).

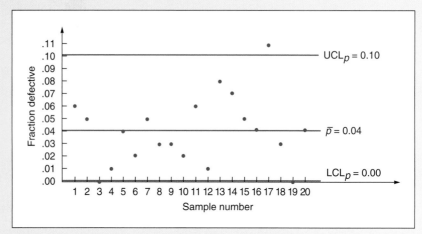

FIGURE S6.6 ■ *p*-Chart for Data Entry for Example S4

The *OM in Action* box, "Unisys Corp.'s Costly Experiment in Health Care Services," provides a real-world followup to Example S4.

c-charts

A quality control chart used to control the number of defects per unit of output.

c-**Charts**   In Example S4, we counted the number of defective records entered. A defective record was one that was not exactly correct because it contained at least one defect. However, a bad record may contain more than one defect. We use *c*-**charts** to control the *number* of defects per unit of output (or per insurance record in the preceding case).

OM IN ACTION

Unisys Corp.'s Costly Experiment in Health Care Services

When Unisys Corp. expanded into the computerized health care service business things looked rosy. It had just beat out Blue Cross/Blue Shield of Florida for an $86-million contract to serve Florida's state employee health-insurance services. Its job was to handle the 215,000 Florida employees' claims processing—a seemingly simple and lucrative growth area for an old-line computer company like Unisys.

But 1 year later the contract was not only torn up, Unisys was fined over $500,000 for not meeting quality standards. Here are two of the measures of quality, both attributes (that is, either "defective" or "not defective") on which the firm was out of control:

1. *Percent of claims processed with errors.* An audit over a 3-month period, by Coopers & Lybrand,

found that Unisys made errors in 8.5% of claims processed. The industry standard is 3.5% "defectives."

2. *Percent of claims processed within 30 days.* For this attribute measure, a "defect" is a processing time longer than the contract's time allowance. In one month's sample, 13% of the claims exceeded the 30-day limit, far above the 5% allowed by the State of Florida.

The Florida contract was a migraine for Unisys, which underestimated the labor-intensiveness of health claims. CEO James Unruh pulled the plug on future ambitions in health care. Meanwhile, the State of Florida's Ron Poppel says, "We really need somebody that's in the insurance business."

Sources: Knight Ridder Tribune Business News (February 7, 2002): 1; and *Business Week* (June 16, 1997): 6.

Testing the stress tolerance on this fabric results in one more observation for the control chart. If the average stress is "out of control," or if a "run" of five averages are above or below the center line, the process needs to be stopped and adjusted.

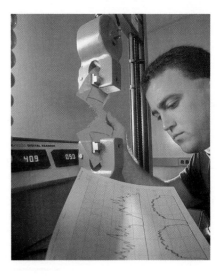

Control charts for defects are helpful for monitoring processes in which a large number of potential errors can occur, but the actual number that do occur is relatively small. Defects may be errors in newspaper words, bad circuits in a microchip, blemishes on a table, or missing pickles on a fast-food hamburger.

The Poisson probability distribution,[3] which has a variance equal to its mean, is the basis for c-charts. Because c is the mean number of defects per unit, the standard deviation is equal to $\sqrt{c}$. To compute 99.73% control limits for $\bar{c}$, we use the formula

$$\text{control limits} = \bar{c} \pm 3\sqrt{\bar{c}} \tag{S6-12}$$

Example S5 shows how to set control limits for a $\bar{c}$-chart.

Example S5

**Excel OM Data File
Ch06SExS5.xla**

Red Top Cab Company receives several complaints per day about the behavior of its drivers. Over a 9-day period (where days are the units of measure), the owner received the following numbers of calls from irate passengers: 3, 0, 8, 9, 6, 7, 4, 9, 8, for a total of 54 complaints.

To compute 99.73% control limits, we take

$$\bar{c} = \frac{54}{9} = 6 \text{ complaints per day}$$

Thus,

$$\text{UCL}_c = \bar{c} + 3\sqrt{\bar{c}} = 6 + 3\sqrt{6} = 6 + 3(2.45) = 13.35$$
$$\text{LCL}_c = \bar{c} - 3\sqrt{\bar{c}} = 6 - 3\sqrt{6} = 6 - 3(2.45) = 0 \leftarrow (\text{since it cannot be negative})$$

After the owner plotted a control chart summarizing these data and posted it prominently in the drivers' locker room, the number of calls received dropped to an average of three per day. Can you explain why this occurred?

Managerial Issues and Control Charts

In an ideal world, there is no need for control charts. Quality is uniform and so high that employees need not waste time and money sampling and monitoring variables and attributes. But because most processes have not reached perfection, managers must make three major decisions regarding control charts.

First, managers must select the points in their process that need SPC. They may ask "Which parts of the job are critical to success" or "Which parts of the job have a tendency to become out of control?"

---

[3]A Poisson probability distribution is a discrete distribution commonly used when the items of interest (in this case, defects) are infrequent and/or occur in time and space.

TABLE S6.2 ■

**Helping You Decide
Which Control Chart
to Use**

USING AN $\bar{x}$-CHART AND AN R-CHART
1. Observations are *variables*, which are usually products measured for size or weight. Examples are the width or length of a wire being cut and the weight of a can of Campbell's soup.
2. Collect 20 to 25 samples of $n = 4$, $n = 5$, or more, each from a stable process, and compute the mean for an $\bar{x}$-chart and the range for an R-Chart.
3. We track samples of n observations each, as in Example S1.

USING A p-CHART
1. Observations are *attributes* that can be categorized as good or bad (or pass–fail, or function–broken), that is, in two states.
2. We deal with fraction, proportion, or percent defectives.
3. There are several samples, with many observations in each. For example, 20 samples of $n = 100$ observations in each, as in Example S4.

USING A c-CHART
1. Observations are *attributes* whose defects per unit of output can be counted.
2. We deal with the number counted, which is a small part of the possible occurrences.
3. Defects may be: number of blemishes on a desk; complaints in a day; crimes in a year; broken seats in a stadium; typos in a chapter of this text; or flaws in a bolt of cloth.

Second, managers need to decide if variable charts (i.e., $\bar{x}$ and R) or attribute charts (i.e., p and c) are appropriate. Variable charts monitor weights or dimensions. Attribute charts are more of a "yes–no" or "go–no go" gauge and tend to be less costly to implement. Table S6.2 can help you understand when to use each of these types of control charts.

Third, the company must set clear and specific SPC policies for employees to follow. For example, should the data entry process be halted if a trend is appearing in percent defective records being keyed? Should an assembly line be stopped if the average length of five successive samples is above the centerline? Figure S6.7 illustrates some of the patterns to look for over time in a process.

A tool called a **run test** is available to help identify the kind of abnormalities in a process that we see in Figure S6.7. In general, a run of 5 points above or below the target or centerline may suggest that an assignable, or nonrandom, variation is present. When this occurs, even though all the points may fall inside the control limits, a flag has been raised. This means the process may not be statistically in control. A variety of run tests are described in books on the subject of quality methods.[4]

Run test

A way to examine the points in a control chart to see if nonrandom variation is present.

FIGURE S6.7 ■

**Patterns to Look for on
Control Charts**

Source: Adapted from Bertrand
L. Hansen, *Quality Control:
Theory and Applications* (1991):
65. Reprinted by permission of
Prentice Hall, Upper Saddle
River, New Jersey.

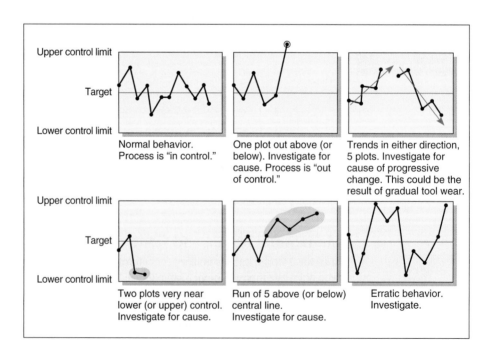

[4]See James R. Thompson and Jacob Koroncki, *Statistical Process Control for Quality Improvement* (New York: Chapman & Hall, 1993); or Fred Aslup and Ricky Watson, *Practical Statistical Process Control* (New York: Van Nostrand Reinhold, 1993).

PROCESS CAPABILITY

Process capability
The ability to meet design specifications.

Statistical process control means we want to keep the process in control. This means that the natural variation of the process must be small (narrow) enough to produce products that meet the standards (quality) required. But a process that is in statistical control may not yield goods or services that meet their *design specifications* (tolerances). The ability of a process to meet design specifications, which are set by engineering design or customer requirements, is called **process capability**. Even though that process may be statistically in control (stable), the output of that process may not conform to specifications.

For example, let's say the time a customer expects to wait for the completion of a lube job at Quik Lube is 12 minutes, with an acceptable tolerance of ±2 minutes. This tolerance gives an upper specification of 14 minutes and a lower specification of 10 minutes. The lube process has to be capable of operating within these design specifications —if not, some customers will not have their requirements met. As a manufacturing example, the tolerances for Harley-Davidson cam gears are extremely low, only 0.0005 inches!

There are two popular measures for quantitatively determining if a process is capable: process capability ratio (C_p) and process capability index (C_{pk}).

Process Capability Ratio (C_p)

For a process to be capable, its values must fall within upper and lower specifications. This typically means the process capability is within ±3 standard deviations from the process mean. Since this range of values is 6 standard deviations, a capable process tolerance, which is the difference between the upper and lower specifications, must be greater than or equal to 6.[5]

C_p
A ratio for determining whether a process meets design specifications.

The process capability ratio, C_p, is computed as:

$$C_p = \frac{\text{Upper Specification} - \text{Lower Specification}}{6\sigma}$$

(S6-13)

A capable process has a C_p of at least 1.0. If the C_p is less than 1.0, the process yields products or services that are outside their allowable tolerance. With a C_p of 1.0, 2.7 parts in 1,000 can be expected to be "out of spec."[6] The higher the process capability ratio, the greater the likelihood the process will be within design specifications. Many firms have chosen a C_p of 1.33 as a target for reducing process variability. This means that only 64 parts per million can be expected to be out of specification.

Example S6

Active Model S6.2

Example S6 is further illustrated in Active Model S6.2 on the CD-ROM.

In a GE insurance claims process, $\bar{x} = 210.0$ minutes, and $\sigma = .516$ minutes.

The design specification to meet customer expectations is 210 ±3 minutes. So the Upper Specification is 213 minutes and the lower specification is 207 minutes.

$$C_p = \frac{\text{Upper Specification} - \text{Lower Specification}}{6\sigma} = \frac{213 - 207}{6(.516)} = 1.938$$

Since a ratio of 1.00 means that 99.73% of a process's outputs are within specifications, this ratio suggests a very capable process, with nonconformance of less than 4 claims per million.

Recall that in Chapter 6 we mentioned the concept of *six sigma* quality, championed by GE and Motorola. This standard equates to a C_p of 2.0, with only 3.4 defective parts per million (very close to zero defects) instead of the 2.7 parts per 1,000 with 3-sigma limits.

While C_p relates to the spread (dispersion) of the process output relative to its tolerance, it does not look at how well the process average is centered on the target value.

---

[5]See GE's *A Pocket Guide of Tools for Quality*, Methuen, MA (1994): 139–143.

[6]This is because a C_p of 1.0 has 99.73% of outputs within specifications. So 1.00 − .9973 = .0027; with 1,000 parts, there are .0027 × 1,000 = 2.7 defects.

For a C_p of 2.0, 99.99966% of outputs are "within spec." So 1.00 − .9999966 = .0000034; with 1 million parts, there are 3.4 defects.

Process Capability Index (C_{pk})

C_{pk}
A proportion of natural variation (3σ) between the center of the process and the nearest specification limit.

The process capability index, C_{pk}, measures the difference between the desired and actual dimensions of goods or services produced.

The formula for C_{pk} is:

$$C_{pk} = \text{minimum of} \left[\frac{\text{Upper Specification Limit} - \overline{X}}{3\sigma}, \frac{\overline{X} - \text{Lower Specification Limit}}{3\sigma} \right] \quad \text{(S6-14)}$$

where $\overline{X}$ = process mean
σ = standard deviation of the process population

When the C_{pk} index equals 1.0, the process variation is centered within the upper and lower specification limits and the process is capable of producing within ±3 standard deviations (fewer than 2,700 defects per million). A C_{pk} of 2.0 means the process is capable of producing fewer than 3.4 defects per million. Figure S6.8 shows the meaning of various measures of C_{pk}, and Example S7 shows an application of C_{pk}.

Example S7

You are the process improvement manager and have developed a new machine to cut insoles for the company's top-of-the-line running shoes. You are excited because the company's goal is no more than 3.4 defects per million and this machine may be the innovation you need. The insoles cannot be more than ±.001 of an inch from the required thickness of .250″. You want to know if you should replace the existing machine, which has a C_{pk} of 1.0. You decide to determine the C_{pk} for the new machine and make a decision on that basis.

Upper Specification Limit = .251 inches

Lower Specification Limit = .249 inches

Mean of the new process $\overline{X}$ = .250 inches.
Estimated standard deviation of the new process = σ = .0005 inches.

$$C_{pk} = \text{minimum of} \left[\frac{\text{Upper Specification Limit} - \overline{X}}{3\sigma}, \frac{\overline{X} - \text{Lower Specification Limit}}{3\sigma} \right]$$

$$C_{pk} = \text{minimum of} \left[\frac{(.251) - .250}{(3).0005}, \frac{.250 - (.249)}{(3).0005} \right]$$

Both calculations result in: $\dfrac{.001}{.0015} = 0.67$.

Because the new machine has a C_{pk} of only 0.67, the new machine should *not* replace the existing machine.

If the mean of the process is not centered on the desired (specified) mean, then the smaller numerator in Equation (S6-14) is used (the minimum of the difference between the upper specification limit and the mean or the Lower Specification Limit and the mean). This application of C_{pk} is shown in Solved Problem S6.4.

When a process is centered between the Upper Specification Limit and the Lower Specification Limit (as was the case in Example S6), the process capability ratio will be the same as the process capability index. However, the C_{pk} index measures the *actual* capability of a process, whether or not its mean is centered between the specification limits. Because in the real world, process distributions are often *not* centered, most companies use C_{pk} to express their expectations to suppliers. Cummins Engine Company, for example, initially required suppliers to use a C_{pk} above 1.33, then worked with suppliers to raise that capability to a C_{pk} above 1.67.

FIGURE S6.8 ■

Meanings of C_{pk} Measures

A C_{pk} index of 1.0 indicates that the process variation is centered within the upper and lower control limits. As the C_{pk} index goes above 1, the process becomes increasingly target-oriented with fewer defects. If the C_{pk} is less than 1.0, the process will not produce within the specified tolerance.

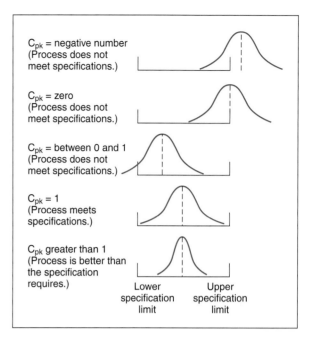

Acceptance sampling
A method of measuring random samples of lots or batches of products against predetermined standards.

Operating characteristic (OC) curve
A graph that describes how well an acceptance plan discriminates between good and bad lots.

Producer's risk
The mistake of having a producer's good lot rejected through sampling.

Consumer's risk
The mistake of a customer's acceptance of a bad lot overlooked through sampling.

ACCEPTANCE SAMPLING[7]

Acceptance sampling is a form of testing that involves taking random samples of "lots," or batches, of finished products and measuring them against predetermined standards. Sampling is more economical than 100% inspection. The quality of the sample is used to judge the quality of all items in the lot. Although both attributes and variables can be inspected by acceptance sampling, attribute inspection is more commonly used, as illustrated in this section.

Acceptance sampling can be applied either when materials arrive at a plant or at final inspection, but it is usually used to control incoming lots of purchased products. A lot of items rejected, based on an unacceptable level of defects found in the sample, can (1) be returned to the supplier or (2) be 100% inspected to cull out all defects, with the cost of this screening usually billed to the supplier. However, acceptance sampling is not a substitute for adequate process controls. In fact, the current approach is to build statistical quality controls at suppliers so that acceptance sampling can be eliminated.

Operating Characteristic Curve

The **operating characteristic (OC) curve** describes how well an acceptance plan discriminates between good and bad lots. A curve pertains to a specific plan—that is, to a combination of *n* (sample size) and *c* (acceptance level). It is intended to show the probability that the plan will accept lots of various quality levels.

With acceptance sampling, two parties are usually involved: the producer of the product and the consumer of the product. In specifying a sampling plan, each party wants to avoid costly mistakes in accepting or rejecting a lot. The producer usually has the responsibility of replacing all defects in the rejected lot or of paying for a new lot to be shipped to the customer. The producer, therefore, wants to avoid the mistake of having a good lot rejected (**producer's risk**). On the other hand, the customer or consumer wants to avoid the mistake of accepting a bad lot because defects found in a lot that has already been accepted are usually the responsibility of the customer (**consumer's risk**). The OC curve shows the features of a particular sampling plan, including the risks of making a wrong decision.[8]

---

[7]**Refer to Tutorial 2 on your CD-ROM for an extended discussion of Acceptance Sampling.**

[8]Note that sampling always runs the danger of leading to an erroneous conclusion. Let us say in this example that the total population under scrutiny is a load of 1,000 computer chips, of which in reality only 30 (or 3%) are defective. This means that we would want to accept the shipment of chips, because 4% is the allowable defect rate. However, if a random sample of $n = 50$ chips were drawn, we could conceivably end up with 0 defects and accept that shipment (that is, it is OK), or we could find all 30 defects in the sample. If the latter happened, we could wrongly conclude that the whole population was 60% defective and reject them all.

FIGURE S6.9 ■

An Operating Characteristic (OC) Curve Showing Producer's and Consumer's Risks

A good lot for this particular acceptance plan has less than or equal to 2% defectives. A bad lot has 7% or more defectives.

Active Model S6.3

Figure S6.9 is further illustrated in Active Model S6.3 on the CD.

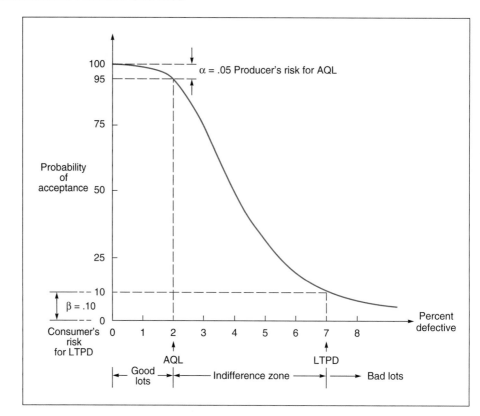

Figure S6.9 can be used to illustrate one sampling plan in more detail. Four concepts are illustrated in this figure.

The **acceptable quality level (AQL)** is the poorest level of quality that we are willing to accept. In other words, we wish to accept lots that have this or better level of quality, but no lower. If an acceptable quality level is 20 defects in a lot of 1,000 items or parts, then AQL is 20/1,000 = 2% defectives.

The **lot tolerance percent defective (LTPD)** is the quality level of a lot that we consider bad. We wish to reject lots that have this or poorer level of quality. If it is agreed that an unacceptable quality level is 70 defects in a lot of 1,000, then the LTPD is 70/1,000 = 7% defective.

To derive a sampling plan, producer and consumer must define not only "good lots" and "bad lots" through the AQL and LTPD, but they must also specify risk levels.

Producer's risk (α) is the probability that a "good" lot will be rejected. This is the risk that a random sample might result in a much higher proportion of defects than the population of all items. A lot with an acceptable quality level of AQL still has an α chance of being rejected. Sampling plans are often designed to have the producer's risk set at $\alpha = .05$, or 5%.

Acceptable quality level (AQL)
The quality level of a lot considered good.

Lot tolerance percent defective (LTPD)
The quality level of a lot considered bad.

This laser tracking device, by Faro Technologies, enables quality control workers to measure and inspect parts and tools during production. The tracker can measure objects from 100 feet away and takes up to 1,000 readings per second.

Type I error
Statistically, the probability of rejecting a good lot.

Type II error
Statistically, the probability of accepting a bad lot.

Average outgoing quality (AOQ)
The percent defective in an average lot of goods inspected through acceptance sampling.

Consumer's risk (β) is the probability that a "bad" lot will be accepted. This is the risk that a random sample might result in a lower proportion of defects than the overall population of items. A common value for consumer's risk in sampling plans is $\beta = .10$, or 10%.

The probability of rejecting a good lot is called a **type I error**. The probability of accepting a bad lot is a **type II error**.

Sampling plans and OC curves may be developed by computer (as seen in the software available with this text), by published tables such as the U.S. Military Standard MIL-STD-105 or Dodge-Romig table, or by calculation, using binomial or Poisson distributions.[9]

Average Outgoing Quality

In most sampling plans, when a lot is rejected, the entire lot is inspected and all defective items replaced. Use of this replacement technique improves the average outgoing quality in terms of percent defective. In fact, given (1) any sampling plan that replaces all defective items encountered and (2) the true incoming percent defective for the lot, it is possible to determine the **average outgoing quality (AOQ)** in percent defective. The equation for AOQ is:

$$AOQ = \frac{(P_d)(P_a)(N - n)}{N} \qquad \text{(S6-15)}$$

where
P_d = true percent defective of the lot
P_a = probability of accepting the lot
N = number of items in the lot
n = number of items in the sample

The maximum value of AOQ corresponds to the highest average percent defective or the lowest average quality for the sampling plan. It is called the *average outgoing quality limit (AOQL)*.

Acceptance sampling is useful for screening incoming lots. When the defective parts are replaced with good parts, acceptance sampling helps to increase the quality of the lots by reducing the outgoing percent defective.

Figure S6.10 compares acceptance sampling, SPC, and C_{pk}. As Figure S6.10 shows, (a) acceptance sampling by definition accepts some bad units, (b) control charts try to keep the process in control, but (c) the C_{pk} index places the focus on improving the process. As operations managers, that is what we want to do—improve the process.

FIGURE S6.10 ■

The Application of Statistical Process Techniques Contributes to the Identification and Systematic Reduction of Process Variability

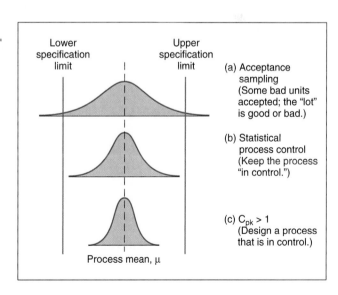

[9]The two most frequently used tables for acceptance plans are *Military Standard Sampling Procedures and Tables for Inspection by Attributes* (MIL-STD-105D) (Washington, DC: U.S. Government Printing Office, 1963); and H. F. Dodge and H. G. Romig, *Sampling Inspection Tables—Single and Double Sampling*, 2nd ed. (New York: John Wiley, 1959).

SUMMARY

Statistical process control is a major statistical tool of quality control. Control charts for SPC help operations managers distinguish between natural and assignable variations. The $\bar{x}$-chart and the R-chart are used for variable sampling, and the p-chart and the c-chart for attribute sampling. The C_{pk} index is a way to express process capability. Operating characteristic (OC) curves facilitate acceptance sampling and provide the manager with tools to evaluate the quality of a production run or shipment.

KEY TERMS

Statistical process control (SPC) *(p. 174)*
Control chart *(p. 174)*
Natural variations *(p. 175)*
Assignable variation *(p. 175)*
$\bar{x}$-chart *(p. 176)*
R-chart *(p. 176)*
Central limit theorem *(p. 177)*
p-charts *(p. 182)*
c-charts *(p. 184)*
Run test *(p. 186)*
Process capability *(p. 187)*

C_p *(p. 187)*
C_{pk} *(p. 188)*
Acceptance sampling *(p. 189)*
Operating characteristic (OC) curve *(p. 189)*
Producer's risk *(p. 189)*
Consumer's risk *(p. 189)*
Acceptable quality level (AQL) *(p. 190)*
Lot tolerance percent defective (LTPD) *(p. 190)*
Type I error *(p. 191)*
Type II error *(p. 191)*
Average outgoing quality (AOQ) *(p. 191)*

USING EXCEL OM FOR SPC

Excel and other spreadsheets are extensively used in industry to maintain control charts. Excel OM's Quality Control module has the ability to develop $\bar{x}$-charts, p-charts, and c-charts. Program S6.1 illustrates Excel OM's spreadsheet approach to computing the $\bar{x}$ control limits for the Oat Flakes company in Example S1. Excel also contains a built-in graphing ability with Chart Wizard.

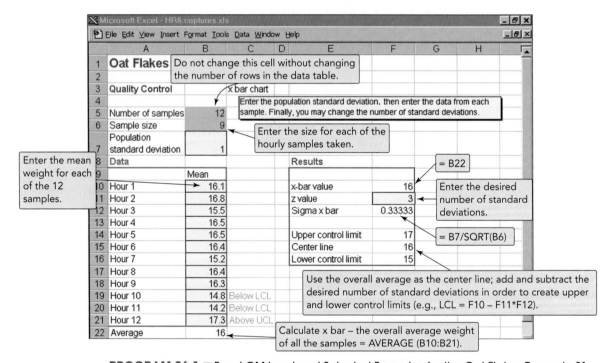

PROGRAM S6.1 ■ Excel OM Input and Selected Formulas for the Oat Flakes Example S1

 # USING POM FOR WINDOWS

POM for Windows' Quality Control module has the ability to compute all of the SPC control charts we introduced in this supplement. It also handles acceptance sampling and operating characteristic curves. See Appendix V for further details.

SOLVED PROBLEMS

Solved Problem S6.1

A manufacturer of precision machine parts produces round shafts for use in the construction of drill presses. The average diameter of a shaft is .56 inch. Inspection samples contain 6 shafts each. The average range of these samples is .006 inch. Determine the upper and lower $\bar{x}$ control chart limits.

SOLUTION

The mean factor A_2 from Table S6.1 where the sample size is 6, is seen to be .483. With this factor, you can obtain the upper and lower control limits:

$$UCL_{\bar{x}} = .56 + (.483)(.006)$$
$$= .56 + .0029$$
$$= .5629 \text{ inches}$$
$$LCL_{\bar{x}} = .56 - .0029$$
$$= .5571 \text{ inches}$$

Solved Problem S6.2

Nocaf Drinks, Inc., a producer of decaffeinated coffee, bottles Nocaf. Each bottle should have a net weight of 4 ounces. The machine that fills the bottles with coffee is new, and the operations manager wants to make sure that it is properly adjusted. Bonnie Crutcher, the operations manager, takes a sample of $n = 8$ bottles and records the average and range in ounces for each sample. The data for several samples is given in the following table. Note that every sample consists of 8 bottles.

SAMPLE	SAMPLE RANGE	SAMPLE AVERAGE	SAMPLE	SAMPLE RANGE	SAMPLE AVERAGE
A	.41	4.00	E	.56	4.17
B	.55	4.16	F	.62	3.93
C	.44	3.99	G	.54	3.98
D	.48	4.00	H	.44	4.01

Is the machine properly adjusted and in control?

SOLUTION

We first find that $\bar{\bar{x}} = 4.03$ and $\bar{R} = .51$. Then, using Table S6.1, we find:

$$UCL_{\bar{x}} = \bar{\bar{x}} + A_2\bar{R} = 4.03 + (.373)(.51) = 4.22$$
$$LCL_{\bar{x}} = \bar{\bar{x}} - A_2\bar{R} = 4.03 - (.373)(.51) = 3.84$$
$$UCL_R = D_4\bar{R} = (1.864)(.51) = .95$$
$$LCL_R = D_3\bar{R} = (.136)(.51) = .07$$

It appears that the process average and range are both in control.

Solved Problem S6.3

Altman Distributors, Inc., fills catalog orders. Among the last 100 orders shipped, the percent of errors was .05. Determine the upper and lower limits for this process for 99.73% confidence.

SOLUTION

$$UCL_p = \bar{p} + 3\sqrt{\frac{\bar{p}(1-\bar{p})}{n}} = .05 + 3\sqrt{\frac{(.05)(1-.05)}{100}}$$
$$= .05 + 3(0.0218) = .1154$$

$$LCL_p = \bar{p} - 3\sqrt{\frac{\bar{p}(1-\bar{p})}{n}} = .05 - 3(.0218)$$
$$= .05 - .0654 = 0 \text{ (because percent defective cannot be negative)}$$

Solved Problem S6.4

Ettlie Engineering has a new catalyst injection system for your countertop production line. Your process engineering department has conducted experiments and determined that the mean is 8.01 grams with a standard deviation of .03. Your specifications are:

$\mu = 8.0$ and $\sigma = .04$, which means an upper specification limit of 8.12 [$= 8.0 + 3(.04)$]
and a lower specification limit of 7.88 [$= 8.0 - 3(.04)$].

What is the C_{pk} performance of the injection system? Using our formula:

$$C_{pk} = \text{minimum of} \left[\frac{\text{Upper Specification Limit} - \overline{X}}{3\sigma}, \frac{\overline{X} - \text{Lower Specification Limit}}{3\sigma} \right]$$

where $\overline{X}$ = process mean
          σ = standard deviation of the process population

$$C_{pk} = \text{minimum of} \left[\frac{8.12 - 8.01}{(3)(.03)}, \frac{8.01 - 7.88}{(3)(.03)} \right]$$

$$\left[\frac{.11}{.09} = 1.22, \frac{.13}{.09} = 1.44 \right]$$

The minimum is 1.22, so the C_{pk} of 1.22 is within specifications, and has an implied error rate of less than 2,700 defects per million.

INTERNET AND STUDENT CD-ROM EXERCISES

Visit our home page or use your student CD-ROM to help with material in this supplement.

 On Our Home Page, www.prenhall.com/heizer

- Self-Tests
- Practice Problems
- Internet Exercises
- Curent Articles and Research
- Virtual Company Tour
- Internet Homework Problems
- Internet Case Study

 On Your Student CD-ROM

- PowerPoint Lecture
- Practice Problems
- Video Clips
- Active Model Exercise
- Excel OM Software
- Excel OM Data Files

ADDITIONAL CASE STUDIES

Internet Case Study: Visit our Web site at www.prenhall.com/heizer **for this free case study:**

- **Green River Chemical Company**: Involves a company that needs to set up a control chart to monitor sulfate content because of customer complaints.

Harvard has selected these Harvard Business School cases to accompany this supplement (textbookcasematch.hbsp.harvard.edu)**:**

- **Deutsche Allgemeinversicherung** (#696–084): A German insurance company tries to adopt *p*-charts to a variety of services it performs.

- **Process Control at Polaroid (A)** (#696–047): This film-production plant moves from traditional QC inspection to worker-based SPC charts.

Process Strategy

Chapter Outline

LEARNING OBJECTIVES

When you complete this chapter you should be able to

IDENTIFY OR DEFINE:

Process focus

Repetitive focus

Product focus

Process reengineering

Service process issues

Environmental issues

DESCRIBE OR EXPLAIN:

Process analysis

Service design

Green manufacturing

Production technology

Mass Customization Provides Dell Computer's Competitive Advantage

Dell Computer started with a single premise. "How can we make the process of buying a computer better?" The answer that its founder, multibillionaire Michael Dell, developed was to undercut other suppliers by selling directly to end customers, thus eliminating a distribution chain whose markup accounted for a high percentage of a PC's price. Dell's slick process has made the company a role model for the computer industry and has allowed Dell to grab first place in sales.

It was no surprise when Michael Dell's $20-billion company reached the number 1 ranking. Dell was only 8 years old when he responded to a magazine ad that promised a fast track to getting a high school diploma. When he founded Dell Computer, at age 19, from his college dorm at the University of Texas, he dreamed of competing

Michael Dell built his first computer in 1983, eight years after the PC was invented. As a freshman at the University of Texas, he built computers in his dorm room.

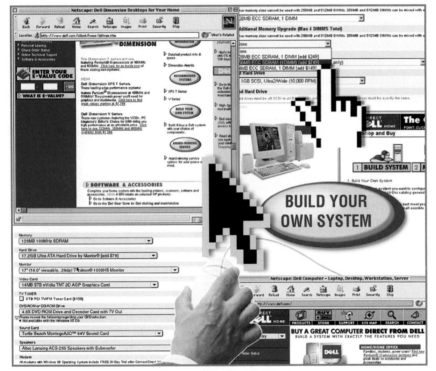

Dell computers are sold over the Internet, then produced and shipped directly to the individual customer. No inventories are kept. Mass customization allows models to change continuously as new technologies become available.

with IBM. Dell bypassed IBM in PC sales in 1999.

Now Dell has embraced the Internet. No comparably sized U.S. company has been as successful in turning the Internet into an everyday tool to enhance productivity. Dell has integrated the Web into every aspect of its business—design, production, sales, and service. Despite a long and diverse global supply chain, Dell operates with just 6 days of inventory, a fraction of its competitors.

Dell has also set standards for quick delivery and mass customization. Dell builds computers rapidly, at low cost, and only when ordered. This process has prevented one of the major prob-

DELL COMPUTER CORP.

Kits of components are made up for each customer. Parts are then delivered as needed, and the final product is assembled by highly trained generalists who put together the entire computer.

lems in the fast changing PC market—outdated PCs. In addition, Dell has gone further to trim inventory. For instance, it takes delivery of components just *minutes* before they are needed. At its new factory in Austin, Texas, a Dell PC can be built, software installed, tested, and packed in 8 hours, down from 10 previously.

How does Dell pull off mass customization? One reason is that instead of investing resources in developing computer parts (like many competitors), it focuses most of its research and development (R&D) on software designed to make the installation and configuration of its PCs fast and simple. Dell's speed impresses many large multinationals who use it as their de facto supplier. After Dell built and shipped 3,700 PCs in 11 days to Delta's reservation centers, Delta executives flew to Austin to throw a party for the factory workers.

Although more than 90% of Dell's personal computer business is custom ordered, the throughput time for each machine is less than 8 hours.

Process strategy
An organization's approach to transform resources into goods and services.

Process focus
A production facility organized around processes to facilitate low-volume, high-variety production.

In Chapter 5, we examined the need for the selection, definition, and design of goods and services. We now turn to their production. A major decision for the operations manager is finding the best way to produce. Let's look at ways to help managers design a process for achieving this goal.

A **process** (or transformation) **strategy** is an organization's approach to transform resources into goods and services. The *objective of a process strategy* is to find a way to produce goods and services that meet customer requirements and product specifications within cost and other managerial constraints. The process selected will have a long-term effect on efficiency and production, as well as the flexibility, cost, and quality of the goods produced. Therefore, much of a firm's strategy is determined at the time of this process decision.

FOUR PROCESS STRATEGIES

Virtually every good or service is made by using some variation of one of four process strategies: (1) process focus, (2) repetitive focus, (3) product focus, and (4) mass customization. Notice the relationship of these four strategies to volume and variety shown in Figure 7.1. Although the figure shows only four strategies, an innovative operations manager can build processes anywhere in the matrix to meet the necessary volume and variety requirements.

Let's look at each of these strategies with an example and a flow diagram. We examine *Standard Register* as a process-focused firm, *Harley-Davidson* as a repetitive producer, *Nucor Steel* as a product-focused operation, and *Dell* as a mass customizer.

Process Focus

Seventy-five percent of all global production is devoted to making *low-volume, high-variety* products in places called "job shops." Such facilities are organized around specific activities or processes. In a factory, these processes might be departments devoted to welding, grinding, and painting. In an office, the processes might be accounts payable, sales, and payroll. In a restaurant, they might be bar, grill, and bakery. Such facilities are **process focused** in terms of equipment, layout, and supervision. They provide a high degree of product flexibility as products move intermittently between processes. Each process is designed to perform a wide variety of activities and handle frequent changes. Consequently, they are also called *intermittent processes*.

These facilities have high variable costs with extremely low utilization of facilities, as low as 5%. This is the case for many restaurants, hospitals, and machine shops. However, some facilities do a little better through the use of innovative equipment, often with electronic controls. With the development of computer numerical-controlled equipment (machines controlled by computer software), it is possible to program machine tools, piece movement, and tool changing, and even to automate placement of the parts on the machine and the movement of materials between machines.

FIGURE 7.1 ■

Process Selected Must Fit with Volume and Variety

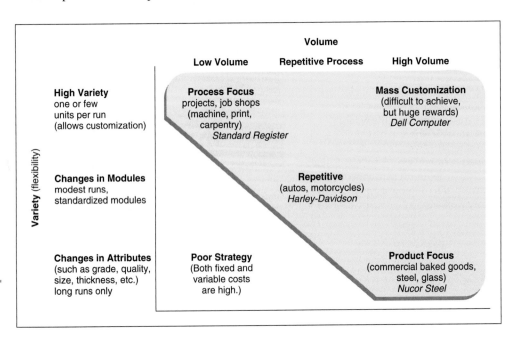

Example 1 shows how Standard Register, a billion-dollar printer and document processor headquartered in Dayton, Ohio, produces paper business forms.

Example 1

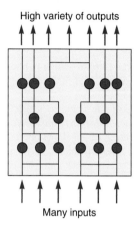

**Process-Focused
(intermittent process)**

High variety of outputs

Many inputs

Job Shop Process Focus at Standard Register

If you've had a pizza delivered to your home recently, there is a good chance that Standard Register printed the order and delivery tag on the box. You probably came in contact with one of Standard's forms this week without knowing it. Thousands of different products are made by the firm, a typical one being a multisheet (3- or 4-layer) business form. Forms used for student college applications, hospital patient admissions, bank drafts, store orders, and job applications are examples. The company has 11 U.S. plants in its Forms Division.

Figure 7.2 is a flow diagram of the entire production process, from order submission to shipment, at Standard's Kirksville, Missouri, plant. This job shop groups people and machines that perform specific activities, such as printing, cutting, or binding, into departments. Entire orders are processed in batches, moving from department to department, rather than in a continuous flow or one at a time.

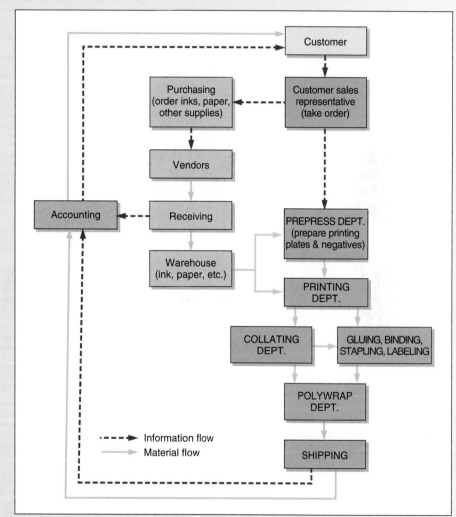

FIGURE 7.2 ■

Flow Diagram of the Production Process at Standard Register's Plant in Kirksville, Missouri

Source: Adapted with permission from J. S. Martinich, *Production and Operations Management* (New York: John Wiley, 1997): 79–87.

The process begins with a sales representative helping the customer design the business form. Once the form is established, the order is transmitted electronically to the Sales Support Department at the manufacturing plant. An order coordinator determines what materials will be needed in production (ink, paper, labels, etc.), computes the production *time* needed, and schedules the job on a particular machine.

The Prepress Department uses computer-aided design (CAD) to convert the product design into printing plates for the presses and then "burns" the image of the form onto an aluminum printing plate. Machine operators in the Printing Department install the plates and inks on their presses and print the forms. After leaving the presses, most products are collated on a machine that places up to 14 copies together, possibly with carbon paper between them. Some products undergo additional processing (for example, gluing, binding, stapling, or labeling). When the forms are completed, most are wrapped in polyethylene before being placed in cartons for shipping. The order is shipped, a "job ticket" is sent to Accounting, and an invoice goes to the customer.

Repetitive Focus

A repetitive process falls between the product and process focuses seen in Figure 7.1. Repetitive processes use modules. Modules are parts or components previously prepared, often in a continuous process.

The **repetitive process** line is the classic assembly line. Widely used in the assembly of virtually all automobiles and household appliances, it has more structure and consequently less flexibility than a process-focused facility.

Fast-food firms are an example of a repetitive process using **modules**. This type of production allows more customizing than a continuous process; modules (for example, meat, cheese, sauce, tomatoes, onions) are assembled to get a quasi-custom product, a cheeseburger. In this manner, the firm obtains both the economic advantages of the continuous model (where many of the modules are prepared) and the custom advantage of the low-volume, high-variety model.

Example 2 shows the Harley-Davidson assembly line. Harley is a repetitive manufacturer located toward the center of Figure 7.1.

Repetitive Manufacturing at Harley-Davidson

Harley-Davidson assembles modules. Most repetitive manufacturers produce on a form of assembly line where the end product can take a variety of shapes depending on the mix of modules. This is the case at Harley, where the modules are motorcycle components and options.

Harley engines are produced in Milwaukee and shipped on a just-in-time basis to the company's York, Pennsylvania, plant. At York, Harley groups parts that require similar processes together into families (see the flow diagram in Figure 7.3). The result is *work cells*. Work cells perform in one location all the operations necessary for the production of specific modules. These work cells feed the assembly line.

Harley-Davidson assembles 2 engine types in 3 displacement sizes for 20 street bike models, which are available in 13 colors and 2 wheel options adding up to 95 total combinations. Harley also produces 4 police and 2 Shriner motorcycles, and offers many custom paint options. This strategy requires that no fewer than 20,000 different pieces be assembled into modules and then into motorcycles.

Repetitive process
A product-oriented production process that uses modules.

Modules
Parts or components of a product previously prepared, often in a continuous process.

Example 2

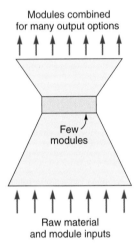

Repetitive Focus

Modules combined for many output options

Few modules

Raw material and module inputs

Video 7.1

Saturn Auto's Mass Production

FIGURE 7.3 ■

Flow Diagram Showing the Production Process at Harley-Davidson's York, Pennsylvania, Assembly Plant

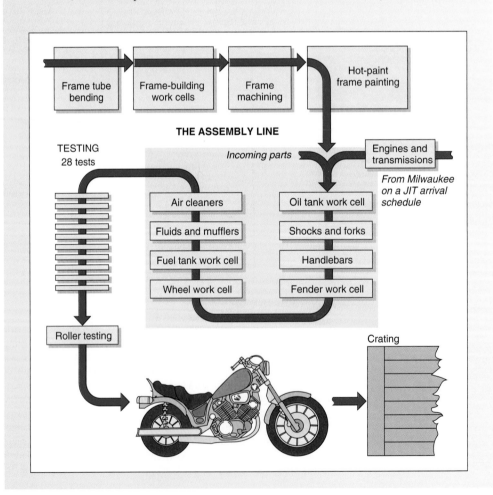

Product Focus

Product focus

A facility organized around products; a product-oriented, high-volume, low-variety process.

High-volume, low-variety processes are **product-focused**. The facilities are organized around *products*. They are also called *continuous processes*, because they have very long, continuous production runs. Products such as glass, paper, tin sheets, lightbulbs, beer, and bolts are made via a continuous process. Some products, such as lightbulbs, are discrete; others, such as rolls of paper, are nondiscrete. Still others, such as repaired hernias at Shouldice Hospital, are services. It is only with standardization and effective quality control that firms have established product-focused facilities. An organization producing the same lightbulb or hot dog bun day after day can organize around a product. Such an organization has an inherent ability to set standards and maintain a given quality, as opposed to an organization that is producing unique products every day, such as a print shop or general-purpose hospital.

A product-focused facility produces high volume and low variety. The specialized nature of the facility requires high fixed cost, but low variable costs reward high facility utilization. The Nucor example follows.

Example 3

Product-Focused (continuous process)

Output variations in size, shape, and packaging

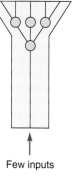

Few inputs

Video 7.2

Wassau Paper's Continuous Work Flow

FIGURE 7.4 ■

A Flow Diagram Showing the Steelmaking Process at Nucor's Crawfordsville, Indiana, Plant

Product-Focused Production at Nucor Steel

Steel is manufactured in a product-oriented facility. Figure 7.4 illustrates Nucor's product-focused flow.

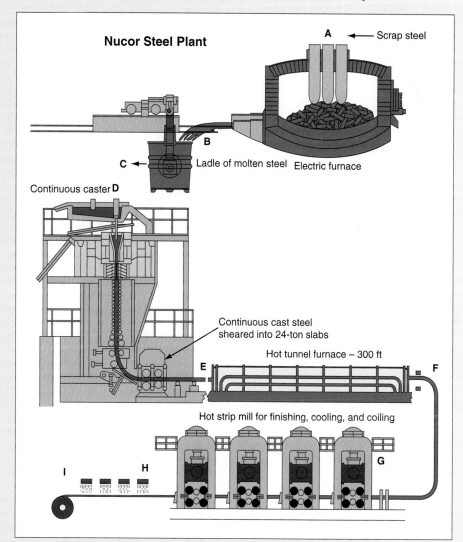

In this process flow diagram, cold scrap steel is first lowered into a furnace that uses an electric arc to melt the steel in 20 seconds (A). Then molten steel pours from the furnace into a preheated ladle (B). The ladle moves on an overhead-track crane to the continuous caster (C). The ladle then opens and steel exits into the caster (D). Shaped steel exits the caster mold as a 2″ × 52″ slab (E). The slab exits the tunnel furnace (F) at a specific temperature needed for rolling. A higher-quality sheet can be produced if the slab temperature is uniform. The steel then enters the rolling mill (G). Water cools the hot-rolled steel before it is coiled (H). The rolled sheet of steel is coiled into rolls of about 25 tons each (I). Finally, a variety of finishing operations can modify the characteristics of the sheet steel to meet customer needs.

Nucor operates 24 hours a day, 6 days a week, with the seventh day reserved for scheduled maintenance.

Mass Customization Focus

Our increasingly wealthy and sophisticated world demands individualized goods and services. Table 7.1 shows the rich variety of goods and services that operations managers are called on to supply. The explosion of variety has taken place in automobiles, movies, breakfast cereals, and a thousand other areas. In spite of this proliferation of products, quality has improved and costs have dropped. Consequently, this wealth of products is available to more people than ever. Operations managers have produced this selection of goods and services through what is known as mass customization. But mass customization is not just about variety, it is about economically making precisely *what* the customer wants *when* the customer wants it.

Mass customization is rapid, low-cost production of goods and services that fulfill increasingly unique customer desires. Mass customization brings us the variety of products traditionally provided by low-volume manufacture (a process focus) at the cost of standardized high-volume (product-focused) production. However, as shown in the upper right section of Figure 7.1, producing to achieve mass customization is a challenge requiring enhanced operational capabilities. The link between sales and production and logistics is much tighter.[1] Operations managers must make imaginative and aggressive use of organizational resources to build agile processes that rapidly and inexpensively produce custom products.

Dell Computer, the focus of the *Global Company Profile* that began this chapter, has demonstrated that the payoff for mass customization can be substantial. More traditional manufacturers include General Motors, which builds six different styles on its Fairfax, Kansas, assembly line. GM adjusts robot welders and other equipment electronically as different models come down the assembly line. Moreover, GM's Cadillac division is now custom manufacturing cars with a 10-day lead time. Not to be outdone, Toyota recently announced delivery of custom-ordered cars in 5 days. Similarly, electronic controls allow designers in the textile industry to rapidly revamp their lines and respond to changes.

The service industry is also moving toward mass customization. For instance, not very many years ago most people had the same telephone service. Now phone service is full of options, like caller ID, call waiting, voice mailboxes, and call forwarding for your specific needs. Insurance companies are adding and tailoring new products with shortened development times to meet the unique needs of their customers. And emusic of California maintains a music sound bite inventory on the

Mass customization

Rapid, low-cost production that caters to constantly changing unique customer desires.

TABLE 7.1 ■

Mass Customization Provides More Choices than Ever[a]

Source: Various; however, much of the data are from the Federal Reserve Bank of Dallas.

ITEM	NUMBER OF CHOICES	
	EARLY 1970S	EARLY 21ST CENTURY
Vehicle models	140	260
Vehicle styles	18	1,212
Bicycle types	8	19
Software titles	0	300,000
Web sites	0	30,727,296[c]
Movie releases	267	458
New book titles	40,530	77,446
Houston TV channels	5	185
Breakfast cereals	160	340
Items (SKUs) in supermarkets	14,000[b]	150,000[d]

[a]Variety available in America; worldwide the variety increases even more.

[b]1989

[c]2002 Net Names International Ltd.

[d]SKUs managed by H. C. Butts grocery chain.

[1]Paul Zipkin, "The Limits of Mass Customization," *MIT Sloan Management Review* (spring 2001): p. 81.

OM IN ACTION

Mass Customization at Borders Books and at Smooth FM Radio

So you want a hard-to-get, high-quality paperback book in 15 minutes? Borders can take care of you—even if you want a book that the store does not carry or have in stock. First, a Borders employee checks the digital database of titles that have been licensed from publishers. If the title is available, a digital file of the book is downloaded to two printers from a central server in Atlanta. One printer makes the book cover and the other the pages. Then the employee puts the two pieces together in a bookbinding machine. A separate machine cuts the book to size. And your book is ready. You get the book you want now, and Borders gets a sale. Books sold this way also avoid both inventory and incoming shipping cost, as well as the cost of returning books that do not sell.

Smooth FM provides a "customized" radio broadcast for Houston, Boston, Milwaukee, Albany, and Jacksonville from its midtown Manhattan station. Here is how it works. During Smooth FM's 40-minute music blocks, an announcer in Manhattan busily records 30-second blocks of local weather and traffic, commercials, promotions, and 5-second station IDs. Then the recorded material is transmitted to the affiliate stations. When the music block is over, the Manhattan announcer hits a button that signals computers at all of the affiliates to simultaneously air the prerecorded "local" segments. Any "national" news or "national" ads can also be added from Manhattan. The result is the economy of mass production *and* a customized product for the local market. Radio people call it "local customization."

Sources: The *Wall Street Journal* (June 1, 1999): B1, B4, and (July 17, 2000): R44; *Computer Networks* (June 2000): 609; and *Computerworld* (June 7, 1999): 6.

Internet that allows customers to select a dozen songs of their choosing and have them made into a custom CD shipped to their door.[2] Similarly, the increasing number of new books and movies each year places demands on operations managers who must build the processes that provide this expanding array of goods and services.

One of the essential ingredients in mass customization is a reliance on modular design. In all of the examples cited, as well as those in the *OM in Action* box, "Mass Customization at Borders Books and at Smooth FM Radio," modular design is the key. However, as Figure 7.5 shows, very effective scheduling and rapid throughput is also required. These three items—imaginative modules, powerful scheduling, and rapid throughput—influence all 10 of the OM decisions, and therefore require excellent operations management. For instance, when mass customization is done well, organizations eliminate the guesswork that comes with sales forecasting and then build to order. This drives down inventories, but increases pressure on scheduling and supply-chain performance. Mass customization is tough, but good organizations are heading there.

FIGURE 7.5 ■

Operations Managers Use Imaginative Modularization, Powerful Scheduling, and Rapid Throughput to Achieve Mass Customization

Video 7.3

Process Strategy at Wheeled Coach Ambulance

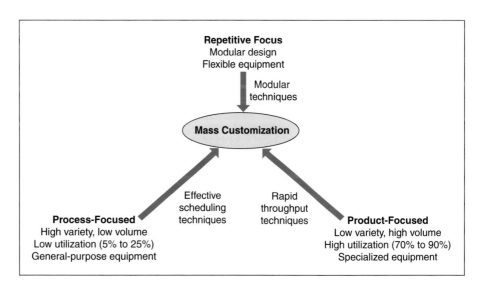

[2]www.emusic.com

Manufactured housing is now 32% of all new homes sold in the U.S. This industry has increased sales as it reduced costs. It did so as it moved production from a process focus to a repetitive focus.

Comparison of Process Choices

The characteristics of the four processes are shown in Table 7.2 and Figure 7.5. Advantages exist across the continuum of processes, and firms may find strategic advantage in any process. Each of the processes, when properly matched to volume and variety, can produce a low-cost advantage. For instance, unit costs will be less in the continuous process case if high volume (and high utilization) exists. However, we do not always use the continuous process (that is, specialized equipment and facilities) because it is too expensive when volumes are low or flexibility is required. A low-volume, unique, highly differentiated good or service is more economical when produced under process focus; this is the way fine-dining restaurants and general purpose hospitals are organized. Just as all four processes, when appropriately selected and well managed, can yield low cost, so too can all four be responsive and produce differentiated products.

Figure 7.5 indicates that equipment utilization in a process-focused facility is often in the range of 5% to 25%. When utilization goes above 15%, moving toward a repetitive or product focus, or

TABLE 7.2 ■ Comparison of the Characteristics of Four Types of Processes

PROCESS FOCUS (LOW VOLUME, HIGH VARIETY) (E.G., STANDARD REGISTER)	REPETITIVE FOCUS (MODULAR) (E.G., HARLEY-DAVIDSON)	PRODUCT FOCUS (HIGH VOLUME, LOW VARIETY) (E.G., NUCOR STEEL)	MASS CUSTOMIZATION (HIGH VOLUME, HIGH VARIETY) (E.G., DELL COMPUTER)
1. Small quantity and large variety of products are produced.	1. Long runs, usually a standardized product with options, are produced from modules.	1. Large quantity and small variety of products are produced.	1. Large quantity and large variety of products are produced.
2. Equipment used is general purpose.	2. Special equipment aids in use of an assembly line.	2. Equipment used is special purpose.	2. Rapid changeover on flexible equipment.
3. Operators are broadly skilled.	3. Employees are modestly trained.	3. Operators are less broadly skilled.	3. Flexible operators are trained for the necessary customization.
4. There are many job instructions because each job changes.	4. Repetitive operations reduce training and changes in job instructions.	4. Work orders and job instructions are few because they are standardized.	4. Custom orders require many job instructions.
5. Raw material inventories are high relative to the value of the product.	5. Just-in-time procurement techniques are used.	5. Raw material inventories are low relative to the value of the product.	5. Raw material inventories are low relative to the value of the product.
6. Work-in-process is high compared to output.	6. Just-in-time inventory techniques are used.	6. Work-in-process inventory is low compared to output.	6. Work-in-process inventory is driven down by JIT, kanban, lean production.
7. Units move slowly through the plant.	7. Movement is measured in hours and days.	7. Swift movement of units through the facility is typical.	7. Goods move swiftly through the facility.
8. Finished goods are usually made to order and not stored.	8. Finished goods are made to frequent forecasts.	8. Finished goods are usually made to a forecast and stored.	8. Finished goods are often made to order.
9. Scheduling orders is complex and concerned with the trade-off between inventory availability, capacity, and customer service.	9. Scheduling is based on building various models from a variety of modules to forecasts.	9. Scheduling is relatively simple and concerned with establishing a rate of output sufficient to meet sales forecasts.	9. Sophisticated scheduling is required to accommodate custom orders.
10. Fixed costs tend to be low and variable costs high.	10. Fixed costs are dependent on flexibility of the facility.	10. Fixed costs tend to be high and variable costs low.	10. Fixed costs tend to be high, but variable costs must be low.
11. Costing, often done by the job, is estimated prior to doing the job, but known only after the job.	11. Costs are usually known because of extensive prior experience.	11. Because fixed costs are high, costs are highly dependent on utilization of capacity.	11. High fixed costs and dynamic variable costs make costing a challenge.

even mass customization, may be advantageous. A cost advantage usually exists by improving utilization, provided the necessary flexibility is maintained. McDonald's started an entirely new industry by moving from process focus to repetitive focus. McDonald's is now trying to add more variety by moving toward mass customization (see the *Global Company Profile* that opens Chapter 9).

Much of what is produced in the world is still produced in very small lots—often as small as one. This is true for legal services, medical services, dental services, and restaurants. An X-ray machine in a dentist's office and much of the equipment in a fine-dining restaurant have low utilization. Hospitals, too, can be expected to be in that range, which would suggest why their costs are considered high. Why such low utilization? In part because excess capacity for peak loads is desirable. Hospital administrators, as well as managers of other service facilities and their patients and customers, expect equipment to be available as needed. Another reason is poor scheduling (although substantial efforts have been made to forecast demand in the service industry) and the resulting imbalance in the use of facilities.

Crossover chart

A chart of costs at the possible volumes for more than one process.

Crossover Charts   The comparison of processes can be further enhanced by looking at the point where the total cost of the processes changes. For instance, Figure 7.6 shows three alternative processes compared on a single chart. Such a chart is sometimes called a **crossover chart**. Process A has the lowest cost for volumes below V_1, process B has the lowest cost between V_1 and V_2, and process C has the lowest cost at volumes above V_2.

Example 4 illustrates how to determine the exact volume where one process become more expensive than another.

FIGURE 7.6 ■

Crossover Charts

Three different processes can be expected to have three different costs. However, at any given volume, only one will have the lowest cost.

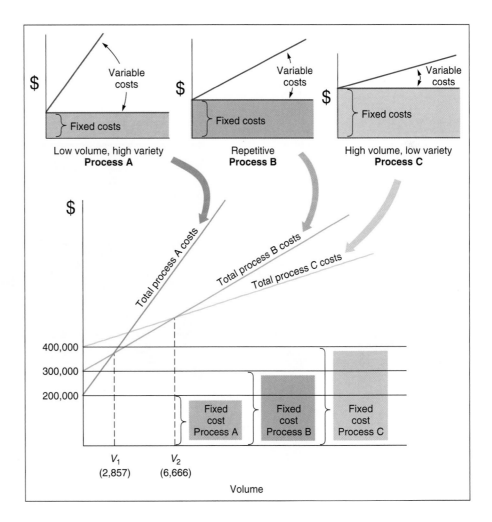

Example 4

Active Model 7.1

Example 4 is further illustrated in Active Model 7.1 on the CD-ROM and in the Exercise located in your Student Lecture Guide.

Kleber Enterprises is evaluating three accounting software products (A, B, and C) to support changes in its internal accounting processes. The resulting processes will have cost structures similar to those shown in Figure 7.6. The costs of the software for these processes are:

	TOTAL FIXED COST	DOLLARS REQUIRED PER ACCOUNTING REPORT
SOFTWARE A	$200,000	$60
SOFTWARE B	$300,000	$25
SOFTWARE C	$400,000	$10

Software A yields a process that is most economical up to V_1, but to exactly what number of reports (volume)? To determine the volume at V_1, we set the cost of software A equal to the cost of software B. V_1 is the unknown volume.

$$200,000 + (60)\ V_1 = 300,000 + (25)\ V_1$$
$$35\ V_1 = 100,000$$
$$V_1 = 2,857$$

This means that software A is most economical from 0 reports to 2,857 reports (V_1).

Similarly, to determine the crossover point for V_2, we set the cost of software B equal to the cost of software C.

$$300,000 + (25)\ V_2 = 400,000 + (10)\ V_2$$
$$15\ V_2 = 100,000$$
$$V_2 = 6,666$$

This means that software B is most economical if the number of reports is between 2,857 (V_1) and 6,666 (V_2), and that software C is most economical if reports exceed 6,666 (V_2).

As you can see, the software chosen is highly dependent on the forecasted volume.

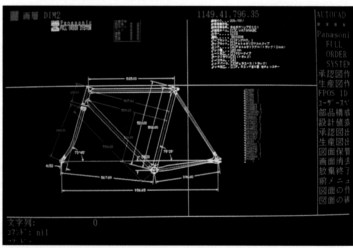

Mass customization improves customer service and provides competitive advantage. National Bicycle's customized bicycle production process begins by defining individual customer needs. The customer mounts the special frame in a bicycle store from which measurements are taken. These custom measurements are then sent to the factory, where CAD software produces a blueprint in about 3 minutes. At the same time, a bar-code label is prepared that will identify bicycle components as they move through production. Time—from beginning to end—is only 3 hours.

Agile organizations are quick and flexible in their response to ever changing customer requirements.

Changing Processes   Changing the production system from one process model to another is difficult and expensive. In some cases, the change may mean starting over. Consider what would be required of a rather simple change—McDonald's adding the flexibility necessary to serve you a charbroiled hamburger. What appears to be rather straightforward will require changes in many of our 10 OM decisions. For instance, changes may be necessary in (1) purchasing (a different quality of meat, perhaps with more fat content, and supplies such as charcoal); (2) quality standards (how long and at what temperature the patty will cook); (3) equipment (the charbroiler); (4) layout (space for the new process and for new exhaust vents); and (5) training. So choosing where to operate on the process strategy continuum may determine the transformation strategy for an extended period. This critical decision must be done right the first time.

PROCESS ANALYSIS AND DESIGN

When analyzing and designing processes to transform resources into goods and services, we ask questions such as:

Each step of your process must add value.

- Is the process designed to achieve competitive advantage in terms of differentiation, response, or low cost?
- Does the process eliminate steps that do not add value?
- Does the process maximize customer value as perceived by the customer?
- Will the process win orders?

A number of tools help us understand the complexities of process design and redesign. They are simply ways of making sense of what happens or must happen in a process. Let's look at four of them: flow diagrams, time-function mapping, process charts, and service blueprinting.

Flow Diagrams

Flow diagram
A drawing used to analyze movement of people or material.

The first tool is the **flow diagram**, which is a schematic or drawing of the movement of material, product, or people. For instance, Figures 7.2, 7.3, and 7.4 showed the processes for Standard Register, Harley-Davidson, and Nucor Steel, respectively. Such diagrams can help understanding, analysis, and communication of a process.

Time-Function Mapping

Time-function mapping (or process mapping)
A flow diagram but with time added on the horizontal axis.

A second tool for process analysis and design is a flow diagram, but with time added on the horizontal axis. Such charts are sometimes called **time-function mapping** or **process mapping**. With time-function mapping, nodes indicate the activities and the arrows indicate the flow direction, with time on the horizontal axis. This type of analysis allows users to identify and eliminate waste such as extra steps, duplication, and delay. Figure 7.7(a, b) shows the use of process mapping before and after process improvement at American National Can Company. In this example, substantial reduction in waiting time and process improvement in order processing contributed to a savings of 46 days.

Process Charts

Process charts
Charts using symbols to analyze the movement of people or material.

The third tool is the *process chart*. **Process charts** use symbols, time, and distance to provide an objective and structured way to analyze and record the activities that make up a process.[3] They allow us to focus on value-added activities. For instance, the process chart shown in Figure 7.8, which includes the present method of hamburger assembly at a fast-food restaurant, includes a value-added line to help us distinguish between value-added activities and waste. Identifying all value-added operations (as opposed to inspection, storage, delay, and transportation, which add no value) allows us to determine the percent of value added to total activities.[4] We can see from the computation at the bottom of Figure 7.8 that the value added in this case is 85.7%. The operations manager's job is to reduce waste and increase the percent of value added. The nonvalue-added items are a waste; they are resources lost to the firm and to society forever.

[3] An additional example of a process chart is shown in Chapter 10, "Human Resources and Job Design."

[4] Waste includes: *inspection* (if the task was done properly, then inspection is unnecessary); *transportation* (movement of material within a process may be a necessary evil, but it adds no value); *delay* (an asset sitting idle and taking up space is waste); *storage* (unless part of a "curing" process, storage is waste).

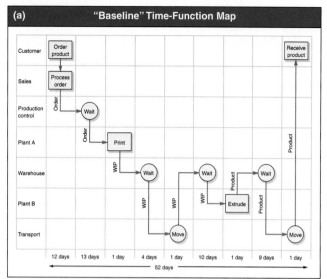
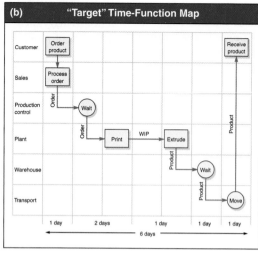

FIGURE 7.7 ■ Time-Function Mapping (Process Mapping) for a Product Requiring Printing and Extruding Operations at American National Can Company

This technique clearly shows that waiting and order processing contributed substantially to the 46 days that can be eliminated in this operation.

Source: Excerpted from "Faster, Better, and Cheaper" by Elaine J. Labach from *Target,* no. 5 (winter 1991): 43 with permission of the Association for Manufacturing Excellence, 380 West Palatine Road, Wheeling, IL 60090-5863, 847/520-3282. **www.ame.org**.

Present Method ☒		PROCESS CHART		Proposed Method ☐

SUBJECT CHARTED *Hamburger Assembly Process*   DATE *1/1/03*

CHART BY *KH*

CHART NO. *1*

DEPARTMENT   SHEET NO. *1* OF *1*

DIST. IN FEET	TIME IN MINS.	CHART SYMBOLS	PROCESS DESCRIPTION
—	—	◯ ⇨ ☐ D ▽	Meat Patty in Storage
1.5	.05	◯ ⇨ ☐ D ▽	Transfer to Broiler
	2.50	◯ ⇨ ☐ D ▽	Broiler
	.05	◯ ⇨ ☐ D ▽	Visual Inspection
1.0	.05	◯ ⇨ ☐ D ▽	Transfer to Rack
	.15	◯ ⇨ ☐ D ▽	Temporary Storage
.5	.10	◯ ⇨ ☐ D ▽	Obtain Buns, Lettuce, etc.
	.20	◯ ⇨ ☐ D ▽	Assemble Order
.5	.05	◯ ⇨ ☐ D ▽	Place in Finish Rack
		◯ ⇨ ☐ D ▽	
3.5	3.15	2 4 1 – 2	TOTALS
Value-added time = Operation time/Total time = (2.50+.20)/3.15 = 85.7%			

◯ = operation; ⇨ = transportation; ☐ = inspection; D = delay; ▽ = storage.

FIGURE 7.8 ■ Process Chart Showing a Hamburger Assembly Process at a Fast-Food Restaurant

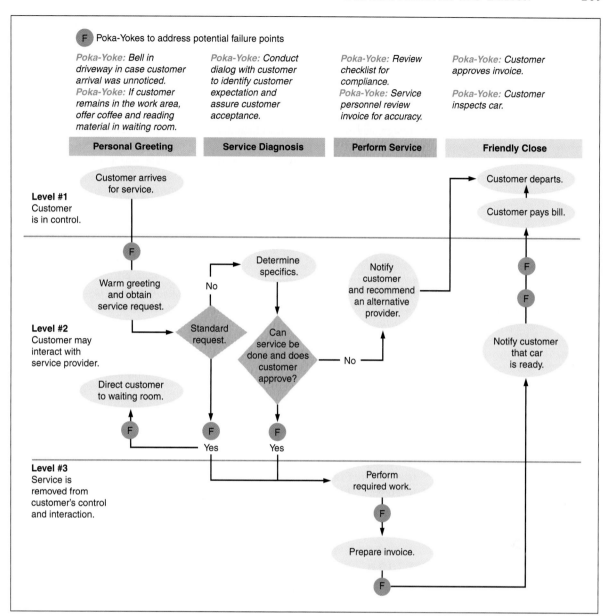

F Poka-Yokes to address potential failure points

Poka-Yoke: Bell in driveway in case customer arrival was unnoticed.
Poka-Yoke: If customer remains in the work area, offer coffee and reading material in waiting room.

Poka-Yoke: Conduct dialog with customer to identify customer expectation and assure customer acceptance.

Poka-Yoke: Review checklist for compliance.
Poka-Yoke: Service personnel review invoice for accuracy.

Poka-Yoke: Customer approves invoice.

Poka-Yoke: Customer inspects car.

| Personal Greeting | Service Diagnosis | Perform Service | Friendly Close |

Level #1
Customer is in control.

Customer arrives for service.

Customer departs.

Customer pays bill.

Level #2
Customer may interact with service provider.

Warm greeting and obtain service request.

Standard request.

Determine specifics.

No

Can service be done and does customer approve?

No

Notify customer and recommend an alternative provider.

Notify customer that car is ready.

Direct customer to waiting room.

Yes

Yes

Level #3
Service is removed from customer's control and interaction.

Perform required work.

Prepare invoice.

FIGURE 7.9 ■ Service Blueprint for Service at Speedy Lube, Inc.

Service Blueprinting

Service blueprinting
A process analysis technique that lends itself to a focus on the customer and the provider's interaction with the customer.

Products with a high service content may warrant use of yet a fourth process technique. **Service blueprinting** is a process analysis technique that focuses on the customer and the provider's interaction with the customer.[5] For instance, the activities at level one of Figure 7.9 are under the control of the customer. In the second level are activities of the service provider interacting with the customer. The third level includes those activities that are performed away from, and not immediately visible to, the customer. Each level suggests different management issues. For instance, the top level may suggest educating the customer or modifying expectations, while the second level may require a focus on personnel selection and training. Finally, the third level lends itself to more typical process innovations. The service blueprint shown in Figure 7.9 also notes potential failure points

---

[5]G. L. Shostack is given credit for the term *service blueprint*. See G. L. Shostack, "Designing Services That Deliver," *Harvard Business Review* 62, no. 1 (January–February, 1984): 133–139.

and shows how poka-yoke techniques can be added to improve quality.[6] The consequences of these failure points can be greatly reduced if identified at the design stage when modifications or appropriate poka-yokes can be included.

Each of these four process analysis tools has its strengths and variations. Flowcharts are a quick way to view the big picture and try to make sense of the entire system. Time-function mapping adds some rigor and a time element to the macro analysis. Process charts are designed to provide a much more detailed view of the process, adding items such as value-added time, delay, distance, storage, and so forth. Service blueprinting, on the other hand, is designed to help us focus on the customer interaction part of the process. Because customer interaction is often an important variable in process design, we now examine some additional aspects of service process design.

SERVICE PROCESS DESIGN

Interaction with the customer often affects process performance adversely. But a service, by its very nature, implies that some interaction and customization is needed. Recognizing that the customer's unique desires tend to play havoc with a process, the more the manager designs the process to accommodate these special requirements, the more effective and efficient the process will be. Notice how well Dell Computer has managed the interface between the customer and the process by using the Internet (see the *Global Company Profile* at the beginning of this chapter). The trick is to find the right combination of cost and customer interaction.

Customer Interaction and Process Design

The four quadrants of Figure 7.10 provide additional insight on how operations managers design service processes to find the best combination of customer interaction and the related customization. The 10 operations decisions we introduced in Chapter 2 are used with a different emphasis in each quadrant. For instance:

- In the upper sections (quadrants) of *mass service* and *professional service*, where *labor content is high*, we expect the manager to focus extensively on human resources. These quadrants require that managers find ways of addressing unique issues that satisfy customers and win orders. This is often done with very personalized services, requiring high labor involvement and therefore significant selection and training issues in the human resources area. This is particularly true in the professional service quadrant.

FIGURE 7.10 ■

Operation Changes within the Service Process Matrix

Source: Adapted from work by Roger Schmenner, "How Can Service Business Survive and Prosper?" *Sloan Management Review* (spring 1986): 21–32. Reprinted by permission of the publisher. Copyright © 1986 by Sloan Management Review Association. All rights reserved.

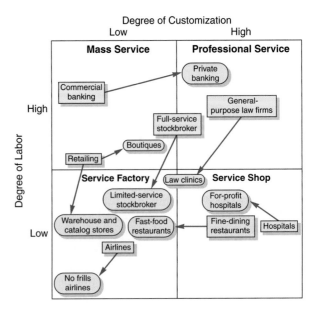

---

[6]For related discussions of poka-yoke in services, see the work of R. B. Chase and D. M. Stewart, "Make Your Service Fail-Safe," *Sloan Management Review* (spring 1994): 34–44.

Restaurants like Darden's Red Lobster are part of the service industry, but they are also the end of a long production line. At the beginning of the line, raw material goes in—at Red Lobster that means 60 million pounds of seafood a year. The seafood is purchased from all over the world. The shrimp arrives in frozen boxes from Ecuador and Thailand at a Red Lobster processing plant in St. Petersburg, Florida. There the shrimp is loaded onto a conveyor belt to be peeled, deveined, cooked, quick-frozen (left), sorted (right), and repacked for ultimate delivery to individual restaurants.

- The quadrants with *low customization* may be able (1) to standardize or restrict some offerings of the service, as do fast-food restaurants, (2) to automate, as have some airlines that have ticket-vending machines, or (3) to remove some services, such as seat assignments, as has Southwest Airlines. Off-loading some aspect of the service through automation may require innovations in process design as well as capital investment. Such is the case with airline ticket vending and bank ATMs. This move to standardization and automation may require added capital expenditure, as well as putting operations managers under pressure to develop new skills for the purchase and maintenance of such equipment. A reduction in a customization capability will require added strength in other areas.
- Because customer feedback is lower in the quadrants with *low customization*, tight control may be required to maintain quality standards.
- Operations with *low labor intensity* may lend themselves particularly well to innovations in process technology and scheduling.

Table 7.3 shows some additional techniques for innovative process design in services. The manager should focus on designing an innovative process that enhances the service. For instance, supermarket *self-service* reduces cost while it allows customers to check for the specific features they want, such as freshness or color. Dell Computer provides another version of self-service by allowing customers to design their own product on the Web. Customers seem to like this, and it is cheaper and faster for Dell.

More Opportunities to Improve Service Processes

Layout   Layout design is an integral part of many service processes, particularly in retailing, dining, and banking. In retailing, layout can provide not only product exposure but also customer education and product enhancement. In restaurants, layout can enhance the dining experience as well as provide an effective flow between bar, kitchen, and dining area. In banks, layout provides security as well as work flow and personal comfort. Because layout is such an integral part of many services, it provides continuing opportunity for winning orders.

Human Resources   Because so many services involve direct interaction with the customer (as the upper quadrants of Figure 7.10 suggest), the human resource issues of recruiting and training can be particularly important ingredients in service processes. Additionally, a committed workforce that exhibits flexibility when schedules are made and is cross-trained to fill in when the process requires less than a full-time person, can have a tremendous impact on overall process performance.

TABLE 7.3 ■

Techniques for
Improving Service
Productivity

STRATEGY	TECHNIQUE	EXAMPLE
Separation	*Structuring service* so customers must go where the service is offered	Bank customers go to a manager to open a new account, to loan officers for loans, and to tellers for deposits
Self-Service	*Self-service* so customers examine, compare, and evaluate at their own pace	Supermarkets and department stores
Postponement	*Customizing at delivery*	Customizing vans at delivery rather than at production
Focus	*Restricting* the offerings	Limited-menu restaurant
Modules	*Modular* selection of service *Modular* production	Investment and insurance selection Prepackaged food modules in restaurants
Automation	*Separating services* that may lend themselves to some type of automation	Automatic teller machines
Scheduling	Precise personnel *scheduling*	Scheduling ticket counter personnel at 15-minute intervals at airlines
Training	*Clarifying the service* options *Explaining how to avoid problems*	Investment counselor, funeral directors After-sale maintenance personnel

SELECTION OF EQUIPMENT AND TECHNOLOGY

Ultimately, the decisions about a particular process require decisions about equipment and technology. Those decisions can be complex because alternative methods of production are present in virtually all operations functions, be they hospitals, restaurants, or manufacturing facilities. Picking the best equipment means understanding the specific industry and available processes and technology. That choice of equipment, be it an X-ray machine for a hospital, a computer-controlled lathe for a factory, or a new computer for an office, requires considering cost, quality, capacity, and flexibility. To make this decision, operations personnel develop documentation that indicates the capacity, size, and tolerances of each option, as well as its maintenance requirements. Any one of these attributes may be the deciding factor regarding selection.

The selection of equipment for a particular type of process can also provide competitive advantage. Many firms, for instance, develop unique machines or techniques within established processes that provide an advantage. This advantage may result in added flexibility in meeting customer requirements, lower cost, or higher quality. Innovations and equipment modification might also allow for a more stable production process that takes less adjustment, maintenance, and operator training. In any case, specialized equipment often provides a way to win orders.

Modern technology also allows operations managers to enlarge the scope of their processes. As a result, an important attribute to look for in new equipment and process selection is flexible equipment. **Flexibility** is the ability to respond with little penalty in time, cost, or customer value. This may mean modular, movable, even cheap equipment. Flexibility may also mean the development of sophisticated electronic equipment, which increasingly provides the rapid changes that mass customization demands. The technological advances that influence OM process strategy are substantial and are discussed next.

Flexibility
The ability to respond with little penalty in time, cost, or customer value.

PRODUCTION TECHNOLOGY

Advances in technology that enhance production and productivity have a wide range of applications in both manufacturing and services. In this section, we introduce nine areas of technology: (1) machine technology, (2) automatic identification systems (AIS), (3) process control, (4) vision systems, (5) robots, (6) automated storage and retrieval systems (ASRSs), (7) automated guided vehicles (AGVs), (8) flexible manufacturing systems (FMSs), and (9) computer-integrated manufacturing (CIM).

Machine Technology

Most of the world's machinery that performs operations such as cutting, drilling, boring, and milling is undergoing tremendous progress in both precision and control. New machinery can

Three critical success factors in the trucking industry are (1) getting shipments to customers promptly (rapid response); (2) keeping trucks busy (capacity utilization); and (3) buying inexpensive fuel (driving down costs). Many firms have now developed devices like the one shown here (on the right) to track location of truck and facilitate communication between drivers and dispatchers. This systems also uses global positioning

satellites (shown on the left), speeds shipment response, maximizes utilization of the truck, and ensures purchase of fuel at the most economical location.

turn out metal components that vary less than a micron—1/76 the width of a human hair. They can accelerate water to three times the speed of sound to cut titanium for surgical tools. Machinery of the twenty-first century is often five times more productive than previous generations while being smaller and using less power. The space and power savings are both significant. And continuing advances in lubricants now allow the use of water-based lubricants rather than oil-based. Using water-based lubricants eliminates hazardous waste—but perhaps even more important, the substitution of water for oil allows shavings to be easily recovered and recycled.

The intelligence now available for the control of new machinery via computer chips allows more complex and precise items to be made faster. Electronic controls increase speed by reducing changeover time, reducing waste (because of fewer mistakes), and enhancing flexibility. Machinery with its own computer and memory is called **computer numerical control (CNC)** machinery.

Advanced versions of such technology are used on Pratt and Whitney's turbine blade plant in Connecticut. The machinery has improved the loading and alignment task so much that Pratt has cut the total time for the grinding process of a turbine blade from 10 days to 2 hours. The new machinery has also contributed to process improvements that mean the blades now travel just 1,800 feet in the plant, down from 8,100 feet. The total throughput time for a turbine blade has been cut from 22 days to 7 days.[7]

Automatic Identification System (AIS)

New equipment, from numerically controlled manufacturing machinery to ATM machines, is controlled by digital electronic signals. Electrons are a great vehicle for transmitting information, but they have a major limitation—most OM data does not start out in bits and bytes. Therefore, operations managers must get the data into an electronic form. Making data digital is done via computer keyboards, bar codes, radio frequencies, optical characters on bank checks, and so forth. These **automatic identification systems (AISs)** help us move data into electronic form where it is easily manipulated. Innovative OM examples are:

- Nurses reducing errors in hospitals by matching bar codes on medication to ID bracelets on patients.
- Radio frequency identification (RFID), which is used for tracking everything from your pet to shipping pallets. Because RFID chips send signals via their own tiny radio antennas, the need for external bar codes and scanning is eliminated.
- Transponders attached to cars allow McDonald's to identify and bill customers who can now zip through the drive-through line without having to stop and pay. The transponders use the same technology that permits motorists to skip stops on some toll roads. McDonald's operations staff is installing antennas that respond to the transponders and estimates that the change speeds up throughput time by 15 seconds.

Computer numerical control (CNC)
Machinery with its own computer and memory.

Automatic identification system (AIS)
A system for transforming data into electronic form, for example, bar codes.

[7]Steve Liesman, "Better Machine Tools Give Manufacturers Newfound Resilience," the *Wall Street Journal* (February 15, 2001): A1, A8.

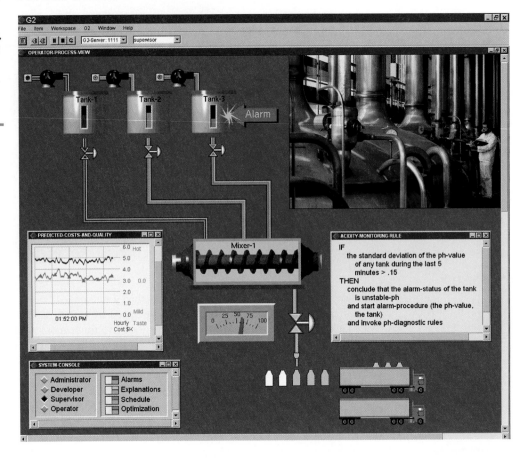

Process Control

Process control
The use of information technology to control a physical process.

Process control is the use of information technology to monitor and control a physical process. For instance, process control is used to measure the moisture content and thickness of paper as it travels over a paper machine at thousands of feet per minute. Process control is also used to determine and control temperatures, pressures, and quantities in petroleum refineries, petrochemical processes, cement plants, steel mills, nuclear reactors, and other product-focused facilities.

Process control systems operate in a number of ways, but the following is typical:

- Sensors—often analog devices—collect data.
- Analog devices read data on some periodic basis, perhaps once a minute or once every second.
- Measurements are translated into digital signals, which are transmitted to a digital computer.
- Computer programs read the file (the digital data) and analyze the data.
- The resulting output may take numerous forms. These include messages on computer consoles or printers, signals to motors to change valve settings, warning lights or horns, statistical process control charts, or schematics as shown in the photo above.

Vision Systems

Vision systems
Using video cameras and computer technology in inspection roles.

Vision systems combine video cameras and computer technology and are often used in inspection roles. Visual inspection is an important task in most food-processing and manufacturing organizations. Moreover, in many applications, visual inspection performed by humans is tedious, mind-numbing, and error-prone. Thus vision systems are widely used when the items being inspected are very similar. For instance, vision systems are used to inspect French fries so that imperfections can be identified as the fries proceed down the production line. Vision systems are used to ensure that sealant is present and in the proper amount on Whirlpool's washing-machine transmissions, and to inspect switch assemblies at the Foster Plant in Des Plaines, Illinois. Vision systems are consistently accurate, do not become bored, and are of modest cost. These systems are vastly superior to individuals trying to perform these tasks.

Robots

When a machine is flexible and has the ability to hold, move, and perhaps "grab" items, we tend to use the word *robot*. **Robots** are mechanical devices that may have a few electronic impulses stored on semiconductor chips that will activate motors and switches. Robots may be used effectively to perform tasks that are especially monotonous or dangerous or those that can be improved by the substitution of mechanical for human effort. Such is the case when consistency, accuracy, speed, strength, or power can be enhanced by the substitution of machines for people. Ford, for example, uses robots to do 98% of the welding on some automobiles.

Automated Storage and Retrieval System (ASRS)

Because of the tremendous labor involved in error-prone warehousing, computer-controlled warehouses have been developed. These systems, known as **automated storage and retrieval systems (ASRSs)**, provide for the automatic placement and withdrawal of parts and products into and from designated places in a warehouse. Such systems are commonly used in distribution facilities of retailers such as Wal-Mart, Tupperware, and Benetton. These systems are also found in inventory and test areas of manufacturing firms.

Automated Guided Vehicle (AGV)

Automated material handling can take the form of monorails, conveyors, robots, or automated guided vehicles. **Automated guided vehicles (AGVs)** are electronically guided and controlled carts used in manufacturing to move parts and equipment. They are also used in offices to move mail and in hospitals and in jails to deliver meals.

Flexible Manufacturing System (FMS)

When a central computer provides instructions to each workstation *and* to the material-handling equipment (which moves material to that station), the system is known as an automated work cell or, more commonly, a **flexible manufacturing system (FMS)**. An FMS is flexible because both the material-handling devices and the machines themselves are controlled by easily changed electronic signals (computer programs). Operators simply load new programs, as necessary, to produce different products. The result is a system that can economically produce low volume but high variety. For example, the Lockheed-Martin facility, near Dallas, efficiently builds one-of-a-kind spare parts for military aircraft. The costs associated with changeover and low utilization have been reduced substantially. FMSs bridge the gap between product-focused and process-focused facilities.

Robots are used not only for labor savings but also, more importantly, for those jobs that are dangerous or monotonous or require consistency, as in the welding on an automobile (which robots do much more effectively than humans).

Advantages of FMS: improved capital utilization, lower direct labor cost, reduced inventory, and consistent quality. Disadvantages of FMS: limited ability to adapt to product changes, substantial preplanning and capital, and tooling and fixture requirements.

Video 7.4

Computer Integrated Manufacturing at Harley-Davidson

Computer-integrated manufacturing (CIM)
A manufacturing system in which CAD, FMS, inventory control, warehousing, and shipping are integrated.

FMSs are not a panacea, however, because the individual components (machines and material-handling devices) have their own physical constraints. For instance, IBM's laptop manufacturing facility in Austin, Texas, can *only* assemble electronic products that fit in the FMS's 2-foot-×-2-foot-×-14-inch space. An FMS also has stringent communication requirements between unique components within it. However, reduced changeover time and more accurate scheduling result in faster throughput and improved utilization. Because there are fewer mistakes, reduced waste also contributes to lowering costs. These features are what operations managers are looking for: flexibility to provide customized products, improved utilization to reduce costs, and improved throughput to improve response.

Computer-Integrated Manufacturing (CIM)

Flexible manufacturing systems can be extended backward electronically into the engineering and inventory control departments and forward to the warehousing and shipping departments. In this way, computer-aided design (CAD) generates the necessary electronic instructions to run a numerically controlled machine. In a computer-integrated manufacturing environment, a design change initiated at a CAD terminal can result in that change being made in the part produced on the shop floor in a matter of minutes. When this capability is integrated with inventory control, warehousing, and shipping as a part of a flexible manufacturing system, the entire system is called **computer-integrated manufacturing (CIM)** (Figure 7.11).

Flexible manufacturing systems and computer-integrated manufacturing are reducing the distinction between low-volume/high-variety and high-volume/low-variety production. Information technology is allowing FMS and CIM to handle increasing variety while expanding to include a growing range of volumes.

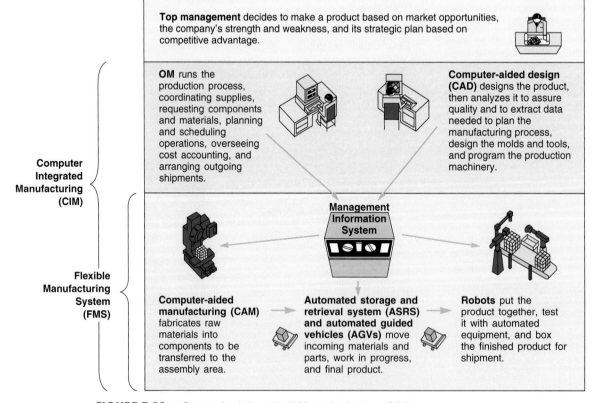

FIGURE 7.11 ■ Computer-Integrated Manufacturing (CIM)

CIM includes computer-aided design (CAD), computer-aided manufacturing (CAM), flexible manufacturing systems (FMSs), automated storage and retrieval systems (ASRSs), and automated guided vehicles (AGVs) to provide an integrated and flexible manufacturing process.

OM IN ACTION

Technology Changes the Hotel Industry

Technology is making a difference in the hotel industry. Hotel owners can now precisely track a maid's time through the use of a security system. When a maid enters a room, a card is inserted that notifies the front-desk computer as to the maid's location. "We can show her a printout of how long she takes to do a room," says one hotel owner.

Security systems also enable guests to use their own credit cards as keys to unlock their doors. There are also other uses for the system. The computer can bar a guest's access to the room after checkout time and automatically control the air conditioning or heat, turning it on at check-in and off at checkout.

And now, with a hand-held infrared unit, housekeeping staff and minibar attendants can check, from the hallway, to see if a room is physically occupied. This both eliminates the embarrassment of having a hotel staffer walk in on a guest *and* improves security for housekeepers.

At Loew's Portofino Bay Hotel at Universal Studios, Orlando, guest smart cards act as credit cards in both the theme park and the hotel, while staff smart cards (programmed for different levels of security access) create an audit trail of employee movement. Starwood Hotels, which runs such properties as Sheraton and Westins, use Casio Pocket PCs to communicate with a hotel wireless network. Now guests can check in and out from any place on the property, such as at their restaurant table after breakfast or lunch.

Sources: The *Wall Street Journal* (September 11, 2002):B2; and *Hotel and Motel Management* (July 3, 2000): 62–63.

TECHNOLOGY IN SERVICES

Just as we have seen rapid advances in technology in the manufacturing sector, so we also find dramatic changes in the service sector. These range from electronic diagnostic equipment at auto repair shops, to blood- and urine-testing equipment in hospitals, to retinal security scanners at airports and high-security facilities. The hospitality industry provides other examples, as discussed in the *OM in Action* box, "Technology Changes the Hotel Industry." As you probably have experienced, you can now authorize payment of your bill from your hotel room via a channel on the room's television set. The labor savings at the registration desk and speedier checkout service provide valuable productivity increases for both the hotel and the customer.

Similarly, Andersen Windows, of Minnesota, has developed user-friendly computer software that enables customers to design their own window specifications. The customer calls up a product information guide, promotion material, a gallery of designs, and a sketch pad to create the designs desired. The software also allows the customer to determine likely energy savings and see a graphic view of their home fitted with the new window.

In retail store, POS terminals now download prices quickly to reflect changing costs or market conditions. For instance, when devaluation struck Mexico, the drugstore chain Farmacias Benavides used such systems to immediately stop the reorders of higher-priced items and to stock up instead on lower-cost generic product lines.

Table 7.4 provides a glimpse of the impact of technology on services. Operations managers in services, as in manufacturing, must be able to evaluate the impact of technology on their firm. This ability requires particular skill when evaluating reliability, investment analysis, human resource requirements, and maintenance/service.

PROCESS REENGINEERING

Process reengineering
The fundamental rethinking and redesign of business processes to bring about dramatic improvements in performance.

Often a firm finds that the initial assumptions of its process are no longer valid. The world is a dynamic place and customer desires, product technology, and product mix change. Consequently, processes are redesigned or, as it is sometimes called, reengineered. **Process reengineering** is the fundamental rethinking and radical redesign of business processes to bring about dramatic improvements in performance.[8] Effective process reengineering relies on reevaluating the purpose

[8]Michael Hammer and Steven Stanton, *The Reengineering Revolution* (New York: HarperCollins, 1995): 3.

TABLE 7.4 ■

Examples of
Technology's Impact on
Services

SERVICE INDUSTRY	EXAMPLE
Financial Services	Debit cards, electronic funds transfer, automatic teller machines, Internet stock trading
Education	Electronic bulletin boards, online journals
Utilities and government	Automated one-man garbage trucks, optical mail and bomb scanners, flood-warning systems.
Restaurants and foods	Wireless orders from waiters to the kitchen, robot butchering, transponders on cars that track sales at drive-throughs.
Communications	Electronic publishing, interactive TV
Hotels	Electronic check-in/checkout, electronic key/lock systems
Wholesale/retail trade	Point-of-sale (POS) terminals, e-commerce, electronic communication between store and supplier, bar-coded data
Transportation	Automatic toll booths, satellite-directed navigation systems
Health care	Online patient-monitoring systems, online medical information systems, robotic surgery
Airlines	Ticketless travel, scheduling, Internet purchases

of the process and questioning both purpose and underlying assumptions. It works only if the basic process and its objectives are reexamined.

Process reengineering also focuses on those activities that cross functional lines. Because managers are often in charge of specific "functions" or specialized areas of responsibility, those activities (processes) that cross from one function or specialty to another may be neglected. Reengineering casts aside all notions of how the process is currently being done and focuses on dramatic improvements in cost, time, and customer value. Any process is a candidate for radical redesign. The process can be a factory layout, a purchasing procedure, or a new way of processing credit applications at IBM, as described in Example 5.

Example 5

The traditional IBM credit application process took many steps. The first step consisted of 14 people answering phones and logging calls from field sales personnel requesting credit for customers. After receiving calls, phone personnel made paper notations that they sent upstairs to credit personnel for credit checks. Then the paper went down the hall to the business practice group where the data were entered into a computer for determination of terms and interest rates. From there, the packet of data went to a clerical group. A week or two after the request, the results of the request were available.

IBM tried to fix the process by keeping a log of each step of every request. Although logging allowed credit personnel to know where in the process the application was, it added a day to the turnaround. Finally, two managers tried a radical approach. They walked a loan request through each step from office to office and found that it took only 90 minutes of actual work. The additional week was spent shuttling the paperwork among departments. This meant that the work along the way was not the problem. Instead, the *process* was at fault. Reengineering resulted in IBM replacing all of its specialists with generalists, called caseworkers, who process applications from start to finish. The firm also developed software that uses the expertise of specialists to support caseworkers. The reengineered process reduced the number of employees and achieved better results. The week-plus turnaround time for a credit request is down to 4 hours. The company now handles 100 times the number of loan requests than it did under the old system.

Source: Adapted from Michael Hammer and James Campy, *Reengineering the Corporation: A Manifesto for Business Revolution* (New York: HarperCollins, 1993).

ENVIRONMENTALLY FRIENDLY PROCESSES

In Chapter 5 we discussed some environment-friendly design techniques; now we introduce some process approaches that address social responsibility and environmental concerns. Many firms have found opportunities in their production processes to reduce the negative impact on the environment. The opportunities range from activities that society perceives as socially responsible to actions that are legally required, such as pollution prevention. These activities include a focus on such issues as efficient use of resources, reduction of waste by-products, emission controls, and recycling.

New products take many forms. This new breed of broiler chickens is featherless. These chickens stay cooler, save on plucking cost, and have less waste (feathers), making them environmentally friendly. But as often happens with new products, the operations manager must adjust the process.

Operations managers can be environmentally sensitive and still achieve a differentiation strategy—and even a low-cost strategy. Here are four examples:

- British cosmetic firm Body Shop has successfully differentiated its products by stressing environmental sensitivity. It pursues a product design, development, and testing strategy that it believes to be socially responsible. This includes environment-friendly ingredients and elimination of animal testing.
- Ben and Jerry's pursues its socially responsible image (and saves $250,000 annually) just by using energy-efficient lighting.
- Standard Register, described in Example 1, produces considerable paper scrap—almost 20 tons of punch holes alone per month—which creates a significant waste issue. But the company developed ways to recycle the paper scrap, as well as aluminum and silver from the plate-making process shown in the flow diagram in Figure 7.2.
- Anheuser-Busch saves $30 million per year in energy and waste-treatment costs by using treated plant wastewater to generate the gas that powers its St. Louis brewery.

Processes can be environmentally friendly and socially responsible while still contributing to profitable strategies.

Shown here is a process developed by Floyd Hammer (right) that converts plastic into weather-resistant park benches, parking lot curbs, and landscaping timbers. Hammer's company, Plastic Recycling Corp., based in Iowa Falls, grows as this new process provides an environmentally friendly product.

SUMMARY

Effective operations managers understand how to use process strategy as a competitive weapon. They select a production process with the necessary quality, flexibility, and cost structure to meet product and volume requirements. They also seek creative ways to combine the low unit cost of high-volume, low-variety manufacturing with the customization available through low-volume, high-variety facilities. Managers use the techniques of lean production and employee participation to encourage the development of efficient equipment and processes. They design their equipment and processes to have capabilities beyond the tolerance required by their customers, while ensuring the flexibility needed for adjustments in technology, features, and volumes.

KEY TERMS

Process strategy *(p. 198)*
Process focus *(p. 198)*
Repetitive process *(p. 200)*
Modules *(p. 200)*
Product focus *(p. 201)*
Mass customization *(p. 202)*
Crossover chart *(p. 205)*
Flow diagram *(p. 207)*
Time-function mapping (or process mapping) *(p. 207)*
Process charts *(p. 207)*
Service blueprinting *(p. 209)*

Flexibility *(p. 212)*
Computer numerical control (CNC) *(p. 213)*
Automatic identification system (AIS) *(p. 213)*
Process control *(p. 214)*
Vision systems *(p. 214)*
Robot *(p. 215)*
Automated storage and retrieval system (ASRS) *(p. 215)*
Automated guided vehicle (AGV) *(p. 215)*
Flexible manufacturing system (FMS) *(p. 215)*
Computer-integrated manufacturing (CIM) *(p. 216)*
Process reengineering *(p. 217)*

SOLVED PROBLEM

Solved Problem 7.1

Clare Copy Shop has a volume of 125,000 black-and-white copies per month. Two salesmen have made presentations to Debbie Clare for machines of equal quality and reliability. The Xemon A has a cost of $2,000 per month and a variable cost of $.03. The other machine (a Camron B) will cost only $1,500 per month but the toner is more expensive, driving the cost per copy up to $.035. If cost and volume are the only considerations, which machine should Clare purchase?

SOLUTION

$$2,000 + .03\,X = 1,500 + .035\,X$$
$$2,000 - 1,500 = .035\,X - .03\,X$$
$$500 = .005\,X$$
$$100,000 = X$$

Because Debbie Clare expects her volume to exceed 100,000 units, she should choose the Xemon A.

INTERNET AND STUDENT CD-ROM EXERCISES

Visit our home page or use your student CD-ROM to help with material in this chapter.

 On Our Home Page, www.prenhall.com/heizer

- Self-Tests
- Practice Problems
- Internet Exercises
- Current Articles and Research
- Virtual company Tour
- Internet Case

 On Your Student CD-ROM

- PowerPoint Lecture
- Practice Problems
- Video Clips and Video Case
- Active Model Exercise

ADDITIONAL CASE STUDIES

Internet Case Study: Visit our Web site at www.prenhall.com/heizer **for this free case study:**

- **Matthew Yachts, Inc.:** Examines a possible process change as the market for yachts changes.

Harvard has selected these Harvard Business School cases to accompany this chapter (textbookcasematch.hbsp.harvard.edu)**:**

- **Massachusetts General Hospital** (#696-015): Describes efforts at Massachusetts General Hospital to reengineer the service delivery process for heart bypass surgery.

- **John Crane UK Ltd.: the CAD/CAM Link** (#692-100): Describes the improvement of manufacturing performance in a job shop.

- **Product Development at Dell** (#699-134): Discusses the new product and process and the management of development risk.

Capacity Planning

Supplement Outline

LEARNING OBJECTIVES

When you complete this supplement you should be able to

IDENTIFY OR DEFINE:

Capacity

Design capacity

Effective capacity

Utilization

DESCRIBE OR EXPLAIN:

Capacity considerations

Net present value analysis

Break-even analysis

Financial considerations

Strategy-driven investments

When designing a concert hall, management hopes that the forecasted capacity (the product mix—opera, symphony, and special events—and the technology needed for these events) is accurate and adequate for operation above the break-even point. However, in many concert halls, even when operating at full capacity, breakeven is not achieved and supplemental funding must be obtained.

CAPACITY

How many concertgoers should a facility seat? How many customers per day should a Hard Rock Cafe be able to service? How many computers should Dell's Nashville plant be able to produce in an 8-hour shift?

After selection of a production process (Chapter 7), we need to determine capacity. **Capacity** is the "throughput," or the number of units a facility can hold, receive, store, or produce in a period of time. The capacity affects a large portion of fixed cost. It also determines if demand will be met or if facilities will be idle. If the facility is too large, portions of it will sit idle and add cost to existing production or clients. If the facility is too small, customers and perhaps entire markets are lost. So determining facility size, with an objective of achieving high levels of utilization and a high return on investment, is critical.

Capacity planning can be viewed in three time horizons. In Figure S7.1 we note that long-range capacity (greater than 1 year) is a function of adding facilities and equipment that has a long lead time. In the intermediate range (3 to 18 months) we can add equipment, personnel, and shifts; we can subcontract; and we can build or use inventory. This is the aggregate planning task. In the short run (usually up to 3 months) we are primarily concerned with scheduling jobs and people, and allocating machinery. It is difficult to modify capacity in the short run; we are using capacity that already exists.

Capacity
The "throughput" or number of units a facility can hold, receive, store, or produce in a period of time.

FIGURE S7.1 ■

Types of Planning over a Time Horizon

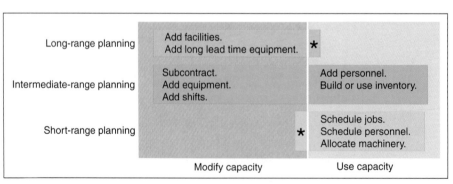

*Limited options exist.

Design and Effective Capacity

Design capacity
The theoretical maximum output of a system in a given period.

Design capacity is the maximum theoretical output of a system in a given period. It is normally expressed as a rate, such as the number of tons of steel that can be produced per week, per month, or per year. For many companies, measuring capacity can be straightforward: It is the maximum number of units produced in a specific time. However, for some organizations, determining capacity can be more difficult. Capacity can be measured in terms of beds (a hospital), active members (a church), or classroom size (a school). Other organizations use total work time available as a measure of overall capacity.

Most organizations operate their facilities at a rate less than the design capacity. They do so because they have found that they can operate more efficiently when their resources are not stretched to the limit. Instead, they expect to operate at perhaps 82% of design capacity. This concept is called effective capacity.

Effective capacity
Capacity a firm can expect to achieve given its product mix, methods of scheduling, maintenance, and standards of quality.

Effective capacity is the capacity a firm *expects* to achieve given the current operating constraints. Effective capacity is often lower than design capacity because the facility may have been designed for an earlier version of the product or a different product mix than is currently being produced.

Two measures of system performance are particularly useful: utilization and efficiency. **Utilization** is simply the percent of *design capacity* actually achieved. **Efficiency** is the percent of *effective capacity* actually achieved. Depending on how facilities are used and managed, it may be difficult or impossible to reach 100% efficiency. Operations managers tend to be evaluated on efficiency. The key to improving efficiency is often found in correcting quality problems and in effective scheduling, training, and maintenance. Utilization and efficiency are computed below:

Utilization
Actual output as a percent of design capacity.

Efficiency
Actual output as a percent of effective capacity.

$$\text{Utilization} = \text{Actual Output/Design Capacity} \qquad \text{(S7-1)}$$

$$\text{Efficiency} = \text{Actual Output/Effective Capacity} \qquad \text{(S7-2)}$$

In Example S1 we determine these values.

Example S1

Active Model S7.1

Example S1 is further illustrated in Active Model S7.1 on your CD-ROM.

Sara James Bakery has a plant for processing breakfast rolls. Last week the facility produced 148,000 rolls. The effective capacity is 175,000 rolls. The production line operates 7 days per week with three 8-hour shifts per day. The line was designed to process a nut-filled, cinnamon-flavored, sugar-coated *Deluxe* roll at the rate of 1,200 per hour. Determine the design capacity, utilization, and efficiency for this plant when producing the *Deluxe* roll.

SOLUTION

Design Capacity = (7 days × 3 shifts × 8 hours) × (1,200 rolls per hour) = 201,600 rolls

Utilization = Actual Output/Design Capacity = 148,000/201,600 = 73.4%

Efficiency = Actual Output/Effective Capacity = 148,000/175,000 = 84.6%

Design capacity, utilization, and efficiency are all important measures for an operations manager. But managers often need to know the expected output of a facility or process. To do this, we use Equation (S7-2) to solve for actual (or in this case, future or expected) output as shown in Equation (S7-3).

$$\text{Actual (or Expected) Output} = (\text{Effective Capacity})(\text{Efficiency}) \qquad \text{(S7-3)}$$

Now with a knowledge of effective capacity and efficiency, a manager can find the expected output of a facility. We do so in Example S2.

Example S2

The manager of Sara James Bakery (see Example S1) now needs to increase production of the increasingly popular *Deluxe* roll. To meet this demand, the operations manager will be adding a second production line. The manager must determine the expected output of this second line for the sales department. Effective capacity on second line is the same as on first line, which is 175,000 *Deluxe* rolls. The first line is operating at an efficiency of 84.6%, as computed in Example S1. But output on the second line will be less than the first line because the crew will be primarily new hires; so the efficiency can be expected to be no more than 75%. What is the expected output?

Expected Output = (Effective Capacity)(Efficiency) = (175,000)(.75) = 131,250 rolls

The sales department should be told the expected output is 131,250 *Deluxe* rolls.

If the expected output is inadequate, additional capacity may be needed. Much of the remainder of this supplement addresses how to effectively and efficiently add that capacity.

Capacity and Strategy

Sustained profits come from building competitive advantage, not just from a good financial return on a specific process. Capacity decisions must be integrated into the organization's mission and strategy. Investments are not to be made as isolated expenditures, but as part of a coordinated plan that will place the firm in an advantageous position.[1] The questions to be asked are "Will these investments eventually win customers?" and "What competitive advantage (such as process flexibility, speed of delivery, improved quality, and so on) do we obtain?"

All of the 10 decisions of operations management we discuss in this text, as well as other organizational elements such as marketing and finance, are impacted by changes in capacity. Change in capacity will have sales and cash flow implications, just as capacity changes have quality, supply chain, human resource, and maintenance implications. All must be considered.

Capacity Considerations

In addition to tight integration of strategy and investments, there are four special considerations for a good capacity decision.

1. **Forecast demand accurately.** An accurate forecast is paramount to the capacity decision. The new product may be Hard Rock Cafe's nightly live music venue that places added demands on the cafe's food service and retail shop, or the product may be a new open-heart surgery capability at Cleveland Clinic, or the new model PT Cruiser at DaimlerChrysler. Whatever the new product, its prospects and the life cycle of existing products, must be determined. Management must know which products are being added and which are being dropped, as well as their expected volumes.

2. **Understand the technology and capacity increments.** The number of initial alternatives may be large, but once the volume is determined, technology decisions may be aided by analysis of cost, human resources required, quality, and reliability. Such a review often reduces the number of alternatives to a few. The technology may dictate the capacity increment. Meeting added demand with a few extra tables in the dining room of a restaurant may not be difficult, but meeting increased demand for a new automobile by adding a new assembly line at BMW may be very difficult—and expensive. But the operations manager is held responsible for the technology and the correct capacity increment.

3. **Find the optimum operating level (volume).** Technology and capacity increments often dictate an optimal size for a facility. A roadside motel may require 50 rooms to be viable. If smaller, the fixed cost is too burdensome; if larger, the facility becomes more than one manager can supervise. A hypothetical optimum for the motel is shown in Figure S7.2. This issue is known as *economies and diseconomies of scale*. GM at one time believed that the optimum auto plant was one with 600 employees. Most businesses have an optimal size—at least

FIGURE S7.2 ■

Economies and Diseconomies of Scale

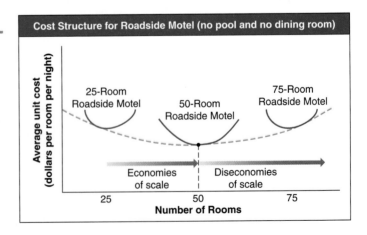

[1]For an excellent discussion on investments that support competitive advantage, see Terry Hill, *Manufacturing Strategy* 2nd ed. (Burr Ridge, IL: Richard D. Irwin, 1994).

until someone comes along with a new business model. For decades, very large integrated steel mills were considered optimal. Then along came Nucor, SMI, and other minimills with a new process and a new business model that changed the optimum size of a steel mill.

4. **Build for change.** In our fast-paced world, change is inevitable. So operations managers build flexibility into the facility and equipment. They evaluate the sensitivity of the decision by testing several revenue projections on both the upside and downside for potential risks. Buildings, and the infrastructure for such things as utilities and parking, can often be built in phases. And buildings and equipment can be designed with modifications in mind to accommodate future changes in product, product mix, and processes.

Rather than strategically manage capacity, managers may tactically manage demand. Here are some techniques for managing demand.

Managing Demand

Even with good forecasting and facilities built to that forecast, there may be a poor match between the actual demand that occurs and available capacity. A poor match may mean demand exceeds capacity or capacity exceeds demand. However, in both cases firms have options.

Demand Exceeds Capacity   When *demand exceeds capacity*, the firm may be able to curtail demand simply by raising prices, scheduling long lead times (which may be inevitable), and discouraging marginally profitable business. However, because inadequate facilities reduce revenue below what is possible, the long-term solution is usually to increase capacity.

Capacity Exceeds Demand   When *capacity exceeds demand*, the firm may want to stimulate demand through price reductions or aggressive marketing, or it may accommodate the market through product changes.

Adjusting to Seasonal Demands   A seasonal or cyclical pattern of demand is another capacity challenge. In such cases, management may find it helpful to offer products with complementary demand patterns—that is, products for which the demand is high for one when low for the other. For example, in Figure S7.3 the firm is adding a line of snowmobile engines to its line of jet ski engines to smooth demand. With appropriate complementing of products, perhaps the utilization of facility, equipment, and personnel can be smoothed.

Tactics for Matching Capacity to Demand   Various tactics for matching capacity to demand exist. Internal changes include adjusting the process to a given volume through:

1. Making staffing changes (increasing or decreasing the number of employees);
2. Adjusting equipment and processes, which might include purchasing additional machinery or selling or leasing out existing equipment;
3. Improving methods to increase throughput; and/or
4. Redesigning the product to facilitate more throughput.

The foregoing tactics can be used to adjust demand to existing facilities. The strategic issue is, of course, how to have a facility of the correct size.

FIGURE S7.3 ■

By Combining Products That Have Complementary Seasonal Patterns, Capacity Can Be Better Utilized

A smoother sales demand contributes to improved scheduling and better human resource strategies.

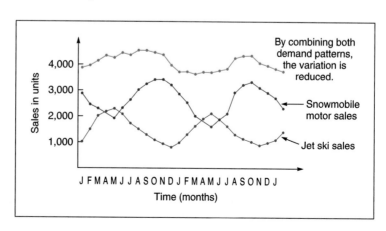

CAPACITY PLANNING

Setting future capacity requirements can be a complicated procedure, one based in large part on future demand. When demand for goods and services can be forecast with a reasonable degree of precision, determining capacity requirements can be straightforward. Determining capacity normally requires two phases. During the first phase, future demand is forecast with traditional models, as we saw in Chapter 4. During the second phase, this forecast is used to determine capacity requirements and the incremental size of each addition to capacity.[2] Interestingly, demand growth is typically gradual in small units, while capacity additions are typically instantaneous in large units. This contradiction often makes the capacity expansion difficult.

Figure S7.4 reveals four approaches to new capacity. As we see in Figure S7.4(a), new capacity is acquired at the beginning of year 1. This capacity will handle increased demand until the beginning of year 2. At the beginning of year 2, new capacity is again acquired, which will allow the organization to stay ahead of demand until the beginning of year 3. This process can be continued indefinitely into the future.

The capacity plan shown in Figure S7.4(a) is only one of an almost limitless number of plans to satisfy future demand. In this figure, new capacity was acquired *incrementally*—at the beginning of year 1 *and* at the beginning of year 2. In Figure S7.4(b), a large increase in capacity is acquired at the beginning of year 1 in order to satisfy expected demand until the beginning of year 3.

FIGURE S7.4 ■

Approaches to
Capacity Expansion

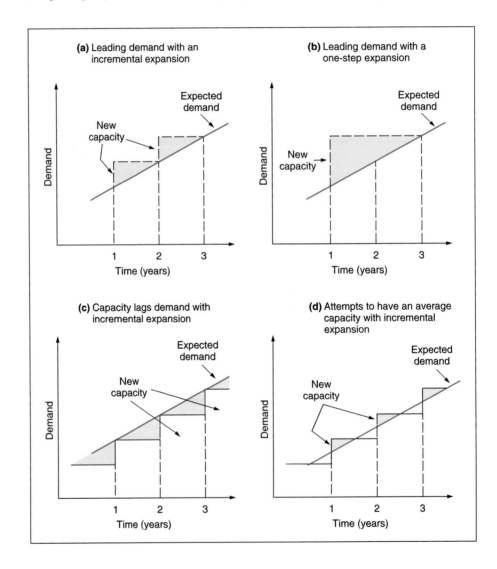

(a) Leading demand with an incremental expansion

(b) Leading demand with a one-step expansion

(c) Capacity lags demand with incremental expansion

(d) Attempts to have an average capacity with incremental expansion

[2]At this point, we make the assumption that management knows the technology and the *type* of facilities to be employed to satisfy future demand requirements—not a minor issue, but beyond the scope of this text.

The capital expenditures for a capacity change can be tremendous. Many companies address this problem by making incremental changes when possible. Others adjust by modifying old equipment or using older equipment even though it may not be as efficient. For instance, managers at family-owned Chelsea Milling Company, makers of Jiffy brand mixes, decided that their company's OM strategy did not support additional capital investment in new equipment. Consequently, when making repairs, modifying equipment, or adjusting for peak loads, they draw on spare, often old, equipment.

The excess capacity provided by plans, Figure S7.4(a) and Figure S7.4(b) gives operations managers flexibility. For instance, in the hotel industry, added capacity in the form of rooms can allow a wider variety of room options and perhaps flexibility in room cleanup schedules. In manufacturing, the excess capacity can be used to do more setups to shorten production runs, driving down inventory. The added capacity may also allow management to build excess inventory and thus delay the capital expenditure and disruption that come with adding new capacity.[3]

Alternatives Figure S7.4(a) and Figure S7.4(b) *lead* capacity—that is, acquire capacity to stay ahead of demand—but Figure S7.4(c) shows an option that *lags* capacity, perhaps using overtime or subcontracting to accommodate excess demand. Figure S7.4(d) attempts to build capacity that is "average," sometimes lagging demand and sometimes leading it.

In some cases, deciding between alternatives can be relatively easy. The total cost of each alternative can be computed and the alternative with the least total cost selected. In other cases, determining the capacity and how to achieve it can be much more complicated. In most cases, numerous subjective factors are difficult to quantify and measure. These factors include technological options; competitor strategies; building restrictions; cost of capital; human resource options; and local, state, and federal laws and regulations.

BREAK-EVEN ANALYSIS

Break-even analysis is a critical tool for determining the capacity a facility must have to achieve profitability. The objective of **break-even analysis** is to find the point, in dollars and units, at which costs equal revenue. This point is the break-even point. Firms must operate above this level to achieve profitability. As shown in Figure S7.5, break-even analysis requires an estimation of fixed costs, variable costs, and revenue.

Fixed costs are costs that continue even if no units are produced. Examples include depreciation, taxes, debt, and mortgage payments. **Variable costs** are those that vary with the volume of units produced. The major components of variable costs are labor and materials. However, other costs, such as the portion of the utilities that varies with volume, are also variable costs. The difference between selling price and variable cost is **contribution.** Only when total contribution exceeds total fixed cost will there be profit.

Another element in break-even analysis is the **revenue function**. In Figure S7.5, revenue begins at the origin and proceeds upward to the right, increasing by the selling price of each unit. Where the revenue function crosses the total cost line (the sum of fixed and variable costs), is the break-even point, with a profit corridor to the right and a loss corridor to the left.

Break-even analysis
A means of finding the point, in dollars and units, at which costs equal revenues.

Fixed costs
Costs that continue even if no units are produced.

Variable costs
Costs that vary with the volume of units produced.

Contribution
The difference between selling price and variable costs.

Revenue function
The function that increases by the selling price of each unit.

---

[3]See related discussion in Sampath Rajagopalan and Jayashankar M. Swaminathan, "Coordinated Production Planning Model with Capacity Expansion and Inventory Management," *Management Science* 47, no. 11, (November 2001): 1562–1580.

FIGURE S7.5 ■

Basic Break-Even Point

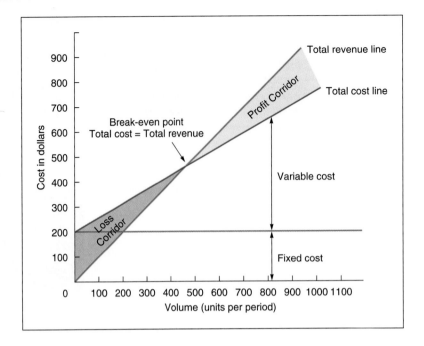

Virtually *no* variable costs
are linear, but we make
that assumption here.

Assumptions   A number of assumptions underlie this basic break-even model. Notably, costs and revenue are shown as straight lines. They are shown to increase linearly—that is, in direct proportion to the volume of units being produced. However, neither fixed costs nor variable costs (nor, for that matter, the revenue function) need be a straight line. For example, fixed costs change as more capital equipment or warehouse space is used; labor costs change with overtime or as marginally skilled workers are employed; the revenue function may change with such factors as volume discounts.

Fixed costs do not remain
constant over all volume;
new warehouses and new
overhead charges result in
step functions in fixed
cost.

Graphic Approach   The first step in the graphic approach to break-even analysis is to define those costs that are fixed and sum them. The fixed costs are drawn as a horizontal line beginning at that dollar amount on the vertical axis. The variable costs are then estimated by an analysis of labor, materials, and other costs connected with the production of each unit. The variable costs are shown as an incrementally increasing cost, originating at the intersection of the fixed cost on the vertical axis and increasing with each change in volume as we move to the right on the volume (or horizontal) axis. Both fixed- and variable-cost information is usually available from a firm's cost accounting department, although an industrial engineering department may also maintain cost information.

Algebraic Approach   The respective formulas for the break-even point in units and dollars are shown below. Let:

BEP_x = Break-even point in units

$BEP_\$$ = Break-even point in dollars

P = Price per unit (after all discounts)

x = Number of units produced

TR = Total revenue = Px

F = Fixed costs

V = Variable costs per unit

TC = Total costs = $F + Vx$

The break-even point occurs where total revenue equals total costs. Therefore

$$TR = TC \quad \text{or} \quad Px = F + Vx$$

Solving for x, we get

$$BEP_x = \frac{F}{P - V}$$

and

$$BEP_\$ = BEP_x P = \frac{F}{P - V} P = \frac{F}{(P - V)/P}$$

$$= \frac{F}{1 - V/P}$$

$$\text{Profit} = TR - TC$$

$$= Px - (F + Vx) = Px - F - Vx$$

$$= (P - V)x - F$$

Using these equations, we can solve directly for break-even point and profitability. The two formulas of particular interest are:

$$\text{Break-even in units} = \frac{\text{Total fixed cost}}{\text{Price} - \text{Variable cost}} \qquad \text{(S7-4)}$$

$$\text{Break-even in dollars} = \frac{\text{Total fixed cost}}{1 - \dfrac{\text{Variable cost}}{\text{Selling price}}} \qquad \text{(S7-5)}$$

Single-Product Case

In Example S3, we determine the break-even point in dollars and units for one product.

Example S3

**Excel OM Data File
Ch07SExS3.xla**

Active Model S7.2

Example S3 is further illustrated in Active Model S7.2 on the CD-ROM and in the Exercise located in your Student Lecture Guide.

Jimmy Stephens, Inc., has fixed costs of $10,000 this period. Direct labor is $1.50 per unit, and material is $.75 per unit. The selling price is $4.00 per unit.

The break-even point in dollars is computed as follows:

$$BEP_\$ = \frac{F}{1 - (V/P)} = \frac{\$10,000}{1 - [(1.50 + .75)/(4.00)]} = \frac{\$10,000}{.4375} = \$22,857.14$$

The break-even point in units is

$$BEP_x = \frac{F}{P - V} = \frac{\$10,000}{4.00 - (1.50 + .75)} = 5,714$$

Note that we use total variable costs (that is, both labor and material).

Paper machines such as the one shown here, at International Paper, require a high capital investment. This investment results in a high fixed cost, but allows production of paper at a very low variable cost. The production manager's job is to maintain utilization above the break-even point to achieve profitability.

Multiproduct Case

Most firms, from manufacturers to restaurants (even fast-food restaurants), have a variety of offerings. Each offering may have a different selling price and variable cost. Utilizing break-even analysis, we modify Equation (S7-5) to reflect the proportion of sales for each product. We do this by "weighting" each product's contribution by its proportion of sales. The formula is then

$$BEP_\$ = \frac{F}{\sum\left[\left(1 - \frac{V_i}{P_i}\right) \times (W_i)\right]}$$ (S7-6)

where V = variable cost per unit
P = price per unit
F = fixed cost
W = percent each product is of total dollar sales
i = each product

Example S4 shows how to determine the break-even point for the multiproduct case.

Example S4

Information for Le Bistro, a French-style deli, follows. Fixed costs are $3,500 per month.

ITEM	PRICE	COST	ANNUAL FORECASTED SALES UNITS
Sandwich	$2.95	$1.25	7,000
Soft drink	.80	.30	7,000
Baked potato	1.55	.47	5,000
Tea	.75	.25	5,000
Salad bar	2.85	1.00	3,000

With a variety of offerings, we proceed with break-even analysis just as in a single-product case, except that we weight each of the products by its proportion of total sales.

MULTIPRODUCT BREAK-EVEN–DETERMINING CONTRIBUTION

1	2	3	4	5	6	7	8
ITEM (i)	SELLING PRICE (P)	VARIABLE COST (V)	(V/P)	1−(V/P)	ANNUAL FORECASTED SALES $	% OF SALES	WEIGHTED CONTRIBUTION (COL. 5 × COL. 7)
Sandwich	$2.95	$1.25	.42	.58	$20,650	.446	.259
Soft drink	.80	.30	.38	.62	5,600	.121	.075
Baked potato	1.55	.47	.30	.70	7,750	.167	.117
Tea	.75	.25	.33	.67	3,750	.081	.054
Salad bar	2.85	1.00	.35	.65	8,550	.185	.120
					$46,300	1.000	.625

For instance, revenue for sandwiches is $20,650 (2.95 × 7,000), which is 44.6% of the total revenue of $46,300. Therefore, the contribution for sandwiches is "weighted" by .446. The weighted contribution is .446 × .58 = .259. In this manner, its *relative* contribution is properly reflected.

Using this approach for each product, we find that the total weighted contribution is .625 for each dollar sales, and the break-even point in dollars is $67,200:

$$BEP_\$ = \frac{F}{\sum\left[\left(1 - \frac{V_i}{P_i}\right) \times (W_i)\right]} = \frac{\$3,500 \times 12}{.625} = \frac{\$42,000}{.625} = \$67,200$$

The information given in this example implies total daily sales (52 weeks at 6 days each) of

$$\frac{\$67,200}{312 \text{ days}} = \$215.38$$

Break-even figures by product provide the manager with added insight as to the realism of his or her sales forecast. They indicate exactly what must be sold each day, as we have done in Example S5.

Example S5

Using the data in Example S4, we take the forecast sandwich sales of 44.6% times the daily break-even of $215.38 divided by the selling price of each sandwich ($2.95). Then sandwich sales must be

$$\frac{.446 \times \$215.38}{\$2.95} = \text{number of sandwiches} = 32.6 \approx 33 \text{ sandwiches each day}$$

Once break-even analysis has been prepared, analyzed, and judged to be reasonable, decisions can be made about the type and capacity of equipment needed. Indeed, a better judgment of the likelihood of success of the enterprise can now be made.

When capacity requirements are subject to significant unknowns, "probabilistic" models may be appropriate. One technique for making successful capacity planning decisions with an uncertain demand is decision theory, including the use of decision trees.

APPLYING DECISION TREES TO CAPACITY DECISIONS

Decision trees require specifying alternatives and various states of nature. For capacity planning situations, the state of nature usually is future demand or market favorability. By assigning probability values to the various states of nature, we can make decisions that maximize the expected value of the alternatives. Example S6 shows how to apply decision trees to a capacity decision.

Example S6

Southern Hospital Supplies, a company that makes hospital gowns, is considering capacity expansion. Its major alternatives are to do nothing, build a small plant, build a medium plant, or build a large plant. The new facility would produce a new type of gown, and currently the potential or marketability for this product is unknown. If a large plant is built and a favorable market exists, a profit of $100,000 could be realized. An unfavorable market would yield a $90,000 loss. However, a medium plant would earn a $60,000 profit with a favorable market. A $10,000 loss would result from an unfavorable market. A small plant, on the other hand, would return $40,000 with favorable market conditions and lose only $5,000 in an unfavorable market. Of course, there is always the option of doing nothing.

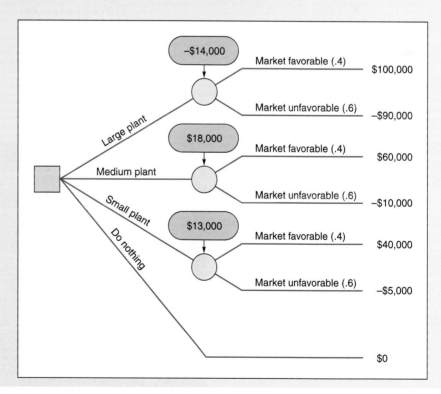

Recent market research indicates that there is a .4 probability of a favorable market, which means that there is also a .6 probability of an unfavorable market. With this information, the alternative that will result in the highest expected monetary value (EMV) can be selected:

$$\text{EMV (large plant)} = (.4)(\$100,000) + (.6)(-\$90,000) = -\$14,000$$

$$\text{EMV (medium plant)} = (.4)(\$60,000) + (.6)(-\$10,000) = +\$18,000$$

$$\text{EMV (small plant)} = (.4)(\$40,000) + (.6)(-\$5,000) = +\$13,000$$

$$\text{EMV (do nothing)} = \$0$$

Based on EMV criteria, Southern should build a medium plant.

STRATEGY-DRIVEN INVESTMENTS

The operations manager may be the one held responsible for return-on-investment (ROI).

Once the strategy implications of potential investments have been considered, traditional investment analysis is appropriate. We introduce the investment aspects of capacity next.

Investment, Variable Cost, and Cash Flow

Capital investment requires cash flow as well as an evaluation of return on investments.

Because capacity and process alternatives exist, so do options regarding capital investment and variable cost. Managers must choose from among different financial options as well as capacity and process alternatives. Analysis should show the capital investment, variable cost, and cash flows as well as net present value for each alternative.

Net Present Value

Net present value
A means of determining the discounted value of a series of future cash receipts.

Determining the discount value of a series of future cash receipts is known as the **net present value** technique. By way of introduction, let us consider the time value of money. Say you invest $100.00 in a bank at 5% for 1 year. Your investment will be worth $100.00 + ($100.00)(.05) = $105.00. If you invest the $105.00 for a second year, it will be worth $105.00 + ($105.00)(.05) = $110.25 at the end of the second year. Of course, we could calculate the future value of $100.00 at 5% for as many years as we wanted by simply extending this analysis. However, there is an easier way to express this relationship mathematically. For the first year:

$$\$105 = \$100(1 + .05)$$

For the second year:

$$\$110.25 = \$105(1 + .05) = \$100(1 + .05)^2$$

In general,

$$F = P(1 + i)^N \tag{S7-7}$$

After the terrorist attacks of September 11, 2001 many airlines found their capacity plans were very wrong. As a result airlines are ferrying their excess capacity to the Mojave Desert for storage. A storage fee accrues each month and a very big capital investment sits idle. Delta Airlines pays a whopping $250,000 per month in lease fees for each Boeing 737 (one of the smaller jets) that sits idle in the desert.

TABLE S7.1 ■

Present Value of $1

YEAR	5%	6%	7%	8%	9%	10%	12%	14%
1	.952	.943	.935	.926	.917	.909	.893	.877
2	.907	.890	.873	.857	.842	.826	.797	.769
3	.864	.840	.816	.794	.772	.751	.712	.675
4	.823	.792	.763	.735	.708	.683	.636	.592
5	.784	.747	.713	.681	.650	.621	.567	.519
6	.746	.705	.666	.630	.596	.564	.507	.456
7	.711	.665	.623	.583	.547	.513	.452	.400
8	.677	.627	.582	.540	.502	.467	.404	.351
9	.645	.592	.544	.500	.460	.424	.361	.308
10	.614	.558	.508	.463	.422	.386	.322	.270
15	.481	.417	.362	.315	.275	.239	.183	.140
20	.377	.312	.258	.215	.178	.149	.104	.073

where F = future value (such as $110.25 or $105)
  P = present value (such as $100.00)
  i = interest rate (such as .05)
  N = number of years (such as 1 year or 2 years)

In most investment decisions, however, we are interested in calculating the present value of a series of future cash receipts. Solving for P, we get

$$P = \frac{F}{(1 + i)^N} \tag{S7-8}$$

When the number of years is not too large, the preceding equation is effective. However, when the number of years, N, is large, the formula is cumbersome. For 20 years, you would have to compute $(1 + i)^{20}$. Without a sophisticated calculator, this computation would be difficult. Interest-rate tables, such as Table S7.1, alleviate this situation. First, let us restate the present value equation:

$$P = \frac{F}{(1 + i)^N} = FX \tag{S7-9}$$

where X = a factor from Table S7.1 defined as = $1/(1 + i)^N$ and F = future value

Thus, all we have to do is find the factor X and multiply it by F to calculate the present value, P. The factors, of course, are a function of the interest rate, i, and the number of years, N. Table S7.1 lists some of these factors.

Equations (S7-8) and (S7-9) are used to determine the present value of one future cash amount, but there are situations in which an investment generates a series of uniform and equal cash amounts. This type of investment is called an *annuity*. For example, an investment might yield $300 per year for 3 years. Of course, you could use Equation (S7-8) three times, for 1, 2, and 3 years, but there is a shorter method. Although there is a formula that can be used to solve for the present value of an annual series of uniform and equal cash flows (an annuity), an easy-to-use table has been developed for this purpose. Like the customary present value computations, this calculation involves a factor. The factors for annuities are in Table S7.2. The basic relationship is

$$S = RX$$

where X = factor from Table S7.2
   S = present value of a series of uniform annual receipts
   R = receipts that are received every year for the life of the investment (the annuity)

The present value of a uniform annual series of amounts is an extension of the present value of a single amount, and thus Table S7.2 can be directly developed from Table S7.1. The factors for any given interest rate in Table S7.2 are nothing more than the cumulative sum of the values in Table

TABLE S7.2 ■

Present Value of an
Annuity of $1

YEAR	5%	6%	7%	8%	9%	10%	12%	14%
1	.952	.943	.935	.926	.917	.909	.893	.877
2	1.859	1.833	1.808	1.783	1.759	1.736	1.690	1.647
3	2.723	2.673	2.624	2.577	2.531	2.487	2.402	2.322
4	3.546	3.465	3.387	3.312	3.240	3.170	3.037	2.914
5	4.329	4.212	4.100	3.993	3.890	3.791	3.605	3.433
6	5.076	4.917	4.766	4.623	4.486	4.355	4.111	3.889
7	5.786	5.582	5.389	5.206	5.033	4.868	4.564	4.288
8	6.463	6.210	5.971	5.747	5.535	5.335	4.968	4.639
9	7.108	6.802	6.515	6.247	5.985	5.759	5.328	4.946
10	7.722	7.360	7.024	6.710	6.418	6.145	5.650	5.216
15	10.380	9.712	9.108	8.559	8.060	7.606	6.811	6.142
20	12.462	11.470	10.594	9.818	9.128	8.514	7.469	6.623

S7.1. In Table S7.1, for example, .952, .907, and .864 are the factors for years 1, 2, and 3 when the interest rate is 5%. The cumulative sum of these factors is 2.723 = .952 + .907 + .864. Now look at the point in Table S7.2 where the interest rate is 5% and the number of years is 3. The factor for the present value of an annuity is 2.723, as you would expect. Table S7.2 can be very helpful in reducing the computations necessary to make financial decisions.

Example S7 shows how to determine the present value of an annuity.

Example S7

River Road Medical Clinic is thinking of investing in a sophisticated new piece of medical equipment. It will generate $7,000 per year in receipts for 5 years. What is the present value of this cash flow? Assume an interest rate of 6%.

$$S = RX = \$7,000(4.212) = \$29,484$$

The factor from Table S7.2 (4.212) was obtained by finding that value when the interest rate is 6% and the number of years is 5. There is another way of looking at this example. If you went to a bank and took a loan for $29,484 today, your payments would be $7,000 per year for 5 years if the bank used an interest rate of 6% compounded yearly. Thus, $29,484 is the present value.

The net present value method is one of the best methods of ranking investment alternatives. The procedure is straightforward: You simply compute the present value of all cash flows for each investment alternative. When deciding among investment alternatives, you pick the investment with the highest net present value. Similarly, when making several investments, those with higher net present values are preferable to investments with lower net present values.

Example S8 shows how to use the net present value to choose between investment alternatives.

Example S8

Quality Plastics, Inc., is considering two different investment alternatives. Investment A has an initial cost of $25,000, and investment B has an initial cost of $26,000. Both investments have a useful life of 4 years. The cash flows for these investments follow. The cost of capital or the interest rate (i) is 8%. (Factors come from Table S7.1).

INVESTMENT A'S CASH FLOW	INVESTMENT B'S CASH FLOW	YEAR	PRESENT VALUE FACTOR AT 8%
$10,000	$9,000	1	.926
9,000	9,000	2	.857
8,000	9,000	3	.794
7,000	9,000	4	.735

To find the present value of the cash flows for each investment, we multiply the present value factor by the cash flow for each investment for each year. The sum of these present value calculations minus the initial investment is the net present value of each investment. The computations appear in the following table:

YEAR	INVESTMENT A'S PRESENT VALUES	INVESTMENT B'S PRESENT VALUES
1	$ 9,260 = (.926)($10,000)	$ 8,334 = (.926)($9,000)
2	7,713 = (.857)($9,000)	7,713 = (.857)($9,000)
3	6,352 = (.794)($8,000)	7,146 = (.794)($9,000)
4	5,145 = (.735)($7,000)	6,615 = (.735)($9,000)
Totals	$28,470	$29,808
Minus initial investment	−25,000	−26,000
Net present value	$ 3,470	$ 3,808

The net present value criterion shows investment B to be more attractive than investment A because it has a higher present value.

In Example S8, it was not necessary to make all of those present value computations for investment B. Because the cash flows are uniform, Table S7.2, the annuity table, gives the present value factor. Of course, we would expect to get the same answer. As you recall, Table S7.2 gives factors for the present value of an annuity. In this example, for payments of $9,000, cost of capital is 8% and the number of years is 4. Looking at Table S7.2 under 8% and 4 years, we find a factor of 3.312. Thus, the present value of this annuity is (3.312)($9,000) = $29,808, the same value as in Example S8.

Although net present value is one of the best approaches to evaluating investment alternatives, it does have its faults. Limitations of the net present value approach include the following:

1. Investments with the same net present value may have significantly different projected lives and different salvage values.
2. Investments with the same net present value may have different cash flows. Different cash flows may make substantial differences in the company's ability to pay its bills.
3. The assumption is that we know future interest rates, which we do not.
4. Payments are always made at the end of the period (week, month, or year), which is not always the case.

SUMMARY

Managers tie equipment selection and capacity decisions to the organization's missions and strategy. They design their equipment and processes to have capabilities beyond the tolerance required by their customers, while ensuring the flexibility needed for adjustments in technology, features, and volumes.

Good forecasting, break-even analysis, decision trees, cash flow, and net present value (NPV) techniques are particularly useful to operations managers when making capacity decisions.

Capacity investments are made effective by ensuring that the investments support a long-term strategy. The criteria for investment decisions are contribution to the overall strategic plan and winning profitable orders, not just return-on-investment. Efficient firms select the correct process and the correct capacity that contributes to their long-term strategy.

KEY TERMS

Capacity (p. 224)
Design capacity (p. 225)
Effective capacity (p. 225)
Utilization (p. 225)
Efficiency (p. 225)
Break-even analysis (p. 229)

Fixed costs (p. 229)
Variable costs (p. 229)
Contribution (p. 229)
Revenue function (p. 229)
Net present value (p. 234)

USING EXCEL OM FOR BREAK-EVEN ANALYSIS

Excel OM's Break-Even Analysis module is illustrated in Program S7.1. Using Jimmy Stephens, Inc., data from Example S3, Program S7.1 shows input data, the Excel formulas used to compute the break-even points, and the solution and graphical output.

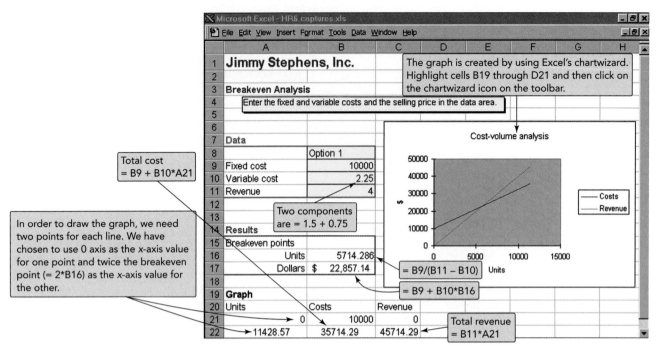

PROGRAM S7.1 ■ Excel OM's Break-Even Analysis, Using Example S3 Data

SOLVED PROBLEMS

Solved Problem S7.1

Sara James Bakery, described earlier in Example S1 and S2, has decided to increase its facilities by adding one additional process line. The firm will have 2 process lines, each working 7 days a week, 3 shifts per day, 8 hours per shift. Effective capacity is now 300,000 rolls. This addition, however, will reduce overall system efficiency to 85%. Compute the expected production with this new effective capacity.

SOLUTION

Expected Production = (Effective Capacity)(Efficiency)

$$= 300,000(.85)$$

$$= 255,000 \text{ rolls per week}$$

Solved Problem S7.2

Marty McDonald has a side business packaging software in Wisconsin. His annual fixed cost is $10,000, direct labor is $3.50 per package, and material is $4.50 per package. The selling price will be $12.50 per package. What is the break-even point in dollars? What is break-even in units?

SOLUTION

$$BEP_\$ = \frac{F}{1-(V/P)} = \frac{\$10,000}{1-(\$8.00/\$12.50)} = \frac{\$10,000}{.36} = \$27,777$$

$$BEP_x = \frac{F}{P-V} = \frac{\$10,000}{\$12.50-\$8.00} = \frac{\$10,000}{\$4.50} = 2,222 \text{ units}$$

Solved Problem S7.3

Your boss, Mr. La Forge, has told you to evaluate the cost of two machines. After some questioning, you are assured that they have the costs shown to the right. Assume:

(a) the life of each machine is 3 years, and

(b) the company thinks it knows how to make 14% on investments no riskier than this one.

	MACHINE A	MACHINE B
Original cost	$13,000	$20,000
Labor cost per year	2,000	3,000
Floor space per year	500	600
Energy (electricity) per year	1,000	900
Maintenance per year	2,500	500
Total annual cost	$ 6,000	$ 5,000
Salvage value	$ 2,000	$ 7,000

Determine via the present value method which machine to purchase.

SOLUTION

		MACHINE A			MACHINE B		
		COLUMN 1	COLUMN 2	COLUMN 3	COLUMN 4	COLUMN 5	COLUMN 6
Now	Expense	1.000	$13,000	$13,000	1.000	$20,000	$20,000
1 yr.	Expense	.877	6,000	5,262	.877	5,000	4,385
2 yr.	Expense	.769	6,000	4,614	.769	5,000	3,845
3 yr.	Expense	.675	6,000	4,050	.675	5,000	3,375
				$26,926			$31,605
3 yr.	Salvage Revenue	.675	$ 2,000	−1,350	.675	$ 7,000	−4,725
				$25,576			$26,880

We use 1.0 for payments with no discount applied against them (that is, when payments are made now, there is no need for a discount). The other values in columns 1 and 4 are from the 14% column and the respective year in Table S7.1 (for example, the intersection of 14% and 1 year is .877, etc.). Columns 3 and 6 are the products of the present value figures times the combined costs. This computation is made for each year and for the salvage value.

The calculation for Machine A for the first year is:

$$.877 \times (\$2,000 + \$500 + \$1,000 + \$2,500) = \$5,262$$

The salvage value of the product is *subtracted* from the summed costs, because it is a receipt of cash. Since the sum of the net costs for Machine B is larger than the sum of the net costs for Machine A, Machine A is the low-cost purchase, and your boss should be so informed.

INTERNET AND STUDENT CD-ROM EXERCISES

Visit our home page or use your student CD-ROM to help with material in this supplement.

 On Our Home Page, www.prenhall.com/heizer

- Self-Tests
- Practice Problems
- Internet Exercises
- Current Articles and Research
- Virtual Company Tour
- Internet Homework Problems
- Internet Cases

 On Your Student CD-ROM

- PowerPoint Lecture
- Practice Problems
- Active Model Exercises
- Excel OM
- Excel OM Data Files

ADDITIONAL CASE STUDIES

Internet Case Studies: Visit our Web site at www.prenhall.com/heizer **for this free case study:**

- **Southwestern University's Food Service:** Requires the development of a multiproduct break-even solution.

Harvard has selected these Harvard Business School cases to accompany this supplement (textbookcasematch.hbsp.harvard.edu)**:**

- **National Cranberry Cooperative** (#688–122): Requires the student to analyze process, bottlenecks, and capacity.
- **Lenzing AG: Expanding in Indonesia** (#796–099): Considers how expansion affects the company's competitive position.
- **Chaparral Steel** (#687–045): Examines a major capacity expansion proposal of Chaparral Steel, a steel minimill.

Location Strategies

Chapter Outline

LEARNING OBJECTIVES

When you complete this chapter you should be able to

IDENTIFY OR DEFINE:

Objective of location strategy

International location issues

Clustering

Geographic information systems

DESCRIBE OR EXPLAIN:

Three methods of solving the location problem:

- Factor-rating method
- Locational break-even analysis
- Center-of-gravity method

Location Provides Competitive Advantage for Federal Express

Overnight-delivery powerhouse Federal Express has believed in the hub concept for its 38-year existence. Even though Fred Smith, founder and CEO, got a C on his college paper proposing a hub for small-package delivery, the idea has proven extremely successful. Starting with a hub in Memphis, Tennessee, the $18-billion firm has added a European hub in Paris and an Asian one in Subic Bay, Philippines. In 1998, a second Asian hub in Taipei, Taiwan, was added as a backup for Subic Bay, should political troubles or weather ever close that facility. Federal Express's fleet of over 600 planes flies into 325 airports worldwide, then delivers to the door with more than 38,000 vans.

Why was Memphis picked as Federal Express's central location? For one thing, it is located in the middle of the U.S. For another, it has very few hours of bad weather closures, perhaps contributing to the firm's excellent flight-safety record.

At the Federal Express hub in Memphis, Tennessee, approximately 100 Federal Express aircraft converge each night around midnight with more than one million documents and packages.

At the preliminary sorting area, packages and documents are sorted and sent to a secondary sorting area. The Memphis facility covers 1.5 million square feet; it is big enough to hold 33 football fields. Packages are sorted and exchanged until 4 A.M.

Packages and documents that have already gone through the primary and secondary sorts are checked by city, state, and zip code. They are then placed in containers that are loaded onto aircraft for delivery to their final destinations in 211 countries.

Each night, except Sunday, Federal Express brings to Memphis packages from throughout the world that are going to cities for which Federal Express does not have direct flights. The central hub permits service to a far greater number of points with fewer aircraft than the traditional City-A-to-City-B system. It also allows Federal Express to match aircraft flights with package loads each night and to reroute flights when load volume requires it, a major cost savings. Moreover, Federal Express also believes that the central hub system helps reduce mishandling and delay in transit because there is total control over the packages from pickup point through delivery.

Federal Express jet departing the Subic Bay Asian hub. Subic Bay's 3,100 employees sort 6,000 boxes and 10,000 documents per hour in the 90,000-square-foot facility located on what used to be a U.S. military base in the Philippines.

THE STRATEGIC IMPORTANCE OF LOCATION

When Federal Express opened its Asian hub in Taiwan in 1998 and doubled its flights to China, it set the stage for its new "round-the-world" flights linking its Paris and Memphis package hubs to Asia. When Mercedes announced its plans to build its first major overseas plant in Vance, Alabama, it completed a year of competition among 170 sites in 30 states and two countries. When Hard Rock Cafe opened in Moscow in late 2002, it ended 3 years of advance preparation of a Russian food supply chain.

One of the most important strategic decisions made by companies like Federal Express, Daimler Chrysler, and Hard Rock is where to locate their operations. The international aspect of these decisions is an indication of the global nature of location decisions. With the opening of the Soviet and Chinese blocs, a great transformation is taking place. World markets have doubled and the global nature of business is accelerating.

Firms throughout the world are using the concepts and techniques of this chapter to address the location decision because location greatly affects both fixed and variable costs. It has a major impact on the overall risk and profit of the company. For instance, depending on the product and type of production or service taking place, transportation costs alone can total as much as 25% of the product's selling price. That is, one-fourth of a firm's total revenue may be needed just to cover freight expenses of the raw materials coming in and finished products going out. Other costs that may be influenced by location include taxes, wages, raw material costs, and rents.

Location and Costs   Because location is such a significant cost driver, the consulting firm McKinsey believes "location ultimately has the power to make (or break) a company's business strategy."[1] Key multinationals in every major industry, from automobiles to cellular phones, now have or are planning a presence in each of their major markets. Motorola, however, has often rejected countries even when costs are lower if infrastructure and education levels cannot support specific production technologies. Location decisions based on a low-cost strategy require careful consideration.

Once management is committed to a specific location, many costs are firmly in place and difficult to reduce. For instance, if a new factory location is in a region with high energy costs, even good management with an outstanding energy strategy is starting at a disadvantage. Management is in a similar bind with its human resource strategy if labor in the selected location is expensive, ill-trained, or has a poor work ethic. Consequently, hard work to determine an optimal facility location is a good investment.

The objective of location strategy is to maximize the benefit of location to a firm.

The location decision often depends on the type of business. For industrial location decisions, the strategy is usually minimizing costs, whereas for retail and professional service organizations, the strategy focuses on maximizing revenue. Warehouse location strategy, however, may be driven by a combination of cost and speed of delivery. In general, the *objective of location strategy* is to maximize the benefit of location to the firm.

Companies make location decisions relatively infrequently, usually because demand has outgrown the current plant's capacity or because of changes in labor productivity, exchange rates, costs, or local attitudes. Companies may also relocate their manufacturing or service facilities because of shifts in demographics and customer demand.

Location options include (1) expanding an existing facility instead of moving, (2) maintaining current sites while adding another facility elsewhere, or (3) closing the existing facility and moving to another location.

FACTORS THAT AFFECT LOCATION DECISIONS

Selecting a facility location is becoming much more complex with the globalization of the workplace. As we saw in Chapter 2, globalization has taken place because of the development of (1) market economics; (2) better international communications; (3) more rapid, reliable travel and shipping; (4) ease of capital flow between countries; and (5) high differences in labor costs. Many firms now consider opening new offices, factories, retail stores, or banks outside their home country. Location decisions transcend national borders. In fact, as Figure 8.1 shows, the sequence of location decisions often begins with choosing a country in which to operate.

Video 8.1

Hard Rock's Location Selection

[1]See Andrew D. Bartness, "The Plant Location Puzzle," *Harvard Business Review* (March–April 1994): 32.

FIGURE 8.1 ■

Some Considerations
and Factors That Affect
Location Decisions

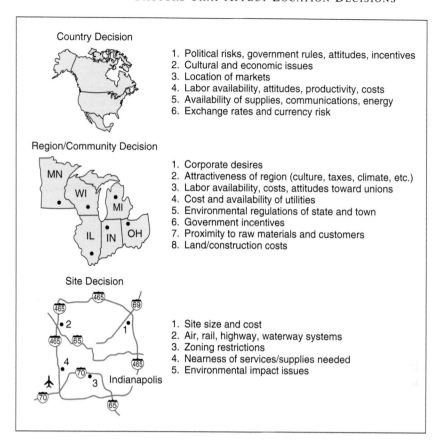

TABLE 8.1 ■

Global Competitiveness
of 75 Selected Countries,
Based on Annual Surveys
of 4,000 Business
Executives

COUNTRY	2001 RANKING
Finland	1
U.S.	2
Netherlands	3
Germany	4
⋮	⋮
Canada	11
⋮	⋮
Japan	15
⋮	⋮
Brazil	30
⋮	⋮
Russia	58
⋮	⋮
Bolivia	75

Source: www.weforum.org,
2002.

One approach to selecting a country is to identify what the parent organization believes are critical success factors (CSFs) needed to achieve competitive advantage. Six possible country CSFs are listed at the top of Figure 8.1. Using such factors (including some negative ones, such as crime) the World Economic Forum annually ranks the global competitiveness of 75 countries (see Table 8.1). Finland landed first in 2001 because of its high rates of saving and investment, openness to trade, quality education, and efficient government.

Once a firm decides which country is best for its location, it focuses on a region of the chosen country and a community. The final step in the location decision process is choosing a specific site within a community. The company must pick the one location that is best suited for shipping and receiving, zoning, utilities, size, and cost. Again, Figure 8.1 summarizes this series of decisions and the factors that affect them.

Besides globalization, a number of other factors affect the location decision. Among these are labor productivity, foreign exchange, culture, changing attitudes toward the industry, and proximity to markets, suppliers, and competitors.

Labor Productivity

When deciding on a location, management may be tempted by an area's low wage rates. However, wage rates cannot be considered by themselves, as Quality Coils, Inc., discovered when it opened its plant in Mexico (see the *OM in Action* box on the next page). Management must also consider productivity.

As discussed in Chapter 1, differences exist in productivity in various countries. What management is really interested in is the combination of productivity and the wage rate. For example, if Quality Coils pays $70 per day with 60 units produced per day in Connecticut, it will spend less on labor than at a Mexican plant that pays $25 per day with a productivity of 20 units per day:

In most cases, it is cheaper to make clothes in Korea, Taiwan, or Hong Kong and ship them to the U.S. than it is to produce them in the U.S. However, final cost is the critical factor and low productivity can negate low cost.

$$\frac{\text{Labor cost per day}}{\text{Productivity (that is, units per day)}} = \text{cost per unit}$$

OM IN ACTION

Quality Coils Pulls The Plug On Mexico

Keith Gibson, president of Quality Coils, Inc., saw the savings of low Mexican wages and headed South. He shut down a factory in Connecticut and opened one in Juarez, where he could pay Mexicans one-third the wage rates he was paying Americans. "All the figures pointed out we should make a killing," says Gibson.

Instead, his company was nearly destroyed. The electromagnetic coil maker regularly lost money during 4 years in Mexico. High absenteeism, low productivity, and problems of long-distance management wore down Gibson until he finally pulled the plug on Juarez.

Moving back to the U.S. and rehiring some of his original workers, Gibson learned, "I can hire one person in Connecticut for what three were doing in Juarez."

When American unions complain that they cannot compete against the low wages in other countries and when the teamster rallies chant "$4 a day/No way!" they overlook several factors. First, productivity in low-wage countries often erases a wage advantage that is not nearly as great as people believe. Second, a host of problems, from poor roads to corrupt governments, run up operating costs. And most importantly, the cost of labor for most U.S. manufacturers is less important than such factors as the skill of the workforce, the quality of transportation, and access to technology.

Sources: Nation's Business (February 1997): 6; and the *Wall Street Journal* (September 15, 1993): A1.

Labor costs in many underdeveloped countries are now one-third of those in developed nations. However, when labor costs are only 15% of manufacturing costs, the difference may not overcome many other disadvantages of low-labor-cost countries.

Case 1: Connecticut plant

$$\frac{\$70 \text{ Wages per day}}{60 \text{ Units produced per day}} = \frac{\$70}{60} = \$1.17 \text{ per unit}$$

Case 2: Juarez, Mexico, plant

$$\frac{\$25 \text{ Wages per day}}{20 \text{ Units produced per day}} = \frac{\$25}{20} = \$1.25 \text{ per unit}$$

Employees with poor training, poor education, or poor work habits may not be a good buy even at low wages. By the same token, employees who cannot or will not always reach their places of work are not much good to the organization, even at low wages. (Labor cost per unit is sometimes called the *labor content* of the product.)

Assembly plants operating along the Mexican side of the border, from Texas to California, are called maquiladoras. Some 2,000 firms and industrial giants, such as General Motors, Zenith, Hitachi, and GE, operate these plants, which were designed to help both sides of the impoverished border region. Over 2 million workers are employed in these cross-border plants. Mexican wages are low, and at current exchange rates, companies don't look to the Far East as they once did.

Exchange Rates and Currency Risk

Although wage rates and productivity may make a country seem economical, unfavorable exchange rates might negate any savings. Sometimes, though, firms can take advantage of a particularly favorable exchange rate by relocating or exporting to a foreign country. However, the values of foreign currencies continually rise and fall in most countries. Such changes could well make what was a good location in 2003 a disastrous one in 2008.

Costs

Tangible costs

Readily identifiable costs that can be measured with some precision.

We can divide location costs into two categories, tangible and intangible. **Tangible costs** are those costs that are readily identifiable and precisely measured. They include utilities, labor, material, taxes, depreciation, and other costs that the accounting department and management can identify. In addition, such costs as transportation of raw materials, transportation of finished goods, and site construction are all factored into the overall cost of a location. Government incentives, as we see in the *OM in Action* box, "How Big Incentives Won Alabama the Auto Industry," certainly impact a location's cost.

Intangible costs

A category of location costs that cannot be easily quantified, such as quality of life and government.

Intangible costs are less easily quantified. They include quality of education, public transportation facilities, community attitudes toward the industry and the company, and quality and attitude of prospective employees. They also include quality-of-life variables, such as climate and sports teams, that may influence personnel recruiting.

Attitudes

Attitudes of national, state, and local governments toward private property, zoning, pollution, and employment stability may be in flux. Governmental attitudes at the time a location decision is made may not be lasting ones. Moreover, management may find that these attitudes can be influenced by their own leadership.

Worker attitudes may also differ from country to country, region to region, and small town to city. Worker views regarding turnover, unions, and absenteeism are all relevant factors. In turn, these attitudes can affect a company's decision whether to make offers to current workers if the firm relocates to a new location. The case study at the end of this chapter, "Southern Recreational Vehicle Company," describes a St. Louis firm that actively chose *not to relocate* any of its workers when it moved to Mississippi.

One of the greatest challenges in a global operations decision is dealing with another country's culture. Cultural variations in punctuality by employees and suppliers make a marked difference in

OM IN ACTION

How Big Incentives Won Alabama the Auto Industry

In 1993, Alabama persuaded Mercedes-Benz to build its first U.S. auto plant in the town of Vance by offering the luxury carmaker $253 million worth of incentives—$169,000 for every job Mercedes promised the state.

Taxpayers considered the deal such a boondoggle that they voted Governor Jim Folsom out of office long before the first Mercedes SUV rolled off the new assembly line in 1997. Today, the deal looks a little more like a bargain—suggesting that the practice of paying millions of taxpayer dollars to lure big employers can sometimes have a big payoff.

Mercedes surpassed its pledge to create 1,500 jobs at the Vance plant with a planned workforce of 4,000 by 2005.

Then in 2001, Honda opened a factory 70 miles east of the Mercedes plant, to build its Odyssey minivan. Toyota Motor Corp.'s new plant near Huntsville started producing engines in 2002. Those two automakers also received incentives.

To cement Alabama's reputation as the South's busiest auto-making center, Hyundai Motor Co. of South Korea picked a site near Montgomery for its first U.S. assembly plant. The factory will begin production in 2005, and will employ 2,000 workers to make 300,000 sedans and SUVs a year.

Is the state giving away more than it gets in return? That's what many economists argue. Other former foes of incentives now argue that manufacturers' arrivals herald "Alabama's new day."

Sources: The Wall Street Journal (April 5, 2002): A1, A24; and *Knight Ridder Tribune Business News* (April 12, 2002): 1.

TABLE 8.2 ■

Ranking Corruption in
Selected Countries
(score of 10 represents a
corruption-free country)

RANK		SCORE
1	Finland	9.7
⋮		⋮
7	Canada	9.0
⋮		⋮
10	United Kingdom	8.7
⋮		⋮
16	U.S.	7.7
⋮		⋮
18	Germany & Israel (tie)	7.3
⋮		⋮
20	Japan	7.1
⋮		⋮
59	China	3.5
⋮		⋮
71	India & Russia (tie)	2.7
⋮		⋮
101	Nigeria	1.6
102	Bangladesh	1.2

Source: Transparency
International's 2002 survey at
www.transparency.org.

Clustering
The location of
competing companies
near each other, often
because of a critical mass
of information, talent,
venture capital, or natural
resources.

production and delivery schedules. Bribery likewise creates substantial economic inefficiency, as well as ethical and legal problems in the global arena. As a result, operations managers face significant challenges when building effective supply chains that include foreign firms. Table 8.2 provides one ranking of corruption in countries around the world.

Proximity to Markets

For many firms it is extremely important to locate near customers. Particularly, service organizations, like drugstores, restaurants, post offices, or barbers, find proximity to market is *the* primary location factor. Manufacturing firms find it useful to be close to customers when transporting finished goods is expensive or difficult (perhaps because they are bulky, heavy, or fragile). In addition, with the trend toward just-in-time production, suppliers want to locate near users to speed deliveries. For a firm like Coca-Cola, whose product's primary ingredient is water, it makes sense to have bottling plants in many cities rather than shipping heavy (and sometimes fragile glass) containers cross country.

Proximity to Suppliers

Firms locate near their raw materials and suppliers because of (1) perishability, (2) transportation costs, or (3) bulk. Bakeries, dairy plants, and frozen seafood processors deal with *perishable* raw materials, so they often locate close to suppliers. Companies dependent on inputs of heavy or bulky raw materials (such as steel producers using coal and iron ore) face expensive inbound *transportation costs*, so transportation costs become a major factor. And goods for which there is a *reduction in bulk* during production (such as lumber mills locating in the Northwest near timber resources) typically need to be near the raw material.

Proximity to Competitors (Clustering)

Companies also like to locate, somewhat surprisingly, near competitors. This tendency, called **clustering**, often occurs when a major resource is found in that region. Such resources include natural resources, information resources, venture capital resources, and talent resources. A natural resource of land and climate motivates wine makers to gravitate to Napa Valley in the U.S. and the Bordeaux region in France. A talent resource entices software firms to cluster in both Silicon Valley and in Boston, with their plentiful supply of bright young graduates from colleges such as Berkeley and Stanford in California and Harvard and MIT in Massachusetts. The venture capital resources available in these areas is an added attraction. Similarly, race car builders from all over the world cluster in the Huntington/North Hampton region of England, where they have found a critical mass of talent and information. Theme parks such as Disney, Universal Studios, and SeaWorld cluster in Orlando, a hot spot for entertainment talent, warm weather, tourists, and inexpensive labor. Even fast-food chains such as McDonald's, Burger King, Wendy's, and Pizza Hut find that locations within 1 mile of each other stimulate sales.

However, Italy may be the true leader when it comes to clustering, with northern zones of that country holding world leadership in such specialties as ceramic tile (Modena), gold jewelry (Vicenza), machine tools (Busto Arsizio), cashmere and wool (Biella), designer eyeglasses (Belluma), and pasta machines (Parma).[2]

METHODS OF EVALUATING LOCATION ALTERNATIVES

Four major methods are used for solving location problems: the factor-rating method, locational break-even analysis, the center-of-gravity method, and the transportation model. This section describes these approaches.

Factor-rating method
A location method that
instills objectivity into the
process of identifying
hard-to-evaluate costs.

The Factor-Rating Method

There are many factors, both qualitative and quantitative, to consider in choosing a location. Some of these factors are more important than others, so managers can use weightings to make the decision process more objective. The **factor-rating method** is popular because a wide variety of factors,

[2]*Financial Times* (November 6, 2001): 11.

TABLE 8.3 ■

Critical Success Factors Affecting Location Selection

Labor costs (including wages, unionization, productivity)
Labor availability (including attitudes, age, distribution, skills)
Proximity to raw materials and suppliers
Proximity to markets
Government fiscal policies (including incentives, taxes, unemployment compensation)
Environmental regulations
Utilities (including gas, electric, water, and their costs)
Site costs (including land, expansion, parking, drainage)
Transportation availability (including rail, air, water, interstate roads)
Quality-of-life issues in the community (including all levels of education, cost of living, health care, sports, cultural activities, transportation, housing, entertainment, religious facilities)
Foreign exchange (including rates, stability)
Quality of government (including stability, honesty, attitudes toward new business—whether overseas or local)

from education to recreation to labor skills, can be objectively included. Table 8.3 lists a few of the many factors that affect location decisions.

The factor-rating method has six steps:

1. Develop a list of relevant factors called *critical success factors* (such as those in Table 8.3).
2. Assign a weight to each factor to reflect its relative importance in the company's objectives.
3. Develop a scale for each factor (for example, 1 to 10 or 1 to 100 points).
4. Have management score each location for each factor, using the scale in step 3.
5. Multiply the score by the weights for each factor and total the score for each location.
6. Make a recommendation based on the maximum point score, considering the results of quantitative approaches as well.

The numbers used in factor weighting can be subjective and the model's results are not "exact" even though this is a quantitative approach.

Example 1

**Excel OM Data File
Ch08Ex1.xla**

Five Flags over Florida, a U.S. chain of 10 family-oriented theme parks, has decided to expand overseas by opening its first park in Europe. The rating sheet in Table 8.4 provides a list of critical success factors that management has decided are important; their weightings and their rating for two possible sites—Dijon, France, and Copenhagen, Denmark—are shown.

TABLE 8.4 ■ Weights, Scores, and Solution

CRITICAL SUCCESS FACTOR	WEIGHT	SCORES (OUT OF 100) FRANCE	DENMARK	WEIGHTED SCORES FRANCE	DENMARK
Labor availability and attitude	.25	70	60	(.25)(70) = 17.5	(.25)(60) = 15.0
People-to-car ratio	.05	50	60	(.05)(50) = 2.5	(.05)(60) = 3.0
Per capita income	.10	85	80	(.10)(85) = 8.5	(.10)(80) = 8.0
Tax structure	.39	75	70	(.39)(75) = 29.3	(.39)(70) = 27.3
Education and health	.21	60	70	(.21)(60) = 12.6	(.21)(70) = 14.7
Totals	1.00			70.4	68.0

Table 8.4 also indicates use of weights to evaluate alternative site locations. Given the option of 100 points assigned to each factor, the French location is preferable. By changing the points or weights slightly for those factors about which there is some doubt, we can analyze the sensitivity of the decision. For instance, we can see that changing the scores for "labor availability and attitude" by 10 points can change the decision.

When a decision is sensitive to minor changes, further analysis of either the weighting or the points assigned may be appropriate. Alternatively, management may conclude that these intangible factors are not the proper criteria on which to base a location decision. Managers therefore place primary weight on the more quantitative aspects of the decision.

Locational Break-Even Analysis

Locational break-even analysis

A cost-volume analysis to make an economic comparison of location alternatives.

Locational break-even analysis is the use of cost-volume analysis to make an economic comparison of location alternatives. By identifying fixed and variable costs and graphing them for each location, we can determine which one provides the lowest cost. Locational break-even analysis can be done mathematically or graphically. The graphic approach has the advantage of providing the range of volume over which each location is preferable.

The three steps to locational break-even analysis are

1. Determine the fixed and variable cost for each location.
2. Plot the costs for each location, with costs on the vertical axis of the graph and annual volume on the horizontal axis.
3. Select the location that has the lowest total cost for the expected production volume.

Example 2

Excel OM Data File
Ch08Ex2.xla

A manufacturer of automobile carburetors is considering three locations—Akron, Bowling Green, and Chicago—for a new plant. Cost studies indicate that fixed costs per year at the sites are $30,000, $60,000, and $110,000, respectively; and variable costs are $75 per unit, $45 per unit, and $25 per unit, respectively. The expected selling price of the carburetors produced is $120. The company wishes to find the most economical location for an expected volume of 2,000 units per year.

For each of the three, we can plot the fixed costs (those at a volume of zero units) and the total cost (fixed costs + variable costs) at the expected volume of output. These lines have been plotted in Figure 8.2.

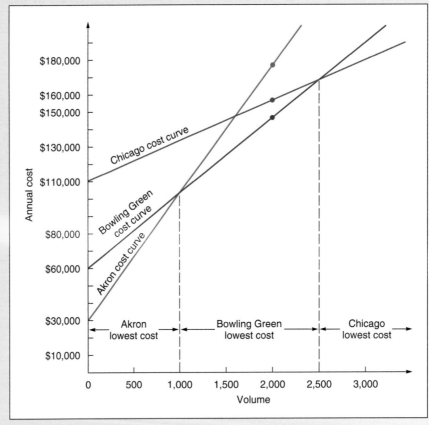

FIGURE 8.2 ■ Crossover Chart for Locational Break-Even Analysis

For Akron,

$$\text{Total Cost} = \$30,000 + \$75(2,000) = \$180,000$$

For Bowling Green,

$$\text{Total Cost} = \$60,000 + \$45(2,000) = \$150,000$$

For Chicago,

$$\text{Total cost} = \$110,000 + \$25(2,000) = \$160,000$$

With an expected volume of 2,000 units per year, Bowling Green provides the lowest cost location. The expected profit is

$$\text{Total revenue} - \text{Total cost} = \$120(2,000) - \$150,000 = \$90,000 \text{ per year}$$

The crossover point for Akron and Bowing Green is

$$30,000 + 75(x) = 60,000 + 45(x)$$
$$30(x) = 30,000$$
$$x = 1,000$$

and the crossover point for Bowling Green and Chicago is

$$60,000 + 45(x) = 110,000 + 25(x)$$
$$20(x) = 50,000$$
$$x = 2,500$$

Thus, for a volume of less than 1,000, Akron would be preferred, and for a volume greater than 2,500, Chicago would yield the greatest profit.

Center-of-Gravity Method

Center-of-gravity method

A mathematical technique used for finding the best location for a single distribution point that services several stores or areas.

The **center-of-gravity method** is a mathematical technique used for finding the location of a distribution center that will minimize distribution costs. The method takes into account the location of markets, the volume of goods shipped to those markets, and shipping costs in finding the best location for a distribution center.[3]

The first step in the center-of-gravity method is to place the locations on a coordinate system. This will be illustrated in Example 3. The origin of the coordinate system and the scale used are arbitrary, just as long as the relative distances are correctly represented. This can be done easily by placing a grid over an ordinary map. The center of gravity is determined by Equations (8-1) and (8-2):

$$x\text{-coordinate of the center of gravity} = \frac{\sum_i d_{ix} Q_i}{\sum_i Q_i} \tag{8-1}$$

$$y\text{-coordinate of the center of gravity} = \frac{\sum_i d_{iy} Q_i}{\sum_i Q_i} \tag{8-2}$$

where d_{ix} = x-coordinate of location i

d_{iy} = y-coordinate of location i

Q_i = Quantity of goods moved to or from location i

Note that Equations (8-1) and (8-2) include the term Q_i, the quantity of supplies transferred to or from location i.

Since the number of containers shipped each month affects cost, distance alone should not be the principal criterion. The center-of-gravity method assumes that cost is directly proportional to both distance and volume shipped. The ideal location is that which minimizes the weighted distance between the warehouse and its retail outlets, where the distance is weighted by the number of containers shipped.

[3]For a discussion of the use of the center-of-gravity method in a warehouse location and consolidation problem, see Charles A. Watts, "Using a Personal Computer to Solve a Warehouse Location/Consolidation Problem," *Production and Inventory Management Journal* (fourth quarter, 2000): 23–28.

Example 3

Active Model 8.1

Example 3 is further illustrated in Active Model 8.1 on the CD-ROM and in the Exercise located in your Student Lecture Guide.

Consider the case of Quain's Discount Department Stores, a chain of four large Target-type outlets. The firm's store locations are in Chicago, Pittsburgh, New York, and Atlanta; they are currently being supplied out of an old and inadequate warehouse in Pittsburgh, the site of the chain's first store. Data on demand rates at each outlet are shown in Table 8.5.

TABLE 8.5 ■ Demand for Quain's Discount Department Stores

STORE LOCATION	NUMBER OF CONTAINERS SHIPPED PER MONTH
Chicago	2,000
Pittsburgh	1,000
New York	1,000
Atlanta	2,000

The firm has decided to find some "central" location in which to build a new warehouse. Its current store locations are shown in Figure 8.3. For example, location 1 is Chicago, and from Table 8.5 and Figure 8.3, we have:

$$d_{1x} = 30$$
$$d_{1y} = 120$$
$$Q_1 = 2,000$$

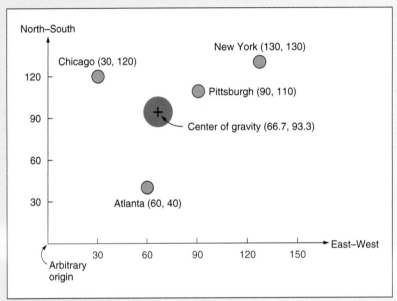

FIGURE 8.3 ■ Coordinate Locations of Four Quain's Department Stores and Center of Gravity

Using the data in Table 8.5 and Figure 8.3 for each of the other cities, in Equations (8-1) and (8-2) we find:

x-coordinate of the center of gravity

$$= \frac{(30)(2000) + (90)(1000) + (130)(1000) + (60)(2000)}{2000 + 1000 + 1000 + 2000} = \frac{400,000}{6,000}$$

$$= 66.7$$

y-coordinate of the center of gravity

$$= \frac{(120)(2000) + (110)(1000) + (130)(1000) + (40)(2000)}{2000 + 1000 + 1000 + 2000} = \frac{560,000}{6,000}$$

$$= 93.3$$

This location (66.7, 93.3) is shown by the crosshair in Figure 8.3. By overlaying a U.S. map on this exhibit, we find this location is near central Ohio. The firm may well wish to consider Columbus, Ohio, or a nearby city as an appropriate location.

FIGURE 8.4 ∎

Worldwide Distribution
of Volkswagens and
Parts

Source: The Economist, Ltd.
Distributed by the *New York
Times*/Special Features.

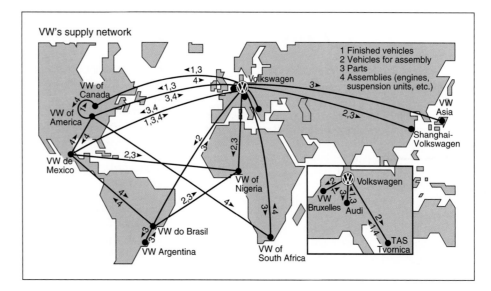

Transportation model
A technique for solving a
class of linear
programming problems.

Transportation Model

The objective of the **transportation model** is to determine the best pattern of shipments from several points of supply (sources) to several points of demand (destinations) so as to minimize total production and transportation costs. Every firm with a network of supply-and-demand points faces such a problem. The complex Volkswagen supply network (shown in Figure 8.4) provides one such illustration. We note in Figure 8.4, for example, that VW de Mexico ships vehicles for assembly and parts to VW of Nigeria and sends assemblies to VW do Brasil, while it receives parts and assemblies from headquarters in Germany.

Although the linear programming (LP) technique can be used to solve this type of problem, more efficient, special-purpose algorithms have been developed for the transportation application. The transportation model finds an initial feasible solution and then makes step-by-step improvement until an optimal solution is reached.

SERVICE LOCATION STRATEGY

It is often desirable to
locate near competition;
large department stores
often attract more
shoppers when
competitors are close by.
The same applies to shoe
stores, fast-food
restaurants, and others.

While the focus in industrial-sector location analysis is on minimizing cost, the focus in the service sector is on maximizing revenue. This is because manufacturing firms find costs tend to vary substantially between locations, while service firms find location often has more impact on revenue than cost. Therefore, for the service firm, a specific location often influences revenue more than it does cost. This means that the location focus for service firms should be on determining the volume of business and revenue. See the *OM in Action* box, "Hospitals Think Location, Location, Location." There are eight major components of volume and revenue for the service firm:

1. Purchasing power of the customer-drawing area.
2. Service and image compatibility with demographics of the customer-drawing area.
3. Competition in the area.
4. Quality of the competition.
5. Uniqueness of the firm's and competitors' locations.
6. Physical qualities of facilities and neighboring businesses.
7. Operating policies of the firm.
8. Quality of management.

Realistic analysis of these factors can provide a reasonable picture of the revenue expected. The techniques used in the service sector include correlation analysis, traffic counts, demographic analysis, purchasing power analysis, the factor-rating method, the center-of-gravity method, and geographic information systems. Table 8.6 provides a summary of location strategies for both service and goods-producing organizations.

OM IN ACTION

Hospitals Think Location, Location, Location

In the competitive medical marketplace of major U.S. cities, the battle for market share—and new patients—is not being waged with scalpels and stethoscopes. Instead, as hospitals build clinics in the neighborhoods where potential customers live, it is being waged with hammers and nails.

New York's Columbia-Presbyterian Medical Center has found that location is becoming fiercely important as managed care systems have begun to take hold. In the new world of fixed reimbursements, shorter hospital stays, and tight rationing of services, Columbia must make its money from volume rather than service fees.

To attract patients, Columbia has opened 23 satellites, ranging from branches in ethnic New York enclaves, to

upscale neighborhoods, to suburban communities, and even to distant cities in Eastern Europe (such as Moscow, Warsaw, Prague, Budapest) that have large American expatriate populations.

Satellites, which typically have no beds and no round-the-clock care, are cheaper to operate than highly specialized hospitals. Babies, for instance, can be delivered at alternative sites for half the main hospital's $9,000 fee.

Marketing its new clinic on Manhattan's elegant Upper East Side, Columbia's ads read "a unique combination of excellence and convenience." For those patients needing overnight accommodation, Dr. William T. Speck, Columbia president, suggests a reservation at the nearby Plaza Hotel. "The food is better," says Dr. Speck.

Sources: Modern Healthcare (June 21, 1999): 30–32; and the *New York Times* (August 22, 1999): 11-1

TABLE 8.6 ■

Location Strategies—Service vs. Goods-Producing Organizations

SERVICE/RETAIL/PROFESSIONAL LOCATION	GOODS-PRODUCING LOCATION
REVENUE FOCUS	**COST FOCUS**

Volume/revenue
    Drawing area; purchasing power
    Competition; advertising/pricing
 	
Physical quality
    Parking/access; security/lighting;
       appearance/image
 	
Cost determinants
    Rent
    Management caliber
    Operation policies (hours, wage rates)
	Tangible costs
	    Transportation cost of raw material
	    Shipment cost of finished goods
	    Energy and utility cost; labor; raw material;
	       taxes, and so on
	
	Intangible and future costs
	    Attitude toward union
	    Quality of life
	    Education expenditures by state
	    Quality of state and local government

TECHNIQUES	**TECHNIQUES**
Regression models to determine importance of various factors Factor-rating method Traffic counts Demographic analysis of drawing area Purchasing power analysis of area Center-of-gravity method Geographic information systems	Transportation method Factor-rating method Locational break-even analysis Crossover charts
ASSUMPTIONS	**ASSUMPTIONS**

Location is a major determinant of revenue
High customer-contact issues are critical
Costs are relatively constant for a given area;
    therefore, the revenue function is critical
	Location is a major determinant of cost
	Most major costs can be identified explicitly
	    for each site
	Low customer contact allows focus on the
	    identifiable costs
	Intangible costs can be evaluated

Even with reduced tax benefits and a saturated hotel market, opportunities still exist when hotel/motel locations are right. Good sites include those near hospitals and medical centers. As medical complexes in metropolitan areas continue to increase, so does the need for hotels to house patients' families. Additionally, medical services such as outpatient care, shorter hospital stays, and more diagnostic tests increase the need for hotels near hospitals.

How Hotel Chains Select Sites

One of the most important decisions in the hospitality industry is location. Hotel chains that pick good sites more accurately and quickly than competitors have a distinct strategic advantage. La Quinta Motor Inns, headquartered in San Antonio, Texas, is a moderately priced chain of 330 inns oriented toward frequent business travelers. To model motel-selection behavior and predict success of a site, La Quinta turned to statistical regression analysis.[4]

The hotel started by testing 35 independent variables, trying to find which of them would have the highest correlation with predicted profitability, the dependent variable. "Competitive" independent variables included the number of hotel rooms in the vicinity and average room rates. "Demand generator" variables were such local attractions as office buildings and hospitals that drew potential customers to a 4-mile-radius trade area. "Demographic" variables, such as local population and unemployment rate, can also affect the success of a hotel. "Market awareness" factors, such as the number of inns in a region, were a fourth category. Finally, "physical characteristics" of the site, such as ease of access or sign visibility, provided the last group of the 35 independent variables.

In the end, the regression model chosen, with a coefficient of determination (r^2) of 51%, included just four predictive variables. They are the *price of the inn*, *median income levels*, the *state population per inn*, and the *location of nearby colleges* (which serves as a proxy for other demand generators). La Quinta then used the regression model to predict profitability and developed a cutoff that gave the best results for predicting success or failure of a site. A spreadsheet is now used to implement the model, which applies the decision rule and suggests "build" or "don't build."

The Telemarketing Industry

Those industries and office activities that require neither face-to-face contact with the customer nor movement of material broaden location options substantially. A case in point is the telemarketing industry and those selling over the Internet, in which our traditional variables (as noted earlier) are

---

[4] Sheryl Kimes and James Fitzsimmons, "Selecting Profitable Hotel Sites at La Quinta Motor Inns," *Interfaces* (March–April 1990): 12–20. Also see the *Wall Street Journal* (July 19, 1995): B1, B5, for a discussion of how Amerihost Inns makes its location decisions.

Where to locate telemarketers? Sixteen states now permit private companies to hire prisoners to pitch products, conduct surveys, or answer hotel/airline reservation systems.

no longer relevant. Where the electronic movement of information is good, the cost and availability of labor may drive the location decision. For instance, Fidelity Investments relocated many of its employees from Boston to Covington, Kentucky. Now employees in the low-cost Covington region connect, by inexpensive fiber-optic phone lines, to their colleagues in the Boston office at a cost of less than a penny per minute. That is less than Fidelity spends on local connections.

The changes in location criteria may also affect a number of other businesses. For instance, states with smaller tax burdens and owners of property in fringe suburbs and scenic rural areas should come out ahead. So should e-mail providers, telecommuting software makers, videoconferencing firms, makers of office electronic equipment, and delivery firms.

Geographic Information Systems

Geographic information systems (GISs) are an important tool to help firms make successful, analytical decisions with regard to location. Retailers, banks, food chains, gas stations, and print shop franchises can all use geographically coded files from a GIS to conduct demographic analyses. By combining population, age, income, traffic flow, and density figures with geography, a retailer can pinpoint the best location for a new store or restaurant.

Here are some of the geographic databases available in many GISs.

- Census data by block, tract, city, county, congressional district, metropolitan area, state, zip code
- Maps of every street, highway, bridge, and tunnel in the U.S.
- Utilities such as electrical, water, and gas lines
- All rivers, mountains, lakes, forests
- All major airports, colleges, hospitals

Airlines, as an example, use GIS to identify airports where ground services are the most effective. This information is then used to help schedule and to decide where to purchase fuel, meals, and other services.

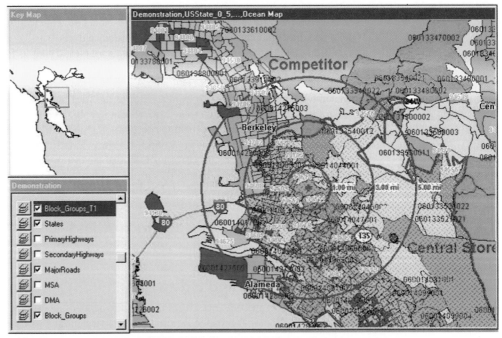

Geographic Information Systems (GIS) are used by a variety of firms to identify target markets by income, ethnicity, product use, age, etc. Here, data from MapInfo helps with competitive analysis. Three concentric rings, each representing various mile radii, were drawn around the competitor's store. This compared to the drive time polygon (shown in red) around the business's own location (red dot) helps to easily visualize competitive overlap.

Commercial office building developers use GIS in the selection of cities for future construction. Building new office space takes several years so developers value the database approach that a GIS can offer. GIS is used to analyze factors that influence the location decisions by addressing five elements for each city: (1) residential areas, (2) retail shops, (3) cultural and entertainment centers, (4) crime incidence, and (5) transportation options. For example, one study of Tampa, Florida, showed that the city's central business district lacks the characteristics to sustain a viable high-demand office market, suggesting the builders should look elsewhere.[5]

Finally Pep Boys, an auto parts retailer headquartered in Philadelphia, has developed models of how GIS technology can be used to identify where the company should locate new stores. It also uses its GIS to decide how many stores are needed to provide proper coverage in a certain geographic area. Pep Boys uses a GIS software product called Atlas GIS (from Strategic Mapping, Inc.). Other similar packages are Hemisphere Solutions (by Unisys Corp.), Map Info (from MapInfo Corp.), Arc/Info (by ESRI), SAS/GIS (by SAS Institute, Inc.), Market Base (by National Decision Systems, Inc.), and MapPoint 2002 (by Microsoft).

To illustrate how extensive some of these GISs can be, consider Microsoft's MapPoint 2002, which includes a comprehensive set of map and demographic data. Its North American maps have more than 6.4 million miles of streets and 1 million points of interest to allow users to locate restaurants, airports, hotels, gas stations, ATMs, museums, campgrounds, and freeway exits. Demographic data includes statistics for population, age, income, education, and housing for 1980, 1990, and 2000, and projections for 2005. These data can be mapped by state, county, city, zip code, or census tract. MapPoint 2002 produces maps that identify business trends; pinpoint market graphics; locate clients, customers, and competitors; and visualize sales performance and product distribution. The European version of MapPoint includes 4.8 million kilometers of roads as well as 300,000 points of interest.[6]

SUMMARY

Location may determine up to 10% of the total cost of an industrial firm. Location is also a critical element in determining revenue for the service, retail, or professional firm. Industrial firms need to consider both tangible and intangible costs. We typically address industrial location problems via a factor-rating method, locational break-even analysis, the center-of-gravity method, and the transportation method of linear programming.

For service, retail, and professional organizations, analysis is typically made of a variety of variables including purchasing power of a drawing area, competition, advertising and promotion, physical qualities of the location, and operating policies of the organization.

KEY TERMS

Tangible costs *(p. 247)*
Intangible costs *(p. 247)*
Clustering *(p. 248)*
Factor-rating method *(p. 248)*

Locational break-even analysis *(p. 250)*
Center-of-gravity method *(p. 251)*
Transportation model *(p. 253)*

USING EXCEL OM TO SOLVE LOCATION PROBLEMS

Our Excel spreadsheet software, Excel OM, may be used to solve Example 1 (with the Factor Rating module), Example 2 (with the Break-Even Analysis module), and Example 3 (with the Center of Gravity module), as well as other location problems. To illustrate the factor-rating method, consider the case of Five Flags over Florida (Example 1), which wishes to expand its corporate presence to Europe. Program 8.1 provides the data inputs for five important factors, including their weights, and ratings on a 1–100 scale (where 100 is the highest rating) for each country. As we see, France is more highly rated, with a 70.4 score versus 68.0 for Denmark.

[5]G. I. Thrall and P. Amos, "Market Evaluation with GIS," *Geo Info System* (November, 1999): 44–49.
[6]*Source:* www.geoplace.com/bg.

PROGRAM 8.1 ■

Excel OM's Factor Rating Module, Including Inputs, Selected Formulas, and Outputs Using Five Flags over Florida Data in Example 1

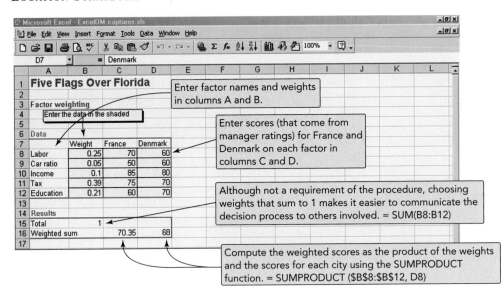

USING POM FOR WINDOWS

Like Excel OM, POM for Windows also includes two different facility location models: the factor-rating method and the center-of-gravity model. For details, refer to Appendix V.

SOLVED PROBLEMS

Solved Problem 8.1

Just as cities and communities can be compared for location selection by the weighted approach model, as we saw earlier in this chapter, so can actual site decisions within those cities. Table 8.7 illustrates four factors of importance to Washington, DC, and the health officials charged with opening that city's first public drug treatment clinic. Of primary concern (and given a weight of 5) was location of the clinic so it would be as accessible as possible to the largest number of patients. Due to a tight budget, the annual lease cost was also of some concern. A suite in the new city Hall, at 14th and U Streets, was highly rated because its rent would be free. An old office building near the downtown bus station received a much lower rating because of its cost. Equally

important as lease cost was the need for confidentiality of patients and, therefore, for a relatively inconspicuous clinic. Finally, because so many of the staff at the clinic would be donating their time, the safety, parking, and accessibility of each site were of concern as well.

Using the factor-rating method, which site is preferred?

SOLUTION

From the three rightmost columns in Table 8.7, the weighted scores are summed. The bus terminal area has a low score and can be excluded from further consideration. The other two sites are virtually identical in total score. The city may now want to consider other factors, including political ones, in selecting between the two remaining sites.

TABLE 8.7 ■ Potential Clinic Sites in Washington, DC

		POTENTIAL LOCATIONS[a]			WEIGHTED SCORES		
FACTOR	IMPORTANCE WEIGHT	HOMELESS SHELTER (2ND AND D, SE)	CITY HALL (14TH AND U, NW)	BUS TERMINAL AREA (7TH AND H, NW)	HOMELESS SHELTER	CITY HALL	BUS TERMINAL AREA
Accessibility for addicts	5	9	7	7	45	35	35
Annual lease cost	3	6	10	3	18	30	9
Inconspicuous	3	5	2	7	15	6	21
Accessibility for health staff	2	3	6	2	6	12	4
				Total scores:	84	83	69

[a]All sites are rated on a 1 to 10 basis, with 10 as the highest score and 1 as the lowest.

Source: From Service Management and Operations, 2/e by Haksever/Render/Russell/Murdick, p. 266. Copyright © 2000. Reprinted by permission of Prentice Hall, Inc., Upper Saddle River, NJ.

Solved Problem 8.2

Chuck Bimmerle is considering opening a new foundry in Denton, Texas; Edwardsville, Illinois; or Fayetteville, Arkansas, to produce high-quality rifle sights. He has assembled the following fixed cost and variable cost data:

| | | PER UNITS COSTS | | |
LOCATION	FIXED COST PER YEAR	MATERIAL	VARIABLE LABOR	OVERHEAD
Denton	$200,000	$ .20	$ .40	$ .40
Edwardsville	$180,000	$ .25	$ .75	$ .75
Fayetteville	$170,000	$1.00	$1.00	$1.00

(a) Graph the total cost lines.

(b) Over what range of annual volume is each facility going to have a competitive advantage?

(c) What is the volume at the intersection of the Edwardsville and Fayetteville cost lines?

SOLUTION

(a) A graph of the total cost lines in shown in Figure 8.5.

(b) Below 8,000 units, the Fayetteville facility will have a competitive advantage (lowest cost); between 8,000 units and 26,666 units, Edwardsville has an advantage; and above 26,666, Denton has the advantage. (We have made the assumption in this problem that other costs—that is, delivery and intangible factors—are constant regardless of the decision.)

(c) From Figure 8.5, we see that the cost line for Fayetteville and the cost line for Edwardsville cross at about 8,000. We can also determine this point with a little algebra:

$$\$180,000 + 1.75Q = \$170,000 + 3.00Q$$
$$\$10,000 = 1.25Q$$
$$8,000 = Q$$

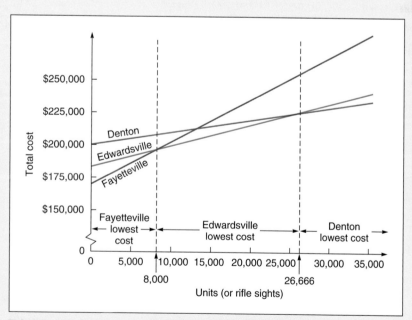

FIGURE 8.5 ■ Graph of Total Cost Lines for Chuck Bimmerle

INTERNET AND STUDENT CD-ROM EXERCISES

Visit our home page or use your student CD-ROM to help with material in this chapter.

 On Our Home Page, www.prenhall.com/heizer

- Self-tests
- Practice Problems
- Internet Exercises
- Current Articles and Research
- Virtual Company Tour
- Internet Homework Problems
- Internet Cases

 On Your Student CD-ROM

- PowerPoint Lecture
- Video Case
- Practice Problems
- ExcelOM
- Excel OM Data Files
- Active Model Exercise

ADDITIONAL CASE STUDIES

Internet Case Studies: Visit our Web site at www.prenhall.com/heizer **for these free case studies:**

- **Consolidated Bottling (A):** Involves finding a centralized location for a quality team to locate its office.
- **Southwestern University's Location Decision:** The university faces three choices in where to locate its football stadium.

Harvard has selected these Harvard Business School case studies to accompany this chapter of our text (textbookcasematch.hbsp.harvard.edu**):**

- **Filene's Basement** (#594-018): This retailer is trying to decide where to add two new stores in its Chicago operation.
- **To Move or Not to Move: Cathy Pacific Airlines** (#HKU-003): Should this airline relocate its data center from Hong Kong to a new country?
- **Wriston Manufacturing** (#698-049): An auto parts producer is trying to decide whether to close one of its Detroit plants.
- **Ellis Manufacturing** (#682-103): This kitchen appliance manufacturer has duplication of resources in its plants.

Layout Strategy

Chapter Outline

LEARNING OBJECTIVES

When you complete this chapter you should be able to

IDENTIFY OR DEFINE:

Fixed-position layout

Process-oriented layout

Work cells

Focused work center

Office layout

Retail layout

Warehouse layout

Product-oriented layout

Assembly-line

DESCRIBE OR EXPLAIN:

How to achieve a good layout for the process facility

How to balance production flow in a repetitive or product-oriented facility

McDonald's Looks for Competitive Advantage with its New High-Tech Kitchen Layout

In its half century of corporate existence, McDonald's has revolutionized the restaurant industry by inventing the limited-menu fast-food restaurant. It has also made five major innovations. The first, the introduction of indoor seating (1950s), was a strategic issue of facility layout, as was the second, drive-through windows (1970s). The third, adding breakfasts to the menu (1980s), was a product strategy. The fourth, adding play areas (1990s), was again a layout decision.

In 2001, McDonald's completed its *fifth* major innovation, and, not surprisingly, it is a new layout to facilitate a mass customization process. This time the corporation banked on the radical redesign of the kitchens in its 13,500 North American outlets. Dubbed the "Made for You" kitchen system, sandwiches are now assembled to order and production levels are controlled by computers. The new layout is intended to both improve the taste of food by ensuring it is always freshly made, and to facilitate the introduction of new products.

Under the new restaurant design, shown in the figure, no food is prepared in advance except the meat patty, which is kept hot in a cabinet. To shorten total production process to 45 seconds, some steps were eliminated and some shortened. For instance, the company developed a toaster that browns buns in 11 seconds instead of half a minute. Bread suppliers had to change the texture of the buns so they could withstand the additional heat. Workers also fig-

The Clock Is Running

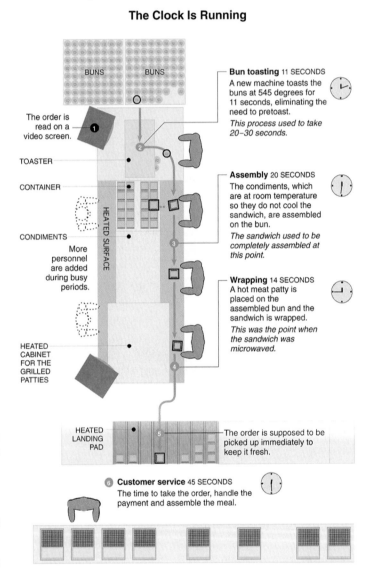

Bun toasting 11 SECONDS
A new machine toasts the buns at 545 degrees for 11 seconds, eliminating the need to pretoast.
This process used to take 20–30 seconds.

Assembly 20 SECONDS
The condiments, which are at room temperature so they do not cool the sandwich, are assembled on the bun.
The sandwich used to be completely assembled at this point.

Wrapping 14 SECONDS
A hot meat patty is placed on the assembled bun and the sandwich is wrapped.
This was the point when the sandwich was microwaved.

The order is read on a video screen.

TOASTER

CONTAINER

CONDIMENTS

More personnel are added during busy periods.

HEATED SURFACE

HEATED CABINET FOR THE GRILLED PATTIES

BUNS   BUNS

HEATED LANDING PAD

The order is supposed to be picked up immediately to keep it fresh.

Customer service 45 SECONDS
The time to take the order, handle the payment and assemble the meal.

N.Y. Times News Service

The redesigned kitchen of a McDonald's in Manhattan. The company hopes the more efficient layout will lift sales.

ured out they could save 2 seconds if condiment containers were repositioned to apply mustard to sandwiches with one motion instead of two.

The payoff for the layout change? McDonald's will save $100 million per year in food costs, largely because only the meat, and no longer the bun or other ingredients, will be discarded when sandwiches do not sell fast enough. The company is banking that with the new layout, new standards of efficiency and happier customers will provide a competitive advantage.

Sources: Nation's Restaurant 35, no. 3 (January 15, 2001): 4–5; *Nation's Restaurant News* 34, no. 42 (October 16, 2000): 1–2; and *Advertising Age* 71, no. 2 (May 15, 2000): 1–2.

The objective of layout
strategy is to develop a
cost-effective layout that
meets the firm's
competitive needs.

THE STRATEGIC IMPORTANCE OF LAYOUT DECISIONS

Layout is one of the key decisions that determines the long-run efficiency of operations. Layout has numerous strategic implications because it establishes an organization's competitive priorities in regard to capacity, processes, flexibility, and cost, as well as quality of work life, customer contact, and image. An effective layout can help an organization achieve a strategy that supports differentiation, low cost, or response. Benetton, for example, supports a *differentiation* strategy by heavy investment in warehouse layouts that contribute to fast, accurate sorting and shipping to its 5,000 outlets. Wal-Mart store layouts support a strategy of *low cost*, as do its warehouse techniques and layouts. Hallmark's office layouts, where many professionals operate in work cells, support *rapid development* of greeting cards. The objective of layout strategy is to develop an economic layout that will meet the firm's competitive requirements. These firms have done so.

In all cases, layout design must consider how to achieve the following:

1. Higher utilization of space, equipment, and people.
2. Improved flow of information, materials, or people.
3. Improved employee morale and safer working conditions.
4. Improved customer/client interaction.
5. Flexibility (whatever the layout is now, it will need to change).

Increasingly, layout designs need to be viewed as dynamic. This means considering small, movable, and flexible equipment. Store displays need to be movable, office desks and partitions modular, and warehouse racks prefabricated. In order to make quick and easy changes in product models and in production rates, operations managers must design flexibility into layouts. To obtain flexibility in layout, managers cross train their workers, maintain equipment, keep investments low, place workstations close together, and use small, movable equipment. In some cases, equipment on wheels is appropriate, in anticipation of the next change in product, process, or volume.

TYPES OF LAYOUT

Layout decisions include the best placement of machines (in production settings), offices and desks (in office settings), or service centers (in settings such as hospitals or department stores). An effective layout facilitates the flow of materials, people, and information within and between areas. To achieve these objectives, a variety of approaches has been developed. We will discuss six of them in this chapter:

1. *Fixed-position layout*—addresses the layout requirements of large, bulky projects such as ships and buildings.
2. *Process-oriented layout*—deals with low-volume, high-variety production (also called "job shop," or intermittent production).
3. *Office layout*—positions workers, their equipment, and spaces/offices to provide for movement of information.
4. *Retail layout*—allocates shelf space and responds to customer behavior.
5. *Warehouse layout*—addresses trade-offs between space and material handling.
6. *Product-oriented layout*—seeks the best personnel and machine utilization in repetitive or continuous production.

Examples for each of these classes of layout problems are noted in Table 9.1.

Because only a few of these six classes can be modeled mathematically, layout and design of physical facilities are still something of an art. However, we do know that a good layout requires determining the following:

1. *Material handling equipment.* Managers must decide about equipment to be used, including conveyors, cranes, automated storage and retrieval systems, and automatic carts to deliver and store material.
2. *Capacity and space requirements.* Only when personnel, machines, and equipment requirements are known can we proceed with layout and provide space for each component. In the case of office work, operations managers must make judgments about the space requirements for each employee. It may be a 6-×-6-foot cubicle plus allowance for hallways, aisles, rest rooms, cafeterias, stairwells, elevators, and so forth, or it may be spacious exec-

TABLE 9.1 ■ Layout Strategies

PROJECT (FIXED POSITION)	JOB SHOP (PROCESS ORIENTED)	OFFICE	RETAIL	WAREHOUSE (STORAGE)	REPETITIVE/ CONTINUOUS (PRODUCT ORIENTED)
		EXAMPLES			
Ingall Ship Building Corp.	Shouldice Hospital	Allstate Insurance	Kroger's Supermarket	Federal-Mogul's warehouse	Sony's TV assembly line
Trump Plaza	Hard Rock Cafes	Microsoft Corp.	Walgreens	The Gap's distribution center	Dodge Caravan minivans
Pittsburgh Airport			Bloomingdales		
		PROBLEM			
Move material to the limited storage areas around the site	Manage varied material flow for each product	Locate workers requiring frequent contact close to one another	Expose customer to high-margin items	Balance low-cost storage with low-cost material handling	Equalize the task time at each workstation

utive offices and conference rooms. Management must also consider allowances for safety requirements that address noise, dust, fumes, temperature, and space around equipment and machines.

3. *Environment and aesthetics.* Layout concerns often require decisions about windows, planters, and height of partitions to facilitate air flow, reduce noise, provide privacy, and so forth.

4. *Flows of information.* Communication is important to any company and must be facilitated by the layout. This issue may require decisions about proximity as well as decisions about open spaces versus half-height dividers versus private offices.

5. *Cost of moving between various work areas.* There may be unique considerations related to moving materials or the importance of certain areas being next to each other. For example, the movement of molten steel is more difficult than the movement of cold steel.

FIXED-POSITION LAYOUT

Fixed-position layout
Addresses the layout requirements of stationary projects or large bulky projects (such as ships or buildings).

In a **fixed-position layout**, the project remains in one place and workers and equipment come to that one work area. Examples of this type of project are a ship, a highway, a bridge, a house, and an oil well.

The techniques for addressing the fixed-position layout are not well developed and are complicated by three factors. First, there is limited space at virtually all sites. Second, at different stages in the construction process, different materials are needed; therefore, different items become critical as the project develops. Third, the volume of materials needed is dynamic. For example, the rate of use of steel panels for the hull of a ship changes as the project progresses.

Different industries handle these problems in different ways. The construction industry usually has a "meeting of the trades" to assign space for various time periods. As suspected, this often yields less than an optimum solution, as the discussion may be more political than analytical. Shipyards, however, have loading areas called "platens" adjacent to the ship, which are loaded by a scheduling department.

Because problems with fixed-position layouts are so difficult to solve well on-site, an alternative strategy is to complete as much of the project as possible off-site. This approach is used in the shipbuilding industry when standard units—say, pipe-holding brackets—are assembled on a nearby assembly line (a product-oriented facility). In an attempt to add efficiency to shipbuilding, Ingall Ship Building Corporation has moved toward product-oriented production when sections of a ship (modules) are similar or when it has a contract to build the same section of several similar ships.[1] Similarly, other shipbuilding firms are experimenting with group technology (see Chapter 5) to

[1]Ingall's 130 Million Dollar Ship Factory," *Shipbuilding and Shipping Record* 115, no. 22 (London: Transport and Technical Publications Ltd.): 25–26.

A house built via traditional fixed-position layout would be constructed on-site, with equipment, materials, and workers brought to the site. However, imaginative OM solutions allow the home pictured here to be built at a much lower cost. The house is built in two movable modules (shown joined here) in a factory, where equipment and material handling are expedited. Prepositioned scaffolding and hoists make the job easier, quicker, and cheaper. The indoor work environment also aids labor productivity, means no weather delays, and eliminates overnight thefts.

group components. As the photo shows, many home builders are moving from a fixed-position layout strategy to one that is more product oriented. About one-third of all new homes in the U.S. are built this way.

PROCESS-ORIENTED LAYOUT

Process-oriented layout
A layout that deals with low-volume, high-variety production; like machines and equipment are grouped together.

The **process-oriented layout** can simultaneously handle a wide variety of products or services. This is the traditional way to support a product differentiation strategy. It is most efficient when making products with different requirements or when handling customers, patients, or clients with different needs. A process-oriented layout is typically the low-volume, high-variety strategy discussed in Chapter 7. In this job-shop environment, each product or each small group of products undergoes a different sequence of operations. A product or small order is produced by moving it from one department to another in the sequence required for that product. A good example of the process-oriented layout is a hospital or clinic. Figure 9.1 illustrates the process for two patients, A and B, at an emergency clinic in Chicago. An inflow of patients, each with his or her own needs, requires routing through admissions, laboratories, operating rooms, radiology, pharmacies, nursing beds, and so on. Equipment, skills, and supervision are organized around these processes.

A big advantage of process-oriented layout is its flexibility in equipment and labor assignments. The breakdown of one machine, for example, need not halt an entire process; work can be transferred to other machines in the department. Process-oriented layout is also especially good for handling the manufacture of parts in small batches, or **job lots**, and for the production of a wide variety of parts in different sizes or forms.

Job lots
Groups or batches of parts processed together.

The disadvantages of process-oriented layout come from the general-purpose use of the equipment. Orders take more time to move through the system because of difficult scheduling, changing setups, and unique material handling. In addition, general-purpose equipment requires high labor skills, and work-in-process inventories are higher because of imbalances in the production process. High labor-skill needs also increase the required level of training and experience, and high work-in-process levels increase capital investment.

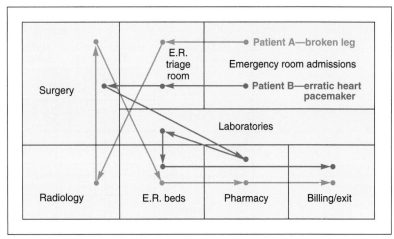

FIGURE 9.1 ■ An Emergency Room Process Layout Showing the Routing of Two Patients

Patient A (broken leg) proceeds (blue arrow) to E.R. triage, to radiology, to surgery, to a bed, to pharmacy, to billing. Patient B (pacemaker problem) moves (purple arrow) to E.R. triage, to surgery, to pharmacy, to lab, to a bed, to billing.

When designing a process layout, the most common tactic is to arrange departments or work centers so as to minimize the costs of material handling. In other words, departments with large flows of parts or people between them should be placed next to one another. Material handling costs in this approach depend on (1) the number of loads (or people) to be moved between two departments during some period of time and (2) the distance-related costs of moving loads (or people) between departments. Cost is assumed to be a function of distance between departments. The objective can be expressed as follows:

$$\text{Minimize cost} = \sum_{i=1}^{n} \sum_{j=1}^{n} X_{ij} C_{ij} \tag{9-1}$$

where n = total number of work centers or departments
i, j = individual departments
X_{ij} = number of loads moved from department i to department j
C_{ij} = cost to move a load between department i and department j

Process-oriented facilities (and fixed-position layouts as well) try to minimize loads or trips, times distance-related costs. The term C_{ij} combines distance and other costs into one factor. We thereby assume not only that the difficulty of movement is equal but also that the pickup and setdown costs are constant. Although they are not always constant, for simplicity's sake we summarize these data (that is, distance, difficulty, and pickup and setdown costs) in this one variable, cost. The best way to understand the steps involved in designing a process layout is to look at an example.

Example 1

Excel OM Data File
Ch09Ex1.xla

Walters Company management wants to arrange the six departments of its factory in a way that will minimize interdepartmental material handling costs. They make an initial assumption (to simplify the problem) that each department is 20 × 20 feet and that the building is 60 feet long and 40 feet wide. The process layout procedure that they follow involves six steps:

Step 1: *Construct a "from-to matrix"* showing the flow of parts or materials from department to department (Figure 9.2).

Step 2: *Determine the space requirements* for each department. (Figure 9.3 shows available plant space.)

Step 3: *Develop an initial schematic diagram* showing the sequence of departments through which parts must move. Try to place departments with a heavy flow of materials or parts next to one another. (See Figure 9.4 on page 269.)

Department	Number of loads per week					
	Assembly (1)	Painting (2)	Machine Shop (3)	Receiving (4)	Shipping (5)	Testing (6)
Assembly (1)		50	100	0	0	20
Painting (2)			30	50	10	0
Machine Shop (3)				20	0	100
Receiving (4)					50	0
Shipping (5)						0
Testing (6)						

FIGURE 9.2 ■ Interdepartmental Flow of Parts

The high flows between 1 and 3, and 3 and 6 are immediately apparent.
Departments 1, 3, and 6, therefore, should be close together.

Room 1	Room 2	Room 3	
Assembly Department (1)	Painting Department (2)	Machine Shop Department (3)	40'
Receiving Department (4)	Shipping Department (5)	Testing Department (6)	
Room 4	Room 5	Room 6	

← ———————————— 60' ————————————→

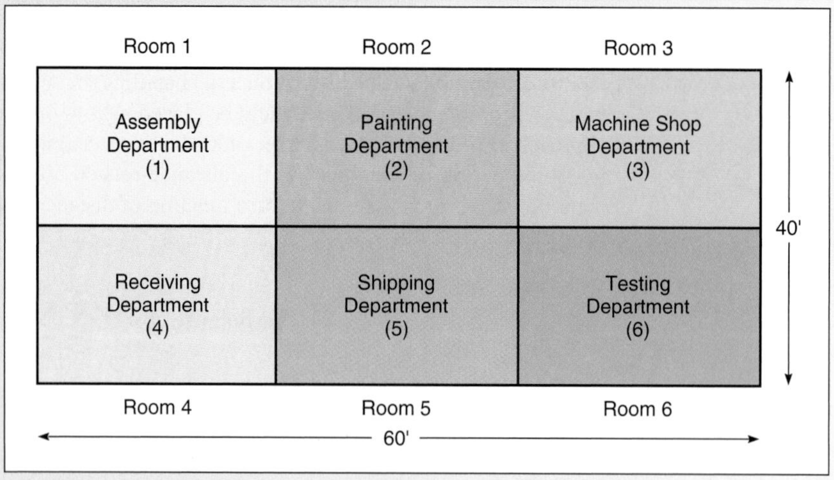

FIGURE 9.3 ■ Building Dimensions and a Possible Department Layout

Step 4: *Determine the cost of this layout by using the material handling cost equation:*

$$\text{Cost} = \sum_{i=1}^{n} \sum_{j=1}^{n} X_{ij} C_{ij}$$

For this problem, Walters Company assumes that a forklift carries all interdepartmental loads. The cost of moving one load between adjacent departments is estimated to be $1. Moving a load between nonadjacent departments costs $2. Looking at Figure 9.2, we thus see that the handling cost between departments 1 and 2 is $50 ($1 × 50 loads), $200 between departments 1 and 3 ($2 × 100 loads), $40 between departments 1 and 6 ($2 × 20 loads), and so on. Rooms that are diagonal to one another, such as 2 and 4, are treated as adjacent. The total cost for the layout shown in Figure 9.4 is:

$$
\begin{aligned}
\text{Cost} = \quad &\$50 \;+\; \$200 \;+\; \$40 \;+\; \$30 \;+\; \$50 \\
&\text{(1 and 2)} \; \text{(1 and 3)} \; \text{(1 and 6)} \; \text{(2 and 3)} \; \text{(2 and 4)} \\[4pt]
&+ \;\; \$10 \;\;+\;\; \$40 \;\;+\; \$100 \;+\; \$50 \\
&\quad \text{(2 and 5)} \; \text{(3 and 4)} \; \text{(3 and 6)} \; \text{(4 and 5)} \\[4pt]
= \;\; &\$570
\end{aligned}
$$

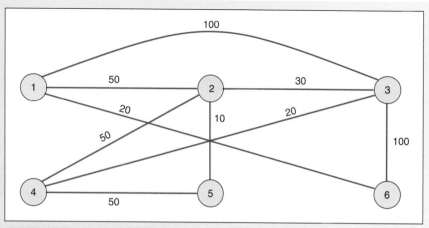

FIGURE 9.4 ■ Interdepartmental Flow Graph Showing Number of Weekly Loads

Step 5: By trial and error (or by a more sophisticated computer program approach that we discuss shortly), *try to improve the layout* pictured in Figure 9.3 to establish a reasonably good arrangement of departments.

By looking at both the flow graph (Figure 9.4) and the cost calculations, we see that placing departments 1 and 3 closer together appears desirable. They currently are nonadjacent, and the high volume of flow between them causes a large handling expense. Looking the situation over, we need to check the effect of shifting departments and possibly raising, instead of lowering, overall costs.

One possibility is to switch departments 1 and 2. This exchange produces a second departmental flow graph (Figure 9.5), which shows a reduction in cost to $480, a savings in material handling of $90.

$$
\begin{aligned}
\text{Cost} = \ &\underset{(1 \text{ and } 2)}{\$50} \ + \ \underset{(1 \text{ and } 3)}{\$100} \ + \ \underset{(1 \text{ and } 6)}{\$20} \ + \ \underset{(2 \text{ and } 3)}{\$60} \ + \ \underset{(2 \text{ and } 4)}{\$50} \\
&+ \ \underset{(2 \text{ and } 5)}{\$10} \ + \ \underset{(3 \text{ and } 4)}{\$40} \ + \ \underset{(3 \text{ and } 6)}{\$100} \ + \ \underset{(4 \text{ and } 5)}{\$50} \\
= \ &\$480
\end{aligned}
$$

This switch, of course, is only one of a large number of possible changes. For a six-department problem, there are actually 720 (or $6! = 6 \times 5 \times 4 \times 3 \times 2 \times 1$) potential arrangements! In layout problems, we seldom find the optimal solution and may have to be satisfied with a "reasonable" one reached after a few trials. Suppose Walters Company is satisfied with the cost figure of $480 and the flow graph of Figure 9.5. The problem may not be solved yet. Often a sixth step is necessary:

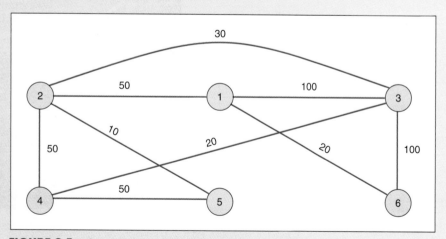

FIGURE 9.5 ■ Second Interdepartmental Flow Graph

Room 1	Room 2	Room 3
Painting Department (2)	Assembly Department (1)	Machine Shop Department (3)
Receiving Department (4)	Shipping Department (5)	Testing Department (6)
Room 4	Room 5	Room 6

FIGURE 9.6 ■ A Feasible Layout for Walters Company

Step 6: *Prepare a detailed plan* arranging the departments to fit the shape of the building and its nonmovable areas (such as the loading dock, washrooms, and stairways). Often this step involves ensuring that the final plan can be accommodated by the electrical system, floor loads, aesthetics, and other factors.

In the case of Walters Company, space requirements are a simple matter (see Figure 9.6).

Computer Software for Process-Oriented Layouts

The graphic approach in Example 1 is fine for small problems. It does not, however, suffice for larger problems. When 20 departments are involved in a layout problem, more than 600 *trillion* different department configurations are possible. Fortunately, computer programs have been written to handle layouts of up to 40 departments. The best-known of these is **CRAFT** (Computerized Relative Allocation of Facilities Technique), a program that produces "good" but not always "optimal" solutions. CRAFT is a search technique that systematically examines alternative departmental rearrangements to reduce total "handling" cost (see Figure 9.7). CRAFT has the added advantage of examining not only load and distance but also a third factor, a difficulty rating.[2]

Other popular process layout packages include the Automated Layout Design program (ALDEP), Computerized Relationship Layout Planning (CORELAP), and Factory Flow.[3] Factory, Flow, illustrated in the photo, is used for optimizing layouts based on material flow distances, fre-

CRAFT
A computer program that systematically examines alternative departmental rearrangements to reduce total material handling cost.

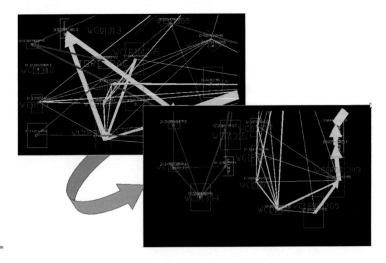

Software programs like Factory Flow, *an E-factory product from EDS, address the complex relationships between routing, material handling equipment, and production volumes. Evaluations of alternatives such as the change from a traditional process oriented layout to two work cells (shown at right) can be made by simply moving equipment with a mouse and recalculating the results.*

[2]Y. A. Bozer, R. R. Meller, and S. J. Erlebacher, "An Improvement-Type Layout Algorithm for Single and Multiple Floor Facilities," *Management Science* 40, no. 7 (1994): 918–933.

[3]See David P. Sly, "Layout Design and Analysis Software," *IIE Solutions* 28, no. 7 (July 1990): 18–25.

FIGURE 9.7 ■ In This Six-Department Outpatient Hospital Example, CRAFT Has Rearranged the Initial Layout (a), with a Cost of $20,100, into the New Layout with a Lower Cost of $14,390 (b). CRAFT does this by systematically testing pairs of departments to see if moving them closer to one another lowers total cost.

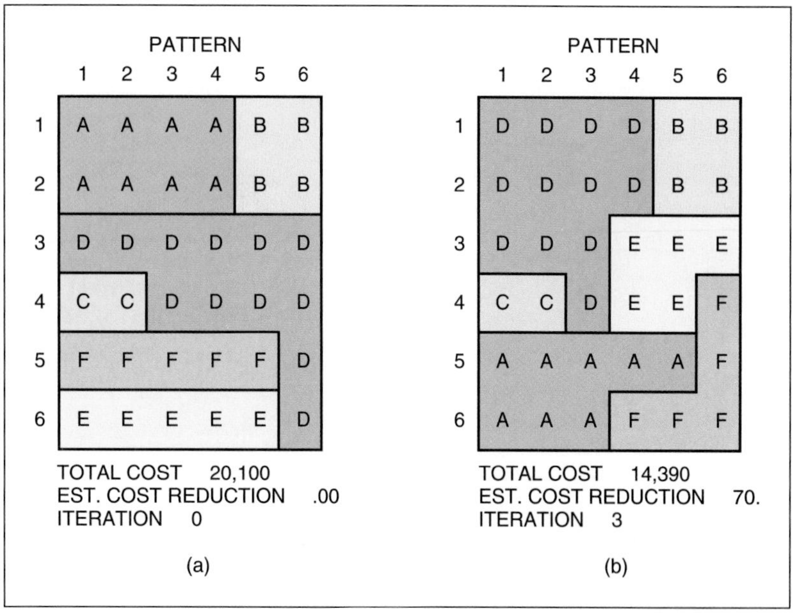

Legend:

☐ A = X-ray/MRI rooms
☐ B = laboratories
☐ C = admissions
☐ D = exam rooms
☐ E = operating rooms
☐ F = recovery rooms

TOTAL COST   20,100
EST. COST REDUCTION   .00
ITERATION   0

(a)

TOTAL COST   14,390
EST. COST REDUCTION   70.
ITERATION   3

(b)

quency, and costs. It employs Auto CAD layout drawings of a factory process; part routing data; and costs, times, and speeds of material handling systems. In one analysis of an existing console assembly plant, Factory Flow was able to reduce material handling costs from $900,000 to $800,000 and to decrease the length of conveyor belts from 3,600 feet to just 700 feet.[4]

Work Cells

Cellular work arrangements are used when volume warrants a special arrangement of machinery and equipment. In a manufacturing environment, group technology identifies products that have similar characteristics and allows not just a particular batch (for example, several units of the same product) but also a family of batches, to be processed in a particular work cell. *Work cells* can be thought of as a special case of process-oriented layout. Although the idea of work cells was first presented by R. E. Flanders in 1925,[5] only with the increasing use of group technology (see Chapter 5) has the technique reasserted itself.

Work cell

A temporary product-oriented arrangement of machines and personnel in what is ordinarily a process-oriented facility.

The **work cell** idea is to reorganize people and machines that would ordinarily be dispersed in various process departments and temporarily arrange them in a small group so that they can focus on making a single product or a group of related products (Figure 9.8). The work cell, therefore, is built around the product. Motorola, for instance, forms work cells to build and test engine control systems for John Deere tractors. These work cells are reconfigured as product design or volume changes. The advantages of work cells are

1. *Reduced work-in-process inventory* because the work cell is set up to provide a balanced flow from machine to machine.
2. *Less floor space* required because less space is needed between machines to accommodate work-in-process inventory.
3. *Reduced raw material and finished goods inventories* because less work-in-process allows more rapid movement of materials through the work cell.
4. *Reduced direct labor cost* because of improved communication between employees, better material flow, and improved scheduling.
5. *Heightened sense of employee participation* in the organization and the product because employees accept the added responsibility of product quality being directly associated with them and their work cell.
6. *Increased use of equipment and machinery* because of better scheduling and faster material flow.
7. *Reduced investment in machinery and equipment* because good facility utilization reduces the number of machines and the amount of equipment and tooling.

[4]"Factory Planning Software," *Industrial Engineering* (December 1993): SS3.

[5]R. E. Flanders, "Design Manufacture and Production Control of a Standard Machine," *Transactions of ASME* 46 (1925).

FIGURE 9.8 ■

Improving Layouts by
Moving to the Work Cell
Concept

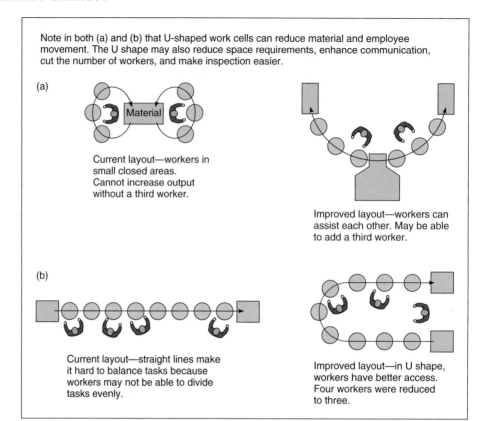

Note in both (a) and (b) that U-shaped work cells can reduce material and employee movement. The U shape may also reduce space requirements, enhance communication, cut the number of workers, and make inspection easier.

(a)

Current layout—workers in small closed areas. Cannot increase output without a third worker.

Improved layout—workers can assist each other. May be able to add a third worker.

(b)

Current layout—straight lines make it hard to balance tasks because workers may not be able to divide tasks evenly.

Improved layout—in U shape, workers have better access. Four workers were reduced to three.

The requirements of cellular production include:

1. Identification of families of products, often through the use of group technology codes or equivalents.
2. A high level of training and flexibility on the part of employees.
3. Either staff support or flexible, imaginative employees to establish work cells initially.
4. Test (poka-yoke) at each station in the cell.

Work cells and assembly lines are sometimes organized in a U shape. U-shaped facilities, as shown in Figure 9.8, have at least five advantages over straight ones: (1) because tasks can be grouped, inspection is often immediate; (2) fewer workers are needed; (3) workers can reach more of the work area; (4) the work area can be more efficiently balanced; and (5) communication is enhanced.

Some 40% of U.S. plants with fewer than 100 employees use some sort of cellular system, whereas 74% of the larger plants surveyed have adopted cellular production methods. Bayside Controls in Queens, N.Y., for example, has in the past decade increased sales from $300,000 per year to $11 million.[6] Much of the gain was attributed to its move to cellular manufacturing. As noted in the *OM in Action* box, Rowe Furniture has had similar success with work cells.

Focused work center
A permanent or semi-permanent product-oriented arrangement of machines and personnel.

Focused factory
A facility designed to produce similar products or components.

The Focused Work Center and the Focused Factory

When a firm has *identified a large family of similar products that have a large and stable demand*, it may organize a focused work center. A **focused work center** moves production from a general-purpose, process-oriented facility to a large work cell that remains part of the present plant. If the focused work center is in a separate facility, it is often called a **focused factory**. A fast-food restaurant is a focused factory—most are easily reconfigured for adjustments to product mix and volume. Burger King, for example, changes the number of personnel and task assignments rather than moving machines and equipment. In this manner, the company balances the assembly line to meet changing production demands. In effect, the "layout" changes numerous times each day.

[6]Stephanie N. Mehta, "Cell Manufacturing Gains Acceptance at Smaller Plants," the *Wall Street Journal* (September 15, 1996): B2.

OM IN ACTION

Work Cells at Rowe Furniture

Many customers dislike buying the standard product. This is particularly true of furniture customers, who usually want a much wider selection than most furniture showrooms can display. Customers really want customization, but they are unhappy waiting months for special orders. So Rowe Furniture Corp. of Salem, Virginia, created a computer network on which customers could order customized combinations of fabrics and styles. This strategy provided the customization, but the real trick was: How could operations people build ordered furniture rapidly and with no increase in cost?

First, Rowe annihilated the old assembly line. Then it formed work cells, each containing teams of workers with the necessary skills—gluers, sewers, staplers, and stuffers. Instead of being scattered along an assembly line, about 3 dozen team members found themselves in work cells. The work cells supported improved communication—perhaps even forced some communication between team members. Cross training followed; gluers began to understand what staplers needed, and stuffers began to understand sewing requirements. Soon, team members realized that they could successfully deal with daily problems and began to develop improved methods. Moreover, both team members and management began to work together to solve problems.

Today the Rowe plant operates at record productivity. "Everybody's a lot happier," says shop worker Sally Huffman.

Sources: Upholstery Design and Management (February 2001): 16–22; and the Wall Street Journal (February 26, 1999): B1.

The term *focused factories* may also refer to facilities that are focused in ways other than by product line or layout. For instance, facilities may be focused in regard to meeting quality, new product introduction, or flexibility requirements.[7]

Focused facilities in manufacturing and in services appear to be better able to stay in tune with their customers, to produce quality products, and to operate at higher margins. This is true whether they are steel mills like SMI, Nucor, or Chaparral, restaurants like McDonald's and Burger King, or a hospital like Shouldice.

Table 9.2 summarizes our discussion of work cells, focused work centers, and focused factories.

TABLE 9.2 ■

Work Cells, Focused Work Centers, and the Focused Factory

WORK CELL	FOCUSED WORK CENTER	FOCUSED FACTORY
A work cell is a temporary product-oriented arrangement of machines and personnel in what is ordinarily a process-oriented facility.	A focused work center is a permanent product-oriented arrangement of machines and personnel in what is ordinarily a process-oriented facility.	A focused factory is a permanent facility to produce a product or component in a product-oriented facility. Many of the focused factories currently being built were originally part of a process-oriented facility.
Example: A job shop with machinery and personnel rearranged to produce 300 unique control panels.	*Example:* Pipe bracket manufacturing at a shipyard.	*Example:* A plant to produce window mechanisms for automobiles.

OFFICE LAYOUT

Office layout
The grouping of workers, their equipment, and spaces/offices to provide for comfort, safety, and movement of information.

The main difference between **office** and factory **layouts** is the importance placed on information. However, in some office environments, just as in manufacturing, production relies on the flow of material. This is the case at Kansas City's Hallmark, which has over half the U.S. greeting card market and produces some 40,000 different cards. In the past, its 700 creative professionals would take up to 2 years to develop a new card. Hallmark's decision to create work cells of artists, writers, lithographers, merchandisers, and accountants, all located in the same area, has resulted in cards prepared in a fraction of the time that the old layout required.

[7]See, for example, Wickham Skinner, "The Focused Factory," *Harvard Business Review* 52, no. 3 (May–June 1974): 113–121.

FIGURE 9.9 ■

Office Relationship
Chart

Source: Adapted from Richard
Muther, *Simplified Systematic
Layout Planning,* 3rd ed.
(Kansas City, Mgt. & Ind'l
Research Publications, 1994).
Used by permission of the
publisher.

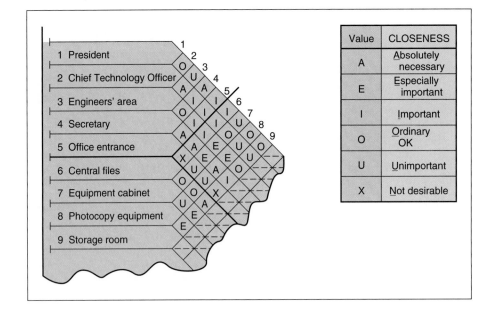

Value	CLOSENESS
A	Absolutely necessary
E	Especially important
I	Important
O	Ordinary OK
U	Unimportant
X	Not desirable

Hallmark's example suggests that maintaining layout flexibility extends to offices as well as factories and remains an important principle of layout design. Just as operations managers build movable, modular equipment to maximize flexibility in the production process, so too should operations managers in office environments. The technological change sweeping manufacturing is also altering the way offices function, making office flexibility a necessity. Consequently, many varieties of modular office equipment that support changing layouts are now available.

Even though the movement of information is increasingly electronic, analysis of office layouts still requires a task-based approach. Managers, therefore, examine both electronic and conventional communication patterns, separation needs, and other conditions affecting employee effectiveness.[8] A useful tool for such an analysis is the *relationship chart* shown in Figure 9.9. This chart, prepared for an office of software engineers, indicates that the chief technology officer must be (1) near the engineers' area, (2) less near the secretary and central files, and (3) not at all near the photocopy or storage room.

General office-area guidelines allot an average of about 100 square feet per person (including corridors). A major executive is allotted about 400 square feet, and a conference room area is based on 25 square feet per person, up to 30 people. In contrast, restaurants provide from 16 square feet to 50 square feet per customer (total kitchen and dining area divided by capacity). By making effective use of the vertical dimension in a workstation, some office designers expand upward instead of outward. This approach keeps each workstation unit (what designers call the "footprint") as small as possible.

These American concepts of space are not universal, however. In the Tokyo office of Toyota, for example, about 110 people work in one large room. As is typical of Japanese offices, they work out in the open, with desks crammed together in clusters called "islands." Islands are arranged in long rows; managers sit at the ends of the rows with subordinates in full view. (When important visitors arrive for meetings, they are ushered into special rooms and do not see these cramped offices.)

On the other hand, some layout considerations are universal (many of which apply to factories as well as to offices). They have to do with working conditions, teamwork, authority, and status. Should all or only part of the work area be air-conditioned? Should all employees use the same entrance, rest rooms, lockers, and cafeteria? As mentioned earlier, layout decisions are part art and part science. Only the science part—which deals with the flow of materials and information—can be analyzed in the same manner as the flow of parts in a process layout.

As a final comment on office layout, we should note two major trends. First, *technology,* such as cellular phones, beepers, faxes, the Internet, home offices, laptop computers, and PDAs, allows increasing layout flexibility by moving information electronically. Second, *virtual companies* (discussed in Chapter 11) create dynamic needs for space and services. These two changes tend to require fewer office employees on-site. For example, when accounting firm Ernst & Young's

Video 9.2

Layout at Service
Organizations

[8]Jacqueline C. Vischer, "Strategic Work-Space Planning," *Sloan Management Review* (fall 1995): 37.

OM IN ACTION

Shopping Mall Layout Meets the Internet

In the 3 months or so since RiverTown Crossings opened its doors in Grandville, Michigan, Nancy McCarty, a 35-year-old mother of six, has rarely ventured beyond just one section of the mall—the one with the Abercrombie Kids, Gap Kids, Gymboree, and other kids' clothing stores. "We can shop this wing and really find everything we need," says Mrs. McCarty.

That's the idea. In an effort to compete with the allure of online shopping, mall owner General Growth Properties Inc. decided on a layout that runs counter to decades of mall retailing wisdom: It clustered competing stores together. The Chicago developer began experimenting with the idea (which has long been popular in many retailing venues, such as New York City's fashion, fur, and diamond districts) 3 years ago. Now, all of its new malls will have clusters.

Worried about a future where shoppers point and click instead of drive and park and walk and wait in line, developers are finally trying to make it more convenient to shop in malls. Some are adding directories that are easier to understand than current mall maps. Others are opting to put anchor department stores closer together—a layout that cuts down on walking and time.

Yet other owners are making malls smaller. Their belief is that many busy consumers, who tend to be relatively affluent, no longer "shop til you drop" but rather "buy on the fly." The Simon De Bartolo Group, the largest owner of shopping centers in the U.S., now wires its malls for kiosks to allow online shopping at tenant stores. This concept links all tenants' Web sites and creates a virtual mall within a mall.

Sources: Real Estate Finance 17, no. 1 (spring 2000): 41–46; and the *Wall Street Journal* (February 8, 2000): B1 and (August 27, 2002): A1, A8.

Chicago office found that 30% to 40% of desks were empty at any given time, the firm developed its new "hoteling programs." Five hundred junior consultants lost their permanent offices; anyone who plans to be in the office (rather than out with clients) for more than half a day books an office through a "concierge," who hangs that consultant's name on the door for the day.

RETAIL LAYOUT

Retail layout
An approach that addresses flow, allocates space, and responds to customer behavior.

Retail layouts are based on the idea that sales and profitability vary directly with customer exposure to products. Thus, most retail operations managers try to expose customers to as many products as possible. (As do shopping malls—see the *OM in Action* box, "Shopping Mall Layout Meets the Internet.") Studies do show that the greater the rate of exposure, the greater the sales and the higher the return on investment. The operations manager can alter *both* with the overall arrangement of the store and the allocation of space to various products within that arrangement.

Five ideas are helpful for determining the overall arrangement of many stores:

1. Locate the high-draw items around the periphery of the store. Thus, we tend to find dairy products on one side of a supermarket and bread and bakery products on another. An example of this tactic is shown in Figure 9.10.
2. Use prominent locations for high-impulse and high-margin items such as housewares, beauty aids, and shampoos.
3. Distribute what are known in the trade as "power items"—items that may dominate a purchasing trip—to both sides of an aisle, and disperse them to increase the viewing of other items.
4. Use end-aisle locations because they have a very high exposure rate.
5. Convey the mission of the store by careful selection in the positioning of the lead-off department. For instance, if prepared foods are part of the mission, position the bakery and deli up front to appeal to convenience-oriented customers.

Once the overall layout of a retail store has been decided, products need to be arranged for sale. Many considerations go into this arrangement. However, the main *objective of retail layout is to maximize profitability per square foot of floor space* (or, in some stores, on linear foot of shelf space). Big-ticket, or expensive, items may yield greater dollar sales, but the profit per square foot may be lower. Computerized programs are available to assist managers in evaluating the profitability of various merchandising plans.

FIGURE 9.10 ■

Store Layout with Dairy and Bread, High-Draw Items, in Different Areas of the Store

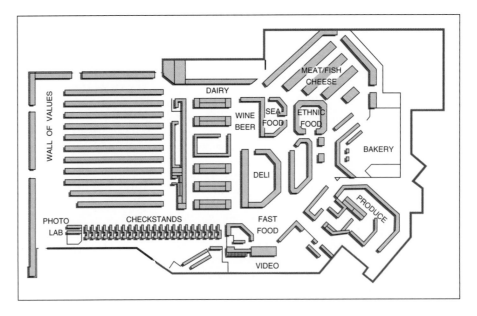

Slotting fees

Fees manufacturers pay to get shelf space for their products.

An additional, and somewhat controversial, issue in retail layout is called slotting. **Slotting fees** are fees manufacturers pay to get their goods on the shelf in a retail store or supermarket chain. The result of massive new-product introductions, retailers can now demand up to $25,000 to place an item in their chain. During the last decade, marketplace economics, consolidations, and technology have provided retailers with this leverage. The competition for shelf space is advanced by POS systems and scanner technology, which improve management and inventory control. Many small firms question the legality and ethics of slotting fees, claiming the fees stifle new products, limit their ability to expand, and cost consumers money.[9]

Servicescapes

Servicescape

The physical surroundings in which a service takes place, and how they affect customers and employees.

Although the main objective of retail layout is to maximize profit, there are other aspects of the service that managers need to consider. Professor Mary Jo Bitner conceived the term **servicescape** to describe the physical surroundings in which the service is delivered and how the surroundings have a humanistic effect on customers and employees.[10] She believes that in order to provide a good service layout, a firm must consider these three elements:

1. *Ambient conditions*, which are background characteristics such as lighting, sound, smell, and temperature. All of these affect workers *and* customers and can affect how much is spent and how long a person stays in the building.
2. *Spatial layout and functionality*, which involve customer circulation path planning, aisle characteristics (such as width, direction, angle, and shelf-spacing), and product grouping.
3. *Signs, symbols, and artifacts*, which are characteristics of building design that carry social significance (such as carpeted areas of a department store that encourage shoppers to slow down and browse).

Examples of each of these three elements of servicescape are

- Ambient conditions   Fine-dining restaurant with linen tablecloths and candlelit atmosphere; Mrs. Field's Cookie bakery smells permeating the shopping mall.
- Layout/functionality   Kroger's long aisles and high shelves.
- Signs, symbols, and artifacts Wal-Mart's greeter at the door; Hard Rock Cafe's wall of guitars.

---

[9]For an interesting discussion of slotting fees, see any of the following: *Forbes* (June 12, 2000): 84–85; the *Wall Street Journal* (September 22, 1999): A23; *Supermarket Business* (1999, Best of Class Issue): 15–18.

[10]Mary Jo Bitner, "Servicescapes: The Impact of Physical Surroundings on Customers and Employees," *Journal of Marketing* 56 (April 1992): 57–71.

A critical element contributing to the bottom line at Hard Rock Cafe is the layout of the cafe and its accompanying retail shop space. The retail space, from 600 to 1,300 square feet in size, is laid out in conjunction with the restaurant area to create the maximum traffic flow before and after eating. Hard Rock treats retail layout like a science—and the payoff is huge. The free-flow layout shown in this London store increases access to product, affords greater visibility, and reduces point-of-sale transaction time. Almost half of a cafe's annual sales is generated from these small retail shops, which have the highest sales per square foot of any retailer in the U.S.

WAREHOUSING AND STORAGE LAYOUTS

Warehouse layout
A design that attempts to minimize total cost by addressing trade-offs between space and material handling.

Automated storage and retrieval systems are reported to improve productivity by an estimated 500% over manual methods.

The objective of **warehouse layout** is to find the optimum trade-off between handling cost and costs associated with warehouse space. Consequently, management's task is to maximize the utilization of the total "cube" of the warehouse—that is, utilize its full volume while maintaining low material handling costs. We define material handling costs as all the costs related to the incoming transport, storage, and outgoing transport of the materials to be warehoused. These costs include equipment, people, material, supervision, insurance, and depreciation. Effective warehouse layouts do, of course, also minimize the damage and spoilage of material within the warehouse.

Management minimizes the sum of the resources spent on finding and moving material plus the deterioration and damage to the material itself. The variety of items stored and the number of items "picked" has direct bearing on the optimum layout. A warehouse storing a few items lends itself to higher density than a warehouse storing a variety of items. Modern warehouse management is, in many instances, an automated procedure using automated storage and retrieval systems (ASRSs).

An important component of warehouse layout is the relationship between the receiving/unloading area and the shipping/loading area. Facility design depends on the type of supplies unloaded, what they are unloaded from (trucks, rail cars, barges, and so on), and where they are unloaded. In some companies, the receiving and shipping facilities, or "docks," as they are called, are even the same area; sometimes they are receiving docks in the morning and shipping docks in the afternoon.

Cross-Docking

Cross-docking
Avoiding the placing of materials or supplies in storage by processing them as they are received for shipment.

Cross-docking means to avoid placing materials or supplies in storage by processing them as they are received. In a manufacturing facility, product is received directly to the assembly line. In a distribution center, labeled and presorted loads arrive at the shipping dock for immediate rerouting, thereby avoiding formal receiving, stocking/storing, and order-selection activities. Because these activities add no value to the product, their elimination is 100% cost savings. Wal-Mart, an early advocate of cross-docking, uses the technique as a major component of its continuing low-cost strategy. With cross-docking, Wal-Mart reduces distribution costs and speeds restocking of stores, thereby improving customer service. Although cross-docking reduces product handling, inventory, and facility costs, it requires both (1) tight scheduling and (2) that shipments received include accu-

The Gap strives for both high quality and low costs. It does so by (1) designing its own clothes, (2) ensuring quality control among its vendors, and (3) maintaining downward pressure on distribution costs. A new automatic distribution center near Baltimore allows The Gap to stock East Coast stores daily rather than only three times a week.

rate product identification, usually with bar codes so they can be promptly moved to the proper shipping dock.

Random Stocking

Automatic identification systems (AISs), usually in the form of bar codes, allow accurate and rapid item identification. When automatic identification systems are combined with effective management information systems, operations managers know the quantity and location of every unit. This information can be used with human operators or with automatic storage and retrieval systems to load units anywhere in the warehouse—randomly. Accurate inventory quantities and locations mean the potential utilization of the whole facility because space does not need to be reserved for certain stock-keeping units (SKUs) or part families. Computerized **random stocking** systems often include the following tasks:

Random stocking
Used in warehousing to locate stock wherever there is an open location. This technique means that space does not need to be allocated to particular items and the facility can be more fully utilized.

1. Maintaining a list of "open" locations.
2. Maintaining accurate records of existing inventory and its locations.
3. Sequencing items on orders to minimize the travel time required to "pick" orders.
4. Combining orders to reduce picking time.
5. Assigning certain items or classes of items, such as high-usage items, to particular warehouse areas so that the total distance traveled within the warehouse is minimized.

Random stocking systems can increase facility utilization and decrease labor cost, but require accurate records.

Customizing

Although we expect warehouses to store as little product as possible and hold it for as short a time as possible, we are now asking warehouses to customize products. Warehouses can be places where value is added through **customizing**. Warehouse customization is a particularly useful way to generate competitive advantage in markets with rapidly changing products. For instance, a warehouse can be a place where computer components are put together, software loaded, and repairs made. Warehouses may also provide customized labeling and packaging for retailers so items arrive ready for display.

Customizing
Using warehousing to add value to the product through component modification, repair, labeling, and packaging.

Increasingly, this type of work goes on adjacent to major airports, in facilities such as the Federal Express terminal in Memphis. Adding value at warehouses adjacent to major airports facilitates overnight delivery. For instance, if your computer terminal has failed, the replacement may be sent to you from such a warehouse for delivery the next morning. When your old terminal arrives back at the warehouse, it is repaired and sent to someone else. These value-added activities at "quasi-warehouses" contribute to strategies of customization, low cost, and rapid response.

REPETITIVE AND PRODUCT-ORIENTED LAYOUT

Video 9.3

Facility Layout at
Wheeled Coach
Ambulances

Product-oriented layouts are organized around products or families of similar high-volume, low-variety products. Repetitive production and continuous production, which are discussed in Chapter 7, use product layouts. The assumptions are

1. Volume is adequate for high equipment utilization.
2. Product demand is stable enough to justify high investment in specialized equipment.
3. Product is standardized or approaching a phase of its life cycle that justifies investment in specialized equipment.
4. Supplies of raw materials and components are adequate and of uniform quality (adequately standardized) to ensure that they will work with the specialized equipment.

Fabrication line

A machine-paced, product-oriented facility for building components.

Assembly line

An approach that puts fabricated parts together at a series of workstations; used in repetitive processes.

Two types of a product-oriented layout are fabrication and assembly lines. The **fabrication line** builds components, such as automobile tires or metal parts for a refrigerator, on a series of machines. An **assembly line** puts the fabricated parts together at a series of workstations. Both are repetitive processes, and in both cases, the line must be "balanced": That is, the time spent to perform work on one machine must equal or "balance" the time spent to perform work on the next machine in the fabrication line, just as the time spent at one workstation by one assembly-line employee must "balance" the time spent at the next workstation by the next employee. The same issues arise when designing the "disassembly lines" of slaughterhouses and automobile makers (see the *OM in Action* box, "Automobile Disassembly Lines: Ecologically Correct").

Fabrication lines tend to be machine-paced and require mechanical and engineering changes to facilitate balancing. Assembly lines, on the other hand, tend to be paced by work tasks assigned to individuals or to workstations. Assembly lines, therefore, can be balanced by moving tasks from one individual to another. The central problem then in product-oriented layout planning, is to balance the output at each workstation on the production line so that it is nearly the same, while obtaining the desired amount of output.

Management's goal is to create a smooth, continuous flow along the assembly line with a minimum of idle time at each workstation. A well-balanced assembly line has the advantage of high personnel and facility utilization and equity between employees' work loads. Some union contracts

---

OM IN ACTION

Automobile Disassembly Lines: Ecologically Correct

Visionaries like Walter Chrysler and Louis Chevrolet could not have imagined the sprawling graveyards of rusting cars and trucks that bear testimony to the automotive culture they helped invent. These days, however, the graveyards are shrinking slightly. "Soon," says Ford's manager of vehicle recycling, "we think people will be buying cars based on how 'green' they are." At BMW, Horst Wolf agrees: "In the long term, all new vehicles will have to be designed in such a way that their materials can be easily reused in the next generation of cars."

In 1990, BMW, sensitive to the political power of Germany's Green Movement, built a pilot "auto disassembly" plant. In the U.S., the company offers $500 toward the purchase of a new-model BMW to anyone bringing a junked BMW to its salvage centers in New York, Los Angeles, or Orlando.

The disassembly line involves removing most of a car's plastic parts and sorting them for recycling. But this is not easy. Disassembly alone might take five people an hour. BMW also had to invent tools to safely puncture and drain fuel tanks with gas in them. Because various plastics are recycled differently, each must be labeled or color-coded. Some types of plastics can be remelted and turned into new parts, such as intake manifolds. Nissan Motor, with disassembly plants in Germany and Japan, now turns 2,000 bumpers a month into air ducts, foot rests, bumper parts, and shipping pallets.

The scrap-metal part of the disassembly line is easier. With shredders and magnets, baseball-sized chunks of metal are sorted after the engines, transmissions, radios, batteries, and exhausts have been removed. Steelmakers have helped over the past 20 years by building minimills that use scrap metal.

The ironic twist for an industry pushed to improve the crashworthiness of its vehicles is that automakers now also need to design cars and trucks that will come apart more easily.

Sources: Professional Engineering (July 11, 2001): 34–36; Ward's Auto World (September 1999): 65; and Futures (June 1998): 425–442.

FIGURE 9.11 ■

An Assembly-Line Layout

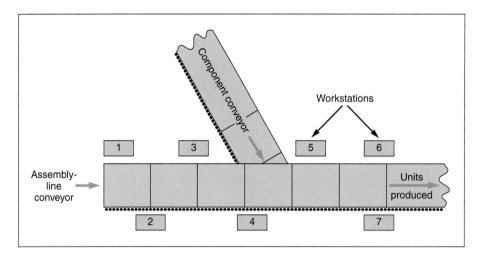

Assembly-line balancing
Obtaining output at each workstation on the production line so delay is minimized.

require that work loads be nearly equal among those on the same assembly line. The term most often used to describe this process is **assembly-line balancing**. Indeed, the *objective of the product-oriented layout is to minimize imbalance in the fabrication or assembly line*.

The main advantages of product-oriented layout are

1. The low variable cost per unit usually associated with high-volume, standardized products.
2. Low material handling costs.
3. Reduced work-in-process inventories.
4. Easier training and supervision.
5. Rapid throughput.

The disadvantages of product layout are

1. High volume is required because of the large investment needed to establish the process.
2. Work stoppage at any one point ties up the whole operation.
3. There is a lack of flexibility when handling a variety of products or production rates.

Product layout can handle only a few products and process designs.

Because the problems of fabrication lines and assembly lines are similar, we focus our discussion on assembly lines. On an assembly line, the product typically moves via automated means, such as a conveyor, through a series of workstations until completed (Figure 9.11). This is the way automobiles are assembled, television sets and ovens are produced, and fast-food hamburgers are made. Product-oriented layouts use more automated and specially designed equipment than do process layouts.

Assembly-Line Balancing

Line balancing is usually undertaken to minimize imbalance between machines or personnel while meeting a required output from the line. In order to produce at a specified rate, management must know the tools, equipment, and work methods used. Then the time requirements for each assembly task (such as drilling a hole, tightening a nut, or spray-painting a part) must be determined. Management also needs to know the *precedence relationship* among the activities—that is, the sequence in which various tasks must be performed. Example 2 shows how to turn these task data into a precedence diagram.

Example 2

We want to develop a precedence diagram for an electrostatic copier that requires a total assembly time of 66 minutes. Table 9.3 and Figure 9.12 give the tasks, assembly times, and sequence requirements for the copier.

TABLE 9.3 ■ Precedence Data

TASK	PERFORMANCE TIME (MINUTES)	TASK MUST FOLLOW TASK LISTED BELOW	
A	10	—	This means that
B	11	A	tasks B and E
C	5	B	cannot be done
D	4	B	until task A has
E	12	A	been completed.
F	3	C, D	
G	7	F	
H	11	E	
I	3	G, H	
	Total time 66		

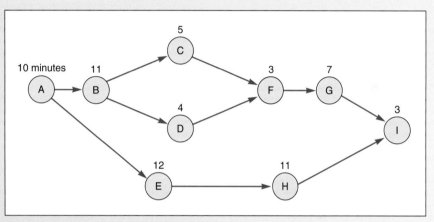

FIGURE 9.12 ■ Precedence Diagram

Once we have constructed a precedence chart summarizing the sequences and performance times, we turn to the job of grouping tasks into job stations so that we can meet the specified production rate. This process involves three steps:

1. Take the units required (demand or production rate) per day and divide it into the productive time available per day (in minutes or seconds). This operation gives us what is called the **cycle time**—namely, the maximum time that the product is available at each workstation if the production rate is to be achieved:

$$\text{Cycle time} = \frac{\text{Production time available per day}}{\text{Units required per day}} \quad (9\text{-}2)$$

Cycle time
The maximum time that the product is available at each workstation.

2. Calculate the theoretical minimum number of workstations. This is the total task-duration time (the time it takes to make the product) divided by the cycle time. Fractions are rounded to the next higher whole number:

$$\text{Minimum number of workstations} = \frac{\sum_{i=1}^{n} \text{Time for task } i}{\text{Cycle time}} \quad (9\text{-}3)$$

where n is the number of assembly tasks.

Some tasks simply cannot be grouped together in one workstation. There may be a variety of physical reasons for this.

3. Balance the line by assigning specific assembly tasks to each workstation. An efficient balance is one that will complete the required assembly, follow the specified sequence,

and keep the idle time at each workstation to a minimum. A formal procedure for doing this is

a. Identify a master list of tasks.
b. Eliminate those tasks that have been assigned.
c. Eliminate those tasks whose precedence relationship has not been satisfied.
d. Eliminate those tasks for which inadequate time is available at the workstation.
e. Use one of the line-balancing "heuristics" described in Table 9.4. The five choices are (1) longest task time, (2) most following tasks, (3) ranked positional weight, (4) shortest task time, and (5) least number of following tasks. You may wish to test several of these **heuristics** to see which generates the "best" solution—that is, the smallest number of workstations and highest efficiency. Remember, however, that although heuristics provide solutions, they do not guarantee an optimal solution.

Heuristic
Problem solving using procedures and rules rather than by mathematical optimization.

TABLE 9.4 ■

Layout Heuristics that may be Used to Assign Tasks to Work Stations in Assembly-Line Balancing

1. *Longest task (operation) time*	From the available tasks, choose the task with the largest (longest) time.
2. *Most following tasks*	From the available tasks, choose the task with the largest number of following tasks.
3. *Ranked positional weight*	From the available tasks, choose the task for which the sum of the times for each following task is longest. (In Example 3 we will see that the ranked positional weight of task C = 5(C) + 3(F) + 7(G) + 3(I) = 18, whereas the ranked positional weight of task D = 4(D) + 3(F) + 7(G) + 3(I) = 17; therefore, C would be chosen first.)
4. *Shortest task (operations) time*	From the available tasks, choose the task with the shortest task time.
5. *Least number of following tasks*	From the available tasks, choose the task with the least number of subsequent tasks.

Example 3 illustrates a simple line-balancing procedure.

Example 3

On the basis of the precedence diagram and activity times given in Example 2, the firm determines that there are 480 productive minutes of work available per day. Furthermore, the production schedule requires that 40 units be completed as output from the assembly line each day. Thus:

$$\text{Cycle time (in minutes)} = \frac{480 \text{ minutes}}{40 \text{ units}}$$

$$= 12 \text{ minutes/unit}$$

$$\text{Minimum number of workstations} = \frac{\text{total task time}}{\text{cycle time}} = \frac{66}{12}$$

$$= 5.5 \text{ or } 6 \text{ stations.}$$

Use the *most following tasks* heuristic to assign jobs to workstations.

Figure 9.13 shows one solution that does not violate the sequence requirements and that groups tasks into six stations. To obtain this solution, activities with the most following tasks were moved into worksta-

FIGURE 9.13 ■

A Six-Station Solution to the Line-Balancing Problem

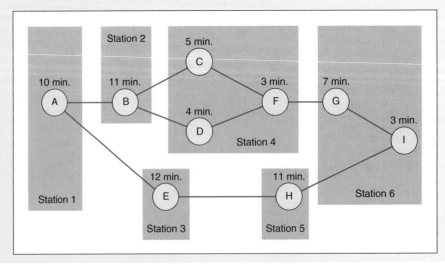

tions to use as much of the available cycle time of 12 minutes as possible. The first workstation consumes 10 minutes and has an idle time of 2 minutes.

   The second workstation uses 11 minutes, and the third consumes the full 12 minutes. The fourth work-station groups three small tasks and balances perfectly at 12 minutes. The fifth has 1 minute of idle time, and the sixth (consisting of tasks G and I) has 2 minutes of idle time per cycle. Total idle time for this solu-tion is 6 minutes per cycle.

The two issues in line balancing are the production rate and efficiency.

We can compute the efficiency of a line balance by dividing the total task time by the product of the number of workstations required times the assigned cycle time:

$$\text{Efficiency} = \frac{\Sigma \text{ task times}}{(\text{actual number of workstations}) \times (\text{assigned cycle time})} \qquad (9\text{-}4)$$

Operations managers compare different levels of efficiency for various numbers of workstations. In this way, the firm can determine the sensitivity of the line to changes in the production rate and workstation assignments.

Example 4

We can calculate the balance efficiency for Example 3 as follows:

$$\text{Efficiency} = \frac{66 \text{ minutes}}{(6 \text{ stations}) \times (12 \text{ minutes})} = \frac{66}{72} = 91.7\%$$

Note that opening a seventh workstation, for whatever reason, would decrease the efficiency of the balance to 78.6%:

$$\text{Efficiency} = \frac{66 \text{ minutes}}{(7 \text{ stations}) \times (12 \text{ minutes})} = 78.6\%$$

Large-scale line-balancing problems, like large process-layout problems, are often solved by com-puters. Several computer programs are available to handle the assignment of workstations on assem-bly lines with 100 (or more) individual work activities. Two computer routines, COMSOAL (Computer Method for Sequencing Operations for Assembly Lines)[11] and ASYBL (General Electric's Assembly Line Configuration program), are widely used in larger problems to evaluate the thousands, or even millions, of possible workstation combinations much more efficiently than could ever be done by hand.

In the case of slaughtering operations, the assembly line is actually a disassembly line. The line-balancing procedures described in this chapter are the same as for an assembly line. The chicken-processing plant shown here must balance the work of several hundred employees. Specialization contributes to efficiency because: (1) one's skills develop with repetition; (2) there is less time lost in changing tools, and (3) specialized tools are developed. The total labor content in each of the chickens processed is a few minutes. How long would it take you to process a chicken by yourself?

[11]G. W. De Puy, "Applying the COMSOAL Computer Heuristic," *Computers & Industrial Engineering* 38, no. 3 (October 2000): 413–422.

SUMMARY

Layouts make a substantial difference in operating efficiency. The six classic layout situations are (1) fixed position, (2) process oriented, (3) office, (4) retail, (5) warehouse, and (6) product oriented. A variety of techniques have been developed in attempts to solve these layout problems. Industrial firms focus on reducing material movement and assembly-line balancing. Retail firms focus on product exposure. Storage layouts focus on the optimum trade-off between storage costs and material handling costs.

Often the variables in the layout problem are so wide-ranging and numerous as to preclude finding an optimal solution. For this reason, layout decisions, although having received substantial research effort, remain something of an art.

KEY TERMS

Fixed-position layout *(p. 265)*
Process-oriented layout *(p. 266)*
Job lots *(p. 266)*
CRAFT *(p. 270)*
Work cell *(p. 271)*
Focused work center *(p. 272)*
Focused factory *(p. 272)*
Office layout *(p. 273)*
Retail layout *(p. 275)*
Slotting fees *(p. 276)*

Servicescape *(p. 276)*
Warehouse layout *(p. 277)*
Cross-docking *(p. 277)*
Random stocking *(p. 278)*
Customizing *(p. 278)*
Fabrication line *(p. 279)*
Assembly line *(p. 279)*
Assembly-line balancing *(p. 280)*
Cycle time *(p. 281)*
Heuristic *(p. 282)*

USING EXCEL OM FOR LAYOUT CALCULATIONS

Excel OM can assist in evaluating a series of room-to-department assignments like we saw for the Walters Company in Example 1. The layout module can generate an optimal solution by enumeration or can compute the "total movement" cost for each layout you wish to examine. As such, it provides a speedy calculator for each flow-distance pairing.

Program 9.1 illustrates our inputs in the top two tables. We first enter department flows, then provide distances between rooms. By entering room assignments on a trial-and-error basis in the upper left of the top table, movement computations take place at the bottom of the screen. Total movement is recalculated each time we try a new room assignment. It turns out that the assignment shown is optimal at 430 movement feet.

PROGRAM 9.1 ■

Using Excel OM's Process Layout Module to Solve the Walters Company Problem in Example 1

Excel OM does not include an Assembly Line Balancing Module.

USING POM FOR WINDOWS FOR LAYOUT DESIGN

POM for Windows' facility layout module can be used to place up to 10 departments in 10 rooms in order to minimize the total distance traveled as a function of the distances between the rooms and the flow between departments. The program performs pair-wise comparisons, exchanging departments until no exchange will reduce the total amount of movement.

POM for Windows' module for line balancing can handle a line with up to 99 tasks, each with up to 6 immediate predecessors. In this program, cycle time can be entered in two ways: (1) either *given*, if known, or (2) *demand* rate can be entered with time available as shown. Five "heuristic rules" may be used: (1) longest operation (task) time, (2) most following tasks, (3) ranked positional weight, (4) shortest operation (task) time, and (5) least number of following tasks. No one rule can guarantee an optimal solution. The default rule is the longest operation time.

Appendix V discusses further details regarding POM for Windows.

SOLVED PROBLEMS

Solved Problem 9.1

The Snow-Bird Hospital is a small emergency-oriented facility located in a popular ski resort area in northern Michigan. Its new administrator, Mary Lord, decides to reorganize the hospital, using the process-layout method she studied in business school. The current layout of Snow-Bird's eight emergency departments is shown in Figure 9.14.

Snow-Bird Hospital Layout

| Entrance/ initial processing | Exam room 1 | Exam room 2 | X-ray | 10' |
| Laboratory tests/EKG | Operating room | Recovery room | Cast-setting room | 10' |

←————————————— 40' —————————————→

FIGURE 9.14 ■ Snow-Bird Hospital Layout

The only physical restriction perceived by Lord is the need to keep the combination entrance/initial processing room in its current location. All other departments or rooms (each 10-feet square) can be moved if layout analysis indicates a move would be beneficial.

First, Lord analyzes records in order to determine the number of trips made by patients between departments in an average month. The data are shown in Figure 9.15. Her objective, Lord decides, is to lay out the rooms so as to minimize the total distance walked by patients who enter for treatment. She writes her objective as

$$\text{Minimize patient movement} = \sum_{i=1}^{8} \sum_{j=1}^{8} X_{ij} C_{ij}$$

where
X_{ij} = number of patients per month (loads or trips) moving from department i to department j
C_{ij} = distance in feet between departments i and j (which, in this case, is the equivalent of cost per load to move between departments)

FIGURE 9.15 ■

Number of Patients Moving between Departments in One Month

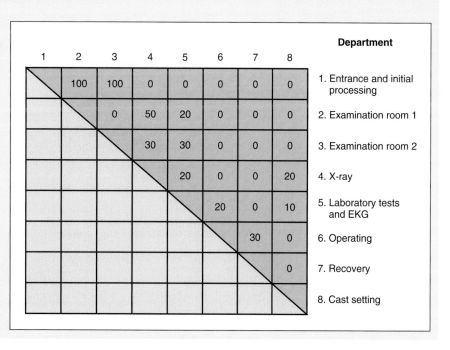

	1	2	3	4	5	6	7	8	Department
		100	100	0	0	0	0	0	1. Entrance and initial processing
			0	50	20	0	0	0	2. Examination room 1
				30	30	0	0	0	3. Examination room 2
					20	0	0	20	4. X-ray
						20	0	10	5. Laboratory tests and EKG
							30	0	6. Operating
								0	7. Recovery
									8. Cast setting

Note that this is only a slight modification of the cost-objective equation shown earlier in the chapter.

Lord assumes that adjacent departments, such as the entrance and examination room 1, have a walking distance of 10 feet. Diagonal departments are also considered adjacent and assigned a distance of 10 feet. Nonadjacent departments, such as the entrance and examination room 2 or the entrance and recovery room, are 20 feet apart, and nonadjacent rooms, such as entrance and X-ray, are 30 feet apart. (Hence, 10 feet is considered 10 units of cost, 20 feet is 20 units of cost, and 30 feet is 30 units of cost.)

Given the above information, redo the layout of Snow-Bird Hospital to improve its efficiency in terms of patient flow.

SOLUTION

First, establish Snow-Bird's current layout, as shown in Figure 9.16. By analyzing Snow-Bird's current layout, patient movement may be computed.

$$
\begin{aligned}
\text{Total movement} = &\quad \underset{\text{1 to 2}}{(100 \times 10')} \quad + \quad \underset{\text{1 to 3}}{(100 \times 20')} \quad + \quad \underset{\text{2 to 4}}{(50 \times 20')} \quad + \quad \underset{\text{2 to 5}}{(20 \times 10')} \\
+ &\quad \underset{\text{3 to 4}}{(30 \times 10')} \quad + \quad \underset{\text{3 to 5}}{(30 \times 20')} \quad + \quad \underset{\text{4 to 5}}{(20 \times 30')} \quad + \quad \underset{\text{4 to 8}}{(20 \times 10')} \\
+ &\quad \underset{\text{5 to 6}}{(20 \times 10')} \quad + \quad \underset{\text{5 to 8}}{(10 \times 30')} \quad + \quad \underset{\text{6 to 7}}{(30 \times 10')} \\
= &\ 1{,}000 + 2{,}000 + 1{,}000 + 200 + 300 + 600 + 600 \\
&+ 200 + 200 + 300 + 300 \\
= &\ 6{,}700 \text{ feet}
\end{aligned}
$$

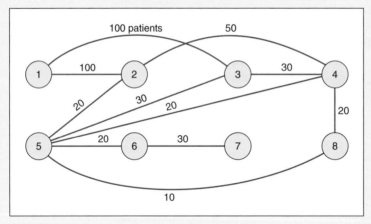

FIGURE 9.16 ■ Current Snow-Bird Patient Flow

Propose a new layout that will reduce the current figure of 6,700 feet. Two useful changes, for example, are to switch rooms 3 and 5 and to interchange rooms 4 and 6. This change would result in the schematic shown in Figure 9.17.

$$
\begin{aligned}
\text{Total movement} = &\quad \underset{\text{1 to 2}}{(100 \times 10')} \quad + \quad \underset{\text{1 to 3}}{(100 \times 10')} \quad + \quad \underset{\text{2 to 4}}{(50 \times 10')} \quad + \quad \underset{\text{2 to 5}}{(20 \times 10')} \\
+ &\quad \underset{\text{3 to 4}}{(30 \times 10')} \quad + \quad \underset{\text{3 to 5}}{(30 \times 20')} \quad + \quad \underset{\text{4 to 5}}{(20 \times 10')} \quad + \quad \underset{\text{4 to 8}}{(20 \times 20')} \\
+ &\quad \underset{\text{5 to 6}}{(20 \times 10')} \quad + \quad \underset{\text{5 to 8}}{(10 \times 10')} \quad + \quad \underset{\text{6 to 7}}{(30 \times 10')} \\
= &\ 1{,}000 + 1{,}000 + 500 + 200 + 300 + 600 + 200 \\
&+ 400 + 200 + 100 + 300 \\
= &\ 4{,}800 \text{ feet}
\end{aligned}
$$

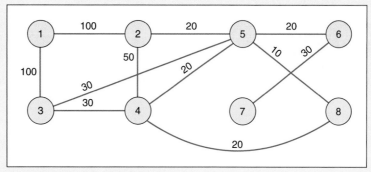

FIGURE 9.17 ■ Improved Layout

Do you see any room for further improvement?

Solved Problem 9.2

The assembly line whose activities are shown in Figure 9.18 has an 8-minute cycle time. Draw the precedence graph and find the minimum possible number of workstations. Then arrange the work activities into workstations so as to balance the line. What is the efficiency of your line balance?

TASK	PERFORMANCE TIME (MINUTES)	TASK MUST FOLLOW THIS TASK
A	5	—
B	3	A
C	4	B
D	3	B
E	6	C
F	1	C
G	4	D, E, F
H	2	G
	28	

FIGURE 9.18 ■

Four-Station Solution to the Line-Balancing Problem

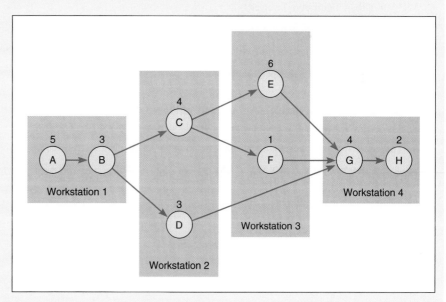

SOLUTION

The theoretical minimum number of workstations is

$$\frac{\sum t_i}{\text{Cycle time}} = \frac{28 \text{ minutes}}{8 \text{ minutes}} = 3.5 \text{ or 4 stations}$$

The precedence graph and one good layout are shown in Figure 9.18.

$$\text{Efficiency} = \frac{\text{total task time}}{(\text{number of workstations}) \times (\text{cycle time})} = \frac{28}{(4)(8)} = 87.5\%$$

INTERNET AND STUDENT CD-ROM EXERCISES

Visit our home page or use your student CD-ROM to help with material in this chapter.

 On Our Home Page, www.prenhall.com/heizer

- Internet Homework Problems
- Internet Cases
- Self-Tests
- Practice Problems
- Internet Exercises
- Current Articles and Research
- Virtual Company Tour

 On Your Student CD-ROM

- PowerPoint Lecture
- Practice Problems
- Video Clips and Video Case
- Active Model Exercise
- ExcelOM
- Excel OM Data Files

ADDITIONAL CASE STUDIES

Internet Case Studies: Visit our Web site at www.prenhall.com/heizer **for these free case studies:**

- **Palm Beach Institute of Sports Medicine**: Deals with all aspects of laying out vacant space for a fitness center.

- **W&G Beer Distributorship**: Involves layout of a warehouse that distributes beer.

- **Microfix, Inc.**: This company needs to balance its PC manufacturing assembly line and deal with sensitivity analysis of time estimates.

- **Des Moines National Bank**: This recently completed building needs to arrange its departments to optimize efficiency.

- **Collier Technical College**: School must decide which of two buildings meets its expansion needs.

Harvard has selected these Harvard Business School cases to accompany this chapter (textbookcasematch.hbsp.harvard.edu**):**

- **Toshiba; Ome Works** (#696-059): Deals with the design of an efficient notebook computer assembly line in the Ome, Japan, factory.

- **Mouawad Bangkok Rare Jewels Manufacturers Co. Ltd. (A)** (#696-056): This small Thai factory faces a challenging production control process.

- **Copeland Corp. (B)** (#686-089): A plant layout must be selected from two alternatives available to this Sidney, Australia, manufacturer.

Chapter 10

Human Resources and Job Design

Chapter Outline

LEARNING OBJECTIVES

When you complete this chapter you should be able to

IDENTIFY OR DEFINE:

Job design

Job specialization

Job expansion

Tools of methods analysis

Ergonomics

Labor standards

Andon

EXPLAIN OR DESCRIBE:

Requirements of good job design

The visual workplace

GLOBAL COMPANY PROFILE:

Human Resources Bring Competitive Advantage to Southwest Airlines

Since its beginning as a Texas airline operating only between San Antonio, Dallas, and Houston, Southwest Airlines has challenged the giants and won. For 30 straight years, in spite of being the little guy fighting both established giants and numerous legal battles, Southwest has been profitable. It has been profitable while other airlines have come and gone. It has been profitable in years when United, Northwest, and USAir lost billions.

What is the critical strategy for this airline with the low-cost advantage? The answer is human resources. Herb Kelleher, the maverick Chairman and former CEO of Southwest, says, "I've tried to create a culture of caring for people in the totality of their lives, not just at work. There is no magic formula. It's like building a giant mosaic—it takes thousands of little pieces. The intangibles are more important than the tangibles. Someone can go out and buy airplanes and ticket counters, but they can't buy our culture, our *esprit de corps.*"

Southwest spends more to recruit and train than any other airline.

Employees are also paid more than the industry average and many receive stock options. These policies may pose a high cost up front, but Southwest finds them effective in the long run. However, that is only part of Southwest's human resource strategy. President Colleen C. Barrett is constantly reinforcing the company's message that employees should be treated like customers and do what is right for the customer. Indeed, before "empowerment" became a management fad, Southwest was doing it. Southwest gives employees freedom from centralized policies

Herb Kelleher, Chairman of Southwest Airlines, works hard at maintaining the Southwest culture. Here Herb is passing out peanuts during a flight.

An innovative operations strategy has allowed Southwest Airlines to grow from a Texas only airline in the 1970s to the nation's fourth largest in 30 years. Southwest consistently ranks at the top of the airline pack in travel surveys and rankings of the most admired companies, and has the lowest rate of complaints filed at the U.S. Department of Transportation.

and teaches them to care. The belief is that if employees know what great service looks like, they will do the right thing.

For instance, to maintain high airplane utilization and high return on assets, Southwest gets planes in and out of the gate in about half of the industry average—20 minutes versus 45 minutes. And, of course, people from all industries come to see how Southwest does it. But as Kelleher observes, "They keep looking for gimmicks, special equipment. It's just a bunch of people knocking themselves out. You have to recognize that people are still most important. How you treat them determines how they treat people on the outside. We have people going around the company all the time doing other people's jobs, but not for cross utilization. We just want everybody to understand what everybody else's problems are."

Sources: The *Wall Street Journal* (August 31, 1999): B1; and *Harvard Business Review* (June 2001): 28–30.

The inhibited personality needs to think twice about working for Southwest Airlines. Here barefoot, Herb Kelleher clings to the tail of a Southwest Airlines jet. Southwest likes to hire people with enthusiasm and sense of humor.

The objective of a human resource strategy is to manage labor and design jobs so people are effectively and efficiently utilized.

Various work cultures, of which Southwest Airlines is just one example, exist all over the world. How are these cultures built and what are the human resource issues for the operations manager? In this chapter, we will examine a variety of human resource issues because organizations do not function without people. Moreover, they do not function well without competent, motivated people. The operations manager's human resource strategy determines the talents and skills available to operations.

As many organizations from Hard Rock Cafe to Lincoln Electric to Southwest Airlines have demonstrated, competitive advantage can be built through human resource strategy. Good human resource strategies are expensive, difficult to achieve, and hard to sustain. However, the payoff potential is substantial because they are hard to copy! So a competitive advantage in this area is particularly beneficial. For these reasons, we now look at the operations manager's human resource options.

HUMAN RESOURCE STRATEGY FOR COMPETITIVE ADVANTAGE

The *objective of a human resource strategy* is to manage labor and design jobs so people are *effectively* and *efficiently utilized*. As we focus on a human resource strategy, we want to ensure that people:

1. Are efficiently utilized within the constraints of other operations management decisions.
2. Have a reasonable quality of work life in an atmosphere of mutual commitment and trust.

By reasonable *quality of work life* we mean a job that is not only reasonably safe and for which the pay is equitable, but that also achieves an appropriate level of both physical and psychological requirements. *Mutual commitment* means that both management and employee strive to meet common objectives. *Mutual trust* is reflected in reasonable, documented employment policies that are honestly and equitably implemented to the satisfaction of both management and employee.[1] When management has a genuine respect for its employees and their contributions to the firm, establishing a reasonable quality of work life and mutual trust is not particularly difficult.

This chapter is devoted to showing how operations managers can achieve an effective human resource strategy, which, as we have suggested in our opening profile of Southwest Airlines, may provide a competitive advantage.

Constraints on Human Resource Strategy

As Figure 10.1 suggests, many decisions made about people are constrained by other decisions. First, the product mix may determine seasonality and stability of employment. Second, technology, equipment, and processes may have implications for safety and job content. Third, the location decision may have an impact on the ambient environment in which the employees work. Finally, layout decisions, such as assembly line versus work cell, influence job content.

Technology decisions impose substantial constraints. For instance, some of the jobs in steel mills are dirty, noisy, and dangerous; slaughterhouse jobs may be stressful and subject workers to stomach-crunching stench; assembly-line jobs are often boring and mind numbing; and high capital expenditures such as those required for manufacturing semiconductor chips may require 24-hour, 7-day-a-week operation in restrictive clothing.

We are not going to change these jobs without making changes in our other strategic decisions. So, the trade-offs necessary to reach a tolerable quality of work life are difficult. Effective managers consider such decisions simultaneously. The result: an effective, efficient system in which both individual and team performance are enhanced through optimum job design.

Acknowledging the constraints imposed on human resource strategy, we now look at three distinct decision areas of human resource strategy: labor planning, job design, and labor standards. The supplement to this chapter expands on the discussion of labor standards and introduces work measurement.

Video 10.1

Human Resources at Hard Rock Cafe

"It's possible to achieve sustainable competitive advantage by how you manage people."
Stanford University
Prof. Jeffrey Pfeffer

---

[1]With increasing frequency we find companies calling their employees *associates*, *individual contributors*, or members of a particular team.

FIGURE 10.1 ■

Constraints on Human
Resource Strategy

*The effective operations
manager understands how
decisions blend together to
constrain the human
resource strategy.*

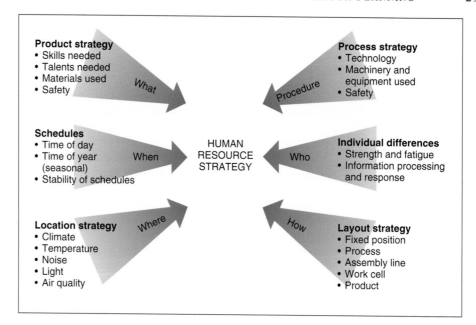

LABOR PLANNING

Labor planning
A means of determining
staffing policies dealing
with employment stability
and work schedules.

Labor planning is determining staffing policies that deal with (1) employment stability and (2) work schedules.

Employment-Stability Policies

Employment stability deals with the number of employees maintained by an organization at any given time. There are two very basic policies for dealing with stability:

1. *Follow demand exactly.* Following demand exactly keeps direct labor costs tied to production, but incurs other costs. These other costs include (a) hiring and termination costs, (b) unemployment insurance, and (c) premium wages to entice personnel to accept unstable employment. This policy tends to treat labor as a variable cost.
2. *Hold employment constant.* Holding employment levels constant maintains a trained workforce and keeps hiring, termination, and unemployment costs to a minimum. However, with employment held constant, employees may not be utilized fully when demand is low, and the firm may not have the human resources it needs when demand is high. This policy tends to treat labor as a fixed cost.

Maintaining a stable workforce may allow a firm to pay lower wages than a firm that follows demand. This savings may provide a competitive advantage. However, firms with highly seasonal work and little control over demand may be best served by a fluctuating workforce. For example, a salmon canner on the Columbia River only processes salmon when the salmon are running. However, the firm may find complementary labor demands in other products or operations, such as making cans and labels or repairing and maintaining facilities.

The above policies are only two of many that can be efficient *and* provide a reasonable quality of work life. Firms must determine policies about employment stability. Employment policies are partly determined by management's view of labor costs—as a variable cost or a fixed cost.

**Standard work
schedule**
Five 8-hour days in the U.S.

Flextime
A system that allows
employees, within limits,
to determine their own
work schedules.

Work Schedules

Although the **standard work schedule** in the U.S. is still five 8-hour days, variations do exist. A currently popular variation is a work schedule called flextime. **Flextime** allows employees, within limits, to determine their own schedules. A flextime policy might allow an employee (with proper notification) to be at work at 8 A.M. plus or minus 2 hours. This policy allows more autonomy and independence on the part of the employee. Some firms have found flextime a low-cost fringe benefit that enhances job satisfaction. The problem from the OM perspective is that much production work requires full staffing for efficient operations. A machine that requires three people cannot run at all if only two show up. Having a waiter show up to serve lunch at 1:30 P.M. rather than 11:30 A.M. is not much help either.

Similarly, some industries find that their process strategies severely constrain their human resource scheduling options. For instance, paper manufacturing, petroleum refining, and power stations require around-the-clock staffing except for maintenance and repair shutdown.

Flexible workweek
A work schedule that deviates from the normal or standard five 8-hour days (such as four 10-hour days).

Another option is the **flexible workweek**. This plan often calls for fewer but longer days, such as four 10-hour days or, as in the case of light assembly plants, 12-hour shifts. Twelve-hour shifts usually mean working 3 days one week and 4 the next. Such shifts are sometimes called *compressed workweeks*. These schedules are viable for many operations functions—as long as suppliers and customers can be accommodated. Firms that have high process start-up times (say, to get a boiler up to operating temperature) find longer workday options particularly appealing. Compressed workweeks have long been common in fire and utility departments, where physical exertion is modest but 24-hour coverage desirable. A recent Gallup survey showed that two-thirds of working adults would prefer toiling four 10-hour days to the standard 5-day schedule. Duke Power Co., Los Angeles County, AT&T, and General Motors are just a few organizations to offer the 4-day week.

Part-time status
When an employee works less than a normal week; less than 32 hours per week often classifies an employee as "part time."

Another option is shorter days rather than longer days. This plan often moves employees to **part-time status**. Such an option is particularly attractive in service industries, where staffing for peak loads is necessary. Banks and restaurants often hire part-time workers. Also, many firms reduce labor costs by reducing fringe benefits for part-time employees.

Job Classifications and Work Rules

Many organizations have strict job classifications and work rules that specify who can do what, when they can do it, and under what conditions they can do it, often as a result of union pressure. These job classifications and work rules restrict employee flexibility on the job, which in turn reduces the flexibility of the operations function. Yet part of an operations manager's task is to manage the unexpected. Therefore, the more flexibility a firm has when staffing and establishing work schedules, the more efficient and responsive it *can* be. This is particularly true in service organizations, where extra capacity often resides in extra or flexible staff. Building morale and meeting staffing requirements that result in an efficient, responsive operation are easier if managers have fewer job classifications and work-rule constraints. If the strategy is to achieve a competitive advantage by responding rapidly to the customer, a flexible workforce may be a prerequisite.[2]

JOB DESIGN

Job design
An approach that specifies the tasks that constitute a job for an individual or a group.

Job design specifies the tasks that constitute a job for an individual or a group. We examine seven components of job design: (1) job specialization, (2) job expansion, (3) psychological components, (4) self-directed teams, (5) motivation and incentive systems, (6) ergonomics and work methods, and (7) the visual workplace.

Labor Specialization

Labor specialization (or job specialization)
The division of labor into unique ("special") tasks.

The importance of job design as a management variable is credited to the eighteenth-century economist Adam Smith.[3] Smith suggested that a division of labor, also known as **labor specialization** (or **job specialization**), would assist in reducing labor costs of multiskilled artisans. This is accomplished in several ways:

1. *Development of dexterity* and faster learning by the employee because of repetition.
2. *Less loss of time* because the employee would not be changing jobs or tools.
3. *Development of specialized tools* and the reduction of investment because each employee has only a few tools needed for a particular task.

The nineteenth-century British mathematician Charles Babbage determined that a fourth consideration was also important for labor efficiency.[4] Because pay tends to follow skill with a rather high correlation, Babbage suggested *paying exactly the wage needed for the particular skill required*. If

---

[2]David M. Upton, "What Really Makes a Factory Flexible?" *Harvard Business Review* (July–August 1995): 74–84.

[3]Adam Smith, *On the Creation of the Wealth of Nations* (London, 1776).

[4]Charles Babbage, *On the Economy of Machinery and Manufacturers* (London: C. Knight, 1832), chapter 18.

the entire job consists of only one skill, then we would pay for only that skill. Otherwise, we would tend to pay for the highest skill contributed by the employee. These four advantages of labor specialization are still valid today.

A classic example of labor specialization is the assembly line. Such a system is often very efficient, although it may require employees to do repetitive, mind-numbing jobs. The wage rate for many of these jobs, however, is very good. Given the relatively high wage rate for the modest skills required in many of these jobs, there is often a large pool of employees from which to choose. This is not an incidental consideration for the manager with responsibility for staffing the operations function. It is estimated that 2% to 3% of the workforce in industrialized nations perform highly specialized, repetitive assembly-line jobs. The traditional way of developing and maintaining worker commitment under labor specialization has been good selection (matching people to the job), good wages, and incentive systems.

From the manager's point of view, a major limitation of specialized jobs is their failure to bring the whole person to the job. Job specialization tends to bring only the employee's manual skills to work. In an increasingly sophisticated knowledge-based society, managers may want employees to bring their mind to work as well.

Job Expansion

In recent years, there has been an effort to improve the quality of work life by moving from labor specialization toward more varied job design. Driving this effort is the theory that variety makes the job "better" and that the employee therefore enjoys a higher quality of work life. This flexibility thus benefits the employee and the organization.

We modify jobs in a variety of ways. The first approach is **job enlargement**, which occurs when we add tasks requiring similar skill to an existing job. **Job rotation** is a version of job enlargement that occurs when the employee is allowed to move from one specialized job to another. Variety has been added to the employee's perspective of the job. Another approach is **job enrichment**, which adds planning and control to the job. An example is to have department store salespeople responsible for ordering, as well as selling, their goods. Job enrichment can be thought of as *vertical expansion*, as opposed to job enlargement, which is *horizontal*. These ideas are shown in Figure 10.2.

A popular extension of job enrichment, **employee empowerment** is the practice of enriching jobs so employees accept responsibility for a variety of decisions normally associated with staff specialists. Empowering employees helps them take "ownership" of their jobs so they have a personal interest in improving performance. (See the *OM in Action* box, "Empowerment at the Ritz-Carlton.")

Job enlargement
The grouping of a variety of tasks about the same skill level; horizontal enlargement.

Job rotation
A system in which an employee is moved from one specialized job to another.

Job enrichment
A method of giving an employee more responsibility that includes some of the planning and control necessary for job accomplishment.

Employee empowerment
Enlarging employee jobs so that the added responsibility and authority is moved to the lowest level possible in the organization.

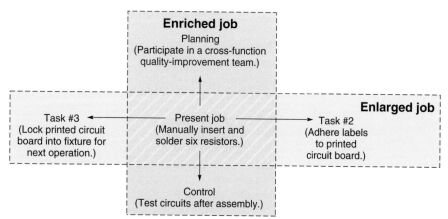

FIGURE 10.2 ■ An Example of Job Enlargement (*horizontal* job expansion) and Job Enrichment (*vertical* job expansion)

The job can be enlarged horizontally by job rotation to tasks 2 and 3 or these tasks can be made a part of the present job. Job enrichment, expanding the job vertically, can occur by adding other types of tasks, such as participation in a quality team (planning) and testing tasks (control).

OM IN ACTION

Empowerment at the Ritz-Carlton

The recently retired president of the Ritz-Carlton hotel chain used to introduce himself to employees with "My name is Horst Schulze. I'm president of this company; I'm very important. (Pause) But so are you. Absolutely. Equally important." This attitude may be the cause of a turnover rate that is less than half the industry average, and may be why the Ritz recently received a Malcolm Baldrige National Quality Award.

Schulze ran the hotel for nearly 20 years with his philosophy of customer service that is embodied in the Gold Standards. The Gold Standards are compressed into a wallet-sized, foldout card that employees are expected to carry at all times. Orientation, training, indeed indoctrination, are extensive with daily review of one of the 20 basics in the Gold Standard. With values like "genuine care and comfort" of guests, "Our job is to make guests feel good so they come back!" and "Ladies and gentlemen serving ladies and gentlemen," Ritz has empowered its employees to take care of the guests.

Speaking of empowerment, Ritz front-desk clerks and sales managers can spend up to $2,000 and $5,000 of company money, respectively—to ensure that guests leave satisfied. For example, when the New York Ritz was overbooked once, 20 guests were sent to another hotel in three limousines packed with champagne and caviar. The cost: $5,000. "The idea was to please guests," says the Ritz manager.

Empowerment also includes taking suggestions of all employees seriously. When a room service waiter proposed the company spend $50,000 to implement a recycling plan, Schulze took a deep breath and then agreed. The idea paid off: Weekly garbage pickups have been reduced and the hotel now sells its paper products rather than paying others to haul it off. The changes have saved $80,000 a year and typify the hotel's reliance on employee suggestions for quality improvement.

Sources: Harvard Business Review (June 2002): 50–62; and the Wall Street Journal (August 30, 1999): B1.

Psychological Components of Job Design

An effective human resources strategy also requires consideration of the psychological components of job design. These components focus on how to design jobs that meet some minimum psychological requirements.

> "We hired workers and human beings came instead."
>
> Max Frisch

Hawthorne Studies   The Hawthorne studies introduced psychology to the workplace. They were conducted in the late 1920s at Western Electric's Hawthorne plant near Chicago. Publication of the findings in 1939[5] showed conclusively that there is a dynamic social system in the workplace. Ironically, these studies were initiated to determine the impact of lighting on productivity. Instead they found the social system and distinct roles played by employees to be more important than the intensity of the lighting. They also found that individual differences may be dominant in what an employee expects from the job and what the employee thinks her or his contribution to the job should be.

Core Job Characteristics   In the eight decades since the Hawthorne studies, substantial research regarding the psychological components of job design has taken place.[6] Hackman and Oldham have incorporated much of that work into five desirable characteristics of job design.[7] Their summary suggests that jobs should include the following characteristics:

1. *Skill variety,* requiring the worker to use a variety of skills and talents.
2. *Job identity,* allowing the worker to perceive the job as a whole and recognize a start and a finish.
3. *Job significance,* providing a sense that the job has impact on the organization and society.
4. *Autonomy,* offering freedom, independence, and discretion.
5. *Feedback,* providing clear, timely information about performance.

---

[5]F. J. Roethlisberger and William J. Dickinson, *Management and the Workers* (New York: John Wiley, 1964, copyright 1939, by the President & Fellows of Harvard College).

[6]See, for instance, the work of Abraham H. Maslow, "A Theory of Human Motivation," *Psychological Review* 50 (1943): 370–396; and Frederick Herzberg, B. Mausner, and B. B. Snyderman, *The Motivation to Work* (New York: John Wiley, 1965).

[7]See "Motivation Through the Design of Work," in *Work Redesign*, eds. Jay Richard Hackman and Greg R. Oldham (Reading, MA: Addison-Wesley, 1980).

FIGURE 10.3 ■

Job Design Continuum

Empowerment can take many forms—planning, scheduling, quality, purchasing, and even hiring authority.

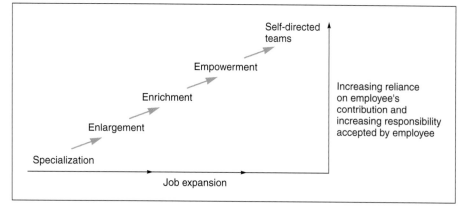

Self-directed team

A group of empowered individuals working together to reach a common goal.

Including these five ingredients in job design is consistent with job enlargement, job enrichment, and employee empowerment. We now want to look at some of the ways in which teams can be used to expand jobs and achieve these five job characteristics.

Self-Directed Teams

Many world-class organizations have adopted teams to foster mutual trust and commitment, and provide the core job characteristics. One team concept of particular note is the **self-directed team:** a group of empowered individuals working together to reach a common goal. These teams may be organized for long-term or short-term objectives. Teams are effective primarily because they can easily provide employee empowerment, ensure core job characteristics, and satisfy many of the psychological needs of individual team members.[8] A job design continuum is shown in Figure 10.3.

Of course, many good job designs *can* provide these psychological needs. Therefore, to maximize team effectiveness, managers do more than just form "teams." For instance, they (1) ensure that those who have a legitimate contribution are on the team, (2) provide management support, (3) ensure the necessary training, and (4) endorse clear objectives and goals. Successful teams should also receive financial and nonfinancial rewards. Finally, managers must recognize that teams may have a life cycle and that achieving an objective may suggest disbanding the team. However, teams may be renewed with a change in members or new assignments.

Teams and other approaches to job expansion should not only improve the quality of work life and job satisfaction but also motivate employees to achieve strategic objectives. Both managers *and* employees need to be committed to achieving strategic objectives. However, employee contribution is fostered in a variety of ways, including organizational climate, supervisory action, *and* job design.

Expanded job designs allow employees to accept more responsibility. For employees who accept this responsibility, we may well expect some enhancement in productivity and product quality. Among the other positive aspects of job expansion are reduced turnover, tardiness, and absenteeism. Managers who expand jobs and build communication systems that elicit suggestions from employees have an added potential for efficiency and flexibility. However, these job designs have a number of limitations.

Limitations of Job Expansion   If job designs that enlarge, enrich, empower, and use teams are so good, why are they not universally used? Let us identify some limitations of expanded job designs:

1. *Higher capital cost.* Job expansion may require facilities that cost more than those with a conventional layout. This extra expenditure must be generated through savings (greater efficiency) or higher prices.
2. *Individual differences.* Some studies indicate that many employees opt for the less complex jobs. In a discussion about improving the quality of work life, we cannot forget the importance of individual differences. Differences in individuals provide latitude for the resourceful operations manager when designing jobs.
3. *Higher wage rates.* People often receive wages for their highest skills, not their lowest. Thus expanded jobs may well require a higher average wage than jobs that are not expanded.

[8]Per H. Engelstad, "Sociotechnical Approach to Problems of Process Control," in *Design of Jobs*, eds. Louis E. Davis and James C. Taylor (Santa Monica: Goodyear Publishing, 1979), 184–205.

4. *Smaller labor pool.* Because expanded jobs require more skill and acceptance of more responsibility, job requirements have increased. Depending upon the availability of labor, this may be a constraint.
5. *Increased accident rates.* Expanded jobs may contribute to a higher accident rate. This indirectly increases wages, insurance costs, and worker's compensation. The alternative may be expanding training and safety budgets.
6. *Current technology may not lend itself to job expansion.* The disassembly jobs at a slaughterhouse and assembly jobs at automobile plants are that way because alternative technologies (if any) are thought to be unacceptable.

These six points provide constraints on job expansion.

In short, job expansion often increases costs. Therefore, for the firm to have a competitive advantage, its savings must be greater than its costs. This is not always the case. The strategic decision may not be an easy one.

Despite the limitations of job expansion, firms are finding ways to make it work. Often the major limitations are not those listed above, but training budgets and the organization's culture, as indicated in Table 10.1. U.S. training budgets must increase. And supervisors must release some control and learn to accept different job responsibilities. Self-directed teams may mean no supervisors on the factory floor. Removing supervisors from the factory floor, as has Harris-Farinon, a world leader in microwave equipment, is often a major culture change. However, Harris-Farinon is setting new performance standards with exactly this type of cultural change.

Service organizations have also reaped substantial advantage from successful human resources strategies. These success stories include Hard Rock Cafe, Southwest Airlines, Ritz-Carlton, Nordstrom, Taco Bell, and Disney. Each has recognized that the creation of value for customers and shareholders begins with the creation of value for employees.[9] Hard Rock goes so far as to give a $10,000 gold Rolex watch to every employee on his or her tenth anniversary with the firm—from the president down to the busboys.

Motivation and Incentive Systems

Our discussion of the psychological components of job design provides insight into the factors that contribute to job satisfaction and motivation. In addition to these psychological factors, there are monetary factors. Money often serves as a psychological as well as financial motivator. Monetary rewards take the form of bonuses, profit and gain sharing, and incentive systems.

Bonuses, typically in cash or stock options, are often used at executive levels to reward management. **Profit-sharing** systems provide some part of the profit for distribution to employees. A variation of profit sharing is **gain sharing**, which rewards employees for improvements made in an organization's performance. The most popular of these is the Scanlon plan, in which any reduction in the cost of labor is shared between management and labor.[10]

The gain-sharing approach used by Panhandle Eastern Corp. of Houston, Texas, allows for employees to receive a bonus of 2% of their salary at year's end if the company earns at least $2.00 per share. When Panhandle earns $2.10 per share, the bonus climbs to 3%. Employees have become much more sensitive about costs since the plan began.

Incentive systems based on individual or group productivity are used throughout the world in a wide variety of applications, including nearly half of the manufacturing firms in America. Production incentives often require employees or crews to produce at or above a predetermined standard. The standard can be based on a "standard time" per task or number of pieces made. Both systems typically guarantee the employee at least a base rate.

With the increasing use of teams, various forms of team-based pay are also being developed. Many are based on traditional pay systems supplemented with some form of bonus or incentive system. However, because many team environments require cross training of enlarged jobs, *knowledge-*

The only thing worse than training an employee and having them go to work somewhere else—is not training them and having them stay!

TABLE 10.1 ■

Average Annual Training Hours per Employee

U.S.	7 hr.
Sweden	170 hr.
Japan	200 hr.

Source: APICS Newsletter.

Bonus
A monetary reward, usually in cash or stock options, given to management.

Profit sharing
A system providing some portion of any profit for distribution to the employees.

Gain sharing
A system of rewards to employees for organizational improvements.

Incentive system
An employee award system based on individual or group productivity.

---

[9]Roger Hallowell, "Southwest Airlines: A Case Study Linking Employees' Needs, Satisfaction, and Organizational Capabilities to Competitive Advantage," *Human Resource Management* (winter 1996): 530.

[10]Fred G. Lesieur and Elbridge S. Puckett, "The Scanlon Plan Has Proved Itself," *Harvard Business Review* 47, no. 5 (September–October 1969): 109–118.

based pay systems have also been developed. Under **knowledge-based** (or skill-based) **pay systems**, a portion of the employee's pay depends on demonstrated knowledge or skills possessed. Knowledge-based pay systems are designed to reward employees for the enlarged scope of their jobs. Some of these pay systems have three dimensions: *horizontal skills* that reflect the variety of tasks the employee can perform; *vertical skills* that reflect the planning and control aspects of the job; and *depth of skills* that reflect quality and productivity. At Wisconsin's Johnsonville Sausage Co., employees receive pay raises *only* by mastering new skills such as scheduling, budgeting, and quality control.

Ergonomics and Work Methods

As mentioned in Chapter 1, Frederick W. Taylor began the era of scientific management in the late 1800s.[11] He and his contemporaries began to examine personnel selection, work methods, labor standards, and motivation.

With the foundation provided by Taylor, we have developed a body of knowledge about people's capabilities and limitations. This knowledge is necessary because humans are hand/eye animals possessing exceptional capabilities and some limitations. Because managers must design jobs that can be done, we now introduce a few of the issues related to people's capabilities and limitations.

Ergonomics   The operations manager is interested in building a good interface between human and machine. Studies of this interface are known as **ergonomics**. Ergonomics means "the study of work." (*Ergon* is the Greek word for *work*.) In the U.S., the term *human factors* is often substituted for the word *ergonomics*. Understanding ergonomics issues helps to improve human performance.

Male and female adults come in limited configurations. Therefore, design of tools and the workplace depends on the study of people to determine what they can and cannot do. Substantial data have been collected that provide basic strength and measurement data needed to design tools and the workplace. The design of the workplace can make the job easier or impossible. Additionally, we now have the ability, through the use of computer modeling, to analyze human motions and efforts.

Let's look briefly at one instance of human measurements: determining the proper height for a writing desk. The desk has an optimum height depending on the size of the individual and the task to be performed. The common height for a writing desk is 29 inches. For typing or data entry at a computer, the surface should be lower. The preferred chair and desk height should result in a very slight angle between the body and arm when the individual is viewed from the front and when the back is straight. This is the critical measurement; it can be achieved via adjustment in either table or chair height.

Ergonomics issues occur in the office as well as in the factory. Here an ergonomics consultant is measuring the angle of a terminal operator's neck. Posture, which is related to desk height, chair height and position, keyboard placement, and computer screen, is an important factor in reducing back and neck pain that can be caused by extended hours at a computer.

[11]Frederick W. Taylor, *Scientific Management* (New York: Harper & Row, 1911): 204.

The Infogrip 'keyboard' has only seven keys, one for each finger and three for the thumb, but replicates all the functions of a traditional QWERTY keyboard. Since finger placement is constant, strain on the hand is reduced. Infogrip, Ventura, CA)

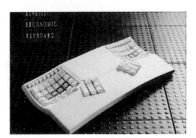

Tests indicate that this keyboard, which more closely fits the natural shape of the hand, is less physically demanding and more comfortable to use than a traditional computer keyboard. (Kinesis Corp., Bellevue, WA)

The "Data-Hand" keyboard allows each hand to rest on its own ergonomically shaped and padded palm support. Five keys surround each fingertip and thumb. (Industrial Innovations, Inc., Scottsdale, AZ)

FIGURE 10.4 ■ Job Design and the Keyboard

Operator Input to Machines   Operator response to machines, be they hand tools, pedals, levers, or buttons, needs to be evaluated. Operations managers need to be sure that operators have the strength, reflexes, perception, and mental capacity to provide necessary control. Such problems as *carpal tunnel syndrome* may result when a tool as simple as a keyboard is poorly designed.[12] The photos in Figure 10.4 indicate recent innovations designed to improve this common tool.

Feedback to Operators   Feedback to operators is provided by sight, sound, and feel; it should not be left to chance. The mishap at the Three Mile Island nuclear facility, America's worst nuclear experience, was in large part the result of poor feedback to the operators about reactor per-

Carpal tunnel syndrome is a wrist disorder that afflicts 23,000 workers annually and costs employers and insurers an average of $30,000 per affected worker. Many of the tools, handles, and computer keyboards now in use put the wrists in an unnatural position. An unnatural position, combined with extensive repetition, may contribute to carpal tunnel syndrome. One of the medical procedures to correct carpal tunnel syndrome is the operation shown here, which reduces the symptoms. The cure, however, lies in the ergonomics of workplace and tool design.

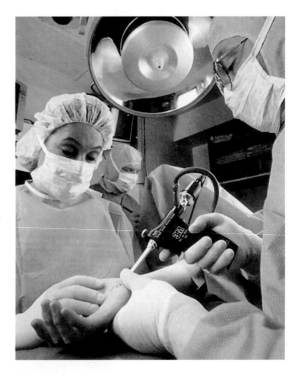

[12]Although carpal tunnel syndrome is routinely referred to as a work-related condition, there is some evidence that an under-lying disease, such as diabetes or arthritis, may be the chief cause. See "Diseases, Not Work, May Be Carpal Tunnel Culprit," *IIE Solutions* (February 1999): 13.

TABLE 10.2 ■

Levels of Illumination
Recommended for
Various Task Conditions

TASK CONDITION	TYPE OF TASK OR AREA	ILLUMINATION LEVEL (FT-C)[a]	TYPE OF ILLUMINATION
Small detail, extreme accuracy	Sewing, inspecting dark materials	100	Overhead ceiling lights and desk lamp
Normal detail, prolonged periods	Reading, parts assembly, general office work	20–50	Overhead ceiling lights
Good contrast, fairly large objects	Recreational facilities	5–10	Overhead ceiling lights
Large objects	Restaurants, stairways, warehouses	2–5	Overhead ceiling lights

[a]FT-C (the footcandle) is a measure of illumination.

Source: C. T. Morgan, J. S. Cook III, A. Chapanis, and M. W. Lund, eds., *Human Engineering Guide to Equipment Design* (New York: McGraw-Hill, 1963).

formance. Nonfunctional groups of large, unclear instruments and inaccessible controls, combined with hundreds of confusing warning lights, contributed to that nuclear failure. Such relatively simple issues make a difference in operator response and, therefore, performance.

The Work Environment   The physical environment in which employees work affects their performance, safety, and quality of work life. Illumination, noise and vibration, temperature, humidity, and air quality are work-environment factors under the control of the organization and the operations manager. The manager must approach them as controllable.

Illumination is necessary, but the proper level depends upon the work being performed. Table 10.2 provides some guidelines. However, other lighting factors are important. These include reflective ability, contrast of the work surface with surroundings, glare, and shadows.

Noise of some form is usually present in the work area, but most employees seem to adjust well. However, high levels of sound will damage hearing. Table 10.3 provides indications of the sound generated by various activities. Extended periods of exposure to decibel levels above 85 dB are permanently damaging. The Occupational Safety and Health Administration (OSHA) requires

TABLE 10.3 ■

Decibel (dB) Levels for Various Sounds (decibel levels are A-weighted sound levels measured with a sound-level meter)

ENVIRONMENT NOISES	COMMON NOISE SOURCES	DECIBELS	
	Jet takeoff (200 ft)	120	
Electric furnace area	Pneumatic hammer	100	Very annoying
Printing press plant	Subway train (20 ft)	90	
	Pneumatic drill (50 ft)	80	Ear protection required if exposed for 8 or more hours
Inside sports car (50 mph)	Vacuum cleaner (10 ft)	70	
Near freeway (auto traffic)	Speech (1 ft)	60	Intrusive
Private business office Light traffic (100 ft)	Large transformer (200 ft)	50	Quiet
Minimum levels, residential areas in Chicago at night		40	
Studio (speech)	Soft whisper (5 ft)	30	Very quiet

Source: Adapted from A. P. G. Peterson and E. E. Gross Jr., *Handbook of Noise Measurement,* 7th ed. (New Concord, MA: General Radio Co.).

Performance during a pit stop makes a difference between winning and losing a race. Activity charts are used to orchestrate the movement of members of a pit crew, an operating room staff, or machine operators in a factory. Solved Problem 10.1 on pages 306–307 shows an activity chart applied to a pit crew.

ear protection above this level if exposure equals or exceeds 8 hours. Even at low levels, noise and vibration can be distracting. Therefore, most managers make substantial effort to reduce noise and vibration through good machine design, enclosures, or segregation of sources of noise and vibration.

Temperature and humidity parameters have been well established. Managers with activities operating outside of the established comfort zone should expect adverse effect on performance.

Methods analysis
Developing work procedures that are safe and produce quality products efficiently.

Methods Analysis   **Methods analysis** focuses on *how* a task is accomplished. Whether controlling a machine or making or assembling components, how a task is done makes a difference in performance, safety, and quality. Using knowledge from ergonomics and methods analysis, methods engineers are charged with ensuring that quality and quantity standards are achieved efficiently and safely. Methods analysis and related techniques are useful in office environments as well as the factory. Methods techniques are used to analyze:

1. Movement of individuals or material. The analysis is performed using *flow diagrams* and *process charts* with varying amounts of detail.
2. Activity of human and machine and crew activity. This analysis is performed using *activity charts* (also known as man-machine charts and crew charts).
3. Body movement (primarily arms and hands). This analysis is performed using *micro-motion charts*.

Flow diagrams
Drawings used to analyze movement of people or material.

Flow diagrams are schematics (drawings) used to investigate movement of people or material. As shown for Britain's Paddy Hopkirk Factory in Figure 10.5, and the *OM in Action* box, "Saving Steps on the B2 Bomber," the flow diagram provides a systematic procedure for looking at long-cycle repetitive tasks. The old method is shown in Figure 10.5(a), and a new method, with improved work flow and requiring less storage and space, is shown in Figure 10.5(b). **Process charts** use symbols, as in Figure 10.5(c), to help us understand the movement of people or material. In this way, movement and delays can be reduced and operations made more efficient. Figure 10.5(c) is a process chart used to supplement the flow diagram shown in Figure 10.5(b).

Process charts
A graphic representation that depicts a sequence of steps for a process.

Activity charts
A way of improving utilization of an operator and a machine or some combination of operators (a crew) and machines.

Activity charts are used to study and improve the utilization of an operator and a machine or some combination of operators (a "crew") and machines. The typical approach is for the analyst to record the present method through direct observation and then propose the improvement on a

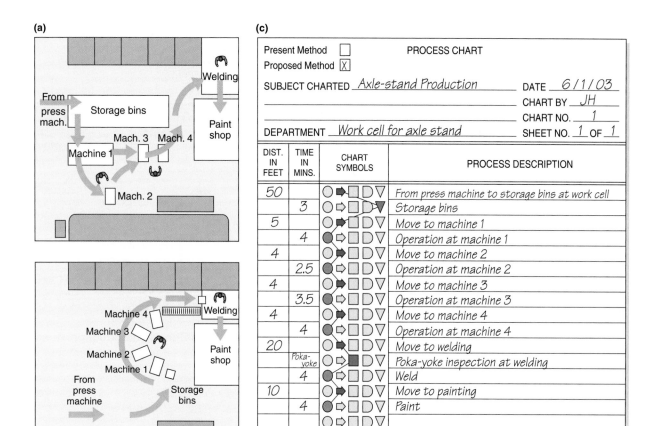

FIGURE 10.5 ■ Flow Diagram of Axle-Stand Production Line at Paddy Hopkirk Factory

(a) Old method; (b) new method; (c) process chart of axle-stand production using Paddy Hopkirk's new method (shown in b).

Process chart (c):

Present Method ☐   PROCESS CHART
Proposed Method ☒

SUBJECT CHARTED _Axle-stand Production_   DATE _6/1/03_
CHART BY _JH_
CHART NO. _1_
DEPARTMENT _Work cell for axle stand_   SHEET NO. _1_ OF _1_

DIST. IN FEET	TIME IN MINS.	CHART SYMBOLS	PROCESS DESCRIPTION
50			From press machine to storage bins at work cell
	3		Storage bins
5			Move to machine 1
	4		Operation at machine 1
4			Move to machine 2
	2.5		Operation at machine 2
4			Move to machine 3
	3.5		Operation at machine 3
4			Move to machine 4
	4		Operation at machine 4
20			Move to welding
	Poka-yoke		Poka-yoke inspection at welding
	4		Weld
10			Move to painting
	4		Paint
97	25		TOTAL

○ = operation; ⇨ = transportation; ☐ = inspection; ◗ = delay; ▽ = storage

OM IN ACTION

Saving Steps on the B2 Bomber

The aerospace industry is noted for making exotic products, but it is also known for doing so in a very expensive way. The historical batch-based processes used in the industry have left a lot of room for improvement. In leading the way, Northrop Grumman analyzed the work flow of a mechanic whose job in the Palmsdale, California, plant was to apply about 70 feet of tape to the B-2

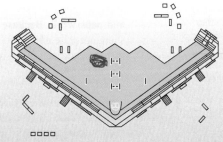

The mechanic's work path is reduced to the small area of purple lines shown here.

Stealth Bomber. The mechanic (see the graphic to the left) walked away from the plane 26 times and took 3 hours just to gather chemicals, hose, gauges, and other material needed just to get ready for the job. By making prepackaged kits for the job, Northrop Grumman cut preparation time to zero and the time to complete the job dropped from 8.4 hours to 1.6 hours (as seen above).

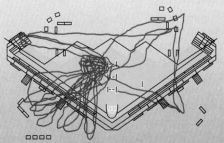

The 26 trips to various workstations to gather the tools and equipment to apply tape to the B2 bomber are shown as purple lines above.

Sources: New York Times (March 9, 1999): C1, C9; and Aviation Week & Space Technology (January 17, 2000): 441.

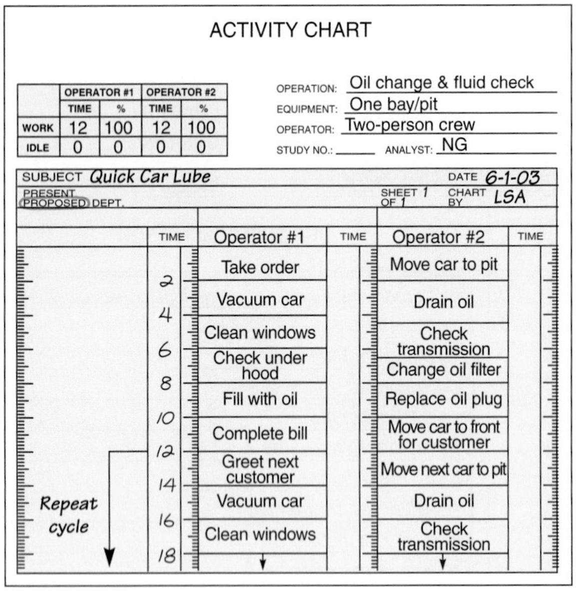

FIGURE 10.6 ■ Activity Chart for Two-Person Crew Doing an Oil Change in 12 Minutes at Quick Car Lube

FIGURE 10.7 ■ Operation Chart (right-hand/left-hand chart) for Bolt-Washer Assembly

Source: Adapted from L. S. Aft, *Productivity Measurement and Improvement* (Upper Saddle River, NJ: Prentice Hall, 1992): 5. Reprinted by permission of Prentice Hall, Inc.

second chart. Figure 10.6 is an activity chart to show a proposed improvement for a two-person crew at Quick Car Lube.

Body movement is analyzed by an **operations chart**. It is designed to show economy of motion by pointing out wasted motion and idle time (delay). The operations chart (also known as *right-hand/left-hand chart*) is shown in Figure 10.7.

Operations chart

A chart depicting right- and left-hand motions.

THE VISUAL WORKPLACE

Visual workplace

Uses a variety of visual communication techniques to rapidly communicate information to stakeholders.

The **visual workplace** uses low-cost visual devices to share information quickly and accurately. Well-designed displays and graphs root out confusion and replace difficult-to-understand printouts and paperwork. Because workplace data change quickly and often, operations managers need to share accurate and up-to-date information. Workplace dynamics, with changing customer requirements, specifications, schedules, and other details on which an enterprise depends, must be rapidly communicated.

Visual systems can include statistical process control (SPC) charts, details of quality, accidents, service levels, delivery performance, costs, cycle time, and such traditional variables as attendance and tardiness. All visual systems should focus on improvement because progress almost always has motivational benefits. An assortment of visual signals and charts is an excellent tool for communication not only among people doing the work, but also among support people, management, visitors, and suppliers. All these stakeholders deserve feedback on the organization. Management reports, if held only in the hands of management, are often useless and perhaps counterproductive. Managers need to think in terms of visual management.

The visual workplace can take many forms. Kanbans are a type of visual signal indicating the need for more production. The 3-minute clocks found in Burger Kings are a type of visual standard indicating the acceptable wait for service. Painted symbols indicating the place for tools are another visual standard to aid housekeeping. Some organizations have found it helpful to have performance standards indicated by hourly quota numbers for all to see. Andon lights are another visual signal. An **andon** is a signal that there is a problem. Andons can be manually initiated by employees when they notice a problem or defect. They can also be triggered automatically when machine performance drops below a certain pace or when the number of cycles indicate that it is time for maintenance. Figure 10.8 shows some visual signals in the workplace.

Andon

Call light that signals problems.

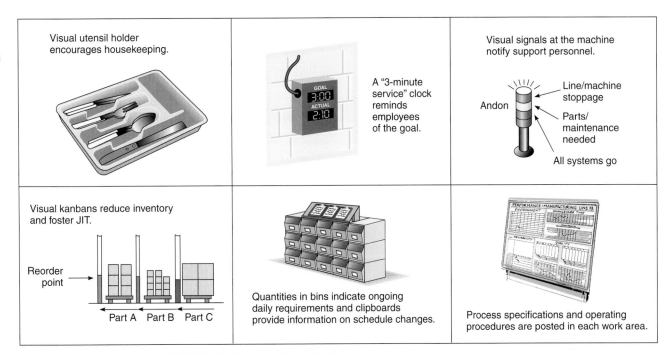

FIGURE 10.8 ■ The Visual Workplace

Visual systems also communicate the larger picture, helping employees to understand the link between their day-to-day activities and the organization's overall performance. At Baldor Electric Co. in Fort Smith, Arkansas, the prior day's closing price of Baldor's stock is posted for all to see. The stock price is to remind employees that a portion of their pay is based on profit sharing and stock options, and to encourage them to keep looking for ways to increase productivity. Similarly, Missouri's Springfield Re Manufacturing Corp. has developed a concept called "open book management," where every employee is trained to understand the importance of financial measures (such as return on equity) and is provided with these measures regularly.

The purpose of the visual workplace is to eliminate non-value-added activities and other forms of waste by making problems, abnormalities, and standards visual. This concept enhances communication and feedback by providing immediate information. The visual workplace needs less supervision because employees understand the standard, see the results, and know what to do.

LABOR STANDARDS

So far in this chapter, we have discussed labor planning and job design. The third requirement of an effective human resource strategy is the establishment of labor standards. Effective manpower planning is dependent on a knowledge of the labor required.

Labor standards are the amount of time required to perform a job or part of a job. Every firm has labor standards, although they may vary from those established via informal methods to those established by professionals. Only when accurate labor standards exist can management know what its labor requirements are, what its costs should be, and what constitutes a fair day's work. Techniques for setting labor standards are presented in the supplement to this chapter.

Labor standards
The amount of time required to perform a job or part of a job.

SUMMARY

Outstanding firms know the importance of an effective and efficient human resource strategy. Often a large percentage of employees and a large part of labor costs are under the direction of OM. Consequently, the operations manager usually has a large role to play in achieving human resource objectives. A prerequisite is to build an environment with mutual respect and commitment and a reasonable quality of work life. Outstanding organizations have designed jobs that use both the mental and physical capabilities of their employees. Regardless of the strategy chosen, the skill with which a firm manages its human resources ultimately determines its success.

KEY TERMS

Labor planning *(p. 293)*
Standard work schedule *(p. 293)*
Flextime *(p. 293)*
Flexible workweek *(p. 294)*
Part-time status *(p. 294)*
Job design *(p. 294)*
Labor specialization (or job specialization) *(p. 294)*
Job enlargement *(p. 295)*
Job rotation *(p. 295)*
Job enrichment *(p. 295)*
Employee empowerment *(p. 295)*
Self-directed team *(p. 297)*
Bonus *(p. 298)*

Profit sharing *(p. 298)*
Gain sharing *(p. 298)*
Incentive system *(p. 298)*
Knowledge-based pay systems *(p. 299)*
Ergonomics *(p. 299)*
Methods analysis *(p. 302)*
Flow diagrams *(p. 302)*
Process charts *(p. 302)*
Activity charts *(p. 302)*
Operations chart *(p. 304)*
Visual workplace *(p. 304)*
Andon *(p. 304)*
Labor standards *(p. 305)*

SOLVED PROBLEM

Solved Problem 10.1

As pit crew manager for Prototype Sports Car, you have just been given the pit-stop rules for next season. You will be allowed only six people over the pit wall at any one time, and one of these must be a designated *fire extinguisher/safety* crewman. This crewman must carry a fire extinguisher and may not service the car. However, the fire extinguisher/safety crewman may also signal the driver where to stop the car in the pit lane and when to leave the pit.

You expect to have air jacks on this year's car. These built-in jacks require only an air hose to make them work. Fuel will also be supplied via a hose, with a second hose used for venting air from the fuel cells. The rate of flow for the fuel hose will be 1 gallon per second. The tank will hold 25 gallons. You expect to have to change all four tires on most pit stops. The length of the races will vary this year, but you expect that the longer races will also require the changing of drivers. Recent stopwatch studies have verified the following times for your experienced crew:

ACTIVITY	TIME IN MINUTES
Install air hose	.075
Remove tire	.125
Mount new tire	.125
Move to air jack hose	.050
Move to rear of car	.050
Help driver	.175
Wipe windshield	.175
Load fuel (per gallon)	.016

Your job is to develop the initial plan for the best way to utilize your six-person pit crew. The six crewmen are identified with letters, as shown in Figure 10.9. You decide to use an activity chart similar to the one shown in Figure 10.6 (page 304) to aid you.

SOLUTION

Your activity chart shows each member of the crew what he or she is to do during each second of the pit stop.

FIGURE 10.9 ■

Position of Car and Six Crewmen (see chart on next page)

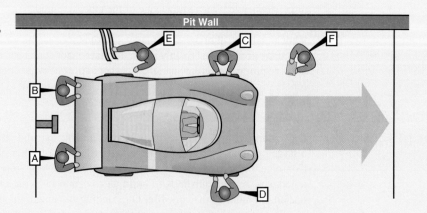

MULTIPLE ACTIVITY CHART

Chart No.:	Sheet No.:	Of:		S U M M A R Y		
PRODUCT:					PRESENT (min.)	PROPOSED	SAVING

CYCLE TIME			
Man			
Machine			
WORKING			
Man			
Machine			
IDLE			
Man			
Machine			
UTILIZATION			
Man			
Machine			

PROCESS: Pit stop for GTO cars

MACHINE(S):

OPERATIVE: **CLOCK NO.:**

CHARTED BY: **DATE:**

TIME (min.)	CREW A	B	C	D	E	F	TIME (min.)	NOTES	TIME (min.)
0.025					Move to car/hoses		0.025	Fire extinguisher/safety crewman F, goes over the pit wall to signal the driver where to stop the car.	0.025
0.050		Install air hose			Gas flows	S	0.050	Crewmen A, C, D move to the car with tires.	0.050
0.075	Remove tire		Remove tire	Remove tire		A	0.075	B places the air jack hose in the connection at the rear of the car.	0.075
0.100		Remove tire				F	0.100	B then returns to pit wall for the fourth tire. E moves to the car with two hoses (one for fuel & one to remove air).	0.100
0.125						E	0.125		0.125
0.150	Mount new tire		Mount new tire	Mount new tire		T	0.150	F is ready with the fire extinguisher. If the driver is to change, the first driver is out in the first 5 seconds.	0.150
0.175						Y	0.175		0.175
0.200							0.200	If there is a driver change, the new driver enters the car.	0.200
0.225						C	0.225		0.225
0.250						R	0.250	A, C, D have their tires mounted. D wipes the windshield with towels from belt.	0.250
0.275	Move to air jack hose	Mount new tire	Help driver	Wipe wind-shield		E	0.275	C helps the driver as necessary with seat belt & ice to cool suit.	0.275
0.300						W	0.300	A removes the air jack hose.	0.300
0.325						M	0.325	B has tire mounted. B moves to the rear of the car.	0.325
0.350		Move to rear			24.5 gal of gas	A	0.350	A removes the air jack hose when B, C, and D signal their tires are mounted.	0.350
0.375						N	0.375	A and B prepare to push car. E (fuel man) disconnects fuel lines. F signals completion of fuel loading.	0.375
0.400	Idle	Idle			24.5 seconds		0.400	F moves to front of car on the pit side and prepares to signal driver when to leave.	0.400
0.425							0.425	A, B, C, and D signal F when they are ready.	0.425
0.450			Idle	Idle			0.450	F signals driver when all is ready. A and B push car out of pit.	0.450
0.475	Push	Push			Idle	Idle	0.475		0.475

FIGURE 10.9 ■ (continued from page 306)

INTERNET AND STUDENT CD-ROM EXERCISES

Visit our home page or use your student CD-ROM to help with material in this chapter.

 On Our Home Page, www.prenhall.com./heizer
- Self-Tests
- Practice Problems
- Internet Exercises
- Current Articles and Research
- Virtual Company Tour

 On Your Student CD-ROM
- Internet Homework Problems
- Internet Case
- PowerPoint Lecture
- Practice Problems
- Video Clip and Video Case

ADDITIONAL CASE STUDIES

Internet Case Study: Visit our Web site at www.prenhall.com/heizer **for this free case study:**

- **Lincoln Electric's Incentive Pay System**: This manufacturer's incentive pay system produces the highest paid factory workers in the world.

Harvard has selected these Harvard Business School cases to accompany this chapter (textbookcasematch.hbsp.harvard.edu):

- **Southwest Airlines: Using Human Resources for Competitive Advantage** (#HR1A): Considers how Southwest Airlines developed a sustainable competitive advantage via human resources.
- **Eli Lilly: The Evista Project** (#699-016): Explores operational realities of two product development teams.
- **PPG: Developing a Self-Directed Workforce** (#693-020): Considers the process of creating a self-directed workforce, including the theory and difficulties.

Work Measurement

Supplement Outline

LEARNING OBJECTIVES

When you complete this supplement you should be able to

IDENTIFY OR DEFINE:

Four ways of establishing labor standards

DESCRIBE OR EXPLAIN:

Requirements for good labor standards

Time study

Predetermined time standards

Work sampling

Each day, in fact 130 times each day, Tim Nelson leans back into a La-Z-Boy recliner, sofa section, or love seat. He is one of 25 inspectors at La-Z-Boy Inc.'s Dayton factory. As Tim leans back into the oversized La-Z-Boy, he inspects for overall comfort; he must sink slightly into the chair, but not too far. Like Goldilocks, the chair must not be too firm or too soft; it must be just right—or it is sent back for restuffing. If it passes the "firm" test, he then rocks back and forth, making certain the chair is properly balanced and moves smoothly. Then Tim checks the footrest, arches his back, and holds the position as if he were taking that Sunday afternoon nap. Hopping to his feet, he does a walk-around visual check; then it is on to the next chair. One down, and 129 to go.[1]

LABOR STANDARDS AND WORK MEASUREMENT

Modern labor standards originated with the works of Fredrick Taylor and Frank and Lillian Gilbreth at the beginning of the twentieth century. At that time, a large proportion of work was manual, and the resulting labor content of products was high. Little was known about what constituted a fair day's work, so managers initiated studies to improve work methods and understand human effort. These efforts continue to this day. Although we are now at the beginning of the twenty-first century and labor costs are often less than 10% of sales, labor standards remain important and continue to play a major role in both service and manufacturing organizations. They are often a beginning point for determining staffing requirements. With over half of the manufacturing plants in America using some form of labor incentive system, good labor standards are a requirement.

Effective operations management requires meaningful standards that can help a firm determine the following:

1. Labor content of items produced (the labor cost).
2. Staffing needs (how many people it will take to meet required production).
3. Cost and time estimates prior to production (to assist in a variety of decisions, from cost estimates to make-or-buy decisions).
4. Crew size and work balance (who does what in a group activity or on an assembly line).
5. Expected production (so that both manager and worker know what constitutes a fair day's work).
6. Basis of wage-incentive plans (what provides a reasonable incentive).
7. Efficiency of employees and supervision (a standard is necessary against which to determine efficiency).

Labor standards exist for telephone operators, auto mechanics, and UPS drivers, as well as many factory workers like Tim Nelson at La-Z-Boy, in the photo above.

[1]Adapted from Jason Strait, "Sitting Down on the Job," *San Antonio Express News* (April 21, 2001): B1.

Properly set labor standards represent the amount of time that it should take an average employee to perform specific job activities under normal working conditions. Labor standards are set in four ways:

1. Historical experience
2. Time studies
3. Predetermined time standards
4. Work sampling

This supplement covers each of these techniques.

HISTORICAL EXPERIENCE

Labor standards can be estimated based on *historical experience*—that is, how many labor-hours were required to do a task the last time it was performed. Historical standards have the advantage of being relatively easy and inexpensive to obtain. They are usually available from employee time cards or production records. However, they are not objective, and we do not know their accuracy, whether they represent a reasonable or a poor work pace, and whether unusual occurrences are included. Because these variables are unknown, their use is not recommended. Instead, time studies, predetermined time standards, and work sampling are preferred.

TIME STUDIES

Time study
Timing a sample of a worker's performance and using it as a basis for setting a standard time.

The classical stopwatch study, or time study, originally proposed by Frederick W. Taylor in 1881, is still the most widely used time-study method.[2] A **time-study** procedure involves timing a sample of a worker's performance and using it to set a standard. A trained and experienced person can establish a standard by following these eight steps:

1. Define the task to be studied (after methods analysis has been conducted).
2. Divide the task into precise elements (parts of a task that often take no more than a few seconds).
3. Decide how many times to measure the task (the number of cycles or samples needed).
4. Time and record elemental times and ratings of performance.
5. Compute the average observed cycle time. The **average observed cycle time** is the arithmetic mean of the times for *each* element measured, adjusted for unusual influence for each element:

Average observed cycle time
The arithmetic mean of the times for each element measured, adjusted for unusual influence for each element.

$$\text{Average observed cycle time} = \frac{\left(\begin{array}{c}\text{sum of the times recorded} \\ \text{to perform each element}\end{array}\right)}{\text{number of cycles observed}} \quad \text{(S10-1)}$$

6. Determine performance rating and then compute the **normal time** for each element.

Normal time
The observed time, adjusted for pace.

$$\text{Normal time} = (\text{average observed cycle time}) \times (\text{performance rating factor}) \quad \text{(S10-2)}$$

The performance rating adjusts the observed time to what a normal worker could expect to accomplish. For example, a normal worker should be able to walk 3 miles per hour. He or she should also be able to deal a deck of 52 cards into 4 equal piles in 30 seconds. A performance rating of 1.05 would indicate that the observed worker performs the task slightly *faster* than average. Numerous videos specify work pace on which professionals agree, and benchmarks have been established by the Society for the Advancement of Management. Performance rating, however, is still something of an art.

7. Add the normal times for each element to develop a total normal time for the task.

Standard time
An adjustment to the total normal time; the adjustment provides allowances for personal needs, unavoidable work delays, and fatigue.

8. Compute the **standard time**. This adjustment to the total normal time provides for allowances such as *personal* needs, unavoidable work *delays*, and worker *fatigue*:

$$\text{Standard time} = \frac{\text{total normal time}}{1 - \text{allowance factor}} \quad \text{(S10-3)}$$

[2]For an illuminating look at the life and influence of Taylor, see Robert Kanigel, *The One Best Way: Frederick Winslow Taylor and the Enigma of Efficiency* (New York: Viking Press, 1997).

1. Constant allowances:
   (A) Personal allowance 5
   (B) Basic fatigue allowance 4
2. Variable allowances:
   (A) Standing allowance 2
   (B) Abnormal position allowance:
       (i) Awkward (bending) 2
       (ii) Very awkward (lying, stretching) 7
   (C) Use of force or muscular energy in
       lifting, pulling, pushing
       Weight lifted (pounds):
       20 3
       40 9
       60 17
   (D) Bad light:
       (i) Well below recommended 2

       (ii) Quite inadequate 5
   (E) Atmospheric conditions (heat and humidity):
       Variable 0–10
   (F) Close attention:
       (i) Fine or exacting 2
       (ii) Very fine or very exacting 5
   (G) Noise level:
       (i) Intermittent—loud 2
       (ii) Intermittent—very loud or high-pitched 5
   (H) Mental strain:
       (i) Complex or wide span of attention 4
       (ii) Very complex 8
   (I) Tediousness:
       (i) Tedious 2
       (ii) Very tedious 5

FIGURE S10.1 ■ **Rest Allowances (in percentage) for Various Classes of Work**

Source: From *Methods, Standards, and Work Design*, 11th ed., by B. W. Niebel and A. Freivalds, (Irwin/McGraw-Hill, 2003).

Personal time allowances are often established in the range of 4% to 7% of total time, depending on nearness to rest rooms, water fountains, and other facilities. *Delay allowances* are often set as a result of the actual studies of the delay that occurs. *Fatigue allowances* are based on our growing knowledge of human energy expenditure under various physical and environmental conditions. A sample set of personal and fatigue allowances is shown in Figure S10.1. Example S1 illustrates the computation of standard time.

Example S1

The time study of a work operation yielded an average observed cycle time of 4.0 minutes. The analyst rated the observed worker at 85%. This means the worker performed at 85% of normal when the study was made. The firm uses a 13% allowance factor. We want to compute the standard time.

SOLUTION

$$\text{Average observed time} = 4.0 \text{ min.}$$

$$\text{Normal time} = (\text{average observed cycle time}) \times (\text{rating factor})$$

$$= (4.0)(.85)$$

$$= 3.4 \text{ min.}$$

$$\text{Standard time} = \frac{\text{normal time}}{1 - \text{allowance factor}} = \frac{3.4}{1 - .13} = \frac{3.4}{.87}$$

$$= 3.9 \text{ min.}$$

Example S2 uses a series of actual stopwatch times for each element.

Example S2

Management Science Associates promotes its management development seminars by mailing thousands of individually composed and typed letters to various firms. A time study has been conducted on the task of preparing letters for mailing. On the basis of the observations below, Management Science Associates wants to develop a time standard for this task. The firm's personal, delay, and fatigue allowance factor is 15%.

	CYCLE OBSERVED (IN MINUTES)					
JOB ELEMENT	1	2	3	4	5	PERFORMANCE RATING
(A) Compose and type letter	8	10	9	21*	11	120%
(B) Type envelope address	2	3	2	1	3	105%
(C) Stuff, stamp, seal, and sort envelopes	2	1	5*	2	1	110%

In many service jobs such as cleaning a Sheraton hotel bathtub, renting a Hertz car, or wrapping a Taco Bell burrito, time and motion studies are effective management tools.

SOLUTION

Once the data have been collected, the procedure is as follows:

1. Delete unusual or nonrecurring observations such as those marked with an asterisk (*). (These might be due to business interruptions, conferences with the boss, or mistakes of an unusual nature; they are not part of the job element, but may be personal or delay time.)

2. Compute the average cycle time for each job element:

$$\text{Average time for A} = \frac{8 + 10 + 9 + 11}{4}$$
$$= 9.5 \text{ min.}$$
$$\text{Average time for B} = \frac{2 + 3 + 2 + 1 + 3}{5}$$
$$= 2.2 \text{ min.}$$
$$\text{Average time for C} = \frac{2 + 1 + 2 + 1}{4}$$
$$= 1.5 \text{ min.}$$

3. Compute the normal time for each job element:

$$\text{Normal time for A} = (\text{average observed time}) \times (\text{rating})$$
$$= (9.5)(1.2)$$
$$= 11.4 \text{ min.}$$
$$\text{Normal time for B} = (2.2)(1.05)$$
$$= 2.31 \text{ min.}$$
$$\text{Normal time for C} = (1.5)(1.10)$$
$$= 1.65 \text{ min.}$$

Note: Normal times are computed for each element because the rating factor may vary for each element, as it did in this case.

4. Add the normal times for each element to find the total normal time (the normal time for the whole job):

$$\text{Total normal time} = 11.40 + 2.31 + 1.65$$
$$= 15.36 \text{ min.}$$

5. Compute the standard time for the job:

$$\text{Standard time} = \frac{\text{total normal time}}{1 - \text{allowance factor}} = \frac{15.36}{1 - .15}$$
$$= 18.07 \text{ min.}$$

Thus, 18.07 minutes is the time standard for this job.

Note: When observed times are not consistent they need to be reviewed. Abnormally short times may be the result of an observational error and are usually discarded. Abnormally long times need to be analyzed to determine if they, too, are an error. However, they may include a seldom occurring but legitimate activity for the element (such as a machine adjustment) or may be personal, delay, or fatigue time.

It is important to let a worker who is going to be observed know about the study in advance in order to avoid misunderstanding or suspicion.

Time study requires a sampling process; so the question of sampling error in the average observed cycle time naturally arises. In statistics, error varies inversely with sample size. So in order to determine just how many cycles we should time, we must consider the variability of each element in the study.

To determine an adequate sample size, three items must be considered:

1. How accurate we want to be (for example, is ±5% of observed cycle time close enough?).
2. The desired level of confidence (for example, the *z* value; is 95% adequate or is 99% required?).
3. How much variation exists within the job elements (for example, if the variation is large, a larger sample will be required).

Sleep Inn® hotels are showing the world that big gains in productivity can be made not only by manufacturers, but in the service industry as well. Designed with labor efficiency in mind, Sleep Inn hotels are staffed with 13% fewer employees than similar budget hotels. Its features include a laundry room that is almost completely automated, round shower stalls that eliminate dirty corners, and closets that have no doors for maids to open and shut.

The formula for finding the appropriate sample size given these three variables is:

$$\text{Required sample size} = n = \left(\frac{zs}{h\bar{x}}\right)^2 \tag{S10-4}$$

where h = accuracy level desired in percent of the job element, expressed as a decimal (5% = .05)

z = number of standard deviations required for desired level of confidence (90% confidence = 1.65; see Table S10.1 or Appendix I for the more common z values)

s = standard deviation of the initial sample

$\bar{x}$ = mean of the initial sample

n = required sample size

TABLE S10.1 ■ Common z Values

DESIRED CONFIDENCE (%)	z VALUE (STANDARD DEVIATION REQUIRED FOR DESIRED LEVEL OF CONFIDENCE)
90.0	1.65
95.0	1.96
95.45	2.00
99.0	2.58
99.73	3.00

We demonstrate with Example S3.

Example S3

Thomas W. Jones Manufacturing Co. has asked you to check a labor standard prepared by a recently terminated analyst. Your first task is to determine the correct sample size. Your accuracy is to be within 5% and your confidence level at 95%. The standard deviation of the sample is 1.0 and the mean 3.00.

SOLUTION

$$h = .05 \qquad \bar{x} = 3.00 \qquad s = 1.0$$

$$z = 1.96 \text{ (from Table S10.1 or Appendix I)}$$

$$n = \left(\frac{zs}{h\bar{x}}\right)^2$$

$$n = \left(\frac{1.96 \times 1.0}{.05 \times 3}\right)^2 = 170.74 \approx 171$$

Therefore, you recommend a sample size of 171.

OM IN ACTION

UPS: The Tightest Ship in the Shipping Business

United Parcel Service (UPS) employs 150,000 people and delivers an average of 9 million packages a day to locations throughout the U.S. and 180 other countries. To achieve its claim of "running the tightest ship in the shipping business," UPS methodically trains its delivery drivers in how to do their jobs as efficiently as possible.

Industrial engineers at UPS have time-studied each driver's route and set standards for each delivery, stop, and pickup. These engineers have recorded every second taken up by stoplights, traffic volume, detours, doorbells, walkways, stairways, and coffee breaks. Even bathroom stops are factored into the standards. All of this information is then fed into company computers to provide detailed time standards for every driver, every day.

To meet their objective of 200 deliveries and pickups each day (versus only 80 at Federal Express), UPS drivers must follow procedures exactly. As they approach a delivery stop, drivers unbuckle their seat belts, honk their horns, and cut their engines. In one seamless motion, they are required to yank up their emergency brakes and push their gearshifts into first. Then they slide to the ground with their clipboards under their right arms and their packages in their left hands. Ignition keys, teeth up, are in their right hands. They walk to the customer's door at the prescribed 3 feet per second and knock first to avoid lost seconds searching for the doorbell. After making the delivery, they do the paperwork on the way back to the truck.

Productivity experts describe UPS as one of the most efficient companies anywhere in applying effective labor standards.

Sources: EBN (May 7, 2001): 70; and Computerworld (June 21, 1999): 16.

Now let's look at two variations of Example S3.

First, if h, the desired accuracy, is expressed as an absolute amount of error (say, 1 minute of error is acceptable), then substitute e for $h\bar{x}$, and the appropriate formula is

$$n = \left(\frac{zs}{e}\right)^2$$ (S10-5)

where e is the absolute amount of acceptable error.

Second, for those cases when s, the standard deviation of the sample, is not provided (which is typically the case outside the classroom), it must be computed. The formula for doing so is given in Equation (S10-6).

$$s = \sqrt{\frac{\sum (x_i - \bar{x})^2}{n-1}} = \sqrt{\frac{\sum (\text{each sample observation} - \bar{x})^2}{\text{number in sample} - 1}}$$ (S10-6)

where x_i = value of each observation
$\bar{x}$ = mean of the observations
n = number of observations in the sample

An example of this computation is provided in Solved Problem S10.3 on pages 320–321.

Although time studies provide accuracy in setting labor standards (see the *OM in Action* box on UPS), they have two disadvantages. First, they require a trained staff of analysts. Second, labor standards cannot be set before tasks are actually performed. This leads us to two alternative work-measurement techniques that we discuss next.

PREDETERMINED TIME STANDARDS

Predetermined time standards
A division of manual work into small basic elements that have established and widely accepted times.

In addition to historical experience and time studies, we can set production standards by using predetermined time standards. **Predetermined time standards** divide manual work into small basic elements that already have established times (based on very large samples of workers). To estimate the time for a particular task, the time factors for each basic element of that task are added together. Developing a comprehensive system of predetermined time standards would be prohibitively expensive for any given firm. Consequently, a number of systems are commercially available. The most common predetermined time standard is *methods time measurement* (MTM), which is a product of the MTM Association.[3]

[3]MTM is really a family of products available from the Methods Time Measurement Association. For example, MTM-HC deals with the healthcare industry, MTM-C handles clerical activities, MTM-M involves microscope activities, MTM-V deals with machine shop tasks, and so on.

FIGURE S10.2 ■

Sample MTM Table for
GET and PLACE Motion

Time values are in TMUs.
Source: Copyrighted by
the MTM Association for
Standards and Research. No
reprint permission without
consent from the MTM
Association, 16–01 Broadway,
Fair Lawn, NJ 07410.

GET and PLACE			DISTANCE RANGE IN IN.	<8	>8 <20	>20 <32
WEIGHT	CONDITIONS OF GET	PLACE ACCURACY	CODE	1	2	3
<2 LBS	EASY	APPROXIMATE	AA	20	35	50
		LOOSE	AB	30	45	60
		TIGHT	AC	40	55	70
	DIFFICULT	APPROXIMATE	AD	20	45	60
		LOOSE	AE	30	55	70
		TIGHT	AF	40	65	80
	HANDFUL	APPROXIMATE	AG	40	65	80
>2 LBS <18 LBS		APPROXIMATE	AH	25	45	55
		LOOSE	AJ	40	65	75
		TIGHT	AK	50	75	85
>18 LBS <45 LBS		APPROXIMATE	AL	90	106	115
		LOOSE	AM	95	120	130
		TIGHT	AN	120	145	160

Therbligs
Basic physical elements
of motion.

**Time measurement
units (TMUs)**
Units for very basic
micromotions in which
1 TMU = .0006 min.
or 100,000 TMUs = 1 hr.

Predetermined time standards are an outgrowth of basic motions called therbligs. The term *therblig* was coined by Frank Gilbreth (*Gilbreth* spelled backwards with the *t* and *h* reversed). **Therbligs** include such activities as select, grasp, position, assemble, reach, hold, rest, and inspect. These activities are stated in terms of **time measurement units (TMUs)**, which are each equal to only .00001 hour, or .0006 minute. MTM values for various therbligs are specified in very detailed tables. Figure S10.2, for example, provides the set of time standards for the motion GET and PLACE. To use GET and PLACE (the most complex motion in the MTM system), one must know what is "gotten," its approximate weight, and where and how far it is supposed to be placed.

Example S4 shows a use of predetermined time standards in setting service labor standards.

Example S4

Pouring a tube specimen in a hospital lab is a repetitive task for which the MTM data in Figure S10.2 may be used to develop standard times. The sample tube is in a rack and the centrifuge tubes in a nearby box. A technician removes the sample tube from the rack, uncaps it, gets the centrifuge tube, pours, and places both tubes in the rack.

The first work element involves getting the tube from the rack. Suppose the conditions for GETTING the tube and PLACING it in front of the technician are

- Weight (less than 2 pounds)
- Conditions of GET (easy)
- Place accuracy (approximate)
- Distance range (8 to 20 inches)

Then the MTM element for this activity is AA2 (as seen from Figure S10.2). The rest of Table S10.2 below is developed from similar MTM tables. Most MTM calculations, by the way, are computerized, so the user need only key in the appropriate MTM codes, such as AA2 in this example.

One of the Gilbreths'
techniques was to use
cameras to record
movement by attaching
lights to an individual's
arms and legs. In that way
they could track the
movement of individuals
while performing various
jobs.

TABLE S10.2 ■ MTM-HC Analysis: Pouring Tube Specimen

ELEMENT DESCRIPTION	ELEMENT	TIME
Get tube from rack	AA2	35
Get stopper, place on counter	AA2	35
Get centrifuge tube, place at sample tube	AD2	45
Pour (3 sec.)	PT	83
Place tubes in rack (simo)	PC2	40
	Total TMU	238

.0006 × 238 = Total standard minutes = .14

Source: A. S. Helms, B. W. Shaw, and C. A. Lindner, "The Development of Laboratory Workload Standards through Computer-Based Work Measurement Technique, Part I," *Journal of Methods-Time Measurement* 12: 43. Used with permission of MTM Association for Standards and Research.

Some firms use a combination of stopwatch studies and predetermined time standards when they are particularly interested in verifying results.

Predetermined time standards have several advantages over direct time studies. First, they may be established in a laboratory environment, where the procedure will not upset actual production activities (which time studies tend to do). Second, because the standard can be set *before* a task is actually performed, it can be used for planning. Third, no performance ratings are necessary. Fourth, unions tend to accept this method as a fair means of setting standards. Finally, predetermined time standards are particularly effective in firms that do substantial numbers of studies of similar tasks. To ensure accurate labor standards, some firms use both time studies and predetermined time standards.

WORK SAMPLING

Work sampling
An estimate, via sampling, of the percent of the time that a worker spends on various tasks.

The fourth method of developing labor or production standards, work sampling, was developed in England by L. Tippet in the 1930s. **Work sampling** estimates the percent of the time that a worker spends on various tasks. It requires random observations to record the activity that a worker is performing. The results are primarily used to determine how employees allocate their time among various activities. Knowledge of this allocation may lead to staffing changes, reassignment of duties, estimates of activity cost, and the setting of delay allowances for labor standards. When work sampling is done to establish delay allowances, it is sometimes called a ratio delay study.

The work-sampling procedure can be summarized in five steps:

1. Take a preliminary sample to obtain an estimate of the parameter value (such as percent of time a worker is busy).
2. Compute the sample size required.
3. Prepare a schedule for observing the worker at appropriate times. The concept of random numbers is used to provide for random observation. For example, let's say we draw the following 5 random numbers from a table: 07, 12, 22, 25, and 49. These can then be used to create an observation schedule of 9:07 A.M., 9:12, 9:22, 9:25, 9:49.
4. Observe and record worker activities.
5. Determine how workers spend their time (usually as a percent).

The cataloger Land's End expects its sales reps to be busy 85% of the time and idle 15%. When the busy ratio hits 90%, the firm believes it is not reaching its goal of high-quality service.

To determine the number of observations required, management must decide on the desired confidence level and accuracy. First, however, the analyst must select a preliminary value for the parameter under study (step 1 above). The choice is usually based on a small sample of perhaps

To reduce the cost of work sampling, Kinemark, a Parsippany, New Jersey, firm, developed this small computer with options for 48 everyday office tasks. As each task is completed, the operator presses the appropriate key. After a week of such reporting, a rather complete picture of what is going on is available.

50 observations. The following formula then gives the sample size for a desired confidence and accuracy:

$$n = \frac{z^2 p(1 - p)}{h^2}$$ (S10-7)

where n = required sample size
   z = standard normal deviate for the desired confidence level
   (z = 1 for 68% confidence, z = 2 for 95.45% confidence, and z = 3 for 99.73% confidence—these values are obtained from Table S10.1 or the Normal Table in Appendix I)
   p = estimated value of sample proportion (of time worker is observed busy or idle)
   h = acceptable error level, in percent

Example S5 shows how to apply this formula.

Example S5

Active Model S10.1

Example S5 is further illustrated in Active Model S10.1 on the CD-ROM and in the Exercise located in your Student Lecture Guide.

The manager of Wilson County's welfare office, Madeline Thimmes, estimates her employees are idle 25% of the time. She would like to take a work sample that is accurate within 3% and wants to have 95.45% confidence in the results.

SOLUTION

In order to determine how many observations should be taken, Madeline applies the following equation:

$$n = \frac{z^2 p(1 - p)}{h^2}$$

where n = required sample size
   z = 2 for 95.45% confidence level
   p = estimate of idle proportion = 25% = .25
   h = acceptable error of 3% = .03

She finds that

$$n = \frac{(2)^2(.25)(.75)}{(.03)^2} = 833 \text{ observations}$$

Thus, 833 observations should be taken. If the percent of idle time observed is not close to 25% as the study progresses, then the number of observations may have to be recalculated and increased or decreased as appropriate.

The focus of work sampling is to determine how workers allocate their time among various activities. This is accomplished by establishing the percent of time individuals spend on these activities rather than the exact amount of time spent on specific tasks. The analyst simply records in a random, nonbiased way the occurrence of each activity. Example S6 shows the procedure for evaluating employees at the state welfare office introduced in Example S5.

Example S6

Madeline Thimmes, the operations manager of Wilson County's state welfare office, wants to be sure her employees have adequate time to provide prompt, helpful service. She believes that service to welfare clients who phone or walk in without an appointment deteriorates rapidly when employees are busy more than 75% of the time. Consequently, she does not want her employees to be occupied with client service activities more than 75% of the time.

The study requires several things: First, based on the calculations in Example S5, 833 observations are needed. Second, observations are to be made in a random, nonbiased way over a period of 2 weeks to ensure a true sample. Third, the analyst must define the activities that are "work." In this case, work is defined as all of the activities that are necessary to take care of the client (filing, meetings, data entry, discussions with the supervisor, etc.). Fourth, personal time is to be included in the 25% of nonwork time. Fifth, the observations are made in a nonintrusive way so as not to distort the normal work patterns. At the end of the 2 weeks, the 833 observations yield the following results:

No. of Observations	Activity
485	On the phone or meeting with a welfare client
126	Idle
62	Personal time
23	Discussions with supervisor
137	Filing, meeting, and computer data entry
833	

The analyst concludes that all but 188 observations (126 idle and 62 personal) are work related. Since 22.6% (= 188/833) is less idle time than Madeline believes necessary to ensure a high client service level, she needs to find a way to reduce current workloads. This could be done through a reassignment of duties or the hiring of additional personnel.

The results of a similar study of salespeople and assembly-line employees are shown in Figure S10.3.

Work sampling offers several advantages over time-study methods. First, because a single observer can observe several workers simultaneously, it is less expensive. Second, observers usually do not require much training, and no timing devices are needed. Third, the study can be temporarily delayed at any time with little impact on the results. Fourth, because work sampling uses instantaneous observations over a long period, the worker has little chance of affecting the study's outcome. Fifth, the procedure is less intrusive and therefore less likely to generate objections.

The disadvantages of work sampling are (1) it does not divide work elements as completely as time studies, (2) it can yield biased or incorrect results if the observer does not follow random routes of travel and observation, and (3) because it is less intrusive, it tends to be less accurate; this is particularly true when cycle times are short.

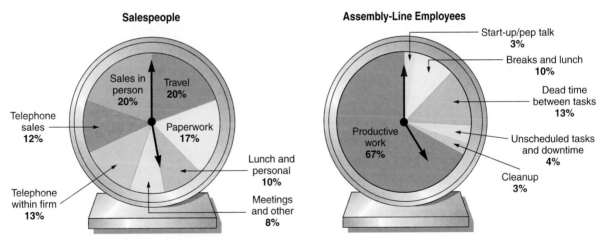

FIGURE S10.3 ■ Work-Sampling Time Studies

These two work-sampling time studies were done to determine what salespeople do at a wholesale electronic distributor (left) and a composite of several auto assembly-line employees (right).
Source: "Memo Busters" by J. H. Hayes from *Forbes,* April 24, 1995. Reprinted by permission of *Forbes Magazine* © 1995 Forbes.

SUMMARY

Labor standards are required for an efficient operations system. They are needed for production planning, labor planning, costing, and evaluating performance. They can also be used as a basis for incentive systems. They are used in both the factory and the office. Standards may be established via historical data, time studies, predetermined time standards, and work sampling.

KEY TERMS

Time study *(p. 311)*
Average observed cycle time *(p. 311)*
Normal time *(p. 311)*
Standard time *(p. 311)*

Predetermined time standards *(p. 315)*
Therbligs *(p. 316)*
Time measurement units (TMUs) *(p. 316)*
Work sampling *(p. 317)*

SOLVED PROBLEMS

Solved Problem S10.1

A work operation consisting of three elements has been subjected to a stopwatch time study. The recorded observations are shown in the following table. By union contract, the allowance time for the operation is personal time 5%, delay 5%, and fatigue 10%. Determine the standard time for the work operation.

JOB ELEMENT	OBSERVATIONS (IN MINUTES)						PERFORMANCE RATING (%)
	1	2	3	4	5	6	
A	.1	.3	.2	.9	.2	.1	90
B	.8	.6	.8	.5	3.2	.7	110
C	.5	.5	.4	.5	.6	.5	80

SOLUTION

First, delete the two observations that appear to be very unusual (.9 minute for job element A and 3.2 minutes for job element B). Then:

$$A\text{'s average observed cycle time} = \frac{.1+.3+.2+.2+.1}{5} = .18 \text{ min.}$$

$$B\text{'s average observed cycle time} = \frac{.8+.6+.8+.5+.7}{5} = .68 \text{ min.}$$

$$C\text{'s average observed cycle time} = \frac{.5+.5+.4+.5+.6+.5}{6} = .50 \text{ min.}$$

$$A\text{'s normal time} = (.18)(.90) = .16 \text{ min.}$$

$$B\text{'s normal time} = (.68)(1.10) = .75 \text{ min.}$$

$$C\text{'s normal time} = (.50)(.80) = .40 \text{ min.}$$

$$\text{Normal time for job} = .16 + .75 + .40 = 1.31 \text{ min.}$$

$$\text{Standard time} = \frac{1.31}{1-.20} = 1.64 \text{ min.}$$

Solved Problem S10.2

The preliminary work sample of an operation indicates the following:

Number of times operator working	60
Number of times operator idle	40
Total number of preliminary observations	100

What is the required sample size for a 99.73% confidence level with ±4% precision?

SOLUTION

$$n = \frac{z^2 p(1-p)}{h^2} = \frac{(3)^2(.6)(.4)}{(.04)^2} = 1{,}350 \text{ sample size}$$

Solved Problem S10.3

Amor Manufacturing Co. of Geneva, Switzerland, has just studied a job in its laboratory in anticipation of releasing the job to the factory for production. The firm wants rather good accuracy for costing and labor forecasting. Specifically, it wants to provide a 99% confidence level and a cycle time that is within 3% of the true value. How many observations should it make? The data collected so far are as follows:

OBSERVATION	TIME
1	1.7
2	1.6
3	1.4
4	1.4
5	1.4

SOLUTION

First, solve for the mean, $\bar{x}$, and the sample standard deviation, s.

$$s = \sqrt{\frac{\sum (\text{each sample observation} - \bar{x})^2}{\text{number in sample} - 1}}$$

OBSERVATION	$\bar{x}_i$	$\bar{x}$	$x_i - \bar{x}$	$(x_i - \bar{x})^2$
1	1.7	1.5	.2	0.04
2	1.6	1.5	.1	0.01
3	1.4	1.5	−.1	0.01
4	1.4	1.5	−.1	0.01
5	1.4	1.5	−.1	0.01
	$\bar{x} = 1.5$			$0.08 = \Sigma(x_i - \bar{x})^2$

$$s = \sqrt{\frac{.08}{n-1}} = \sqrt{\frac{.08}{4}} = .141$$

Then, solve for $n = \left(\dfrac{zs}{h\bar{x}}\right)^2 = \left[\dfrac{(2.58)(.141)}{(.03)(1.5)}\right]^2 = 65.3$

where $\bar{x} = 1.5$

$s = .141$

$z = 2.58$

$h = .03$

Therefore, you recommend 65 observations.

Solved Problem S10.4

At Maggard Micro Manufacturing, Inc., workers press semiconductors into predrilled slots on printed-circuit boards. The elemental motions for normal time used by the company are as follows:

Reach 6 inches for semiconductors	10.5 TMU
Grasp the semiconductor	8.0 TMU
Move semiconductor to printed-circuit board	9.5 TMU
Position semiconductor	20.1 TMU
Press semiconductor into slots	20.3 TMU
Move board aside	15.8 TMU

(Each time measurement unit is equal to .0006 min.) Determine the normal time for this operation in minutes and in seconds.

SOLUTION

Add the time measurement units:

$$10.5 + 8.0 + 9.5 + 20.1 + 20.3 + 15.8 = 84.2$$

Time in minutes = $(84.2)(.0006 \text{ min.}) = .05052$ min.

Time in seconds = $(.05052)(60 \text{ sec.}) = 3.0312$ sec.

Solved Problem S10.5

To obtain the random sample needed for work sampling, a manager divides a typical workday into 480 minutes. Using a random-number table to decide what time to go to an area to sample work occurrences, the manager records observations on a tally sheet like the following:

STATUS	TALLY
Productively working	⫴⫴⫴ I
Idle	IIII

SOLUTION

In this case, the supervisor made 20 observations and found that employees were working 80% of the time. So, out of 480 minutes in an office workday, 20%, or 96 minutes, was idle time, and 384 minutes was productive. Note that this procedure describes what a worker *is* doing, not necessarily what he or she *should* be doing.

INTERNET AND STUDENT CD-ROM EXERCISES

Visit our home page or use your student CD-ROM to help with material in this supplement.

On Our Home Page, www.prenhall.com/heizer

- Self-Tests
- Practice Problems
- Internet Exercises
- Current Articles and Research
- Internet Homework Problems
- Internet Cases

On Your Student CD-ROM

- PowerPoint Lecture
- Practice Problems
- Active Model Exercise

ADDITIONAL CASE STUDIES

Internet Case Studies: Visit our Web site at www.prenhall.com/heizer **for these free case studies:**

- **Chicago Southern Hospital**: Examines the requirements for a work-sampling plan for nurses.

- **Telephone Operator Standards at AT&T**: Examines the implications of work standards for telephone operators at AT&T.

Harvard has selected this Harvard Business School case to accompany this supplement
(textbookcasematch.hbsp.harvard.edu**):**

- **Lincoln Electric** (#376-028): Discusses the compensation system and company culture at this welding equipment manufacturer.

Supply-Chain Management

Chapter Outline

LEARNING OBJECTIVES

When you complete this chapter you should be able to

IDENTIFY OR DEFINE:

Supply-chain management

Purchasing

Outsourcing

E-procurement

Materials management

Keiretsu

Virtual companies

DESCRIBE OR EXPLAIN:

Supply-chain strategies

Approaches to negotiations

GLOBAL COMPANY PROFILE:

Volkswagen's Radical Experiment in Supply-Chain Management

In its new Brazilian plant 100 miles northwest of Rio de Janeiro, Volkswagen is radically altering its supply chain. With this experimental truck factory, Volkswagen is betting that it has found a system that will reduce the number of defective parts, cut labor costs, and improve efficiency. Because VW's potential market is small, this is a relatively small plant, with scheduled production of only 100 trucks per day with only 1,000 workers. However, only 200 of the 1,000 work for Volkswagen. The VW employees are responsible for overall quality, marketing, research, and design. The other 800, who work for suppliers such as Rockwell International, Cummins Engines, Delga Automotiva, Remon, and VDO, do the assembly work. Volkswagen's innovative supply chain will, it hopes, improve quality and drive down costs, as each subcontractor accepts responsibility for its units and worker compensation. With this strategy, Volkswagen subcontractors accept more of the direct costs and risks.

As the schematic shows, at the first stop in the assembly process, workers from Iochpe-Maxion mount the gas tank, transmission lines, and steering blocks. As the chassis moves down the line, employees from Rockwell mount axles and brakes. Then workers from Remon put on wheels and adjust tire pressure. The MWM/Cummins team installs the engine and transmission. Truck cabs, produced by the Brazilian firm Delga Automotiva, are painted by Eisenmann, and then finished and upholstered by VDO, both of Germany. Volkswagen employees do an evaluation of the final truck.

Because technology and economic efficiency demand specialization, many firms, like Volkswagen, are increasing their commitment to outsourcing and supply-chain integration. At this

Volkswagen's major suppliers are assigned space in the VW plant, but supply their own components, supplies, and workers. Workers from various suppliers build the truck as it moves down the assembly line. Volkswagen personnel inspect.

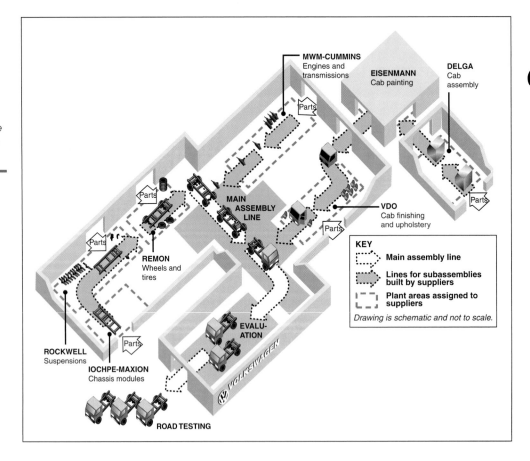

MWM-CUMMINS
Engines and transmissions

EISENMANN
Cab painting

DELGA
Cab assembly

Parts

MAIN ASSEMBLY LINE

VDO
Cab finishing and upholstery

Parts

Parts

REMON
Wheels and tires

Parts

EVALU-ATION

VOLKSWAGEN

ROCKWELL
Suspensions

Parts

IOCHPE-MAXION
Chassis modules

ROAD TESTING

KEY

- Main assembly line
- Lines for subassemblies built by suppliers
- Plant areas assigned to suppliers

Drawing is schematic and not to scale.

Volkswagen plant, however, VW is buying not only the materials but also labor and the related services. Suppliers are integrated tightly into VW's own network, right down to the assembly work in the plant.

Because purchase costs in the auto industry exceed 60% of the sales dollar, even modest reductions in these costs could make Volkswagen's payoff substantial. The results are not in yet, but VW is already trying a similar approach in plants in Buenos Aires, Argentina, and with Skoda, in the Czech Republic. Volkswagen's new level of integration in supply-chain management may be the wave of the future.

Remon workers attach the wheels as other parts of the truck are assembled simultaneously.

Nearly finished trucks move down the assembly line. The plant produces 100 trucks a day.

Supply-chain management

Management of activities that procure materials and services, transforming them into intermediate goods and final products, and delivering the products through a distribution system.

Most firms, like VW, spend over 50 percent of their sales dollars on purchases. Because such a high percentage of an organization's costs are determined by purchasing, relationships with suppliers are increasingly integrated and long-term. Joint efforts that improve innovation, speed design, and reduce costs are common. Such efforts can dramatically improve both partners' competitiveness. This changing focus places added emphasis on procurement and supplier relationships which must be managed. The discipline that manages these relationships is known as *supply-chain management*.

THE STRATEGIC IMPORTANCE OF THE SUPPLY CHAIN

Supply-chain management is the integration of the activities that procure materials and services, transform them into intermediate goods and final products, and deliver them to customers. These activities include purchasing and outsourcing activities, plus many other functions that are important to the relationship with suppliers and distributors. As Figure 11.1 suggests, supply-chain management includes determining (1) transportation vendors, (2) credit and cash transfers, (3) suppliers, (4) distributors and banks, (5) accounts payable and receivable, (6) warehousing and inventory levels, (7) order fulfillment, and (8) sharing customer, forecasting, and production information. The objective is to build a chain of suppliers that focuses on maximizing value to the ultimate customer. Activities of supply-chain managers cut across accounting, finance, marketing, and the operations discipline.

As firms strive to increase their competitiveness via product customization, high quality, cost reductions, and speed-to-market, they place added emphasis on the supply chain. The key to effective supply-chain management is to make the suppliers "partners" in the firm's strategy to satisfy an ever changing marketplace.

To ensure that the supply chain supports the firm's strategy, we need to consider the supply-chain issues shown in Table 11.1. Just as the OM function supports the firm's overall strategy, the supply chain is designed to support the OM strategy. Strategies of low cost or rapid response demand different things from a supply chain than a strategy of differentiation. For instance, a low-cost strategy, as Table 11.1 indicates, requires that we select suppliers based primarily on cost. Such suppliers should have the ability to design low-cost products that meet the functional requirements, minimize inventory, and drive down lead times. The firm must achieve integration of its selected strategy up and down the supply chain, and expect that strategy to be different for different products and change as products move through their life cycle.

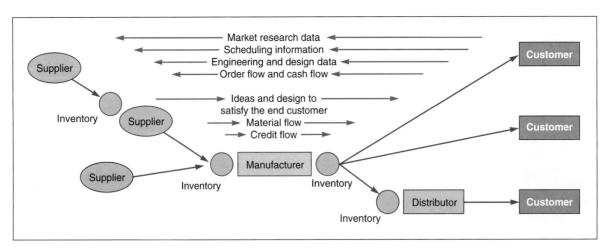

FIGURE 11.1 ■ The Supply Chain

The supply chain includes all the interactions between suppliers, manufacturers, distributors, and customers. The chain includes transportation, scheduling information, cash and credit transfers, as well as ideas, designs, and material transfers.

TABLE 11.1 ■ How the Supply Chain Decisions Impact Strategy

	LOW-COST STRATEGY	**RESPONSE STRATEGY**	**DIFFERENTIATION STRATEGY**
Supplier's Goal	Supply demand at lowest possible cost (e.g., Emerson Electric, Taco Bell)	Respond quickly to changing requirements and demand to minimize stockouts (e.g., Dell Computers)	Share market research; jointly develop products and options (e.g., Benetton)
Primary Selection Criteria	Select primarily for cost	Select primarily for capacity, speed, and flexibility	Select primarily for product development skills
Process Characteristics	Maintain high average utilization	Invest in excess capacity and flexible processes	Modular processes that lend themselves to mass customization
Inventory Characteristics	Minimize inventory throughout the chain to hold down costs	Develop responsive system, with buffer stocks positioned to ensure supply	Minimize inventory in the chain to avoid obsolescence
Lead-Time Characteristics	Shorten lead time as long as it does not increase costs	Invest aggressively to reduce production lead time	Invest aggressively to reduce development lead time
Product-Design Characteristics	Maximize performance and minimize cost	Use product designs that lead to low setup time and rapid production ramp-up	Use modular design to postpone product differentiation for as long as possible

See related table and discussion in Marshall L. Fisher, "What Is the Right Supply Chain for Your Product?" *Harvard Business Review* (March–April 1997): 105.

The objective of global supply-chain management is to build a chain of suppliers that focus on maximizing value to the ultimate customer.

Global Supply-Chain Issues

When companies enter growing global markets such as Eastern Europe, China, South America, or even Mexico, expanding their supply chains becomes a strategic challenge. Quality production in those areas may be a challenge, just as distribution systems may be less reliable, suggesting higher inventory levels than would be needed in one's home country. Also, tariffs and quotas may block nonlocal companies from doing business. Moreover, both political and currency risk remain high in much of the world.[1]

So the development of a successful strategic plan for supply chain management requires innovative planning and careful research. Supply chains in a global environment must be

1. Flexible enough to react to sudden changes in parts availability, distribution or shipping channels, import duties, and currency rates.
2. Able to use the latest computer and transmission technologies to schedule and manage the shipment of parts in and finished products out.
3. Staffed with local specialists to handle duties, trade, freight, customs, and political issues.

McDonald's planned for a global supply-chain challenge 6 years in advance of its opening in Russia. Creating a $60-million "food town," it developed independently owned supply plants in Moscow to keep its transportation costs and handling times low and its quality and customer-service levels high. Every component in this food chain—meat plant, chicken plant, bakery, fish plant, and lettuce plant—is closely monitored to make sure that all the system's links are strong.

Firms like Ford and Boeing also face global procurement decisions. Ford's Mercury has only 227 suppliers worldwide, a small number compared to the 700 involved in previous models. Ford has set a trend to develop a global network of *fewer* suppliers that provide the lowest cost and highest quality regardless of home country. So global is the production of the Boeing 777 that officials proclaim that "The Chinese now make so many Boeing parts, that when Boeing planes fly to China, they are going home."[2] The *OM in Action* box, "A Rose is a Rose, But Only if It Is Fresh," details the global supply chain that ends with your local forist.

[1]Note the devaluation of the Mexican peso in 1992, the Thai bhat and the Malaysian ringgit in 1997, and the Argentine peso in 2002, as well as armed conflicts in about two dozen countries at any given time.

[2]*New York Times* (October 21, 1997): A21.

OM IN ACTION

A Rose Is a Rose, But Only if It Is Fresh

How fast can supply chains be? How good can supply chains be? Supply chains for food and flowers must be fast and they must be good. When the food-supply chain has a problem, the best that can happen is the customer does not get fed on time; the worst that happens is the customer gets food poisoning and dies. In the floral industry, the timing and temperature are also critical. Indeed, flowers are the most perishable agricultural item—even more so than fish. Flowers not only need to move fast, but they must also be kept cool, at a constant temperature of 33 to 36 degrees. And they must be provided preservative-treated water while in transit. Roses are especially delicate, fragile, and perishable.

Seventy percent of the roses sold in the U.S. market arrive by air from rural Columbia and Ecuador. Roses move through this supply chain via an intricate but fast transportation network. This network stretches from Ecuadorian growers who cut, grade, bundle, pack and ship, to importers who make the deal, to the U.S. Department of Agriculture personnel who quarantine and inspect for insects, diseases, and parasites, to U.S. Customs agents who inspect and approve, to facilitators who provide clearance and labeling, to wholesalers who distribute, to retailers who arrange and sell, and finally to the customer. Each and every minute the product is deteriorating, but when roses meet the scrutinizing eye of a final recipient, every delicate petal is critical. The time and temperature sensitivity of perishables like roses has added sophistication and refined new standards in the supply chain. The result is improved quality and reduced losses. After all, when it's Valentine's Day, what good is a shipment of roses that arrives wilted or late? This is a difficult supply chain; only an excellent one will get the job done.

Sources: IIE Solutions (February 2002): 26–32; and *World Trade* (March 1999): 70–72.

TABLE 11.2 ■

Supply Chain Costs as a Percent of Sales

INDUSTRY	% PURCHASED
All industry	52
Automobile	67
Food	60
Lumber	61
Paper	55
Petroleum	79
Transportation	62

SUPPLY-CHAIN ECONOMICS

The supply chain receives such attention because it is an integral part of a firm's strategy and the most costly activity in most firms. For both goods and services, supply-chain costs as a percent of sales are often substantial (see Table 11.2). Because such a huge portion of revenue is devoted to the supply chain, an effective strategy is vital. The supply chain provides a major opportunity to reduce costs and increase contribution margins.

Table 11.3 illustrates the amount of leverage available to the operations manager through the supply chain. Firms spending 50% of their sales dollar in the supply chain and having a net profit of 6% would require $3.57 worth of sales in order to equal the savings that accrues to the company from a $1 savings in procurement. These numbers indicate the strong role that procurement can play in profitability.

TABLE 11.3 ■ Dollars of Additional Sales Needed to Equal $1 Saved through the Supply Chain[a]

PERCENT NET PROFIT OF FIRM	PERCENT OF SALES SPENT IN THE SUPPLY CHAIN						
	30%	40%	50%	60%	70%	80%	90%
2	$2.78	$3.23	$3.85	$4.76	$6.25	$9.09	$16.67
4	$2.70	$3.13	$3.70	$4.55	$5.88	$8.33	$14.29
6	$2.63	$3.03	$3.57	$4.35	$5.56	$7.69	$12.50
8	$2.56	$2.94	$3.45	$4.17	$5.26	$7.14	$11.11
10	$2.50	$2.86	$3.33	$4.00	$5.00	$6.67	$10.00

[a]The required increase in sales assumes that 50% of the costs other than purchases are variable and that 1/2 of the remaining (less profit) are fixed. Therefore, at sales of $100 (50% purchases and 2% margin), $50 are purchases, $24 are other variable costs, $24 are fixed costs, and $2 profit. Increasing sales by $3.85 yields the following:

Purchases at 50%	$ 51.93
Other Variable Costs	24.92
Fixed Cost	24.00
Profit	3.00
	$103.85

Through $3.85 of additional sales, we have increased profit by $1, from $2 to $3. The same increase in margin could have been obtained by reducing supply chain costs by $1.

Example 1

> The Goodwin Company spends 50% of its sales dollar in the supply chain. The firm has a net profit of 4%. Of the remaining 46%, 23% is fixed and the remaining 23% is variable. From Table 11.3, we see that the dollar value of sales needed to generate the same profit that results from $1 of supply chain savings would be $3.70.

Make-or-Buy Decisions

Make-or-buy decision
Choosing between producing a component or a service in-house or purchasing it from an outside source.

A wholesaler or retailer buys everything that it sells; a manufacturing operation hardly ever does. Manufacturers, restaurants, and assemblers of products buy components and subassemblies that go into final products. As we saw in Chapter 5, choosing products and services that can be advantageously obtained *externally* as opposed to produced *internally* is known as the **make-or-buy decision**. Supply-chain personnel evaluate alternative suppliers and provide current, accurate, complete data relevant to the buy alternative. Table 11.4 lists a variety of considerations in the make-or-buy decision. Regardless of the decision, supply-chain performance should be reviewed periodically. Vendor competence and costs change, as do a firm's own strategy, production capabilities, and costs.

Outsourcing

Outsourcing
Transferring a firm's activities that have traditionally been internal to external suppliers.

Outsourcing transfers some of what are traditional internal activities and resources of a firm to outside vendors, making it slightly different than the traditional make-or-buy decision. Outsourcing is part of the continuing trend toward utilizing the efficiency that comes with specialization.[3] The vendor performing the outsourced service is an expert in that particular specialty, and the outsourcing firm can focus on its critical success factors—its core competencies.

With outsourcing there is no tangible product and no transfer of title. The contracting firm usually provides the resources necessary for accomplishing the activities. The resources transferred to the supplying firm may include facilities, people, and equipment. Many firms now outsource their information-technology requirements, accounting work, legal functions, and even product assembly. Electronic Data Systems (EDS) provides information technology outsourcing for many firms, including Delphi Automotive and Nextel. Similarly, Automatic Data Processing (ADP) provides payroll services for thousands of firms. And much of IBM's computer assembly work has been outsourced to a specialist in electronic assembly, Solectron.

TABLE 11.4 ■

Considerations for the Make-or-Buy Decision

REASONS FOR MAKING	REASONS FOR BUYING
1. Maintain core competence	1. Frees management to deal with its primary business
2. Lower production cost	2. Lower acquisition cost
3. Unsuitable suppliers	3. Preserve supplier commitment
4. Assure adequate supply (quantity or delivery)	4. Obtain technical or management ability
5. Utilize surplus labor or facilities and make a marginal contribution	5. Inadequate capacity
6. Obtain desired quality	6. Reduce inventory costs
7. Remove supplier collusion	7. Ensure alternative sources
8. Obtain unique item that would entail a prohibitive commitment for a supplier	8. Inadequate managerial or technical resources
9. Protect personnel from a layoff	9. Reciprocity
10. Protect proprietary design or quality	10. Item is protected by a patent or trade secret
11. Increase or maintain size of the company (management preference)	

[3]See related discussions in Richard C. Insinga and Michael J. Werle, "Linking Outsourcing to Business Strategy," *The Academy of Management Executive* 14, no. 4 (November 2000): 58–70; and James Brian Quinn, "Outsourcing Innovation: The New Engine of Growth," *MIT Sloan Management Review* (summer 2000): 13–27.

Supply-Chain
Management at
Regal Marine

SUPPLY-CHAIN STRATEGIES

For goods and services to be obtained from outside sources, the firm must decide on a supply-chain strategy. One such strategy is the approach of *negotiating with many suppliers* and playing one supplier against another. A second strategy is to develop *long-term, "partnering"* relationships with a few suppliers to satisfy the end customer. A third strategy is *vertical integration*, where firms may decide to use vertical backward integration by actually buying the supplier. A fourth variation is a combination of few suppliers and vertical integration, known as a *keiretsu*. In a *keiretsu*, *suppliers become part of a company coalition.* Finally, a fifth strategy is to develop virtual companies *that use suppliers on an as-needed basis.* We will now discuss each of these strategies.

Many Suppliers

With the many-supplier strategy, the supplier responds to the demands and specifications of a "request for quotation," with the order usually going to the low bidder. This is a common strategy when products are commodities. This strategy plays one supplier against another and places the burden of meeting the buyer's demands on the supplier. Suppliers aggressively compete with one another. Although many approaches to negotiations can be used with this strategy, long-term "partnering" relationships are not the goal. This approach holds the supplier responsible for maintaining the necessary technology, expertise, and forecasting abilities, as well as cost, quality, and delivery competencies.

Few Suppliers

Integrating suppliers, production, and distribution requires that operations be as agile as possible.

A strategy of few suppliers implies that rather than looking for short-term attributes, such as low cost, a buyer is better off forming a long-term relationship with a few dedicated suppliers. Long-term suppliers are more likely to understand the broad objectives of the procuring firm and the end customer. Using few suppliers can create value by allowing suppliers to have economies of scale and a learning curve that yields both lower transaction costs and lower production costs.

Few suppliers, each with a large commitment to the buyer, may also be more willing to participate in JIT systems, as well as provide design innovations and technological expertise. Many firms have moved aggressively to incorporate suppliers into their supply systems. DaimlerChrysler, for one, now seeks to choose suppliers even before parts are designed. Motorola also evaluates suppliers on rigorous criteria, but in many instances has eliminated traditional supplier bidding, placing added emphasis on quality and reliability. On occasion these relationships yield contracts that extend through the product's life cycle. The expectation is that both the purchaser and supplier collaborate, becoming more efficient and reducing prices over time. The natural outcome of such relationships is fewer suppliers, but those that remain have long-term relationships.

About 100 years ago, Henry Ford surrounded himself with reliable suppliers, many on his own property, making his assembly operation close to self-sufficient.

Service companies like Marks and Spencer, a British retailer, have also demonstrated that cooperation with suppliers can yield cost savings for customers and suppliers alike. This strategy has resulted in suppliers that develop new products, winning customers for Marks and Spencer and the supplier. The move toward tight integration of the suppliers and purchasers is occurring in both manufacturing and services.

Like all strategies, a downside exists. With few suppliers, the cost of changing partners is huge, so both buyer and supplier run the risk of becoming captives of the other. Poor supplier performance is only one risk the purchaser faces. The purchaser must also be concerned about trade secrets and suppliers that make other alliances or venture out on their own. This happened when the U.S. Schwinn Bicycle Co., needing additional capacity, taught Taiwan's Giant Manufacturing Company to make and sell bicycles. Giant Manufacturing is now the largest bicycle manufacturer in the world, and Schwinn was acquired by Pacific Cycle LLC out of bankruptcy.

Vertical Integration

Vertical integration
Developing the ability to produce goods or services previously purchased or actually buying a supplier or a distributor.

Purchasing can be extended to take the form of vertical integration. By **vertical integration**, we mean developing the ability to produce goods or services previously purchased or actually buying a supplier or a distributor. As shown in Figure 11.2, vertical integration can take the form of *forward* or *backward integration*.

Backward integration suggests a firm purchase its suppliers, as in the case of Ford Motor Company deciding to manufacture its own car radios. Forward integration, on the other hand, suggests that a

FIGURE 11.2 ■

Vertical Integration Can
Be Forward or Backward

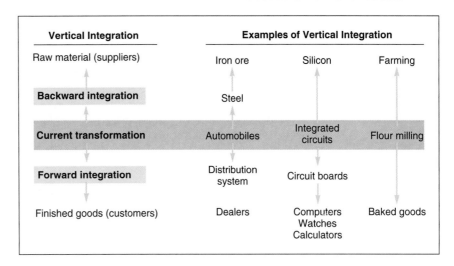

manufacturer of components make the finished product. An example is Texas Instruments, a manufacturer of integrated circuits that also makes calculators and computers containing integrated circuits.

Vertical integration can offer a strategic opportunity for the operations manager. For firms with the necessary capital, managerial talent, and required demand, vertical integration may provide substantial opportunities for cost reduction. Other advantages in inventory reduction and scheduling can accrue to the company that effectively manages vertical integration or close, mutually beneficial relationships with suppliers.

Because purchased items represent such a large part of the costs of sales, it is obvious why so many organizations find interest in vertical integration. Vertical integration can yield cost reduction, quality adherence, and timely delivery. Vertical integration appears to work best when the organization has large market share or the management talent to operate an acquired vendor successfully.[4] However, backward integration may be particularly dangerous for firms in industries undergoing technological change if management cannot keep abreast of those changes or invest the financial resources necessary for the next wave of technology.

Sanford Corporation is one of America's largest producers of highlighters and markers. Sanford is vertically integrated, making its own inks—a strategy that gives it a research, development, quality, and product-flexibility advantage.

---

[4]Robert D. Buzzell, "Is Vertical Integration Profitable?" *Harvard Business Review* 61, no. 1 (January–February 1983): 92–102.

Keiretsu Networks

Keiretsu
A Japanese term to describe suppliers who become part of a company coalition.

Many large Japanese manufacturers have found a middle ground between purchasing from few suppliers and vertical integration. These manufacturers are often financial supporters of suppliers through ownership or loans. The supplier then becomes part of a company coalition known as a *keiretsu*. Members of the *keiretsu* are assured long-term relationships and are therefore expected to function as partners, providing technical expertise and stable quality production to the manufacturer. Members of the *keiretsu* can also have suppliers further down the chain, making second- and even third-tier suppliers part of the coalition.

Virtual Companies

Virtual companies
Companies that rely on a variety of supplier relationships to provide services on demand. Also known as hollow corporations or network companies.

As noted before, the limitations to vertical integration are severe. Our technological society continually demands more specialization, which further complicates vertical integration. Moreover, a firm that has a department or division of its own for everything may be too bureaucratic to be world-class. So rather than letting vertical integration lock an organization into businesses that it may not understand or be able to manage, another approach is to find good flexible suppliers. **Virtual companies** rely on a variety of supplier relationships to provide services on demand. Virtual companies have fluid, moving organizational boundaries that allow them to create a unique enterprise to meet changing market demands. Suppliers may provide a variety of services that include doing the payroll, hiring personnel, designing products, providing consulting services, manufacturing components, conducting tests, or distributing products. The relationships may be short-term or long-term and may include true partners, collaborators, or simply able suppliers and subcontractors. Whatever the formal relationship, the result can be exceptionally lean performance. The advantages of virtual companies include specialized management expertise, low capital investment, flexibility, and speed. The result is efficiency.

The apparel business provides a *traditional* example of virtual organizations. The designers of clothes seldom manufacture their designs; rather, they license the manufacture. The manufacturer may then rent a loft, lease sewing machines, and contract for labor. The result is an organization that has low overhead, remains flexible, and can respond rapidly to the market.

A *contemporary* example is the semiconductor industry, exemplified by Visioneer in Palo Alto. This California firm subcontracts almost everything: Software is written by several partners, hardware is manufactured by a subcontractor in Silicon Valley, printed circuit boards are made in Singapore, and plastic cases are made in Boston, where units are also tested and packed for shipment. In the virtual company, the purchasing function is demanding and dynamic.

Each company makes its own judgment about the appropriate degree of vertical integration. Jaguar has changed its approach to vertical integration. In the past, Jaguar made virtually every part it could, even some simple items such as washers. However, Jaguar now focuses on those items that make a car unique: the body, engine, and suspension. Outside suppliers with their own capabilities, expertise, and efficiencies provide most other components.

MANAGING THE SUPPLY CHAIN

As managers move toward integration of the supply chain, substantial efficiencies are possible. The cycle of materials—as they flow from suppliers, to production, to warehousing, to distribution, to the customer—is taking place between separate and often very independent organizations. Therefore, there are significant management issues that may result in serious inefficiencies. Success begins with mutual agreement on goals, followed by mutual trust, and continues with compatible organizational cultures.

Mutual Agreement on Goals   An integrated supply chain requires more than just agreement on the contractual terms of a buy/sell relationship. Partners in the chain must appreciate that the only entity that puts money into a supply chain is the end customer. Therefore, establishing a mutual understanding of the mission, strategy, and goals of participating organizations is essential. The integrated supply chain is about adding economic value and maximizing the total content of the product.

The supplier must be treated as an extension of the company.

Trust   Trust is critical to an effective and efficient supply chain. Members of the chain must enter into a relationship that shares information—a relationship built on mutual trust. Supplier relationships are more likely to be successful if risk and cost savings are shared—and activities such as end-customer research, sales analysis, forecasting, and production planning are joint activities.

Compatible Organizational Cultures   A positive relationship between the purchasing and supplying organizations that comes with compatible organizational cultures can be a real advantage in making a supply chain hum. A champion within one of the two firms promotes both formal and informal contacts, and those contacts contribute to the alignment of the organizational cultures, further strengthening the relationship.

The operations manager is dealing with a supply chain that is made up of independent specialists, each trying to satisfy its own customers at a profit. This leads to actions that may not optimize the entire chain. On the other hand, the supply chain is replete with opportunities to reduce waste and enhance value. We now look at some of the significant issues and opportunities.

Issues in an Integrated Supply Chain

Three issues complicate development of an efficient, integrated supply chain: local optimization, incentives, and large lots.

Local Optimization   Members of the chain are inclined to focus on maximizing local profit or minimizing immediate cost based on their limited knowledge. Slight upturns in demand are overcompensated for because no one wants to be caught short. Similarly, slight downturns are overcompensated for because no one wants to be caught holding excess inventory. So fluctuations are magnified. For instance, a pasta distributor does not want to run out of pasta for his retail customers; the natural response to an extra large order is to compensate with an even larger order to the manufacturer on the assumption that sales are picking up. Neither the distributor nor the manufacturer knows that the retailer had a major one-time promotion that moved a lot of pasta. This is exactly the issue that complicated the implementation of efficient distribution at the Italian pasta maker Barilla.

Incentives (Sales Incentives, Quantity Discounts, Quotas, and Promotions)
Incentives push merchandise into the chain for sales that have not occurred. This generates fluctuations that are ultimately expensive to all members of the chain.

Large Lots   There is often a bias toward large lots because large lots tend to reduce unit costs. The logistics manager wants to ship large lots, preferably in full trucks, and the production manager wants long production runs. Both actions drive down unit cost, but fail to reflect actual sales.

These three common occurrences (local optimization, incentives, and large lots) contribute to distortions of information about what is really going on in the supply chain. A well-running supply system needs to be based on accurate information about how many products are truly being pulled through the chain. The inaccurate information is unintentional, but it results in distortions and fluctuations in the supply chain and causes what is known as the bullwhip effect.

Bullwhip effect

The increasing fluctuation in orders that often occurs as orders move through the supply chain.

The **bullwhip effect** occurs as orders are relayed from retailers, to wholesalers, to manufacturers, with fluctuations increasing at each step in the sequence. The "bullwhip" fluctuations in the supply chain increase the costs associated with inventory, transportation, shipping, and receiving, while decreasing customer service and profitability. Procter & Gamble found that, although the use of Pampers diapers was steady and the retail-store orders had little fluctuation, as orders moved through the supply chain, fluctuations increased. By the time orders were initiated for raw material, the variability was substantial.[5] Similar behavior has been observed and documented at many companies, including Campbell Soup, Hewlett Packard, and Applied Materials.[6] A number of opportunities exist for reducing the bullwhip effect and improving opportunities in the supply chain. These are discussed in the following section.

Opportunities in an Integrated Supply Chain

Opportunities for effective management in the supply chain include the following 10 items.

Pull data

Accurate sales data that initiates transactions to "pull" product through the supply chain.

Accurate "Pull" Data   Generate accurate **pull data** by sharing (1) point-of-sales (POS) information, so that each member of the chain can schedule effectively, and (2) computer-assisted ordering (CAO). This implies using POS systems that collect sales data and then adjusting that data for market factors, inventory on hand, and outstanding orders. Then a net order is sent directly to the supplier who is responsible for maintaining the finished-goods inventory.

Lot Size Reduction   Reduce lot sizes by aggressive management. This may include: (1) developing economical shipments of less than truckload lots; (2) providing discounts based on total annual volume rather than size of individual shipments; and (3) reducing the cost of ordering through techniques such as standing orders and various forms of electronic purchasing.

Single stage control of replenishment

Fixing responsibility for monitoring and managing inventory for the retailer.

Single Stage Control of Replenishment   **Single stage control of replenishment** means designating a member in the chain as responsible for monitoring and managing inventory in the supply chain based on the "pull" from the end user. This approach removes distorted information and multiple forecasts that create the bullwhip effect. Control may be in the hands of:

- A sophisticated retailer who understands demand patterns. How Wal-Mart does this for some of its inventory with radio frequency (RF) tags is shown in the *OM in Action* box, "RF Tags: Keeping the Shelves Stocked."
- A distributor who manages the inventory for a particular distribution area. Distributors who handle grocery items, beer, and soft drinks may do this.
- A manufacturer that has a well-managed distribution system, such as Procter & Gamble.

Vendor managed inventory (VMI)

Supplier maintains material for the buyer, often delivering directly to the buyer's using department.

Vendor Managed Inventory   **Vendor managed inventory (VMI)** means the use of a local supplier (usually a distributor) to maintain inventory for the manufacturer or retailer. The supplier delivers directly to the purchaser's using department rather than to a receiving dock or stockroom. If the supplier can maintain the stock of inventory for a variety of customers who use the same product or whose differences are very minor (say, at the packaging stage), then there should be a net savings. These systems work without the immediate direction of the purchaser.

Postponement

Delaying any modifications or customization to the product as long as possible in the production process.

Postponement   **Postponement** withholds any modification or customization to the product (keeping it generic) as long as possible.[7] For instance, after analyzing the supply chain for its printers, Hewlett-Packard (H-P) determined that if the printer's power supply was moved out of the printer itself and into a power cord, H-P could ship the basic printer anywhere in the world. H-P modified the printer, its power cord, its packaging, and its documentation so that only the power cord and documentation needed to be added at the final distribution point. This modification allowed the firm to manufacture and hold centralized inventories of the generic printer for shipment

---

[5]Hau L. Lee, V. Padmanabhan, and W. Whang, "The Bullwhip Effect in Supply Chains," *MIT Sloan Management Review* (spring 1997): 93–106; and Dinah Greek, "Whip Hand," *Professional Engineering* (May 24, 2000): 43.

[6]Robert Ristelhueber, "Supply Chain Strategies—Applied Materials Seek to Snap Bullwhip Effect," *EBN* (January 22, 2001): 61.

[7]Hau L. Lee and Corey Billington, "The Evolution of Supply-Chain-Management Models and Practice at Hewlett-Packard," *Interfaces* 25, no. 5 (September–October 1995): 42–63.

OM IN ACTION

Radio Frequency Tags: Keeping the Shelves Stocked

The supply chain works smoothly when sales are steady, but it often breaks down when confronted by a sudden surge in demand. Radio Frequency ID (or RFID) tags could change that by providing real-time information about what's happening on store shelves. Here's how the system works.

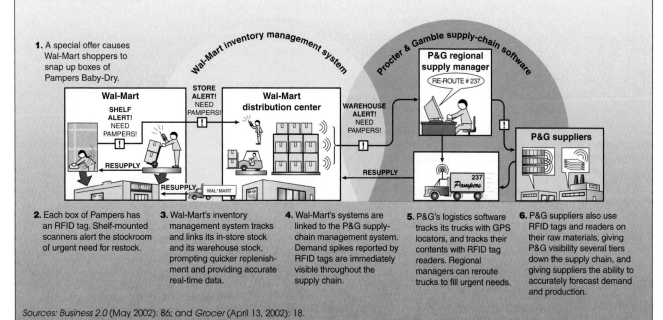

1. A special offer causes Wal-Mart shoppers to snap up boxes of Pampers Baby-Dry.

2. Each box of Pampers has an RFID tag. Shelf-mounted scanners alert the stockroom of urgent need for restock.

3. Wal-Mart's inventory management system tracks and links its in-store stock and its warehouse stock, prompting quicker replenishment and providing accurate real-time data.

4. Wal-Mart's systems are linked to the P&G supply-chain management system. Demand spikes reported by RFID tags are immediately visible throughout the supply chain.

5. P&G's logistics software tracks its trucks with GPS locators, and tracks their contents with RFID tag readers. Regional managers can reroute trucks to fill urgent needs.

6. P&G suppliers also use RFID tags and readers on their raw materials, giving P&G visibility several tiers down the supply chain, and giving suppliers the ability to accurately forecast demand and production.

Sources: Business 2.0 (May 2002): 86; and *Grocer* (April 13, 2002): 18.

---

as demand changed. Only the unique power system and documentation had to be held in each country. This understanding of the entire supply chain reduced both risk and investment in inventory.[8]

Channel assembly
Postpones final assembly of a product so the distribution channel can assemble it.

Channel Assembly Channel assembly is a variation of postponement. **Channel assembly** sends individual components and modules, rather than finished products, to the distributor. The distributor then assembles, tests, and ships. Channel assembly treats distributors more as manufacturing partners than as distributors. This technique has proven successful in industries where products are undergoing rapid change, such as personal computers. With this strategy, finished goods inventory is reduced because units are built to a shorter, more accurate forecast. Consequently, market response is better, with lower investment—a nice combination.

Drop shipping
Shipping directly from the supplier to the end consumer, rather than from the seller, saving both time and reshipping costs.

Drop Shipping and Special Packaging **Drop shipping** means the supplier will ship directly to the end consumer, rather than to the seller, saving both time and reshipping costs. Other cost-saving measures include the use of special packaging, labels, and optimal placement of labels and bar codes on containers. The final location down to the department and number of units in each shipping container can also be indicated. Substantial savings can be obtained through management techniques such as these. Some of these techniques can be of particular benefit to wholesalers and retailers by reducing shrinkage (lost, damaged, or stolen merchandise) and handling cost.

For instance, Dell Computer has decided that its core competence is not in stocking peripherals, but in assembling PCs. So if you order a PC from Dell, with a printer and perhaps other components, the computer comes from Dell, but the printer and many of the other components will be drop shipped from the manufacturer.

---

[8]M. Eric Johnson and Tom Davis, "Gaining an Edge with Supply Chain Management," *APICS: The Performance Advantage* 5, no. 12 (December 1995): 26–31.

Blanket order

A long-term purchase commitment to a supplier for items that are to be delivered against short-term releases to ship.

Blanket Orders   Blanket orders are unfilled orders with a vendor.[9] A **blanket order** is a contract to purchase certain items from the vendor. It is not an authorization to ship anything. Shipment is made only on receipt of an agreed-on document, perhaps a shipping requisition or shipment release.

Standardization

Reducing the number of variations in materials and components as an aid to cost reduction.

Standardization   The purchasing department should make special efforts to increase levels of **standardization**: That is, rather than obtaining a variety of similar components with labeling, coloring, packaging, or perhaps even slightly different engineering specifications, the purchasing agent should try to have those components standardized.

Electronic Ordering and Funds Transfer   Electronic ordering and funds transfer reduces paper transactions. Paper transactions consist of a purchase order, a purchase release, a receiving document, authorization to pay an invoice (which is matched with the approved receiving report), and finally the issuance of a check. Purchasing departments can reduce this barrage of paperwork by electronic ordering, acceptance of all parts as 100% good, and electronic funds transfer to pay for units received. Not only can electronic ordering reduce paperwork, but it also speeds up the traditionally long procurement cycle.

Electronic data interchange (EDI)

A standardized data-transmittal format for computerized communications between organizations.

Transactions between firms often use electronic data interchange. **Electronic data interchange (EDI)** is a standardized data-transmittal format for computerized communications between organizations. EDI provides data transfer for virtually any business application, including purchasing. Under EDI, for instance, data for a purchase order, such as order date, due date, quantity, part number, purchase order number, address, and so forth, are fitted into the standard EDI format.

An extension of EDI is **Advanced Shipping Notice (ASN)**, which is a shipping notice delivered directly from vendor to purchaser. When the vendor is ready to ship, shipping labels are printed and the advanced shipping notice is created and transmitted to the purchaser. While both manufacturing and retail establishments use this technique, the Internet's ease of use and lower cost are replacing both EDI and ASN in their current form.

Advanced Shipping Notice (ASN)

A shipping notice delivered directly from vendor to purchaser.

INTERNET PURCHASING

Internet purchasing (e-procurement)

Order releases communicated over the Internet or approved vendor catalogs available on the Internet for use by employees of the purchasing firm.

State-of-the-art supply-chain systems combine many of the above techniques within automated purchasing systems. **Internet purchasing** or as it is sometimes called, *e-procurement*, takes two forms. First, Internet purchasing may just imply that the Internet is used to communicate order releases to suppliers. This would occur for those items for which a blanket purchase order exists. In this application, the Internet replaces more traditional electronic data interchange (EDI) with an order release to the supplier via the Internet. Secondly, for nonstandard items, for which there is no blanket order, catalogs and ordering procedures enhance the communication features of the Internet. In this application of Internet purchasing, long-term master agreements with approved vendors result in catalogs being put online for the buying organization's use. In such systems, ordering lead time is reduced and purchasing costs are controlled. As an added benefit, Internet purchasing lends itself to comparison shopping, rapid ordering, and reduction of inventory. (The supplement to this chapter, "E-Commerce and Operations Management," discusses these issues further.)

Suppliers like e-procurement systems because online selling means they are getting closer to their customers. The seller's margins may also improve as total cycle time (time from order to delivery) is cut. As an added sweetener, the capital investment for e-procurement systems is low.

Internet purchasing may be a part of an integrated Enterprise Resource Planning (ERP) system with Internet communication between units of the supply chain. In such systems the "order release" not only tells the shipper to ship, but also updates the appropriate portions of the ERP system. Other, less integrated, Internet purchasing systems may not be part of a fully integrated ERP system, but even in these, purchases are usually posted automatically to the financial and distribution system of the purchasing firm, thus reducing internal transaction costs. Internet purchasing, as discussed in the *OM in Action* box, "Purchasing Managers Live and Die by the Net," is a radical change.

---

[9]Unfilled orders are also referred to as "open" orders or "incomplete" orders.

OM IN ACTION

Purchasing Managers Live and Die by the Net

Will procurement executives fade into the woodwork, replaced partly by online companies and partly by executives schooled in finance and technology rather than the fine points of negotiations? "Finance people and design engineers will do the negotiating, while third-party e-procurement companies comb the Net for new suppliers," says professor Safwan Masri of Columbia University. "Four years from now, chief procurement officers will have been replaced by chief trading officers with Internet skills who coordinate buying and selling on the Web," adds Ray Graham, president of Casbah Corp.

Companies such as United Technologies have indeed farmed out much of their buying to third parties. The firm has moved engineers and finance executives into its new supply side management function. And Siemens, the German manufacturing giant, just turned part of its purchasing department into a wholly owned subsidiary. The subsidiary buys products for Siemens and sells its procurement expertise to outsiders. "We're telling other companies they can use us; they don't need their own purchasing people," says Peter Ochel, vice president for e-procurement.

The big question is the risk to long-term relationships. Suppliers have long tried to hold onto big clients by establishing strong relationships with their procurement people. Reverse auctions or blanket-buying exchanges could wipe that out.

Sources: New York Times (April 8, 2000): B1, B4, and (February 28, 2001): 7.

San Diego State University, for example, uses software to check order status, receive invoices and acknowledgements, and generate activity reports. The supply catalog content is real time and supplier managed. Purchasing transactions are integrated with the university's financial software, which is supplied by Oracle. Texas Instruments has installed a similar system to drive down its purchasing costs while improving item availability. TI employees worldwide now order directly from their desktops, and the transaction data are interfaced with SAP's Enterprise Resource Planning software. (SAP and ERP are treated in more detail in Chapter 14.)

VENDOR SELECTION

Vendor selection
A decision regarding from whom to buy goods or services.

For those goods and services a firm buys, vendors must be selected. **Vendor selection** considers numerous factors, such as strategic fit, vendor competence, delivery, and quality performance. Because a firm may have some competence in all areas and exceptional competence in only a few, selection can be challenging. We now examine vendor selection as a three-stage process. Those stages are (1) vendor evaluation, (2) vendor development, and (3) negotiations.

Vendor Evaluation

The first stage, *vendor evaluation*, involves finding potential vendors and determining the likelihood of their becoming good suppliers. This phase requires the development of evaluation criteria such as those in Example 2. Both the criteria and the weights selected depend on the supply-chain strategy to be achieved. (Refer to Table 11.1 shown earlier in the chapter.)

Example 2

Hau Lee, president of Creative Toys in Palo Alto, is interested in evaluating suppliers who will work with him to make nontoxic, environmentally friendly paints and dies for his line of children's toys. This is a critical strategic element of his supply chain, and he desires a firm that will contribute to his product. He begins his analysis of one potential supplier, Faber Paint and Dye.

SOLUTION

Hau first reviews the supplier differentiation attributes in Table 11.1 and develops the list of selection criteria shown on the top of the next page.[10] He then assigns the weights shown to help him perform an objective review of potential vendors. His staff assigns the scores shown and computes the total weighted score.

[10]A discussion of vendor selection criteria can be found in Chapter 8 of Robert Monczka, Robert Trent, and Robert Handfield, *Purchasing and Supply Chain Management*, 2nd ed. (Cincinnati: Southwestern, 2002); and Chapter 5 of Judith A. Stimson, *Supplier Selection* (Miami: PT Publications, Inc., 1998).

CRITERIA	WEIGHTS	SCORES (1–5) (5 HIGHEST)	WEIGHT × SCORE
Engineering/research/innovation skills	.20	5	1.0
Production process capability (flexibility/technical assistance)	.15	4	.6
Distribution/delivery capability	.05	4	.2
Quality systems and performance	.10	2	.2
Facilities/location	.05	2	.1
Financial and managerial strength (stability and cost structure)	.15	4	.6
Information systems capability (e-commerce, Internet)	.10	2	.2
Integrity (environmental compliance/ethics)	.20	5	1.0
	1.00		3.9 Total

Faber Paint and Dye receives an overall score of 3.9. Hau now has a basis for comparison to other potential vendors, selecting the one with the highest overall rating.

The selection of competent suppliers is critical. If good suppliers are not selected, then all other supply-chain efforts are wasted. As firms move toward fewer longer-term suppliers, the issues of financial strength, quality, management, research, technical ability, and potential for a close long-term relationship play an increasingly important role. These attributes should be noted in the evaluation process.

Vendor Development

The second stage is *vendor development*. Assuming a firm wants to proceed with a particular vendor, how does it integrate this supplier into its system? The buyer makes sure the vendor has an appreciation of quality requirements, engineering changes, schedules and delivery, the purchaser's payment system, and procurement policies. *Vendor development* may include everything from training, to engineering and production help, to procedures for information transfer. Procurement policies also need to be established. Those might address issues such as percent of business done with any one supplier or with minority businesses.

Negotiations

Negotiation strategies
Approaches taken by supply-chain personnel to develop contractual relationships with suppliers.

Regardless of the supply-chain strategy adopted, negotiations regarding the critical elements of the contractual relationship must take place. These negotiations often focus on quality, delivery, payment, and cost. We will look at three classic types of **negotiation strategies**: the cost-based model, the market-based price model, and competitive bidding.

Cost-Based Price Model   The *cost-based price model* requires that the supplier open its books to the purchaser. The contract price is then based on time and materials or on a fixed cost with an escalation clause to accommodate changes in the vendor's labor and materials cost.

Market-Based Price Model   In the market-based price model, price is based on a published, auction, or index price. Many commodities (agriculture products, paper, metal, etc.) are priced this way. Paperboard prices, for instance, are available via the *Official Board Markets* weekly publication (www.advanstar.com/subscribe/).[11] Nonferrous metal prices are quoted in *Platt's Metals Week* (www.platts.com/plattsmetals/) and other metals at www.metalworld.com.

Competitive Bidding   When suppliers are not willing to discuss costs or where near-perfect markets do not exist, competitive bidding is often appropriate. Infrequent work (such as construction, tooling, and dies) is usually purchased based on a bid. Bidding may take place via mail, fax, or an Internet auction. Competitive bidding is the typical policy in many firms for the majority of their purchases. Bidding policies usually require that the purchasing agent have several potential suppliers of the product (or its equivalent) and quotations from each. The major disadvantage of this method, as mentioned earlier, is that the development of long-term relations between buyer and seller are hin-

[11]The "yellow sheet" is the commonly used name of the *Official Board Markets*, published by *Magazines for Industry*, Chicago. It contains announced paperboard prices for containerboard and boxboard.

dered. Competitive bidding may effectively determine initial cost. However, it may also make difficult the communication and performance that are vital for engineering changes, quality, and delivery.

Yet a fourth approach is *to combine one or more* of the preceding negotiation techniques. The supplier and purchaser may agree on review of certain cost data, accept some form of market data for raw material costs, or agree that the supplier will "remain competitive." In any case, a good supplier relationship is one in which both partners have established a degree of mutual trust and a belief in the competence of each other.

> Negotiations should not be viewed as a win/lose game; it can be a win/win game.

LOGISTICS MANAGEMENT

> **Logistics management**
> An approach that seeks efficiency of operations through the integration of all material acquisition, movement, and storage activities.

Procurement activities may be combined with various shipping, warehousing, and inventory activities to form a logistics system. The purpose of **logistics management** is to obtain efficiency of operations through the integration of all material acquisition, movement, and storage activities. When transportation and inventory costs are substantial on both the input and output sides of the production process, an emphasis on logistics may be appropriate. The potential for competitive advantage is found via both reduced costs and improved customer service.

Firms recognize that the distribution of goods to and from their facilities can represent as much as 25% of the cost of products. In addition, the total distribution cost in the U.S. is over 10% of the gross national product (GNP). Because of this high cost, firms constantly evaluate their means of distribution. Five major means of distribution are trucking, railroads, airfreight, waterways, and pipelines.

Distribution Systems

Trucking   The vast majority of manufactured goods moves by truck. The flexibility of shipping by truck is only one of its many advantages. Companies that have adopted JIT programs in recent years have put increased pressure on truckers to pick up and deliver on time, with no damage, with paperwork in order, and at low cost. Trucking firms are increasingly using computers to monitor weather, find the most effective route, reduce fuel cost, and analyze the most efficient way to unload. UPS chairman James Kelly describes his company as "a global conveyor belt," arranging JIT delivery of materials and data from anywhere in the world.[12]

Railroads   Railroads in the U.S. employ 250,000 people and ship 90% of all coal, 67% of autos, 68% of paper products, and about one-half of all food, lumber, and chemicals. Containerization has made intermodal shipping of truck trailers on railroad flat cars, often piggybacked as double-decks, a popular means of distribution. Over 4 million trailer loads are moved in the U.S. each year by rail. With the growth of JIT, however, rail transport has been the biggest loser because small-batch manufacture requires frequent, smaller shipments that are likely to move via truck or air.

Rapid response to customer requirements can make airfreight the carrier of choice even for heavy, bulky items such as the turbine shown here, as well as the preferred carrier for the international shipment of many lightweight, high-value items.

[12]"Overnight, Everything Changed for FedEx" the *Wall Street Journal* (November 4, 1999): A16.

OM IN ACTION

DHL's Role in the Supply Chain

It's the dead of night at DHL International's air express hub in Brussels, yet the massive building is alive with busy forklifts and sorting workers. The boxes going on and off the DHL plane range from Dell computers and Cisco routers to Caterpillar mufflers and Komatsu hydraulic pumps. Sun Microsystems computers from California are earmarked for Finland; CD-ROMS from Teac's plant in Malaysia are destined for Bulgaria.

The door-to-door movement of time-sensitive packages is the key to e-commerce, JIT, short product-life cycles, mass customization, reduced inventories, and the entire global supply chain. Global supply chains are in continuous motion, which is ideal for the air express industry.

With a decentralized network covering 227 countries and territories (more than are in the U.N.), DHL is a true multinational. The Brussels headquarters has only 450 of the company's 60,000 employees, but includes 26 nationalities.

DHL has assembled an extensive global network of express logistics centers for strategic goods. In its Brussels logistics center, for instance, DHL upgrades, repairs, and configures Fijitsu computers, InFocus projectors, and Johnson & Johnson medical equipment. It stores and provides parts for EMC and Hewlett-Packard and replaces Nokia and Philips phones. "If something breaks down on a Thursday at 4 o'clock, the relevant warehouse knows at 4:05, and the part is on a DHL plane at 7 or 8 that evening," says Robert Kuijpers, DHL International's CEO.

Sources: EBN (February 25, 2002): 27, and *Forbes* (October 18, 1999): 120–124.

Airfreight   Airfreight represents only about 1% of tonnage shipped in the U.S. However, the recent proliferation of airfreight carriers such as Federal Express, UPS, and DHL makes it the fastest growing mode of shipping. Clearly, for national and international movement of lightweight items such as medical and emergency supplies, flowers, fruits, and electronic components, airfreight offers speed and reliability. See the *OM in Action* box, "DHL's Role in the Supply Chain."

Waterways   Waterways are one of the nation's oldest means of freight transportation, dating back to construction of the Erie Canal in 1817. Included in U.S. waterways are the nation's rivers, canals, the Great Lakes, coastlines, and oceans connecting to other countries. The usual cargo on waterways is bulky, low-value cargo such as iron ore, grains, cement, coal, chemicals, limestone, and petroleum products. This distribution system is important when shipping cost is more important than speed.

Pipelines   Pipelines are an important form of transporting crude oil, natural gas, and other petroleum and chemical products. An amazing 90% of the state of Alaska's budget is derived from the 1.5 million barrels of oil pumped daily through the pipeline at Prudhoe Bay.

Cost of Shipping Alternatives

The longer a product is in transit, the longer the firm has its money invested. But faster shipping is usually more expensive than slow shipping. A simple way to obtain some insight into this trade-off is to evaluate carrying cost against shipping options. We do this in Example 3.

Example 3

A shipment of new connectors for semiconductors needs to go from San Jose to Singapore for assembly. The value of the connectors is $1,750.00 and holding cost is 40% per year. One airfreight carrier can ship the connectors 1 day faster than its competitor, at an extra cost of $20.00.

First we determine the daily holding cost.

$$\text{Daily cost of holding the product} = (\text{annual holding cost} \times \text{product value})/365$$
$$= (.40 \times \$1,750.00)/365$$
$$= \$1.92$$

Since the cost of saving one day is $20.00, which is much more than the daily holding cost of $1.92, we decide on the less costly of the carriers and take the extra day to make the shipment. This saves $18.08 ($20.00 − $1.92).

Note: The solution becomes radically different if the 1-day delay in getting the connectors to Singapore delays delivery (making a customer mad) or delays payment of a $150,000.00 final product. (Even 1 day's interest on $150,000.00 or a mad customer makes a savings of $18.08 insignificant.)

Example 3 only looks at holding costs versus shipping cost. For the operations or logistics manager there are many other considerations, including coordinating shipments to maintain a schedule, getting a new product to market, and keeping a customer happy.[13] There is no reason why estimates of these other costs cannot be added to the estimate of daily holding cost. Determining the impact and cost of these many other considerations is what makes the evaluation of shipping alternatives interesting.

BENCHMARKING SUPPLY-CHAIN MANAGEMENT

For most companies the percent of revenue spent on labor is going down, but the percent spent in the supply chain is going up.

As Table 11.5 shows, well-managed supply-chain relationships result in firms setting world-class benchmarks. Benchmark firms have driven down costs, lead times, late deliveries, and shortages, all while improving quality. Effective supply-chain management provides a competitive advantage by aiding firms in their response to a demanding global marketplace. Wal-Mart, for example, has developed a competitive edge through effective supply-chain management. With its own fleet of 2,000 trucks, 19 distribution centers, and a satellite communication system, Wal-Mart (with the help of its suppliers) replenishes store shelves an average of twice per week. Competitors resupply every other week. Economical and speedy resupply means high levels of product availability and reductions in inventory investment.

TABLE 11.5 ■

Supply-Chain
Performance
Compared

	TYPICAL FIRMS	BENCHMARK FIRMS
Administrative costs as percent of purchases	3.3%	.8%
Lead time (weeks)	15	8
Time spent placing an order	42 minutes	15 minutes
Percentage of late deliveries	33%	2%
Percentage of rejected material	1.5%	.0001%
Number of shortages per year	400	4

Source: Adapted from a McKinsey & Company report.

SUMMARY

A substantial portion of the cost and quality of the products of many firms, including most manufacturing, restaurant, wholesale, and retail firms, is determined by how efficiently they manage the supply chain. Supply-chain management provides a great opportunity for firms to develop a competitive advantage, often using e-commerce. Supply-chain management is an approach to working with suppliers that includes not only purchasing but also a comprehensive approach to developing maximum value from the supply chain. Five supply-chain strategies have been identified. They are (1) many suppliers, (2) few suppliers, (3) vertical integration, (4) *keiretsu* networks, and (5) virtual companies. Leading companies determine the right supply-chain strategy and often develop a logistics management organization to ensure effective warehousing and distribution.

KEY TERMS

Supply-chain management *(p. 326)*
Make-or-buy decision *(p. 329)*
Outsourcing *(p. 329)*
Vertical integration *(p. 330)*
Keiretsu (p. 332)
Virtual companies *(p. 332)*
Bullwhip effect *(p. 334)*
Pull data *(p. 334)*
Single stage control of replenishment *(p. 334)*
Vendor managed inventory (VMI) *(p. 334)*
Postponement *(p. 334)*

Channel assembly *(p. 335)*
Drop shipping *(p. 335)*
Blanket order *(p. 336)*
Standardization *(p. 336)*
Electronic data interchange (EDI) *(p. 336)*
Advanced Shipping Notice (ASN) *(p. 336)*
Internet purchasing (e-procurement) *(p. 336)*
Vendor selection *(p. 337)*
Negotiation strategies *(p. 338)*
Logistics management *(p. 339)*

[13]The cost of an unhappy customer can be equated to the stockout cost discussed in Chapter 12, "Inventory Management."

INTERNET AND STUDENT CD-ROM EXERCISES

Visit our home page or use your student CD-ROM to help with material in this chapter.

 On Our Home Page, www.prenhall.com/heizer

- Self-Tests
- Practice Problems
- Internet Exercises
- Current Articles and Research
- Virtual Company Tour
- Internet Homework Problem
- Internet Cases

 On Your Student CD-ROM

- PowerPoint Lecture
- Practice Problems
- Video Clip and Video Case

ADDITIONAL CASE STUDIES

Internet Case Studies: Visit our Web site at www.prenhall.com/heizer **for these free case studies:**

- **AT&T Buys a Printer**: The company needs to select printers from one of five suppliers.
- **Blue and Gray, Inc.**: This firm must decide whether to produce a part in-house or go to an outside vendor.
- **Factory Enterprises, Inc.**: The company is considering the advantages of a supply-chain management concept.
- **Thomas Manufacturing Company**: This firm is considering radical changes in its supply-chain/purchasing practices.

Harvard has selected these Harvard Business School cases to accompany this chapter (textbookcasematch.hbsp.harvard.edu**):**

- **Supply Chain Management at World Co. Ltd.** (#601-072): Illustrates the value of response times and how response times can be reduced.
- **Ford Motor Co.: Supply Chain Strategy** (#699-198): Evaluation of whether Ford should "virtually integrate" on the Dell Computer model.
- **Sport Obermeyer Ltd.** (#695-022): Examines how to match supply with demand for products with high demand uncertainty.
- **Barilla SpA (A)** (#694-046): Allows students to analyze how a company can implement a continuous replenishment system.

Supplement 11

E-Commerce and Operations Management

Supplement Outline

LEARNING OBJECTIVES

When you complete this supplement you should be able to

IDENTIFY OR DEFINE:

- E-commerce
- E-business
- Online catalogs
- Outsourcing
- E-procurement

DESCRIBE OR EXPLAIN:

- How E-commerce is changing the supply chain
- Online auctions
- Pass-through warehouses

Here a FreeMarkets team monitors an online auction from the firm's Global Market Operations Center in Pittsburgh. FreeMarkets provides support for the entire global sourcing process, including software, supplier development, competitive negotiations, and savings implementation. Online bidding through electronic commerce leads to greater cost savings than more traditional procurement.

THE INTERNET

Internet
An international computer network connecting people and organizations around the world.

The **Internet** is a revolutionary development for managing a firm's operations. This international computer network connects hundreds of millions of companies and people around the world. While the Internet's impact on our lives is only in its infancy, the impact on OM is already significant. Internet technology enables integration of traditional internal information systems as well as enhancement of communication between organizations. Internet-based systems tie together global design, manufacturing, delivery, sales, and after-services activities.

The Internet has reshaped how business thinks about delivering value to its customer, interacting with suppliers, and managing its employees. A prime benefit is speed, with managers able to make decisions with better information much more quickly than in the past.[1] Here are just a few examples of its applications:

- Customers visiting www.dell.com can configure, price, and order computer systems 24 hours a day, 7 days a week. They can get current order status and delivery information and have online access to the same technical reference materials used by Dell telephone-support teams.
- Lufthansa Airlines and Boeing have complete maintenance information on the Web for use by worldwide service personnel. Boeing also permits online access to most of its current engineering information.
- Integrated Technologies Ltd., a British manufacturer of medical diagnostic equipment, exchanges 3-D design models in real time with clients in Europe through a password-secured site on the Internet. This practice allows customers to not only review the technical aspects of products, but also to do more sophisticated analysis—like simulations of stress, or flow through a valve.
- Multinational robotics manufacturer NSK.RHP has built a Web-enabled factory (calling it a "cyberfactory") in which all machining centers are linked to the Internet. Machine operators use Microsoft Explorer's Internet search engine to access plant setup information and operating procedures, to take training, and to leave shift-to-shift-messages.
- Fast-food restaurants, like Burger King, are installing remote tracking systems. Managers can now check the time clock, review cash-register sales, or monitor refrigerator temperatures over the Internet. (See the *OM in Action* box, "Internet Keeps Burger King Manager in the Know.")

Intranet
An in-house Internet.

In a similar vein, Hallmark cards uses an in-house Internet, known as an **intranet**, to view images of previous popular cards, share artwork, and even route new cards to production. Hallmark is joining those business processes together so that there is never a handoff until design goes to manufacturing, where a plate is created for the printing press.

---

[1]The potential of the Internet/manufacturing interface is examined on the following Internet locations: http://www.isr.umd.edu and http://iac.dtic.mil.

OM IN ACTION

Internet Keeps Burger King Manager in the Know

Paul Bobo tracks Burger King performance over the Internet. With Burger Kings in Florida and Georgia, Mr. Bobo can now log onto the Internet and see how sales are going, double-check temperatures of the freezer and refrigerator, verify the fryer temperature, and view activity inside the restaurant. With electronic feeds from the cash register, wireless thermometers placed directly in food, and sensors on doors and mechanical equipment, the information on all three restaurants is at his fingertips. He can now make informed decisions even from his home. The system can also alert Mr. Bobo of any trouble through a computer, pager, cell phone, or other hand-held device such as a personal digital assistant (PDA). Thousands of fast-food franchises have adopted the technology.

Restaurants provide a signal from the point-of-sale terminals that updates sales information after every transaction. This allows managers to access real-time data at any time of the day or night. A slow drive-through dragging down sales or any change in the products or product mix is easily noted, allowing timely adjustments made to incoming orders. If sales slow on a certain day, a manager can call and send excess personnel home, shaving labor costs. The Internet-connected software is flexible enough that information sent to managers can be sorted or changed to meet individual preferences. Another feature is the use of Web cams, so activity in the restaurant can be monitored. If an employee is doing homework rather than working, is out of uniform, or is dipping into the cash register, the Web cam can pick it up.

Not only is the data real time, but it also removes the nightly grind of preparing the restaurant's sales report, allowing managers more time to focus on problem solving, working with employees, and building a better business.

Sources: The *Wall Street Journal* (August 30, 2001): B6; Apigent Solutions, Inc. (**www.apigent.com**); and *Franchising World* (July/August 2001): 14–16.

In-house use, technical collaboration, and transfer of information to and from the customer are making the Internet a powerful operations tool. Detailed global accessibility to engineering data/drawings, to inventory and suppliers, to ordering and order status, and to procedure and documentation are the new tools of the Internet age.

The Internet is proving a tremendous vehicle for OM change. The range of applications for the Internet seems limited only by our imagination and creativity. This high-speed network that spans most of the globe is available at a very reasonable cost, and its use is growing daily, with well over 30 million domains registered worldwide.[2]

Are companies that use the Internet more efficient? The answer is *yes*, as you will see throughout this supplement.

ELECTRONIC COMMERCE

E-commerce

The use of computer networks, primarily the Internet, to buy and sell products, services, and information.

E-commerce (or its synonym, e-business) is the use of computer networks, primarily the Internet, to buy and sell products, services, and information. The result of e-commerce is a great range of fast, low-cost electronic services. Although e-commerce implies information between businesses, the technology is equally applicable between business and consumers and indeed between consumers themselves. The business applications are evident across business activities, from tracking consumer behavior in marketing functions, to collaboration on product design in production functions, to speeding transactions in accounting functions. Former IBM chairman Louis Gerstner believes e-commerce is a whole new way of doing business and describes it as ". . . all about cycle time, speed, globalization, enhanced productivity, reaching new customers, and sharing knowledge across institutions for competitive advantage."[3] We begin with some definitions within e-commerce.

---

[2]Net Names International, Ltd; www.domainstats.com.

[3]E. Turban et al., *Electronic Commerce* (Upper Saddle River, NJ: Prentice Hall, 2000): 5.

E-commerce Definitions

Within the popular term e-commerce, four definitions are frequently used. They are based on the type of transaction taking place and are:

- *Business-to-business (B2B)*. This implies that both sides of the transaction are businesses, nonprofit organizations, or governments.
- *Business-to-consumer (B2C)*. These are e-commerce transactions in which buyers are individual consumers.
- *Consumer-to-consumer (C2C)*. Here consumers sell directly to each other by electronic classified advertisements or auction sites.
- *Consumer-to-business (C2B)*. In this category individuals sell services or goods to businesses.

These four types of transactions are shown in Figure S11.1.

Our focus in this supplement is business-to-business e-commerce. The B2B segment of e-commerce has grown to over $1 trillion in the U.S. and constitutes about 80% of the e-commerce market. Table S11.1 lists the types of data we can expect to find in B2B applications.

ECONOMICS OF E-COMMERCE

E-commerce is revolutionizing operations management because it reduces costs so effectively. It reduces costs by improving communication and disseminating economically valuable information. The new middleman driving down transaction costs is the e-commerce provider. This middleman is cheaper and faster than the traditional broker. E-commerce increases economic efficiencies by matching buyers and sellers. It facilitates the exchange of information, goods, and services. These added efficiencies reduce costs for everyone; they also reduce barriers to entry. E-commerce opens both large and small organizations to economies not previously available. Perfect information is a big contributor to efficiency, and e-commerce is moving us a bit closer to what economists call perfect markets.

- Product—drawings, specifications, video, or simulation demonstrations, prices.
- Production Processes—capacities, commitments, product plans.
- Transportation—carrier availability, lead times, costs.
- Inventory—inventory tracking, levels, costs, and location.
- Suppliers—product catalog, quality history, lead times, terms, and conditions.
- Supply Chain Alliances—key contact, partners' roles and responsibilities, schedules.
- Supply Chain Process and Performance—process descriptions, performance measures such as quality and delivery.
- Competitor—benchmarking, product offerings, market share.
- Sales and Marketing—point of sale (POS) data entry, promotions, pricing, discounts.
- Customer—sales history and forecasts.
- Costs—market indexes, auction results.

In addition, the time constraints inherent in many transactions all but disappear. The firm or individual at the other end of the transaction need not always be immediately available. The convenience to both parties is improved because cheap electronic storage is built into e-commerce systems. Information, transactions, and creativity in the way we communicate have never been easier or cheaper.

Honeywell's Consumer Products Group, for example, used to have 28 employees who took orders by phone or fax. As 4,000 corporate customers shifted to online ordering, the employees were reassigned to other jobs such as outside sales, increasing labor productivity enormously.[4] This is but one of the benefits of e-commerce listed in Table S11.2.

TABLE S11.2 ■

Benefits and Limitations
of E-Commerce

Benefits of E-Commerce
- Improved, lower-cost information that makes buyers and sellers more knowledgeable has an inherent power to drive down costs.
- Lower entry costs increase information sharing.
- Available 24 hours a day, virtually any place in the world, enabling convenient transactions for those concerned.
- Availability expands the market for both buyers and sellers.
- Decreases the cost of creating, processing, distributing, storing, and retrieving paper-based information.
- Reduces the cost of communication.
- Richer communication than traditional paper and telephone communication because of video clips, voice, and demonstrations.
- Fast delivery of digitized products such as drawings, documents, and software.
- Increased flexibility of locations. (That is, it allows some processes to be located anywhere electronic communication can be established, and allows people to shop and work from home.)

Limitations of E-Commerce
- Lack of system security, reliability, and standards.
- Lack of privacy.
- Insufficient band width; some transactions are still rather slow.
- Integrating e-commerce software with existing software and databases is still a challenge.
- Lack of trust in (1) unknowns about the integrity of those on the other end of a transaction, (2) integrity of the transaction itself, and (3) electronic money that is only bits and bytes.

Sources: For related discussions see **www.capsresearch.org**; S. Chopra and P. Meindl, *Supply Chain Management* (Upper Saddle River, NJ: Prentice Hall, 2001); or D. Neef, *e-Procurement* (Upper Saddle River, NJ: Prentice Hall, 2001).

PRODUCT DESIGN

Shorter life cycles require faster product development cycles and lead to time-based competition (a topic we addressed in Chapter 5). E-commerce is accelerating time-based competition even more. However, the operations manager is finding that e-commerce collaboration in product and process design by virtual teams may not only be cheaper, but may also yield better and quicker decisions. Members of product teams in different locations can now easily share knowledge at low cost. At General Electric, for example, engineers in 100 countries now share ideas and information, and work on projects simultaneously. Based on 600 projects completed so far using the Web, GE estimates the time it takes to develop a new product has been cut 20%.

Operations managers are also addressing this acceleration by managing product data over the Internet. New communication and collaboration tools allow engineering changes and configuration management to extend to the supply chain. Accurate data to suppliers, subcontractors, and strategic partners become more important with globalization and extended supply chains. The complexity of managing product development and product definition increases as design responsibilities shift away from a central team to dispersed product development teams worldwide. E-commerce, with the rapid transfer of specifications, 3-dimensional drawings, and speedy collaboration, eases the task.

General Motors, as an example, is tying thousands of suppliers into its electronic engineering and design network. The Web-based network will let GM's suppliers work online in real time with

---

[4]*New York Times* (June 7, 2000): E1, E18.

its designers, creating and editing 3-D CAD models. In the past, suppliers worked from static blueprints and engineering schematics and waited for printed updates. Now they get real-time updates to designs online.[5]

E-PROCUREMENT

E-procurement

Purchases and order releases communicated over the Internet or to approved vendor online catalogs.

Online catalog

Electronic (Internet) presentation of products that were traditionally presented in paper catalogs.

Modern procurement is often e-procurement. **E-procurement**, as we saw in Chapter 11, is purchasing or order release communicated over the Internet or via approved online vendor catalogs.

Online Catalogs

Online catalogs are information about products in electronic form via the Internet. They are quickly improving cost comparison and bidding processes. These electronic catalogs can enrich traditional catalogs by incorporating voice and video clips, much as does the CD-ROM that accompanies this text. Online catalogs are available in three versions: (1) those provided by vendors, (2) those developed by intermediaries, and (3) those provided by buyers. We now discuss each.

Online Catalog Provided by Vendor   Among those catalogs provided by vendors is that of W. W. Grainger. W. W. Grainger (www.Grainger.com) is probably the world's largest seller of MRO items (items for maintenance, repair, and operations). Grainger needed to take care of frequent, relatively low dollar, purchases by buyers looking for very specific ways to fill their needs. Rather than treat its Web site as just another toll-free line, Grainger rethought its business model and moved its 4,000-page catalog online. The online catalog is integrated into the company's sales and service agenda, making it easy to use and informative, with relevant pricing and product availability. Customized versions reflect discounts applicable to each customer. Moreover, the system takes orders 24 hours a day rather than just when Grainger's stores are open. Sales personnel are rewarded for online sales just as they are for telephone or over-the-counter sales. These catalogs are a win–win for the operations manager (the customer) and for Grainger. Operations personnel find the online catalog easier to use and available whenever they need it. Moreover, the average order size is up almost 50%.[6]

Online catalogs are often available on every employee's desktop computer. Once approved and established, each employee can do his or her own purchasing. In many process industries, maintenance, repair, and operations items account for a substantial portion of the sales dollar. Many of these purchases are individually small dollar value, and as such fail to receive the attention of other "normal" purchases. The result is a huge inefficiency. Online e-commerce provides an opportunity for substantial savings; plus, paper trails related to ordering become less-expensive electronic trails. Operations managers obtain convenience while purchasing department costs decrease.

"... e-procurement ... integrates supply chains between different buyers and sellers, and makes a company's supply chain a key competitive advantage."

Robert Derocher
Deloitte Consulting

Online Catalog Provided by Intermediaries   Intermediaries are companies that run a site where business buyers and sellers can meet. Typical of these is ProcureNet (www.procurenet.com). ProcureNet has combined 30 seller sites with over 100,000 parts for the electronics industry. Qualified buyers can place orders with selling companies. Boeing Aircraft, for example, maintains an intermediary site for its customers. On this Web site, called Boeing PART (Part Analysis and Requirements Tracking), 500 Boeing customers can place orders for replacement parts, many of which are drop shipped from suppliers. The cost is significantly less than traditional faxes, telephone calls, and purchase orders.

Online Exchange Provided by Buyer   The three auto giants, GM, Ford, and DaimlerChrysler, created the first major online Internet trading exchange, called Covisint, to buy virtually everything from paper clips to stamping presses to contract manufacturing. (See the *OM in Action* box, "The Face of Covisint's Online Exchange.") Virtually every other industry quickly followed. A few of these online bazaars are shown in Table S11.3.

The first exchange listed in Table S11.3, the Global Health Care Exchange, is typical of the others. Its model is shown in Figure S11.2. In a traditional hospital, the current supply chain (on the left) starts in the purchasing department. Shelves of catalogs, filled with information and prices that could be years out of date, line the walls. Orders flow by fax or phone to thousands of distributors

---

[5]*Informationweek.com* (April 15, 2002): 24.

[6]Michael E. Porter, "Strategy and the Internet," *Harvard Business Review* 79, no. 3 (March, 2001): 63–78.

OM IN ACTION

The Face of Covisint's Online Exchange

Imagine being 24 years old and responsible for a $10-million order for car-axle assemblies. That's what Aaron Gillum does every day. He is what is known as an auction engineer for Covisint, the B2B online exchange created in 2000 by the Big Three Automakers—Ford, GM, and DaimlerChrysler—and later joined by Nissan, Renault, and Peugeot. Covisint's role is to move part purchasing online to save costs and speed up the buying cycle.

Auction engineers like Mr. Gillum (who works only on purchases for DaimlerChrysler) handle almost $100 billion in auto-parts orders per year. In its biggest auction ever, Covisint conducted a 4-day auction for DaimlerChrysler in which the automaker purchased $2.6 billion in parts.

In the old days, when fax machines were the cutting edge of technology, an auto-supply auction could take from several days to several weeks. The auto manufacturer would invite several suppliers to bid by mail or fax. Of course, bidders had no idea what competitors were bidding, and purchasing managers often asked for a second round of bids. Now, an auction can be wrapped up in as little as 10 minutes, although the average time is 45 minutes. Once an auction starts, suppliers can place their bids and instantly see what others are bidding, so they can adjust their own price. It is not unusual for an auto manufacturer to save 40% of the opening price on an item.

Sources: New York Times (February 11, 2002): R126; and *Automotive News* (April 8, 2002): 6, and (January 14, 2002): 33.

and manufacturers. Some hospitals and suppliers are linked electronically, but these still usually require manual intervention. At the other end of the chain are the makers or distributors of everything from heart valves to toilet paper. The new online exchange, on the right, puts downward pressure on price and improves transaction efficiency.

TABLE S11.3 ■

Internet Trading Exchanges

Health care products—set up by Johnson & Johnson, GE Medical Systems, Baxter International, Abbott Laboratories, and Medtronic Inc; called the Global Health Care Exchange (ghx.com).
Defense and aerospace products—created by Boeing, Raytheon, Lockheed-Martin, and Britain's BAE Systems; called the Aerospace and Defense Industry Trading Exchange (exostar.com).
Food, beverage, consumer products—set up by 49 leading food and beverage firms; called Transora (transora.com).
Retail goods—set up by Sears and France's Carrefour; called Global Net Xchange for retailers (gnx.com).
Steel and metal products—such as New View Technologies (exchange.e-steel.com) and Metal-Site (metalsite.com).
Hotels—created by Marriott and Hyatt, and later joined by Fairmont, Six Continents, and Club Corp: called Avendra (avendra.com) buys for 2,800 hotels.

Sources: New York Times (February 11, 2002): R16; and the *Wall Street Journal* (May 23, 2002): D8.

FIGURE S11.2 ■

The Medical Supply Chain Goes Online

An estimated $11 billion of waste exists in the current medical supply chain. Efforts are underway to streamline systems and reduce costs with online exchanges.
Sources: Adapted from the *Wall Street Journal* (February 28, 2000): B4; and Chase H&Q industry study.

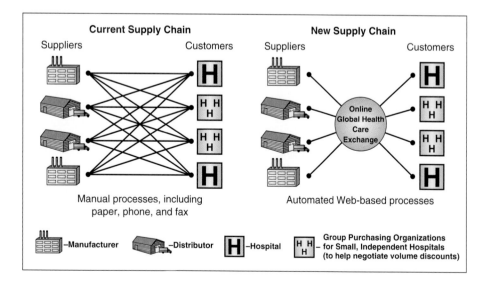

RFQs and Bid Packaging

The cost of preparing requests for quotes (RFQs) can be substantial; consequently, e-commerce has found another area ripe for improvement. General Electric, for example, has been able to make major advances in this aspect of its procurement process. Purchasing personnel now have access to an extensive database of vendor, delivery, and quality data. With this extensive history, supplier selection for obtaining quotes has improved. Electronic files containing engineering drawings are also available. The combination allows purchasing agents to attach electronic copies of the necessary drawings to RFQs and send the entire electronic-encrypted package to vendors in a matter of hours rather than days. The system is both faster, by about 3 weeks, and less expensive.

Internet Outsourcing

Internet outsourcing
The transfer of an organization's activities that have traditionally been internal to Internet suppliers.

Creative organizations are proving the Internet's versatility by supplying business processes such as payroll, accounting, and human resource services via the Internet. **Internet outsourcing** transfers an organization's activities that have traditionally been internal to Internet suppliers. Firms that want to outsource their noncore human resource function can find organizations such as Employease (Employease.com), which will provide the service via the Internet. Other companies handle employee benefits (www.OnlineBenefits.com). These and similar firms duplicate some or all of an internal human resource function on a Web site. With various restrictions, supervisors, employees, and human resource personnel have access to appropriate information. Other outsourcing possibilities include travel (TheTrip.com), document management (CyLex.com), and shipping (FedEx.com).

Online Auctions

Online auction sites can also be maintained by sellers, buyers, or intermediaries. General Motors' approach to selling excess steel is to post it on the Web and expect that its own suppliers who need steel to buy it from GM. This is a forerunner of business-to-business (B2B) auctions that every industry can be expected to pursue. Operations managers find online auctions a fertile area for disposing of excess raw material and discontinued or excess inventory. FreeMarkets, as discussed in the *OM in Action* box, maintains an intermediary auction site. Online auctions lower entry barriers and simultaneously increase the potential number of customers.

OM IN ACTION

FreeMarkets' B2B Model for E-procurement

The consumer side of Internet shopping is rising, with no end in sight. But the B2B side of purchasing is where the action is. Glen Meakem, CEO of FreeMarkets Online (www.freemarkets.com), a Pittsburgh industrial e-commerce company, says that live online bidding and better-informed purchasing decisions lead to greater cost savings than does more traditional purchasing. Meakem notes that buyers save an average of 16% through online auctions.

Consider this case: United Technologies Corp. needed suppliers to make $24 million worth of circuit boards. FreeMarkets evaluated 1,000 potential suppliers and invited 50 especially qualified ones to bid. Three hours of online competitive bidding was conducted. FreeMarkets divided the job into 12 lots, each of which was put up to bid. At 8 A.M., the first lot, valued at $2.25 million, was placed on the Net. The first bid was $2.25 million, which was seen by all. Minutes later, another bidder placed a $2-million bid. Further bidders reduced the price again. Minutes before the bid closed on the first lot, at 8:45 A.M., the 42nd bidder placed a $1.1-million bid. When it all ended, the bids for all 12 lots totaled $18 million (about 35% savings to United Technologies).

FreeMarkets then analyzed the bidders, recommended the winners, and collected its fees (up to 2.5% commission) from both buyer and suppliers.

Sources: Internet Week (February 12, 2001): 11; the Wall Street Journal (January 5, 2000): B1; and New York Times (June 7, 2000): E1, E18.

E-commerce is supported by bar-code tracking of shipments. At each step of a journey, from initial pick-up to final destination, bar-codes are read and stored (left). Within seconds, this tracking information is available online to customers worldwide (right).

The key for auction firms such as FreeMarkets is to find and build a huge base of potential bidders—indeed, half of FreeMarket's employees spend their time not running electronic auctions, but qualifying new suppliers. For the operations manager, the implications of this approach to procurement are significant—and the supply chain now requires a new set of skills.

From our discussion of supply chains in Chapter 11, you may recall that many firms spend over half of their sales dollar on purchases. Preliminary estimates of 10% savings in procurement through these e-commerce options may be conservative. Recently, Honeywell's avionics group saved $400,000, or 19.5%, on the purchase of $1.7 million in parts. Carrier Corp. saved 16% on air-conditioner motors, using Asian suppliers. Sun Microsystems claims savings of over $1 billion a year using its in-house reverse auction system (called Dynamic Bidding): The firm now spends 1 hour pricing out items that used to take weeks or months to negotiate.[7]

INVENTORY TRACKING

FedEx's pioneering efforts at tracking packages from pickup to delivery have shown the way for operations managers to do the same for their shipments and inventory. (See the *Global Company Profile* discussing FedEx's worldwide network in Chapter 8). Surely if FedEx can track millions of documents each day worldwide, operations managers in other firms can also do so. The tools of e-commerce, including the discipline of data collection, bar code technology, radio frequency, and electronic communications to track inventory in transit, on the shop floor, and in the warehouse are now perfected and available to the resourceful operations manager.

Ford has hired UPS to track vehicles as they move from factory to dealers. Tracking cars and trucks has been an embarrassingly inexact science for years. Ford's tracking system is expected to track more than 4 million Ford cars and trucks each year. Using bar codes and the Internet, dealers are able to log onto the Web site and find out exactly where the vehicles they have ordered are in the distribution system. As operations managers move to an era of mass customization, with each customer ordering exactly the car he or she wants, customers will expect to know where their car is and exactly when they can pick it up. The Internet and e-commerce can provide this service and do so economically.

INVENTORY REDUCTION

Warehousing for E-commerce

The new e-commerce warehouse is not run by the producer, but by the logistics vendor. As the *OM in Action* box "In E-Commerce, 'Pass-Through Facilities' Replace Warehouses" suggests, it is less a warehouse than a "pass-through facility." Working with United Parcel Service, Nike, Inc. uses

[7]G. Verga, "Reverse Auctions," *APICS—The Performance Advantage* (January 2002): 28–31.

OM IN ACTION

In E-Commerce, "Pass-Through Facilities" Replace Warehouses

The two buildings spread themselves over 160,000 square feet—more than two football fields—on the south side of Portland, Oregon's airport. As an answer to a vexing issue in e-commerce, namely, actually delivering the goods ordered, these "pass-through facilities" (formerly called warehouses) help make the product available. For e-commerce retailers, selling an item online is the easy part; getting it quickly to the customer is a lot trickier.

A pass-through facility is conceived less as a holding area than as a transportation hub. The massive, automated conveyers and storage equipment are intended to speed orders in and out. Targeted by corporations that deal with high volume and quick delivery, customers include "integrators" such as UPS, Emery, DHL, and Fed Ex, as well as freight forwarders, logistics companies, and airlines.

The buildings in a pass-through facility are configured to make it easy for loading equipment to maneuver. With direct access to taxiways, four 747s and 68 large trucks are able to load and unload at the Portland buildings at the same time. "The closer you can get your product to the air cargo center, the quicker your delivery time. The only way e-commerce is going to compete effectively with local retailers is if a customer logs on and orders something, and they can get it in a day," says Steven Bradford, VP of the facility's developer, Trammell-Crow.

Sources: New York Times (January 23, 2000): B-3; *Transportation & Distribution* (January 2000): 42–52; and *Consulting-Specifying Engineer* (June 2000): 30–34.

---

such a facility in Louisville, Kentucky, to handle online orders. And FedEx's warehouse park next to the airport in Memphis can receive an order after a store closes for the evening and locate, package, and ship the merchandise that night. Delivery is guaranteed by 10 A.M. the next day.

Just-in-Time Delivery for E-Commerce

Just-in-time systems in manufacturing (see Chapter 16) are based on the premise that parts and materials will be delivered exactly on time. Electronic commerce can support this goal by coordinating the supplier's inventory system with the service capabilities of the delivery firm.

FedEx has a short but successful history of using the Internet for online tracking in the world of e-commerce. In 1996 the firm launched FedEx InterNetShip, which within 18 months had 75,000 customers. A FedEx.com customer today can compute shipping costs, print labels, adjust invoices, and track package status all on the same Web site. (FedEx, by the way, saves $3 for each inquiry made via the Web compared to a phone call). FedEx also plays a core role in other firms' logistics processes. In some cases, FedEx runs the server for retailer Web sites. In other cases, such as for Dell Computer, it operates warehouses that pick, pack, test, and assemble products, then handle delivery and even customs clearance. FedEx's B2B service, called "Virtual Order," integrates different companies' Web catalogs and customer orders for Dell. FedEx then fulfills orders and delivers them via its fleet of trucks and planes. FedEx is effectively demonstrating that an e-commerce service company can economically manage complex transactions for other companies.

SCHEDULING AND LOGISTICS IMPROVEMENTS

Coordinated Pickup and Delivery

FedEx maintains a unified view of the data residing in different parts of its network so it can better track and coordinate orders for its end customer. This led the firm to the new model of coordinated pickup and delivery. It works like this: Cisco and FedEx established an alliance in which FedEx picks up and delivers Cisco components where needed, when needed. E-commerce allows FedEx to know where each piece of Cisco's shipments is headed and when it will be ready for shipment. FedEx then delivers the items precisely when and where they are needed for assembly and installation. FedEx merges the orders in transit. The components never go to a warehouse. The economies are found in the reduction of in-transit inventory and having the components present when needed—no sooner and no later. The amount of time items are in the distribution system is reduced, as is the quantity of items in the system. These techniques are shrinking delivery and installation time while reducing costs.

Logistics Cost Reduction

Recent data indicate that the motor carrier industry averages a capacity utilization of only 50%. That underutilized space costs the U.S. economy over $31 billion per year. To improve logistics efficiency, Schneider National established a Web site (Schneider Connection at www.schneider.com) that lets shippers and truckers match up to use some of this idle capacity. Shippers may pick from thousands of approved North American carriers that have registered with Schneider Connection. The opportunity for operations managers to use e-commerce technology to reduce logistics cost is substantial.

SUMMARY

E-commerce is revolutionizing the way operations managers achieve greater efficiencies. Economical collaboration can improve decision making and reduce costs. Cost reduction can occur in transaction processing, purchasing efficiencies, inventory reduction, scheduling, and logistics. The opportunities are amazing. Getting up to e-commerce speed is not an option. Stragglers won't just be left behind—they will be eliminated. Operations personnel who use e-commerce to their advantage will overpower their rivals.

KEY TERMS

Internet (p. 344)
Intranet (p. 344)
E-commerce (p. 345)

E-procurement (p. 348)
Online catalog (p. 348)
Internet outsourcing (p. 350)

INTERNET AND STUDENT CD-ROM EXERCISES

Visit our home page or use your student CD-ROM to help with material in this supplement.

 On Our Home Page, www.prenhall.com/heizer

- Self-Tests
- Internet Case Studies
- Internet Exercises
- Current Articles and Research

 On Your Student CD-ROM

- PowerPoint Lecture

ADDITIONAL CASE STUDIES

Internet Case Studies: Visit our Web site at www.prenhall.com/heizer **for these free case studies:**

- **Cisco's E-Commerce Connection**: Discusses the impact of the Internet on Cisco's operations functions.

- **Fruit of the Loom Tries E-Commerce**: Examines how Fruit of the Loom integrated its distributors into e-commerce.

These Harvard Business School cases accompany this supplement (textbookcasematch.hbsp.harvard.edu)**:**

- **H. E. Butt Grocery Co: The New Digital Strategy (A)** (#300-106): Examines how this company's supply chain has moved to e-commerce.

- **Webvan** (#602-037): Discusses the processes by which Webvan delivers groceries to customers' homes.

- **Cisco Systems: Building Leading Internet Capabilities** (#301-133): Cisco's efforts to broaden its Internet use are explored.

Inventory Management

Chapter Outline

LEARNING OBJECTIVES

When you complete this chapter you should be able to

IDENTIFY AND DEFINE:

ABC analysis

Record accuracy

Cycle counting

Independent and dependent demand

Holding, ordering, and setup costs

DESCRIBE OR EXPLAIN:

The functions of inventory and basic inventory models

GLOBAL COMPANY PROFILE:

Inventory Management Provides Competitive Advantage at Amazon.com

When Jeff Bezos opened his revolutionary business in 1995, Amazon.com was intended to be a "virtual" retailer—no inventory, no warehouses, no overhead—just a bunch of computers taking orders and authorizing others to fill them. Things clearly didn't work out that way. Now Amazon stocks millions of items of inventory, amid hundreds of thousands of bins on metal shelves, in warehouses around the country that have twice the floor space of the Empire State Building.

Precisely managing this massive inventory has forced Amazon into becoming a world-class leader in warehouse management and automation. This profile shows what goes on behind the scenes. When you place an order at Amazon.com, you are not only doing business with an Internet company, you are doing business with a company that obtains competitive advantage through inventory management.

Sources: New York Times (January 21, 2002): C-3; *Time* (December 27, 1999): 68–73; and the *Wall Street Journal* (November 22, 2002): A1, A6.

1. You order three items, and a computer in Seattle takes charge. A computer assigns your order—a book, a game, and a digital camera—to one of Amazon's massive U.S. distribution centers, such as the 800,000 square foot facility in McDonough, Georgia.
2. The "flow meister" in McDonough receives your order (right). *She determines which workers go where to fill your order.*

3. Rows of red lights show which products are ordered (left). *Workers move from bulb to bulb, retrieving an item from the shelf above and pressing a button that resets the light. This is known as a "pick-to-light" system. This system doubles the picking speed of manual operators and drops the error rate to nearly zero.*
4. Your items are put into crates on moving belts (below). *Each item goes into a large green crate that contains many customers' orders. When full, the crates ride a series of conveyor belts that winds more than 10 miles through the plant at a constant speed of 2.9 feet per second. The bar code on each item is scanned 15 times, by machines and by many of the 600 workers. The goal is to reduce errors to zero—returns are very expensive.*

5. *All three items converge in a chute, and then inside a box.* All of the crates arrive at a central point where bar codes are matched with order numbers to determine who gets what. Your three items end up in a 3-foot-wide chute—one of several thousand—and are placed into a cardboard box with a new bar code that identifies your order. Picking is sequenced to reduce operator travel.

6. *Any gifts you've chosen are wrapped by hand* (left). Amazon trains an elite group of gift wrappers, each of whom processes 30 packages an hour.

7. *The box is packed, taped, weighed, and labeled before leaving the warehouse in a truck* (left). The McDonough plant was designed to ship as many as 200,000 pieces a day. About 60% of orders are shipped via the U.S. Postal Service; nearly everything else goes through United Parcel Service.

8. *Your order arrives at your doorstep.* Within a week, your order is delivered.

As Amazon.com wells knows, inventory is one of the most expensive assets of many companies, representing as much as 50 percent of total invested capital. Operations managers around the globe have long recognized that good inventory management is crucial. On the one hand, a firm can reduce costs by reducing inventory. On the other hand, production may stop and customers become dissatisfied when an item is out of stock. Thus, companies must strike a balance between inventory investment and customer service. You can never achieve a low-cost strategy without good inventory management.

All organizations have some type of inventory planning and control system. A bank has methods to control its inventory of cash. A hospital has methods to control blood supplies and pharmaceuticals. Government agencies, schools, and, of course, virtually every manufacturing and production organization are concerned with inventory planning and control.

In cases of physical products, the organization must determine whether to produce goods or to purchase them. Once this decision has been made, the next step is to forecast demand, as discussed in Chapter 4. Then operations managers determine the inventory necessary to service that demand. In this chapter, we discuss the functions, types, and management of inventory. We then address two basic inventory issues: how much to order and when to order.

FUNCTIONS OF INVENTORY

Inventory can serve several functions that add flexibility to a firm's operations. The four functions of inventory are

1. To *"decouple" or separate various parts of the production process*. For example, if a firm's supplies fluctuate, extra inventory may be necessary to decouple the production process from suppliers.
2. To *decouple the firm from fluctuations in demand* and *provide a stock of goods that will provide a selection for customers*. Such inventories are typical in retail establishments.
3. To *take advantage of quantity discounts*, because purchases in larger quantities may reduce the cost of goods or their delivery.
4. To *hedge against inflation* and upward price changes.

Types of Inventory

Raw material inventory
Materials that are usually purchased but have yet to enter the manufacturing process.

Work-in-process inventory (WIP)
Products or components that are no longer raw material but have yet to become finished products.

To accommodate the functions of inventory, firms maintain four types of inventories: (1) raw material inventory, (2) work-in-process inventory, (3) maintenance/repair/operating supply (MRO) inventory, and (4) finished goods inventory.

Raw material inventory has been purchased but not processed. This inventory can be used to decouple (i.e., separate) suppliers from the production process. However, the preferred approach is to eliminate supplier variability in quality, quantity, or delivery time so that separation is not needed. **Work-in-process (WIP) inventory** is components or raw material that have undergone some change but are not completed. WIP exists because of the time it takes for a product to be made (called *cycle time*). Reducing cycle time reduces inventory. Often this task is not difficult: During most of the time a product is "being made," it is in fact sitting idle. As Figure 12.1 shows, actual work time or "run" time is a small portion of the material flow time, perhaps as low as 5%.

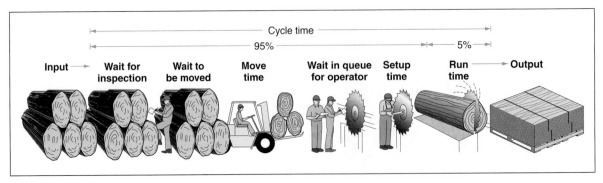

FIGURE 12.1 ■ The Material Flow Cycle

Most of the time that work is in-process (95% of the cycle time) is not productive time.

MRO
Maintenance, repair, and operating materials.

Finished goods inventory
An end item ready to be sold, but still an asset on the company's books.

MROs are inventories devoted to **maintenance/repair/operating** supplies necessary to keep machinery and processes productive. They exist because the need and timing for maintenance and repair of some equipment are unknown. Although the demand for MRO inventories is often a function of maintenance schedules, other unscheduled MRO demands must be anticipated. **Finished goods inventory** is completed product awaiting shipment. Finished goods may be inventoried because future customer demands are unknown.

INVENTORY MANAGEMENT

Operations managers establish systems for managing inventory. In this section, we briefly examine two ingredients of such systems: (1) how inventory items can be classified (called *ABC analysis*) and (2) how accurate inventory records can be maintained. We will then look at inventory control in the service sector.

ABC Analysis

ABC analysis
A method for dividing on-hand inventory into three classifications based on annual dollar volume.

ABC analysis divides on-hand inventory into three classifications on the basis of annual dollar volume.[1] ABC analysis is an inventory application of what is known as the Pareto principle. The Pareto principle states that there are a "critical few and trivial many."[2] The idea is to establish inventory policies that focus resources on the *few critical* inventory parts and not the many trivial ones. It is not realistic to monitor inexpensive items with the same intensity as very expensive items.

To determine annual dollar volume for ABC analysis, we measure the *annual demand* of each inventory item times the *cost per unit*. Class A items are those on which the annual dollar volume is high. Although such items may represent only about 15% of the total inventory items, they represent 70% to 80% of the total dollar usage. *Class B* items are those inventory items of medium annual dollar volume. These items may represent about 30% of inventory items and 15% to 25% of the total value. Those with low annual dollar volume are *Class C*, which may represent only 5% of the annual dollar volume but about 55% of the total inventory items.

Graphically, the inventory of many organizations would appear as presented in Figure 12.2. An example of the use of ABC analysis is shown in Example 1.

Example 1

Excel OM Data File Ch12Ex1.xla

The breakdown into A, B, C categories is not a hard-and-fast rule. The objective is just to try to separate the "important" from the "unimportant."

Silicon Chips, Inc., maker of superfast DRAM chips, has organized its 10 inventory items on an annual dollar-volume basis. Shown below are the items (identified by stock number), their annual demand, unit cost, annual dollar volume, and the percentage of the total represented by each item. In the table below, we show these items grouped into ABC classifications:

ABC CALCULATION

ITEM STOCK NUMBER	PERCENT OF NUMBER OF ITEMS STOCKED	ANNUAL VOLUME (UNITS)	×	UNIT COST	=	ANNUAL DOLLAR VOLUME	PERCENT OF ANNUAL DOLLAR VOLUME		CLASS
#10286	20%	1,000		$ 90.00		$ 90,000	38.8%	72%	A
#11526		500		154.00		77,000	33.2%		A
#12760	30%	1,550		17.00		26,350	11.3%	23%	B
#10867		350		42.86		15,001	6.4%		B
#10500		1,000		12.50		12,500	5.4%		B
#12572	50%	600		$14.17		8,502	3.7%	5%	C
#14075		2,000		.60		1,200	.5%		C
#01036		100		8.50		850	.4%		C
#01307		1,200		.42		504	.2%		C
#10572		250		.60		150	.1%		C
		8,550				$232,057	100.0%		

[1]H. Ford Dickie is given credit for developing this technique. See H. Ford Dickie, *Modern Manufacturing* (formerly *Factory Management and Maintenance*) (July 1951).

[2]After Vilfredo Pareto, nineteenth-century Italian economist.

FIGURE 12.2 ■

Graphic Representation
of ABC Analysis

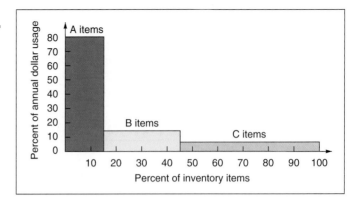

Most automated inventory
management systems
include ABC analysis.

Criteria other than annual dollar volume can determine item classification. For instance, anticipated engineering changes, delivery problems, quality problems, or high unit cost may dictate upgrading items to a higher classification. The advantage of dividing inventory items into classes allows policies and controls to be established for each class.

Policies that may be based on ABC analysis include the following:

1. Purchasing resources expended on supplier development should be much higher for individual A items than for C items.
2. A items, as opposed to B and C items, should have tighter physical inventory control; perhaps they belong in a more secure area, and perhaps the accuracy of inventory records for A items should be verified more frequently.
3. Forecasting A items may warrant more care than forecasting other items.

Better forecasting, physical control, supplier reliability, and an ultimate reduction in safety stock can all result from appropriate inventory management policies. ABC analysis guides the development of those policies.

Record Accuracy

Good inventory policies are meaningless if management does not know what inventory is on hand. Accuracy of records is a critical ingredient in production and inventory systems. Record accuracy allows organizations to focus on those items that are needed, rather than settling for being sure that "some of everything" is in inventory. Only when an organization can determine accurately what it has on hand can it make precise decisions about ordering, scheduling, and shipping.

To ensure accuracy, incoming and outgoing record keeping must be good, as must be stockroom security. A well-organized stockroom will have limited access, good housekeeping, and storage areas that hold fixed amounts of inventory. Bins, shelf space, and parts will be labeled accurately. The U.S. Marines' approach to improved inventory record accuracy is discussed in the *OM in Action* box, "What the Marines Learned about Inventory from Wal-Mart."

Video 12.1

Inventory Control at
Wheeled Coach
Ambulance

Cycle Counting

Cycle counting
A continuing reconciliation
of inventory with inventory
records.

Even though an organization may have made substantial efforts to record inventory accurately, these records must be verified through a continuing audit. Such audits are known as **cycle counting**. Historically, many firms performed annual physical inventories. This practice often meant shutting down the facility and having inexperienced people count parts and material. Inventory records should instead be verified via cycle counting. Cycle counting uses inventory classifications developed through ABC analysis. With cycle counting procedures, items are counted, records are verified, and inaccuracies are periodically documented. The cause of inaccuracies is then traced and appropriate remedial action taken to ensure integrity of the inventory system. A items will be counted frequently, perhaps once a month; B items will be counted less frequently, perhaps once a quarter; and C items will be counted perhaps once every 6 months. Example 2 illustrates how to compute the number of items of each classification to be counted each day.

OM IN ACTION

What the Marines Learned about Inventory from Wal-Mart

The U.S. Marine Corps knew it had inventory problems. A few years ago, when a soldier at Camp Pendleton, near San Diego, put in an order for a spare part, it took him a week to get it—from the other side of the base. Worse, the force had 207 computer systems worldwide. Called the "Rats' Nest" by marine techies, most systems didn't even talk to each other.

To execute a victory over uncontrolled supplies, the corps studied Wal-Mart, Caterpillar, Inc., and UPS. "We're in the middle of a revolution," says General Gary McKissock. McKissock aims to reduce inventory for the corps by half, saving $200 million, and to shift 2,000 marines from inventory detail to the battlefield.

By replacing inventory with information, the corps won't have to stockpile tons of supplies near the battlefield, like it did during the Gulf War, only to find it couldn't keep track of what was in containers. Then there was the marine policy requiring a 60-day supply of everything. McKissock figured out there was no need to overstock commodity items, like office supplies, that can be obtained anywhere. And with advice from the private sector, the marines have been upgrading warehouses, adding wireless scanners for real-time inventory placement and tracking. Now, if containers need to be sent into a war zone, they will have radio frequency transponders which, when scanned, will link to a database detailing what's inside.

Sources: Business Week (December 24, 2001): 24; and Federal Computer Week (December 11, 2000): 9.

Example 2

Cole's Trucks, Inc., a builder of high-quality refuse trucks, has about 5,000 items in its inventory. After hiring Matt Clark, a bright young OM student, for the summer, the firm determined that it has 500 A items, 1,750 B items, and 2,750 C items. Company policy is to count all A items every month (every 20 working days), all B items every quarter (every 60 working days), and all C items every 6 months (every 120 working days). How many items should be counted each day?

ITEM CLASS	QUANTITY	CYCLE COUNTING POLICY	NUMBER OF ITEMS COUNTED PER DAY
A	500	Each month (20 working days)	500/20 = 25/day
B	1,750	Each quarter (60 working days)	1,750/60 = 29/day
C	2,750	Every 6 months (120 working days)	2,750/120 = 23/day
			77/day

Seventy-seven items are counted each day.

In Example 2, the particular items to be cycle-counted can be sequentially or randomly selected each day. Another option is to cycle-count items when they are reordered.

Cycle counting also has the following advantages:

1. Eliminates the shutdown and interruption of production necessary for annual physical inventories.
2. Eliminates annual inventory adjustments.
3. Trained personnel audit the accuracy of inventory.
4. Allows the cause of the errors to be identified and remedial action to be taken.
5. Maintains accurate inventory records.

Control of Service Inventories

Shrinkage
Retail inventory that is unaccounted for between receipt and sale.

Pilferage
A small amount of theft.

Management of service inventories deserves special consideration. Although we may think of services as not having inventory, that is not the case. For instance, extensive inventory is held in wholesale and retail businesses, making inventory management crucial and often a factor in a manager's advancement. In the food-service business, for example, control of inventory can make the difference between success and failure. Moreover, inventory that is in transit or idle in a warehouse is lost value. Similarly, inventory damaged or stolen prior to sale is a loss. In retailing, inventory that is unaccounted for between receipt and time of sale is known as **shrinkage**. Shrinkage occurs from damage and theft as well as sloppy paperwork. Inventory theft is also known as **pilferage**. Retail inventory losses of 1% of sales is considered good, with losses in many stores exceeding 3%. Because the impact on profitability is substantial, inventory accuracy and control are critical. Applicable techniques include the following:

1. Good personnel selection, training, and discipline. These are never easy, but very necessary in food-service, wholesale, and retail operations, where employees have access to directly consumable merchandise.
2. Tight control of incoming shipments. This task is being addressed by many firms through the use of bar-code and radio frequency ID systems that read every incoming shipment and automatically check tallies against purchase orders. When properly designed, these systems are very hard to defeat. Each item has its own stock keeping unit (SKU), pronounced "skew."
3. Effective control of all goods leaving the facility. This job is done with bar codes on items being shipped, magnetic strips on merchandise, or via direct observation. Direct observation can be personnel stationed at exits (as at Costco and Sam's Club wholesale stores) and in potentially high-loss areas or can take the form of one-way mirrors and video surveillance.

Successful retail operations require very good store-level control with accurate inventory in its proper location. One recent study found that consumers and clerks could not find 16% of the items at one of the U.S.'s largest retailers—not because the items were out-of-stock, but because they were misplaced (in a backroom, a storage area, or on the wrong aisle). By the researcher's estimates, major retailers lose 10% to 25% of overall profits due to poor or inaccurate inventory records.[3]

INVENTORY MODELS

We now examine a variety of inventory models and the costs associated with them.

Independent versus Dependent Demand

Inventory control models assume that demand for an item is either independent of or dependent on the demand for other items. For example, the demand for refrigerators is *independent* of the demand for toaster ovens. However, the demand for toaster oven components is *dependent* on the requirements of toaster ovens.

This chapter focuses on managing inventory where demand is *independent*. Chapter 14 presents *dependent* demand management.

Holding, Ordering, and Setup Costs

Holding cost
The cost to keep or carry inventory in stock.

Holding costs are the costs associated with holding or "carrying" inventory over time. Therefore, holding costs also include obsolescence and costs related to storage, such as insurance, extra staffing, and interest payments. Table 12.1 shows the kinds of costs that need to be evaluated to

TABLE 12.1 ■

Determining Inventory Holding Costs

CATEGORY	COST (AND RANGE) AS A PERCENT OF INVENTORY VALUE
Housing costs (building rent or depreciation, operating cost, taxes, insurance)	6% (3–10%)
Material handling costs (equipment lease or depreciation, power, operating cost)	3% (1–3.5%)
Labor cost	3% (3–5%)
Investment costs (borrowing costs, taxes, and insurance on inventory)	11% (6–24%)
Pilferage, scrap, and obsolescence	3% (2–5%)
Overall carrying cost	**26%**

Note: All numbers are approximate, as they vary substantially depending on the nature of the business, location, and current interest rates. Any inventory holding cost of less than 15% is suspect, but annual inventory holding costs often approach 40% of the value of inventory.

[3]A. Raman, N. DeHoratius, and Z. Ton, "Execution: The Missing Link in Retail Operations," *California Management Review* 43, no. 3 (spring 2001): 136–141.

Ordering cost
The cost of the ordering process.

Setup cost
The cost to prepare a machine or process for production.

Setup time
The time required to prepare a machine or process for production.

determine holding costs. Many firms fail to include all of the inventory holding costs. Consequently, inventory holding costs are often understated.

Ordering cost includes costs of supplies, forms, order processing, clerical support, and so forth. When orders are being manufactured, ordering costs also exist, but they are a part of what is called setup costs. **Setup cost** is the cost to prepare a machine or process for manufacturing an order. This includes time and labor to clean and change tools or holders. Operations managers can lower ordering costs by reducing setup costs and by using such efficient procedures as electronic ordering and payment.

In many environments, setup cost is highly correlated with **setup time**. Setups usually require a substantial amount of work prior to setup actually being performed at the work center. With proper planning much of the preparation required by a setup can be done prior to shutting down the machine or process. Setup times can thus be reduced substantially. Machines and processes that traditionally have taken hours to set up are now being set up in less than a minute by the more imaginative world-class manufacturers. As we shall see later in this chapter, reducing setup times is an excellent way to reduce inventory investment and to improve productivity.

INVENTORY MODELS FOR INDEPENDENT DEMAND

In this section, we introduce three inventory models that address two important questions: *when to order* and *how much to order*. These *independent* demand models are

1. Basic economic order quantity (EOQ) model.
2. Production order quantity model.
3. Quantity discount model.

The Basic Economic Order Quantity (EOQ) Model

Economic order quantity (EOQ) model
A widely used technique for inventory control.

The **economic order quantity (EOQ) model** is one of the oldest and most commonly known inventory-control techniques.[4] This technique is relatively easy to use but is based on several assumptions:

1. Demand is known, constant, and independent.
2. Lead time—that is, the time between placement and receipt of the order—is known and constant.
3. Receipt of inventory is instantaneous and complete. In other words, the inventory from an order arrives in one batch at one time.
4. Quantity discounts are not possible.
5. The only variable costs are the cost of setting up or placing an order (setup cost) and the cost of holding or storing inventory over time (holding or carrying cost). These costs were discussed in the previous section.
6. Stockouts (shortages) can be completely avoided if orders are placed at the right time.

With seasonal surges in demand, retailers and suppliers may rely on large inventories. Warehouses are often full in November in preparation of the holiday season. This can mean huge holding costs.

---

[4]The research on EOQ dates back to 1915; see Ford W. Harris, *Operations and Cost* (Chicago: A. W. Shaw, 1915).

FIGURE 12.3 ■

Inventory Usage over Time

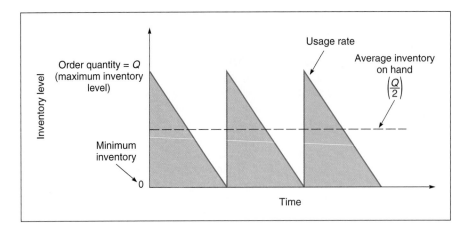

With these assumptions, the graph of inventory usage over time has a sawtooth shape, as in Figure 12.3. In Figure 12.3, Q represents the amount that is ordered. If this amount is 500 dresses, all 500 dresses arrive at one time (when an order is received). Thus, the inventory level jumps from 0 to 500 dresses. In general, an inventory level increases from 0 to Q units when an order arrives.

Because demand is constant over time, inventory drops at a uniform rate over time. (Refer to the sloped lines in Figure 12.3.) When the inventory level reaches 0 each time, the new order is placed and received, and the inventory level again jumps to Q units (represented by the vertical lines). This process continues indefinitely over time.

Minimizing Costs

The objective of most inventory models is to minimize total costs. With the assumptions just given, significant costs are setup (or ordering) cost and holding (or carrying) cost. All other costs, such as the cost of the inventory itself, are constant. Thus, if we minimize the sum of setup and holding costs, we will also be minimizing total costs. To help you visualize this, in Figure 12.4 we graph total costs as a function of the order quantity, Q. The optimal order size, Q^*, will be the quantity that minimizes the total costs. As the quantity ordered increases, the total number of orders placed per year will decrease. Thus, as the quantity ordered increases, the annual setup or ordering cost will decrease. But as the order quantity increases, the holding cost will increase due to the larger average inventories that are maintained.

As we can see in Figure 12.4, a reduction in either holding or setup cost will reduce the total cost curve. A reduction in setup cost curve also reduces the optimal order quantity (lot size). In addition, smaller lot sizes have a positive impact on quality and production flexibility. At Toshiba, the $40-billion Japanese conglomerate, workers can make as few as 10 laptop computers before changing mod-

FIGURE 12.4 ■

Total Cost as a Function of Order Quantity

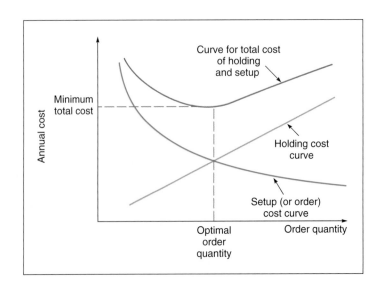

els. This lot-size flexibility has allowed Toshiba to move toward a "build-to-order" mass customization system, an important ability in an industry that has product life cycles measured in months, not years.[5]

You should note that in Figure 12.4, the optimal order quantity occurs at the point where the ordering-cost curve and the carrying-cost curve intersect. This was not by chance. With the EOQ model, the optimal order quantity will occur at a point where the total setup cost is equal to the total holding cost.[6] We use this fact to develop equations that solve directly for Q^*. The necessary steps are:

1. Develop an expression for setup or ordering cost.
2. Develop an expression for holding cost.
3. Set setup cost equal to holding cost.
4. Solve the equation for the optimal order quantity.

Using the following variables, we can determine setup and holding costs and solve for Q^*:

$$Q = \text{Number of pieces per order}$$
$$Q^* = \text{Optimum number of pieces per order (EOQ)}$$
$$D = \text{Annual demand in units for the inventory item}$$
$$S = \text{Setup or ordering cost for each order}$$
$$H = \text{Holding or carrying cost per unit per year}$$

1. Annual setup cost = (Number of orders placed per year) × (Setup or order cost per order)

$$= \left(\frac{\text{Annual demand}}{\text{Number of units in each order}}\right) \text{(Setup or order cost per order)}$$

$$= \left(\frac{D}{Q}\right)(S)$$

$$= \frac{D}{Q}S$$

2. Annual holding cost = (Average inventory level) × (Holding cost per unit per year)

$$= \left(\frac{\text{Order quantity}}{2}\right) \text{(Holding cost per unit per year)}$$

$$= \left(\frac{Q}{2}\right)(H)$$

$$= \frac{Q}{2}H$$

3. Optimal order quantity is found when annual setup cost equals annual holding cost, namely,

$$\frac{D}{Q}S = \frac{Q}{2}H$$

4. To solve for Q^*, simply cross multiply terms and isolate Q on the left of the equal sign.

$$2DS = Q^2H$$

$$Q^2 = \frac{2DS}{H}$$

$$Q^* = \sqrt{\frac{2DS}{H}} \tag{12-1}$$

[5]R. Anthony Inman, "The Impact of Lot-Size Reduction on Quality," *Production and Inventory Management Journal* 35, no. 1 (first quarter 1994): 5–8; and *International Journal of Production Economics* (August 1, 1996): 37–46.

[6]This is the case when holding costs are linear and begin at the origin—that is, when inventory costs do not decline (or increase) as inventory volume increases and all holding costs are in small increments. Additionally, there is probably some learning each time a setup (or order) is executed—a fact that lowers subsequent setup costs. Consequently, the EOQ model is probably a special case. However, we abide by the conventional wisdom that this model is a reasonable approximation.

Now that we have derived equations for the optimal order quantity, Q^*, it is possible to solve inventory problems directly, as in Example 3.

Example 3

Excel OM Data File
Ch12Ex3.xla

Sharp, Inc., a company that markets painless hypodermic needles to hospitals, would like to reduce its inventory cost by determining the optimal number of hypodermic needles to obtain per order. The annual demand is 1,000 units; the setup or ordering cost is $10 per order; and the holding cost per unit per year is $.50. Using these figures, we can calculate the optimal number of units per order:

$$Q^* = \sqrt{\frac{2DS}{H}}$$

$$Q^* = \sqrt{\frac{2(1,000)(10)}{0.50}} = \sqrt{40,000} = 200 \text{ units}$$

We can also determine the expected number of orders placed during the year (N) and the expected time between orders (T) as follows:

$$\text{Expected number of orders} = N = \frac{\text{Demand}}{\text{Order quantity}} = \frac{D}{Q^*} \tag{12-2}$$

$$\text{Expected time between orders} = T = \frac{\text{Number of working days per year}}{N} \tag{12-3}$$

Example 4 illustrates this concept.

Example 4

Using the data from Sharp, Inc., in Example 3, and assuming a 250-day working year, we find the number of orders (N) and the expected time between orders (T) as:

$$N = \frac{\text{Demand}}{\text{Order quantity}}$$

$$= \frac{1,000}{200} = 5 \text{ orders per year}$$

$$T = \frac{\text{Number of working days per year}}{\text{Expected number of orders}}$$

$$= \frac{250 \text{ working days per year}}{5 \text{ orders}} = 50 \text{ days between orders}$$

Active Model 12.1

Examples 3, 4, and 5 are further illustrated in Active Model 12.1 on your CD-ROM and in the Exercise located in your Student Lecture Guide.

As mentioned earlier in this section, the total annual variable inventory cost is the sum of setup and holding costs:

$$\text{Total annual cost} = \text{Setup cost} + \text{Holding cost} \tag{12-4}$$

In terms of the variables in the model, we can express the total cost TC as

$$TC = \frac{D}{Q}S + \frac{Q}{2}H \tag{12-5}$$

Example 5 shows how to use this formula.

Example 5

Again using the Sharp, Inc., data from Examples 3 and 4, we determine that the total annual inventory costs are

$$TC = \frac{D}{Q}S + \frac{Q}{2}H$$

$$= \frac{1,000}{200}(\$10) + \frac{200}{2}(\$.50)$$

$$= (5)(\$10) + (100)(\$.50)$$

$$= \$50 + \$50 = \$100$$

This store takes 4 weeks to get an order for Levis 501 jeans filled by the manufacturer. If the store sells 10 pairs of size 30–32 Levis a week, the store manager could set up two containers, keep 40 pairs of jeans in the second container, and place an order whenever the first container is empty. This would be a fixed-quantity reordering system. It is also called a "two-bin" system and is an example of a very elementary, but effective, approach to inventory management.

The total inventory cost expression may also be written to include the actual cost of the material purchased. If we assume that the annual demand and the price per hypodermic needle are known values (for example, 1,000 hypodermics per year at $P = \$10$) and total annual cost should include purchase cost, then Equation (12-5) becomes

$$TC = \frac{D}{Q} S + \frac{Q}{2} H + PD$$

Because material cost does not depend on the particular order policy, we still incur an annual material cost of $D \times P = (1{,}000)(\$10) = \$10{,}000$. (Later in this chapter we will discuss the case in which this may not be true—namely, when a quantity discount is available.)[7]

Robust

A model that gives satisfactory answers even with substantial variation in its parameters.

Robust Model   A benefit of the EOQ model is that it is robust. By **robust** we mean that it gives satisfactory answers even with substantial variation in its parameters. As we have observed, determining accurate ordering costs and holding costs for inventory is often difficult. Consequently, a robust model is advantageous. Total cost of the EOQ changes little in the neighborhood of the minimum. The curve is very shallow. This means that variations in setup costs, holding costs, demand, or even EOQ make relatively modest differences in total cost. Example 6 shows the robustness of EOQ.

Example 6

If management in the Sharp, Inc., examples underestimated total annual demand by 50% (say demand is actually 1,500 needles rather than 1,000 needles) while using the same Q, the annual inventory cost increases only $25 ($100 versus $125), or 25%. Here is why.

If demand in Example 5 is actually 1,500 needles rather than 1,000, but management uses an order quantity of $Q = 200$ (when it should be $Q = 244.9$ based on $D = 1{,}500$), the sum of holding and ordering cost increases 25%:

$$\text{Annual cost} = \frac{D}{Q} S + \frac{Q}{2} H$$

$$= \frac{1{,}500}{200} (\$10) + \frac{200}{2} (\$.50)$$

$$= \$75 + \$50 = \$125$$

---

[7]The formula for the Economic Order Quantity (Q^*) can also be determined by finding where the total cost curve is at a minimum (i.e., where the slope of the total cost curve is zero). Using calculus, we set the derivative of the total cost with respect to Q^* equal to 0.

The calculations for finding the minimum of $TC = \frac{D}{Q} S + \frac{Q}{2} H + PD$

are: $\dfrac{d(TC)}{dQ} = \left(\dfrac{-DS}{Q^2} \right) + \dfrac{H}{2} + 0 = 0$

Thus, $Q^* = \sqrt{\dfrac{2DS}{H}}$

However, had we known that the demand was for 1,500 with an EOQ of 244.9 units, we would have spent $122.48, as shown below:

$$\text{Annual cost} = \frac{1,500}{244.9}\,(\$10) + \frac{244.9}{2}\,(\$.50)$$
$$= 6.125\,(\$10) + 122.45\,(\$.50)$$
$$= \$61.24 + \$61.24 = \$122.48$$

Note that the expenditure of $125.00, made with an estimate of demand that was substantially wrong, is only 2% ($2.52/$122.48) higher than we would have paid had we known the actual demand and ordered accordingly.

We may conclude that the EOQ is indeed robust and that significant errors do not cost us very much. This attribute of the EOQ model is most convenient because our ability to accurately forecast demand, holding cost, and ordering cost is limited.

Reorder Points

Now that we have decided *how much* to order, we will look at the second inventory question, *when* to order. Simple inventory models assume that receipt of an order is instantaneous. In other words, they assume (1) that a firm will place an order when the inventory level for that particular item reaches zero, and (2) that it will receive the ordered items immediately. However, the time between placement and receipt of an order, called **lead time**, or delivery time, can be as short as a few hours or as long as months. Thus, the when-to-order decision is usually expressed in terms of a **reorder point (ROP)**—the inventory level at which an order should be placed (see Figure 12.5).

The reorder point (ROP) is given as

$$\text{ROP} = (\text{Demand per day})(\text{Lead time for a new order in days})$$
$$= d \times L \tag{12-6}$$

This equation for ROP *assumes that demand during lead time and lead time itself are constant.* When this is not the case, extra stock, often called **safety stock**, should be added.

The demand per day, d, is found by dividing the annual demand, D, by the number of working days in a year:

$$d = \frac{D}{\text{Number of working days in a year}}$$

Computing the reorder point is demonstrated in Example 7.

Lead time
In purchasing systems, the time between placing an order and receiving it; in production systems, it is the wait, move, queue, setup, and run times for each component produced.

Reorder point (ROP)
The inventory level (point) at which action is taken to replenish the stocked item.

Safety stock
Extra stock to allow for uneven demand; a buffer.

Each order may require a change in the way a machine or process is set up. Reducing setup time usually means a reduction in setup cost; and reductions in setup costs make smaller batches (lots) economical to produce. Increasingly, set up (and operation) is performed by computer-controlled machines operating from previously written programs.

FIGURE 12.5 ■

The Reorder Point (ROP) Curve

Q* is the optimum order quantity, and lead time represents the time between placing and receiving an order.

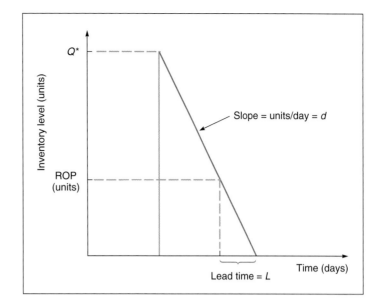

Example 7

Electronic Assembler, Inc., has a demand for 8,000 VCRs per year. The firm operates a 250-day working year. On average, delivery of an order takes 3 working days. We calculate the reorder point as

$$d = \frac{D}{\text{Number of working days in a year}} = \frac{8,000}{250}$$

$$= 32 \text{ units}$$

$$\text{ROP} = \text{Reorder point} = d \times L = 32 \text{ units per day} \times 3 \text{ days}$$

$$= 96 \text{ units}$$

Thus, when inventory stock drops to 96, an order should be placed. The order will arrive 3 days later, just as the firm's stock is depleted.

Safety stock is especially important in firms whose raw material deliveries may be uniquely unreliable. See the *OM in Action* box, "Inventory Modeling at the Philippines' San Miguel Corp."

OM IN ACTION

Inventory Safety Stock at the Philippines' San Miguel Corporation

At the San Miguel Corporation (SMC), which produces and distributes more than 300 products to every corner of the Philippine archipelago, raw material accounts for about 10% of total assets. The significant amount of money tied up in inventory encouraged the company's operations managers to develop a series of cost-minimizing inventory models.

One major SMC product, ice cream, uses dairy and cheese curd imported from Australia, New Zealand, and Europe. The normal mode of delivery is sea, and delivery frequencies are determined by supplier schedules, not San Miguel's. Stockouts, however, are avoidable through airfreight expediting. SMC's inventory model for ice cream balances ordering, carrying, and stockout costs while considering delivery frequency constraints and minimum order quantities. Modifications in delivery mode indicated that current safety stocks of 30 to 51 days could be cut in half for dairy and cheese curd. Even with the increased use of expensive airfreight, SMC saved $170,000 per year through the new policy.

Another SMC product, beer, consists of three major ingredients: malt, hops, and chemicals. Since these ingredients are characterized by low expediting costs and high unit costs, inventory modeling pointed to optimal policies that reduced safety stock levels, saving another $180,000 per year.

Sources: Businessworld (May 24, 2000): 1; and OR/MS Today (April 1999): 44–45.

Production Order Quantity Model

In the previous inventory model, we assumed that the entire inventory order was received at one time. There are times, however, when the firm may receive its inventory over a period of time. Such cases require a different model, one that does not require the instantaneous-receipt assumption. This model is applicable under two situations: (1) when inventory continuously flows or builds up over a period of time after an order has been placed or (2) when units are produced and sold simultaneously. Under these circumstances, we take into account daily production (or inventory-flow) rate and daily demand rate. Figure 12.6 shows inventory levels as a function of time.

Production order quantity model

An economic order quantity technique applied to production orders.

Because this model is especially suitable for the production environment, it is commonly called the **production order quantity model**. It is useful when inventory continuously builds up over time and traditional economic order quantity assumptions are valid. We derive this model by setting ordering or setup costs equal to holding costs and solving for optimal order size, Q^*. Using the following symbols, we can determine the expression for annual inventory holding cost for the production order quantity model:

$$Q = \text{Number of pieces per order}$$
$$H = \text{Holding cost per unit per year}$$
$$p = \text{Daily production rate}$$
$$d = \text{Daily demand rate, or usage rate}$$
$$t = \text{Length of the production run in days}$$

1. $$\left(\begin{array}{c}\text{Annual inventory}\\\text{holding cost}\end{array}\right) = (\text{Average inventory level}) \times \left(\begin{array}{c}\text{Holding cost}\\\text{per unit per year}\end{array}\right)$$

2. $$\left(\begin{array}{c}\text{Average inventory}\\\text{level}\end{array}\right) = (\text{Maximum inventory level})/2$$

3. $$\left(\begin{array}{c}\text{Maximum}\\\text{inventory level}\end{array}\right) = \left(\begin{array}{c}\text{Total produced during}\\\text{the production run}\end{array}\right) - \left(\begin{array}{c}\text{Total used during}\\\text{the production run}\end{array}\right)$$
$$= pt - dt$$

However, Q = total produced = pt, and thus $t = Q/p$. Therefore,

$$\text{Maximum inventory level} = p\left(\frac{Q}{p}\right) - d\left(\frac{Q}{p}\right)$$

$$= Q - \frac{d}{p}Q$$

$$= Q\left(1 - \frac{d}{p}\right)$$

4. Annual inventory holding cost (or simply holding cost) =

$$\frac{\text{Maximum inventory level}}{2}(H) = \frac{Q}{2}\left[1 - \left(\frac{d}{p}\right)\right]H$$

FIGURE 12.6 ■

Change in Inventory Levels over Time for the Production Model

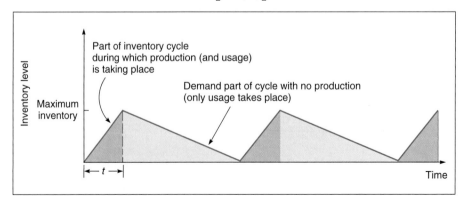

Intermec Technologies uses bar code readers to automate inventory control at its production and distribution facilities. Bar coding makes the data-collection process more accurate as well as faster and cheaper. With rapidly obtained data, shipments can be checked against production records and sales invoices to verify inventory accuracy and reduce losses. The scanning device shown is linked to the central computer by wireless data transmission.

A major difference between the production order model and the basic EOQ model is the annual holding cost, which is reduced in the production order quantity model.

Using this expression for holding cost and the expression for setup cost developed in the basic EOQ model, we solve for the optimal number of pieces per order by equating setup cost and holding cost:

$$\text{Setup cost} = (D/Q)S$$

$$\text{Holding cost} = \tfrac{1}{2} HQ[1 - (d/p)]$$

Set ordering cost equal to holding cost to obtain Q_p^*:

$$\frac{D}{Q} S = \tfrac{1}{2} HQ[1 - (d/p)]$$

$$Q^2 = \frac{2DS}{H[1 - (d/p)]}$$

$$Q_p^* = \sqrt{\frac{2DS}{H[1 - (d/p)]}}$$

(12-7)

In Example 8, we use the above equation, Q_p^*, to solve for the optimum order or production quantity when inventory is consumed as it is produced.

Example 8

**Excel Data OM File
Ch12Ex8.xla**

Active Model 12.2

Example 8 is further illustrated in Active Model 12.2 on the CD-ROM.

Nathan Manufacturing, Inc., makes and sells specialty hubcaps for the retail automobile aftermarket. Nathan's forecast for its wire-wheel hubcap is 1,000 units next year, with an average daily demand of 4 units. However, the production process is most efficient at 8 units per day. So the company produces 8 per day but uses only 4 per day. Given the following values, solve for the optimum number of units per order. (*Note:* This plant schedules production of this hubcap only as needed, during the 250 days per year the shop operates.)

$$\text{Annual demand} = D = 1,000 \text{ units}$$

$$\text{Setup costs} = S = \$10$$

$$\text{Holding cost} = H = \$0.50 \text{ per unit per year}$$

$$\text{Daily production rate} = p = 8 \text{ units daily}$$

$$\text{Daily demand rate} = d = 4 \text{ units daily}$$

$$Q_p^* = \sqrt{\frac{2DS}{H[1 - (d/p)]}}$$

$$Q_p^* = \sqrt{\frac{2(1,000)(10)}{0.50[1 - (4/8)]}}$$

$$= \sqrt{\frac{20,000}{0.50(1/2)}} = \sqrt{80,000}$$

$$= 282.8 \text{ hubcaps or } 283 \text{ hubcaps}$$

OM IN ACTION

Inventory Accuracy at Milton Bradley

Milton Bradley, a division of Hasbro, Inc., has been manufacturing toys for more than 100 years. Founded by Milton Bradley in 1860, the company started by making a lithograph of Abraham Lincoln. Using his printing skills, Bradley developed games, including the Game of Life, Chutes and Ladders, Candy Land, Scrabble, and Lite Brite. Today, the company produces hundreds of games, requiring billions of plastic parts.

Once Milton Bradley has determined the optimal quantities for each production run, it must make them and assemble them as a part of the proper game. Some games require literally hundreds of plastic parts, including spinners, hotels, people, animals, cars, and so on. According to Gary Brennan, director of manufacturing, getting the right number of pieces to the right toys and production lines is the most important issue for the credibility of the company. Some orders can require 20,000 or more perfectly assembled games delivered to their warehouses in a matter of days.

Games with the incorrect number of parts and pieces can result in some very unhappy customers. It is also time-consuming and expensive for Milton Bradley to supply the extra parts or have toys or games returned. And if shortages are found during the assembly stage, the entire production run can be stopped until the problem is corrected. Counting parts by hand or machine is not always accurate. As a result, Milton Bradley now weighs pieces and completed games to determine if the correct number of parts have been included. If the weight is not exact, there is a problem that is resolved before shipment. Using highly accurate digital scales, Milton Bradley is now able to get the right parts in the right game at the right time. Without this simple innovation, the most sophisticated production schedule is meaningless.

Sources: The *Wall Street Journal* (April 15, 1999): B1; and *Plastics World* (March 1997): 22–26.

You may want to compare this solution with the answer in Example 3. Eliminating the instantaneous-receipt assumption, where $p = 8$ and $d = 4$, has resulted in an increase in Q^* from 200 in Example 3 to 283. This increase in Q^* occurred because holding cost dropped from \$.50 to ($\$.50 \times \frac{1}{2}$), making a larger order quantity optimal. Also note that

$$d = 4 = \frac{D}{\text{Number of days the plant is in operation}} = \frac{1,000}{250}$$

We can also calculate Q_p^* when *annual* data are available. When annual data are used, we can express Q_p^* as

$$Q_p^* = \sqrt{\frac{2DS}{H\left(1 - \dfrac{\text{annual demand rate}}{\text{annual production rate}}\right)}} \qquad (12\text{-}8)$$

Quantity Discount Models

Quantity discount
A reduced price for items purchased in large quantities.

To increase sales, many companies offer quantity discounts to their customers. A **quantity discount** is simply a reduced price (P) for an item when it is purchased in larger quantities. It is not uncommon to have a discount schedule with several discounts for large orders. A typical quantity discount schedule appears in Table 12.2. As can be seen in the table, the normal price of the item is \$5. When 1,000 to 1,999 units are ordered at one time, the price per unit drops to \$4.80; when the quantity ordered at one time is 2,000 units or more, the price is \$4.75 per unit. As always, management must

TABLE 12.2 ■

A Quantity Discount Schedule

DISCOUNT NUMBER	DISCOUNT QUANTITY	DISCOUNT (%)	DISCOUNT PRICE (P)
1	0 to 999	no discount	\$5.00
2	1,000 to 1,999	4	\$4.80
3	2,000 and over	5	\$4.75

decide when and how much to order. However, with an opportunity to save money on quantity discounts, how does the operations manager make these decisions?

As with other inventory models discussed so far, the overall objective is to minimize total cost. Because the unit cost for the third discount in Table 12.2 is the lowest, you might be tempted to order 2,000 units or more merely to take advantage of the lower product cost. Placing an order for that quantity, however, even with the greatest discount price, might not minimize total inventory cost. Granted, as discount quantity goes up, the product cost goes down. However, holding cost increases because orders are larger. Thus the major trade-off when considering quantity discounts is between *reduced product cost* and *increased holding cost*. When we include the cost of the product, the equation for the total annual inventory cost can be calculated as follows:

$$\text{Total cost} = \text{Setup cost} + \text{Holding cost} + \text{Product cost}$$

or

$$TC = \frac{D}{Q}S + \frac{QH}{2} + PD \tag{12-9}$$

where Q = Quantity ordered
D = Annual demand in units
S = Ordering or setup cost per order or per setup
P = Price per unit
H = Holding cost per unit per year

Now, we have to determine the quantity that will minimize the total annual inventory cost. Because there are several discounts, this process involves four steps:

Step 1: For each discount, calculate a value for optimal order size Q^*, using the following equation:

$$Q^* = \sqrt{\frac{2DS}{IP}} \tag{12-10}$$

Note that the holding cost is IP instead of H. Because the price of the item is a factor in annual holding cost, we cannot assume that the holding cost is a constant when the price per unit changes for each quantity discount. Thus, it is common to express the holding cost (I) as a percentage of unit price (P) instead of as a constant cost per unit per year, H.

Step 2: For any discount, if the order quantity is too low to qualify for the discount, adjust the order quantity upward to the *lowest* quantity that will qualify for the discount. For example, if Q^* for discount 2 in Table 12.2 were 500 units, you would adjust this value up to 1,000 units. Look at the second discount in Table 12.2. Order quantities between 1,000 and 1,999 will qualify for the 4% discount. Thus, if Q^* is below 1,000 units, we will adjust the order quantity up to 1,000 units.

> Don't forget to adjust order quantity upward if the quantity is too low to qualify for the discount.

The reasoning for step 2 may not be obvious. If the order quantity, Q^*, is below the range that will qualify for a discount, a quantity within this range may still result in the lowest total cost.

As shown in Figure 12.7, the total cost curve is broken into three different total cost curves. There is a total cost curve for the first ($0 \le Q \le 999$), second ($1,000 \le Q \le 1,999$), and third ($Q \ge 2,000$) discount. Look at the total cost (TC) curve for discount 2. Q^* for discount 2 is less than the allowable discount range, which is from 1,000 to 1,999 units. As the figure shows, the lowest allowable quantity in this range, which is 1,000 units, is the quantity that minimizes total cost. Thus, the second step is needed to ensure that we do not discard an order quantity that may indeed produce the minimum cost. Note that an order quantity computed in step 1 that is *greater* than the range that would qualify it for a discount may be discarded.

Step 3: Using the preceding total cost equation compute a total cost for every Q^* determined in steps 1 and 2. If you had to adjust Q^* upward because it was below the allowable quantity range, be sure to use the adjusted value for Q^*.

Step 4: Select the Q^* that has the lowest total cost, as computed in step 3. It will be the quantity that will minimize the total inventory cost.

Let us see how this procedure can be applied with an example.

FIGURE 12.7 ■

Total Cost Curve for the
Quantity Discount
Model

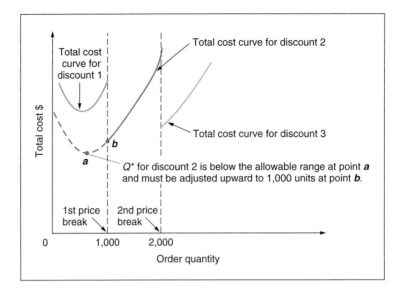

Example 9

**Excel OM Data File
Ch12Ex9.xla**

Wohl's Discount Store stocks toy race cars. Recently, the store has been given a quantity discount schedule for these cars. This quantity schedule was shown in Table 12.2. Thus, the normal cost for the toy race cars is $5.00. For orders between 1,000 and 1,999 units, the unit cost drops to $4.80; for orders of 2,000 or more units, the unit cost is only $4.75. Furthermore, ordering cost is $49.00 per order, annual demand is 5,000 race cars, and inventory carrying charge, as a percentage of cost, I, is 20%, or .2. What order quantity will minimize the total inventory cost?

The first step is to compute Q^* for every discount in Table 12.2. This is done as follows:

$$Q_1^* = \sqrt{\frac{2(5,000)(49)}{(.2)(5.00)}} = 700 \text{ cars order}$$

$$Q_2^* = \sqrt{\frac{2(5,000)(49)}{(.2)(4.80)}} = 714 \text{ cars order}$$

$$Q_3^* = \sqrt{\frac{2(5,000)(49)}{(.2)(4.75)}} = 718 \text{ cars order}$$

The second step is to adjust upward those values of Q^* that are below the allowable discount range. Since Q_1^* is between 0 and 999, it need not be adjusted. Because Q_2^* is below the allowable range of 1,000 to 1,999, it must be adjusted to 1,000 units. The same is true for Q_3^*: It must be adjusted to 2,000 units. After this step, the following order quantities must be tested in the total cost equation:

$$Q_1^* = 700$$

$$Q_2^* = 1,000 \text{---adjusted}$$

$$Q_3^* = 2,000 \text{---adjusted}$$

The third step is to use the total cost equation and compute a total cost for each order quantity. This step is taken with the aid of Table 12.3, which presents the computations for each level of discount introduced in Table 12.2.

TABLE 12.3 ■ Total Cost Computations for Wohl's Discount Store

DISCOUNT NUMBER	UNIT PRICE	ORDER QUANTITY	ANNUAL PRODUCT COST	ANNUAL ORDERING COST	ANNUAL HOLDING COST	TOTAL
1	$5.00	700	$25,000	$350	$350	$25,700
2	$4.80	1,000	$24,000	$245	$480	$24,725
3	$4.75	2,000	$23,750	$122.50	$950	$24,822.50

The fourth step is to select that order quantity with the lowest total cost. Looking at Table 12.3, you can see that an order quantity of 1,000 toy race cars will minimize the total cost. You should see, however, that the total cost for ordering 2,000 cars is only slightly greater than the total cost for ordering 1,000 cars. Thus, if the third discount cost is lowered to $4.65, for example, then this quantity might be the one that minimizes total inventory cost.

PROBABILISTIC MODELS WITH CONSTANT LEAD TIME

Probabilistic model
A statistical model applicable when product demand or any other variable is not known, but can be specified by means of a probability distribution.

All of the inventory models we have discussed so far make the assumption that demand for a product is constant and certain. We now relax this assumption. The following inventory models apply when product demand is not known but can be specified by means of a probability distribution. These types of models are called **probabilistic models**.

An important concern of management is maintaining an adequate service level in the face of uncertain demand. The **service level** is the *complement* of the probability of a stockout. For instance, if the probability of a stockout is 0.05, then the service level is .95. Uncertain demand raises the possibility of a stockout. One method of reducing stockouts is to hold extra units in inventory. As we noted, such inventory is usually referred to as safety stock. It involves adding a number of units as a buffer to the reorder point. As you recall from our previous discussion:

Service level
The complement of the probability of a stockout.

$$\text{Reorder point} = \text{ROP} = d \times L$$

where $d = $ Daily demand
             $L = $ Order lead time, or number of working days it takes to deliver an order

The inclusion of safety stock (ss) changes the expression to

$$\text{ROP} = d \times L + ss \qquad (12\text{-}11)$$

The amount of safety stock maintained depends on the cost of incurring a stockout and the cost of holding the extra inventory. Annual stockout cost is computed as follows:

$$\begin{aligned} \text{Annual stockout costs} = \text{the sum of the units short} \times \text{the probability} \\ \times \text{ the stockout cost/unit} \times \text{the number of orders per year} \end{aligned} \qquad (12\text{-}12)$$

Example 10 illustrates this concept.

Example 10

David Rivera Optical has determined that its reorder point for eyeglass frames is 50 ($d \times L$) units. Its carrying cost per frame per year is $5, and stockout (or lost sale) cost is $40 per frame. The store has experienced the following probability distribution for inventory demand during the reorder period. The optimum number of orders per year is six.

	NUMBER OF UNITS	PROBABILITY
	30	.2
	40	.2
ROP →	50	.3
	60	.2
	70	.1
		1.0

How much safety stock should David Rivera keep on hand?

SOLUTION

The objective is to find the amount of safety stock that minimizes the sum of the additional inventory holding costs and stockout costs. The annual holding cost is simply the holding cost per unit multiplied by the units added to the ROP. For example, a safety stock of 20 frames, which implies that the new ROP, with safety stock, is 70 (= 50 + 20), raises the annual carrying cost by $5(20) = $100.

However, computing annual stockout cost is more interesting. For any level of safety stock, stockout cost is the expected cost of stocking out. We can compute it, as in Equation (12.12), by multiplying the number of frames short by the probability of demand at that level, by the stockout cost, by the number of times per year the stockout can occur (which in our case is the number of orders per year). Then we add stockout costs for each possible stockout level for a given ROP. For zero safety stock, for example, a shortage of 10 frames will occur if demand is 60, and a shortage of 20 frames will occur if the demand is 70. Thus the stockout costs for zero safety stock are

$$(10 \text{ frames short}) \, (.2) \, (\$40 \text{ per stockout}) \, (6 \text{ possible stockouts per year})$$
$$+ (20 \text{ frames short}) \, (.1) \, (\$40) \, (6) = \$960$$

The following table summarizes the total costs for each alternative:

SAFETY STOCK	ADDITIONAL HOLDING COST	STOCKOUT COST		TOTAL COST
20	(20) ($5) = $100		$ 0	$100
10	(10) ($5) = $ 50	(10) (.1) ($40) (6)	= $240	$290
0	$ 0	(10) (.2) ($40) (6) + (20) (.1) ($40) (6) = $960		$960

The safety stock with the lowest total cost is 20 frames. Therefore, this safety stock changes the reorder point to 50 + 20 = 70 frames.

When it is difficult or impossible to determine the cost of being out of stock, a manager may decide to follow a policy of keeping enough safety stock on hand to meet a prescribed customer service level. For instance, Figure 12.8 shows the use of safety stock when demand (for hospital resuscitation kits) is probabilistic. We see that the safety stock in Figure 12.8 is 16.5 units, and the reorder point is also increased by 16.5.

The manager may want to define the service level as meeting 95% of the demand (or, conversely, having stockouts only 5% of the time). Assuming that demand during lead time (the reorder period) follows a normal curve, only the mean and standard deviation are needed to define the inventory requirements for any given service level. Sales data are usually adequate for computing the mean and standard deviation. In the following example we use a normal curve with a known mean (μ) and

FIGURE 12.8 ■

Probabilistic Demand for a Hospital Item

Expected number of kits needed during lead time is 350, but for a 95% service level, the reorder point should be raised to 366.5.

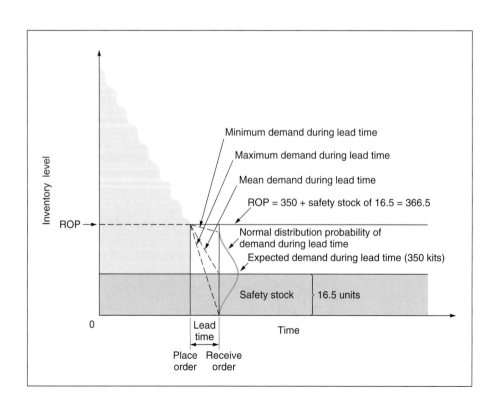

standard deviation (σ) to determine the reorder point and safety stock necessary for a 95% service level. We use the following formula:

$$\text{ROP} = \text{expected demand during lead time} + Z\sigma \qquad (12\text{-}13)$$

where Z = number of standard deviations
               σ = standard deviation of lead time demand

Example 11

Memphis Regional Hospital stocks a "code blue" resuscitation kit that has a normally distributed demand during the reorder period. The mean (average) demand during the reorder period is 350 kits, and the standard deviation is 10 kits. The hospital administrator wants to follow a policy that results in stockouts occurring only 5% of the time.

(a) What is the appropriate value of Z? (b) How much safety stock should the hospital maintain? (c) What reorder point should be used? The following figure may help you visualize the example:

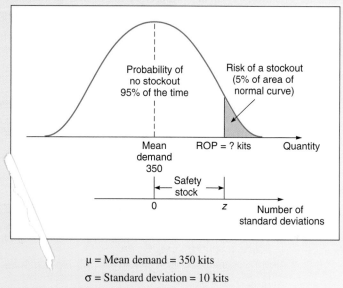

μ = Mean demand = 350 kits

σ = Standard deviation = 10 kits

Z = Number of standard normal deviates

The cost of the inventory policy increases dramatically with an increase in service levels. Indeed, inventory costs increase exponentially as service level increases

SOLUTION

a.   We use the properties of a standardized normal curve to get a Z value for an area under the normal curve of .95 (or $1 - .05$). Using a normal table (see Appendix I), we find a Z value of 1.65 standard deviations from the mean.

b.   Safety stock $= x - \mu$

Because $Z = \dfrac{x - \mu}{\sigma}$

Then   Safety stock $= Z\sigma$ (12-14)

Solving for safety stock, as in Equation (12-14), gives

$$\text{Safety stock} = 1.65(10) = 16.5 \text{ kits}$$

This is the situation illustrated in Figure 12.8.

c.   The reorder point is

$$\text{ROP} = \text{expected demand during lead time} + \text{safety stock}$$
$$= 350 \text{ kits} + 16.5 \text{ kits of safety stock} = 366.5, \text{ or } 367 \text{ kits}$$

Equations (12-13) and (12-14) assume that both an estimate of expected demand during lead times and its standard deviation are available. When data on lead time demand are *not* at hand, these formulas cannot be applied and we need to determine if: (a) demand is variable and lead time is constant; or (b) only lead time is variable; or (c) both demand and lead time are variable. For each of these situations, a different formula is needed to compute ROP.[8]

FIXED-PERIOD (P) SYSTEMS

Fixed-quantity (Q) system

An EOQ ordering system, with the same order amount each time.

Perpetual inventory system

A system that keeps track of each withdrawal or addition to inventory continuously, so records are always current.

Fixed-period (P) system

A system in which inventory orders are made at regular time intervals.

The inventory models that we have considered so far are **fixed-quantity,** or **Q systems**. That is, the same fixed amount is added to inventory every time an order for an item is placed. We saw that orders are event-triggered. When inventory decreases to the reorder point (ROP), a new order for Q units is placed.

To use the fixed-quantity model, inventory must be continuously monitored. This is called a **perpetual inventory system**. Every time an item is added to or withdrawn from inventory, records must be updated to make sure the ROP has not been reached.

In a **fixed-period,** or **P system**, on the other hand, inventory is ordered at the end of a given period. Then, and only then, is on-hand inventory counted. Only the amount necessary to bring total inventory up to a prespecified target level is ordered. Figure 12.9 illustrates this concept.

Fixed-period systems have several of the same assumptions as the basic EOQ fixed-quantity system:

- The only relevant costs are the ordering and holding costs.
- Lead times are known and constant.
- Items are independent of one another.

The downward-sloped line in Figure 12.9 again represents on-hand inventory. But now, when the time between orders (*P*) passes, we place an order to raise inventory up to the target value (*T*). The amount ordered during the first period may be Q_1, the second period Q_2, and so on. The Q_i value is the difference between current on-hand inventory and the target inventory level. Example 12 illustrates how much to reorder in a simple P system.

FIGURE 12.9 ■

Inventory Level in a Fixed-Period (P) System

Various amounts (Q_1, Q_2, Q_3, etc.) are ordered at regular time intervals (P) based on the quantity necessary to bring inventory up to the target maximum (T).

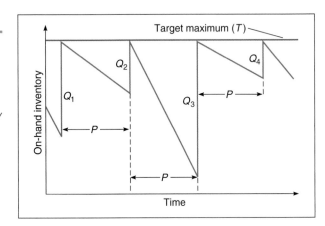

Example 12

Hard Rock London has a back order for three leather bomber jackets in its retail shop. There are no jackets in stock, none are expected from earlier orders, and it is time to place an order. The target value is 50 jackets. How many bomber jackets should be ordered?

SOLUTION

Order amount (*Q*) = Target (*T*) − On-hand inventory − Earlier
orders not yet received + Back orders = 50 − 0 − 0 + 3 = 53 jackets

The advantage of the fixed-period system is that there is no physical count of inventory items after an item is withdrawn—this occurs only when the time for the next review comes up. This procedure is also convenient administratively, especially if inventory control is only one of several duties of an employee.

[8](a) If *only* the *demand* (*d*) is variable, then ROP = *average* daily demand × lead time in days + $Z\sigma_{dLT}$

where σ_{dLT} = standard deviation of demand per day = $\sqrt{\text{lead time}}\ \sigma_d$

(b) If *only lead time* is variable, then ROP = daily demand × *average* lead time in days + $Zd\sigma_{LT}$

(c) If *both* are variable, then ROP = average daily demand × average lead time + $Z\ \sqrt{\text{average lead time} \times \sigma_d^2 + \bar{d}^2\sigma_{LT}^2}$

A fixed-period system is appropriate when vendors make routine (that is, at fixed-time interval) visits to customers to take fresh orders or when purchasers want to combine orders to save ordering and transportation costs (therefore, they will have the same review period for similar inventory items). For example, a vending machine company may come to refill its machines every Tuesday.

The disadvantage of the P system is that because there is no tally of inventory during the review period, there is the possibility of a stockout during this time. This scenario is possible if a large order draws the inventory level down to zero right after an order is placed. Therefore, a higher level of safety stock (as compared to a fixed-quantity system) needs to be maintained to provide protection against stockout during both the time between reviews and the lead time.

SUMMARY

Inventory investment: your company's largest asset.

Inventory represents a major investment for many firms. This investment is often larger than it should be because firms find it easier to have "just-in-case" inventory rather than "just-in-time" inventory. Inventories are of four types:

1. Raw material and purchased components.
2. Work-in-process.
3. Maintenance, repair, and operating (MRO).
4. Finished goods.

In this chapter, we discussed independent inventory, ABC analysis, record accuracy, cycle counting, and inventory models used to control independent demands. The EOQ model, production order quantity model, and quantity discount model can all be solved using Excel OM or POM for Windows software. A summary of the inventory models presented in this chapter is shown in Table 12.4.

TABLE 12.4 ■

Models for Independent Demand Summarized

Q = Number of pieces per order
EOQ = Optimum order quantity ($Q*$)
D = Annual demand in units
S = Setup or ordering cost for each order
H = Holding or carrying cost per unit per year in dollars
p = Daily production rate
d = Daily demand rate

P = Price
I = Annual inventory carrying cost as a percentage of price
μ = Mean demand
σ = Standard deviation
x = Mean demand + Safety stock
Z = Standardized value under the normal curve

EOQ:

$$Q^* = \sqrt{\frac{2DS}{H}} \qquad (12\text{-}1)$$

Quantity discount EOQ model:

$$Q^* = \sqrt{\frac{2DS}{IP}} \qquad (12\text{-}10)$$

EOQ production order quantity model:

$$Q_p^* = \sqrt{\frac{2DS}{H[1 - (d/p)]}} \qquad (12\text{-}7)$$

Probability model:

$$\text{Safety stock} = Z\sigma = x - \mu \qquad (12\text{-}14)$$

Total cost for the EOQ and quantity discount EOQ models:

TC = Total cost

= Setup cost + Holding cost + Product cost

$$= \frac{D}{Q}S + \frac{Q}{2}H + PD \qquad (12\text{-}9)$$

KEY TERMS

Raw material inventory *(p. 358)*
Work-in-process inventory (WIP) *(p. 358)*
MRO *(p. 359)*
Finished goods inventory *(p. 359)*
ABC analysis *(p. 359)*
Cycle counting *(p. 360)*
Shrinkage *(p. 361)*
Pilferage *(p. 361)*
Holding cost *(p. 362)*
Ordering cost *(p. 363)*
Setup cost *(p. 363)*
Setup time *(p. 363)*

Economic order quantity (EOQ) model *(p. 363)*
Robust *(p. 367)*
Lead time *(p. 368)*
Reorder point (ROP) *(p. 368)*
Safety stock *(p. 368)*
Production order quantity model *(p. 370)*
Quantity discount *(p. 372)*
Probabilistic models *(p. 375)*
Service level *(p. 375)*
Fixed-quantity (Q) system *(p. 378)*
Perpetual inventory system *(p. 378)*
Fixed-period (P) system *(p. 378)*

USING EXCEL OM FOR INVENTORY

Excel OM allows us to easily model inventory problems ranging from ABC analysis, to the basic EOQ model, to the production model, to quantity discount situations. Two of these models are illustrated in this section.

Program 12.1 shows the input data, selected formulas, and results for an ABC analysis, using data from Example 1 (p. 359). After the data are entered, we use the *Data* and *Sort* Excel commands to rank the items from largest to smallest dollar volumes.

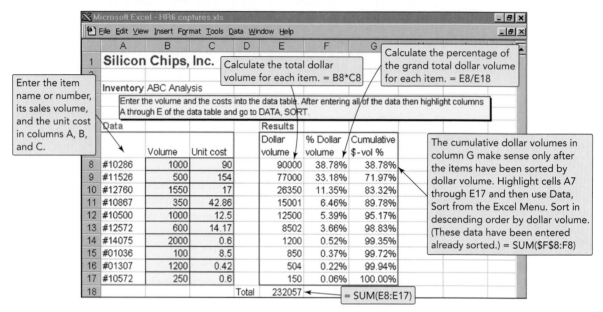

PROGRAM 12.1 ■ Using Excel OM for an ABC Analysis, with Data from Example 1

We illustrate the production inventory model in Program 12.2 using the data from Example 8 (p. 371). Input data, selected Excel formulas, output (including an optional graph of order quantity versus cost) all appear in Program 12.2.

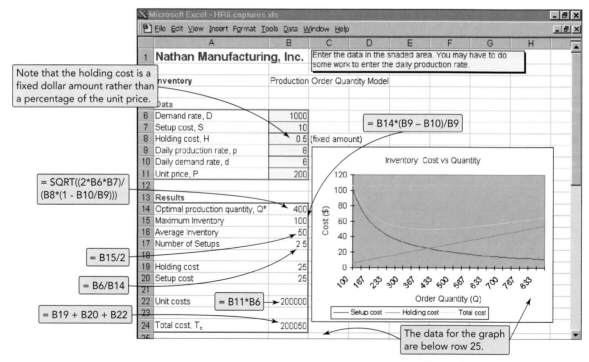

PROGRAM 12.2 ■ Using Excel OM for a Production Model, with Data from Example 8

USING POM FOR WINDOWS TO SOLVE INVENTORY PROBLEMS

The POM for Windows inventory module can solve the entire EOQ family of problems, as well as ABC inventory management. Please refer to Appendix V for further details.

SOLVED PROBLEMS

Solved Problem 12.1

The Warren W. Fisher Computer Corporation purchases 8,000 transistors each year as components in minicomputers. The unit cost of each transistor is $10, and the cost of carrying one transistor in inventory for a year is $3. Ordering cost is $30 per order.

What are (a) the optimal order quantity, (b) the expected number of orders placed each year, and (c) the expected time between orders? Assume that Fisher operates a 200-day working year.

SOLUTION

(a) $Q^* = \sqrt{\dfrac{2DS}{H}} = \sqrt{\dfrac{2(8,000)(30)}{3}} = 400$ units

(b) $N = \dfrac{D}{Q^*} = \dfrac{8,000}{400} = 20$ orders

(c) Time between orders $= T = \dfrac{\text{Number of working days}}{N} = \dfrac{200}{20} = 10$ working days

Hence, an order for 400 transistors is placed every 10 days. Presumably, then, 20 orders are placed each year.

Solved Problem 12.2

Annual demand for notebook binders at Salinas' Stationery Shop is 10,000 units. Teresita Salinas operates her business 300 days per year and finds that deliveries from her supplier generally take working days. Calculate the reorder point for the notebook binders

SOLUTION

$$L = 5 \text{ days}$$

$$d = \dfrac{10,000}{300} = 33.3 \text{ units per day}$$

$$\text{ROP} = d \times L = (33.3 \text{ units per day})(5 \text{ days})$$

$$= 166.7 \text{ units}$$

Thus, Teresita should reorder when her stock reaches 167 units.

Solved Problem 12.3

Leonard Presby, Inc., has an annual demand rate of 1,000 units but can produce at an average production rate of 2,000 units. Setup cost is $10; carrying cost is $1. What is the optimal number of units to be produced each time?

SOLUTION

$$Q^* = \sqrt{\dfrac{2DS}{H\left(1 - \dfrac{\text{annual demand rate}}{\text{annual production rate}}\right)}} = \sqrt{\dfrac{2(1,000)(10)}{1[1 - (1,000/2,000)]}}$$

$$= \sqrt{\dfrac{20,000}{1/2}} = \sqrt{40,000} = 200 \text{ units}$$

Solved Problem 12.4

What safety stock should Ron Satterfield Corporation maintain if mean sales are 80 during the reorder period, the standard deviation is 7, and Ron can tolerate stockouts 10% of the time?

SOLUTION

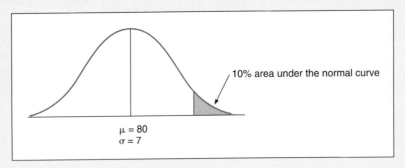

From Appendix I, Z at an area of .9 (or 1 − .10) = 1.28

$$Z = \frac{x - \mu}{\sigma} = \frac{ss}{\sigma}$$

$$ss = Z\sigma$$

$$= 1.28(7) = 8.96 \text{ units, or } 9 \text{ units}$$

INTERNET AND STUDENT CD-ROM EXERCISES

Visit our home page or use your student CD-ROM to help with material in this chapter.

 On Our Home Page, www.prenhall.com/heizer

- Self-Tests
- Practice Problems
- Internet Exercises
- Current Articles and Research
- Virtual Company Tour
- Internet Homework Problems
- Internet Cases

 On Your Student CD-ROM

- PowerPoint Lecture
- Practice Problems
- Video Clip and Video Case
- Active Model Exercises
- Excel OM
- Excel OM Example Data Files

ADDITIONAL CASE STUDIES

Internet Case Studies: Visit our Web site at www.prenhall.com/heizer **for these free case studies:**

- **Martin-Pullin Bicycle Corp.:** A forecasting and EOQ case.
- **Professional Video Management**: This firm faces a quantity discount decision.
- **Western Ranchman Outfitters**: Involves using EOQ with an unreliable supplier of jeans.
- **LaPlace Power and Light**: This utility company is evaluating its current inventory policies.

Harvard has selected these Harvard Business School cases to accompany this chapter
(textbookcasematch.hbsp.harvard.edu)**:**

- **Pioneer Hi-Bred International, Inc.** (#898-238): Deals with the challenges in managing inventory in a large, complex agribusiness firm.
- **L.L. Bean, Inc.: Item Forecasting and Inventory** (#893-003): The firm must balance costs of understocking and overstocking when demand for catalog items is uncertain.
- **Blanchard Importing and Distribution Co., Inc.** (#673-033): Illustrates two main types of errors resulting from the use of EOQ models.

Aggregate Planning

Chapter Outline

LEARNING OBJECTIVES

When you complete this chapter you should be able to

IDENTIFY OR DEFINE:

Aggregate planning

Tactical scheduling

Graphic technique for aggregate planning

Mathematical techniques for planning

DESCRIBE OR EXPLAIN:

How to do aggregate planning

How service firms develop aggregate plans

Aggregate Planning Provides a Competitive Advantage at Anheuser-Busch

Anheuser-Busch produces close to 40% of the beer consumed in the U.S. The company achieves efficiency at such volume by doing an excellent job of matching capacity to demand.

Matching capacity and demand in the intermediate term (3 to 18 months) is the heart of aggregate planning. Anheuser-Busch matches fluctuating demand by brand to specific plant, labor, and inventory capacity. Meticulous cleaning between batches, effective maintenance, and efficient employee and facility scheduling contribute to high facility utilization, a major factor in all high capital investment facilities.

Beer is made in a product-focused facility—one that produces high volume and low variety. Product-focused production processes usually require high fixed cost but typically have the benefit of low variable costs. Maintaining high use of such facilities is critical because high capital costs require high use to be competitive. Performance above the break-even point requires high use, and downtime is disastrous.

Beer production can be divided into four stages. The first stage is the selection and assurance of raw material delivery and quality. The second stage is the actual brewing process from milling to aging. The third stage is packaging into the wide variety of containers desired by the market.

Shown are brew kettles in which wort, later to become beer, is boiled and hops are added for the flavor and bitter character they impart.

The fourth and final stage is distribution, which includes temperature-controlled delivery and storage. Each stage has its resource limitations. Developing the aggregate plan to make it all work is demanding.

Effective aggregate planning is a major ingredient in competitive advantage at Anheuser-Busch.

In the brewhouse control room, process control uses computers to monitor the starting-cellar process, where wort is in its final stage of preparation before being fermented into beer.

The canning line imprints on each can: a code that identifies the day, year, and 15-minute period of production; the plant at which the product was brewed and packaged; and the production line used. This system allows any quality-control problems to be tracked and corrected.

**Aggregate planning
(or aggregate
scheduling)**
An approach to
determine the quantity
and timing of production
for the intermediate future
(usually 3 to 18 months
ahead).

Scheduling decisions
Making plans that match
production to changes in
demand.

Manufacturers like Anheuser-Busch, GE, and Yamaha face tough decisions when trying to schedule products like beer, air conditioners, and jet skis, the demand for which is heavily dependent on seasonal variation. If the firms increase output and a summer is warmer than usual, they stand to increase sales and market share. However, if the summer is cool, they may be stuck with expensive unsold product. Developing plans that minimize costs connected with such forecasts is one of the main functions of an operations manager.

Aggregate planning (also known as **aggregate scheduling**) is concerned with determining the quantity and timing of production for the intermediate future, often from 3 to 18 months ahead. Operations managers try to determine the best way to meet forecasted demand by adjusting production rates, labor levels, inventory levels, overtime work, subcontracting rates, and other controllable variables. Usually, *the objective of aggregate planning is to minimize cost over the planning period.* However, other strategic issues may be more important than low cost. These strategies may be to smooth employment levels, to drive down inventory levels, or to meet a high level of service.

For manufacturers, the aggregate schedule ties the firm's strategic goals to production plans, but for service organizations, the aggregate schedule ties strategic goals to workforce schedules.

Four things are needed for aggregate planning:

- A logical overall unit for measuring sales and output, such as air-conditioning units at GE or cases of beer at Anheuser-Busch.
- A forecast of demand for a reasonable intermediate planning period in these aggregate terms.
- A method for determining the costs that we discuss in this chapter.
- A model that combines forecasts and costs so that scheduling decisions can be made for the planning period.

In this chapter we describe the aggregate planning decision, show how the aggregate plan fits into the overall planning process, and describe several techniques that managers use when developing an aggregate plan. We stress both manufacturing and service-sector firms.

THE PLANNING PROCESS

In Chapter 4, we saw that demand forecasting can address short-, medium-, and long-range problems. Long-range forecasts help managers deal with capacity and strategic issues and are the responsibility of top management (see Figure 13.1). Top management formulates policy-related questions, such as facility location and expansion, new product development, research funding, and investment over a period of several years.

Medium-range planning begins once long-term capacity decisions are made. This is the job of the operations manager. **Scheduling decisions** address the problem of matching productivity to fluctuating demands. These plans need to be consistent with top management's long-range strategy and work within the resources allocated by earlier strategic decisions. Medium- (or "intermediate-") range planning is accomplished by building an aggregate production plan.

Short-range planning may extend up to a year but is usually less than 3 months. This plan is also the responsibility of operations personnel, who work with supervisors and foremen to "disaggregate" the intermediate plan into weekly, daily, and hourly schedules. Tactics for dealing with short-term planning involve loading, sequencing, expediting, and dispatching, which are discussed in Chapter 15.

Figure 13.1 illustrates the time horizons and features for short-, intermediate-, and long-range planning.

THE NATURE OF AGGREGATE PLANNING

As the term *aggregate* implies, an aggregate plan means combining appropriate resources into general, or overall, terms. Given demand forecast, facility capacity, inventory levels, workforce size, and related inputs, the planner has to select the rate of output for a facility over the next 3 to 18 months. The plan can be for manufacturing firms such as Anheuser-Busch and Whirlpool, hospitals, colleges, or Prentice Hall, the company that published this textbook.

Take, for a manufacturing example, IBM or Hewlett-Packard, each of which produces different models of microcomputers. They make (1) laptops, (2) desktops, (3) notebook computers, and (4)

FIGURE 13.1 ■

Planning Tasks and
Responsibilities

If top management does
a poor or inconsistent job
of long-term planning,
problems will develop
that make the aggregate
planner's job very tough.

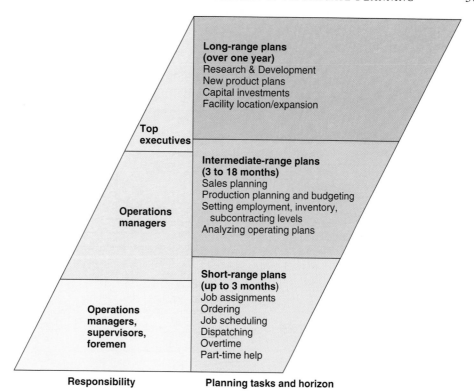

Top executives

**Long-range plans
(over one year)**
Research & Development
New product plans
Capital investments
Facility location/expansion

Operations managers

**Intermediate-range plans
(3 to 18 months)**
Sales planning
Production planning and budgeting
Setting employment, inventory,
 subcontracting levels
Analyzing operating plans

**Operations managers,
supervisors,
foremen**

**Short-range plans
(up to 3 months)**
Job assignments
Ordering
Job scheduling
Dispatching
Overtime
Part-time help

Responsibility   **Planning tasks and horizon**

advanced technology machines with high-speed chips. For each month in the upcoming three quarters, the aggregate plan for IBM or Hewlett-Packard might have the following output (in units of production) for this "family" of microcomputers:

QUARTER 1			QUARTER 2			QUARTER 3		
Jan.	Feb.	March	April	May	June	July	Aug.	Sept.
150,000	120,000	110,000	100,000	130,000	150,000	180,000	150,000	140,000

Note that the plan looks at production *in the aggregate*, not on a product-by-product breakdown. Likewise, an aggregate plan for GM tells the auto manufacturer how many cars to make, but not how many should be two-door versus four-door or red versus green. It tells Nucor Steel how many tons of steel to produce, but does not differentiate grades of steel.

In the service sector, consider Computrain, a company that provides microcomputer training for managers. The firm offers courses on spreadsheets, graphics, databases, word processing, and the Internet, and employs several instructors to meet the demand for its services from business and government. Demand for training tends to be very low near holiday seasons and during summer, when many people take their vacations. To meet the fluctuating needs for courses, the company can hire and lay off instructors, advertise to increase demand in slow seasons, or subcontract its work to other training agencies during peak periods. Again, aggregate planning makes decisions about intermediate-range capacity, not specific courses or instructors.

Disaggregation
The process of breaking
the aggregate plan into
greater detail.

**Master production
schedule**
A timetable that specifies
what is to be made and
when.

Aggregate planning is part of a larger production planning system. Therefore, understanding the interfaces between the plan and several internal and external factors is useful. Figure 13.2 shows that not only does the operations manager receive input from the marketing department's demand forecast, but must also deal with financial data, personnel, capacity, and availability of raw materials. In a manufacturing environment, the process of breaking the aggregate plan down into greater detail is called **disaggregation**. Disaggregation results in a **master production schedule**, which provides input to material requirements planning (MRP) systems. The master production schedule addresses the purchasing or production of parts or components needed to make final products (see Chapter 14). Detailed work schedules for people and priority scheduling for products result as the final step of the production planning system (and are discussed in Chapter 15).

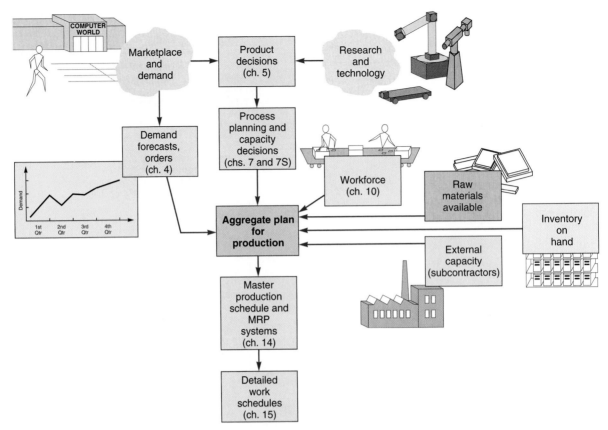

FIGURE 13.2 ■ Relationships of the Aggregate Plan

AGGREGATE PLANNING STRATEGIES

When generating an aggregate plan, the operations manager must answer several questions:

1. Should inventories be used to absorb changes in demand during the planning period?
2. Should changes be accommodated by varying the size of the workforce?
3. Should part-timers be used, or should overtime and idle time absorb fluctuations?
4. Should subcontractors be used on fluctuating orders so a stable workforce can be maintained?
5. Should prices or other factors be changed to influence demand?

All of these are legitimate planning strategies. They involve the manipulation of inventory, production rates, labor levels, capacity, and other controllable variables. We will now examine eight options in more detail. The first five are called *capacity options* because they do not try to change demand but attempt to absorb the fluctuations in it. The last three are *demand options* through which firms try to smooth out changes in the demand pattern over the planning period.

Capacity Options

A firm can choose from the following basic capacity (production) options:

Aggregate planning in the real world involves a lot of trial and error.

1. *Changing inventory levels.* Managers can increase inventory during periods of low demand to meet high demand in future periods. If we select this strategy, costs associated with storage, insurance, handling, obsolescence, pilferage, and capital invested will increase. (These costs typically range from 15% to 40% of the value of an item annually.) On the other hand, when the firm enters a period of increasing demand, shortages can result in lost sales due to potentially longer lead times and poorer customer service.
2. *Varying workforce size by hiring or layoffs.* One way to meet demand is to hire or lay off production workers to match production rates. However, often new employees need to be

Federal Express's huge aircraft fleet is used to near capacity for nighttime delivery of packages but is 100% idle during the daytime. In an attempt to better utilize their capacity (and leverage its assets), Federal Express considered two services with opposite or countercyclical demand patterns to its nighttime service—commuter passenger service and passenger charter service. However, after a thorough analysis, the 12% to 13% return on investment was judged insufficient for the risks involved. Facing the same issues, though, UPS decided to begin a charter airline that operates on weekends.

trained, and the average productivity drops temporarily as they are absorbed into the firm. Layoffs or firings, of course, lower the morale of all workers and can lead to lower productivity.

3. *Varying production rates through overtime or idle time.* It is sometimes possible to keep a constant workforce while varying working hours, cutting back the number of hours worked when demand is low and increasing them when it rises. Yet when demand is on a large upswing, there is a limit on how much overtime is realistic. Overtime pay requires more money, and too much overtime can wear workers down to the point that overall productivity drops off. Overtime also implies the increased overhead needed to keep a facility open. On the other hand, when there is a period of decreased demand, the company must somehow absorb workers' idle time—usually a difficult process.

4. *Subcontracting.* A firm can acquire temporary capacity by subcontracting work during peak demand periods. Subcontracting, however, has several pitfalls. First, it may be costly; second, it risks opening your client's door to a competitor. Third, it is often hard to find the perfect subcontract supplier, one who always delivers the quality product on time.

5. *Using part-time workers.* Especially in the service sector, part-time workers can fill unskilled labor needs. This practice is common in restaurants, retail stores, and supermarkets. The *OM in Action* box on the next page, describing Federal Express and United Parcel Service, provides two views of this strategy.

Demand Options

The basic demand options are the following:

1. *Influencing demand.* When demand is low, a company can try to increase demand through advertising, promotion, personal selling, and price cuts. Airlines and hotels have long offered weekend discounts and off-season rates; telephone companies charge less at night; some colleges give discounts to senior citizens; and air conditioners are least expensive in winter. However, even special advertising, promotions, selling, and pricing are not always able to balance demand with production capacity.

OM IN ACTION

A Tale of Two Delivery Services

Federal Express and United Parcel Service are direct competitors in package delivery. Both firms are successful, but they approach aggregate planning quite differently.

Managers at Federal Express use a large number of part-time employees in their huge package-sorting facility. This Memphis facility is designed and staffed to sort over a million envelopes and packages in a short 4-hour shift during the middle of the night. Federal Express found that college students provide a good source of labor. These high-energy part-timers help meet peak demands, and the firm believes that full-timers could not be effectively utilized for a full 8-hour shift.

At UPS's package-sorting hub, managers are also faced with the decision whether to staff with mostly full-time or part-time employees. UPS chose the mostly full-time approach. The firm also researches job designs and work processes thoroughly, hoping to provide a high level of job satisfaction and a strong sense of teamwork. Hours at UPS are long, the work is hard, and UPS generates some complaints about its demanding levels of productivity. Yet when openings occur, UPS has never had a shortage of job applicants.

Sources: The Wall Street Journal *(March 31, 2000): 1, and (September 14, 1999): A1.*

Negative inventory means we owe units to customers. We either lose sales or back order to make it up.

2. *Back ordering during high-demand periods.* Back orders are orders for goods or services that a firm accepts but is unable (either on purpose or by chance) to fill at the moment. If customers are willing to wait without loss of their goodwill or order, back ordering is a possible strategy. Many firms back order, but the approach often results in lost sales.

3. *Counterseasonal product and service mixing.* A widely used active smoothing technique among manufacturers is to develop a product mix of counterseasonal items. Examples include companies that make both furnaces and air conditioners or lawn mowers and snow-

John Deere and Company, the "granddaddy" of farm equipment manufacturers, uses sales incentives to smooth demand. During the fall and winter off-seasons, sales are boosted with price cuts and other incentives. About 70% of Deere's big machines are ordered in advance of seasonal use—about double the industry rate. Incentives hurt margins, but Deere keeps its market share and controls costs by producing more steadily all year long. Similarly, in service businesses like L.L. Bean, some customers are offered free shipping on orders placed before the Christmas rush.

blowers. However, companies that follow this approach may find themselves involved in products or services beyond their area of expertise or beyond their target market.

These eight options, along with their advantages and disadvantages, are summarized in Table 13.1.

Mixing Options to Develop a Plan

Although each of the five capacity options and three demand options might produce an effective aggregate schedule, some combination of capacity options and demand options may be better.

Many manufacturers assume that the use of the demand options has been fully explored by the marketing department and those reasonable options incorporated into the demand forecast. The operations manager then builds the aggregate plan based on that forecast. However, using the five capacity options at his command, the operations manager still has a multitude of possible plans. These plans can embody, at one extreme, a *chase strategy* and, at the other, a *level-scheduling strategy*. They may, of course, fall somewhere in between.

Chase strategy

Sets production equal to forecasted demand.

Chase Strategy   A **chase strategy** attempts to achieve output rates for each period that match the demand forecast for that period. This strategy can be accomplished in a variety of ways. For example, the operations manager can vary workforce levels by hiring or laying off or can vary production by means of overtime, idle time, part-time employees, or subcontracting. Many service organizations favor the chase strategy because the inventory option is difficult or impossible to adopt. Industries that have moved toward a chase strategy include education, hospitality, and construction.

Level scheduling

Maintaining a constant output rate, production rate, or workforce level over the planning horizon.

Level Strategy   A level strategy (or **level scheduling**) is an aggregate plan in which daily production is uniform from period to period. Firms like Toyota and Nissan keep production at uniform levels and may (1) let the finished goods inventory go up or down to buffer the difference between demand and production or (2) find alternative work for employees. Their philosophy is that a stable workforce leads to a better-quality product, less turnover and absenteeism, and more employee commitment to corporate goals. Other hidden savings include employees who are more experienced, easier scheduling and supervision, and fewer dramatic startups and shutdowns. Level scheduling works well when demand is reasonably stable.

TABLE 13.1 ■ Aggregate Planning Options: Advantages and Disadvantages

OPTION	ADVANTAGES	DISADVANTAGES	SOME COMMENTS
Changing inventory levels.	Changes in human resources are gradual or none; no abrupt production changes.	Inventory holding costs may increase. Shortages may result in lost sales.	Applies mainly to production, not service, operations.
Varying workforce size by hiring or layoffs.	Avoids the costs of other alternatives.	Hiring, layoff, and training costs may be significant.	Used where size of labor pool is large.
Varying production rates through overtime or idle time.	Matches seasonal fluctuations without hiring/training costs.	Overtime premiums; tired workers; may not meet demand.	Allows flexibility within the aggregate plan.
Subcontracting.	Permits flexibility and smoothing of the firm's output.	Loss of quality control; reduced profits; loss of future business.	Applies mainly in production settings.
Using part-time workers.	Is less costly and more flexible than full-time workers.	High turnover/training costs; quality suffers; scheduling difficult.	Good for unskilled jobs in areas with large temporary labor pools.
Influencing demand.	Tries to use excess capacity. Discounts draw new customers.	Uncertainty in demand. Hard to match demand to supply exactly.	Creates marketing ideas. Overbooking used in some businesses.
Back ordering during high-demand periods.	May avoid overtime. Keeps capacity constant.	Customer must be willing to wait, but goodwill is lost.	Many companies back order.
Counterseasonal product and service mixing.	Fully utilizes resources; allows stable workforce.	May require skills or equipment outside firm's areas of expertise.	Risky finding products or services with opposite demand patterns.

METHODS FOR AGGREGATE PLANNING

Mixed strategy
A planning strategy that uses two or more controllable variables to set a feasible production plan.

Mixed plans are more complex than single or "pure" ones but typically yield a better strategy.

For most firms, neither a chase strategy nor a level strategy is likely to prove ideal, so a combination of the eight options (called a **mixed strategy**) must be investigated to achieve minimum cost. However, because there are a huge number of possible mixed strategies, managers find that aggregate planning can be a challenging task. Finding the one "optimal" plan is not always possible. Indeed, some companies have no formal aggregate planning process: They use the same plan from year to year, making adjustments up or down just enough to fit the new annual demand. This method certainly does not provide much flexibility, and if the original plan was suboptimal, the entire production process will be locked into suboptimal performance.

In this section, we introduce several techniques that operations managers use to develop more useful and appropriate aggregate plans. They range from the widely used charting (or graphical) method to a series of more formal mathematical approaches, including the transportation method of linear programming.

Graphical and Charting Methods

Graphical and charting techniques
Aggregate planning techniques that work with a few variables at a time to allow planners to compare projected demand with existing capacity.

Graphical and charting techniques are popular because they are easy to understand and use. Basically, these plans work with a few variables at a time to allow planners to compare projected demand with existing capacity. They are trial-and-error approaches that do not guarantee an optimal production plan, but they require only limited computations and can be performed by clerical staff. Following are the five steps in the graphical method:

1. Determine the demand in each period.
2. Determine capacity for regular time, overtime, and subcontracting each period.
3. Find labor costs, hiring and layoff costs, and inventory holding costs.
4. Consider company policy that may apply to the workers or to stock levels.
5. Develop alternative plans and examine their total costs.

These steps are illustrated in Examples 1 to 4.

Example 1

A Juarez, Mexico, manufacturer of roofing supplies has developed monthly forecasts for an important product and presented the 6-month period January to June in Table 13.2.

TABLE 13.2 ■

MONTH	EXPECTED DEMAND	PRODUCTION DAYS	DEMAND PER DAY (COMPUTED)
Jan.	900	22	41
Feb.	700	18	39
Mar.	800	21	38
Apr.	1,200	21	57
May	1,500	22	68
June	1,100	20	55
	6,200	124	

The demand per day is computed by simply dividing the expected demand by the number of production or working days each month.

To illustrate the nature of the aggregate planning problem, the firm also draws a graph (Figure 13.3) that charts daily demand each month. The dotted line across the chart represents the production rate required to meet average demand over the 6-month period. It is computed as follows:

$$\text{Average requirement} = \frac{\text{Total expected demand}}{\text{Number of production days}} = \frac{6,200}{124} = 50 \text{ units per day}$$

Note that in the first 3 months, expected demand is lower than average, while expected demand in April, May, and June is above average.

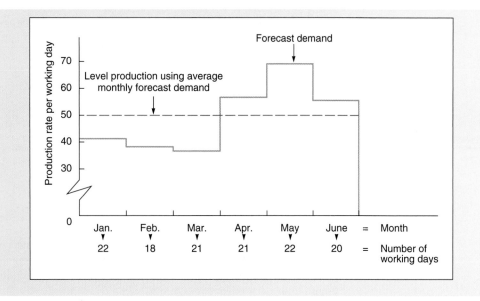

FIGURE 13.3 ■

Graph of Forecast
and Average Forecast
Demand

The graph in Figure 13.3 illustrates how the forecast differs from the average demand. Some strategies for meeting the forecast were listed earlier. The firm, for example, might staff in order to yield a production rate that meets *average* demand (as indicated by the dashed line). Or it might produce a steady rate of, say, 30 units and then subcontract excess demand to other roofing suppliers. Other plans might combine overtime work with subcontracting to absorb demand. Examples 2 to 4 illustrate three possible strategies.

Example 2

**Excel OM Data File
Ch13Ex2.xla**

Active Model 13.1

Example 2 is further illustrated in Active Model 13.1 on the CD-ROM and in the Exercise located in your Student Lecture Guide.

One possible strategy (call it plan 1) for the manufacturer described in Example 1 is to maintain a constant workforce throughout the 6-month period. A second (plan 2) is to maintain a constant workforce at a level necessary to meet the lowest demand month (March) and to meet all demand above this level by subcontracting. Both plan 1 and plan 2 have level production and are, therefore, called *level strategies*. Plan 3 is to hire and lay off workers as needed to produce exact monthly requirements—*a chase strategy*. Table 13.3 provides cost information necessary for analyzing these three alternatives:

TABLE 13.3 ■ Cost Information

Inventory carrying cost	$ 5 per unit per month
Subcontracting cost per unit	$ 10 per unit
Average pay rate	$ 5 per hour ($40 per day)
Overtime pay rate	$ 7 per hour (above 8 hours per day)
Labor-hours to produce a unit	1.6 hours per unit
Cost of increasing daily production rate (hiring and training)	$300 per unit
Cost of decreasing daily production rate (layoffs)	$600 per unit

Analysis of Plan 1. When analyzing this approach, which assumes that 50 units are produced per day, we have a constant workforce, no overtime or idle time, no safety stock, and no subcontractors. The firm accumulates inventory during the slack period of demand, January through March, and depletes it during the higher-demand warm season, April through June. We assume beginning inventory = 0 and planned ending inventory = 0:

MONTH	PRODUCTION AT 50 UNITS PER DAY	DEMAND FORECAST	MONTHLY INVENTORY CHANGE	ENDING INVENTORY
Jan.	1,100	900	+200	200
Feb.	900	700	+200	400
Mar.	1,050	800	+250	650
Apr.	1,050	1,200	−150	500
May	1,100	1,500	−400	100
June	1,000	1,100	−100	0
				1,850

Total units of inventory carried over from one month to the next month = 1,850 units

Workforce required to produce 50 units per day = 10 workers

Because each unit requires 1.6 labor-hours to produce, each worker can make 5 units in an 8-hour day. Thus to produce 50 units, 10 workers are needed.

The costs of plan 1 are computed as follows:

COSTS		CALCULATIONS
Inventory carrying	$ 9,250	(= 1,850 units carried × $5 per unit)
Regular-time labor	49,600	(= 10 workers × $40 per day × 124 days)
Other costs (overtime, hiring, layoffs, subcontracting)	0	
Total cost	$58,850	

The graph for Example 2 was shown in Figure 13.3. Some planners prefer a *cumulative* graph to display visually how the forecast deviates from the average requirements. Note that both the level production line and the forecast line produce the same total production. Such a graph is provided in Figure 13.4.

FIGURE 13.4 ■

Cumulative Graph for Plan 1

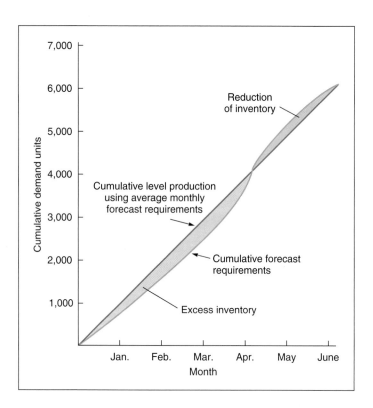

Example 3

Analysis of Plan 2. Although a constant workforce is also maintained in plan 2, it is set low enough to meet demand only in March, the lowest month. To produce 38 units per day in-house, 7.6 workers are needed. (You can think of this as 7 full-time workers and 1 part-timer.) *All* other demand is met by subcontracting. Subcontracting is thus required in every other month. No inventory holding costs are incurred in plan 2.

Because 6,200 units are required during the aggregate plan period, we must compute how many can be made by the firm and how many must be subcontracted:

$$\text{In-house production} = 38 \text{ units per day} \times 124 \text{ production days}$$

$$= 4,712 \text{ units}$$

$$\text{Subcontract units} = 6,200 - 4,712 = 1,488 \text{ units}$$

The costs of plan 2 are computed as follows:

COSTS		CALCULATIONS
Regular-time labor	$37,696	(= 7.6 workers × $40 per day × 124 days)
Subcontracting	14,880	(= 1,488 units × $10 per unit)
Total cost	$52,576	

Example 4

Analysis of Plan 3. The final strategy, plan 3, involves varying the workforce size by hiring and firing as necessary. The production rate will equal the demand, and there is no change in production from the previous month, December. Table 13.4 shows the calculations and the total cost of plan 3. Recall that it costs $600 per unit produced to reduce production from the previous month's daily level and $300 per unit change to increase the daily rate of production through hirings:

TABLE 13.4 ■ Cost Computations for Plan 3

MONTH	FORECAST (UNITS)	DAILY PRODUCTION RATE	BASIC PRODUCTION COST (DEMAND × 1.6 HRS PER UNIT × $5 PER HR)	EXTRA COST OF INCREASING PRODUCTION (HIRING COST)	EXTRA COST OF DECREASING PRODUCTION (LAYOFF COST)	TOTAL COST
Jan.	900	41	$ 7,200	—	—	$ 7,200
Feb.	700	39	5,600	—	$1,200 (= 2 × $600)	6,800
Mar.	800	38	6,400	—	$ 600 (= 1 × $600)	7,000
Apr.	1,200	57	9,600	$5,700 (= 19 × $300)	—	15,300
May	1,500	68	12,000	$3,300 (= 11 × $300)	—	15,300
June	1,100	55	8,800	—	$7,800 (= 13 × $600)	$16,600
			$49,600	$9,000	$9,600	$68,200

So the total cost, including production, hiring, and layoff for plan 3 is $68,200.

The final step in the graphical method is to compare the costs of each proposed plan and to select the approach with the least total cost. A summary analysis is provided in Table 13.5. We see that because plan 2 has the lowest cost, it is the best of the three options.

TABLE 13.5 ■

Comparison of the Three Plans

COST	PLAN 1 (CONSTANT WORKFORCE OF 10 WORKERS)	PLAN 2 (WORKFORCE OF 7.6 WORKERS PLUS SUBCONTRACT)	PLAN 3 (HIRING AND LAYOFFS TO MEET DEMAND)
Inventory carrying	$ 9,250	$ 0	$ 0
Regular labor	49,600	37,696	49,600
Overtime labor	0	0	0
Hiring	0	0	9,000
Layoffs	0	0	9,600
Subcontracting	0	14,880	0
Total cost	$58,850	$52,576	$68,200

Of course, many other feasible strategies can be considered in a problem like this, including combinations that use some overtime. Although charting and graphing is a popular management tool, its help is in evaluating strategies, not generating them. To generate strategies, a systematic approach that considers all costs and produces an effective solution is needed.

Mathematical Approaches to Planning

This section briefly describes some of the mathematical approaches to aggregate planning that have been developed over the past 50 years.

**Transportation method
of linear programming**
A way of solving for the
optimal solution to an
aggregate planning
problem.

The Transportation Method of Linear Programming   When an aggregate planning problem is viewed as one of allocating operating capacity to meet forecasted demand, it can be formulated in a linear programming format. The **transportation method of linear programming** is not a trial-and-error approach like charting, but rather produces an optimal plan for minimizing costs. It is also flexible in that it can specify regular and overtime production in each time period, the number of units to be subcontracted, extra shifts, and the inventory carryover from period to period.

In Example 5, the supply consists of on-hand inventory and units produced by regular time, overtime, and subcontracting. Costs, in the upper right-hand corner of each cell of the matrix in Table 13.7, relate to units produced in a given period or units carried in inventory from an earlier period.

Example 5

**Excel OM Data File
Ch13Ex5.xla**

Farnsworth Tire Company developed data that relate to production, demand, capacity, and cost at its West Virginia plant. These data are shown in Table 13.6:

TABLE 13.6 ■ Farnsworth's Production, Demand, Capacity, and Cost Data

	SALES PERIOD		
	MAR.	APR.	MAY
Demand	800	1,000	750
Capacity:			
Regular	700	700	700
Overtime	50	50	50
Subcontracting	150	150	130
Beginning inventory	100 tires		

COSTS	
Regular time	$40 per tire
Overtime	$50 per tire
Subcontract	$70 per tire
Carrying cost	$ 2 per tire per month

Table 13.7 illustrates the structure of the transportation table and an initial feasible solution.

When setting up and analyzing this table, you should note the following:

1. Carrying costs are $2/tire per month. Tires produced in 1 period and held for 1 month will have a $2 higher cost. Because holding cost is linear, 2 months' holdover costs $4. So when you move across a row from left to right, regular time, overtime, and subcontracting costs are lowest when output is used the same period it is produced. If goods are made in one period and carried over to the next, holding costs are incurred.
2. Transportation problems require that supply equals demand; so, a dummy column called "unused capacity" has been added. Costs of not using capacity are zero.
3. Because back ordering is not a viable alternative for this particular company, no production is possible in those cells that represent production in a period to satisfy demand in a past period (i.e., those periods with an "X"). If back ordering *is* allowed, costs of expediting, loss of goodwill, and loss of sales revenues are summed to estimate backorder cost.
4. Quantities in each column of Table 13.7 designate the levels of inventory needed to meet demand requirements. Demand of 800 tires in March is met by using 100 tires from beginning inventory and 700 tires from regular time.
5. In general, to complete the table, allocate as much production as you can to a cell with the smallest cost without exceeding the unused capacity in that row or demand in that column. If there is still some demand left in that row, allocate as much as you can to the next-lowest-cost cell. You then repeat this process for periods 2 and 3 (and beyond, if necessary). When you are finished, the sum of all your entries in a row must equal the total row capacity, and the sum of all entries in a column must equal the demand for that period. (This step can be accomplished by the transportation method or by using POM for Windows or Excel OM software.)

Try to confirm that the cost of this initial solution is $105,900. The initial solution is not optimal, however. See if you can find the production schedule that yields the least cost (which turns out to be $105,700) using software or by hand.

TABLE 13.7 ■ Farnsworth's Transportation Table[a]

	SUPPLY FROM	DEMAND FOR — Period 1 (Mar.)	Period 2 (Apr.)	Period 3 (May)	Unused Capacity (dummy)	TOTAL CAPACITY AVAILABLE (supply)
	Beginning inventory	0 / 100	2	4	0	100
Period 1	Regular time	40 / 700	42	44	0	700
	Overtime	50	52 / 50	54	0	50
	Subcontract	70	72 / 150	74	0	150
Period 2	Regular time	×	40 / 700	42	0	700
	Overtime	×	50 / 50	52	0	50
	Subcontract	×	70 / 50	72	0 / 100	150
Period 3	Regular time	×	×	40 / 700	0	700
	Overtime	×	×	50 / 50	0	50
	Subcontract	×	×	70 / 70	0 / 130	130
	TOTAL DEMAND	800	1,000	750	230	2,780

[a]Cells with an x indicate that back orders are not used at Farnsworth. When using Excel OM or POM for Windows to solve, you must insert a *very* high cost (e.g., 9999) in each cell that is not used for production.

The transportation method of linear programming described in the above example was originally formulated by E. H. Bowman in 1956.[1] Although it works well in analyzing the effects of holding inventories, using overtime, and subcontracting, it does not work when nonlinear or negative factors are introduced. So, when other factors such as hiring and layoffs are introduced, the more general method of linear programming must be used.

Management coefficients model

A formal planning model built around a manager's experience and performance.

Management Coefficients Model Bowman's **management coefficients model**[2] builds a formal decision model around a manager's experience and performance. The assumption is that the manager's past performance is pretty good, so it can be used as a basis for future decisions. The technique uses a regression analysis of past production decisions made by managers. The regression line provides the relationship between variables (such as demand and labor) for future decisions. According to Bowman, managers' deficiencies are mostly inconsistencies in decision making.

Other Models Two additional aggregate planning models are the linear decision rule and simulation. The *linear decision rule (LDR)* attempts to specify an optimum production rate and workforce level over a specific period. It minimizes the total costs of payroll, hiring, layoffs, overtime, and inventory through a series of quadratic cost curves.[3]

[1]See E. H. Bowman, "Production Planning by the Transportation Method of Linear Programming," *Operations Research* 4, no. 1 (February 1956): 100–103.

[2]E. H. Bowman, "Consistency and Optimality in Managerial Decision Making," *Management Science* 9, no. 2 (January 1963): 310–321.

[3]Because LDR was developed by Charles C. Holt, Franco Modigliani, John F. Muth, and Nobel prize winner Herbert Simon, it is popularly known as the HMMS rule. For details, see C. C. Holt et al., *Production Planning, Inventories, and Work Force* (Upper Saddle River, NJ: Prentice Hall, 1960).

A computer model called *scheduling by simulation* uses a search procedure to look for the minimum-cost combination of values for workforce size and production rate.[4]

Comparison of Aggregate Planning Methods

Although these mathematical models have been found by researchers to work well under certain conditions, and linear programming has found some acceptance in industry, the fact is that most sophisticated planning models are not widely used. Why? Perhaps it reflects the average manager's attitude about what he or she views as overly complex models. Like all of us, planners like to understand how and why the models on which they are basing important decisions work. Additionally, operations managers need to make decisions quickly based on the changing dynamics of the workplace—and building good models is time-consuming. This may explain why the simpler charting and graphical approach is more generally accepted.

Table 13.8 highlights some of the main features of charting, transportation, and management coefficients planning models.

TABLE 13.8 ■

Summary of Three Major Aggregate Planning Methods

TECHNIQUE	SOLUTION APPROACHES	IMPORTANT ASPECTS
Graphical/charting methods	Trial and error	Simple to understand and easy to use. Many solutions; one chosen may not be optimal.
Transportation method of linear programming	Optimization	LP software available; permits sensitivity analysis and new constraints; linear functions may not be realistic.
Management coefficients model	Heuristic	Simple, easy to implement; tries to mimic manager's decision process; uses regression.

AGGREGATE PLANNING IN SERVICES

Some service organizations conduct aggregate planning in exactly the same way as we did in Examples 1 through 5 in this chapter, but with demand management taking a more active role. Because most services pursue *combinations* of the eight capacity and demand options discussed earlier, they usually formulate mixed aggregate planning strategies. In actuality, in such industries as banking, trucking, and fast foods, aggregate planning may be easier than in manufacturing.

Controlling the cost of labor in service firms is critical.[5] It involves the following:

1. Close scheduling of labor-hours to assure quick response to customer demand.
2. Some form of on-call labor resource that can be added or deleted to meet unexpected demand.
3. Flexibility of individual worker skills that permits reallocation of available labor.
4. Individual worker flexibility in rate of output or hours of work to meet expanded demand.

These options may seem demanding, but they are not unusual in service industries, in which labor is the primary aggregate planning vehicle. For instance:

- Excess capacity is used to provide study and planning time by real estate and auto salespersons.
- Police and fire departments have provisions for calling in off-duty personnel for major emergencies. Where the emergency is extended, police or fire personnel may work longer hours and extra shifts.
- When business is unexpectedly light, restaurants and retail stores send personnel home early.
- Supermarket stock clerks work cash registers when checkout lines become too lengthy.
- Experienced waitresses increase their pace and efficiency of service as crowds of customers arrive.

Approaches to aggregate planning differ by the type of service provided. Here we discuss five service scenarios.

[4]R. C. Vergin, "Production Scheduling under Seasonal Demand," *Journal of Industrial Engineering* 17, no. 5 (May 1966): 260–266.

[5]Glenn Bassett, *Operations Management for Service Industries* (Westport, CT: Quorum Books, 1992): 77.

Restaurants

In a business with a highly variable demand, such as a restaurant, aggregate scheduling is directed toward (1) smoothing the production rate and (2) finding the size of the workforce to be employed. The general approach usually requires building very modest levels of inventory during slack periods and depleting inventory during peak periods, but using labor to accommodate most of the changes in demand. Because this situation is very similar to those found in manufacturing, traditional aggregate planning methods may be applied to services as well. One difference that should be noted is that even modest amounts of inventory may be perishable. In addition, the relevant units of time may be much smaller than in manufacturing. For example, in fast-food restaurants, peak and slack periods may be measured in hours and the "product" may be inventoried for as little as 10 minutes.

Hospitals

Hospitals face aggregate planning problems in allocating money, staff, and supplies to meet the demands of patients. Michigan's Henry Ford Hospital, for example, plans for bed capacity and personnel needs in light of a patient-load forecast developed by moving averages. The necessary labor focus of its aggregate plan has led to the creation of a new floating staff pool serving each nursing pod.[6]

National Chains of Small Service Firms

With the advent of national chains of small service businesses such as funeral homes, quick lube outlets, photocopy/printing centers, and computer centers, the question of aggregate planning versus independent planning at each business establishment becomes an issue. Both output and purchasing may be centrally planned when demand can be influenced through special promotions. This approach to aggregate scheduling is advantageous because it reduces purchasing and advertising costs and helps manage cash flow at independent sites.

Miscellaneous Services

Most "miscellaneous" services—financial, transportation, and many communication and recreation services—provide intangible output. Aggregate planning for these services deals mainly with planning for human resource requirements and managing demand. The twofold goal is to level demand peaks and to design methods for fully utilizing labor resources during low-demand periods. Example 6 illustrates such a plan for a legal firm.

Example 6

Aggregate Planning in a Law Firm

Klasson and Avalon, a medium-size Tampa law firm of 32 legal professionals, has developed a 3-month forecast for 5 categories of legal business it anticipates (see Table 13.9). Assuming a 40-hour workweek and that 100% of each lawyer's hours are billed, about 500 billable hours are available from each lawyer this fiscal quarter. Hours of billable time are forecast and accumulated for the quarter by the 5 categories of skill (column 1), then divided by 500 to provide a count of lawyers needed to cover the estimated business. Between 30 and 39 lawyers will be needed to cover the variations in level of business between worst and best levels of demand. (For example, best-case scenario of 19,500 total hours, divided by 500 hours per lawyer, equals 39 lawyers needed.)

Because all 32 lawyers at Klasson and Avalon are qualified to perform basic legal research, this skill area has maximum scheduling flexibility (column 6). The most highly skilled (and capacity-constrained) categories are trial work and corporate law. In these areas, the firm's best-case forecast just barely covers trial work with 3.6 lawyers needed (see column 5) and 4 qualified (column 6). Meanwhile, corporate law is short 1 full person. Overtime can be used to cover the excess this quarter, but as business expands, it might be necessary to hire or develop talent in both of these areas. Real estate and criminal practice are adequately covered by available staff, as long as other needs do not use their excess capacity.

[6]G. Buxey, "Production Planning for Seasonal Demand," *International Journal of Operations and Production Management* 13, no. 7 (1993): 4–21.

TABLE 13.9 ■ Labor Allocation at Klasson and Avalon, Attorneys-at-Law; Forecasts for Coming Quarter (1 lawyer = 500 hours of labor)

(1) CATEGORY OF LEGAL BUSINESS	LABOR HOURS REQUIRED			CAPACITY CONSTRAINTS	
	(2) BEST CASE (HOURS)	(3) LIKELY CASE (HOURS)	(4) WORST CASE (HOURS)	(5) MAXIMUM DEMAND IN PEOPLE	(6) NUMBER OF QUALIFIED PERSONNEL
Trial work	1,800	1,500	1,200	3.6	4
Legal research	4,500	4,000	3,500	9.0	32
Corporate law	8,000	7,000	6,500	16.0	15
Real estate law	1,700	1,500	1,300	3.4	6
Criminal law	3,500	3,000	2,500	7.0	12
Total hours	19,500	17,000	15,000		
Lawyers needed	39	34	30		

With its current legal staff of 32, Klasson and Avalon's best-case forecast will increase the workload by 20% (assuming no new hires). This represents one extra day of work per lawyer per week. The worst-case scenario will result in about a 6% underutilization of talent. For both these scenarios, the firm has determined that available staff will provide adequate service.

Source: Adapted from Glenn Bassett, *Operations Management for Service Industries* (Westport, CT: Quorum Books, 1992): 110.

Airline Industry

Airlines and auto-rental firms also have unique aggregate scheduling problems. Consider an airline that has its headquarters in New York, two hub sites in cities such as Atlanta and Dallas, and 150 offices in airports throughout the country. Aggregate planning consists of tables or schedules for (1) number of flights in and out of each hub; (2) number of flights on all routes; (3) number of passengers to be serviced on all flights; and (4) number of air personnel and ground personnel required at each hub and airport.

This planning is considerably more complex than aggregate planning for a single site or even for a number of independent sites. Scheduling decisions for airlines also include determining the seats to be allocated to various fare classes. Techniques for doing this allocation are called yield, or revenue, management, our next topic.

YIELD MANAGEMENT

Yield (or revenue) management
Capacity decisions that determine the allocation of classes of resources in order to maximize profit or yield.

Yield (or revenue) management is the aggregate planning process of allocating resources to customers at prices that will maximize yield or revenue. Its use dates back to the 1980s when American Airlines' reservation system (called SABRE) allowed the airline to alter ticket prices, in real time and on any route, based on demand information. If it looked like demand for expensive seats was low, more discounted seats were offered. If demand for full-fare seats was high, the number of discounted seats was reduced.

American Airlines' success in yield management spawned many other companies and industries to adopt the concept. Yield management in the hotel industry began in the late 1980s at Marriott International, which now claims an additional $400 million a year in profit from its management of revenue. The competing Omni hotel chain uses software that performs more than 100,000 calculations every night at each facility. The Dallas Omni, for example, now charges its highest rates (about $199) on weekdays, but heavily discounts (to as low as $59) on weekends. Its sister hotel in San Antonio, which is in a more tourist-oriented destination, reverses this rating scheme, with better deals for its consumers on weekdays. The *OM in Action* box, "Yield Management at Hertz," describes this practice in the rental car industry.

OM IN ACTION

Yield Management at Hertz

For over 90 years, Hertz has been renting standard cars for a fixed amount per day. During the past two decades, however, a significant increase in demand has derived from airline travelers flying for business purposes. As the auto-rental market has changed and matured, Hertz has offered more options, including allowing customers to pick up and drop off in different locations. This option has resulted in excess capacity in some cities and shortages in others.

These shortages and overages alerted Hertz to the need for a yield management system similar to those used in the airline industry. The system is used to set prices, regulate the movement, and ultimately determine the availability of cars at each location. Through research Hertz found that different city locations peak on different days of the week. So cars are moved to peak-demand locations from locations where the demand is low. By altering both the price and quantity of cars at various locations, Hertz has been able to increase "yield" and boost revenue.

The yield management system is primarily used by regional and local managers to better deal with changes in demand in the U.S. market. Hertz's plan to go global with the system, however, faces major challenges in foreign countries, where restrictions against moving empty cars across national borders are common.

Sources: Cornell Hotel and Restaurant Quarterly (December 2001): 33–46; and the *Wall Street Journal* (March 3, 2000): W-4.

Organizations that have *perishable inventory*, such as airlines, hotels, car rental agencies, cruise lines, and even electrical utilities, have the following shared characteristics that make yield management of interest:[7]

1. Service or product can be sold in advance of consumption.
2. Demand fluctuates.
3. Capacity is relatively fixed.
4. Demand can be segmented.
5. Variable costs are low and fixed costs are high.

Example 7 illustrates how yield management works in a hotel.

Example 7

The Cleveland Downtown Inn is a 100-room hotel that has historically charged one set price for its rooms, $150 per night. The variable cost of a room being occupied is low. Management believes the cleaning, air-conditioning, and incidental costs of soap, shampoo, and so forth, are $15 per room per night. Sales average 50 rooms per night. Figure 13.5 illustrates the current pricing scheme. Net sales are $6,750 per night with a single price point.

FIGURE 13.5 ■

Hotel Sets Only One Price Level

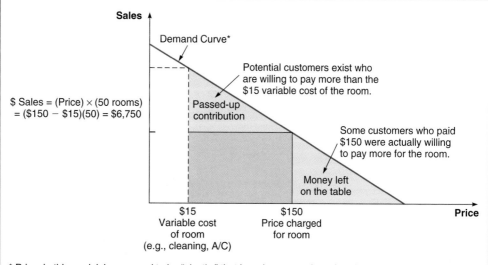

$ Sales = (Price) × (50 rooms)
= ($150 − $15)(50) = $6,750

* Price, in this model, is assumed to be "elastic," that is, sales respond to price changes. A change in demand is caused by a change in price. A product is said to be "inelastic" if a higher price does not affect demand.

---

[7]R. Oberwetter, "Revenue Management," *OR/MS Today* (June 2001): 41–44.

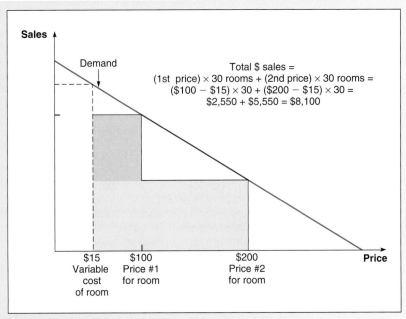

FIGURE 13.6 ■ Hotel with Two Price Levels

We note in Figure 13.5, however, that some guests would have been willing to spend more than $150 per room—"money left on the table." Others would be willing to pay more than the variable cost of $15, but less than $150—"passed-up contribution."

In Figure 13.6, the Inn decides to set *two* price levels. It estimates that 30 rooms per night can be sold at $100, and another 30 rooms at $200, using yield management software that is widely available. Total profit is now $8,100 ($2,550 from $100 rooms and $5,550 from $200 rooms). It may be that even more price levels are called for at Cleveland Downtown Inn.

Industries traditionally associated with revenue management operate in quadrant 2 of Figure 13.7.[8] They are able to apply variable pricing for their product and control product use or availability (number of airline seats or hotel rooms sold at economy rate). On the other hand, movie theaters, arenas, or performing arts centers (quadrant 1) have less pricing flexibility, but still use time

FIGURE 13.7 ■

Yield Management Matrix

Industries in quadrant 2 are traditionally associated with revenue management.
Source: Adapted from S. Kimes and K. McGuire, "Function Space Revenue Management," *Cornell Hotel and Restaurant Administration Quarterly* 42, no. 6 (December 2001): 33–46.

		Price	
		Tend to be fixed	**Tend to be variable**
Duration of use	**Predictable use**	Quadrant 1: Movies Stadiums/arenas Convention centers Hotel meeting space	Quadrant 2: Hotels Airlines Rental cars Cruise lines
	Unpredictable use	Quadrant 3: Restaurants Golf courses Internet service providers	Quadrant 4: Continuing care hospitals

[8]S. E. Kimes and K. A. McGuire, "Function Space Revenue Management," *Cornell Hotel and Restaurant Administration Quarterly* 42, no. 6 (December 2001): 33–46.

(evening or matinee) and location (orchestra, side, or balcony) to manage revenue. In both cases, management has control over the amount of the resource used—the duration of the resource—such as a seat for 2 hours.

In the lower half of Figure 13.7, the manager's job is more difficult because the duration of the use of the resource is less controllable. However, with imagination, managers are using excess capacity even for these industries. For instance, the golf course may sell less desirable tee times at a reduced rate, and the restaurant may have an "early bird" special to generate business before the usual dinner hour.

To make yield management work, the company needs to manage three issues:

1. Multiple pricing structures must be feasible and appear logical (and preferably fair) to the customer. Such justification may take various forms, for example, first-class seats on an airline or the preferred starting time at a golf course.
2. Forecasts of the use and duration of the use. How many economy seats should be available? How much will customers pay for a room with an ocean view?
3. Changes in demand. This means managing the increased use as more capacity is sold. It also means dealing with issues that occur because the pricing structure may not seem logical and fair to all customers. Finally, it means managing new issues, such as overbooking because the forecast was not perfect.

SUMMARY

Aggregate planning provides companies with a necessary weapon to help capture market shares in the global economy. The aggregate plan provides both manufacturing and service firms the ability to respond to changing customer demands while still producing at low-cost and high-quality levels.

The aggregate schedule sets levels of inventory, production, subcontracting, and employment over an intermediate time range, usually 3 to 18 months. This chapter describes several aggregate planning techniques, ranging from the popular charting approach to a variety of mathematical models such as linear programming.

The aggregate plan is an important responsibility of an operations manager and a key to efficient production. Output from the aggregate schedule leads to a more detailed master production schedule, which is the basis for disaggregation, job scheduling, and MRP systems.

Aggregate plans for manufacturing firms and service systems are similar. Restaurants, airlines, and hotels are all service systems that employ aggregate plans, and have an opportunity to implement yield management. But regardless of the industry or planning method, the most important issue is the implementation of the plan. In this respect, managers appear to be more comfortable with faster, less complex, and less mathematical approaches to planning.

KEY TERMS

Aggregate planning (or aggregate scheduling) *(p. 388)*
Scheduling decisions *(p. 388)*
Disaggregation *(p. 389)*
Master production schedule *(p. 389)*
Chase strategy *(p. 393)*
Level scheduling *(p. 393)*

Mixed strategy *(p. 394)*
Graphical and charting techniques *(p. 394)*
Transportation method of linear programming *(p. 398)*
Management coefficients model *(p. 399)*
Yield (or revenue) management *(p. 402)*

USING EXCEL OM FOR AGGREGATE PLANNING

Excel OM's Aggregate Planning module is illustrated in Program 13.1. Again using data from Example 2, Program 13.1 provides input and some of the formulas used to compute the costs of regular time, overtime, subcontracting, holding, shortage, and increase or decrease in production. The user must provide the production plan for Excel OM to analyze.

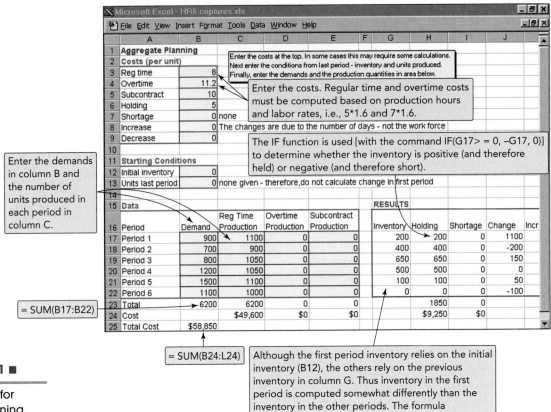

Enter the demands in column B and the number of units produced in each period in column C.

= SUM(B17:B22)

= SUM(B24:L24)

Although the first period inventory relies on the initial inventory (B12), the others rely on the previous inventory in column G. Thus inventory in the first period is computed somewhat differently than the inventory in the other periods. The formula for G22 is = G21 + SUM(C22:E22) − B22.

PROGRAM 13.1 ■

Using Excel OM for
Aggregate Planning,
with Example 2 Data

USING POM FOR WINDOWS FOR AGGREGATE PLANNING

POM for Windows' Aggregate Planning module performs aggregate or production planning for up to 12 time periods. Given a set of demands for future periods, you can try various plans to determine the lowest-cost plan based on holding, shortage, production, and changeover costs. Four methods are available for planning. More help is available on each after you choose the method. See Appendix V for further details.

SOLVED PROBLEMS

Solved Problem 13.1

The roofing manufacturer described in Examples 1 to 4 of this chapter wishes to consider yet a fourth planning strategy (plan 4). This one maintains a constant workforce of eight people and uses overtime whenever necessary to meet demand. Use the cost information found in Table 13.3 on page 395. Again, assume beginning and ending inventories are equal to zero.

SOLUTION

Employ eight workers and use overtime when necessary. Note that carrying costs will be encountered in this plan.

MONTH	PRODUCTION AT 40 UNITS PER DAY	BEGINNING-OF-MONTH INVENTORY	FORECAST DEMAND THIS MONTH	OVERTIME PRODUCTION NEEDED	ENDING INVENTORY
Jan.	880	—	900	20 units	0 units
Feb.	720	0	700	0 units	20 units
Mar.	840	20	800	0 units	60 units
Apr.	840	60	1,200	300 units	0 units
May	880	0	1,500	620 units	0 units
June	800	0	1,100	300 units	0 units
				1,240 units	80 units

Carrying cost totals = 80 units × $5/unit/month = $400

Regular pay:

8 workers × $40/day × 124 days = $39,680

To produce 1,240 units at overtime rate (of $7/hour) requires 1,984 hours.

Overtime pay = $7/hour × 1,984 hours = $13,888

Plan 4

COSTS (WORKFORCE OF 8 PLUS OVERTIME)

Carrying cost	$ 400	(80 units carried × $5/unit)
Regular labor	39,680	(8 workers × $40/day × 124 days)
Overtime	13,888	(1,984 hours × $7/hour)
Hiring or firing	0	
Subcontracting	0	
Total costs	$53,968	

Solved Problem 13.2

A Dover, Delaware, plant has developed the accompanying supply, demand, cost, and inventory data. The firm has a constant workforce and meets all of its demand. Allocate production capacity to satisfy demand at a minimum cost. What is the cost of this plan?

Supply Capacity Available (in units)

PERIOD	REGULAR TIME	OVERTIME	SUBCONTRACT
1	300	50	200
2	400	50	200
3	450	50	200

Demand Forecast

PERIOD	DEMAND (UNITS)
1	450
2	550
3	750

Other Data

Initial inventory	50 units
Regular-time cost per unit	$50
Overtime cost per unit	$65
Subcontract cost per unit	$80
Carrying cost per unit per period	$ 1
Back order cost per unit per period	$ 4

SOLUTION

SUPPLY FROM		Period 1	Period 2	Period 3	Unused Capacity (dummy)	TOTAL CAPACITY AVAILABLE (supply)
Beginning inventory		0 / 50	1	2	0	50
Period 1	Regular time	50 / 300	51	52	0	300
	Overtime	65 / 50	66	67	0	50
	Subcontract	80 / 50	81	82	0 / 150	200
Period 2	Regular time	54	50 / 400	51	0	400
	Overtime	69	65 / 50	66	0	50
	Subcontract	84	80 / 100	81 / 50	0 / 50	200
Period 3	Regular time	58	54	50 / 450	0	450
	Overtime	73	69	65 / 50	0	50
	Subcontract	88	84	80 / 200	0	200
TOTAL DEMAND		450	550	750	200	1,950

Cost of plan:

Period 1:	50($0) + 300($50) + 50($65) + 50($80)	= $22,250
Period 2:	400($50) + 50($65) + 100($80)	= $31,250
Period 3:	50($81) + 450($50) + 50($65) + 200($80)	= $45,800
	Total cost	$99,300

INTERNET AND STUDENT CD-ROM EXERCISES

Visit our home page or use your student CD-ROM to help with material in this chapter.

 On Our Home Page, www.prenhall.com/heizer

- Self-Tests
- Practice Problems
- Internet Exercises
- Current Articles and Research
- Virtual Company Tour
- Internet Homework Problems
- Internet Case

 On Your Student CD-ROM

- PowerPoint Lecture
- Practice Problems
- Active Model Exercise
- ExcelOM
- Excel OM Example Data File

ADDITIONAL CASE STUDIES

Internet Case Studies: Visit our Web site at www.prenhall.com/heizer **for this free case study:**

- **Cornell Glass**: Involves setting a production schedule for an auto glass producer.

Harvard has selected these Harvard Business School cases to accompany this chapter (textbookcasematch.hbsp.harvard.edu)**:**

- **MacPherson Refrigeration Ltd.** (#93 D021): Students need to evaluate three aggregate production plans for the company's products.

- **Sport Obermeyer Ltd.** (#695-022): This Asian skiwear company has to match supply with demand for products with uncertain demand and a globally dispersed supply chain.

- **Chaircraft Corp.** (#689-082): Illustrates effective production planning in a multistage process affected by seasonal demand.

Chapter 14

Material Requirements Planning (MRP) and ERP

Chapter Outline

LEARNING OBJECTIVES

When you complete this chapter you should be able to

IDENTIFY OR DEFINE:

Planning bills and kits

Phantom bills

Low-level coding

Lot sizing

DESCRIBE OR EXPLAIN:

Material requirements planning

Distribution requirements planning

Enterprise resource planning

How ERP works

Advantages and disadvantages of ERP systems

MRP Provides a Competitive Advantage for Collins Industries

Collins Industries, headquartered in Hutchinson, Kansas, is the largest manufacturer of ambulances in the world. The $200-million firm is an international competitor that sells more than 20% of its vehicles to markets outside the U.S. In its largest ambulance subsidiary (named Wheeled Coach), located in Winter Park, Florida, vehicles are produced on assembly lines (i.e., a repetitive process). Twelve major ambulance designs are assembled at the Florida plant, and they use 18,000 different inventory items, including 6,000 manufactured parts and 12,000 purchased parts.

This variety of products and the nature of the process demand good material requirements planning. Effective use of an MRP system requires accurate bills of material and inventory records. The Collins system, which uses MAPICS DB software on an IBM AS400 minicomputer, provides daily updates and has reduced inventory by over 30% in just 2 years.

Collins insists that four key tasks be performed properly. First, the material plan must meet both the requirements of the master schedule and the capabilities of the production facility. Second, the plan must be executed as designed.

Third, effective "time-phased" material deliveries, consignment inventories, and a constant review of purchase methods reduce inventory investment. Finally, Collins maintains excellent record integrity. Record accuracy is recognized as a fundamental ingredient of its successful MRP program. Collins's cycle counters are charged with material audits that not only correct errors, but also investigate and correct problems.

Collins Industries uses MRP as the catalyst for low inventory, high quality, tight schedules, and accurate records. Collins has found competitive advantage via MRP.

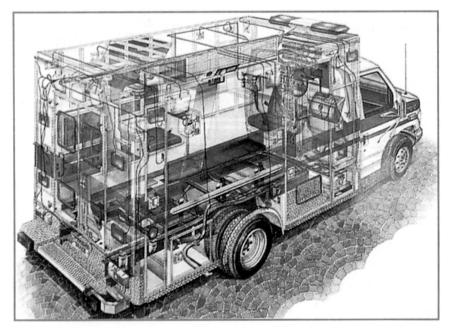

This cutaway of one ambulance interior indicates the complexity of the product, which for some rural locations may be the equivalent of a hospital emergency room in miniature. To complicate production, virtually every ambulance is custom-ordered. This customization necessitates precise orders, excellent bills of materials, exceptional inventory control from supplier to assembly, and an MRP system that works.

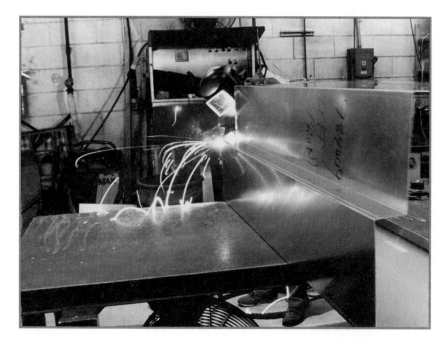

Collins uses work cells to feed the assembly line. It maintains a complete carpentry shop (to provide interior cabinetry), a paint shop (to prepare, paint, and detail each vehicle), an electrical shop (to provide for the complex electronics in a modern ambulance), an upholstery shop (to make interior seats and benches), and as shown here, a metal fabrication shop (to construct the shell of the ambulance).

On six parallel lines, ambulances move forward each day to the next workstation. The MRP system makes certain that just the materials needed at each station arrive overnight for assembly the next day.

411

Material requirements planning (MRP)
A dependent demand technique that uses bill-of-material, inventory, expected receipts, and a master production schedule to determine material requirements.

Collins Industries and many other firms have found important benefits in MRP. These benefits include (1) better response to customer orders as the result of improved adherence to schedules, (2) faster response to market changes, (3) improved utilization of facilities and labor, and (4) reduced inventory levels. Better response to customer orders and to the market wins orders and market share. Better utilization of facilities and labor yields higher productivity and return on investment. Less inventory frees up capital and floor space for other uses. These benefits are the result of a strategic decision to use a *dependent* inventory scheduling system. Demand for every component of an ambulance is dependent.

By *dependent demand*, we mean the demand for one item is related to the demand for another item. Consider the Ford Explorer. Ford's demand for auto tires and radiators depends on the production of Explorers. Four tires and one radiator go into each finished Explorer. Demand for items is *dependent* when the relationship between the items can be determined. Therefore, once management receives an order or makes a forecast of the demand for the final product, quantities required for all components can be computed, because all components are dependent items. The Boeing Aircraft operations manager who schedules production of one plane per week, for example, knows the requirements down to the last rivet. For any product, all components of that product are dependent demand items. *More generally, for any item for which a schedule can be established, dependent techniques should be used.*

When their requirements are met, dependent models are preferable to the EOQ models described in Chapter 12.[1] Dependency exists for all component parts, subassemblies, and supplies once a master schedule is known. Dependent models are better not only for manufacturers and distributors but also for a wide variety of firms from restaurants to hospitals.[2] The dependent technique used in a production environment is called **material requirements planning (MRP)**.

Because MRP provides such a clean structure for dependent demand, it has evolved as the basis for Enterprise Resource Planning (ERP). ERP is an information system for identifying and planning the enterprise-wide resources needed to take, make, ship, and account for customer orders. We will discuss ERP in the latter part of this chapter.

DEPENDENT INVENTORY MODEL REQUIREMENTS

Effective use of dependent inventory models requires that the operations manager know the following:

1. Master production schedule (what is to be made and when).
2. Specifications or bill of material (materials and parts required to make the product).
3. Inventory availability (what is in stock).
4. Purchase orders outstanding (what is on order).
5. Lead times (how long it takes to get various components).

We now discuss each of these requirements in the context of material requirements planning (MRP).

Master Production Schedule

Master production schedule (MPS)
A timetable that specifies what is to be made and when.

A **master production schedule (MPS)** specifies what is to be made (i.e., the number of finished products or items) and when. The schedule must be in accordance with a production plan. The production plan sets the overall level of output in broad terms (for example, product families, standard hours, or dollar volume). The plan also includes a variety of inputs, including financial plans, customer demand, engineering capabilities, labor availability, inventory fluctuations, supplier performance, and other considerations. Each of these inputs contributes in its own way to the production plan, as shown in Figure 14.1.

As the planning process moves from the production plan to execution, each of the lower-level plans must be feasible. When one is not, feedback to the next higher level is used to make the necessary adjustment. One of the major strengths of MRP is its ability to determine precisely the feasibility of a schedule within capacity constraints. This planning process can yield excellent

---

[1]The inventory models (EOQ) discussed in Chapter 12 assumed that the demand for one item was independent of the demand for another item. For example, EOQ assumes the demand for refrigerator parts is *independent* of the demand for refrigerators and that demand is constant.

[2]Aleda V. Roth and Roland van Dierdonck, "Hospital Resource Planning: Concepts, Feasibility, and Framework," *Production and Operations Management* (winter 1995): 2–29.

FIGURE 14.1 ■

The Planning Process

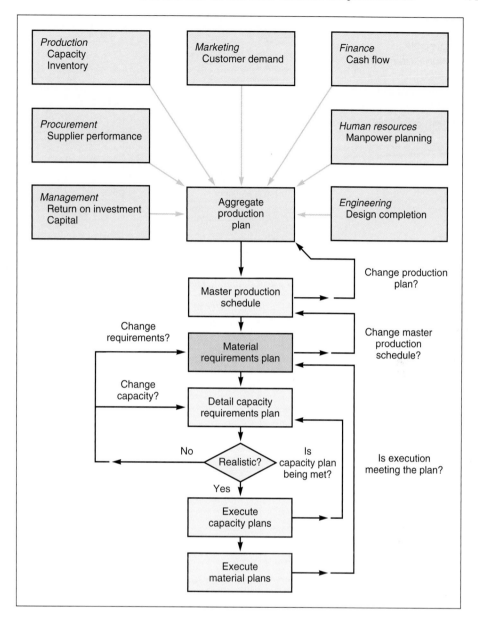

Regardless of the complexity of the planning process, the production plan and its derivative, the master production schedule, must be developed.

The master production schedule is derived from the aggregate schedule.

Video 14.1

MRP at Wheeled Coach Ambulances

results. The production plan sets the upper and lower bounds on the master production schedule. The result of this production planning process is the master production schedule.

The master production schedule tells us what is required to satisfy demand and meet the production plan. This schedule establishes what items to make and when: It *disaggregates* the aggregate production plan. While the *aggregate production plan* (as discussed in Chapter 13) is established in gross terms such as families of products or tons of steel, the *master production schedule* is established in terms of specific products. Figure 14.2 shows the master schedules for three stereo models that flow from the aggregate production plan for a family of stereo amplifiers.

Managers must adhere to the schedule for a reasonable length of time (usually a major portion of the production cycle—the time it takes to produce a product). Many organizations establish a master production schedule and establish a policy of not changing ("fixing") the near-term portion of the plan. This near-term portion of the plan is then referred to as the "fixed," "firm," or "frozen" schedule. The Wheeled Coach division of Collins Industries, the subject of the *Global Company Profile* for this chapter, fixes the last 14 days of its schedule. Only changes beyond the fixed schedule are permitted. The schedule then becomes a "rolling" production schedule. For example, a fixed 7-week plan has an additional week added to it as each week is completed so a 7-week fixed sched-

FIGURE 14.2 ■

The Aggregate
Production Plan Provides
the Basis for
Development of the
Detailed Master
Production Schedule

Months	January				February				
Aggregate Production Plan (Shows the total quantity of amplifiers)	1,500				1,200				
Weeks	1	2	3	4	5	6	7	8	
Master Production Schedule (Shows the specific type and quantity of amplifier to be produced)									
240 watt amplifier	100		100		100		100		
150 watt amplifier		500		500		450		450	
75 watt amplifier			300			100			

ule is maintained. Note that the master production schedule is a statement of *what is to be produced*, not a forecast of demand. The master schedule can be expressed in any of the following terms:

1. A *customer order in a job shop* (make-to-order) company.
2. *Modules in a repetitive* (assemble-to-stock) company.
3. An *end item in a continuous* (make-to-stock) company.

This relationship of the master production schedule to the processes is shown in Figure 14.3.

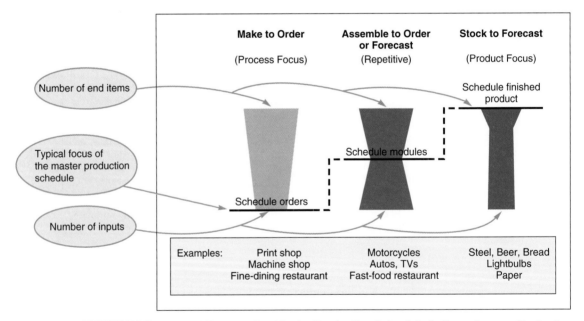

FIGURE 14.3 ■ Typical Focus of the Master Production Schedule in Three Process Strategies

A master production schedule for two of Nancy's Specialty Foods' products, crabmeat quiche and spinach quiche, might look like Table 14.1.

TABLE 14.1 ■

Master Production
Schedule for Crabmeat
Quiche and Spinach
Quiche at Nancy's
Specialty Foods

GROSS REQUIREMENTS FOR CRABMEAT QUICHE									
Day	6	7	8	9	10	11	12	13	14	and so on
Amount	50		100	47	60		110	75		

GROSS REQUIREMENTS FOR SPINACH QUICHE									
Day	7	8	9	10	11	12	13	14	15	16	and so on
Amount	100	200	150			60	75		100		

Bills of Material

Defining what goes into a product may seem simple, but it can be difficult in practice. As we noted in Chapter 5, to aid this process, manufactured items are defined via a bill of material. A **bill of material (BOM)** is a list of quantities of components, ingredients, and materials required to make a product. Individual drawings describe not only physical dimensions but also any special processing as well as the raw material from which each part is made. Nancy's Specialty Foods has a recipe for quiche, specifying ingredients and quantities, just as Collins Industries has a full set of drawings for an ambulance. Both are bills of material (although we call one a recipe and they do vary somewhat in scope).

Because there is often a rush to get a new product to market, however, drawings and bills of material may be incomplete or even nonexistent. Moreover, complete drawings and BOM (as well as other forms of specifications) often contain errors in dimensions, quantities, or countless other areas. When errors are identified, engineering change notices (ECNs) are created, further complicating the process. An *engineering change notice* is a change or correction to an engineering drawing or bill of material.

One way a bill of material defines a product is by providing a product structure. Example 1 shows how to develop the product structure and "explode" it to reveal the requirements for each component. A bill of material for item A in Example 1 consists of items B and C. Items above any level are called *parents*; items below any level are called *components* or *children*.

Example 1

Excel Om Data File
Ch14Ex1.xla

Speaker Kits, Inc., packages high-fidelity components for mail order. Components for the top-of-the-line speaker kit, "Awesome" (A), include 2 standard 12-inch speaker kits (Bs) and 3 speaker kits with amp-boosters (Cs).

Each B consists of 2 speakers (Ds) and 2 shipping boxes each with an installation kit (E). Each of the three 300-watt stereo kits (Cs) has 2 speaker boosters (Fs) and 2 installation kits (Es). Each speaker booster (F) includes 2 speakers (Ds) and 1 amp-booster (G). The total for each Awesome is 4 standard 12-inch speakers and twelve 12-inch speakers with the amp-booster. (Most purchasers require hearing aids within 2 years, and at least one court case is pending because of structural damage to a men's dormitory.) As we can see, the demand for B, C, D, E, F, and G is completely dependent on the master production schedule for A—the Awesome speaker kits. Given this information, we can construct the following product structure:

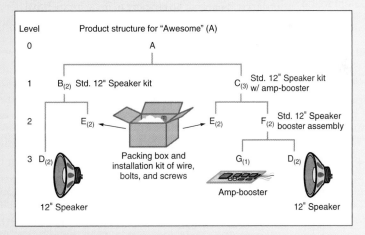

This structure has four levels: 0, 1, 2, and 3. There are four parents: A, B, C, and F. Each parent item has at least one level below it. Items B, C, D, E, F, and G are components because each item has at least one level above it. In this structure, B, C, and F are both parents and components. The number in parentheses indicates how many units of that particular item are needed to make the item immediately above it. Thus, $B_{(2)}$ means that it takes two units of B for every unit of A, and $F_{(2)}$ means that it takes two units of F for every unit of C.

Once we have developed the product structure, we can determine the number of units of each item required to satisfy demand for a new order of 50 Awesome speaker kits. This information is displayed below:

Part B:	$2 \times$ number of As =	$(2)(50)$ =	100
Part C:	$3 \times$ number of As =	$(3)(50)$ =	150
Part D:	$2 \times$ number of Bs + $2 \times$ number of Fs =	$(2)(100) + (2)(300)$ =	800

Part E: $2 \times$ number of Bs $+ 2 \times$ number of Cs $= (2)(100) + (2)(150) = 500$

Part F: $2 \times$ number of Cs $=$ $\qquad$ $(2)(150) =$ $\qquad$ 300

Part G: $1 \times$ number of Fs $=$ $\qquad$ $(1)(300) =$ $\qquad$ 300

Thus, for 50 units of A, we will need 100 units of B, 150 units of C, 800 units of D, 500 units of E, 300 units of F, and 300 units of G.

Bills of material not only specify requirements but also are useful for costing, and they can serve as a list of items to be issued to production or assembly personnel. When bills of material are used in this way, they are usually called *pick lists*.

Modular bills

Bills of material organized by major subassemblies or by product options.

Modular Bills   Bills of material may be organized around product modules (see Chapter 5). *Modules* are not final products to be sold, but are components that can be produced and assembled into units. They are often major components of the final product or product options. Bills of material for modules are called **modular bills**. Bills of material are sometimes organized as modules (rather than as part of a final product) because production scheduling and production are often facilitated by organizing around relatively few modules rather than a multitude of final assemblies. For instance, a firm may make 138,000 different final products but have only 40 modules that are mixed and matched to produce those 138,000 final products. The firm builds an aggregate production plan and prepares its master production schedule for the 40 modules, not the 138,000 configurations of the final product. This approach allows the MPS to be prepared for a reasonable number of items (the narrow portion of the middle graphic in Figure 14.3) and to postpone assembly. The 40 modules can then be configured for specific orders at final assembly.

Planning bills (or kits)

A material grouping created in order to assign an artificial parent to the bill of material.

Planning Bills and Phantom Bills   Two other special kinds of bills of material are planning bills and phantom bills. **Planning bills** are created in order to assign an artificial parent to the bill of material. Such bills are used (1) when we want to group subassemblies so the number of items to be scheduled is reduced and (2) when we want to issue "kits" to the production department. For instance, it may not be efficient to issue inexpensive items such as washers and cotter pins with each of numerous subassemblies, so we call this a *kit* and generate a planning bill. The planning bill specifies the *kit* to be issued. Consequently, a planning bill may also be known as **kitted material** or **kit**. **Phantom bills of material** are bills of material for components, usually subassemblies, that exist only temporarily. These components go directly into another assembly and are never inventoried. Therefore, components of phantom bills of material are coded to receive special treatment; lead times are zero, and they are handled as an integral part of their parent item. An example is a transmission shaft with gears and bearings assembly that is placed directly into a transmission.

Phantom bills of material

Bills of material for components, usually assemblies, that exist only temporarily; they are never inventoried.

Low-level coding

A number that identifies items at the lowest level at which they occur.

Low-Level Coding   Low-level coding of an item in a BOM is necessary when identical items exist at various levels in the BOM. **Low-level coding** means that the item is coded at the lowest level at which it occurs. For example, item D in Example 1 is coded at the lowest level at which it is used. Item D could be coded as part of B and occur at level 2. However, because D is also part of F, and F is level 2, item D becomes a level-3 item. Low-level coding is a convention to allow easy computing of the requirements of an item. When the BOM has thousands of items or when requirements are frequently recomputed, the ease and speed of computation become a major concern.

Low-level coding ensures that an item is always at the lowest level of usage.

Accurate Inventory Records

As we saw in Chapter 12, knowledge of what is in stock is the result of good inventory management. Good inventory management is an absolute necessity for an MRP system to work. If the firm has not yet achieved at least 99% record accuracy, then material requirements planning will not work.

Purchase Orders Outstanding

Knowledge of outstanding orders should exist as a by-product of well-managed purchasing and inventory-control departments. When purchase orders are executed, records of those orders and their scheduled delivery dates must be available to production personnel. Only with good purchasing data can managers prepare good production plans and effectively execute an MRP system.

For manufacturers like Harley-Davidson, which produces a large number of end products from a relatively small number of options, modular bills of material provide an effective solution.

Lead Times for Each Component

Lead time

In purchasing systems, the time between recognition of the need for an order and receiving it; in production systems, it is the order, wait, move, queue, setup, and run times for each component produced.

Once managers determine when products are needed, they determine when to acquire them. The time required to acquire (that is, purchase, produce, or assemble) an item is known as **lead time**. Lead time for a manufactured item consists of *move*, *setup*, and *assembly* or *run times* for each component. For a purchased item, the lead time includes the time between recognition of need for an order and when it is available for production.

When the bill of material for Awesome speaker kits (As), in Example 1, is turned on its side and modified by adding lead times for each component (see Table 14.2), we then have a *time-phased product structure*. Time in this structure is shown on the horizontal axis of Figure 14.4 with item A due for completion in week 8. Each component is then offset to accommodate lead times.

TABLE 14.2 ■ Lead Times for Awesome Speaker Kits (As)

COMPONENT	LEAD TIME
A	1 week
B	2 weeks
C	1 week
D	1 week
E	2 weeks
F	3 weeks
G	2 weeks

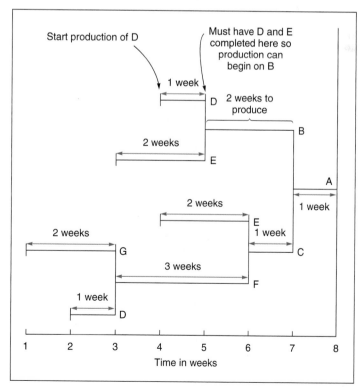

FIGURE 14.4 ■ Time-Phased Product Structure

FIGURE 14.5 ■

Structure of the MRP
System

MRP software programs
are popular because
many organizations face
dependent demand
situations.

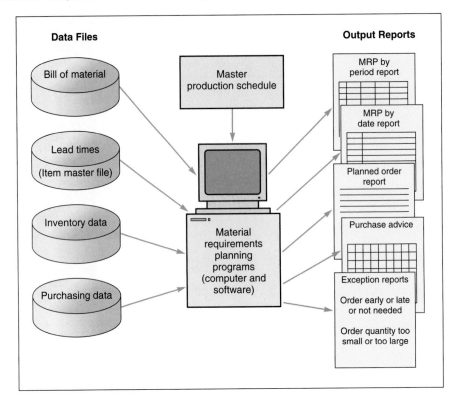

**Gross material
requirements plan**

A schedule that shows
the total demand for an
item (prior to subtraction
of on-hand inventory
and scheduled receipts)
and (1) when it must be
ordered from suppliers,
or (2) when production
must be started to meet
its demand by a
particular date.

MRP STRUCTURE

Although most MRP systems are computerized, the MRP procedure is straightforward and can be done by hand. A master production schedule, a bill of material, inventory and purchase records, and lead times for each item are the ingredients of a material requirements planning system (see Figure 14.5).

Once these ingredients are available and accurate, the next step is to construct a gross material requirements plan. The **gross material requirements plan** is a schedule, as shown in Example 2. It combines a master production schedule (that requires one unit of A in week 8) and the time-phased schedule (Figure 14.4). It shows when an item must be ordered from suppliers if there is no inventory on hand or when the production of an item must be started in order to satisfy demand for the finished product by a particular date.

Example 2

Each Awesome speaker kit (item A of Example 1) requires all the items in the product structure for A. Lead times are shown in Table 14.2. Using this information, we construct the gross material requirements plan and draw up a production schedule that will satisfy the demand of 50 units of A by week 8. The result is shown in Table 14.3.

You can interpret the gross material requirements shown in Table 14.3 as follows: If you want 50 units of A at week 8, you must start assembling A in week 7. Thus, in week 7, you will need 100 units of B and 150 units of C. These two items take 2 weeks and 1 week, respectively, to produce. Production of B, therefore, should start in week 5, and production of C should start in week 6 (lead time subtracted from the required date for these items). Working backward, we can perform the same computations for all of the other items. Because D and E are used in two different places in Awesome speaker kits, there are two entries in each data record.

The material requirements plan shows when production of each item should begin and end in order to have 50 units of A at week 8.

TABLE 14.3 ■ Gross Material Requirements Plan for 50 Awesome Speaker Kits (As)

				WEEK					
	1	2	3	4	5	6	7	8	LEAD TIME
A. Required date								50	
Order release date							50		1 week
B. Required date							100		
Order release date					100				2 weeks
C. Required date							150		
Order release date						150			1 week
E. Required date					200	300			
Order release date			200	300					2 weeks
F. Required date						300			
Order release date			300						3 weeks
D. Required date				600	200				
Order release date		600		200					1 week
G. Required date			300						
Order release date	300								2 weeks

So far, we have considered *gross material requirements*, which assumes that there is no inventory on hand. When there is inventory on hand, we prepare a *net requirements plan*. When considering on-hand inventory, we must realize that many items in inventory contain subassemblies or parts. If the gross requirement for Awesome speaker kits (As) is 100 and there are 20 of those speakers on hand, the net requirement for Awesome speaker kits (As) is 80 (that is, 100 − 20). However, each Awesome speaker kit on hand contains 2 Bs. As a result, the requirement for Bs drops by 40 Bs (20 A kits on hand × 2 Bs per A). Therefore, if inventory is on hand for a parent item, the requirements for the parent item and all its components decrease because each Awesome kit contains the components for lower-level items. Example 3 shows how to create a net requirements plan.

Example 3

Active Model 14.1

Examples 1–3 are further illustrated in Active Model 14.1 on the CD-ROM and in the Exercise located in your Student Lecture Guide.

Net material requirements
The result of adjusting gross requirements for inventory on hand and scheduled receipts.

Planned order receipt
The quantity planned to be received at a future date.

Planned order release
The scheduled date for an order to be released.

In Example 1, we developed a product structure from a bill of material, and in Example 2, we developed a gross requirements plan. Given the following on-hand inventory, we now construct a net requirements plan.

ITEM	ON HAND
A	10
B	15
C	20
D	10
E	10
F	5
G	0

A **net material requirements** plan includes gross requirements, on-hand inventory, net requirements, planned order receipt, and planned order release for each item. We begin with A and work backward through the components. Shown in the chart on page 420 is the net material requirements plan for product A.

Constructing a net requirements plan is similar to constructing a gross requirements plan. Starting with item A, we work backward to determine net requirements for all items. To do these computations, we refer to the product structure, on-hand inventory, and lead times. The gross requirement for A is 50 units in week 8. Ten items are on hand; therefore, the net requirements and the scheduled **planned order receipt** are both 40 items in week 8. Because of the 1-week lead time, the **planned order release** is 40 items in week 7 (see the arrow connecting the order receipt and order release). Referring to week 7 and the product structure in Example 1, we can see that 80 (2 × 40) items of B and 120 (3 × 40) items of C are required in week 7 in order to have a total for 50 items of A in week 8. The letter A to the right of the gross figure for items B and C was generated as a result of the demand for the parent, A. Performing the same type of analysis for B and C yields the net requirements for D, E, F, and G. Note the on-hand inventory in row E in week 6 is zero. It is zero because the on-hand inventory (10 units) was used to make B in week 5. By the same token, the inventory for D was used to make F.

Lot Size	Lead Time (weeks)	On Hand	Safety Stock	Allocated	Low-Level Code	Item Identification		Week 1	Week 2	Week 3	Week 4	Week 5	Week 6	Week 7	Week 8
Lot-for-Lot	1	10	—	—	0	A	Gross Requirements								50
							Scheduled Receipts								
							Projected On Hand (10)	10	10	10	10	10	10	10	10
							Net Requirements								40
							Planned Order Receipts								40
							Planned Order Releases							40	
Lot-for-Lot	2	15	—	—	1	B	Gross Requirements							80^A	
							Scheduled Receipts								
							Projected On Hand (15)	15	15	15	15	15	15	15	
							Net Requirements							65	
							Planned Order Receipts							65	
							Planned Order Releases					65			
Lot-for-Lot	1	20	—	—	1	C	Gross Requirements							120^A	
							Scheduled Receipts								
							Projected On Hand (20)	20	20	20	20	20	20	20	
							Net Requirements							100	
							Planned Order Receipts							100	
							Planned Order Releases						100		
Lot-for-Lot	2	10	—	—	2	E	Gross Requirements						130^B	200^C	
							Scheduled Receipts								
							Projected On Hand (10)	10	10	10	10	10			
							Net Requirements						120	200	
							Planned Order Receipts						120	200	
							Planned Order Releases				120	200			
Lot-for-Lot	3	5	—	—	2	F	Gross Requirements							200^C	
							Scheduled Receipts								
							Projected On Hand (5)	5	5	5	5	5	5		
							Net Requirements							195	
							Planned Order Receipts							195	
							Planned Order Releases				195				
Lot-for-Lot	1	10	—	—	3	D	Gross Requirements					390^F		130^B	
							Scheduled Receipts								
							Projected On Hand (10)	10	10	10					
							Net Requirements					380		130	
							Planned Order Receipts					380		130	
							Planned Order Releases				380		130		
Lot-for-Lot	2	0	—	—	3	G	Gross Requirements					195^F			
							Scheduled Receipts								
							Projected On Hand					0			
							Net Requirements					195			
							Planned Order Receipts					195			
							Planned Order Releases			195					

Net Material Requirements Plan for Product A   *Note that the superscript is the source of the demand.*

Examples 2 and 3 considered only product A, the Awesome speaker kit, and its completion only in week 8. Fifty units of A were required in week 8. Normally, however, there is a demand for many products over time. For each product, management must prepare a master production schedule (as we saw earlier in Table 14.1). Scheduled production of each product is added to the master schedule and ultimately to the net material requirements plan. Figure 14.6 shows how several product schedules, including requirements for components sold directly, can contribute to one gross material requirements plan.

Most inventory systems also note the number of units in inventory that have been assigned to specific future production but not yet used or issued from the stockroom. Such items are often referred to as *allocated* items. Allocated items increase requirements and may then be included in an MRP planning sheet, as shown in Figure 14.7.

FIGURE 14.6 ■

Several Schedules
Contributing to a Gross
Requirements Schedule
for B

*One "B" is in each A and
one "B" is in each S;
additionally, 10 Bs sold
directly are scheduled in
week 1 and 10 more that
are sold directly are
scheduled in week 2.*

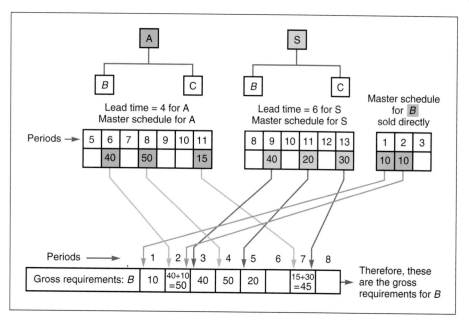

FIGURE 14.7 ■ Sample MRP Planning Sheet for Item Z

The allocated quantity has the effect of increasing the requirements (or, alternatively, reducing the quantity on hand). The logic, then, of a net requirements MRP is

$$\underbrace{\left[\left(\begin{array}{c}\text{gross}\\\text{requirements}\end{array}\right)+\left(\text{allocations}\right)\right]}_{\text{total requirements}}-\underbrace{\left[\left(\begin{array}{c}\text{on}\\\text{hand}\end{array}\right)+\left(\begin{array}{c}\text{scheduled}\\\text{receipts}\end{array}\right)\right]}_{\text{available inventory}}=\begin{array}{c}\text{net}\\\text{requirements}\end{array}$$

MRP MANAGEMENT

The material requirements plan is not static. And since MRP systems increasingly are integrated with just-in-time (JIT) techniques, we will now discuss these two issues.

MRP Dynamics

Bills of material and material requirements plans are altered as changes in design, schedules, and production processes occur. Additionally, changes occur in material requirements whenever the master production schedule is modified. Regardless of the cause of any changes, the MRP model can be manipulated to reflect them. In this manner, an up-to-date requirements schedule is possible.

Due to the changes that occur in MRP data, it is not uncommon to recompute MRP requirements about once a week. Conveniently, a central strength of MRP is its timely and accurate *replanning*

System nervousness
Frequent changes in the
MRP system.

capability. However, many firms find they do not want to respond to minor scheduling or quantity changes even if they are aware of them. These frequent changes generate what is called **system nervousness** and can create havoc in purchasing and production departments if implemented. Consequently, OM personnel reduce such nervousness by evaluating the need and impact of changes prior to disseminating requests to other departments. Two tools are particularly helpful when trying to reduce MRP system nervousness.

Time fences
A way of allowing a
segment of the master
schedule to be
designated as "not
to be rescheduled."

The first is time fences. **Time fences** allow a segment of the master schedule to be designated as "not to be rescheduled." This segment of the master schedule is thus not changed during the periodic regeneration of schedules. The second tool is pegging. **Pegging** means tracing upward in the BOM from the component to the parent item. By pegging upward, the production planner can determine the cause for the requirement and make a judgment about the necessity for a change in the schedule.

Pegging
In material requirements
planning systems, tracing
upward in the bill of
material (BOM) from the
component to the
parent item.

With MRP, the operations manager *can* react to the dynamics of the real world. How frequently the manager wishes to impose those changes on the firm requires professional judgment. Moreover, if the nervousness is caused by legitimate changes, then the proper response may be to investigate the production environment—not adjust via MRP.[3]

MRP and JIT

MRP is a planning and scheduling technique with *fixed* lead times, while just-in-time (JIT) is a way to move material expeditiously. Fixed lead times can be a limitation. For instance, the lead time to produce 50 units may vary substantially from the lead time to produce 1 unit. This limitation complicates the marriage of JIT and MRP. In many respects, however, an MRP system combined with JIT provides the best of both worlds. MRP provides a good master schedule and an accurate picture of requirements, and JIT reduces work-in-process inventory. Let's look at two approaches for integrating the two systems: small buckets and balanced flow.

Small Bucket Approach MRP is an excellent tool for resource and scheduling management in process-focused facilities, that is, in job shops. Such facilities include machine shops, hospitals, and restaurants, where lead times are relatively stable and poor balance between work centers is expected. Schedules are often driven by work orders, and lot sizes are the exploded bill-of-material size. In these enterprises, MRP can be integrated with JIT through the following steps.

Buckets
Time units in a material
requirements planning
(MRP) system.

Step 1: Reduce MRP "buckets" from weekly to daily to perhaps hourly. **Buckets** are time units in an MRP system. Although the examples in this chapter have used weekly *time buckets*, many firms now use daily or even fraction-of-a-day time buckets. Some systems use a **bucketless system** in which all time-phased data have dates attached rather than defined time periods or buckets.

Bucketless system
Time-phased data are
referenced using dated
records rather than
defined time periods,
or buckets.

Step 2: The planned receipts that are part of a firm's planned orders in an MRP system are communicated to the work areas for production purposes and used to sequence production.

Step 3: Inventory is moved through the plant on a JIT basis.

Step 4: As products are completed, they are moved into inventory (typically finished goods inventory) in the normal way. Receipt of these products into inventory reduces the quantities required for subsequent planned orders in the MRP system.

Back flush
A system to reduce
inventory balances by
deducting everything in
the bill of material on
completion of the unit.

Step 5: A system known as *back flush* is used to reduce inventory balances. **Back flushing** uses the bill of materials to deduct component quantities from inventory as each unit is completed.

The focus in these facilities becomes one of maintaining schedules. Nissan achieves success with this approach by computer communication links to suppliers. These schedules are confirmed, updated, or changed every 15 to 20 minutes. Suppliers provide deliveries 4 to 16 times per day. Master schedule performance is 99% on time, as measured every hour. On-time delivery from suppliers is 99.9% and for manufactured piece parts, 99.5%.

---

[3]Jay H. Heizer, "The Production Manager Can Be a Good Guy in the Factory with a Future," *APICS—The Performance Advantage* (July 1994): 30–34.

Balanced Flow Approach MRP supports the planning and scheduling necessary for repetitive operations, such as the assembly lines at Harley-Davidson, Whirlpool, and a thousand other places. In these environments, the planning portion of MRP is combined with JIT execution. The JIT portion uses kanbans, visual signals, and reliable suppliers to pull the material through the facility. In these systems, execution is achieved by maintaining a carefully balanced flow of material to assembly areas with small lot sizes.[4]

LOT-SIZING TECHNIQUES

An MRP system is an excellent way to determine production schedules and net requirements. However, whenever we have a net requirement, a decision must be made about *how much* to order. This decision is called a **lot-sizing decision**. There are a variety of ways to determine lot sizes in an MRP system; commercial MRP software usually includes the choice of several lot-sizing techniques. We will now review a few of them.

Lot-for-Lot In Example 3, we used a lot-sizing technique known as **lot-for-lot**, which produced exactly what was required. This decision is consistent with the objective of an MRP system, which is to meet the requirements of *dependent* demand. Thus, an MRP system should produce units only as needed, with no safety stock and no anticipation of further orders. When frequent orders are economical and just-in-time inventory techniques implemented, lot-for-lot can be very efficient. However, when setup costs are significant or management has been unable to implement JIT, lot-for-lot can be expensive. Example 4 uses the lot-for-lot criteria and determines cost for 10 weeks of demand.

> **Lot-sizing decision**
> The process of, or techniques used in, determining lot size.

> **Lot-for-lot**
> A lot-sizing technique that generates exactly what was required to meet the plan.

Example 4

Speaker Kits, Inc., wants to compute its ordering and carrying cost of inventory on lot-for-lot criteria. Speaker Kits has determined that, for the 12-inch speaker/booster assembly, setup cost is $100 and holding cost is $1 per period. The production schedule, as reflected in net requirements for assemblies, is as follows:

MRP LOT-SIZING PROBLEM: LOT-FOR-LOT TECHNIQUE

		1	2	3	4	5	6	7	8	9	10
Gross requirements		35	30	40	0	10	40	30	0	30	55
Scheduled receipts											
Projected on hand	35	35	0	0	0	0	0	0	0	0	0
Net requirements		0	30	40	0	10	40	30	0	30	55
Planned order receipts			30	40		10	40	30		30	55
Planned order releases		30	40		10	40	30		30	55	

Holding costs = $1/unit/week; setup cost = $100; gross requirements average per week = 27; lead time = 1 week.

Shown above is the lot-sizing solution using the lot-for-lot technique and its cost. The holding cost is zero, but seven separate setups (one associated with each order) yield a total cost of $700.

Economic Order Quantity As discussed in Chapter 12, EOQ can be used as a lot-sizing technique. But as we indicated there, EOQ is preferable when *relatively constant* independent demand exists, not when we *know* the demand. EOQ is a statistical technique using averages (such as average demand for a year) whereas our MRP procedure assumes *known* (dependent) demand reflected in a master production schedule. Operations managers should take advantage of demand information when it is known, rather than assuming a constant demand. EOQ is examined in Example 5.

> MRP is preferable when demand is *dependent*. Statistical techniques such as EOQ may be preferable when demand is *independent*.

---

[4]For a related discussion, see Sylvain Landry, Claude R. Duguay, Sylvain Chausse, and Jean-Luc Themens, "Integrating MRP, Kanban, and Bar-Coding Systems to Achieve JIT Procurement," *Production and Inventory Management Journal* (first quarter 1997): 8–12.

This Nissan line in Smyrna, Tennessee, has little inventory because Nissan schedules to a razor's edge. At Nissan, MRP helps to reduce inventory to world-class standards. World-class automobile assembly requires that purchased parts have a turnover of slightly more than once a day and that overall turnover approaches 150 times per year.

Example 5

With a setup cost of $100 and a holding cost per week of $1, Speaker Kits, Inc., examines its cost with lot sizes based on an EOQ criteria. Using the same requirements as in Example 4, the net requirements and lot sizes follow:

MRP LOT-SIZING PROBLEM: EOQ TECHNIQUE

		1	2	3	4	5	6	7	8	9	10
Gross requirements		35	30	40	0	10	40	30	0	30	55
Scheduled receipts											
Projected on hand	35	35	0	43	3	3	66	26	69	69	39
Net requirements		0	30	0	0	7	0	4	0	0	16
Planned order receipts			73			73		73			73
Planned order releases		73			73		73			73	

Holding costs = $1/unit/week; setup cost = $100; gross requirements average per week = 27; lead time = 1 week.

Ten-week usage equals gross requirement of 270 units; therefore, weekly usage equals 27, and 52 weeks (annual usage) equals 1,404 units. From Chapter 12, the EOQ model is

$$Q^* = \sqrt{\frac{2DS}{H}}$$

where D = annual usage = 1,404
      S = setup cost = $100
      H = holding (carrying) cost, on an annual basis per unit
        = $1 × 52 weeks = $52

$$Q^* = 73 \text{ units}$$

Setups = 1,404/73 = 19 per year

Setup cost = 19 × $100 = $1,900

Holding cost = $\frac{73}{2}$ × ($1 × 52 weeks) = $1,898

Setup cost + holding cost = $1,900 + 1,898 = $3,798

The EOQ solution yields a computed 10-week cost of $730 [$3,798 × (10 weeks/52 weeks) = $730].

Notice that actual holding cost will vary from the computed $730, depending on the rate of actual usage. From the preceding table, we can see that in our 10-week example, costs really are $400 for four setups, plus a holding cost of 318 units at $1 per week for a total of $718. Because usage was not constant, the actual computed cost was in fact less than the theoretical EOQ ($730), but more than the lot-for-lot rule ($700). If any stockouts had occurred, these costs too would need to be added to our actual EOQ of $718.

Part period balancing (PPB)
An inventory ordering technique that balances setup and holding costs by changing the lot size to reflect requirements of the next lot size in the future.

Economic part period (EPP)
That period of time when the ratio of setup cost to holding cost is equal.

Part Period Balancing Part period balancing (PPB) is a more dynamic approach to balance setup and holding cost.[5] PPB uses additional information by changing the lot size to reflect requirements of the next lot size in the future. PPB attempts to balance setup and holding cost for known demands. Part period balancing develops an **economic part period (EPP)**, which is the ratio of setup cost to holding cost. For our Speaker Kits example, EPP = $100/$1 = 100 units. Therefore, holding 100 units for one period would cost $100, exactly the cost of one setup. Similarly, holding 50 units for two periods also costs $100 (2 periods × $1 × 50 units). PPB merely adds requirements until the number of part periods approximates the EPP—in this case, 100. Example 6 shows the application of part period balancing.

Example 6

Once again, Speaker Kits, Inc., computes the costs associated with a lot size by using a $100 setup cost and a $1 holding cost. This time, however, part period balancing is used. The data are shown in the following table:

PPB CALCULATIONS

PERIODS COMBINED	TRIAL LOT SIZE (CUMULATIVE NET REQUIREMENTS)	PART PERIODS	Costs		
			SETUP	HOLDING	TOTAL
2	30	0	40 units held for 1 period = $40		
2, 3	70	$40 = 40 \times 1$	10 units held for 3 periods = $30		
2, 3, 4	70	40			
2, 3, 4, 5	80	$70 = 40 \times 1 + 10 \times 3$	100 +	70	= 170
2, 3, 4, 5, 6	120	$230 = 40 \times 1 + 10 \times 3 + 40 \times 4$			
(Therefore, combine periods 2 through 5; 70 is as close to our EPP of 100 as we are going to get.)					
6	40	0			
6, 7	70	$30 = 30 \times 1$			
6, 7, 8	70	$30 = 30 \times 1 + 0 \times 2$			
6, 7, 8, 9	100	$120 = 30 \times 1 + 30 \times 3$	100 +	120	= 220
(Therefore, combine periods 6 through 9; 120 is as close to our EPP of 100 as we are going to get.)					
10	55	0	100 +	0	= 100
			300 +	190	= 490

MRP LOT-SIZING PROBLEM: PPB TECHNIQUE

		1	2	3	4	5	6	7	8	9	10
Gross requirements		35	30	40	0	10	40	30	0	30	55
Scheduled receipts											
Projected on hand	35	35	0	50	10	10	0	60	30	30	0
Net requirements		0	30	0	0	0	40	0	0	0	55
Planned order receipts			80				100				55
Planned order releases		80				100			55		

Holding costs = $1/unit/week; setup cost = $100; gross requirements average per week = 27; lead time = 1 week.

EPP is 100 (setup cost divided by holding cost = $100/$1). The first lot is to cover periods 1, 2, 3, 4, and 5 and is 80.

The total costs are $490, with setup costs totaling $300 and holding costs totaling $190.

[5]J. J. DeMatteis, "An Economic Lot-Sizing Technique: The Part-Period Algorithms," *IBM Systems Journal* 7 (1968): 30–38.

Many MRP programs, such as Resource Manager for Excel and DB, *are commercially available.*
Resource Manager*'s initial menu screen is shown here. A demo program is available for student use at*
www.usersolutions.com.

Wagner-Whitin procedure

A technique for lot-size computation that assumes a finite time horizon beyond which there are no additional net requirements to arrive at an ordering strategy.

Wagner-Whitin Algorithm   The **Wagner-Whitin procedure** is a dynamic programming model that adds some complexity to the lot-size computation. It assumes a finite time horizon beyond which there are no additional net requirements. It does, however, provide good results.[6] The technique is seldom used in practice, but this may change with increasing understanding and software sophistication.

Lot-Sizing Summary   In the three Speaker Kits lot-sizing examples, we found the following costs:

Lot-for-lot	$700
EOQ	$730
Part period balancing	$490

These examples should not, however, lead operations personnel to hasty conclusions about the preferred lot-sizing technique. In theory, new lot sizes should be computed whenever there is a schedule or lot-size change anywhere in the MRP hierarchy. However, in practice, such changes cause the instability and system nervousness referred to earlier in this chapter. Consequently, such frequent changes are not made. This means that all lot sizes are wrong because the production system cannot respond to frequent changes.

In general, the lot-for-lot approach should be used whenever economical. Lot-for-lot is the goal. Lots can be modified as necessary for scrap allowances, process constraints (for example, a heat-treating process may require a lot of a given size), or raw material purchase lots (for example, a truckload of chemicals may be available in only one lot size). However, caution should be exercised prior to any modification of lot size because the modification can cause substantial distortion of actual requirements at lower levels in the MRP hierarchy. When setup costs are significant and demand is reasonably smooth, part period balancing (PPB), Wagner-Whitin, or even EOQ should provide satisfactory results. Too much concern with lot sizing yields false accuracy because of MRP dynamics. A correct lot size can be determined only after the fact, based on what actually happened in terms of requirements.[7]

[6]See James M. Fordyce and Francis M. Webster, "The Wagner-Whitin Algorithm Made Simple," *Production and Inventory Management* (second quarter 1984): 21–27. This article provides as straightforward an explanation of the Wagner-Whitin technique as the authors have found. The Wagner-Whitin Algorithm yields a cost of $455 for the data in Examples 4, 5, and 6.

[7]See discussions by Joseph Orlicky, *Material Requirements Planning* (New York: McGraw-Hill, 1975): 136–137; and G. Nandakumar, "Lot-Sizing Techniques in a Multiproduct Multilevel Environment," *Production and Inventory Management* 26 (first quarter 1985): 46–54.

EXTENSIONS OF MRP

Recent years have seen the development of a number of extensions of MRP. In this section, we review three of them.

Closed-Loop MRP

Closed-loop material requirements planning implies an MRP system that provides feedback to scheduling from the inventory control system. Specifically, a **closed-loop MRP system** provides information to the capacity plan, master production schedule, and ultimately to the production plan (as shown in Figure 14.8). Virtually all commercial MRP systems are closed-loop.

Capacity Planning

In keeping with the definition of closed-loop MRP, feedback about workload is obtained from each work center. **Load reports** show the resource requirements in a work center for all work currently assigned to the work center, all work planned, and expected orders. Figure 14.9(a) shows that the initial load in the milling center exceeds capacity in weeks 4 and 6. Closed-loop MRP systems allow production planners to move the work between time periods in order to smooth the load or at least bring it within capacity. (This is the "Capacity Planning" side of Figure 14.8.) The closed-loop MRP system can then reschedule all items in the net requirements plan (see Figure 14.9[b]).

FIGURE 14.8 ■

Closed-Loop Material Requirements Planning

Source: Adapted from *Capacity Planning and Control Study Guide* (Alexandria, VA: American Production and Inventory Control Society). Reprinted by permission.

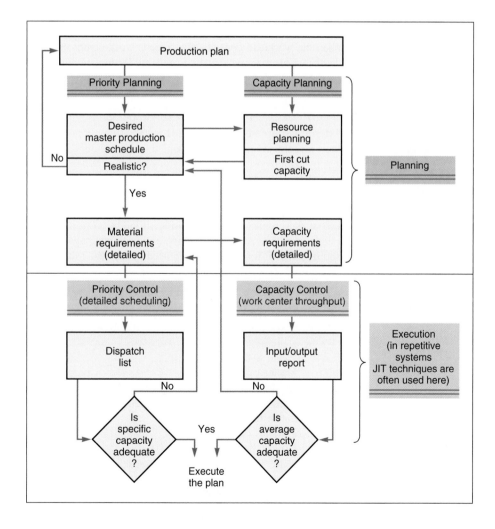

FIGURE 14.9 ■

(a) Initial Resource Requirements Profile for a Milling Center (b) Smoothed Resource Requirements Profile for a Milling Center

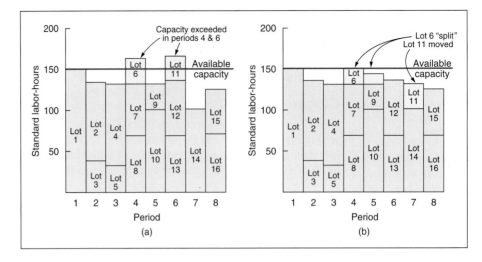

Tactics for smoothing the load and minimizing the impact of changed lead time include the following:

1. *Overlapping,* which reduces the lead time, sends pieces to the second operation before the entire lot is completed on the first operation.
2. *Operations splitting* sends the lot to two different machines for the same operation. This involves an additional setup, but results in shorter throughput times, because only part of the lot is processed on each machine.
3. *Lot splitting* involves breaking up the order and running part of it ahead of schedule.

When the workload consistently exceeds work-center capacity, the tactics just discussed are not adequate. This may mean adding capacity. Options include adding capacity via personnel, machinery, overtime, or subcontracting.

Material Requirements Planning II (MRP II)

Material requirements planning II

A system that allows, with MRP in place, inventory data to be augmented by other resource variables; in this case, MRP becomes *material resource planning*.

Material requirements planning II is an extremely powerful technique. Once a firm has MRP in place, inventory data can be augmented by labor-hours, by material cost (rather than material quantity), by capital cost, or by virtually any resource. When MRP is used this way, it is usually referred to as **MRP II**, and *resource* is usually substituted for *requirements*. MRP then stands for material *resource* planning.

For instance, so far in our discussion of MRP, we have scheduled units (quantities). However, each of these units requires resources in addition to its components. Those additional resources include labor-hours, machine-hours, and accounts payable (cash). Each of these resources can be used in an MRP format just as we used quantities. Table 14.4 shows how to determine the labor-

TABLE 14.4 ■

Material Resource Planning (MRP II)

By utilizing the logic of MRP, resources such as labor, machine-hours, and cost can be accurately determined and scheduled. Weekly demand for labor, machine-hours, and payables for 100 units are shown.

		WEEK		
	5	6	7	8
A. Units (lead time 1 week)				100
Labor: 10 hours each				1,000
Machine: 2 hours each				200
Payable: $0 each				0
B. Units (lead time 2 weeks, 2 each required)			200	
Labor: 10 hours each			2,000	
Machine: 2 hours each			400	
Payable: Raw material at $5 each			1,000	
C. Units (lead time 4 weeks, 3 each required)	300			
Labor: 2 hours each	600			
Machine: 1 hour each	300			
Payable: Raw material at $10 each	3,000			

hours, machine-hours, and cash that a sample master production schedule will require in each period. These requirements are then compared with the respective capacity (that is, labor-hours, machine-hours, cash, etc.), so operations managers can make schedules that will work.

To aid the functioning of MRP II, most MRP II computer programs are tied into other computer files that provide data to the MRP system or receive data from the MRP system. Purchasing, production scheduling, capacity planning, and warehouse management are a few examples of this data integration.

MRP IN SERVICES

The demand for many services or service items is classified as dependent demand when it is directly related to or derived from the demand for other services. For example, in a restaurant where bread and vegetables are included in meals ordered, the demand for bread and vegetables is dependent on the demand for meals. The meal is an end item and the bread and vegetables are component items.

Figure 14.10 shows a bill of material (b) and accompanying product-structure tree (a) for veal picante, a top-selling entrée in a New Orleans restaurant. Note that the various components of veal picante (that is, veal, sauce, and linguini) are prepared by different kitchen personnel (see part [a] of Figure 14.10). These preparations also require different amounts of time to complete. Figure 14.10(c) shows a bill of labor for the veal dish. It lists the operations to be performed, the order of operations, and the labor requirements for each operation (types of labor and labor-hours).

FIGURE 14.10 ■

Product Structure Tree, Bill of Material, and Bill of Labor for Veal Picante

Source: Adapted from John G. Wacker, "Effective Planning and Cost Control for Restaurants," *Production and Inventory Management* (first quarter 1985): 60. Reprinted by permission of American Production and Inventory Control Society.

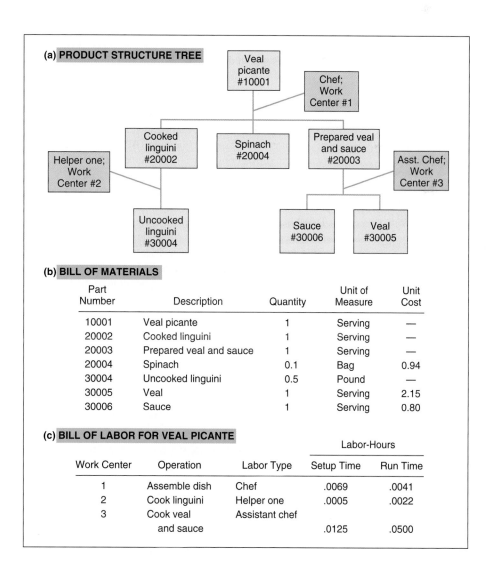

(a) PRODUCT STRUCTURE TREE

(b) BILL OF MATERIALS

Part Number	Description	Quantity	Unit of Measure	Unit Cost
10001	Veal picante	1	Serving	—
20002	Cooked linguini	1	Serving	—
20003	Prepared veal and sauce	1	Serving	—
20004	Spinach	0.1	Bag	0.94
30004	Uncooked linguini	0.5	Pound	—
30005	Veal	1	Serving	2.15
30006	Sauce	1	Serving	0.80

(c) BILL OF LABOR FOR VEAL PICANTE

Work Center	Operation	Labor Type	Labor-Hours Setup Time	Run Time
1	Assemble dish	Chef	.0069	.0041
2	Cook linguini	Helper one	.0005	.0022
3	Cook veal and sauce	Assistant chef	.0125	.0500

MRP is also applied in hospitals, especially when dealing with surgeries that require equipment, materials, and supplies. Houston's Park Plaza Hospital and many hospital suppliers, for example, use the technique to improve the scheduling and management of expensive surgical inventory.

DISTRIBUTION RESOURCE PLANNING (DRP)

Distribution resource planning (DRP)

A time-phased stock-replenishment plan for all levels of a distribution network.

When dependent techniques are used in the supply chain, they are called distribution resource planning (DRP). **Distribution resource planning (DRP)** is a time-phased stock-replenishment plan for all levels of the supply chain. DRP procedures and logic are analogous to MRP. DRP requires the following:

1. Gross requirements, which are the same as expected demand or sales forecasts.
2. Minimum levels of inventory to meet customer-service levels.
3. Accurate lead time.
4. Definition of the distribution structure.

With DRP, expected demand becomes gross requirements. Net requirements are determined by allocating available inventory to gross requirements. The DRP procedure starts with the forecast at the retail level (or the most distant point of the distribution network being supplied). All other levels are computed. As is the case with MRP, inventory is then reviewed with an aim to satisfying demand. So that stock will arrive when it is needed, net requirements are offset by the necessary lead time. A planned order release quantity becomes the gross requirement at the next level down the distribution chain.

DRP *pulls* inventory through the system. Pulls are initiated by the top or retail level ordering more stock. Allocations are made to the top level from available inventory and production after being adjusted to obtain shipping economies. The goal of the DRP system is small and frequent replenishment within the bounds of economical ordering and shipping.

ENTERPRISE RESOURCE PLANNING (ERP)

Enterprise Resource Planning (ERP)

An information system for identifying and planning the enterprise-wide resources needed to take, make, ship, and account for customer orders.

Advances in MRP II systems that tie customers and suppliers to MRP II have led to the development of Enterprise Resource Planning (ERP) systems. **Enterprise Resource Planning (ERP)** is software that allows companies to: (1) automate and integrate many of their business processes, (2) share a common database and business practices throughout the enterprise, and (3) produce information in real time. A schematic showing some of these relationships for a manufacturing firm appears in Figure 14.11.

The objective of an ERP system is to coordinate a firm's whole business, from supplier evaluation to customer invoicing. This objective is seldom achieved, but ERP systems are evolving as umbrella systems that tie together a variety of specialized systems. This is accomplished by using a centralized database to assist the flow of information between business functions. Exactly what is tied together, and how, varies on a case-by-case basis.[8] In addition to the traditional components of MRP, ERP systems usually provide financial and human resource (HR) management information. ERP systems also include:

- *Supply Chain Management (SCM)* software to support sophisticated vendor communication, e-commerce, and those activities necessary for efficient warehousing and logistics. The idea is to tie operations (MRP) to procurement, to materials management, and to suppliers, providing the tools necessary for evaluation of all four.
- *Customer Relationship Management (CRM)* software for the incoming side of the business. CRM is designed to aid analysis of sales, target the most profitable customers, and manage the sales force.

Besides these five modules (MRP, finance, HR, SCM, and CRM), many other options are usually available from vendors of ERP software. These vendors have built modules to provide a variety of "solution" packages that are mixed and matched to individual company needs. Indeed, the trick to these large database and integrated ERP systems is to develop interfaces that allow file access to the databases. SAP, a large ERP vendor, has developed about a thousand *business application-*

[8]*Midrange MRP* (March 2000): 10–11, 18–22, 24–26.

FIGURE 14.11 ■

MRP and ERP
Information Flows,
Showing Customer
Relationship
Management
(CRM), Supply
Chain Management
(SCM), and
Finance/Accounting

*Other functions such as
human resources are often
also included in ERP
systems.*

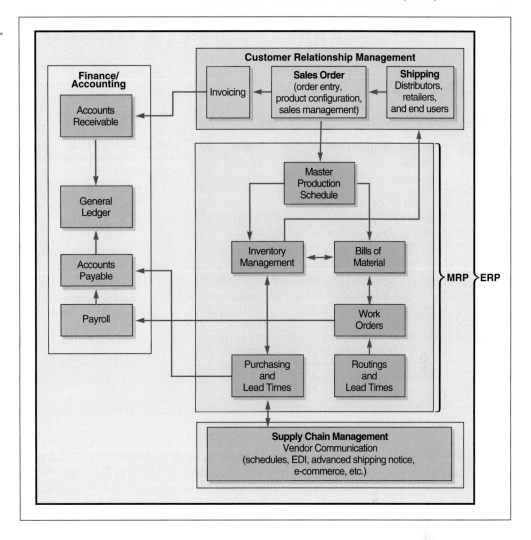

programming interfaces (BAPIs) to access its database. Similarly, other ERP vendors have designed the systems to facilitate integration with third-party software. The demand for interfaces to ERP systems is so large that a new software industry has developed to write the interfaces. This new category of programs is sometimes called *middleware* or *enterprise application integration* (EAI) software. These interfaces allow the expansion of ERP systems so they can integrate with other systems, such as warehouse management, logistics exchanges, electronic catalogs, quality management, and product life cycle management. It is this potential for integration with other systems, including the rich supply of third-party software offerings, that makes ERP so enticing.

In addition to data integration, ERP software promises reduced transaction costs and speed and accuracy of information. A strategic emphasis on just-in-time systems and tying suppliers and distributors more closely to the firm drives the desire for enterprise-wide integration.

In an ERP system, data are entered one time only into a common, complete, and consistent database shared by all applications. For example, when a Nike salesperson enters an order into his ERP system for 20,000 pairs of sneakers for Foot Locker, the data are instantly available on the manufacturing floor. Production crews start filling the order if it is not in stock, accounting prints Foot Locker's invoice, and shipping notifies the Foot Locker of the future delivery date. The salesperson, or even the customer, can check the progress of the order at any point. This is all accomplished using the same data and common applications. To reach this consistency, however, the data fields must be defined identically across the entire enterprise. In Nike's case, this means integrating operations at production sites from Vietnam to China to Mexico, at business units across the globe, in many currencies, and with reports in a variety of languages. The *OM in Action* box, "Managing Benetton with ERP Software," provides an example of how ERP software helps integrate company operations.

OM IN ACTION

Managing Benetton with ERP Software

Thanks to ERP, the Italian sportswear company Benetton can probably claim to have the world's fastest factory and the most efficient distribution in the garment industry. Located in Ponzano, Italy, Benetton makes and ships 50 million pieces of clothing each year. That is 30,000 boxes every day—boxes that must be filled with exactly the items ordered going to the correct store of the 5,000 Benetton outlets in 60 countries. This highly automated distribution center uses only 19 people. Without ERP, 400 people would be needed.

Here is how ERP software works:

1. **Ordering.** A salesperson in the south Boston store finds that she is running out of a best-selling blue sweater. Using a laptop PC, her local Benetton sales agent taps into the ERP sales module.
2. **Availability.** ERP's inventory software simultaneously forwards the order to the mainframe in Italy and finds that half the order can be filled immediately from the Italian warehouse. The rest will be manufactured and shipped in 4 weeks.

3. **Production.** Because the blue sweater was originally created by computer-aided design (CAD), ERP manufacturing software passes the specifications to a knitting machine. The knitting machine makes the sweaters.
4. **Warehousing.** The blue sweaters are boxed with a bar code addressed to the Boston store and placed in one of the 300,000 slots in the Italian warehouse. A robot flies by, reading bar codes, picks out any and all boxes ready for the Boston store, and loads them for shipment.
5. **Order tracking.** The Boston salesperson logs onto the ERP system through the Internet and sees that the sweater (and other items) are completed and being shipped.
6. **Planning.** Based on data from ERP's forecasting and financial modules, Benetton's chief buyer decides that blue sweaters are in high demand and quite profitable. She decides to add three new hues.

Sources: MIT Sloan Management Review (fall 2001): 46–53; and *Newsweek* (September 17, 2001): 36.

Useful ERP Web Links
The Baan Company:
www.baan.com
PeopleSoft:
www.peoplesoft.com
Oracle:
www.oracle.com
SAP:
www.sap.com
JD Edwards:
www.jdedwards.com

Most ERP applications are customized products. The major vendors, SAP AG (a German firm), Baan (from the Netherlands), and J. D. Edwards, People Soft, and Oracle (all of the U.S.), begin with modules customized for specific industries. For example, the Monsanto Company, Dow Chemical, and DuPont, all global chemical firms, have purchased SAP's R/2 (mainframe based) and R/3 (client/server based) product for their industry. The software is then customized to meet their individual needs. Although the vendors build the software to keep the customization process simple, most companies spend up to five times the cost of the software to customize. With ERP programs costing a minimum of $300,000 for a very small company to hundreds of millions of dollars for a global giant like General Motors or Coca-Cola, it is easy to see that ERP systems are expensive. As the *OM in Action* box notes, Nestlé, too, found nothing easy about ERP.

ERP is possible because of advances in hardware and software that have taken place in recent years. The ERP program is designed to take advantage of *client/server networks* with software designated either as a client (one who requests a service) or as a server (one who provides a service). Client/server software is also flexible enough to run on a PC, workstation, or a mainframe and be linked via local area networks. With the rapid growth of the Internet and e-commerce, ERP software vendors include Internet options to complement client/server technology.

Advantages and Disadvantages of ERP Systems

We have alluded to some of the pluses and minuses of ERP. Here is a more complete list of both.

Advantages:
1. Provides integration of the supply-chain, production, and administrative process.
2. Creates commonality of databases.
3. Can incorporate improved, reengineered, "best processes."
4. Increases communication and collaboration between business units and sites.
5. Has a software database that is off-the-shelf coding.
6. May provide a strategic advantage over competitors.

OM IN ACTION

There Is Nothing Easy about ERP

In 2000, the Switzerland-based consumer food giant, Nestlé SA, signed a $200-million contract with SAP for an ERP system. To this $200 million, Nestlé added $80 million for consulting and maintenance. And this is in addition to $500 million for hardware and software as part of a data center overhaul. Jeri Dunn, CIO of Nestlé USA, counsels that successful implementation is dependent on changing business processes and achieving universal "buy-in." Then, and only then, can an organization focus on installing the software. With many autonomous divisions and 200 operating companies and subsidiaries in 80 countries, the challenge of changing the processes and obtaining "buy-in" was substantial.

Standardizing processes is difficult, fraught with dead ends and costly mistakes. Nestlé had 28 points of customer order entry, multiple purchasing systems, and no idea how much volume was being done with a particular vendor; every factory did purchasing on its own with its own specifications. Nestlé USA was paying 29 different prices for vanilla—to the same vendor!

The newly established common databases and business processes led to consistent data and more trustworthy demand forecasts for the many Nestlé products. Nestlé now forecasts down to the level of the distribution center. This improved forecasting allows the company to reduce inventory and the related transportation expenses that occur when too much of a product is sent to one place while there is a shortage in another. The supply chain improvements accounted for much of Nestlé's $325 million in savings.

ERP projects are notorious for taking a long time and a lot of money, and this one is no exception, but after almost 3 years, the last modules of Nestlé's system were installed—and Nestlé is ready to call this installation a success.

Sources: CIO (May 15, 2002): 62–70; and *Information Week* (June 26, 2000): 185.

Disadvantages:

1. Is very expensive to purchase, and even more costly to customize.
2. Implementation may require major changes in the company and its processes.
3. Is so complex that many companies cannot adjust to it.
4. Involves an ongoing process for implementation, which may never be completed.
5. Expertise in ERP is limited, with staffing an ongoing problem.

ERP in the Service Sector

Efficient consumer response (ECR)
Supply chain management systems in the grocery industry; they tie sales to buying, to inventory, to logistics, and to production.

ERP has not penetrated the service sector as much as it has manufacturing. However, ERP vendors have developed a series of service modules for such markets as health care, government, retail stores, and financial services. Springer-Miller Systems, for example, has created an ERP package for the hotel market with software that handles all front- and back-office functions. This system integrates tasks such as maintaining guest histories, booking room and dinner reservations, scheduling golf tee times, and managing multiple properties in a chain.[9] J. D. Edwards's "One World" package combines ERP with supply-chain management (at a price of $13 million) to coordinate airline meal preparation.[10] In the grocery industry, these supply chain systems are known as *efficient consumer response* (ECR) systems. As is the case in manufacturing, **efficient consumer response** systems tie sales to buying, to inventory, to logistics, and to production.

SUMMARY

Material requirements planning (MRP) is the preferred way to schedule production and inventory when demand is dependent. For MRP to work, management must have a master schedule, precise requirements for all components, accurate inventory and purchasing records, and accurate lead times. Distribution resource planning (DRP) is a time-phased stock-replacement technique for supply chains based on MRP procedures and logic.

Production should often be lot-for-lot in an MRP system, and replenishment orders in a DRP system should be small and frequent, given the constraints of ordering and transportation costs.

---

[9] J. R. Gordon and S. R. Gordon, *Information Systems: A Management Approach*, 2nd ed. (Fort Worth: Dryden Press, 1999): 151.

[10] C. Stedman, "Archive Food Vendor Seeks Savings on Production," *Computerworld* (June 14, 1999): 43.

Both MRP and DRP, when properly implemented, can contribute in a major way to reduction in inventory while improving customer-service levels. These techniques allow the operations manager to schedule and replenish stock on a "need-to-order" basis rather than simply a "time-to-order" basis.

The continuing development of MRP systems has led to the integration of production data with a variety of other activities, including the supply chain and sales. As a result, we now have integrated database-oriented Enterprise Resource Management (ERP) systems. These expensive and difficult to install ERP systems, when successful, support strategies of differentiation, response, and cost leadership.

KEY TERMS

Material requirements planning (MRP) *(p. 412)*
Master production schedule (MPS) *(p. 412)*
Bill of material (BOM) *(p. 415)*
Modular bills *(p. 416)*
Planning bills (or kits) *(p. 416)*
Phantom bills of material *(p. 416)*
Low-level coding *(p. 416)*
Lead time *(p. 417)*
Gross material requirements plan *(p. 418)*
Net material requirements *(p. 419)*
Planned order receipt *(p. 419)*
Planned order release *(p. 419)*
System nervousness *(p. 422)*
Time fences *(p. 422)*
Pegging *(p. 422)*

Buckets *(p. 422)*
Bucketless system *(p. 422)*
Back flush *(p. 422)*
Lot-sizing decision *(p. 423)*
Lot-for-lot *(p. 423)*
Part period balancing (PPB) *(p. 425)*
Economic part period (EPP) *(p. 425)*
Wagner-Whitin procedure *(p. 426)*
Closed-loop MRP system *(p. 427)*
Load report *(p. 427)*
Material requirements planning II (MRP II) *(p. 428)*
Distribution resource planning (DRP) *(p. 430)*
Enterprise Resource Planning (ERP) *(p. 430)*
Efficient consumer response (ECR) *(p. 433)*

USING EXCEL OM TO SOLVE MRP PROBLEMS

Using Excel OM's MRP module requires the careful entry of several pieces of data. The initial MRP screen is where we enter (1) the total number of occurrences of items in the BOM (including the top item), (2) what we want the BOM items to be called (i.e., Item no., Part, etc.), (3) total number of periods to be scheduled, and (4) what we want the periods called (i.e., days, weeks, etc.).

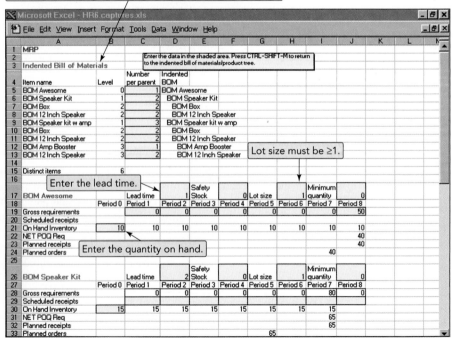

PROGRAM 14.1 ■

Using Excel OM's MRP Module to Solve Examples 1, 2, and 3

Excel OM's second MRP screen provides the data entry for an indented bill of material. Here we enter (1) the name of each item in the BOM, (2) the quantity of that item in the assembly, and (3) the correct indent (i.e., parent/child relationship) for each item. The indentations are critical as they provide the logic for the BOM explosion. The indentations should follow the logic of the product structure tree with indents for each assembly item in that assembly.

Excel OM's third MRP screen repeats the indented BOM and provides the standard MRP tableau for entries. This is shown in Program 14.1 using the data from Examples 1, 2, and 3.

USING POM FOR WINDOWS TO SOLVE MRP PROBLEMS

POM for Windows' MRP module can also solve Examples 1 to 3. Up to 18 periods can be analyzed. Here are the inputs required:

1. *Item names.* The item names are entered in the left column. The same item name will appear in more than one row if the item is used by two parent items. Each item must follow its parents.
2. *Item level.* The level in the indented BOM must be given here. The item *cannot* be placed at a level more than one below the item immediately above.
3. *Lead time.* The lead time for an item is entered here. The default is 1 week.
4. *Number per parent.* The number of units of this subassembly needed for its parent is entered here. The default is one.
5. *On hand.* List current inventory on hand once, even if the subassembly is listed twice.
6. *Lot size.* The lot size can be specified here. A 0 or 1 will perform lot-for-lot ordering. If another number is placed here, then all orders for that item will be in integer multiples of that number.
7. *Demands.* The demands are entered in the end item row in the period in which the items are demanded.
8. *Scheduled receipts.* If units are scheduled to be received in the future, they should be listed in the appropriate time period (column) and item (row). (An entry here in level 1 is a demand; all other levels are receipts.)

Further details regarding POM for Windows are seen in Appendix V.

SOLVED PROBLEMS

Solved Problem 14.1

Determine the low-level coding and the quantity of each component necessary to produce 10 units of an assembly we will call Alpha. The product structure and quantities of each component needed for each assembly are noted in parenthesis.

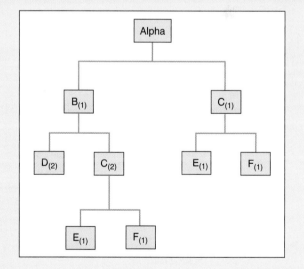

SOLUTION

Redraw the product structure with low-level coding. Then multiply down the structure until the requirements of each branch are determined. Then add across the structure until the total for each is determined.

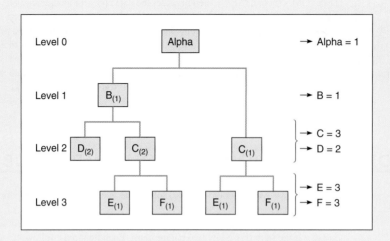

Es required for left branch:

$$(1_{\text{alpha}} \times 1_B \times 2_C \times 1_E) = 2$$

and Es required for right branch:

$$(1_{\text{alpha}} \times 1_C \times 1_E) = \underline{1}$$
$$3 \text{ Es required in total}$$

Then "explode" the requirement by multiplying each by 10, as shown in the following table:

LEVEL	ITEM	QUANTITY PER UNIT	TOTAL REQUIREMENTS FOR 10 ALPHA
0	Alpha	1	10
1	B	1	10
2	C	3	30
2	D	2	20
3	E	3	30
3	F	3	30

Solved Problem 14.2

Using the product structure for Alpha in Solved Problem 14.1, and the lead times, quantity on hand, and master production schecule shown below, prepare a net MRP table for Alphas.

ITEM	LEAD TIME	QTY ON HAND
Alpha	1	10
B	2	20
C	3	0
D	1	100
E	1	10
F	1	50

Master Production Schedule for Alpha

PERIOD	6	7	8	9	10	11	12	13
Gross requirements			50			50		100

SOLUTION

See the chart on the next page.

Lot Size	Lead Time (# of Periods)	On Hand	Safety Stock	Allocated	Low-Level Code	Item ID		1	2	3	4	5	6	7	8	9	10	11	12	13
Lot-for-Lot	1	10	—	—	0	Alpha (A)	Gross Requirements								50			50		100
							Scheduled Receipts													
							Projected On Hand 10													
							Net Requirements								40			50		100
							Planned Order Receipts								40			50		100
							Planned Order Releases							40			50		100	
Lot-for-Lot	2	20	—	—	1	B	Gross Requirements							40(A)			50(A)		100(A)	
							Scheduled Receipts													
							Projected On Hand 20													
							Net Requirements							20			50		100	
							Planned Order Receipts							20			50		100	
							Planned Order Releases					20			50		100			
Lot-for-Lot	3	0	—	—	2	C	Gross Requirements					40(B)		40(A)	100(B)		200(B) + 50(A)		100(A)	
							Scheduled Receipts													
							Projected On Hand 0													
							Net Requirements					40		40	100		250		100	
							Planned Order Receipts					40		40	100		250		100	
							Planned Order Releases		40		40	100		250		100				
Lot-for-Lot	1	100	—	—	2	D	Gross Requirements					40(B)			100(B)		200(B)			
							Scheduled Receipts													
							Projected On Hand 100					60			0		0			
							Net Requirements					0			40		200			
							Planned Order Receipts								40		200			
							Planned Order Releases				0			40		200				
Lot-for-Lot	1	10	—	—	3	E	Gross Requirements		40(C)		40(C)	100(C)		250(C)		100(C)				
							Scheduled Receipts													
							Projected On Hand 10													
							Net Requirements		30		40	100		250		100				
							Planned Order Receipts		30		40	100		250		100				
							Planned Order Releases	30		40	100		250		100					
Lot-for-Lot	1	50	—	—	3	F	Gross Requirements		40(C)		40(C)	100(C)		250(C)		100(C)				
							Scheduled Receipts													
							Projected On Hand 50		10		–									
							Net Requirements		0		30	100		250		100				
							Planned Order Receipts				30	100		250		100				
							Planned Order Releases			30	100		250		100					

Net Material Requirements Planning Sheet for Alpha

The letter in parentheses (A) is the source of the demand.

437

INTERNET AND STUDENT CD-ROM EXERCISES

Visit our home page or use your student CD-ROM to help with material in this chapter.

 On Our Home Page, www.prenhall.com/heizer

- Self-Tests
- Practice Problems
- Internet Exercises
- Current Articles and Research
- Virtual Company Tour
- Internet Homework Problems
- Internet Cases

 On Your Student CD-ROM

- PowerPoint Lecture
- Practice Problems
- Video Clip and Video Case
- Active Model Exercise
- ExcelOM
- Excel OM Example Data File

ADDITIONAL CASE STUDIES

Internet Case Studies: Visit our Web site at www.prenhall.com/heizer **for these free case studies:**

- **Auto Parts, Inc.**: Distributor of automobile replacement parts has major MRP problems.

- **Ruch Manufacturing**: Truck manufacturer seeks to revise its lot sizing policy.

Harvard has selected these Harvard Business School cases to accompany this chapter
(textbookcasematch.hbsp.harvard.edu)**:**

- **Digital Equipment Corp.: The Endpoint Model** (#688-059): Implementation of an MRP II system to reduce cycle time of orders.

- **Tektronix, Inc.: Global ERP Implementation** (#699-043): Tektronix's implementation of an ERP system in its three global business divisions.

- **Vardelay Industries, Inc.** (#697-037): Discusses ERP and related issues of process reengineering, standardization, and change management.

Short-Term Scheduling

Chapter Outline

LEARNING OBJECTIVES

When you complete this chapter you should be able to

IDENTIFY OR DEFINE:

Gantt charts

Assignment method

Sequencing rules

Johnson's rule

Bottlenecks

DESCRIBE OR EXPLAIN:

Scheduling

Sequencing

Shop loading

Theory of constraints

GLOBAL COMPANY PROFILE:

Scheduling Airplanes When Weather Is the Enemy

Throughout the ordeals of tornadoes, ice storms, and snowstorms, airlines across the globe struggle to cope with delays, cancellations, and furious passengers. Close to 10% of Delta Airlines' flights are disrupted in a typical year, half because of weather; the cost is $440 million in lost revenue, overtime pay, and food and lodging vouchers.

Now Delta is taking the sting out of the scheduling nightmares that come from weather-related problems with its recently opened $33-million high-tech nerve center adjacent to the Atlanta Airport. From computers to telecommunications systems to deicers, Delta's new Operations Control Center more quickly notifies customers of schedule

4 A.M.
FORECAST:
Rain with a chance of light snow for Atlanta.

ACTION:
Discuss status of planes and possible need for cancellations.

10 A.M.
FORECAST:
Freezing rain after 5 P.M.

ACTION:
Ready deicing trucks; develop plans to cancel 50% to 80% of flights after 6 P.M.

1:30 P.M.
FORECAST:
Rain changing to snow.

ACTION:
Cancel half the flights from 6 P.M. to 10 A.M.; notify passengers and reroute planes.

5 P.M.
FORECAST:
Less snow than expected.

ACTION:
Continue calling passengers and arrange alternate flights.

10 P.M.
FORECAST:
Snow tapering off.

ACTION:
Find hotels for 1,600 passengers stranded by the storm.

Here is what Delta officials had to do one December day when a storm bore down on Atlanta.

To improve flight rescheduling efforts, Delta employees monitor giant screens that display meterological charts, weather patterns, and maps of Delta flights at its Operations Control Center in Atlanta.

changes, reroutes flights, and gets jets into the air.

With earlier access to information, the center's staff of 18 pores over streams of data transmitted by computers. Using mathematical scheduling models described in this chapter, Delta decides on schedule and route changes. Its software, called the Inconvenienced Passenger Rebooking System, notifies passengers of cancellations or delays and even books them onto rival airlines if necessary. With 100,000 passengers flying into and out of Atlanta every day, Delta estimates its new scheduling efforts save $35 million a year.

Sources: U.S.A. Today (February 18, 2002): B-01, and (January 12, 2001): A14; and *New York Times* (January 21, 1997): C1, C20.

In an effort to maintain schedules, Delta Airlines uses elaborate equipment as shown here for ice removal.

THE STRATEGIC IMPORTANCE OF SHORT-TERM SCHEDULING

Delta Airlines doesn't just schedule its 500-plus aircraft every day. It also schedules over 20,000 pilots and flight attendants to accommodate passengers who wish to reach their destinations. This schedule, based on huge computer programs, plays a major role in satisfying customers. Delta finds competitive advantage with its flexibility for last-minute adjustments to demand and weather disruptions.

Manufacturing firms also make schedules that match production to customer demands. Lockheed-Martin's Dallas plant schedules machines, tools, and people to make aircraft parts. Lockheed's mainframe computer downloads schedules for parts production into a flexible machining system (FMS) in which a manager makes the final scheduling decision. The FMS allows parts of many sizes or shapes to be made, in any order, without disrupting production. This versatility in scheduling results in parts ready on a just-in-time basis, with low setup times, little work-in-process, and high machine utilization. Efficient scheduling is how companies like Lockheed-Martin meet due dates promised to customers and face time-based competition.

The strategic importance of scheduling is clear:

1. By scheduling effectively, companies use assets more effectively and create greater capacity per dollar invested, which, in turn, *lowers cost*.
2. This added capacity and related flexibility provides *faster delivery* and, therefore, better customer service.
3. Good scheduling is a competitive advantage because it contributes to *dependable* delivery.

SCHEDULING ISSUES

Scheduling deals with the timing of operations. Table 15.1 illustrates scheduling decisions faced in five organizations: a hospital, a college, a manufacturer, a restaurant, and an airline. As you can see from Figure 15.1, scheduling begins with *capacity* planning, which involves facility and equipment acquisition (discussed in Chapter 7 and its Supplement). In the aggregate planning stage (Chapter 13), decisions regarding the *use* of facilities, inventory, people, and outside contractors are made. Then the master schedule breaks down the aggregate plan and develops an *overall* schedule for outputs. Short-term schedules then translate capacity decisions, intermediate planning, and master schedules into job sequences and specific assignments of personnel, materials, and machinery. In this chapter, we describe the narrow issue of scheduling goods and services in the *short run* (that is, on a weekly, daily, or hourly basis).

Japan's Nippon Steel maintains its world-class operation by automating the scheduling of people, machines, and tools through its hot-roller control room. Computerized scheduling software helps managers monitor production.

TABLE 15.1 ■

Scheduling Decisions

Video 15.1 Scheduling at Hard Rock

ORGANIZATION	MANAGERS MUST SCHEDULE THE FOLLOWING
Mount Sinai Hospital	Operating room use
	Patient admissions
	Nursing, security, maintenance staffs
	Outpatient treatments
Indiana University	Classrooms and audiovisual equipment
	Student and instructor schedules
	Graduate and undergraduate courses
Lockheed-Martin factory	Production of goods
	Purchases of materials
	Workers
Hard Rock Cafe	Chef, waiters, bartenders
	Delivery of fresh foods
	Entertainers
	Opening of dining areas
Delta Airlines	Maintenance of aircraft
	Departure timetables
	Flight crews, catering, gate, and ticketing personnel

Forward and Backward Scheduling

Scheduling involves assigning due dates to specific jobs, but many jobs compete simultaneously for the same resources. To help address the difficulties inherent in scheduling, we can categorize scheduling techniques as (1) forward scheduling and (2) backward scheduling.

Forward scheduling

A schedule that begins as soon as the requirements are known.

Forward scheduling starts the schedule as soon as the job requirements are known. Forward scheduling is used in a variety of organizations such as hospitals, clinics, fine-dining restaurants, and machine tool manufacturers. In these facilities, jobs are performed to customer order, and delivery is often requested as soon as possible. Forward scheduling is usually designed to produce a schedule that can be accomplished even if it means not meeting the due date. In many instances, forward scheduling causes a buildup of work-in-process inventory.

FIGURE 15.1 ■

The Relationship between Capacity Planning, Aggregate Planning, Master Schedule, and Short-Term Scheduling

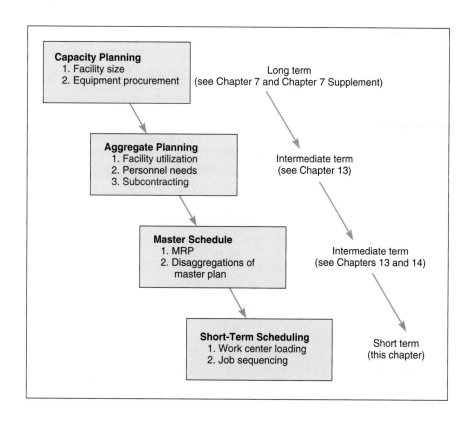

Backward scheduling
Scheduling that begins with the due date and schedules the final operation first and the other job steps in reverse order.

Forward scheduling works well in firms whose suppliers are usually behind in meeting schedules.

Backward scheduling begins with the due date, scheduling the *final* operation first. Steps in the job are then scheduled, one at a time, in reverse order. By subtracting the lead time for each item, the start time is obtained. However, the resources necessary to accomplish the schedule may not exist. Backward scheduling is used in many manufacturing environments, as well as service environments such as catering a banquet or scheduling surgery. In practice, a combination of forward and backward scheduling is often used to find a reasonable trade-off between what can be achieved and customer due dates.

Machine breakdowns, absenteeism, quality problems, shortages, and other factors further complicate scheduling. (See the *OM in Action* box, "Scheduling Workers Who Fall Asleep on the Job Is Not Easy.") Consequently, assignment of a date does not ensure that the work will be performed according to the schedule. Many specialized techniques have been developed to aid us in preparing reliable schedules.

Scheduling Criteria

The correct scheduling technique depends on the volume of orders, the nature of operations, and the overall complexity of jobs, as well as the importance placed on each of four criteria. Those four criteria are

1. *Minimize completion time.* This criterion is evaluated by determining the average completion time per job.
2. *Maximize utilization.* This is evaluated by determining the percent of the time the facility is utilized.
3. *Minimize work-in-process (WIP) inventory.* This is evaluated by determining the average number of jobs in the system. The relationship between the number of jobs in the system and WIP inventory will be high. Therefore, the fewer the number of jobs that are in the system, the lower the inventory.
4. *Minimize customer waiting time.* This is evaluated by determining the average number of late days.

These four criteria are used in this chapter, as they are in industry, to evaluate scheduling performance. Additionally, good scheduling approaches should be simple, clear, easily understood, easy to carry out, flexible, and realistic. Given these considerations, *the objective of scheduling is to optimize the use of resources so that production objectives are met.* In this chapter, we examine scheduling in process-focused (intermittent) production, repetitive production, and the service sector.

OM IN ACTION

Scheduling Workers Who Fall Asleep on the Job Is Not Easy

Unable to cope with a constantly changing work schedule, an operator at a big oil refinery dozes off in the middle of the night—and inadvertently dumps thousands of gallons of chemicals into a nearby river.

A similar story holds for pilots. Their inconsistent schedules often force them to snooze in the cockpit in order to get enough sleep. "There have been times I've been so sleepy, I'm nodding off as we're taxiing to get into take-off position," says a Federal Express pilot. "I've fallen asleep reading checklists. I've fallen asleep in the middle of a word."

An estimated 20 million people in the U.S. work in industries that maintain round-the-clock schedules. In interviews with researchers, employees from the graveyard shift report tales of seeing sleeping assembly-line workers fall off their stools, batches of defective parts sliding past dozing inspectors, and exhausted forklift operators crashing into walls. "It's kind of too ugly. How can you admit that your nuclear power plant operators regularly fall asleep on the job?" says a Harvard researcher.

Scheduling is a major problem in firms with late shifts. Some companies, but far from all, are taking steps to deal with schedule-related sleep problems among workers. Dow Chemical, Detroit Edison, Pennzoil, and Exxon, for instance, are giving all workers several days off between shift changes. The Philadelphia Police Department is now using fewer and less random schedule changes and reports a 40% decline in officers' on-the-job auto accidents.

As more is learned about the economic toll of constant schedule changes, companies will find they cannot afford to continue ignoring the problem. As one researcher says, "Megabucks are involved, and, sometimes, lives."

Sources: OH&S Canada (January–February 2002): 46–47; Nursing Management (2001): 54; and the Wall Street Journal (August 9, 1999): A4.

SCHEDULING PROCESS-FOCUSED WORK CENTERS

Process-focused facilities (also known as *intermittent* or *job-shop facilities*)[1] are high-variety, low-volume systems commonly found in manufacturing and service organizations. Examples include auto paint shops, printers, surgical centers, and fancy French restaurants. These are production systems in which products are made to order. Items made under this system usually differ considerably in terms of materials used, order of processing, processing requirements, time of processing, and setup requirements. Because of these differences, scheduling can be complex. To run a facility in a balanced and efficient manner, the manager needs a production planning and control system. This system should

1. Schedule incoming orders without violating capacity constraints of individual work centers.
2. Check the availability of tools and materials before releasing an order to a department.
3. Establish due dates for each job and check progress against need dates and order lead times.
4. Check work in progress as jobs move through the shop.
5. Provide feedback on plant and production activities.
6. Provide work efficiency statistics and monitor operator times for payroll and labor distribution analyses.

Whether the scheduling system is manual or automated, it must be accurate and relevant. This means it requires a production database with both planning and control files.[2] Three types of **planning files** are

1. An *item master file*, which contains information about each component the firm produces or purchases.
2. A *routing file*, which indicates each component's flow through the shop.
3. A *work-center master file*, which contains information about the work center, such as capacity and efficiency.

Control files track the actual progress made against the plan for each work order.

LOADING JOBS IN WORK CENTERS

Loading means the assignment of jobs to work or processing centers. Operations managers assign jobs to work centers so that costs, idle time, or completion times are kept to a minimum. Loading work centers takes two forms.[3] One is oriented to capacity; the second is related to assigning specific jobs to work centers.

First, we examine loading from the perspective of capacity via a technique known as input–output control. Then, we present two approaches used for loading: *Gantt charts* and the *assignment method* of linear programming.

Input–Output Control

Many firms have difficulty scheduling (that is, achieving effective throughput) because they overload the production processes. This often occurs because they do not know actual performance in the work centers. Effective scheduling depends on matching the schedule to performance. Lack of knowledge about capacity and performance causes reduced throughput.

Input–output control is a technique that allows operations personnel to manage facility work flows. If the work is arriving faster than it is being processed, we are overloading the facility and a backlog develops. Overloading causes crowding in the facility, leading to inefficiencies and quality problems. If the work is arriving at a slower rate than jobs are being performed, we are underloading the facility and the work center may run out of work. Underloading the facility results in idle capacity and wasted resources. Example 1 shows the use of input–output controls.

---

[1]Much of the literature on scheduling is about manufacturing; therefore, the traditional term *job-shop scheduling* is often used.

[2]For an expanded discussion, see *APICS Study Aid—Detailed Scheduling and Planning* (Alexandria, VA: American Production and Inventory Control Society).

[3]Note that this discussion can apply to work centers that might be called a "shop" in a manufacturing firm, a "ward" in a hospital, or a "department" in an office or large kitchen.

Example 1

Figure 15.2 shows the planned capacity for the DNC Milling work center for 5 weeks (weeks 6/6 through 7/4). The planned input is 280 standard hours per week. The actual input is close to this figure, varying between 250 and 285. Output is scheduled at 320 standard hours, which is the assumed capacity. A backlog of 300 hours (not shown in the figure) exists in the work center. However, actual output (270 hours) is substantially less than planned. Therefore, neither the input plan nor the output plan is being achieved. Indeed, the backlog of work in this work center has actually increased by 5 hours by week 6/27. This increases work-in-process inventory, complicating the scheduling task and indicating the need for manager action.

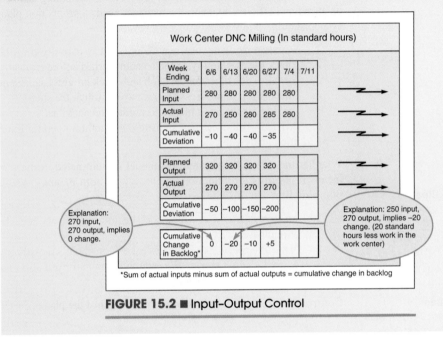

FIGURE 15.2 ■ Input–Output Control

The options available to operations personnel to manage facility work flow include the following:

1. Correcting performances.
2. Increasing capacity.
3. Increasing or reducing input to the work center by (a) routing work to or from other work centers, (b) increasing or decreasing subcontracting, (c) producing less (or producing more).

Producing less is not a popular solution for many managers, but the advantages can be substantial. First, customer-service level may improve because units may be produced on time. Second, efficiency may actually improve because of less work-in-process cluttering the work center and adding to overhead costs. Third, quality may improve because less work-in-process hides fewer problems.

Gantt Charts

Gantt charts

Planning charts used to schedule resources and allocate time.

Gantt charts are visual aids that are useful in loading and scheduling. The name is derived from Henry Gantt, who developed them in the late 1800s. The charts show the use of resources, such as work centers and labor.

When used in *loading*, Gantt charts show the loading and idle times of several departments, machines, or facilities. They display the relative workloads in the system so that the manager knows what adjustments are appropriate. For example, when one work center becomes overloaded, employees from a low-load center can be transferred temporarily to increase the workforce. Or if waiting jobs can be processed at different work centers, some jobs at high-load centers can be transferred to low-load centers. Versatile equipment may also be transferred among centers. Example 2 illustrates a simple Gantt load chart.

Example 2

A New Orleans washing machine manufacturer accepts special orders for machines to be used in such unique facilities as submarines, hospitals, and large industrial laundries. The production of each machine requires varying tasks and durations. Figure 15.3 shows the load chart for the week of March 8.

The four work centers process several jobs during the week. This particular chart indicates that the metalworks and painting centers are completely loaded for the entire week. The mechanical and electronic centers have some idle time scattered during the week. We also note that the metalworks center is unavailable on Tuesday and the painting center is unavailable on Thursday, perhaps for preventive maintenance.

FIGURE 15.3 ■

Gantt Load Chart for the Week of March 8

Work Center \ Day	Monday	Tuesday	Wednesday	Thursday	Friday
Metalworks	Job 349	✕		Job 350	
Mechanical		Job 349		Job 408	
Electronics	Job 408			Job 349	
Painting	Job 295		Job 408	✕	Job 349

☐ Processing   ☐ Unscheduled   ✕ Center not available (for example, maintenance time, repairs, shortages)

The Gantt *load chart* does have a major limitation: It does not account for production variability such as unexpected breakdowns or human errors that require reworking a job. Consequently, the chart must also be updated regularly to account for new jobs and revised time estimates.

A Gantt *schedule chart* is used to monitor jobs in progress.[4] It indicates which jobs are on schedule and which are ahead of or behind schedule. In practice, many versions of the chart are found. The schedule chart in Example 3 places jobs in progress on the vertical axis and time on the horizontal axis.

Example 3

First Printing and Copy Center in Winter Park, Florida, uses the Gantt chart in Figure 15.4 to show the scheduling of three orders, jobs A, B, and C. Each pair of brackets on the time axis denotes the estimated starting and finishing of a job enclosed within it. The solid bars reflect the actual status or progress of the job. Job A, for example, is about one-half day behind schedule at the end of day 5. Job B was completed after equipment maintenance. Job C is ahead of schedule.

FIGURE 15.4 ■

Gantt Scheduling Chart for Jobs A, B, and C at a Printing Firm

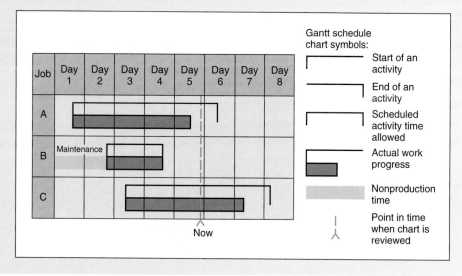

[4]Gantt charts are also used for project scheduling and were noted in Chapter 3, "Project Management."

Assignment Method

Assignment method
A special class of linear programming models that involves assigning tasks or jobs to resources.

The **assignment method** involves assigning tasks or jobs to resources. Examples include assigning jobs to machines, contracts to bidders, people to projects, and salespeople to territories. The objective is most often to minimize total costs or time required to perform the tasks at hand. One important characteristic of assignment problems is that only one job (or worker) is assigned to one machine (or project).

Each assignment problem uses a table. The numbers in the table will be the costs or times associated with each particular assignment. For example, if First Printing and Copy Center has three available typesetters (A, B, and C) and three new jobs to be completed, its table might appear as follows. The dollar entries represent the firm's estimate of what it will cost for each job to be completed by each typesetter.

	TYPESETTER		
JOB	**A**	**B**	**C**
R-34	$11	$14	$ 6
S-66	$ 8	$10	$11
T-50	$ 9	$12	$ 7

The assignment method involves adding and subtracting appropriate numbers in the table in order to find the lowest *opportunity cost*[5] for each assignment. There are four steps to follow:

1. Subtract the smallest number in each row from every number in that row and then, from the resulting matrix, subtract the smallest number in each column from every number in that column. This step has the effect of reducing the numbers in the table until a series of zeros, meaning *zero opportunity costs*, appear. Even though the numbers change, this reduced problem is equivalent to the original one, and the same solution will be optimal.

2. Draw the minimum number of vertical and horizontal straight lines necessary to cover all zeros in the table. If the number of lines equals either the number of rows or the number of columns in the table, then we can make an optimal assignment (see step 4). If the number of lines is less than the number of rows or columns, we proceed to step 3.

3. Subtract the smallest number not covered by a line from every other uncovered number. Add the same number to any number(s) lying at the intersection of any two lines. Do not change the value of the numbers which are covered by only one line. Return to step 2 and continue until an optimal assignment is possible.

4. Optimal assignments will always be at zero locations in the table. One systematic way of making a valid assignment is first to select a row or column that contains only one zero square. We can make an assignment to that square and then draw lines through its row and column. From the uncovered rows and columns, we choose another row or column in which there is only one zero square. We make that assignment and continue the procedure until we have assigned each person or machine to one task.

Example 4 shows how to use the assignment method.

Example 4

**Excel OM Data File
Ch15Ex4.xla**

The cost table shown earlier in this section is repeated here. We find the minimum total cost assignment of jobs to typesetters by applying steps 1 through 4.

TYPESETTER JOB	A	B	C
R-34	$11	$14	$ 6
S-66	$ 8	$10	$11
T-50	$ 9	$12	$ 7

[5]Opportunity costs are those profits foregone or not obtained.

Step 1a: Using the previous table, subtract the smallest number in each row from every number in the row. The result is shown below (left).

TYPESETTER / JOB	A	B	C
R-34	5	8	0
S-66	0	2	3
T-50	2	5	0

TYPESETTER / JOB	A	B	C
R-34	5	6	0
S-66	0	0	3
T-50	2	3	0

Step 1b: Using the above (left) table, subtract the smallest number in each column from every number in the column. The result is shown above (right).

Step 2: Draw the minimum number of vertical and horizontal straight lines needed to cover all zeros. Because two lines suffice, the solution is not optimal.

TYPESETTER / JOB	A	B	C
R-34	5	6	0
S-66	0	0	3
T-50	②	3	0

Smallest uncovered number

Step 3: Subtract the smallest uncovered number (2 in this table) from every other uncovered number and add it to numbers at the intersection of two lines.

TYPESETTER / JOB	A	B	C
R-34	3	4	0
S-66	0	0	5
T-50	0	1	0

Return to step 2. Cover the zeros with straight lines again.

TYPESETTER / JOB	A	B	C
R-34	3	4	0
S-66	0	0	5
T-50	0	1	0

Because three lines are necessary, an optimal assignment can be made (see step 4). Assign R-34 to person C, S-66 to person B, and T-50 to person A. Referring to the original cost table, we see that:

$$\text{Minimum cost} = \$6 + \$10 + \$9 = \$25$$

Note: If we had assigned S-66 to typesetter A, we could not assign T-50 to a zero location.

Some assignment problems entail *maximizing* profit, effectiveness, or payoff of an assignment of people to tasks or of jobs to machines. It is easy to obtain an equivalent minimization problem by converting every number in the table to an *opportunity loss*. To convert a maximizing problem to an

The problem of scheduling American League umpiring crews from one series of games to the next is complicated by many restrictions on travel, ranging from coast-to-coast time changes, airline flight schedules, and night games running late. The league strives to achieve these two conflicting objectives: (1) balance crew assignments relatively evenly among all teams over the course of a season, and (2) minimize travel costs. Using the assignment problem formulation, the time it takes the league to generate a schedule has been significantly decreased, and the quality of the schedule has improved.

Sequencing
Determining the order in which jobs should be done at each work center.

equivalent minimization problem, we subtract every number in the original payoff table from the largest single number in that table. We then proceed to step 1 of the four-step assignment method. It turns out that minimizing the opportunity loss produces the same assignment solution as the original maximization problem.

Priority rules
Rules that are used to determine the sequence of jobs in process-oriented facilities.

SEQUENCING JOBS IN WORK CENTERS

Scheduling provides a basis for assigning jobs to work centers. *Loading* is a capacity-control technique that highlights overloads and underloads. **Sequencing** specifies the order in which jobs should be done at each center. For example, suppose that 10 patients are assigned to a medical clinic for treatment. In what order should they be treated? Should the first patient to be served be the one who arrived first or the one who needs emergency treatment? Sequencing methods provide such detailed information. These methods are referred to as priority rules for dispatching jobs to work centers.

First come, first served (FCFS)
Jobs are completed in the order they arrived.

Priority Rules for Dispatching Jobs

Priority rules provide guidelines for the sequence in which jobs should be worked. The rules are especially applicable for process-focused facilities such as clinics, print shops, and manufacturing job shops. We will examine a few of the most popular priority rules. Priority rules try to minimize completion time, number of jobs in the system, and job lateness, while maximizing facility utilization.

Shortest processing time (SPT)
Jobs with the shortest processing times are assigned first.

The most popular priority rules are

Earliest due date (EDD)
Earliest due date jobs are performed next.

- FCFS: **First come, first served.** The first job to arrive at a work center is processed first.
- SPT: **Shortest processing time.** The shortest jobs are handled first and completed.
- EDD: **Earliest due date.** The job with the earliest due date is selected first.
- LPT: **Longest processing time.** The longer, bigger jobs are often very important and are selected first.

Longest processing time (LPT)
Jobs with the longest processing time are completed next.

Example 5 compares these rules.

Example 5

Five architectural rendering jobs are waiting to be assigned at Ajax, Tarney and Barnes Architects. Their work (processing) times and due dates are given in the following table. We want to determine the sequence of processing according to (1) FCFS, (2) SPT, (3) EDD, and (4) LPT rules. Jobs were assigned a letter in the order they arrived.

JOB	JOB WORK (PROCESSING) TIME (DAYS)	JOB DUE DATE (DAYS)
A	6	8
B	2	6
C	8	18
D	3	15
E	9	23

Active Model 15.1

Example 5 is further illustrated in Active Model 15.1 on the CD-ROM and in the Exercise located in your Student Lecture Guide.

1. The *FCFS* sequence shown in the next table is simply A-B-C-D-E. The "flow time" in the system for this sequence measures the time each job spends waiting plus time being processed. Job B, for example, waits 6 days while job A is being processed, then takes 2 more days of operation time itself; so it will be completed in 8 days—which is 2 days later than its due date.

JOB SEQUENCE	JOB WORK (PROCESSING) TIME	FLOW TIME	JOB DUE DATE	JOB LATENESS
A	6	6	8	0
B	2	8	6	2
C	8	16	18	0
D	3	19	15	4
E	9	28	23	5
	28	77		11

The first-come, first-served rule results in the following measures of effectiveness:

a. Average completion time $= \dfrac{\text{Sum of total flow time}}{\text{Number of jobs}}$

$$= \frac{77 \text{ days}}{5} = 15.4 \text{ days}$$

b. Utilization $= \dfrac{\text{Total job work (processing) time}}{\text{Sum of total flow time}}$

$$= \frac{28}{77} = 36.4\%$$

c. Average number of jobs in the system $= \dfrac{\text{Sum of total flow time}}{\text{Total job work (processing) time}}$

$$= \frac{77 \text{ days}}{28 \text{ days}} = 2.75 \text{ jobs}$$

Excel OM Data File Ch15Ex5.xla

d. Average job lateness $= \dfrac{\text{Total late days}}{\text{Number of jobs}} = \dfrac{11}{5} = 2.2 \text{ days}$

2. The *SPT* rule shown in the next table results in the sequence B-D-A-C-E. Orders are sequenced according to processing time, with the highest priority given to the shortest job.

JOB SEQUENCE	JOB WORK (PROCESSING) TIME	FLOW TIME	JOB DUE DATE	JOB LATENESS
B	2	2	6	0
D	3	5	15	0
A	6	11	8	3
C	8	19	18	1
E	9	28	23	5
	28	65		9

Measurements of effectiveness for SPT are

a. Average completion time = $\dfrac{65}{5}$ = 13 days

b. Utilization = $\dfrac{28}{65}$ = 43.1%

c. Average number of jobs in the system = $\dfrac{65}{28}$ = 2.32 jobs

d. Average job lateness = $\dfrac{9}{5}$ = 1.8 days

3. The *EDD* rule shown in the next table gives the sequence B-A-D-C-E. Note that jobs are ordered by earliest due date first.

JOB SEQUENCE	JOB WORK (PROCESSING) TIME	FLOW TIME	JOB DUE DATE	JOB LATENESS
B	2	2	6	0
A	6	8	8	0
D	3	11	15	0
C	8	19	18	1
E	9	28	23	5
	28	68		6

Measurements of effectiveness for EDD are

a. Average completion time = $\dfrac{68}{5}$ = 13.6 days

b. Utilization = $\dfrac{28}{68}$ = 41.2%

c. Average number of jobs in the system = $\dfrac{68}{28}$ = 2.43 jobs

d. Average job lateness = $\dfrac{6}{5}$ = 1.2 days

4. The *LPT* rule shown in the next table results in the order E-C-A-D-B.

JOB SEQUENCE	JOB WORK (PROCESSING) TIME	FLOW TIME	JOB DUE DATE	JOB LATENESS
E	9	9	23	0
C	8	17	18	0
A	6	23	8	15
D	3	26	15	11
B	2	28	6	22
	28	103		48

Measures of effectiveness for LPT are

a. Average completion time = $\dfrac{103}{5}$ = 20.6 days

b. Utilization = $\dfrac{28}{103}$ = 27.2%

c. Average number of jobs in the system = $\dfrac{103}{28}$ = 3.68 jobs

d. Average job lateness = $\dfrac{48}{5}$ = 9.6 days

The results of these four rules are summarized in the following table:

RULE	AVERAGE COMPLETION TIME (DAYS)	UTILIZATION (%)	AVERAGE NUMBER OF JOBS IN SYSTEM	AVERAGE LATENESS (DAYS)
FCFS	15.4	36.4	2.75	2.2
SPT	13.0	43.1	2.32	1.8
EDD	13.6	41.2	2.43	1.2
LPT	20.6	27.2	3.68	9.6

Your doctor may use a first-come, first-served priority rule satisfactorily. However, such a rule may be less than optimal for this emergency room. What priority rule might be best, and why? What priority rule is often used on the TV programs M*A*S*H and E.R.?

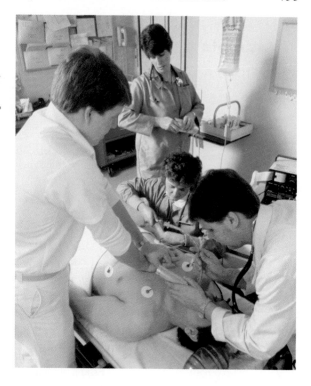

As we can see in Example 5, LPT is the least effective measurement of sequencing for the Ajax, Tarney and Barnes firm. SPT is superior in three measures and EDD in the fourth (average lateness). This is typically true in the real world also. We find that no one sequencing rule always excels on all criteria. Experience indicates the following:

The results of a dispatching rule change depending on how full the facility is.

1. Shortest processing time is generally the best technique for minimizing job flow and minimizing the average number of jobs in the system. Its chief disadvantage is that long-duration jobs may be continuously pushed back in priority in favor of short-duration jobs. Customers may view this dimly, and a periodic adjustment for longer jobs must be made.
2. First come, first served does not score well on most criteria (but neither does it score particularly poorly). It has the advantage, however, of appearing fair to customers, which is important in service systems.
3. Earliest due date minimizes maximum tardiness, which may be necessary for jobs that have a very heavy penalty after a certain date. In general, EDD works well when lateness is an issue.

Critical Ratio

Critical ratio (CR)
A sequencing rule that is an index number computed by dividing the time remaining until due date by the work time remaining.

Another type of sequencing rule is the critical ratio. The **critical ratio (CR)** is an index number computed by dividing the time remaining until due date by the work time remaining. As opposed to the priority rules, critical ratio is dynamic and easily updated. It tends to perform better than FCFS, SPT, EDD, or LPT on the average job-lateness criterion.

The critical ratio gives priority to jobs that must be done to keep shipping on schedule. A job with a low critical ratio (less than 1.0) is one that is falling behind schedule. If CR is exactly 1.0, the job is on schedule. A CR greater than 1.0 means the job is ahead of schedule and has some slack.

The formula for critical ratio is

$$CR = \frac{\text{Time remaining}}{\text{Workdays remaining}} = \frac{\text{Due date} - \text{Today's date}}{\text{Work (lead) time remaining}}$$

Example 6 shows how to use the critical ratio.

Example 6

Today is day 25 on Zyco Medical Testing Laboratories' production schedule. Three jobs are on order, as indicated here:

Job	Due Date	Workdays Remaining
A	30	4
B	28	5
C	27	2

We compute the critical ratios, using the formula for CR.

Job	Critical Ratio	Priority Order
A	$(30 - 25)/4 = 1.25$	3
B	$(28 - 25)/5 = .60$	1
C	$(27 - 25)/2 = 1.00$	2

Job B has a critical ratio of less than 1, meaning it will be late unless expedited. Thus, it has the highest priority. Job C is on time and job A has some slack. Once Job B has been completed, we would recompute the critical ratios for Jobs A and C to determine whether their priorities have changed.

In most production scheduling systems, the critical-ratio rule can help do the following:

1. Determine the status of a specific job.
2. Establish relative priority among jobs on a common basis.
3. Relate both stock and make-to-order jobs on a common basis.
4. Adjust priorities (and revise schedules) automatically for changes in both demand and job progress.
5. Dynamically track job progress.

Sequencing *N* Jobs on Two Machines: Johnson's Rule

The next step in complexity is the case in which *N* jobs (where *N* is 2 or more) must go through two different machines or work centers in the same order. This is called the *N*/2 problem.

Johnson's rule
An approach that minimizes processing time for sequencing a group of jobs through two work centers while minimizing total idle time in the work centers.

Johnson's rule can be used to minimize the processing time for sequencing a group of jobs through two work centers.[6] It also minimizes total idle time on the machines. *Johnson's rule* involves four steps:

1. All jobs are to be listed, and the time that each requires on a machine is to be shown.
2. Select the job with the shortest activity time. If the shortest time lies with the first machine, the job is scheduled first. If the shortest time lies with the second machine, schedule the job last. Ties in activity times can be broken arbitrarily.
3. Once a job is scheduled, eliminate it.
4. Apply steps 2 and 3 to the remaining jobs, working toward the center of the sequence.

Example 7 shows how to apply Johnson's rule.

Example 7

Five specialty jobs at a Fredonia, New York, tool and die shop must be processed through two work centers (drill press and lathe). The time for processing each job follows:

	WORK (PROCESSING) TIME FOR JOBS (IN HOURS)	
Job	Work Center 1 (Drill Press)	Work Center 2 (Lathe)
A	5	2
B	3	6
C	8	4
D	10	7
E	7	12

[6]S. M. Johnson, "Optimal Two and Three Stage Production Schedules with Set-Up Times Included," *Naval Research Logistics Quarterly* 1, no. 1 (March 1954): 61–68.

1.   We wish to set the sequence that will minimize the total processing time for the five jobs. The job with the shortest processing time is A, in work center 2 (with a time of 2 hours). Because it is at the second center, schedule A last. Eliminate it from consideration.

				A

2.   Job B has the next shortest time (3 hours). Because that time is at the first work center, we schedule it first and eliminate it from consideration.

B				A

3.   The next shortest time is job C (4 hours) on the second machine. Therefore, it is placed as late as possible.

B			C	A

4.   There is a tie (at 7 hours) for the shortest remaining job. We can place E, which was on the first work center, first. Then D is placed in the last sequencing position.

B	E	D	C	A

The sequential times are

Work center 1	3	7	10	8	5
Work center 2	6	12	7	4	2

The time-phased flow of this job sequence is best illustrated graphically:

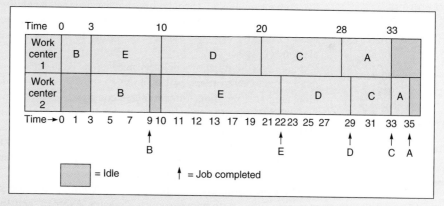

Thus, the five jobs are completed in 35 hours. The second work center will wait 3 hours for its first job, and it will also wait 1 hour after completing job B.

LIMITATIONS OF RULE-BASED DISPATCHING SYSTEMS

Scheduling can be complex to perform and still yield poor results—not a very fruitful combination. Even with sophisticated rules, good scheduling is very difficult.

The scheduling techniques just discussed are rule-based techniques, but rule-based systems have a number of limitations. Among those:

1.   Scheduling is dynamic; therefore, rules need to be revised to adjust to changes in process, equipment, product mix, and so forth.
2.   Rules do not look upstream or downstream; idle resources and bottleneck resources in other departments may not be recognized.
3.   Rules do not look beyond due dates. For instance, two orders may have the same due date. One order involves restocking a distributor and the other is a custom order that will shut down the customer's factory if not completed. Both may have the same due date, but clearly the custom order is more important.

FIGURE 15.5 ■

Most Finite Scheduling
Systems Generate a
Gantt Chart That Can Be
Manipulated by the User
on a Computer Screen

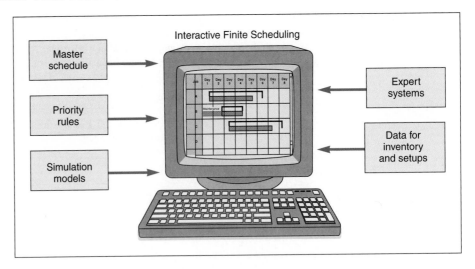

Despite these limitations, schedulers often use sequencing rules such as SPT, EDD, or critical ratio. They apply these methods periodically at each work center and then the scheduler modifies the sequence to deal with a multitude of real-world variables. They may do this manually or with finite scheduling software.

FINITE SCHEDULING

Finite scheduling
Computerized short-term scheduling that overcomes the disadvantage of rule-based systems by providing the user with graphical interactive computing.

Short-term scheduling is becoming integrated with finite scheduling. **Finite scheduling** overcomes the disadvantages of rule-based systems by providing the scheduler with graphical interactive computing. Finite schedules are characterized by the ability of the scheduler to make schedule changes based on up-to-the-minute information. These schedules are often displayed in Gantt chart form. The scheduler has the flexibility to handle any situation, including order, labor, or machine changes.

Finite scheduling allows delivery needs to be balanced against efficiency based on today's conditions and today's orders, not according to some predefined rule. Many of the current finite scheduling computer programs offer resource constraint features, a multitude of rules, and the ability of the scheduler to work interactively with the scheduling system to create a realistic schedule. These systems may also combine an "expert system" and simulation techniques and allow the scheduler to assign costs to various options. Finite scheduling leaves it up to the scheduler to determine what constitutes a "good" schedule. Figure 15.5 illustrates a computer screen with a Gantt chart, which is the result of a variety of data, rules, and models.

There are about 100 different finite scheduling software packages at the disposal of manufacturing firms, with names such as Preactor, Asprova, and Jobplan. These are currently used at over 60% of U.S. plants. Not all systems are successful, however. Plants that use fully integrated finite-scheduling systems, with electronic inputs, and that regenerate their schedules daily (as opposed to once a week) are more likely to benefit from this software than other firms.[7]

Theory of constraints (TOC)
That body of knowledge that deals with anything that limits an organization's ability to achieve its goals.

THEORY OF CONSTRAINTS

Managers need to identify the operations that constrain output because it is throughput—that is, units processed through the facility and sold—that makes the difference. This has led to the use of the term *theory of constraints*. The **theory of constraints (TOC)** is that body of knowledge that deals with anything that limits an organization's ability to achieve its goals. Constraints can be

[7]R. L. LaForge and C. W. Craighead, "Computer-Based Scheduling in Manufacturing Firms," *Production and Inventory Management Journal* 41, no. 1 (first quarter 2000): 29.

OM IN ACTION

Banking and the Theory of Constraints (TOC)

When a midwestern U.S. bank identified its weakest link as the mortgage department, with a home-loan processing time of over a month, it turned to the principles of TOC to reduce the average loan time. A cross functional mortgage improvement team of eight people employed the five steps outlined in the text. Using flow-charting, the team discovered that it was taking too long to (1) conduct property appraisals and surveys, and (2) verify applicant employment. So the first step of TOC was to identify these two constraints.

The second step in TOC was to develop a plan to reduce the time taken for employment verification and for conducting appraisals and surveys. The team learned that it could reduce employment verification to 2 weeks by having the loan officer request the last 2 years of W-2 forms and last month's pay stub. It found similar solutions to reducing survey/appraisal time.

As a third step, it had personnel refocus their resources so the two constraints could be performed at a higher level of efficiency. The result is decreased operating expense and inventory (money, in this banking example) and increased throughput.

The fourth TOC step required that employees support the earlier steps by focusing on the two time constraints. The bank also placed a higher priority on verification so that constraint could be overcome.

Finally, the bank began to look for new constraints once the first ones were overcome. Like all continuing improvement efforts, the process starts over before complacency sets in.

Sources: Decision Support Systems (March 2001): 451–468; *The Banker's Magazine* (January–February 1997): 53–59; and *Bank Systems and Technology* (September 1999): S10.

physical (such as process or personnel availability, raw materials, or supplies) or nonphysical (such as procedures, morale, training). Recognizing and managing these limitations through a five-step process is the basis of the theory of constraints:

Step 1: Identify the constraints.

Step 2: Develop a plan for overcoming the identified constraints.

Step 3: Focus resources on accomplishing step 2.

Step 4: Reduce the effects of the constraints by off-loading work or by expanding capability. Make sure that the constraints are recognized by all those who can have impact on them.

Step 5: Once one set of constraints is overcome, go back to step 1 and identify new constraints.

The *OM in Action* box, "Banking and the Theory of Constraints (TOC)," illustrates these five steps and shows that TOC is used in services as well as manufacturing.

Dr. Eliyahu Goldratt, a physicist, popularized the theory of constraints in a book he wrote with Jeff Cox, called *The Goal: A Process of Ongoing Improvement.*[8]

BOTTLENECK WORK CENTERS

Bottleneck
An operation that limits output in the production sequence.

Bottleneck work centers are constraints that limit the output of production. Bottlenecks have less capacity than the prior or following work centers. They constrain throughput. Bottlenecks are a common occurrence because even well-designed systems are seldom balanced for very long. Changing products, product mixes, and volumes often create multiple and shifting bottlenecks. Consequently, bottleneck work centers occur in nearly all process-focused facilities, from hospitals and restaurants to factories. Successful operations managers deal with bottlenecks by increasing the bottleneck's capacity, rerouting work, changing lot size, changing work sequence, or accepting idleness at other workstations. Substantial research has been done on the bottleneck issue.

[8]Eliyahu M. Goldratt and Jeff Cox, *The Goal: A Process of Ongoing Improvement* (Croton-on-Hudson, NY: North River Press, 1986). For more discussion of the constraints, see D. Nave, "How to Compare Six Sigma, Lean, and the Theory of Constraints," *Quality Progress* 35, no. 3 (March 2002): 73–79; and L. Cheng, "Line Balancing vs. Theory of Constraints," *IIE Solutions* 34, no. 4 (April 2002): 30–33.

Scheduling limited resources, such as this Boeing 747 simulator at New Zealand Airlines, is a critical job. Training and refreshing of pilots must occur at specific intervals. Sophisticated optimization models schedule pilots' flying time as well as their time at the simulator.

To increase throughput, the bottleneck constraint must be maximized by imaginative management, well-trained employees, and a well-maintained process. Several techniques for dealing with the bottleneck are available. They include:

1. Increasing capacity of the constraint. This may require a capital investment or more people and take a while to implement.
2. Ensuring that well-trained and cross-trained employees are available to operate and maintain the work center causing the constraint.
3. Developing alternative routings, processing procedures, or subcontractors.
4. Moving inspections and tests to a position just before the bottleneck. This approach has the advantage of rejecting any potential defects before they enter the bottleneck.
5. Scheduling throughput to match the capacity of the bottleneck. This may mean scheduling less work at the work center.

REPETITIVE MANUFACTURING

The scheduling goals defined at the beginning of this chapter are also appropriate for repetitive production. You may recall from Chapter 7 that repetitive producers make standard products from modules. Repetitive producers want to satisfy customer demands, lower inventory investment, reduce the batch (or lot) size, and utilize equipment and processes. The way to move toward these goals is

Level material use
The use of frequent, high-quality, small lot sizes that contribute to just-in-time production.

to move to a level material-use schedule. **Level material use** means frequent, high-quality, small lot sizes that contribute to just-in-time production. This is exactly what world-class producers such as Harley-Davidson and John Deere do. The advantages of level material use are

1. Lower inventory levels, which releases capital for other uses.
2. Faster product throughput (that is, shorter lead times).
3. Improved component quality and hence improved product quality.
4. Reduced floor-space requirements.
5. Improved communication among employees because they are closer together (which can result in improved teamwork and *esprit de corps*).
6. Smoother production process because large lots have not "hidden" the problems.

Suppose a repetitive producer runs large monthly batches. With a level material-use schedule, management would move toward shortening this monthly cycle to a weekly, daily, or even hourly cycle.

One way to develop a level material-use schedule is to first determine the minimum lot size that will keep the production process moving. This is illustrated in the next chapter, "Just-in-Time and Lean Production Systems."

SCHEDULING FOR SERVICES

Scheduling service systems differs from scheduling manufacturing systems in several ways. First, in manufacturing, the scheduling emphasis is on materials; in services, it is on staffing levels. Second, service systems seldom store inventories. Third, services are labor-intensive, and the demand for this labor can be highly variable.

Service systems try to match fluctuating customer demand with the capability to meet that demand. In some businesses, such as doctors' and lawyers' offices, an *appointment system* is the schedule. In retail shops, a post office, or a fast-food restaurant, a *first-come, first-served* rule for serving customers may suffice. Scheduling in these businesses is handled by bringing in extra workers, often part-timers, to help during peak periods. *Reservations systems* work well in rental car agencies, hotels, and some restaurants as a means of minimizing customer waiting time and avoiding disappointment over unfilled service.

Hospitals
A hospital is an example of a service facility that may use a scheduling system every bit as complex as one found in a job shop. Hospitals seldom use a machine shop priority system such as first come, first served (FCFS) for treating emergency patients. However, they do schedule products (such as surgeries) just like a factory, even though finished goods inventories cannot be kept and capacities must meet wide variations in demand.

Banks
Cross training of the workforce in a bank allows loan officers and other managers to provide short-term help for tellers if there is a surge in demand. Banks also employ part-time personnel to provide a variable capacity.

Airlines
Airlines face two constraints in scheduling flight crews: (1) a complex set of FAA work-time limitations and (2) union contracts that guarantee crew pay for some number of hours each day or each trip. Airline planners must build crew schedules that meet or exceed crews' pay guarantees. Planners must also make efficient use of their other expensive resource: aircraft. These schedules are typically built using linear programming models. The *OM in Action* box, "Scheduling Aircraft Turnaround," details how very short-term schedules (20 minutes) can help an airline become more efficient.

OM IN ACTION

Scheduling Aircraft Turnaround

Airlines that face increasingly difficult financial futures have recently discovered the importance of efficient scheduling of ground turnaround activities for flights. For some low-cost, point-to-point carriers like Southwest Airlines, scheduling turnarounds in 20 minutes has been standard policy for years. Yet for others, like Continental, United, and US Airways, the approach is new. This figure illustrates how US Airways deals with speedier schedules. Now its planes average seven trips a day, instead of six, meaning the carrier can sell tens of thousands more seats a day.

Sources: US Airways, Boeing, and *Aviation Week & Space Technology* (January 29, 2001): 50.

US Airways is cutting the turnaround time on commercial flights from the current 45 minutes to 20 minutes for Boeing 737s. Below is a list of procedures that must be completed before the flight can depart:

① Ticket agent takes flight plan to pilot, who loads information into aircraft computer. About 130 passengers disembark from the plane.

② Workers clean trash cans, seat pockets, lavatories, etc.

③ Catering personnel board plane and replenish supply of drinks and ice.

④ A fuel truck loads up to 5,300 gallons of fuel into aircraft's wings.

⑤ Baggage crews unload up to 4,000 pounds of luggage and 2,000 pounds of freight. "Runners" rush the luggage to baggage claim area in terminal.

⑥ Ramp agents, who help park aircraft upon arrival, "push" plane back away from gate.

24/7 Operations   Emergency hot lines, police/fire departments, telephone operations, and mail-order businesses (such as L.L. Bean) schedule employees 24 hours a day, 7 days a week. To allow management flexibility in staffing, sometimes part-time workers can be employed. This provides both benefits (in using odd shift lengths or matching anticipated workloads) and difficulties (from the large number of possible alternatives in terms of days off, lunch hour times, rest periods, starting times). Most companies use computerized scheduling systems to cope with these complexities.[9]

Scheduling Service Employees with Cyclical Scheduling

A number of techniques and algorithms exist for scheduling service-sector employees such as nurses, restaurant staff, tellers, and retail sales clerks. Managers, who can spend 20 hours per month on employee schedules, often consider a fairly long planning period (say, 6 weeks). Then they try to set a timely and efficient schedule that keeps personnel happy. Although there are several ways of tackling this problem, one approach that is both workable yet simple is *cyclical scheduling*.[10]

Cyclical scheduling has seven steps:

1. Plan a schedule equal in weeks to the number of people being scheduled.
2. Determine how many of each of the least desirable off-shifts must be covered each week.
3. Begin the schedule for one employee by scheduling the days off during the planning cycle (at a rate of 2 days per week on the average).
4. Assign off-shifts for that first employee using step 2. Here is an example of one person's 42-day schedule, where X is the day off, D is the day shift, and E is the evening shift:

S	M	T	W	T	F	S		S	M	T	W	T	F	S		S	M	T	W	T	F	S		S	M	T	W	T	F	S		S	M	T	W	T	F	S		S	M	T	W	T	F	S
E	E	E	E	E	X	X		X	X	E	E	E	E	E		E	D	X	D	D	D	D		D	D	X	X	D	E	X		X	E	E	E	E	E	X		D	E	D	D	X	X	E

5. Repeat this pattern for each of the other employees, but offsetting each one by 1 week from the previous one.
6. Allow each employee to pick his or her "slot" or "line" in order of seniority.
7. Mandate that any changes from a chosen schedule are strictly between the personnel wanting to switch.

Using this approach at Colorado General Hospital, the head nurse saved an average of 10 to 15 hours a month and found these advantages: (1) No computer was needed, (2) the nurses were happy with the schedule, (3) the cycles could be changed during different seasons (to accommodate avid skiers), and (4) recruiting was easier because of predictability and flexibility.[11]

To manage her hundreds of retail cookie outlets, Debbi Fields decided to capture her experience in a scheduling system that every store could access at any time. Her software takes advantage of headquarters' expertise in scheduling minimum-wage employees, the predominant counter help. It draws up a work schedule, including breaks, creates a full-day projection of the amount of dough to be processed, and charts progress and sales on an hourly basis. The scheduling system even tells staff when to cut back production and start offering free samples to passing customers.

[9]See, for example, Gary M. Thompson, "Assigning Telephone Operators to Shifts at New Brunswick Telephone Company," *Interfaces* 27 no. 4 (July–August 1997): 1–11; and B. Andrews and S. Cunningham, "L.L. Bean Improves Call Center Forecasting," *Interfaces* 25, no. 6 (November–December 1995): 1–13.

[10]For more details, see C. Haksever, B. Render, and R. Russell, *Service Management and Operations*, 2nd ed. (Upper Saddle River, NJ: Prentice Hall, 2000).

[11]With chronic staffing shortages in the nursing field, scheduling has taken a new turn. To help recruit and retain, hospitals have taken to a technique called "self-scheduling." See R. Hung, "A Note on Nurse Self-Scheduling," *Nursing Economics* (January–February 2002): 37–39.

SUMMARY

Scheduling involves the timing of operations to achieve the efficient movement of units through a system. This chapter addressed the issues of short-term scheduling in process-focused, repetitive, and service environments. We saw that process-focused facilities are production systems in which products are made to order and that scheduling tasks in them can become complex. Several aspects and approaches to scheduling, loading, and sequencing of jobs were introduced. These ranged from Gantt charts and the assignment methods of scheduling to a series of priority rules, the critical-ratio rule, and Johnson's rule for sequencing. We also examined the theory of constraints and the concept of bottlenecks.

Service systems generally differ from manufacturing systems. This leads to the use of appointment systems, first-come, first-served systems, and reservation systems, as well as to heuristics and linear programming approaches for servicing customers.

KEY TERMS

Forward scheduling *(p. 443)*
Backward scheduling *(p. 444)*
Planning files *(p. 445)*
Control files *(p. 445)*
Loading *(p. 445)*
Input–output control *(p. 445)*
Gantt charts *(p. 446)*
Assignment method *(p. 448)*
Sequencing *(p. 450)*
Priority rules *(p. 450)*

First come, first served (FCFS) *(p. 450)*
Shortest processing time (SPT) *(p. 450)*
Earliest due date (EDD) *(p. 450)*
Longest processing time (LPT) *(p. 450)*
Critical ratio (CR) *(p. 453)*
Johnson's rule *(p. 454)*
Finite scheduling *(p. 456)*
Theory of constraints (TOC) *(p. 456)*
Bottleneck *(p. 457)*
Level material use *(p. 458)*

USING EXCEL OM FOR SHORT-TERM SCHEDULING

Excel OM has two modules that help solve short-term scheduling problems: Assignment and Job Shop Scheduling. The Assignment module is illustrated in Programs 15.1 and 15.2. The input screen, using the Example 4 data, appears first, as Program 15.1. Once the data are all entered, we choose the Tools command, followed by the Solver command. Excel's Solver uses linear programming to optimize assignment problems. The constraints are also shown in Program 15.1. We then select the Solve command and the solution appears in Program 15.2.

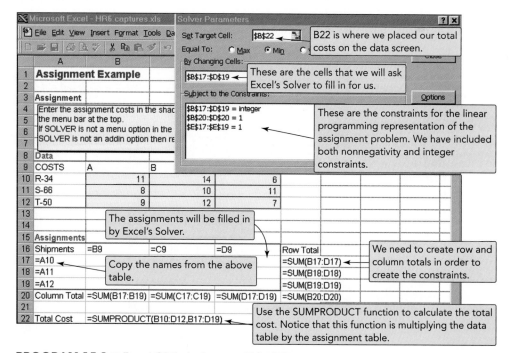

PROGRAM 15.1 ■ Excel OM's Assignment Module

After entering the problem data in the yellow area, select Tools, then Solver.

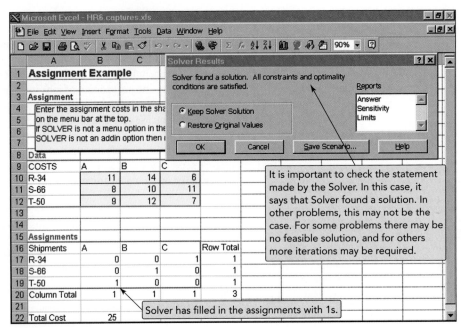

PROGRAM 15.2 ■ Excel OM Output Screen for Assignment Problem Described in Program 15.1

Excel OM's Job Shop Scheduling module is illustrated in Program 15.3. Program 15.3 uses Example 5's data. Because jobs are listed in the sequence in which they arrived (see column A), the results are for the FCFS rule. Program 15.3 also shows some of the formulas (columns E, F, G) used in the calculations.

To solve with the SPT rule, we need four intermediate steps: (1) Select (that is, screen) the data in columns A, B, C for all jobs; (2) invoke the Data command; (3) invoke the Sort command; and (4) sort by Time (column B) in *ascending* order. To solve for EDD, step 4 changes to sort by Due Date (column C) in *ascending* order. Finally, for an LPT solution, step 4 becomes sort by Due Date (column C) in *descending* order.

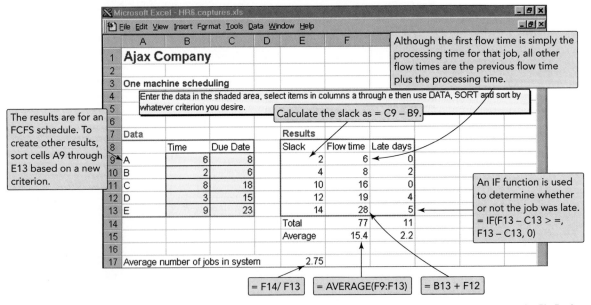

PROGRAM 15.3 ■ Excel OM's Job Shop Scheduling Module Applied to Example 5's Data

USING POM FOR WINDOWS TO SOLVE SCHEDULING PROBLEMS

POM for Windows can handle both categories of scheduling problems we see in this chapter. Its Assignment module is used to solve the traditional one-to-one assignment problem of people to tasks, machines to jobs, and so on. Its Job Shop Scheduling module can solve a one- or two-machine job shop problem. Available priority rules include SPT, FCFS, EDD, and LPT. Each can be examined in turn once the data are all entered. Refer to Appendix V for specifics regarding POM for Windows.

SOLVED PROBLEMS

Solved Problem 15.1

King Finance Corporation, headquartered in New York, wants to assign three recently hired college graduates, Julie Jones, Al Smith, and Pat Wilson, to regional offices. However, the firm also has an opening in New York and would send one of the three there if it were more economical than a move to Omaha, Dallas, or Miami. It will cost $1,000 to relocate Jones to New York, $800 to relocate Smith there, and $1,500 to move Wilson. What is the optimal assignment of personnel to offices?

OFFICE / HIREE	OMAHA	MIAMI	DALLAS
Jones	$800	$1,100	$1,200
Smith	$500	$1,600	$1,300
Wilson	$500	$1,000	$2,300

SOLUTION

(a) The cost table has a fourth column to represent New York. To "balance" the problem, we add a "dummy" row (person) with a zero relocation cost to each city.

OFFICE / HIREE	OMAHA	MIAMI	DALLAS	NEW YORK
Jones	$800	$1,100	$1,200	$1,000
Smith	$500	$1,600	$1,300	$ 800
Wilson	$500	$1,000	$2,300	$1,500
Dummy	0	0	0	0

(b) Subtract the smallest number in each row and cover all zeros (column subtraction will give the same numbers and therefore is not necessary):

OFFICE / HIREE	OMAHA	MIAMI	DALLAS	NEW YORK
Jones	0	300	400	200
Smith	0	1,100	800	300
Wilson	0	500	1,800	1,000
Dummy	0	0	0	0

(c) Subtract the smallest uncovered number (200) from all uncovered numbers, and add it to each square where two lines intersect. Then cover all zeros:

OFFICE / HIREE	OMAHA	MIAMI	DALLAS	NEW YORK
Jones	0	100	200	0
Smith	0	900	600	100
Wilson	0	300	1,600	800
Dummy	200	0	0	0

(d) Subtract the smallest uncovered number (100) from all uncovered numbers, and add it to each square where two lines intersect. Then cover all zeros:

OFFICE / HIREE	OMAHA	MIAMI	DALLAS	NEW YORK
Jones	0	0	100	0
Smith	0	800	500	100
Wilson	0	200	1,500	800
Dummy	300	0	0	100

(e) Subtract the smallest uncovered number (100) from all uncovered numbers, add it to squares where two lines intersect, and cover all zeros:

OFFICE / HIREE	OMAHA	MIAMI	DALLAS	NEW YORK
Jones	100	0	100	0
Smith	0	700	400	0
Wilson	0	100	1,400	700
Dummy	400	0	0	100

(f) Because it takes four lines to cover all zeros, an optimal assignment can be made at zero squares. We assign

Wilson to Omaha
Jones to Miami
Dummy (no one) to Dallas
Smith to New York

$$\text{Cost} = \$0 + \$500 + \$800 + \$1,100$$
$$= \$2,400$$

Solved Problem 15.2

A defense contractor in Dallas has six jobs awaiting processing. Processing time and due dates are given in the table to the right. Assume that jobs arrive in the order shown. Set the processing sequence according to FCFS and evaluate.

Job	Job Processing Time (days)	Job Due Date (days)
A	6	22
B	12	14
C	14	30
D	2	18
E	10	25
F	4	34

Solution

FCFS has the sequence A-B-C-D-E-F.

Job Sequence	Job Processing Time	Flow Time	Due Date	Job Lateness
A	6	6	22	0
B	12	18	14	4
C	14	32	30	2
D	2	34	18	16
E	10	44	25	19
F	4	48	34	14
	48	182		55

1. Average completion time = 182/6 = 30.33 days
2. Average number of jobs in system = 182/48 = 3.79 jobs
3. Average job lateness = 55/6 = 9.16 days
4. Utilization = 48/182 = 26.4%

Solved Problem 15.3

The Dallas firm in Solved Problem 15.2 also wants to consider job sequencing by the SPT priority rule. Apply SPT to the same data and provide a recommendation.

Solution

SPT has the sequence D-F-A-E-B-C.

Job Sequence	Job Processing Time	Flow Time	Due Date	Job Lateness
D	2	2	18	0
F	4	6	34	0
A	6	12	22	0
E	10	22	25	0
B	12	34	14	20
C	14	48	30	18
	48	124		38

1. Average completion time = 124/6 = 20.67 days
2. Average number of jobs in system = 124/48 = 2.58 jobs
3. Average job lateness = 38/6 = 6.33 days
4. Utilization = 48/124 = 38.7%

SPT is superior to FCFS in this case on all four measures. If we were to also analyze EDD, we would, however, find its average job lateness to be lowest at 5.5 days. SPT is a good recommendation. SPT's major disadvantage is that it makes long jobs wait, sometimes for a long time.

Solved Problem 15.4

Use Johnson's rule to find the optimum sequence for processing the jobs shown on the right through two work centers. Times at each center are in hours.

JOB	WORK CENTER 1	WORK CENTER 2
A	6	12
B	3	7
C	18	9
D	15	14
E	16	8
F	10	15

SOLUTION

B	A	F	D	C	E

The sequential times are

Work center 1	3	6	10	15	18	16
Work center 2	7	12	15	14	9	8

Solved Problem 15.5

Illustrate the throughput time and idle time at the two work centers in Solved Problem 15.4 by constructing a time-phased chart.

SOLUTION

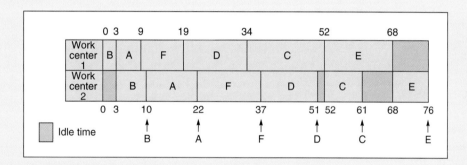

INTERNET AND STUDENT CD-ROM EXERCISES

Visit our home page or use your student CD-ROM to help with material in this chapter.

 On Our Home Page, www.prenhall.com/heizer

- Internet Homework Problems
- Internet Cases
- Self-Tests
- Practice Problems
- Internet Exercises
- Current Articles and Research
- Virtual Company Tour

 On Your Student CD-ROM

- PowerPoint Lecture
- Practice Problems
- Video Clip and Video Case
- Active Mode Exercise
- ExcelOM
- Excel OM Example Data Files

ADDITIONAL CASE STUDIES

Internet Case Study: Visit our Web site at www.prenhall.com/heizer **for this free case study:**

- **Old Oregon Wood Store**: Involves finding the best assignment of workers to the task of manufacturing tables.

Harvard has selected these Harvard Business School cases to accompany this chapter (textbookcasematch.hbsp.harvard.edu)**:**

- **The Patient Care Delivery Model at Massachusetts General Hospital** (#699–154): Examines the implementation of a new patient care delivery model.

- **Southern Pulp and Paper** (#696-103): Describes a paper mill whose poorly scheduled paper machines are a bottleneck in the operation.

Just-in-Time and Lean Production Systems

LEARNING OBJECTIVES

*When you complete this chapter you
should be able to*

IDENTIFY OR DEFINE:

Types of waste

Variability

Kanban

DESCRIBE OR EXPLAIN:

Just-in-time philosophy

Pull systems

Push systems

The goals of JIT partnerships

Lean production

Just-in-Time (JIT) Provides Competitive Advantage at Green Gear

Green Gear Cycling, Inc., of Eugene, Oregon, designs and manufactures a high performance travel bicycle, known as Bike Friday. The name is a take-off on Robinson Crusoe's "man Friday," who was always there when needed. Bike Friday is a bike in a suitcase—always there when you need it. This unique line of folding suitcase travel bicycles is built-to-order. Green Gear's goal, from its inception in 1992, has been to produce a high-quality custom bike rapidly and economically. This goal suggested a mass customization strategy requiring fast throughput, low inventory, work cells, and elimination of machine setups. It also meant adopting the best practices in operations management with a major focus on just-in-time (JIT) and supply-chain management.

Green Gear has carefully integrated just-in-time manufacturing and continuous improvement into its culture and processes. Inventory resides in individual containers at their point of use. Each container is labeled and includes a kanban card to trigger reordering.

Work cells make extensive use of visual signaling and explicit labeling of all inventory, tooling, and equipment. Each item is kept in a specific location, which facilitates cross training and assignment of employees to various cells. Effective work cell design translates to low inventory with work-in-process inventory of one bike per cell.

GREEN GEAR CYCLING

Through collaboration with suppliers, Green Gear has developed and implemented JIT deliveries that contribute to minimal inventory levels. And by storing inventory at the point of use and aggressively developing internal systems that support small reorder quantities, the firm has been able to drive inventory down and push quality up. Dedicated machinery and single application jigs also contribute to the small reorder quantities of component items. This success with JIT is instrumental in allowing Green Gear's high quality, low inventory system to work. Such systems are known as "kanban" systems and often use a simple signal such as a card, rather than a formal order, to signal the need for more parts.

For competitive as well as efficiency reasons, managers at Green Gear want to maintain a total throughput time, from raw tubing to completed bicycle, of less than 1 day. A manufacturing layout with such a high throughput requires minimizing or eliminating setups. The result is two flow lines, one for tandem bikes and one for single bikes. The seven work cells on these two lines are fed components from three support cells. The three support cells supply subassemblies, powder coating, and wheels. They receive orders via kanban cards. Each work cell's throughput time is balanced to match each of the others. These well-designed work cells contribute to rapid product throughput at Green Gear with little work-in-process.

Each bike is built to size, configured for the customer purchasing it, and shipped immediately on completion. So this build-to-order JIT system requires little raw material and little work-in-process inventory and *no* finished goods inventory. Supplier collaboration, creative work cells, elimination of setups, and exceptional quality contribute to low inventory and aid Green Gear in its continuing effort to speed bikes through the plant with a lot size of one.

Each Bike Friday is custom built to order from 14 base models, totaling 67 preconfigured bills, multiple paint colors, and additional sizing options. The total number of possible Bike Friday combinations exceeds 211,000. But lot size is one. And it fits in a suitcase.

Just-in-time (JIT)

A philosophy of continuous and forced problem solving that drives out waste.

Lean production

A way to eliminate waste through a focus on exactly what the customer wants.

TEN OM STRATEGY DECISIONS

Design of Goods and Services

Managing Quality

Process Strategy

Location Strategies

Layout Strategies

Human Resources

Supply-Chain Management

Inventory Management

Independent Demand

Dependent Demand

JIT & Lean Production

Scheduling

Maintenance

Variability

Any deviation from the optimum process that delivers perfect product on time, every time.

Pull system

A JIT concept that results in material being produced only when requested and moved to where it is needed just as it is needed.

As shown in the *Global Company Profile*, just-in-time (JIT) contributes to an efficient operation at Green Gear Cycling. In this chapter we discuss JIT as a philosophy of continuing improvement that drives out waste and supports lean organizations.

JUST-IN-TIME AND LEAN PRODUCTION

Just-in-time is a philosophy of continuous and forced problem solving that supports lean production. **Lean production** supplies the customer with exactly what the customer wants when the customer wants it, without waste, through continuous improvement. Lean production is driven by the "pull" of the customer's order. JIT is a key ingredient of lean production. When implemented as a comprehensive manufacturing strategy, JIT and lean production sustain competitive advantage and result in greater overall returns.[1]

With JIT, supplies and components are "pulled" through a system to arrive *where* they are needed *when* they are needed. When good units do not arrive just as needed, a "problem" has been identified. This makes JIT an excellent tool to help operations managers add value by driving out waste and unwanted variability. Because there is no excess inventory or excess time in a JIT system, costs associated with unneeded inventory are eliminated and throughput improved. Consequently, the benefits of JIT are particularly helpful in supporting strategies of rapid response and low cost.

Because elimination of *waste* and *variability* and the concept of *"pulling"* materials are fundamental to both JIT and lean production, we will briefly discuss them in this section. We will then introduce applications of JIT with suppliers, layout, inventory, scheduling, quality, and employee empowerment. Then we will review some of the distinguishing features of lean production and look at JIT applied to services.

Waste Reduction   When we talk about waste in the production of goods or services, we are describing *anything that does not add value*. Products being *stored, inspected,* or *delayed, products waiting in queues,* and *defective products* do not add value; they are 100% waste. Moreover, any activity that does not add value to a product *from the customer's perspective* is waste. JIT speeds up throughput, allowing faster delivery times and reducing work-in-process. Reducing work-in-process releases assets in inventory for other, more productive, purposes.

Variability Reduction   To achieve just-in-time material movement, managers *reduce variability caused by both internal and external factors*. **Variability** is any deviation from the optimum process that delivers perfect product on time, every time. Inventory hides variability—a polite word for problems. The less variability in the system, the less waste in the system. Most variability is caused by tolerating waste or by poor management. Variability occurs because:

1. Employees, machines, and suppliers produce units that do not conform to standards, are late, or are not the proper quantity.
2. Engineering drawings or specifications are inaccurate.
3. Production personnel try to produce before drawings or specifications are complete.
4. Customer demands are unknown.

Variability can often go unseen when inventory exists. This is why JIT is so effective. The JIT philosophy of continuous improvement removes variability. The removal of variability allows us to move good materials just-in-time for use. JIT reduces material throughout the supply chain. It helps us focus on adding value at each stage. Table 16.1 outlines the contributions of JIT; we discuss each of these concepts in this chapter.

Pull versus Push   The concept behind JIT is that of a **pull system**: a system that *pulls* a unit to where it is needed just as it is needed. A pull system uses signals to request production and delivery from stations upstream to the station that has production capacity available. The pull concept is used both within the immediate production process and with suppliers. By *pulling* material through the system in very small lots just as it is needed, the cushion of inventory that hides problems is

---

[1]Research suggests that the more JIT is comprehensive in breadth and depth, the greater overall returns will be. See Rosemary R. Fullerton and Cheryl S. McWatters, "The Production Performance Benefits from JIT Implementation," *Journal of Operations Management* 19, no. 1 (January 2001): 81–96.

TABLE 16.1 ■

JIT Contributes to
Competitive Advantage

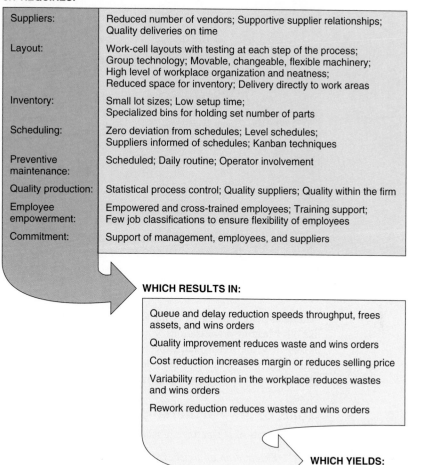

JIT REQUIRES:

Suppliers:	Reduced number of vendors; Supportive supplier relationships; Quality deliveries on time
Layout:	Work-cell layouts with testing at each step of the process; Group technology; Movable, changeable, flexible machinery; High level of workplace organization and neatness; Reduced space for inventory; Delivery directly to work areas
Inventory:	Small lot sizes; Low setup time; Specialized bins for holding set number of parts
Scheduling:	Zero deviation from schedules; Level schedules; Suppliers informed of schedules; Kanban techniques
Preventive maintenance:	Scheduled; Daily routine; Operator involvement
Quality production:	Statistical process control; Quality suppliers; Quality within the firm
Employee empowerment:	Empowered and cross-trained employees; Training support; Few job classifications to ensure flexibility of employees
Commitment:	Support of management, employees, and suppliers

WHICH RESULTS IN:

Queue and delay reduction speeds throughput, frees assets, and wins orders

Quality improvement reduces waste and wins orders

Cost reduction increases margin or reduces selling price

Variability reduction in the workplace reduces wastes and wins orders

Rework reduction reduces wastes and wins orders

WHICH YIELDS:

Faster response to the customer at lower cost and higher quality—

A Competitive Advantage

Manufacturing cycle time
The time between the arrival of raw materials and the shipping of finished products.

Push system
A system that pushes materials into downstream workstations regardless of their timeliness or availability of resources to perform the work.

JIT partnerships
Partnerships of suppliers and purchasers that remove waste and drive down costs for mutual benefits.

removed, problems become evident, and continuous improvement is emphasized. Removing the cushion of inventory also reduces both investment in inventory and manufacturing cycle time.

Manufacturing cycle time is the time between the arrival of raw materials and the shipping of finished products. For example, at Northern Telecom, a phone-system manufacturer, materials are pulled directly from qualified suppliers to the assembly line. This effort reduced Northern's receiving segment of manufacturing cycle time from 3 weeks to just 4 hours, the incoming inspection staff from 47 to 24, and problems on the shop floor caused by defective materials by 97%.

Many firms still move material through their facilities in a "push" fashion. A **push system** dumps orders on the next downstream workstation regardless of timeliness and resource availability. Push systems are the antithesis of JIT.

SUPPLIERS

Incoming material is often delayed at the shipper, in transit, at receiving departments, and at incoming inspection. Similarly, finished goods are often stored or held at warehouses prior to shipment to distributors or customers. Because holding inventory is wasteful, JIT partnerships are directed toward reducing such waste.

JIT partnerships exist when supplier and purchaser work together with a mutual goal of removing waste and driving down costs. Such relationships are critical for successful JIT. Every *moment*

Many services have adopted JIT techniques as a normal part of their business. Most restaurants, and certainly all fine-dining restaurants, expect and receive JIT deliveries. Both buyer and supplier expect fresh, high-quality produce delivered without fail just when it is needed. The system doesn't work any other way.

material is held, some process that adds value should be occurring. To ensure this is the case, Xerox, like other leading organizations, views the supplier as an extension of its own organization. Because of this view, the Xerox staff expects suppliers to be as fully committed to improvement as Xerox. This relationship requires a high degree of openness[2] by both supplier and purchaser. Table 16.2 shows the characteristics of JIT partnerships.

Goals of JIT Partnerships

The four goals of JIT partnerships are:

1. *Elimination of unnecessary activities.* With good suppliers, for instance, receiving activity and incoming-inspection activity are unnecessary under JIT.
2. *Elimination of in-plant inventory.* JIT delivers materials where and when needed. Raw material inventory is necessary only if there is reason to believe that suppliers are undependable. Likewise, parts or components should be delivered in small lots directly to the using department as needed.
3. *Elimination of in-transit inventory.* General Motors once estimated that at any given time, over one-half of its inventory is in transit. Modern purchasing departments are now addressing in-transit inventory reduction by encouraging suppliers and prospective suppliers to locate near manufacturing plants and provide frequent small shipments. The shorter the flow of material in the resource pipeline, the less inventory. Inventory can also be reduced by a technique known as *consignment.* **Consignment inventory** (see the *OM in Action* box. "Lean Production at Cessna Aircraft"), a variation of vendor managed inventory (Chapter 11), means the supplier maintains the title to the inventory until it is used. For instance, an assembly plant may find a hardware supplier that is willing to locate its warehouse where the user currently has its stockroom. In this manner, when hardware is needed, it is no farther than the stockroom, and the supplier can ship to other, perhaps smaller, purchasers from the "stockroom."
4. *Elimination of poor suppliers.* When a firm reduces the number of suppliers, it increases long-term commitments. To obtain improved quality and reliability, vendors and purchasers have mutual understanding and trust. Achieving deliveries only when needed and in the exact quantities needed also requires *perfect quality*—or as it is also known, *zero defects.* Of course, *both* the supplier and the delivery system must be excellent.

Consignment inventory
An arrangement in which the supplier maintains title to the inventory until it is used.

[2]J. Douglas Blocher, Charles W. Lackey, and Vincent A. Mabert, "From JIT Purchasing to Supplier Partnerships at Xerox," *Target* 9, no. 3 (May–June 1993): 12–18.

TABLE 16.2 ■

Characteristics of JIT Partnerships

To get JIT to work, the purchasing agent must communicate the company's goal to the supplier. This includes delivery, packaging, lot sizes, quality, and so on.

SUPPLIERS

Few suppliers
Nearby suppliers
Repeat business with same suppliers
Analysis and support to enable desirable suppliers to become or to stay price competitive
Competitive bidding mostly limited to new purchases
Buyer resists vertical integration and subsequent wipeout of supplier business
Suppliers encouraged to extend JIT buying to their suppliers (second- and third-tier suppliers)

QUANTITIES

Steady output rate
Frequent deliveries in small-lot quantities
Long-term contract agreements
Minimal or no paperwork to release orders (use EDI or Internet)
Delivery quantities fixed for whole contract term
Little or no permissible overage or underage
Suppliers package in exact quantities
Suppliers reduce their production lot sizes

QUALITY

Minimal product specifications imposed on supplier
Help suppliers meet quality requirements
Close relationships between buyers' and suppliers' quality assurance people
Suppliers use poka-yoke and process control charts instead of lot-sampling inspection

SHIPPING

Scheduling of inbound freight
Gain control by use of company-owned or contract shipping and warehousing
Use of advanced shipping notice (ASN)

Source: Adapted from M. Schniederjans, *Topics in Just in Time,* Boston: Allyn & Bacon (1993); *International Journal of Operations and Production Management* 10, no. 4 (1990): 31–41; *Production and Inventory Management Journal* 29, no. 3 (1988): 45–50; and *California Management Review* 26, no. 1 (1983).

OM IN ACTION

Lean Production at Cessna Aircraft

When Cessna Aircraft opened its new plant in Independence, Kansas, in 1996, it saw the opportunity to switch from a craftwork mentality producing small single-engine planes to a lean manufacturing system. In doing so, Cessna adopted three lean manufacturing practices.

First, Cessna set up consignment- and vendor-managed inventories with several of its suppliers. Honeywell, for example, maintains a 30-day supply of avionic parts on-site. Other vendors were encouraged to use a nearby warehouse to keep parts that could then be delivered daily to the production line.

Second, Cessna managers committed to a philosophy of cross training in which team members learn the duties of other team members and can shift across

assembly lines as needed. To develop these technical skills, Cessna brought in retired assembly-line workers to mentor and teach new employees. Employees were taught to work as a team and to assume responsibility for their team's quality.

Third, the company used group technology (Chapter 5) and manufacturing cells (Chapter 9) to move away from a batch process which resulted in large inventories and unsold planes. Now, Cessna pulls product through its plant only when a specific order is placed.

These long term commitments to manufacturing efficiency are part of the lean production system that has made Cessna an industry leader, with about half the market in general aviation planes.

Sources: Fortune (May 1, 2000): 1222B–1222Z; *Manufacturing Systems, Supply Chain Yearbook: 2002* (2000): 46–48; and *New York Times* (June 7, 2000): E-18.

Concerns of Suppliers

To establish JIT partnerships, several supplier concerns must be addressed. The supplier concerns include:

1. *Desire for diversification.* Many suppliers do not want to tie themselves to long-term contracts with one customer. The suppliers' perception is that they reduce their risk if they have a variety of customers.
2. *Poor customer scheduling.* Many suppliers have little faith in the purchaser's ability to reduce orders to a smooth, coordinated schedule.
3. *Engineering changes.* Frequent engineering changes, with inadequate lead time for suppliers to carry out tooling and process changes, play havoc with JIT.
4. *Quality assurance.* Production with zero defects is not considered realistic by many suppliers.
5. *Small lot sizes.* Suppliers often have processes designed for large lot sizes and see frequent delivery to the customer in small lots as a way to transfer holding costs to suppliers.
6. *Proximity.* Depending on the customer's location, frequent supplier delivery of small lots may be seen as economically prohibitive.

For those who remain skeptical of JIT partnerships, we would point out that virtually every restaurant in the world practices JIT, and with little staff support. Many restaurants order food for the next day in the middle of the night for delivery the next morning. They are ordering just *what* is needed, for delivery *when* it is needed, from reliable suppliers.

JIT LAYOUT

JIT layouts reduce another kind of waste—movement. The movement of material on a factory floor (or paper in an office) does not add value. Consequently, we want flexible layouts that reduce the movement of both people and material. JIT layouts move material directly to the location where needed. For instance, an assembly line should be designed with delivery points next to the line so material need not be delivered first to a receiving department elsewhere in the plant, then moved again. This is what VF Corporation's Wrangler Division in Greensboro, North Carolina, did. Now denim is delivered directly to the line. When a layout reduces distance, the firm also saves space and eliminates potential areas for unwanted inventory. Table 16.3 provides a list of layout tactics.

TABLE 16.3 ■
Layout Tactics

Build work cells for
   families of products
Minimize distance
Design little space for
   inventory
Improve employee
   communication
Use poka-yoke devices
Build flexible or movable
   equipment
Cross train workers to add
   flexibility

Distance Reduction

Reducing distance is a major contribution of work cells, work centers, and focused factories (see Chapter 9). The days of long production lines and huge economic lots, with goods passing through monumental, single-operation machines, are gone. Now firms use work cells, often arranged in a U shape, containing several machines performing different operations. These work cells are often based on group technology codes (as discussed in Chapter 5). Group technology codes help us identify components with similar characteristics so we can group them into families. Once families are identified, work cells are built for them. The result can be thought of as a small product-oriented facility where the "product" is actually a group of similar products—a family of products. The cells produce one good unit at a time, and ideally they produce the units *only* after a customer orders them.

Increased Flexibility

Modern work cells are designed so they can be easily rearranged to adapt to changes in volume, product improvements, or even new designs. Almost nothing in these new departments is bolted down. This same concept of layout flexibility applies to office environments. Not only are most office furniture and equipment movable, but so are office walls, computer connections, and telecommunications. Equipment is modular. Layout flexibility aids the changes that result from product *and* process improvements that are inevitable with a philosophy of continuous improvement.

In a JIT system, each worker is inspecting the part as it comes to him or her. Each worker knows that the part must be good before it goes on to the next "customer."

Impact on Employees

Employees working together are cross trained so they can bring flexibility and efficiency to the work cell. JIT layouts allow employees to work together so they can tell each other about problems and opportunities for improvement. When layouts provide for sequential operations, feedback can be immediate. Defects are waste. When workers produce units one at a time, they test each product or component at each subsequent production stage. Machines in work cells with self-testing "poka-yoke" functions detect defects and stop automatically when they occur. Before JIT, defective products were replaced from inventory. Because surplus inventory is not kept in JIT facilities, there are no such buffers. Getting it right the first time is critical.

Reduced Space and Inventory

Because JIT layouts reduce travel distance, they also reduce inventory by removing space for inventory. When there is little space, inventory must be moved in very small lots or even single units. Units are always moving because there is no storage. For instance, each month Security Pacific Corporation's focused facility sorts 7 million checks, processes 5 million statements, and mails 190,000 customer statements. With a JIT layout, mail processing time has been reduced by 33%, salary costs by tens of thousands of dollars per year, floor space by 50%, and in-process waiting lines by 75% to 90%. Storage, including shelves and drawers, has been removed.

INVENTORY

Inventories in production and distribution systems often exist "just in case" something goes wrong. That is, they are used just in case some variation from the production plan occurs. The "extra" inventory is then used to cover variations or problems. Effective inventory tactics require "just in time," not "just in case." **Just-in-time inventory** is the minimum inventory necessary to keep a perfect system running. With just-in-time inventory, the exact amount of goods arrives at the moment it is needed, not a minute before or a minute after. The *OM in Action* box, "Let's Try Zero Inventory," suggests it can be done. Some useful JIT inventory tactics are shown in Table 16.4 and discussed in more detail in the following sections.

Reduce Variability

The idea behind JIT is to eliminate inventory that hides variability in the production system. This concept is illustrated in Figure 16.1, which shows a lake full of rocks. The water in the lake represents inventory flow, and the rocks represent problems such as late deliveries, machine breakdowns, and poor personnel performance. The water level in the lake hides variability and problems. Because inventory hides problems, they are hard to find.

TABLE 16.4 ■
JIT Inventory Tactics

Use a pull system to move inventory

Reduce lot size

Develop just-in-time delivery systems with suppliers

Deliver directly to point of use

Perform to schedule

Reduce setup time

Use group technology

Just-in-time inventory
The minimum inventory necessary to keep a perfect system running.

OM IN ACTION

Let's Try Zero Inventory

Just-in-time tactics are being incorporated in manufacturing to improve quality, drive down inventory investment, and reduce other costs. However, JIT is also established practice in restaurants, where customers expect it, and a necessity in the produce business, where there is little choice. Pacific Pre-Cut Produce, a $14-million fruit and vegetable processing company in Tracy, California, holds inventory to zero. Buyers are in action in the wee hours of the morning. At 6 A.M., produce production crews show up. Orders for very specific cuts and mixtures of fruit and vegetable salads and stir-fry ingredients for supermarkets, restaurants, and institutional kitchens pour in from 8 A.M. until 4 P.M. Shipping begins at 10 P.M. and continues until the last order is filled and loaded at 5 A.M. the next morning. Inventories are once again zero and

things are relatively quiet for an hour or so; then the routine starts again. Pacific Pre-Cut Produce has accomplished a complete cycle of purchase, manufacture, and shipping in about 24 hours.

VP Bob Borzone calls the process the ultimate in mass-customization. "We buy everything as a bulk commodity, then slice and dice it to fit the exact requirements of the end user. There are 20 different stir-fry mixes. Some customers want the snow peas clipped on both ends, some just on one. Some want only red bell peppers in the mix, some only yellow. You tailor the product to the customer's requirements. You're trying to satisfy the need of a lot of end users, and each restaurant and retailer wants to look different."

Sources: Inbound Logistics (August 1997): 26–32; and *Restaurant Business* (June 10, 1992): 16–20.

Video 16.1

Sailing through the
Problems of Excess
Inventory

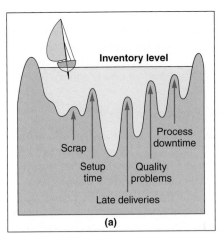

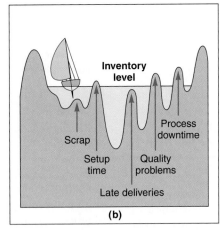

FIGURE 16.1 ■ Inventory Has Two Costs, One for Holding the Inventory and the Second for the Problems it Hides; Just as Water in a Lake Hides the Rocks

"Inventory is evil."

Shigeo Shingo

Reduce Inventory

Operations managers move toward JIT by first removing inventory. Reducing inventory uncovers the "rocks" in Figure 16.1(a) that represent the variability and problems currently being tolerated. With reduced inventory, management chips away at the exposed problems until the lake is clear. After the lake is clear, managers make additional cuts in inventory and continue to chip away at the next level of exposed problems (see Figure 16.1[b]). Ultimately, there will be virtually no inventory and no problems (variability).

Shigeo Shingo, codeveloper of the Toyota JIT system, says, "Inventory is evil." He is not far from the truth. If inventory itself is not evil, it hides evil at great cost.

Reduce Lot Sizes

Just-in-time has also come to mean elimination of waste by reducing investment in inventory. The key to JIT is producing good product in small lot sizes. Reducing the size of batches can be a major help in reducing inventory and inventory costs. As we saw in Chapter 12, when inventory usage is constant, the average inventory level is the sum of the maximum inventory plus the minimum inventory divided by two. Figure 16.2 shows that lowering the order size increases the number of orders but drops inventory levels.

Ideally, in a JIT environment, order size is one and single units are being pulled from one adjacent process to another. More realistically, analysis of the process, transportation time, and contain-

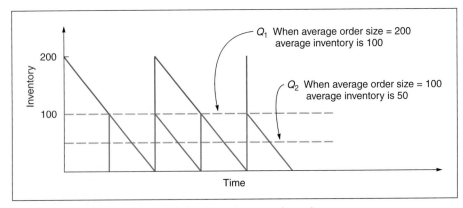

FIGURE 16.2 ■ Frequent Orders Reduce Average Inventory

A lower order size increases the number of orders and total ordering cost, but reduces average inventory and total holding cost.

ers used for transport are considered when determining lot size. Such analysis typically results in a small lot size but a lot size larger than one. Once a lot size has been determined, the EOQ production order quantity model can be modified to determine the desired setup time. We saw in Chapter 12 that the production order quantity model takes the form:

$$Q^* = \sqrt{\frac{2DS}{H[1-(d/p)]}}$$ (16-1)

where D = Annual demand
S = Setup cost
H = Holding cost
d = Daily demand
p = Daily production

Example 1 shows how Crate Furniture, Inc., a firm that produces rustic furniture, moves toward a reduced lot size.

Example 1

Crate Furniture's production analyst, Aleda Roth, determined that a 2-hour production cycle would be acceptable between two departments. Further, she concluded that a setup time that would accommodate the 2-hour cycle time should be achieved. Roth developed the following data and procedure to determine optimum setup time analytically:

D = Annual demand = 400,000 units

d = Daily demand = 400,000 per 250 days = 1,600 units per day

p = Daily production rate = 4,000 units per day

Q = EOQ desired = 400 (which is the 2-hour demand; that is, 1,600 per day per four 2-hour periods

H = Holding cost = $20 per unit per year

S = Setup cost (to be determined)

Roth determines that the cost, on an hourly basis, of setting up equipment is $30. Further, she computes that the setup cost per setup should be

$$Q = \sqrt{\frac{2DS}{H(1-d/p)}}$$

$$Q^2 = \frac{2DS}{H(1-d/p)}$$

$$S = \frac{(Q^2)(H)(1-d/p)}{2D}$$

$$S = \frac{(400)^2(20)(1-1,600/4,000)}{2(400,000)}$$

$$= \frac{(3,200,000)(0.6)}{800,000} = \$2.40$$

Setup time = $2.40/(hourly labor rate)

= $2.40/($30 per hour)

= 0.08 hour, or 4.8 minutes

Now, rather than producing components in large lots, Crate Furniture can produce in a 2-hour cycle with the advantage of an inventory turnover of four *per day.*

Only two changes need to be made for small-lot material flow to work. First, we need to improve material handling and work flow. With short production cycles, there can be very little wait time. Improving material handling is usually easy and straightforward. The second change is more challenging, and that is a radical reduction in setup times. We discuss setup reduction next.

FIGURE 16.3 ■

Lower Setup Costs Will Lower Total Cost

More frequent orders require reducing setup costs; otherwise, inventory costs will rise. As the setup costs are lowered (from S_1 to S_2), inventory costs also fall (from T_1 and T_2).

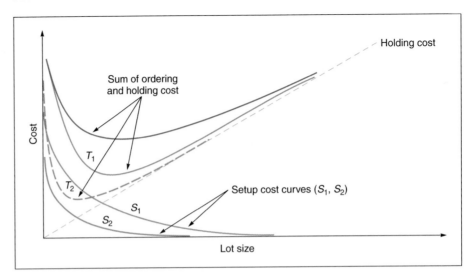

Reduce Setup Costs

Both inventory and the cost of holding it go down as the inventory-reorder quantity and the maximum inventory level drops. However, because inventory requires incurring an ordering or setup cost that must be applied to the units produced, managers tend to purchase (or produce) large orders. With large orders, each unit purchased or ordered absorbs only a small part of the setup cost. Consequently, the way to drive down lot sizes *and* reduce average inventory is to reduce setup cost, which in turn lowers the optimum order size.

The effect of reduced setup costs on total cost and lot size is shown in Figure 16.3. Moreover, smaller lot sizes hide fewer problems. In many environments, setup cost is highly correlated with setup time. In a manufacturing facility, setups usually require a substantial amount of work prior to actually being accomplished at a work center. Much of the preparation required by a setup can be done prior to shutting down the machine or process. Setup times can be reduced substantially, as shown in Figure 16.4. For instance, in Kodak's Guadalajara, Mexico, plant a team reduced the setup time to change a bearing from 12 hours to 6 minutes![3] This is the kind of progress that is typical of world-class manufacturers.

Reduced lot sizes must be accompanied by reduced setup times; otherwise, the setup cost must be assigned to fewer units.

FIGURE 16.4 ■

Steps to Reduce Setup Times

Reduced setup times are a major JIT component.

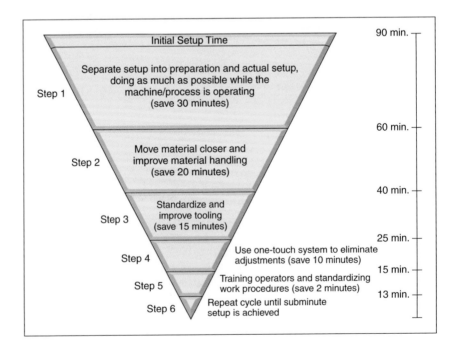

[3]Frank Carguello and Marty Levin, "Excellence at Work in Guadalajara, Mexico, Operation," *Target* 15, no. 3 (third quarter 1999): 51–53.

Just as setup costs can be reduced at a machine in a factory, setup time can also be reduced during the process of getting the order ready. It does little good to drive down factory setup time from hours to minutes if orders are going to take 2 weeks to process or "set up" in the office. This is exactly what happens in organizations that forget that JIT concepts have applications in offices as well as the factory. Reducing setup time (and cost) is an excellent way to reduce inventory investment and to improve productivity.

TABLE 16.5 ■
JIT Scheduling Tactics

Communicate schedules
  to suppliers
Make level schedules
Freeze part of the
  schedule
Perform to schedule
Seek one-piece-make and
  one-piece-move
Eliminate waste
Produce in small lots
Use kanbans
Make each operation
  produce a perfect part

Level schedules
Scheduling products
so that each day's
production meets the
demand for that day.

SCHEDULING

Effective schedules, communicated both within the organization and to outside suppliers, support JIT. Better scheduling also improves the ability to meet customer orders, drives down inventory by allowing smaller lot sizes, and reduces work-in-process. For instance, Ford Motor Company now ties some suppliers to its final assembly schedule. Ford communicates its schedules to bumper manufacturer Polycon Industries from the Ford Oakville production control system. The scheduling system describes the style and color of the bumper needed for each vehicle moving down the final assembly line. The scheduling system transmits the information to portable terminals carried by Polycon warehouse personnel who load the bumpers onto conveyors leading to the loading dock. The bumpers are then trucked 50 miles to the Ford plant. Total time is 4 hours.[4] Table 16.5 suggests several items that can contribute to achieving these goals, but two techniques (in addition to communicating schedules) are paramount. They are *level schedules* and *kanban*.

Level Schedules

Level schedules process frequent small batches rather than a few large batches. Because this technique schedules many small lots that are always changing, it has on occasion been called "jelly bean" scheduling. Figure 16.5 contrasts a traditional large-lot approach using large batches with a JIT level schedule using many small batches. The operations manager's task is to make and move small lots so the level schedule is economical. This requires success with the issues discussed in this chapter that allow small lots. As lots get smaller, the constraints may change and become increasingly challenging. At some point, processing a unit or two may not be feasible. The constraint may be the way units are sold and shipped (four to a carton), or an expensive paint changeover (on an automobile assembly line), or the proper number of units in a sterilizer (for a food-canning line).

The scheduler may find that *freezing* the portion of the schedule closest to due dates allows the production system to function and the schedule to be met. Freezing means not allowing changes to be part of the schedule. Operations managers expect the schedule to be achieved with no deviations from the schedule.

Kanban

One way to achieve small lot sizes is to move inventory through the shop only as needed rather than *pushing* it on to the next workstation whether or not the personnel there are ready for it. As noted earlier, when inventory is moved only as needed, it is referred to as a *pull* system, and the ideal lot size is one. The Japanese call this system *kanban*,

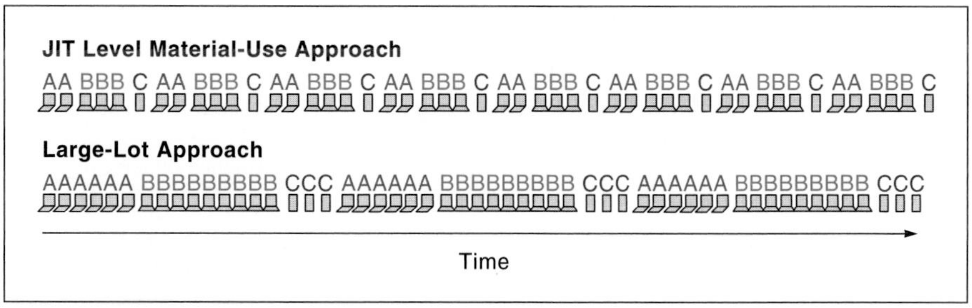

FIGURE 16.5 ■ Scheduling Small Lots of Parts A, B, and C Increases Flexibility to Meet Customer Demand and Reduces Inventory

The JIT approach to scheduling produces just as many of each model per time period as the large-lot approach, provided setup times are lowered.

---

[4]Mike Ngo and Paul Szucs, "Four Hours," *APICS—The Performance Advantage* 6, no. 1 (January 1996): 30–32.

A kanban need not be as formal as signal lights or empty carts. The cook in a fast-food restaurant knows that when six cars are in line, eight meat patties and six orders of French fries should be cooking.

Kanban

The Japanese word for card that has come to mean "signal"; a kanban system moves parts through production via a "pull" from a signal.

Kanban is a Japanese word for *card*. In their effort to reduce inventory, the Japanese use systems that "pull" inventory through work centers. They often use a "card" to signal the need for another container of material—hence the name *kanban*. *The card is the authorization for the next container of material to be produced.* Typically, a kanban signal exists for each container of items to be obtained. An order for the container is then initiated by each kanban and "pulled" from the producing department or supplier. A sequence of kanbans "pulls" the material through the plant.

The system has been modified in many facilities so that, even though it is called a *kanban*, the card itself does not exist. In some cases, an empty position on the floor is sufficient indication that the next container is needed. In other cases, some sort of signal, such as a flag or rag (Figure 16.6) alerts that it is time for the next container.

When there is visual contact between producer and user, the process works like this:

1. The user removes a standard size container of parts from a small storage area, as shown in Figure 16.6.
2. The signal at the storage area is seen by the producing department as authorization to replenish the using department or storage area. Because there is an optimum lot size, the producing department may make several containers at a time.

Figure 16.7 shows how a kanban works, pulling units as needed through successive phases of production. This system is similar to the resupply that occurs in your neighborhood supermarket: The customer buys; the stock clerk observes the shelf or receives notice from the end-of-day sales list and restocks. When the limited supply, if any, in the store's storage is depleted, a "pull" signal is sent to the warehouse, distributor, or manufacturer for resupply, usually that night. The complicating factor in a manufacturing firm is the need for the actual manufacturing (production) to take place.

FIGURE 16.6 ■

Diagram of Outbound Stockpoint with Warning-Signal Marker

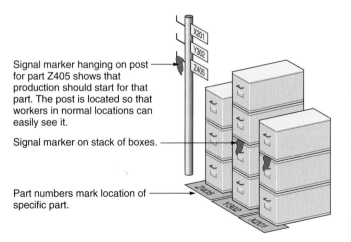

Signal marker hanging on post for part Z405 shows that production should start for that part. The post is located so that workers in normal locations can easily see it.

Signal marker on stack of boxes.

Part numbers mark location of specific part.

Several additional points regarding kanbans may be helpful:

- When the producer and user are not in visual contact, a card can be used; otherwise, a light or flag or empty spot on the floor may be adequate.
- Because a pull station may require several resupply components, several kanban pull techniques can be used for different products at the same pull station.
- Usually, each card controls a specific quantity or parts, although multiple card systems are used if the producing work cell produces several components or if the lot size is different from the move size.

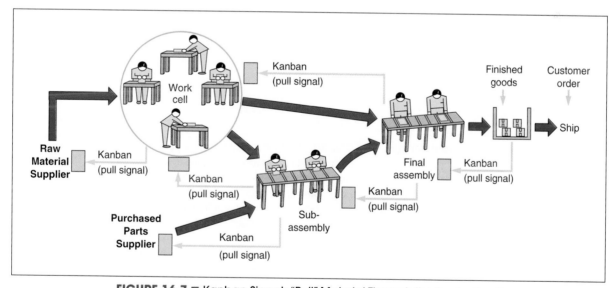

FIGURE 16.7 ■ Kanban Signals "Pull" Material Through the Production Process

As a customer "pulls" an order from finished goods, a signal (card) is sent to the final assembly area. The final assembly area produces and resupplies finished goods. When final assembly needs components, it sends a signal to its suppliers, a subassembly area and a work cell. These areas supply final assembly. The work cell, in turn, sends a signal to the raw material supplier, and the subassembly area notifies the work cell and purchased parts supplier of a requirement.

- In an MRP system (see Chapter 14) the schedule can be thought of as a "build" authorization and the kanban as a type of "pull" system that initiates the actual production.
- The kanban cards provide a direct control (limit) on the amount of work-in-process between cells.
- If there is an immediate storage area, a two-card system may be used—one card circulates between user and storage area, and the other circulates between the storage area and the producing area.

Determining the Number of Kanban Cards or Containers   The number of kanban cards, or containers, in a JIT system sets the amount of authorized inventory. To determine the number of containers moving back and forth between the using area and the producing areas, management first sets the size of each container. This is done by computing the lot size, using a model such as the production order quantity model (discussed in Chapter 12 and shown again on page 477 in Equation [16-1]). Setting the number of containers involves knowing (1) lead time needed to produce a container of parts and (2) the amount of safety stock needed to account for variability or uncertainty in the system. The number of kanban cards is computed as follows:

$$\text{Number of kanbans (containers)} = \frac{\text{Demand during lead time} + \text{Safety stock}}{\text{Size of container}}$$

Example 2 illustrates how to calculate the number of kanbans needed.

Example 2

Hobbs Bakery produces short runs of cakes that are shipped to grocery stores. The owner, Ken Hobbs, wants to try to reduce inventory by changing to a kanban system. He has developed the following data and asked you to finish the project by telling him the number of kanbans (containers) needed.

Daily demand = 500 cakes
Production lead time = 2 days
Safety stock = $\frac{1}{2}$ day
Container size (determined on a production order size EOQ basis) = 250 cakes

Solution

Demand during lead time (= lead time × daily demand = 2 days × 500 cakes =) 1,000

Safety Stock = 250

$$\text{Number of kanbans (containers) needed} =$$
$$\frac{\text{Demand during lead time} + \text{Safety stock}}{\text{Container size}} = \frac{1,000 + 250}{250} = 5$$

Advantages of Kanban   Containers are typically very small, usually a matter of a few hours' worth of production. Such a system requires tight schedules. Small quantities must be produced several times a day. The process must run smoothly with little variability in quality of lead time because any shortage has an almost immediate impact on the entire system. Kanban places added emphasis on meeting schedules, reducing the time and cost required by setups, and economical material handling.

Whether it is called kanban or something else, the advantages of small inventory and *pulling* material through the plant only when needed are significant. For instance, small batches allow only a very limited amount of faulty or delayed material. Problems are immediately evident. Numerous aspects of inventory are bad; only one aspect—availability—is good. Among the bad aspects are poor quality, obsolescence, damage, occupied space, committed assets, increased insurance, increased material handling, and increased accidents. Kanban systems put downward pressure on all of these negative aspects of inventory.

In-plant kanban systems often use standardized, reusable containers that protect the specific quantities to be moved. Such containers are also desirable in the supply chain. Standardized containers reduce weight and disposal costs, generate less wasted space in trailers, and require less labor to pack, unpack, and prepare items.

Manufacturers' inventory/sales ratio was substantially lower in the last recession than in earlier recessions, thanks in large part to JIT inventories.

The New United Motor Manufacturing (NUMMI) plant in Fremont, California, which builds the Toyota Corolla and the GM Prizm, is a joint venture between Toyota and General Motors. The plant was designed as a just-in-time (JIT) facility. Management even moved a water tower to ensure that new loading docks would facilitate JIT arrivals and JIT movement of parts within the plant. This plant, like most JIT facilities, also empowers employees so they can stop the entire production line by pulling the overhead cord if any quality problems are spotted.

TABLE 16.6 ■
JIT Quality Tactics

Use statistical process control
Empower employees
Build fail-safe methods (poka-yoke, checklists, etc.)
Provide immediate feedback

QUALITY

The relationship between JIT and quality is a strong one.[5] They are related in three ways. First, JIT cuts the cost of obtaining good quality. This saving occurs because scrap, rework, inventory investment, and damage costs are buried in inventory. JIT forces down inventory; therefore, fewer bad units are produced and fewer units must be reworked. In short, whereas inventory *hides* bad quality, JIT immediately *exposes* it.

Second, JIT improves quality. As JIT shrinks queues and lead time, it keeps evidence of errors fresh and limits the number of potential sources of error. In effect, JIT creates an early warning system for quality problems so that fewer bad units are produced and feedback is immediate. This advantage can accrue both within the firm and with goods received from outside vendors.

Finally, better quality means fewer buffers are needed and, therefore, a better, easier-to-employ JIT system can exist. Often the purpose of keeping inventory is to protect against unreliable quality. If consistent quality exists, JIT allows firms to reduce all costs associated with inventory. Table 16.6 suggests some requirements for quality in a JIT environment.

EMPLOYEE EMPOWERMENT

Whereas some JIT techniques require policy and strategy decisions, many are part of the purview of empowered employees. Empowered employees can bring their involvement to bear on most of the daily operations issues that are so much a part of a just-in-time philosophy. This means that those tasks that have traditionally been assigned to staff can move to empowered employees.

Employee empowerment follows the management adage that no one knows the job better than those who do it. Firms not only train and cross train, but need to take full advantage of that investment by enriching jobs.[6] Aided by aggressive cross training and few job classifications, firms can engage the mental as well as physical capacities of employees in the challenging task of improving the workplace.

JIT's philosophy of continuous improvement gives employees the opportunity to enrich their jobs and their lives. When empowerment is managed successfully, companies gain from mutual commitment and respect on the part of both employees and management.

LEAN PRODUCTION

Lean production can be thought of as the end result of a well-run OM function. The major difference between JIT and lean production is that JIT is a philosophy of continuing improvement with an *internal* focus, while lean production begins *externally* with a focus on the customer. Understanding what the customer wants and ensuring customer input and feedback are starting

[5]See related discussion in Barbara B. Flynn, Sadao Sakakibara, and Roger G. Schroeder, "Relationship Between JIT and TQM: Practices and Performance," *Academy of Management Journal* 38, no. 5 (1995): 1325–1360.

[6]Richard J. Schonberger, "Human Resource Management Lessons from a Decade of Total Quality Management and Reengineering," *California Management Review* (summer 1994): 109–123.

OM IN ACTION

Dell's Lean Production

Dell's 200,000-square-foot computer assembly plant in Austin, Texas, is a showcase of efficient manufacturing. Dell's JIT and lean production practices allowed it to wring a half billion dollars out of manufacturing costs last year. Both Dell's suppliers and in-house purchasing personnel evaluate inventories on an hour-by-hour basis to hold work-in-process (WIP) to a minimum. Six-person teams assemble 18 computers each hour with parts that arrive via an overhead conveyor system. If the cell has a problem, the parts are instantly shifted to another cell, avoiding stops that are common in traditional assembly lines.

Dell's lean production practices begin with instantaneous customer feedback. Because of its direct sales model, Dell is the first to know of changes in the marketplace. Dell has been so successful at lean production and knowing its customer that over the last 2 years production has increased by a third. Over the same period, lean practices have cut manufacturing space in half.

Robots are now being tested to shave seconds from the time required to load computers into cartons. Additional seconds are saved by combining the downloading of software and computer testing into one step. Dell keeps product design under constant review, simplifying components, speeding assembly, and saving even more seconds. Time saved improves throughput, adds capacity, and contributes to flexibility in the production process. The added throughput, capacity, and flexibility allow Dell to respond to the sudden and frequent shifts in demand that characterize the PC market.

Dell has tripled production per square foot over the past 5 years and expects to triple it again over the next 5 years. The firm's reputation is such that CEO Michael Dell is counseling the U.S. automotive industry in lean manufacturing techniques. Dell's model of lean production is showing the way.

Sources: Forbes (June 10, 2002): 110; and *Infotech Update* (July–August, 2001): 6.

---

points for lean production. Lean production means identifying customer value by analyzing all of the activities required to produce the product, and then optimizing the entire process from the view of the customer. The manager finds what creates value for the customer and what does not.

Toyota Production System (TPS)
Developed by Toyota Motor Company, TPS is the forerunner of lean production concepts, emphasizing employee learning and empowerment.

Lean production is sometimes called the **Toyota Production System (TPS)**, with Toyota Motor Company's Eiji Toyoda and Taiichi Ohno given credit for its approach and innovations.[7] If there is any distinction between JIT, lean production, and TPS it is that JIT emphasizes continuous improvement, lean production emphasizes understanding the customer, and TPS emphasizes employee learning and empowerment in an assembly line environment. In practice, there is little difference and the terms are often used interchangeably.

The transition to lean production is difficult. Building an organizational culture where learning and continuous improvement are the norm is a challenge. However, we find that organizations that focus on JIT, quality, and employee empowerment are often lean producers. Such firms drive out activities that do not add value in the eyes of the customer: They include leaders like Toyota, United Parcel Service, and Dell Computer. Dells' exceptional performance is noted in the *OM in Action* box, "Dell's Lean Production." These lean producers adopt a philosophy of minimizing waste by striving for perfection through continuous learning, creativity, and teamwork. Success requires the full commitment and involvement of all employees and of the company's suppliers. The rewards reaped by lean producers are spectacular. Lean producers often become benchmark performers. They share the following attributes:

- *Use just-in-time techniques* to eliminate virtually all inventory.
- *Build systems that help employees* produce a perfect part every time.
- *Reduce space requirements* by minimizing the distance a part travels.
- *Develop close relationships with suppliers*, helping them to understand their needs and their customers' needs.
- *Educate suppliers* to accept responsibility for helping meet customer needs.
- *Eliminate all but value-added activities*. Material handling, inspection, inventory, and rework jobs are among the likely targets because these do not add value to the product.
- *Develop the workforce* by constantly improving job design, training, employee participation and commitment, and teamwork.
- *Make jobs more challenging*, pushing responsibility to the lowest level possible.
- *Reduce the number of job classes* and build worker flexibility.

---

[7]Eiji Toyoda was part of the founding family of Toyota Motor Company and Taiichi Ohno was production manager.

Lean producers set their sights on perfection: no bad parts, no inventory, and only value-added activities. Traditional producers have limited goals, accepting, for instance, the production of some defective parts and some inventory.

JIT IN SERVICES

All of the JIT techniques for dealing with suppliers, layout, inventory, and scheduling are used in services.

Suppliers   As we have noted, virtually every restaurant deals with its suppliers on a JIT basis. Those that do not are usually unsuccessful. The waste is too evident—food spoils and customers complain.

Layouts   JIT layouts are required in restaurant kitchens, where cold food must be served cold and hot food hot. McDonald's for example, has reconfigured its kitchen layout at great expense (see the *Global Company Profile*, Chapter 9) to drive seconds out of the production process, thereby speeding delivery to customers. With the new process, McDonald's can produce made-to-order hamburgers in 45 seconds. Layouts also make a difference in airline baggage claim, where customers expect their bags just-in-time.

In a hospital with JIT, suppliers bring ready-to-use supplies directly to storage areas, nurses' stations, and operating rooms. Only a 24-hour reserve is maintained.

Inventory   Every stockbroker drives inventory down to nearly zero. Most sell and buy orders occur on a JIT basis because an unexecuted sell or buy order is not acceptable to most clients. A broker may be in serious trouble if left holding an unexecuted trade. Similarly, McDonald's maintains a finished goods inventory of only 10 minutes; after that, it is thrown away. Hospitals also endorse JIT inventory and low safety stocks, even for such critical supplies as pharmaceuticals, by developing community networks as backup systems. In this manner, if one pharmacy runs out of a needed drug, a member of the network can supply it until the next day's shipment arrives.[8]

Scheduling   At airline ticket counters, the focus of a JIT system is customer demand, but rather than being satisfied by the inventory of a tangible product, that demand must be satisfied by personnel. Through elaborate scheduling, airline ticket-counter personnel show up just-in-time to satisfy customer demand, and they provide the service on a JIT basis. In other words, personnel are scheduled, rather than "things" inventoried. Personnel schedules are critical. At a beauty salon, the focus is only slightly different: The customer is scheduled to assure JIT service. Similarly, at McDonald's, as at most fast-food restaurants, scheduling of personnel is down to 15-minute increments based on precise forecasting of demand. Additionally, production is done in small lots to ensure that fresh, hot hamburgers are delivered just-in-time. In short, both personnel and production are scheduled on a JIT basis to meet specific demand.

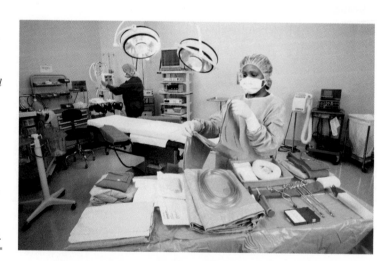

JIT takes on an unusual form in an operating room. Baxter International, like many other hospital suppliers, provides surgical supplies for hospitals on a JIT basis. One, it delivers prepackaged surgical supplies based on hospital operating schedules, and, two, the surgical packages themselves are prepared so supplies are available in the sequence in which they will be used during surgery.

[8]Daniel Whitson, "Applying Just-in-Time Systems in Health Care," *Industrial Engineering Solutions* (August 1997): 33–37.

Notice that in all three of these examples—the airline ticket counter, the beauty salon, and McDonald's—scheduling is a key ingredient in effective JIT. Excellent forecasts drive those schedules. Those forecasts may be very elaborate, with seasonal, daily, and even hourly components in the case of the airline ticket counter (holiday sales, flight time, etc.), seasonal and weekly components at the beauty salon (holidays and Fridays creating special problems), or down to a few minutes at McDonald's.

In order to deliver goods and services to customers under continuously changing demand, suppliers need to be reliable, inventories lean, cycle times short, and schedules nimble. These issues are currently being managed with great success in many firms regardless of their products. JIT techniques are widely used in both goods-producing and service-producing firms; they just look different.

SUMMARY

JIT and lean production are philosophies of continuous improvement. Lean production begins with a focus on customer desires, but both concepts focus on driving all waste out of the production process. Because waste is found in anything that does not add value, JIT and lean organizations are adding value more efficiently than other firms. Waste occurs when defects are produced within the production process or by outside suppliers. JIT and lean production attack wasted space because of a less-than-optimal layout; they attack wasted time because of poor scheduling; they attack waste in idle inventory; they attack waste from poorly maintained machinery and equipment. The expectation is that committed, empowered employees work with committed management and suppliers to build systems that respond to customers with ever lower cost and higher quality.

KEY TERMS

Just-in-time (JIT) *(p. 470)*
Lean production *(p. 470)*
Variability *(p. 470)*
Pull system *(p. 470)*
Manufacturing cycle time *(p. 471)*
Push system *(p. 471)*

JIT partnerships *(p. 471)*
Consignment inventory *(p. 472)*
Just-in-time inventory *(p. 475)*
Level schedules *(p. 479)*
Kanban *(p. 480)*
Toyota Production System (TPS) *(p. 484)*

SOLVED PROBLEM

Solved Problem 16.1

Krupp Refrigeration, Inc., is trying to reduce inventory and wants you to install a kanban system for compressors on one of its assembly lines. Determine the size of the kanban and the number of kanbans (containers) needed.

Setup cost = $10

Annual holding cost per compressor = $100

Daily production = 200 compressors

Annual usage = 25,000 (50 weeks × 5 days each
                        × daily usage of 100 compressors

Lead time = 3 days

Safety stock = $\frac{1}{2}$ day's production of compressors

SOLUTION

First, we must determine kanban container size. To do this, we determine the production order quantity (see discussion in Chapter 12 or Equation [16-1]), which determines our kanban size:

$$Q_p = \sqrt{\frac{2DS}{H\left(1 - \dfrac{d}{p}\right)}} = \sqrt{\frac{2(25,000)(10)}{H\left(1 - \dfrac{d}{p}\right)}} = \sqrt{\frac{500,000}{100\left(1 - \dfrac{100}{200}\right)}} = \sqrt{\frac{500,000}{50}}$$

$$= \sqrt{10,000} = 100 \text{ compressors}$$

Then we determine the number of kanbans:

$$\text{Demand during lead time} = 300\ (= 3\ \text{days} \times \text{daily usage of } 100)$$

$$\text{Safety stock} = 100\ (= \tfrac{1}{2}\ \text{day's production} \times 200)$$

$$\text{Number of kanbans} = \frac{\text{Demand during lead time} + \text{Safety stock}}{\text{Size of container}}$$

$$= \frac{300 + 100}{100} = \frac{400}{100} = 4\ \text{containers}$$

INTERNET AND STUDENT CD-ROM EXERCISES

Visit our home page or use your student CD-ROM to help with material in this chapter.

 On Our Home Page www.prenhall.com/heizer

- Self-Tests
- Practice Problems
- Internet Exercises
- Current Articles and Research
- Virtual Company Tour
- Internet Homework Problems

 On Your Student CD-ROM

- PowerPoint Lecture
- Practice Problems
- Video Clips
- ExcelOM

ADDITIONAL CASE STUDIES

Harvard has selected these Harvard Business School cases to accompany this text (textbookcasematch.hbsp.harvard.edu):

- **Johnson Controls Automotive Systems Group: The Georgetown Kentucky Plant** (#693-086): Examines the challenge of JIT with growing variation and a change from JIT delivery to JIT assembly.

- **Injex Industries** (#697-003): Examines supplier concerns as Injex provides components to a single, demanding customer on a JIT basis.

Maintenance and Reliability

Chapter Outline

LEARNING OBJECTIVES

When you complete this chapter you should be able to

IDENTIFY OR DEFINE:

Maintenance

Mean time between failures

Redundancy

Preventive maintenance

Breakdown maintenance

Infant mortality

DESCRIBE OR EXPLAIN:

How to measure system reliability

How to improve maintenance

How to evaluate maintenance performance

Maintenance and Reliability Are the Critical Success Factors for NASA's Space Shuttles

Space Shuttle Atlantis *is rolled from its hangar to the launch pad.*

Atlantis's *main engine is installed in the Orbiter Processing facility.*

From miles away, a space shuttle looks gleaming white on the launch pad. Yet up close, in the hangars where the three NASA shuttles, *Endeavor*, *Atlantis*, and *Discovery*, spend most of their lives, a shuttle can show its true colors: moldy green; burnt brown; grungy gray; sooty black.

In one Kennedy Space Center hangar, *Atlantis* sits with its guts spread out. Its three engines (each the size of a Volkswagen) are detached and in another shop for maintenance. It has a gaping hole in its nose because its front jets are sitting on the floor. With millions of miles on its odometer, *Atlantis* is like a used car. However, NASA has no plans to retire this multibillion-dollar work-horse. *Atlantis* is expected to make a dozen more voyages as a global science lab, with its hold rented out to dozens of nations for scientific experiments and satellite launches.

Such a plan requires world-class reliability. It also requires maintenance. Indeed it means about 600 computer-generated maintenance jobs, each with hundreds of tasks, during the 3-month turnaround between flights. There are platforms to install, engine inspections, turbo-pump checks, tile reworks, lube-oil drainings, drag-chute removal and reinstallation. More than 100 men and women work behind the scenes to maintain *Atlantis's* long-standing reputa-tion of reliability. In the memory of each are the January 1986 *Challenger* explo-sion and the February 2003 *Columbia* disaster. Despite the tragic losses, shut-tle program manager Ron Dittemore still believes, "the vehicles are kept in pristine shape."

A tunnel leading into the cargo bay space lab is inspected.

THE STRATEGIC IMPORTANCE OF MAINTENANCE AND RELIABILITY

Managers at NASA must avoid the undesirable results of a shuttle that fails. The results of failure can be disruptive, inconvenient, wasteful, and expensive in dollars and in lives. Machine and product failures can have far-reaching effects on an organization's operation, reputation, and profitability. In complex, highly mechanized plants, an out-of-tolerance process or a machine breakdown may result in idle employees and facilities, loss of customers and goodwill, and profits turning into losses. In an office, the failure of a generator, an air-conditioning system, or a computer may halt operations. A good maintenance and reliability strategy protects both a firm's performance and its investment.

The objective of maintenance and reliability is to maintain the capability of the system while controlling costs. A good maintenance system drives out system variability. Systems must be designed and maintained to reach expected performance and quality standards. **Maintenance** includes all activities involved in keeping a system's equipment in working order. **Reliability** is the probability that a machine part or product will function properly for a specified time under stated conditions.

Two firms that recognize the strategic importance of dedicated maintenance are Walt Disney Company and United Parcel Service. Disney World, in Florida, is intolerant of failures or breakdowns. Disney's reputation makes it not only one of the most popular vacation destinations in the world but also a mecca for benchmarking teams that want to study its maintenance and reliability practices.

Likewise, UPS's famed maintenance strategy keeps its delivery vehicles operating and looking as good as new for 20 years or more. The UPS program involves dedicated drivers who operate the same truck every day and dedicated mechanics who maintain the same group of vehicles. Drivers and mechanics are both responsible for the performance of a vehicle and stay closely in touch with each other.

The interdependency of operator, machine, and mechanic is a hallmark of successful maintenance and reliability. As Figure 17.1 illustrates, it is not only good maintenance and reliability procedures that make Disney and UPS successful, but the involvement of their employees as well.

In this chapter, we examine four important tactics for improving the reliability and maintenance not only of products and equipment but also of the systems that produce them. The four tactics are organized around reliability and maintenance.

The reliability tactics are

1. Improving individual components.
2. Providing redundancy.

The maintenance tactics are

1. Implementing or improving preventive maintenance.
2. Increasing repair capabilities or speed.

Maintenance

All activities involved in keeping a system's equipment in working order.

Reliability

The probability that a machine part or product will function properly for a specified time under stated conditions.

FIGURE 17.1 ■

Good Maintenance and Reliability Strategy Requires Employee Involvement and Good Procedures

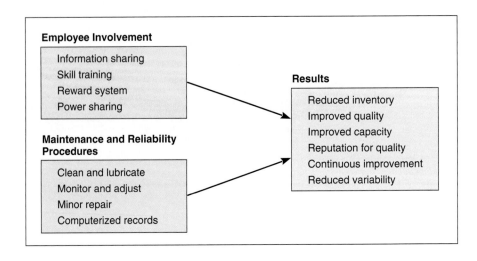

The operations manager must drive out variability: Designing for reliability and managing for maintenance are crucial ingredients for doing so.

RELIABILITY

Systems are composed of a series of individual interrelated components, each performing a specific job. If any *one* component fails to perform, for whatever reason, the overall system (for example, an airplane or machine) can fail.

Improving Individual Components

Because failures do occur in the real world, understanding their occurrence is an important reliability concept. We will now examine the impact of failure in a series. Figure 17.2 shows that as the number of components in a *series* increases, the reliability of the whole system declines very quickly. A system of $n = 50$ interacting parts, each of which has a 99.5% reliability, has an overall reliability of 78%. If the system or machine has 100 interacting parts, each with an individual reliability of 99.5%, the overall reliability will be only about 60%!

To measure reliability in a system in which each individual part or component may have its own unique rate of reliability, we cannot use the reliability curve in Figure 17.2. However, the method of computing system reliability (R_s) is simple. It consists of finding the product of individual reliabilities as follows:

$$R_s = R_1 \times R_2 \times R_3 \times \ldots \times R_n \tag{17-1}$$

where R_1 = reliability of component 1
         R_2 = reliability of component 2

and so on.

Equation (17-1) assumes that the reliability of an individual component does not depend on the reliability of other components (that is, each component is independent). Additionally, in this equation as in most reliability discussions, reliabilities are presented as *probabilities*, Thus, a .90 reliability means that the unit will perform as intended 90% of the time. It also means that it will fail $1 - .90 = .10 = 10\%$ of the time. We can use this method to evaluate the reliability of a service or a product, such as the one we examine in Example 1.

FIGURE 17.2 ■

Overall System Reliability as a Function of Number of Components and Component Reliability with Components in a Series

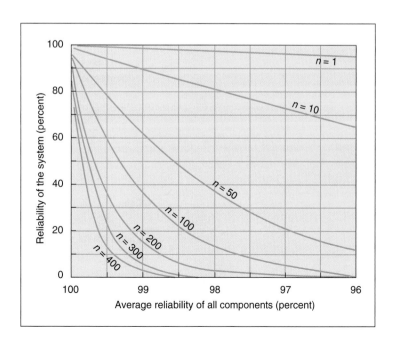

Example 1

Active Model 17.1

Example 1 is further illustrated in Active Model 17.1 on your CD-ROM.

The National Bank of Greeley, Colorado, processes loan applications through three clerks set up in series:

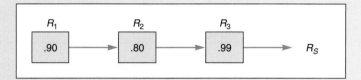

If the clerks have reliabilities of .90, .80, .99, then the reliability of the loan process is

$$R_s = R_1 \times R_2 \times R_3 = (.90)(.80)(.99) = .713, \text{ or } 71.3\%$$

Component reliability is often a design or specification issue for which engineering design personnel may be responsible. However, supply chain personnel may be able to improve components of systems by staying abreast of suppliers' products and research efforts. Supply chain personnel can also contribute directly to the evaluation of supplier performance.

The basic unit of measure for reliability is the *product failure rate* (FR). Firms producing high-technology equipment often provide failure-rate data on their products. As shown in Equations (17-2) and (17-3), the failure rate measures the percentage of failures among the total number of products tested, FR(%), or a number of failures during a period of time, FR(*N*):

$$\text{FR}(\%) = \frac{\text{Number of failures}}{\text{Number of units tested}} \times 100\% \tag{17-2}$$

$$\text{FR}(N) = \frac{\text{Number of failures}}{\text{Number of unit-hours of operating time}} \tag{17-3}$$

Mean time between failures (MTBF)
The expected time between a repair and the next failure of a component, machine, process, or product.

Perhaps the most common term in reliability analysis is the **mean time between failures (MTBF)**, which is the reciprocal of FR(*N*):

$$\text{MTBF} = \frac{1}{\text{FR}(N)} \tag{17-4}$$

In Example 2, we compute the percentage of failure FR(%), number of failures FR(*N*), and mean time between failures (MTBF).

Example 2

Twenty air-conditioning systems designed for use by astronauts in NASA space shuttles were operated for 1,000 hours at NASA's Huntsville, Alabama, test facility. Two of the systems failed during the test—one after 200 hours and the other after 600 hours. To compute the percentage of failures, we use the following equation:

$$\text{FR}(\%) = \frac{\text{Number of failures}}{\text{Number of units tested}} = \frac{2}{20}(100\%) = 10\%$$

Next we compute the number of failures per operating hour:

$$\text{FR}(N) = \frac{\text{Number of failures}}{\text{Operating time}}$$

where

$$\text{Total time} = (1,000 \text{ hr})(20 \text{ units})$$
$$= 20,000 \text{ unit-hr}$$
$$\text{Nonoperating time} = 800 \text{ hr for 1st failure} + 400 \text{ hr for 2nd failure}$$
$$= 1,200 \text{ unit-hr}$$
$$\text{Operating time} = \text{Total time} - \text{Nonoperating time}$$
$$\text{FR}(N) = \frac{2}{20,000 - 1,200} = \frac{2}{18,800}$$
$$= .000106 \text{ failure/unit-hr}$$

and because $\text{MTBF} = \dfrac{1}{\text{FR}(N)}$

$$\text{MTBF} = \frac{1}{.000106} = 9,434 \text{ hr}$$

If the typical space shuttle trip lasts 60 days, NASA may be interested in the failure rate per trip:

$$\text{Failure rate} = (\text{failures/unit-hr})(24 \text{ hr/day})(60 \text{ days/trip})$$
$$= (.000106)(24)(60)$$
$$= .152 \text{ failure/trip}$$

Because the failure rate recorded in Example 2 is probably too high, NASA will have to either increase the reliability of individual components, and thus of the system, or else install several backup air-conditioning units on each space shuttle. Backup units provide redundancy.

Providing Redundancy

Redundancy
The use of components in parallel to raise reliabilities.

To increase the reliability of systems, **redundancy** is added. The technique here is to "back up" components with additional components. This is known as putting units in parallel and is a standard operations management tactic, as noted in the *OM in Action* box, "Tomcat F-14 Pilots Love Redundancy." Redundancy is provided to ensure that if one component fails, the system has recourse to another. For instance, say that reliability of a component is .80 and we back it up with another component with reliability of .80. The resulting reliability is the probability of the first component working plus the probability of the backup (or parallel) component working multiplied by the probability of needing the backup component $(1 - .8 = .2)$. Therefore:

$$
\begin{pmatrix} \text{Probability} \\ \text{of first} \\ \text{component} \\ \text{working} \end{pmatrix}
+
\left[\begin{pmatrix} \text{Probability} \\ \text{of second} \\ \text{component} \\ \text{working} \end{pmatrix} \times \begin{pmatrix} \text{Proability} \\ \text{of needing} \\ \text{second} \\ \text{component} \end{pmatrix} \right] =
$$

$$\quad (.8) \quad + \quad [(.8) \quad \times \quad (1 - .8)] \quad = .8 + .16 = .96$$

OM IN ACTION

Tomcat F-14 Pilots Love Redundancy

In a world that accepts software with bugs and computer systems that crash, it is worth remembering that some computer systems operate without fail. Where are these systems? They are in fighter jets, the space shuttle, nuclear power plants, and flood-control systems. These systems are all extraordinarily reliable, even though they depend heavily on software. Such systems are all about redundancy—they have their own software and their own processors—and use most of their cycles to perform internal quality checks.

The Tomcat F-14's variable-wing geometry allows it to fly very fast and to slow down quickly when landing on an aircraft carrier. The calculations to determine the correct wing position as air speed changes are determined by software and dedicated processors. The processors run in tandem so multiple calculations verify outgoing signals.

Only 10% of the F-14's software is used to fly the plane; 40% is used to do automatic testing and verification; the remaining 50% is redundancy. Highly reliable systems work because the design includes self-checking and redundancy. These redundant systems find potential problems and correct them before a failure can occur. If you are a Tomcat F-14 pilot, you love redundancy.

Source: Information.com (April 1, 2002): 34.

Example 3 shows how redundancy can improve the reliability of the loan process presented in Example 1.

Example 3

Active Model 17.2

Example 3 is further illustrated in Active Model 17.2 on the CD-ROM and in the Exercise located in your Student Lecture Guide.

The National Bank is disturbed that its loan-application process has a reliability of only .713 (see Example 1). Therefore, the bank decides to provide redundancy for the two least reliable clerks. This procedure results in the system shown below:

$$R_1 \qquad R_2 \qquad R_3$$

$$\boxed{0.90} \qquad \boxed{0.8}$$
$$\downarrow \qquad \downarrow$$
$$\boxed{0.90} \rightarrow \boxed{0.8} \rightarrow \boxed{0.99} = [.9 + .9(1-.9)] \times [.8 + .8(1-.8)] \times .99$$
$$= [.9 + (.9)(.1)] \times [.8 + (.8)(.2)] \times .99$$
$$= .99 \times .96 \times .99 = .94$$

By providing redundancy for two clerks, National Bank has increased reliability of the loan process from .713 to .94.

MAINTENANCE

There are two types of maintenance: preventive maintenance and breakdown maintenance. **Preventive maintenance** involves performing routine inspections and servicing and keeping facilities in good repair. These activities are intended to build a system that will find potential failures and make changes or repairs that will prevent failure. Preventive maintenance is much more than just keeping machinery and equipment running. It also involves designing technical and human systems that will keep the productive process working within tolerance; it allows the system to perform. The emphasis of preventive maintenance is on understanding the process and keeping it working without interruption. **Breakdown maintenance** occurs when equipment fails and must be repaired on an emergency or priority basis.

Implementing Preventive Maintenance

Preventive maintenance implies that we can determine when a system needs service or will need repair. Therefore, to perform preventive maintenance, we must know when a system requires service or when it is likely to fail. Failures occur at different rates during the life of a product. A high initial failure rate, known as **infant mortality**, may exist for many products.[1] This is why many electronic firms "burn in" their products prior to shipment: That is to say, they execute a variety of tests (such as a full wash cycle at Maytag) to detect "start-up" problems prior to shipment. Firms may also provide 90-day warranties. We should note that many infant mortality failures are not product failures per se, but rather failure due to improper use. This fact points up the importance in many industries of operations management's building an after-sales service system that includes installing and training.

Once the product, machine, or process "settles in," a study can be made of the MTBF (mean time between failure) distribution. Such distributions often follow a normal curve. When these distributions exhibit small standard deviations, then we know we have a candidate for preventive maintenance, even if the maintenance is expensive.[2]

Once our firm has a candidate for preventive maintenance, we want to determine *when* preventive maintenance is economical. Typically, the more expensive the maintenance, the narrower must be the MTBF distribution (that is, have a small standard deviation). Additionally, if the process is no more expensive to repair when it breaks down than the cost of preventive maintenance, perhaps we should let the process break down and then do the repair. However, the consequence of the breakdown must be fully considered. Even some relatively minor breakdowns have catastrophic conse-

Preventive maintenance
A plan that involves routine inspections, servicing, and keeping facilities in good repair to prevent failure.

Breakdown maintenance
Remedial maintenance that occurs when equipment fails and must be repaired on an emergency or priority basis.

Infant mortality
The failure rate early in the life of a product or process.

---

[1] Infant mortality failures often follow a negative exponential distribution.

[2] See, for example, the work of P. M. Morse, *Queues, Inventories, and Maintenance* (New York: John Wiley, 1958): 161–168; and J. Michael Brock, John R. Michael, and David Morganstein, "Using Statistical Thinking to Solve Maintenance Problems," *Quality Progress* (May 1989): 55–60.

Preventive maintenance is critical to the Orlando Utilities Commission (OUC), a Central Florida electric utility company. Its coal-fired unit requires that maintenance personnel perform about 12,000 repair and preventive maintenance tasks a year. These are scheduled daily by a computerized maintenance program. An unexpected forced outage can cost OUC from $250,000 to $500,000 per day. The value of preventive maintenance was illustrated by the first overhaul of a new generator, which revealed a cracked rotor blade that could have destroyed a $27-million piece of equipment.

quences. At the other extreme, preventive maintenance costs may be so incidental that preventive maintenance is appropriate even if the MTBF distribution is rather flat (that is, it has a large standard deviation). In any event, consistent with job enrichment practices, machine operators must be held responsible for preventive maintenance of their own equipment and tools.

With good reporting techniques, firms can maintain records of individual processes, machines, or equipment. Such records can provide a profile of both the kinds of maintenance required and the timing of maintenance needed. Maintaining equipment history is an important part of a preventive maintenance system, as is a record of the time and cost to make the repair. Such records can also contribute to similar information about the family of equipment as well as suppliers.

Record keeping is of such importance that most good maintenance systems are now computerized. Figure 17.3 shows the major components of such a system with files to be maintained on the left and reports generated on the right.

FIGURE 17.3 ■

A Computerized
Maintenance System

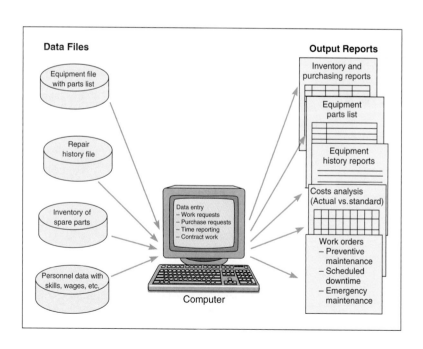

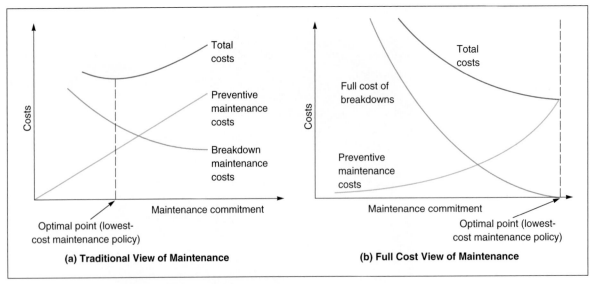

FIGURE 17.4 ■ Maintenance Costs

Figure 17.4(a) shows a traditional view of the relationship between preventive maintenance and breakdown maintenance. Under this view, operations managers consider a *balance* between the two costs. On the one hand, allocating more resources to preventive maintenance will reduce the number of breakdowns. At some point, however, the decrease in breakdown maintenance costs may be less than the increase in preventive maintenance costs. At this point, the total cost curve will begin to rise. Beyond this optimal point, the firm will be better off waiting for breakdowns to occur and repairing them when they do.

Unfortunately, cost curves such as in Figure 17.4(a) seldom consider the *full costs of a breakdown*. Many costs are ignored because they are not *directly* related to the immediate breakdown. For instance, the cost of inventory maintained to compensate for downtime is not typically considered. Moreover, downtime can have a devastating effect on morale: Employees may begin to believe that performance to standard and maintaining equipment are not important. Finally, downtime adversely affects delivery schedules, destroying customer relations and future sales. When the full impact of breakdowns is considered, Figure 17.4(b) may be a better representation of maintenance costs. In Figure 17.4(b), total costs are at a minimum when the system does not break down.

Assuming that all potential costs associated with downtime have been identified, the operations staff can compute the optimal level of maintenance activity on a theoretical basis. Such analysis, of course, also requires accurate historical data on maintenance costs, breakdown probabilities, and repair times. Example 4 shows how to compare preventive and breakdown maintenance costs in order to select the least expensive maintenance policy.

Example 4

Huntsman and Associates is a CPA firm specializing in payroll preparation. The firm has been successful in automating much of its work, using high-speed printers for check processing and report preparation. The computerized approach, however, has problems. Over the past 20 months, the printers have broken down at the rate indicated in the following table:

NUMBER OF BREAKDOWNS	NUMBER OF MONTHS THAT BREAKDOWNS OCCURRED
0	2
1	8
2	6
3	4
	Total: 20

Each time the printers break down, Huntsman estimates that it loses an average of $300 in time and service expenses. One alternative is to purchase a service contract for preventive maintenance. Even if Huntsman contracts for preventive maintenance, there will still be breakdowns, *averaging* one breakdown per month. The price for this service is $150 per month. To decide whether Huntsman should contract for preventive maintenance, we will follow a 4-step approach:

Step 1: Compute the *expected number* of breakdowns (based on past history) if the firm continues as is, without the service contract.

Step 2: Compute the expected breakdown cost per month with no preventive maintenance contract.

Step 3: Compute the cost of preventive maintenance.

Step 4: Compare the two options and select the one that will cost less.

1.

NUMBER OF BREAKDOWNS	FREQUENCY	NUMBER OF BREAKDOWNS	FREQUENCY
0	2/20 = .1	2	6/20 = 0.3
1	8/20 = .4	3	4/20 = 0.2

$$\begin{pmatrix} \text{Expected number} \\ \text{of breakdowns} \end{pmatrix} = \sum \left[\begin{pmatrix} \text{Number of} \\ \text{breakdowns} \end{pmatrix} \times \begin{pmatrix} \text{Corresponding} \\ \text{frequency} \end{pmatrix} \right]$$

$$= (0)(.1) + (1)(.4) + (2)(.3) + (3)(.2)$$

$$= 0 + .4 + .6 + .6$$

$$= 1.6 \text{ breakdowns/month}$$

2. $\text{Expected breakdown cost} = \begin{pmatrix} \text{Expected number} \\ \text{of breakdowns} \end{pmatrix} \times \begin{pmatrix} \text{Cost per} \\ \text{breakdown} \end{pmatrix}$

$$= (1.6)(\$300)$$

$$= \$480/\text{month}$$

3. $\begin{pmatrix} \text{Preventive} \\ \text{maintenance cost} \end{pmatrix} = \begin{pmatrix} \text{Cost of expected} \\ \text{breakdowns if service} \\ \text{contract signed} \end{pmatrix} + \begin{pmatrix} \text{Cost of} \\ \text{service contract} \end{pmatrix}$

$$= (1 \text{ breakdown/month})(\$300) + \$150/\text{month}$$

$$= \$450/\text{month}$$

4. Because it is less expensive overall to hire a maintenance service firm ($450) than to not do so ($480), Huntsman should hire the service firm.

Through variations of the technique shown in Example 4, operations managers can examine maintenance policies.

Increasing Repair Capabilities

Because reliability and preventive maintenance are seldom perfect, most firms opt for some level of repair capability. Enlarging or improving repair facilities can get the system back in operation faster. A good maintenance facility should have these six features:

1. Well-trained personnel.
2. Adequate resources.
3. Ability to establish a repair plan and priorities.[3]
4. Ability and authority to do material planning.
5. Ability to identify the cause of breakdowns.
6. Ability to design ways to extend MTBF.

However, not all repairs can be done in the firm's facility. Managers must, therefore, decide where repairs are to be performed. Figure 17.5 shows some of the options and how they rate in terms of speed, cost, and competence. Consistent with the advantages of employee empowerment, a strong case can be made for employees' maintaining their own equipment. This approach, however, may

[3]You may recall from our discussion of network planning in Chapter 3 that DuPont developed the critical path method (CPM) to improve the scheduling of maintenance projects.

FIGURE 17.5 ■

The Operations
Manager Must
Determine How
Maintenance Will Be
Performed

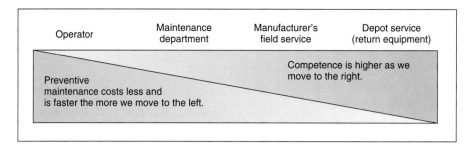

also be the weakest link in the repair chain because not every employee can be trained in all aspects of equipment repair. Moving to the right in Figure 17.5 may improve the competence of the repair work, but it also increases cost as it may entail expensive off-site repair with corresponding increases in replacement time and shipping.

However, preventive maintenance policies and techniques must include an emphasis on employees accepting responsibility for the maintenance they are capable of doing. Employee maintenance may be only of the "clean, check, and observe" variety, but if each operator performs those activities within his or her capability, the manager has made a step toward both employee empowerment and maintaining system performance.

TOTAL PRODUCTIVE MAINTENANCE

Total productive maintenance (TPM)
Combines total quality management with strategic view of maintenance from process and equipment design to preventive maintenance.

Many firms have moved to bring total quality management concepts to the practice of preventive maintenance with an approach known as **total productive maintenance (TPM)**. It involves the concept of reducing variability through employee involvement and excellent maintenance records. In addition, total productive maintenance includes:

- Designing machines that are reliable, easy to operate, and easy to maintain.
- Emphasizing total cost of ownership when purchasing machines, so that service and maintenance are included in the cost.
- Developing preventive maintenance plans that utilize the best practices of operators, maintenance departments, and depot service.
- Training workers to operate and maintain their own machines.

High utilization of facilities, tight scheduling, low inventory, and consistent quality demand reliability.[4] Total productive maintenance is the key to reducing variability and improving reliability.

TECHNIQUES FOR ESTABLISHING MAINTENANCE POLICIES

Two other OM techniques have proven beneficial to effective maintenance: simulation and expert systems.

Simulation   Because of the complexity of some maintenance decisions, computer simulation is a good tool for evaluating the impact of various policies. For instance, operations personnel can decide whether to add more staff by determining the trade-offs between machine downtime costs and the costs of additional labor.[5] Management can also simulate the replacement of parts that have

[4]This conclusion is supported by a number of studies; see, for example, recent work by Kathleen E. McKone, Roger G. Schroeder, and Kristy O. Cua, "The Impact of Total Productive Maintenance Practices on Manufacturing Performance," *Journal of Operations Management* 19, no. 1 (January 2001): 39–58.

[5]Christian Striffler, Walton Hancock, and Ron Turkett, "Maintenance Staffs: Size Them Right," *IIE Solutions* 32, no. 12 (December 2000): 33–38.

not yet failed as a way of preventing future breakdowns. Simulation via physical models can also be useful. For example, a physical model can vibrate an airplane to simulate thousands of hours of flight time in order to evaluate maintenance needs.

Expert Systems   OM managers use expert systems (that is, computer programs that mimic human logic) to assist staff in isolating and repairing various faults in machinery and equipment. For instance, General Electric's DELTA system asks a series of detailed questions that aid the user in identifying a problem. DuPont uses expert systems to monitor equipment and to train repair personnel.

SUMMARY

Operations managers focus on design improvements and backup components to improve reliability. Reliability improvements also can be obtained through the use of preventive maintenance and excellent repair facilities.

Some firms use automated sensors and other controls to warn when production machinery is about to fail or is becoming damaged by heat, vibration, or fluid leaks. The goal of such procedures is not only to avoid failures but also to perform preventive maintenance before machines are damaged.

Finally, many firms give employees a sense of "ownership" of their equipment. When workers repair or do preventive maintenance on their own machines, breakdowns are less common. Well-trained and empowered employees ensure reliable systems through preventive maintenance. In turn, reliable, well-maintained equipment not only provides higher utilization but also improves quality and performance to schedule. Top firms build and maintain systems so that customers can count on products and services that are produced to specifications and on time.

KEY TERMS

Maintenance *(p. 492)*
Reliability *(p. 492)*
Mean time between failures (MTBF) *(p. 494)*
Redundancy *(p. 495)*

Preventive maintenance *(p. 496)*
Breakdown maintenance *(p. 496)*
Infant mortality *(p. 496)*
Total productive maintenance (TPM) *(p. 500)*

USING POM FOR WINDOWS TO SOLVE RELIABILITY PROBLEMS

POM for Windows' Reliability module allows us to enter (1) number of systems (components) in the series (1 through 10); (2) number of backup, or parallel, components (1 through 12); and (3) component reliability for both series and parallel data. Refer to Appendix V for details.

SOLVED PROBLEMS

Solved Problem 17.1

The semiconductor used in the Sullivan Wrist Calculator has 5 parts, each of which has its own reliability rate. Component 1 has a reliability of .90; component 2, .95; component 3, .98; component 4, .90; and component 5, .99. What is the reliability of one semiconductor?

SOLUTION

$$\text{Semiconductor reliability, } R_s = R_1 \times R_2 \times R_3 \times R_4 \times R_5$$
$$= (.90)(.95)(.98)(.90)(.99)$$
$$= .7466$$

Solved Problem 17.2

A recent engineering change at Sullivan Wrist Calculator places a backup component in each of the two least reliable transistor circuits. The new circuit will look like the following:

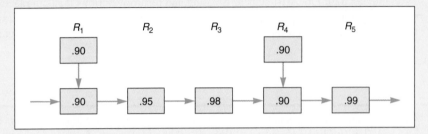

What is the reliability of the new system?

SOLUTION

$$\text{Reliability} = [.9 + (1 - .9) \times .9] \times .95 \times .98 \times [.9 + (1 - .9) \times .9] \times .99$$
$$= [.9 + .09] \times .95 \times .98 \times [.9 + .09] \times .99$$
$$= .99 \times .95 \times .98 \times .99 \times .99$$
$$= .903$$

INTERNET AND STUDENT CD-ROM EXERCISES

Visit our home page or use your student CD-ROM to help with material in this chapter.

 On Our Home Page, www.prenhall.com/heizer

- Self-Tests
- Practice Problems
- Internet Exercises
- Current Articles and Research
- Virtual Company Tour
- Internet Homework Problems
- Internet Case

On Your Student CD-ROM

- PowerPoint Lecture
- Practice Problems
- Active Model Exercise

ADDITIONAL CASE STUDIES

Internet Case Study: Visit our Web site at www.prenhall.com/heizer **for this free case study:**

- **Cartak's Department Store**: Requires the evaluation of the impact of an additional invoice verifier.

Harvard has selected these Harvard Business School cases to accompany this chapter (textbookcasematch.hbsp.harvard.edu)**:**

- **The Dana–Farber Cancer Institute** (#699-025): Examines organizational and process characteristics that may have contributed to a medical error.

- **Workplace Safety at Alcoa (A)** (#692-042): Looks at the challenge facing the manager of a large aluminum manufacturing plant in its drive for improved safety.

- **A Brush with AIDS (A)** (#394-058): Ethical dilemma when needles penetrate container walls.

Decision-Making Tools

Module Outline

LEARNING OBJECTIVES

When you complete this module you should be able to

IDENTIFY OR DEFINE:

Decision trees and decision tables

Highest monetary value

Expected value of perfect information

Sequential decisions

DESCRIBE OR EXPLAIN:

Decision making under risk

Decision making under uncertainty

Decision making under certainty

The wildcatter's decision was a tough one. Which of his new Kentucky lease areas—Blair East or Blair West—should he drill for oil? A wrong decision in this type of wildcat oil drilling could mean the difference between success and bankruptcy for the company. Talk about decision making under uncertainty and pressure! But using a decision tree, Tomco Oil President Thomas E. Blair identified 74 different options, each with its own potential net profit. What had begun as an overwhelming number of geological, engineering, economic, and political factors now became much clearer. Says Blair, "Decision tree analysis provided us with a systematic way of planning these decisions and clearer insight into the numerous and varied financial outcomes that are possible."[1]

"The business executive is by profession a decision maker. Uncertainty is his opponent. Overcoming it is his mission."

John McDonald

Operations managers are decision makers. To achieve the goals of their organizations, managers must understand how decisions are made and know which decision-making tools to use. To a great extent, the success or failure of both people and companies depends on the quality of their decisions. Bill Gates, who developed the DOS and Windows operating systems, became chairman of the most powerful software firm in the world (Microsoft) and a billionaire. In contrast, the Firestone manager who headed the team that designed the flawed tires (which caused so many accidents with Ford Explorers in the late 1990s) is not working there anymore.

THE DECISION PROCESS IN OPERATIONS

What makes the difference between a good decision and a bad decision? A "good" decision—one that uses analytic decision making—is based on logic and considers all available data and possible alternatives. It also follows these six steps:

1. Clearly define the problem and the factors that influence it.
2. Develop specific and measurable objectives.
3. Develop a model—that is, a relationship between objectives and variables (which are measurable quantities).
4. Evaluate each alternative solution based on its merits and drawbacks.
5. Select the best alternative.
6. Implement the decision and set a timetable for completion.

Throughout this book, we introduce a broad range of mathematical models and tools to help operations managers make better decisions. Effective operations depend on careful decision making. Fortunately, there are a whole variety of analytic tools to help make these decisions. This module

---

[1]J. Hosseini, "Decision Analysis and Its Application in the Choice between Two Wildcat Ventures," *Interfaces*, Vol. 16, No. 2. Reprinted by permission, INFORMS, 901 Elkridge Landing Road, Suite 400, Linthicum, Maryland 21090 USA.

"Management means,
in the last analysis, the
substitution of thought
for brawn and muscle,
of knowledge for folklore
and tradition, and of
cooperation for force."

Peter Drucker

introduces two of them—decision tables and decision trees. They are used in a wide number of OM situations, ranging from new-product analysis (Chapter 5), to equipment selection (Chapter 7), to location planning (Chapter 8), to scheduling (Chapter 15), and to maintenance planning (Chapter 17).

FUNDAMENTALS OF DECISION MAKING

Regardless of the complexity of a decision or the sophistication of the technique used to analyze it, all decision makers are faced with alternatives and "states of nature." The following notation will be used in this module:

1. Terms:
    a. *Alternative*—a course of action or strategy that may be chosen by a decision maker (for example, not carrying an umbrella tomorrow).
    b. *State of nature*—an occurrence or a situation over which the decision maker has little or no control (for example, tomorrow's weather).
2. Symbols used in a decision tree:
    a. □—decision node from which one of several alternatives may be selected.
    b. ○—a state-of-nature node out of which one state of nature will occur.

To present a manager's decision alternatives, we can develop *decision trees* using the above symbols. When constructing a decision tree, we must be sure that all alternatives and states of nature are in their correct and logical places and that we include *all* possible alternatives and states of nature.

Example A1

Getz Products Company is investigating the possibility of producing and marketing backyard storage sheds. Undertaking this project would require the construction of either a large or a small manufacturing plant. The market for the product produced—storage sheds—could be either favorable or unfavorable. Getz, of course, has the option of not developing the new product line at all. A decision tree for this situation is presented in Figure A.1.

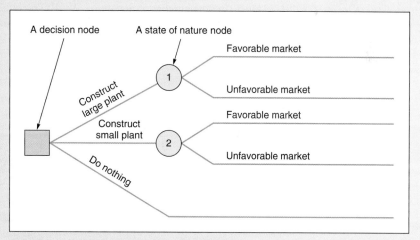

FIGURE A.1 ■ Getz Products Decision Tree

DECISION TABLES

Decision table

A tabular means of analyzing decision alternatives and states of nature.

We may also develop a decision or payoff table to help Getz Products define its alternatives. For any alternative and a particular state of nature, there is a *consequence* or *outcome*, which is usually expressed as a monetary value. This is called a *conditional value*. Note that all of the alternatives in Example A2 are listed down the left side of the table, that states of nature (outcomes) are listed across the top, and that conditional values (payoffs) are in the body of the **decision table**.

Example A2

We construct a decision table for Getz Products (Table A.1), including conditional values based on the following information. With a favorable market, a large facility would give Getz Products a net profit of $200,000. If the market is unfavorable, a $180,000 net loss would occur. A small plant would result in a net profit of $100,000 in a favorable market, but a net loss of $20,000 would be encountered if the market is unfavorable.

The toughest part of decision tables is getting the data to analyze.

TABLE A.1 ■ Decision Table with Conditional Values for Getz Products

	STATES OF NATURE	
ALTERNATIVES	FAVORABLE MARKET	UNFAVORABLE MARKET
Construct large plant	$200,000	−$180,000
Construct small plant	$100,000	−$ 20,000
Do nothing	$ 0	$ 0

In Examples A3 and A4, we see how to use decision tables.

Decision Making Under Uncertainty

When there is complete *uncertainty* as to which state of nature in a decision table may occur (that is, when we cannot even assess probabilities for each possible outcome), we rely on three decision methods:

Maximax
A criterion that finds an alternative that maximizes the maximum outcome.

Maximin
A criterion that finds an alternative that maximizes the minimum outcome.

Equally likely
A criterion that assigns equal probability to each state of nature.

1. **Maximax**—this method finds an alternative that *max*imizes the *max*imum outcome for every alternative. First, we find the maximum outcome within every alternative, and then we pick the alternative with the maximum number. Because this decision criterion locates the alternative with the *highest* possible *gain*, it has been called an "optimistic" decision criterion.

2. **Maximin**—this method finds the alternative that *max*imizes the *min*imum outcome for every alternative. First, we find the minimum outcome within every alternative, and then we pick the alternative with the maximum number. Because this decision criterion locates the alternative that has the *least* possible *loss*, it has been called a "pessimistic" decision criterion.

3. **Equally likely**—this method finds the alternative with the highest average outcome. First, we calculate the average outcome for every alternative, which is the sum of all outcomes divided by the number of outcomes. We then pick the alternative with the maximum number. The equally likely approach assumes that each state of nature is equally likely to occur.

Example A3 applies each of these approaches to the Getz Products Company.

Example A3

There are optimistic decision makers ("maximax") and pessimistic ones ("maximin").

Given Getz's decision table of Example A2, determine the maximax, maximin, and equally likely decision criteria (see Table A.2).

TABLE A.2 ■ Decision Table for Decision Making under Uncertainty

	STATES OF NATURE				
ALTERNATIVES	FAVORABLE MARKET	UNFAVORABLE MARKET	MAXIMUM IN ROW	MINIMUM IN ROW	ROW AVERAGE
Construct large plant	$200,000	−$180,000	$200,000 ←	−$180,000	$10,000
Construct small plant	$100,000	−$20,000	$100,000	−$20,000	$40,000 ←
Do nothing	$ 0	$ 0	$ 0	$ 0 ←	$ 0
			Maximax ┘	Maximin ┘	Equally likely ┘

1. The maximax choice is to construct a large plant. This is the *max*imum of the *max*imum number within each row, or alternative.
2. The maximin choice is to do nothing. This is the *max*imum of the *min*imum number within each row, or alternative.
3. The equally likely choice is to construct a small plant. This is the maximum of the average outcome of each alternative. This approach assumes that all outcomes for any alternative are *equally likely*.

Decision Making Under Risk

Expected monetary value (EMV)
The expected payout or value of a variable that has different possible states of nature, each with an associated probability.

Decision making under risk, a more common occurrence, relies on probabilities. Several possible states of nature may occur, each with an assumed probability. The states of nature must be mutually exclusive and collectively exhaustive and their probabilities must sum to 1.[2] Given a decision table with conditional values and probability assessments for all states of nature, we can determine the **expected monetary value (EMV)** for each alternative. This figure represents the expected value or *mean* return for each alternative *if we could repeat the decision a large number of times*.

The EMV for an alternative is the sum of all possible payoffs from the alternative, each weighted by the probability of that payoff occurring.

Even when probabilities are available, maximax and maximin are of value in that they present best case–worst case planning scenarios.

$$\begin{aligned}
\text{EMV (Alternative } i) = {} & \text{(Payoff of 1st state of nature)} \\
& \times \text{ (Probability of 1st state of nature)} \\
+ {} & \text{(Payoff of 2nd state of nature)} \\
& \times \text{ (Probability of 2nd state of nature)} \\
+ \cdots + {} & \text{ (Payoff of last state of nature)} \\
& \times \text{ (Probability of last state of nature)}
\end{aligned}$$

Example A4 illustrates how to compute the maximum EMV.

Example A4

**Excel OM Data File
ModAEx4.xla**

Getz Products operations manager believes that the probability of a favorable market is exactly the same as that of an unfavorable market; that is, each state of nature has a .50 chance of occurring. We can now determine the EMV for each alternative (see Table A.3):

1. $\text{EMV}(A_1) = (.5)(\$200,000) + (.5)(-\$180,000) = \$10,000$
2. $\text{EMV}(A_2) = (.5)(\$100,000) + (.5)(-\$20,000) = \$40,000$
3. $\text{EMV}(A_3) = (.5)(\$0) + (.5)(\$0) = \$0$

The maximum EMV is seen in alternative A_2. Thus, according to the EMV decision criterion, we would build the small facility.

TABLE A.3 ■ Decision Table for Getz Products

	STATES OF NATURE	
ALTERNATIVES	**FAVORABLE MARKET**	**UNFAVORABLE MARKET**
Construct large plant (A_1)	$200,000	−$180,000
Construct small plant (A_2)	$100,000	−$ 20,000
Do nothing (A_3)	$ 0	$ 0
Probabilities	.50	.50

Decision Making Under Certainty

Now suppose that the Getz operations manager has been approached by a marketing research firm that proposes to help him make the decision about whether to build the plant to produce storage sheds. The marketing researchers claim that their technical analysis will tell Getz with certainty whether the market is favorable for the proposed product. In other words, it will change Getz's environment from one of decision making *under risk* to one of decision making *under certainty*. This information could prevent Getz from making a very expensive mistake. The marketing research firm would charge Getz $65,000 for the information. What would you recommend? Should the operations manager hire the firm to make the study? Even if the information from the study is perfectly accurate, is it worth $65,000? What might it be worth? Although some of these questions are difficult to answer, determining the value of such *perfect information* can be very useful. It places an upper bound on what you would be willing to spend on information, such as that being sold by a marketing consultant. This is the concept of the expected value of perfect information, which we now introduce.

EVPI places an upper limit on what you should pay for information.

---

[2]To review these and other statistical terms, refer to the CD-ROM Tutorial 1, "Statistical Review for Managers."

Expected Value of Perfect Information (EVPI)

Expected value of perfect information (EVPI)
The difference between the payoff under certainty and the payoff under risk.

If a manager were able to determine which state of nature would occur, then he or she would know which decision to make. Once a manager knows which decision to make, the payoff increases because the payoff is now a certainty, not a probability. Because the payoff will increase with knowledge of which state of nature will occur, this knowledge has value. Therefore, we now look at how to determine the value of this information. We call this difference between the payoff under certainty and the payoff under risk the **expected value of perfect information (EVPI)**.

$$\text{EVPI} = \text{Expected value under certainty} - \text{Maximum EMV}$$

Expected value under certainty
The expected (average) return if perfect information is available.

To find the EVPI, we must first compute the **expected value under certainty**, which is the expected (average) return, if we have perfect information before a decision has to be made. In order to calculate this value, we choose the best alternative for each state of nature and multiply its payoff times the probability of occurrence of that state of nature.

$$
\begin{aligned}
\text{Expected value under certainty} = {} & \text{(Best outcome or consequence for 1st state of nature)} \\
& \times \text{(Probability of 1st state of nature)} \\
& + \text{(Best outcome for 2nd state of nature)} \\
& \times \text{(Probability of 2nd state of nature)} \\
& + \cdots + \text{(Best outcome for last state of nature)} \\
& \times \text{(Probability of last state of nature)}
\end{aligned}
$$

We will use the data and decision table from Example A4 to examine the expected value of perfect information. We do so in Example A5.

Example A5

By referring back to Table A.3, the Getz operations manager can calculate the maximum that he would pay for information—that is, the expected value of perfect information, or EVPI. He follows a two-stage process. First, the expected value under certainty is computed. Then, using this information, EVPI is calculated. The procedure is outlined as follows:

1. The best outcome for the state of nature "favorable market" is "build a large facility" with a payoff of $200,000. The best outcome for the state of nature "unfavorable market" is "do nothing" with a payoff of $0. Expected value under certainty = ($200,000)(0.50) + ($0)(0.50) = $100,000. Thus, if we had perfect information, we would expect (on the average) $100,000 if the decision could be repeated many times.
2. The maximum EMV is $40,000, which is the expected outcome without perfect information. Thus:

$$
\begin{aligned}
\text{EVPI} &= \text{Expected value under certainty} - \text{Maximum EMV} \\
&= \$100,000 - \$40,000 = \$60,000
\end{aligned}
$$

In other words, the *most* Getz should be willing to pay for perfect information is $60,000. This conclusion, of course, is again based on the assumption that the probability of each state of nature is 0.50.

DECISION TREES

Decisions that lend themselves to display in a decision table also lend themselves to display in a decision tree. We will, therefore, analyze some decisions using decision trees. Although the use of a decision table is convenient in problems having one set of decisions and one set of states of nature, many problems include *sequential* decisions and states of nature. When there are two or more sequential decisions and later decisions are based on the outcome of prior ones, the decision tree approach becomes appropriate. A **decision tree** is a graphic display of the decision process that indicates decision alternatives, states of nature and their respective probabilities, and payoffs for each combination of decision alternative and state of nature.

Decision tree
A graphical means of analyzing decision alternatives and states of nature.

Expected monetary value (EMV) is the most commonly used criterion for decision tree analysis. One of the first steps in such analysis is to graph the decision tree and to specify the monetary consequences of all outcomes for a particular problem.

Decision tree software is a relatively new advance that permits users to solve decision-analysis problems with flexibility, power, and ease. Programs such as DPL, Tree Plan, and Supertree allow decision problems to be analyzed with less effort and in greater depth than ever before. Full-color presentations of the options open to managers always have impact. In this photo, wildcat drilling options are explored with DPL, a product of Applied Decision Analysis.

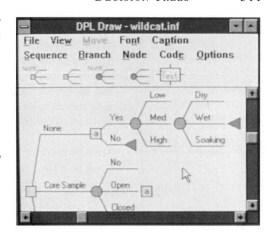

Analyzing problems with *decision trees* involves five steps:

1. Define the problem.
2. Structure or draw the decision tree.
3. Assign probabilities to the states of nature.
4. Estimate payoffs for each possible combination of decision alternatives and states of nature.
5. Solve the problem by computing expected monetary values (EMV) for each state-of-nature node. This is done by working *backward*—that is, by starting at the right of the tree and working back to decision nodes on the left.

Example A6

A completed and solved decision tree for Getz Products is presented in Figure A.2. Note that the payoffs are placed at the right-hand side of each of the tree's branches. The probabilities (first used by Getz in Example A4) are placed in parentheses next to each state of nature. The expected monetary values for each state-of-nature node are then calculated and placed by their respective nodes. The EMV of the first node is $10,000. This represents the branch from the decision node to "construct a large plant." The EMV for node 2, to "construct a small plant," is $40,000. The option of "doing nothing" has, of course, a payoff of $0. The branch leaving the decision node leading to the state of nature node with the highest EMV will be chosen. In Getz's case, a small plant should be built.

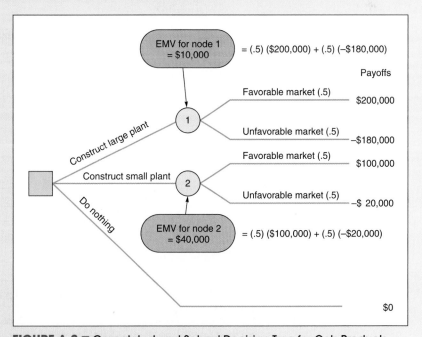

FIGURE A.2 ■ Completed and Solved Decision Tree for Getz Products

A More Complex Decision Tree

There is a widespread use of decision trees beyond OM. Managers often appreciate a graphical display of a tough problem.

When a *sequence* of decisions must be made, decision trees are much more powerful tools than are decision tables. Let's say that Getz Products has two decisions to make, with the second decision dependent on the outcome of the first. Before deciding about building a new plant, Getz has the option of conducting its own marketing research survey, at a cost of $10,000. The information from this survey could help it decide whether to build a large plant, to build a small plant, or to not build at all. Getz recognizes that although such a survey will not provide it with *perfect* information, it may be extremely helpful.

Getz's new decision tree is represented in Figure A.3 of Example A7. Take a careful look at this more complex tree. Note that *all possible outcomes and alternatives* are included in their logical sequence. This procedure is one of the strengths of using decision trees. The manager is forced to examine all possible outcomes, including unfavorable ones. He or she is also forced to make decisions in a logical, sequential manner.

Example A7

Examining the tree in Figure A.3, we see that Getz's first decision point is whether to conduct the $10,000 market survey. If it chooses not to do the study (the lower part of the tree), it can either build a large plant, a small plant, or no plant. This is Getz's second decision point. If the decision is to build, the market will be either favorable (.50 probability) or unfavorable (also .50 probability). The payoffs for each of the possible consequences are listed along the right-hand side. As a matter of fact, this lower portion of Getz's tree is *identical* to the simpler decision tree shown in Figure A.2.

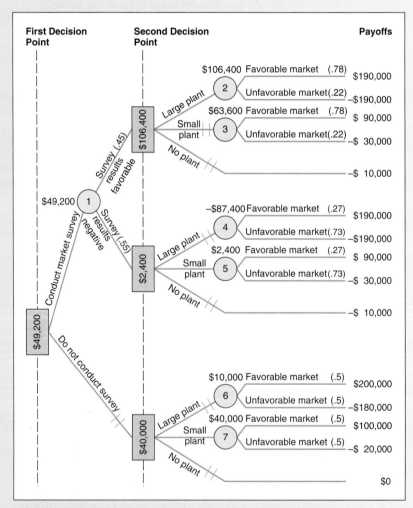

FIGURE A.3 ■ Getz Products Decision Tree with Probabilities and EMVs Shown

The short parallel lines mean "prune" that branch, as it is less favorable than another available option and may be dropped.

You can reduce complexity by viewing and solving a number of smaller trees—start at the end branches of a large one. Take one decision at a time.

The upper part of Figure A.3 reflects the decision to conduct the market survey. State-of-nature node number 1 has 2 branches coming out of it. Let us say there is a 45% chance that the survey results will indicate a favorable market for the storage sheds. We also note that the probability is .55 that the survey results will be negative.

The rest of the probabilities shown in parentheses in Figure A.3 are all *conditional* probabilities. For example, .78 is the probability of a favorable market for the sheds given a favorable result from the market survey. Of course, you would expect to find a high probability of a favorable market given that the research indicated that the market was good. Don't forget, though: There is a chance that Getz's $10,000 market survey did not result in perfect or even reliable information. Any market research study is subject to error. In this case, there remains a 22% chance that the market for sheds will be unfavorable given positive survey results.

Likewise, we note that there is a 27% chance that the market for sheds will be favorable given negative survey results. The probability is much higher, .73, that the market will actually be unfavorable given a negative survey.

Finally, when we look to the payoff column in Figure A.3, we see that $10,000—the cost of the marketing study—has been subtracted from each of the top 10 tree branches. Thus, a large plant constructed in a favorable market would normally net a $200,000 profit. Yet because the market study was conducted, this figure is reduced by $10,000. In the unfavorable case, the loss of $180,000 would increase to $190,000. Similarly, conducting the survey and building *no plant* now results in a −$10,000 payoff.

With all probabilities and payoffs specified, we can start calculating the expected monetary value of each branch. We begin at the end or right-hand side of the decision tree and work back toward the origin. When we finish, the best decision will be known.

1. Given favorable survey results,

$$\text{EMV (node 2)} = (.78)(\$190,000) + (.22)(-\$190,000) = \$106,400$$
$$\text{EMV (node 3)} = (.78)(\$90,000) + (.22)(-\$30,000) = \$63,600$$

The EMV of no plant in this case is −$10,000. Thus, if the survey results are favorable, a large plant should be built.

2. Given negative survey results,

$$\text{EMV (node 4)} = (.27)(\$190,000) + (.73)(-\$190,000) = -\$87,400$$
$$\text{EMV (node 5)} = (.27)(\$90,000) + (.73)(-\$30,000) = \$2,400$$

The EMV of no plant is again −$10,000 for this branch. Thus, given a negative survey result, Getz should build a small plant with an expected value of $2,400.

3. Continuing on the upper part of the tree and moving backward, we compute the expected value of conducting the market survey.

$$\text{EMV(node 1)} = (.45)(\$106,400) + (.55)(\$2,400) = \$49,200$$

4. If the market survey is *not* conducted.

$$\text{EMV (node 6)} = (.50)(\$200,000) + (.50)(-\$180,000) = \$10,000$$
$$\text{EMV (node 7)} = (.50)(\$100,000) + (.50)(-\$20,000) = \$40,000$$

The EMV of no plant is $0. Thus, building a small plant is the best choice, given the marketing research is not performed.

5. Because the expected monetary value of conducting the survey is $49,200—versus an EMV of $40,000 for not conducting the study—the best choice is to *seek marketing information*. If the survey results are favorable, Getz should build the large plant; if they are unfavorable, it should build the small plant.

SUMMARY

This module examines two of the most widely used decision techniques—decision tables and decision trees. These techniques are especially useful for making decisions under risk. Investments in research and development, plant and equipment, and even new buildings and structures can be analyzed with these decision models. Problems in inventory control, aggregate planning, maintenance, scheduling, and production control are just a few other decision table and decision tree applications.

KEY TERMS

Decision table *(p. 507)*
Maximax *(p. 508)*
Maximin *(p. 508)*
Equally likely *(p. 508)*
Expected monetary value (EMV) *(p. 509)*

Expected value of perfect information (EVPI) *(p. 510)*
Expected value under certainty *(p. 510)*
Decision tree *(p. 510)*

✕ USING EXCEL OM FOR DECISION MODELS

Excel OM allows decision makers to evaluate decisions quickly and to perform sensitivity analysis on the results. Program A.1 uses the Getz data to illustrate input, output, and selected formulas needed to compute the EMV and EVPI values.

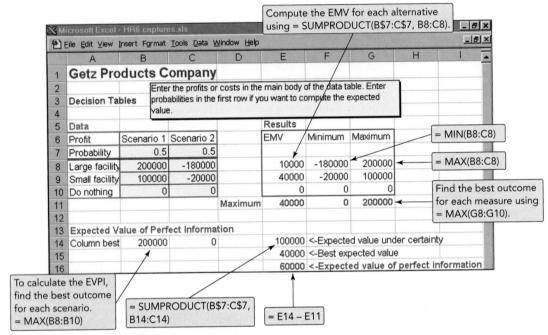

PROGRAM A.1 ■ Using Excel OM to Compute EMV and Other Measures for Getz

USING POM FOR WINDOWS

POM for Windows can be used to calculate all of the information described in the decision tables and decision trees in this module. For details on how to use this software, please refer to Appendix V.

SOLVED PROBLEMS

Solved Problem A.1

Stella Yan Hua is considering the possibility of opening a small dress shop on Fairbanks Avenue, a few blocks from the university. She has located a good mall that attracts students. Her options are to open a small shop, a medium-sized shop, or no shop at all. The market for a dress shop can be good, average, or bad. The probabilities for these three possibilities are .2 for a good market, .5 for an average market, and .3 for a bad market. The net profit or loss for the medium-sized or small shops for the various market conditions are given in the following table. Building no shop at all yields no loss and no gain. What do you recommend?

ALTERNATIVES	GOOD MARKET ($)	AVERAGE MARKET ($)	BAD MARKET ($)
Small shop	75,000	25,000	−40,000
Medium-sized shop	100,000	35,000	−60,000
No shop	0	0	0
Probabilities	.20	.50	.30

SOLUTION

The problem can be solved by computing the expected monetary value (EMV) for each alternative.

$$\text{EMV (Small shop)} = (.2)(\$75,000) + (.5)(\$25,000) + (.3)(-\$40,000) = \$15,500$$
$$\text{EMV (Medium-sized shop)} = (.2)(\$100,000) + (.5)(\$35,000) + (.3)(-\$60,000) = \$19,500$$
$$\text{EMV (No shop)} = (.2)(\$0) + (.5)(\$0) + (.3)(\$0) = \$0$$

As you can see, the best decision is to build the medium-sized shop. The EMV for this alternative is $19,500.

Solved Problem A.2

Daily demand for cases of Tidy Bowl cleaner at Ravinder Nath's Supermarket has always been 5, 6, or 7 cases. Develop a decision tree that illustrates her decision alternatives as to whether to stock 5, 6, or 7 cases.

SOLUTION

The decision tree is shown in Figure A.4.

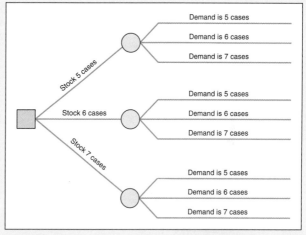

FIGURE A.4 ■ Demand at Ravinder Nath's Supermarket

INTERNET AND STUDENT CD-ROM EXERCISES

Visit our home page or use your student CD-ROM to help with this material in this module.

 On Our Home Page, www.prenhall.com/heizer

- Self Tests
- Practice Problems
- Internet Exercises
- Current Articles and Research
- Internet Homework Problems
- Internet Cases

On Your Student CD-ROM

- PowerPoint Lecture
- Practice Problems
- ExcelOM
- Excel OM Example Data File

ADDITIONAL CASE STUDIES

See our Internet homepage at www.prenhall.com/heizer **for these additional free case studies:**

- **Arctic, Inc.:** A refrigeration company has several major options with regard to capacity and expansion.
- **Toledo Leather Company:** This firm is trying to select new equipment based on potential costs.

Quantitative Module B

Linear Programming

Module Outline

LEARNING OBJECTIVES

When you complete this module you should be able to

IDENTIFY OR DEFINE:

- Objective function
- Constraints
- Feasible region
- Iso-profit/iso-cost methods
- Corner-point solution
- Shadow price

DESCRIBE OR EXPLAIN:

- How to formulate linear models
- Graphical method of linear programming
- How to interpret sensitivity analysis

The storm front closed in quickly on Chicago's O'Hare Airport, shutting it without warning. The heavy thunderstorms, lightning, and poor visibility sent American Airlines passengers and ground crew scurrying. Because American Airlines uses linear programming (LP) to schedule flights, hotels, crews, and refueling, LP has a direct impact on profitability. As the president of AA's Decision Technology Group says, "Finding fast solutions to LP problems is essential. If we get a major weather disruption at one of the hubs, such as Dallas or Chicago, then a lot of flights may get canceled, which means we have a lot of crews and airplanes in the wrong places. What we need is a way to put that whole operation back together again." LP is the tool that helps airlines such as American unsnarl and cope with this weather mess.

Linear Programming (LP)

A mathematical technique designed to help operations managers plan and make decisions relative to the trade-offs necessary to allocate resources.

Many operations management decisions involve trying to make the most effective use of an organization's resources. Resources typically include machinery (such as planes, in the case of an airline), labor (such as pilots), money, time, and raw materials (such as jet fuel). These resources may be used to produce products (such as machines, furniture, food, or clothing) or services (such as airline schedules, advertising policies, or investment decisions). **Linear programming (LP)** is a widely used mathematical technique designed to help operations managers plan and make the decisions necessary to allocate resources.

A few examples of problems in which LP has been successfully applied in operations management are

1.   Scheduling school buses to *minimize* the total distance traveled when carrying students.
2.   Allocating police patrol units to high crime areas in order to *minimize* response time to 911 calls.
3.   Scheduling tellers at banks so that needs are met during each hour of the day while *minimizing* the total cost of labor.
4.   Selecting the product mix in a factory to make best use of machine- and labor-hours available while *maximizing* the firm's profit.
5.   Picking blends of raw materials in feed mills to produce finished feed combinations at *minimum* cost.
6.   Determining the distribution system that will *minimize* total shipping cost from several warehouses to various market locations.
7.   Developing a production schedule that will satisfy future demands for a firm's product and at the same time *minimize* total production and inventory costs.
8.   Allocating space for a tenant mix in a new shopping mall so as to *maximize* revenues to the leasing company. (See the *OM in Action* box, "Using LP to Select Tenants in a Shopping Mall.")

OM IN ACTION

Using LP to Select Tenants in a Shopping Mall

Homart Development Company is one of the largest shopping-center developers in the U.S. When starting a new center, Homart produces a tentative floor plan, or "footprint," for the mall. This plan outlines sizes, shapes, and spaces for large department stores. Leasing agreements are reached with the two or three major department stores that will become anchor stores in the mall. The anchor stores are able to negotiate highly favorable occupancy agreements. Homart's profits come primarily from the rent paid by the nonanchor tenants—the smaller stores that lease space along the aisles of the mall. The decision as to allocating space to potential tenants is, therefore, crucial to the success of the investment.

The tenant mix describes the desired stores in the mall by their size, general location, and type of merchandise or service provided. For example, the mix might specify two small jewelry stores in a central section of the mall and a medium-size shoe store and a large restaurant in one of the side aisles. In the past, Homart developed a plan for tenant mix using "rules of thumb" developed over years of experience in mall development.

Now, to improve its bottom line in an increasingly competitive marketplace, Homart treats the tenant-mix problem as an LP model. First, the model assumes that tenants can be classified into categories according to the type of merchandise or service they provide. Second, the model assumes that for each store type, store sizes can be estimated by distinct category. For example, a small jewelry store is said to contain about 700 square feet and a large one about 2,200 square feet. The tenant-mix model is a powerful tool for enhancing Homart's mall planning and leasing activities.

Sources: Chain Store Age (March 2000):191–192; *Business World* (March 18, 2002):1; and *Interfaces* (March–April 1988):1–9.

REQUIREMENTS OF A LINEAR PROGRAMMING PROBLEM

All LP problems have four properties in common:

Objective function
A mathematical expression in linear programming that maximizes or minimizes some quantity (often profit or cost, but any goal may be used).

1. LP problems seek to *maximize* or *minimize* some quantity (usually profit or cost). We refer to this property as the **objective function** of an LP problem. The major objective of a typical firm is to maximize dollar profits in the long run. In the case of a trucking or airline distribution system, the objective might be to minimize shipping costs.
2. The presence of restrictions, or **constraints**, limits the degree to which we can pursue our objective. For example, deciding how many units of each product in a firm's product line to manufacture is restricted by available labor and machinery. We want, therefore, to maximize or minimize a quantity (the objective function) subject to limited resources (the constraints).

Constraints
Restrictions that limit the degree to which a manager can pursue an objective.

3. There must be *alternative courses of action* to choose from. For example, if a company produces three different products, management may use LP to decide how to allocate among them its limited production resources (of labor, machinery, and so on). If there were no alternatives to select from, we would not need LP.
4. The objective and constraints in linear programming problems must be expressed in terms of *linear equations* or inequalities.

FORMULATING LINEAR PROGRAMMING PROBLEMS

One of the most common linear programming applications is the *product-mix problem*. Two or more products are usually produced using limited resources. The company would like to determine how many units of each product it should produce in order to maximize overall profit given its limited resources. Let's look at an example.

Active Model B.1

This example is further illustrated in Active model B.1 on the CD-ROM and in the Exercise located in your Student Lecture Guide.

Shader Electronics Example

The Shader Electronics Company produces two products: (1) the Shader Walkman, a portable AM/FM cassette player, and (2) the Shader Watch-TV, a wristwatch-size black-and-white television. The production process for each product is similar in that both require a certain number of hours of electronic work and a certain number of labor-hours in the assembly department. Each Walkman takes 4 hours of electronic work and 2 hours in the assembly shop. Each Watch-TV requires 3 hours in electronics and 1 hour in assembly. During the current production period, 240 hours of electronic time are available and 100 hours of assembly department time are available. Each Walkman sold yields a profit of $7; each Watch-TV produced may be sold for a $5 profit.

Shader's problem is to determine the best possible combination of Walkmans and Watch-TVs to manufacture in order to reach the maximum profit. This product-mix situation can be formulated as a linear programming problem.

TABLE B.1 ■

Shader Electronics
Company Problem
Data

| | HOURS REQUIRED TO PRODUCE 1 UNIT | | |
DEPARTMENT	WALKMANS (X_1)	WATCH-TVs (X_2)	AVAILABLE HOURS THIS WEEK
Electronic	4	3	240
Assembly	2	1	100
Profit per unit	$7	$5	

We begin by summarizing the information needed to formulate and solve this problem (see Table B.1). Further, let's introduce some simple notation for use in the objective function and constraints. Let

We name the decision variables X_1 and X_2 here but point out that any notation (such as WM and WT) would be fine as well.

$$X_1 = \text{number of Walkmans to be produced}$$
$$X_2 = \text{number of Watch-TVs to be produced}$$

Now we can create the LP *objective function* in terms of X_1 and X_2:

$$\text{Maximize profit} = \$7X_1 + \$5X_2$$

Our next step is to develop mathematical relationships to describe the two constraints in this problem. One general relationship is that the amount of a resource used is to be less than or equal to ($\leq$) the amount of resource *available*.

First constraint: Electronic time used is $\leq$ Electronic time available.

$$4X_1 + 3X_2 \leq 240 \text{ (hours of electronic time)}$$

Second constraint: Assembly time used is $\leq$ Assembly time available.

$$2X_1 + 1X_2 \leq 100 \text{ (hours of assembly time)}$$

Both of these constraints represent production capacity restrictions and, of course, affect the total profit. For example, Shader Electronics cannot produce 70 Walkmans during the production period because if $X_1 = 70$, both constraints will be violated. It also cannot make $X_1 = 50$ Walkmans and $X_2 = 10$ Watch-TVs. This constraint brings out another important aspect of linear programming: That is, certain interactions will exist between variables. The more units of one product that a firm produces, the fewer it can make of other products.

GRAPHICAL SOLUTION TO A LINEAR PROGRAMMING PROBLEM

Graphical solution approach
A means of plotting a solution to a two-variable problem on a graph.

The easiest way to solve a small LP problem such as that of the Shader Electronics Company is the **graphical solution approach**. The graphical procedure can be used only when there are two decision variables (such as number of Walkmans to produce, X_1, and number of Watch-TVs to produce, X_2). When there are more than two variables, it is *not* possible to plot the solution on a two-dimensional graph; we then must turn to more complex approaches described later in this module.

Graphical Representation of Constraints

Decision variables
Choices available to a decision maker.

In order to find the optimal solution to a linear programming problem, we must first identify a set, or region, of feasible solutions. The first step in doing so is to plot the problem's constraints on a graph.

The variable X_1 (Walkmans, in our example) is usually plotted as the horizontal axis of the graph, and the variable X_2 (Watch-TVs) is plotted as the vertical axis. The complete problem may be restated as:

$$\text{Maximize profit} = \$7X_1 + \$5X_2$$

Subject to the constraints

$$4X_1 + 3X_2 \leq 240 \ (\textit{electronics constraint})$$

These two constraints are also called the nonnegativity constraints.

$$2X_1 + 1X_2 \leq 100 \ (\textit{assembly constraint})$$

$$X_1 \geq 0 \ (\textit{number of Walkmans produced is greater than or equal to } 0)$$

$$X_2 \geq 0 \ (\textit{number of Watch-TVs produced is greater than or equal to } 0)$$

The first step in graphing the constraints of the problem is to convert the constraint *inequalities* into *equalities* (or equations).

$$\text{Constraint A:} \quad 4X_1 + 3X_2 = 240$$

$$\text{Constraint B:} \quad 2X_1 + 1X_2 = 100$$

The equation for constraint A is plotted in Figure B.1 and for constraint B in Figure B.2.

To plot the line in Figure B.1, all we need to do is to find the points at which the line $4X_1 + 3X_2 = 240$ intersects the X_1 and X_2 axes. When $X_1 = 0$ (the location where the line touches the X_2 axis), it implies that $3X_2 = 240$ and that $X_2 = 80$. Likewise, when $X_2 = 0$, we see that $4X_1 = 240$ and that $X_1 = 60$. Thus, constraint A is bounded by the line running from $(X_1 = 0, X_2 = 80)$ to $(X_1 = 60, X_2 = 0)$. The shaded area represents all points that satisfy the original *inequality*.

Constraint B is illustrated similarly in Figure B.2. When $X_1 = 0$, then $X_2 = 100$; and when $X_2 = 0$, then $X_1 = 50$. Constraint B, then, is bounded by the line between $(X_1 = 0, X_2 = 100)$ and $(X_1 = 50, X_2 = 0)$. The shaded area represents the original inequality.

Figure B.3 shows both constraints together. The shaded region is the part that satisfies both restrictions. The shaded region in Figure B.3 is called the *area of feasible solutions*, or simply the **feasible region**. This region must satisfy *all* conditions specified by the program's constraints and is thus the region where all constraints overlap. Any point in the region would be a *feasible solution* to the Shader Electronics Company problem. Any point outside the shaded area would represent an *infeasible solution*. Hence, it would be feasible to manufacture 30 Walkmans

Feasible region
The set of all feasible combinations of decision variables.

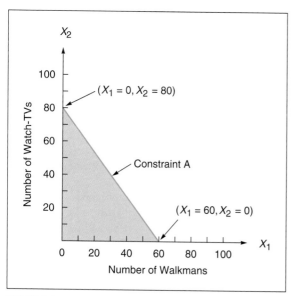

FIGURE B.1 ■ Constraint A

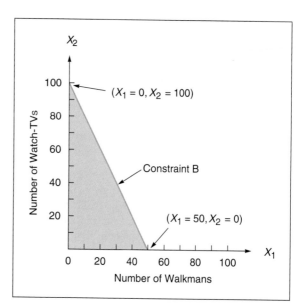

FIGURE B.2 ■ Constraint B

FIGURE B.3 ■

Feasible Solution Region
for the Shader Electronics
Company Problem

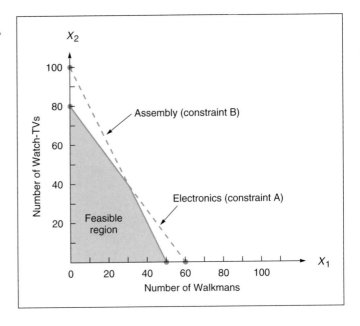

and 20 Watch-TVs ($X_1 = 30$, $X_2 = 20$), but it would violate the constraints to produce 70 Walkmans and 40 Watch-TVs. This can be seen by plotting these points on the graph of Figure B.3.

Iso-Profit Line Solution Method

Now that the feasible region has been graphed, we can proceed to find the *optimal* solution to the problem. The optimal solution is the point lying in the feasible region that produces the highest profit.

Once the feasible region has been established, several approaches can be taken in solving for the optimal solution. The speediest one to apply is called the **iso-profit line method**.[1]

Iso-profit line method
An approach to solving a linear programming maximization problem graphically.

We start by letting profits equal some arbitrary but small dollar amount. For the Shader Electronics problem, we may choose a profit of $210. This is a profit level that can easily be obtained without violating either of the two constraints. The objective function can be written as $210 = 7X_1 + 5X_2$.

This expression is just the equation of a line; we call it an *iso-profit line*. It represents all combinations (of X_1, X_2) that would yield a total profit of $210. To plot the profit line, we proceed exactly as we did to plot a constraint line. First, let $X_1 = 0$ and solve for the point at which the line crosses the X_2 axis:

$$210 = \$7(0) + \$5X_2$$
$$X_2 = 42 \text{ Watch-TVs}$$

Then let $X_2 = 0$ and solve for X_1:

$$210 = \$7X_1 + \$5(0)$$
$$X_1 = 30 \text{ Walkmans}$$

We can now connect these two points with a straight line. This profit line is illustrated in Figure B.4. All points on the line represent feasible solutions that produce a profit of $210.

We see, however, that the iso-profit line for $210 does not produce the highest possible profit to the firm. In Figure B.5, we try graphing two more lines, each yielding a higher profit. The middle equation, $280 = \$7X_1 + \$5X_2$, was plotted in the same fashion as the lower line. When $X_1 = 0$,

$$280 = \$7(0) + \$5X_2$$
$$X_2 = 56 \text{ Watch-TVs}$$

---

[1]*Iso* means "equal" or "similar." Thus, an iso-profit line represents a line with all profits the same, in this case $210.

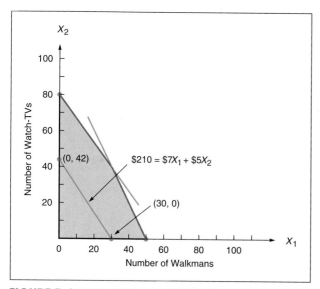

FIGURE B.4 ■ A Profit Line of $210 Plotted for the Shader Electronics Company

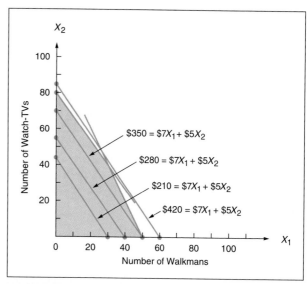

FIGURE B.5 ■ Four Iso-Profit Lines Plotted for the Shader Electronics Company

When $X_2 = 0$,

$$\$280 = \$7X_1 + \$5(0)$$
$$X_1 = 40 \text{ Walkmans}$$

Again, any combination of Walkmans (X_1) and Watch-TVs (X_2) on this iso-profit line will produce a total profit of $280.

Note that the third line generates a profit of $350, even more of an improvement. The farther we move from the 0 origin, the higher our profit will be. Another important point to note is that these iso-profit lines are parallel. We now have two clues as to how to find the optimal solution to the original problem. We can draw a series of parallel profit lines (by carefully moving our ruler in a plane parallel to the first profit line). The highest profit line that still touches some point of the feasible region will pinpoint the optimal solution. Notice that the fourth line ($420) is too high to count because it does not touch the feasible region.

The highest possible iso-profit line is illustrated in Figure B.6. It touches the tip of the feasible region at the corner point ($X_1 = 30$, $X_2 = 40$) and yields a profit of $410.

FIGURE B.6 ■

Optimal Solution for the Shader Electronics Problem

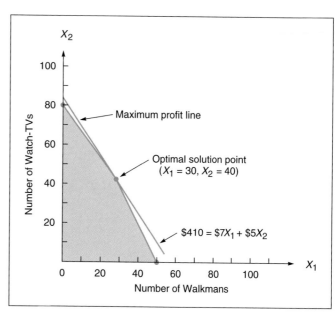

Corner-Point Solution Method

Corner-point method
A method for solving graphical linear programming problems.

A second approach to solving linear programming problems employs the **corner-point method**. This technique is simpler in concept than the iso-profit line approach, but it involves looking at the profit at every corner point of the feasible region.

The mathematical theory behind linear programming states that an optimal solution to any problem (that is, the values of X_1, X_2 that yield the maximum profit) will lie at a *corner point*, or *extreme point*, of the feasible region. Hence, it is necessary to find only the values of the variables at each corner; the maximum profit or optimal solution will lie at one (or more) of them.

Once again we can see (in Figure B.7) that the feasible region for the Shader Electronics Company problem is a four-sided polygon with four corner, or extreme, points. These points are labeled ①, ②, ③ and ④ on the graph. To find the (X_1, X_2) values producing the maximum profit, we find out what the coordinates of each corner point are, then determine and compare their profit levels.

Point ①: $(X_1 = 0, X_2 = 0)$ Profit $7(0) + $5(0) = $0
Point ②: $(X_1 = 0, X_2 = 80)$ Profit $7(0) + $5(80) = $400
Point ④: $(X_1 = 50, X_2 = 0)$ Profit $7(50) + $5(0) = $350

We skipped corner point ③ momentarily because in order to find its coordinates *accurately*, we will have to solve for the intersection of the two constraint lines. As you may recall from algebra, we can apply the method of *simultaneous equations* to the two constraint equations:

$$4X_1 + 3X_2 = 240 \quad \text{(electronics time)}$$
$$2X_1 + 1X_2 = 100 \quad \text{(assembly time)}$$

To solve these equations simultaneously, we multiply the second equation by -2:

$$-2(2X_1 + 1X_2 = 100) = -4X_1 - 2X_2 = -200$$

and then add it to the first equation:

$$
\begin{aligned}
+4X_1 + 3X_2 &= 240 \\
-4X_1 - 2X_2 &= -200 \\
\hline
+1X_2 &= 40
\end{aligned}
$$

or

$$X_2 = 40$$

FIGURE B.7 ■

The Four Corner Points of the Feasible Region

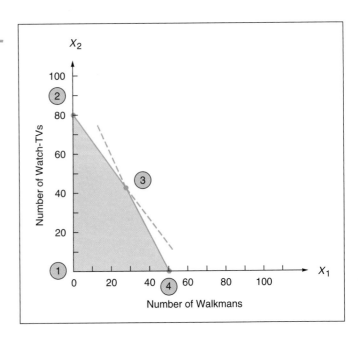

Number of Walkmans

Doing this has enabled us to eliminate one variable, X_1, and to solve for X_2. We can now substitute 40 for X_2 in either of the original equations and solve for X_1. Let us use the first equation. When $X_2 = 40$, then

$$4X_1 + 3(40) = 240$$
$$4X_1 + 120 = 240$$

or

$$4X_1 = 120$$
$$X_1 = 30$$

Although the values for X_1 and X_2 are integers for Shader Electronics, this will not always be the case.

Thus, point ③ has the coordinates $(X_1 = 30, X_2 = 40)$. We can compute its profit level to complete the analysis:

Point ③: $(X_1 = 30, X_2 = 40)$ Profit $= \$7(30) + \$5(40) = \$410$

Because point ③ produces the highest profit of any corner point, the product mix of $X_1 = 30$ Walkmans and $X_2 = 40$ Watch-TVs is the optimal solution to the Shader Electronics problem. This solution will yield a profit of \$410 per production period; it is the same solution we obtained using the iso-profit line method.

SENSITIVITY ANALYSIS

Parameter
Numerical value that is given in a model.

Operations managers are usually interested in more than the optimal solution to an LP problem. In addition to knowing the value of each decision variable (the X_is) and the value of the objective function, they want to know how sensitive these answers are to input **parameter** changes. For example, what happens if the coefficients of the objective function are not exact, or if they change by 10% or 15%? What happens if right-hand-side values of the constraints change? Because solutions are based on the assumption that input parameters are constant, the subject of sensitivity analysis comes into play. **Sensitivity analysis**, or postoptimality analysis, is the study of how sensitive solutions are to parameter changes.

Sensitivity analysis
An analysis that projects how much a solution might change if there were changes in the variables or input data.

There are two approaches to determining just how sensitive an optimal solution is to changes. The first is simply a trial-and-error approach. This approach usually involves resolving the entire problem, preferably by computer, each time one input data item or parameter is changed. It can take a long time to test a series of possible changes in this way.

The approach we prefer is the analytic postoptimality method. After an LP problem has been solved, we determine a range of changes in problem parameters that will not affect the optimal solution or change the variables in the solution. This is done without resolving the whole problem. LP software, such as Excel's Solver or POM for Windows, has this capability. Let us examine several scenarios relating to the Shader Electronics example.

Program B.1 is part of the Excel Solver computer-generated output available to help a decision maker know whether a solution is relatively insensitive to reasonable changes in one or more of the parameters of the problem. (The complete computer run for these data, including input and full output, is illustrated in Programs B.2 and B.3 later in this module.)

Sensitivity Report

The *Sensitivity Report* has two parts: Adjustable Cells and Constraints.

The Excel *Sensitivity Report* for the Shader Electronics example in Program B.1 has two distinct components: (1) a table titled Adjustable Cells and (2) a table titled Constraints. These tables permit us to answer several what-if questions regarding the problem solution.

We are analyzing only one change at a time.

It is important to note that while using the information in the sensitivity report to answer what-if questions, we assume that we are considering a change to only a *single* input data value. That is, the sensitivity information does not always apply to simultaneous changes in several input data values.

The *Adjustable Cells* table presents information regarding the impact of changes to the objective function coefficients (i.e., the unit profits of \$7 and \$5) on the optimal solution. The *Constraints* table presents information related to the impact of changes in constraint right-hand-side (RHS) values (i.e., the 240 hours and 100 hours) on the optimal solution. Although different LP software packages may format and present these tables differently, the programs all provide essentially the same information.

PROGRAM B.1 ■

Sensitivity Analysis for
Shader Electronics Using
Excel's Solver

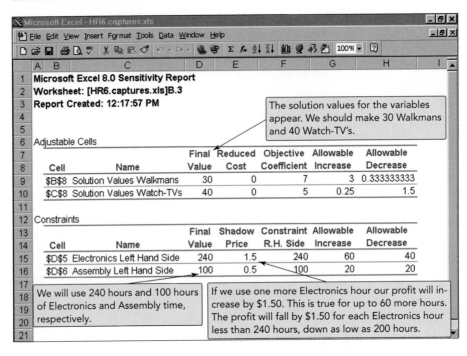

Changes in the Resources or Right-Hand-Side Values

The right-hand-side values of the constraints often represent resources available to the firm. The resources could be labor-hours or machine time or perhaps money or production materials available. In the Shader Electronics example, the two resources are hours available of electronics time and hours of assembly time. If additional hours were available, a higher total profit could be realized. How much should the company be willing to pay for additional hours? Is it profitable to have some additional electronics hours? Should we be willing to pay for more assembly time? Sensitivity analysis about these resources will help us answer these questions.

If the size of the feasible region increases, the optimal objective function value could improve.

If the right-hand side of a constraint is changed, the feasible region will change (unless the constraint is redundant), and often the optimal solution will change. In the Shader example, there were 100 hours of assembly time available each week and the maximum possible profit was $410. If the available assembly hours are *increased* to 110 hours, the new optimal solution seen in Figure B.8(a) is (45,20) and the profit is $415. Thus, the extra 10 hours of time resulted in an increase in profit of $5 or $0.50 per hour. If the hours were *decreased* to 90 hours as shown in Figure B.8(b), the new optimal solution is (15,60) and the profit is $405. Thus, reducing the hours by 10 results in a decrease in profit of $5 or $0.50 per hour. This $0.50 per hour change in profit that resulted from a change in the hours available is called the shadow price, or **dual** value. The **shadow price** for a constraint is the improvement in the objective function value that results from a one-unit increase in the right-hand side of the constraint.

Shadow price (or dual)
The value of one additional unit of a scarce resource in LP.

Validity Range for the Shadow Price   Given that Shader Electronics' profit increases by $0.50 for each additional hour of assembly time, does it mean that Shader can do this indefinitely, essentially earning infinite profit? Clearly, this is illogical. How far can Shader increase its assembly time availability and still earn an extra $0.50 profit per hour? That is, for what level of increase in the RHS value of the assembly time constraint is the shadow price of $0.50 valid?

The shadow price is valid only as long as the change in the RHS is within the Allowable Increase and Allowable Decrease values.

The shadow price of $0.50 is valid as long as the available assembly time stays in a range within which all current corner points continue to exist. The information to compute the upper and lower limits of this range is given by the entries labeled Allowable Increase and Allowable Decrease in the *Sensitivity Report* in Program B.1. In Shader's case, these values show that the shadow price of $0.50 for assembly time availability is valid for an increase of up to 20 hours from the current value and a decrease of up to 20 hours. That is, the available assembly time can range from a low of 80 (= 100 − 20) to a high of 120 (= 100 +20) for the shadow price of $0.50 to be valid. Note that the allowable decrease implies that for each hour of assembly time that Shader loses (up to 20 hours), its profit decreases by $0.50.

(a)

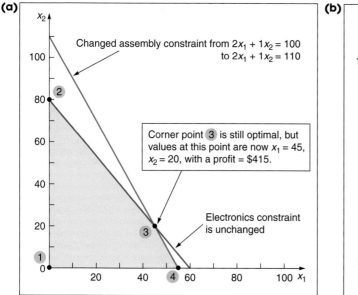

(b)

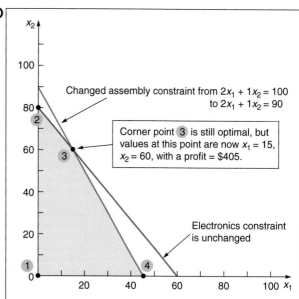

FIGURE B.8 ■ Shader Electronics Sensitivity Analysis on Right-Hand-Side (RHS) Resources

Changes in the Objective Function Coefficient

Let us now focus on the information provided in Program B.1 titled Adjustable Cells. Each row in the Adjustable Cells table contains information regarding a decision variable (i.e., Walkmans or Watch-TVs) in the LP model.

Allowable Ranges for Objective Function Coefficients

As the unit profit contribution of either product changes, the slope of the iso-profit lines we saw earlier in Figure B.5 changes. The size of the feasible region, however, remains the same. That is, the locations of the corner points do not change.

There is an allowable decrease and an allowable increase for each objective function coefficient over which the current optimal solution remains optimal.

The limits to which the profit coefficient of Walkmans or Watch-TVs can be changed without affecting the optimality of the current solution is revealed by the values in the Allowable Increase and Allowable Decrease columns of the *Sensitivity Report* in Program B.1. The allowable increase in the objective function coefficient for Watch-TVs is only $0.25. In contrast, the allowable decrease is $1.50. Hence, if the unit profit of Watch-TVs drops to $4 (i.e., a decrease of $1 from the current value of $5), it is still optimal to produce 30 Walkmans and 40 Watch-TVs. The total profit will drop to $370 (from $410) because each Watch-TV now yields lesser profit (of $1 per unit). However, if the unit profit drops below $3.50 per Watch-TV (i.e., a decrease of more than $1.50 from the current $5 profit), the current solution is no longer optimal. The LP problem would then have to be resolved using Solver, or other software, to find the new optimal corner point.

A new corner point becomes optimal if an objective function coefficient is decreased or increased too much.

SOLVING MINIMIZATION PROBLEMS

Many linear programming problems involve *minimizing* an objective such as cost instead of maximizing a profit function. A restaurant, for example, may wish to develop a work schedule to meet staffing needs while minimizing the total number of employees. Also, a manufacturer may seek to distribute its products from several factories to its many regional warehouses in such a way as to minimize total shipping costs.

Iso-cost

An approach to solving a linear programming minimization problem graphically.

Minimization problems can be solved graphically by first setting up the feasible solution region and then using either the corner-point method or an **iso-cost** line approach (which is analogous to the iso-profit approach in maximization problems) to find the values of X_1 and X_2 that yield the minimum cost.

Example B1 shows how to solve a minimization problem.

Example B1

Cohen Chemicals, Inc., produces two types of photo-developing fluids. The first, a black-and-white picture chemical, costs Cohen \$2,500 per ton to produce. The second, a color photo chemical, costs \$3,000 per ton.

Based on an analysis of current inventory levels and outstanding orders, Cohen's production manager has specified that at least 30 tons of the black-and-white chemical and at least 20 tons of the color chemical must be produced during the next month. In addition, the manager notes that an existing inventory of a highly perishable raw material needed in both chemicals must be used within 30 days. In order to avoid wasting the expensive raw material, Cohen must produce a total of at least 60 tons of the photo chemicals in the next month.

We may formulate this information as a minimization LP problem. Let

$$X_1 = \text{number of tons of black-and-white picture chemical produced}$$
$$X_2 = \text{number of tons of color picture chemical produced}$$

Subject to:

$$X_1 \geq 30 \text{ tons of black-and-white chemical}$$
$$X_2 \geq 20 \text{ tons of color chemical}$$
$$X_1 + X_2 \geq 60 \text{ tons total}$$
$$X_1, X_2 \geq \$0 \text{ nonnegativity requirements}$$

To solve the Cohen Chemicals problem graphically, we construct the problem's feasible region, shown in Figure B.9.

FIGURE B.9 ■

Cohen Chemicals' Feasible Region

The area is not bounded to the right in a minimization problem as it is in a maximization problem.

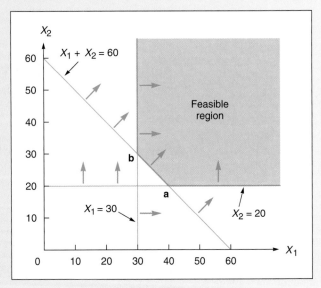

Minimization problems are often unbounded outward (that is, on the right side and on the top), but this characteristic causes no problem in solving them. As long as they are bounded inward (on the left side and the bottom), we can establish corner points. The optimal solution will lie at one of the corners.

In this case, there are only two corner points, **a** and **b**, in Figure B.9. It is easy to determine that at point **a**, $X_1 = 40$ and $X_2 = 20$, and that at point **b**, $X_1 = 30$ and $X_2 = 30$. The optimal solution is found at the point yielding the lowest total cost.

Thus

$$\text{Total cost at } \mathbf{a} = 2,500X_1 + 3,000X_2$$
$$= 2,500(40) + 3,000(20)$$
$$= \$160,000$$
$$\text{Total cost at } \mathbf{b} = 2,500X_1 + 3,000X_2$$
$$= 2,500(30) + 3,000(30)$$
$$= \$165,000$$

The lowest cost to Cohen Chemicals is at point **a**. Hence the operations manager should produce 40 tons of the black-and-white chemical and 20 tons of the color chemical.

OM IN ACTION

Scheduling Planes at Delta Airlines with LP

It has been said that an airline seat is the most perishable commodity in the world. Each time an airliner takes off with an empty seat, a revenue opportunity is lost forever. For Delta Airlines, which flies over 2,500 domestic flight legs per day using about 450 aircraft of 10 different models, its schedule is the very heartbeat of the airline.

One flight leg for Delta might consist of a Boeing 757 jet assigned to fly at 6:21 A.M. from Atlanta to arrive in Boston at 8:45 A.M. Delta's problem, like that of every competitor, is to match airplanes such as 747s, 757s, or 767s to flight legs such as Atlanta-Boston and to fill seats with paying passengers. Recent advances in linear programming algorithms and computer hardware have made it possible to solve optimization problems of this scope for the first time. Delta calls its huge LP model "Coldstart" and runs the model every day. Delta is the first airline to solve a problem of this scope.

The typical size of a daily Coldstart model is about 40,000 constraints and 60,000 variables. The constraints include aircraft availability, balancing arrivals and departures at airports, aircraft maintenance needs, and so on. Coldstart's objective is to minimize a combination of operating costs and lost passenger revenue, called "spill costs."

The savings from the model so far have been phenomenal, estimated at $220,000 per day over Delta's earlier schedule planning tool, which was nicknamed "Warmstart." Delta expects to save $300 million over the next 3 years through this use of linear programming.

Sources: R. Subramanian et al., *Interfaces* 24, 1 (January–February 1994):104–120; Peter R. Horner, *OR/MS Today* 22, 4 (August 1995):14–15.

LINEAR PROGRAMMING APPLICATIONS

The foregoing examples each contained just two variables (X_1 and X_2). Most real-world problems contain many more variables, however. Let's use the principles already developed to formulate a few more-complex problems. The practice you will get by "paraphrasing" the following LP situations should help develop your skills for applying linear programming to other common operations situations.

Production-Mix Example

Example B2 involves another *production-mix* decision. Limited resources must be allocated among various products that a firm produces. The firm's overall objective is to manufacture the selected products in such quantities as to maximize total profits.

Example B2

Failsafe Electronics Corporation primarily manufactures four highly technical products, which it supplies to aerospace firms that hold NASA contracts. Each of the products must pass through the following departments before they are shipped: wiring, drilling, assembly, and inspection. The time requirements in each department (in hours) for each unit produced and its corresponding profit value are summarized in this table:

	DEPARTMENT				
PRODUCT	WIRING	DRILLING	ASSEMBLY	INSPECTION	UNIT PROFIT
XJ201	.5	3	2	.5	$ 9
XM897	1.5	1	4	1.0	$12
TR29	1.5	2	1	.5	$15
BR788	1.0	3	2	.5	$11

The production time available in each department each month and the minimum monthly production requirement to fulfill contracts are as follows:

DEPARTMENT	CAPACITY (IN HOURS)	PRODUCT	MINIMUM PRODUCTION LEVEL
Wiring	1,500	XJ201	150
Drilling	2,350	XM897	100
Assembly	2,600	TR29	300
Inspection	1,200	BR788	400

The production manager has the responsibility of specifying production levels for each product for the coming month. Let

$$X_1 = \text{number of units of XJ201 produced}$$
$$X_2 = \text{number of units of XM897 produced}$$
$$X_3 = \text{number of units of TR29 produced}$$
$$X_4 = \text{number of units of BR788 produced}$$

$$\text{Maximize profit} = 9X_1 + 12X_2 + 15X_3 + 11X_4$$

$$\begin{aligned}
\text{subject to} \quad .5X_1 + 1.5X_2 + 1.5X_3 + 1X_4 &\leq 1{,}500 \text{ hours of wiring available}\\
3X_1 + 1X_2 + 2X_3 + 3X_4 &\leq 2{,}350 \text{ hours of drilling available}\\
2X_1 + 4X_2 + 1X_3 + 2X_4 &\leq 2{,}600 \text{ hours of assembly available}\\
.5X_1 + 1X_2 + .5X_3 + .5X_4 &\leq 1{,}200 \text{ hours of inspection}\\
X_1 &\geq 150 \text{ units of XJ201}\\
X_2 &\geq 100 \text{ units of XM897}\\
X_3 &\geq 300 \text{ units of TR29}\\
X_4 &\geq 400 \text{ units of BR788}\\
X_1, X_2, X_3, X_4 &\geq 0
\end{aligned}$$

Diet Problem Example

Example B3 illustrates the *diet problem*, which was originally used by hospitals to determine the most economical diet for patients. Known in agricultural applications as the *feed-mix problem*, the diet problem involves specifying a food or feed ingredient combination that will satisfy stated nutritional requirements at a minimum cost level.

Example B3

The Feed 'N Ship feedlot fattens cattle for local farmers and ships them to meat markets in Kansas City and Omaha. The owners of the feedlot seek to determine the amounts of cattle feed to buy in order to satisfy minimum nutritional standards and, at the same time, minimize total feed costs.

Each grain stock contains different amounts of four nutritional ingredients: A, B, C, and D. Here are the ingredient contents of each grain, in *ounces per pound of grain*.

	FEED		
INGREDIENT	STOCK X	STOCK Y	STOCK Z
A	3 oz.	2 oz.	4 oz.
B	2 oz.	3 oz.	1 oz.
C	1 oz.	0 oz.	2 oz.
D	6 oz.	8 oz.	4 oz.

The cost per pound of grains X, Y, and Z is $.02, $.04, and $.025, respectively. The minimum requirement per cow per month is 64 ounces of ingredient A, 80 ounces of ingredient B, 16 ounces of ingredient C, and 128 ounces of ingredient D.

The feedlot faces one additional restriction—it can obtain only 500 pounds of stock Z per month from the feed supplier, regardless of its need. Because there are usually 100 cows at the Feed 'N Ship feedlot at any given time, this constraint limits the amount of stock Z for use in the feed of each cow to no more than 5 pounds, or 80 ounces, per month. Let

$$X_1 = \text{number of pounds of stock X purchased per cow each month}$$
$$X_2 = \text{number of pounds of stock Y purchased per cow each month}$$
$$X_3 = \text{number of pounds of stock Z purchased per cow each month}$$

Minimum cost $= .02X_1 + .04X_2 + .025X_3$ subject to

$$\text{Ingredient A requirement: } 3X_1 + 2X_2 + 4X_3 \geq 64$$
$$\text{Ingredient B requirement: } 2X_1 + 3X_2 + 1X_3 \geq 80$$
$$\text{Ingredient C requirement: } 1X_1 + 0X_2 + 2X_3 \geq 16$$
$$\text{Ingredient D requirement: } 6X_1 + 8X_2 + 4X_3 \geq 128$$
$$\text{Stock Z limitation: } \qquad\quad X_3 \leq 80$$
$$X_1, X_2, X_3 \geq 0$$

The cheapest solution is to purchase 40 pounds of grain X_1, at a cost of $0.80 per cow.

Production Scheduling Example

One of the most important areas of linear programming application is *production scheduling*. Solving a production scheduling problem allows the production manager to set an efficient, low-cost production schedule for a product over several production periods. Basically, the problem resembles the common product-mix model for each period in the future. Production levels must allow the firm to meet demand for its product within manpower and inventory limitations. The objective is either to maximize profit or to minimize the total cost (of production plus inventory).

Example B4

The T. E. Callarman Appliance Company is thinking of manufacturing and selling trash compactors on an experimental basis over the next 6 months. The manufacturing costs and selling prices of the compactors are projected to vary from month to month. Table B.2 gives these forecast costs and prices.

TABLE B.2 ■ Manufacturing Costs and Selling Prices

MONTH	MANUFACTURING COST	SELLING PRICE (DURING MONTH)
July	$60	—
August	$60	$80
September	$50	$60
October	$60	$70
November	$70	$80
December	—	$90

All compactors manufactured during any month are shipped out in one large load at the end of that month. The firm can sell as many as 300 units per month, but its operation is limited by the size of its warehouse, which holds a maximum of 100 compactors.

Callarman's operations manager, Richard Deckro, needs to determine the number of compactors to manufacture and sell each month in order to maximize the firm's profit. Callarman has no compactors on hand at the beginning of July and wishes to have no compactors on hand at the end of the test period in December.

To formulate this LP problem, Deckro lets

$$X_1, X_2, X_3, X_4, X_5, X_6 = \text{number of units } \textit{manufactured} \text{ during July (first month),}$$
$$\text{August (second month), etc.}$$
$$Y_1, Y_2, Y_3, Y_4, Y_5, Y_6 = \text{number of units } \textit{sold} \text{ during July, August, etc.}$$

He notes that because the company starts with no compactors (and because it takes 1 month to gear up and ship out the first batch), it cannot sell any units in July (that is, $Y_1 = 0$). Also, because it wants zero inventory at the end of the year, manufacture during the month of December must be zero (that is, $X_6 = 0$).

Profit for Callarman Appliances is sales minus manufacture cost. Hence, Deckro's objective function is

$$\text{Maximize profit} = 80Y_2 + 60Y_3 + 70Y_4 + 80Y_5 + 90Y_6$$
$$- (60X_1 + 60X_2 + 50X_3 + 60X_4 + 70X_5)$$

The first part of this expression is the sales price times the units sold each month. The second part is the manufacture cost, namely, the costs from Table B.2 times the units manufactured.

To set up the constraints, Deckro needs to introduce a new set of variables: $I_1, I_2, I_3, I_4, I_5, I_6$. These represent the inventory at the end of a month (after all sales have been made and after the amount produced during the month has been stocked in the warehouse). Thus

$$\begin{matrix} \text{Inventory at} \\ \text{end of this} \\ \text{month} \end{matrix} = \begin{matrix} \text{Inventory at} \\ \text{end of} \\ \text{previous month} \end{matrix} + \begin{matrix} \text{Current} \\ \text{month's} \\ \text{production} \end{matrix} - \begin{matrix} \text{This month's} \\ \text{sales} \end{matrix}$$

For July, this is $I_1 = X_1$ because there is neither previous inventory nor sales. For August,

$$I_2 = I_1 + X_2 - Y_2$$

Constraints for the remaining months are as follows:

$$\begin{aligned} \text{September:} \quad & I_3 = I_2 + X_3 - Y_3 \\ \text{October:} \quad & I_4 = I_3 + X_4 - Y_4 \\ \text{November:} \quad & I_5 = I_4 + X_5 - Y_5 \\ \text{December:} \quad & I_6 = I_5 - Y_6 \end{aligned}$$

Constraints for the storage capacity are

$$I_1 \le 100, \quad I_2 \le 100, \quad I_3 \le 100, \quad I_4 \le 100, \quad I_5 \le 100$$

and $I_6 = 0$ (in order to end up with zero inventory at the end of December).
Constraints for demand are

$$\text{all} \quad Y_i \le 300$$

The final solution yields a profit of $19,000 with

$$\begin{aligned} & X_1 = 100, X_2 = 200, X_3 = 400, X_4 = 300, X_5 = 300, X_6 = 0, \\ & Y_1 = 100, Y_2 = 300, Y_3 = 300, Y_4 = 300, Y_5 = 300, Y_6 = 100, \\ & I_1 = 100, \quad I_2 = 0, \quad I_3 = 100, \quad I_4 = 100, \quad I_5 = 100, \quad I_6 = 0 \end{aligned}$$

Alternative optimal solutions exist.

Labor Scheduling Example

Labor scheduling problems address staffing needs over a specific time period. They are especially useful when managers have some flexibility in assigning workers to jobs that require overlapping or interchangeable talents. Large banks and hospitals frequently use LP to tackle their labor scheduling. Example B5 describes how one bank uses LP to schedule tellers.

Example B5

Arlington Bank of Commerce and Industry is a busy bank that has requirements for between 10 and 18 tellers depending on the time of day. Lunchtime, from noon to 2 P.M., is usually heaviest. The table below indicates the workers needed at various hours that the bank is open.

TIME PERIOD	NUMBER OF TELLERS REQUIRED	TIME PERIOD	NUMBER OF TELLERS REQUIRED
9 A.M.–10 A.M.	10	1 P.M.–2 P.M.	18
10 A.M.–11 A.M.	12	2 P.M.–3 P.M.	17
11 A.M.–Noon	14	3 P.M.–4 P.M.	15
Noon–1 P.M.	16	4 P.M.–5 P.M.	10

The bank now employs 12 full-time tellers, but many people are on its roster of available part-time employees. A part-time employee must put in exactly 4 hours per day, but can start anytime between 9 A.M. and 1 P.M. Part-timers are a fairly inexpensive labor pool because no retirement or lunch benefits are provided them. Full-timers, on the other hand, work from 9 A.M. to 5 P.M. but are allowed 1 hour for lunch. (Half the full-timers eat at 11 A.M., the other half at noon.) Full-timers thus provide 35 hours per week of productive labor time.

By corporate policy, the bank limits part-time hours to a maximum of 50% of the day's total requirement.

Part-timers earn $6 per hour (or $24 per day) on average, whereas full-timers earn $75 per day in salary and benefits on average. The bank would like to set a schedule that would minimize its total manpower costs. It will release 1 or more of its full-time tellers if it is profitable to do so.

We can let

$$F = \text{Full-time tellers}$$

$$P_1 = \text{Part-timers starting at 9 A.M. (leaving at 1 P.M.)}$$

$$P_2 = \text{Part-timers starting at 10 A.M. (leaving at 2 P.M.)}$$

$$P_3 = \text{Part-timers starting at 11 A.M. (leaving at 3 P.M.)}$$

$$P_4 = \text{Part-timers starting at noon (leaving at 4 P.M.)}$$

$$P_5 = \text{Part-timers starting at 1 P.M. (leaving at 5 P.M.)}$$

Objective function:

$$\text{Minimize total daily manpower cost} = \$75F + \$24(P_1 + P_2 + P_3 + P_4 + P_5)$$

Constraints: For each hour, the available labor-hours must be at least equal to the required labor-hours.

$$
\begin{aligned}
F + P_1 &\geq 10 && (\textit{9 A.M. to 10 A.M. needs}) \\
F + P_1 + P_2 &\geq 12 && (\textit{10 A.M. to 11 A.M. needs}) \\
\tfrac{1}{2}F + P_1 + P_2 + P_3 &\geq 14 && (\textit{11 A.M. to noon needs}) \\
\tfrac{1}{2}F + P_1 + P_2 + P_3 + P_4 &\geq 16 && (\textit{noon to 1 P.M. needs}) \\
F + P_2 + P_3 + P_4 + P_5 &\geq 18 && (\textit{1 P.M. to 2 P.M. needs}) \\
F + P_3 + P_4 + P_5 &\geq 17 && (\textit{2 P.M. to 3 P.M. needs}) \\
F + P_4 + P_5 &\geq 15 && (\textit{3 P.M. to 4 P.M. needs}) \\
F + P_5 &\geq 10 && (\textit{4 P.M. to 5 P.M. needs})
\end{aligned}
$$

Only 12 full-time tellers are available, so

$$F \leq 12$$

Part-time worker-hours cannot exceed 50% of total hours required each day, which is the sum of the tellers needed each hour.

$$4(P_1 + P_2 + P_3 + P_4 + P_5) \leq .50(10 + 12 + 14 + 16 + 18 + 17 + 15 + 10)$$

or

$$4P_1 + 4P_2 + 4P_3 + 4P_4 + 4P_5 \leq 0.50(112)$$
$$F, P_1, P_2, P_3, P_4, P_5 \geq 0$$

There are two alternative optimal schedules that Arlington Bank can follow. The first is to employ only 10 full-time tellers ($F = 10$) and to start 7 part-timers at 10 A.M. ($P_2 = 7$), 2 part-timers at 11 A.M. and noon ($P_3 = 2$ and $P_4 = 2$), and 3 part-timers at 1 P.M. ($P_5 = 3$). No part-timers would begin at 9 A.M.

The second solution also employs 10 full-time tellers, but starts 6 part-timers at 9 A.M. ($P_1 = 6$), 1 part-timer at 10 A.M. ($P_2 = 1$), 2 part-timers at 11 A.M. and noon ($P_3 = 2$ and $P_4 = 2$), and 3 part-timers at 1 P.M. ($P_5 = 3$). The cost of either of these two policies is $1,086 per day.

THE SIMPLEX METHOD OF LP

Simplex method

An algorithm developed by Dantzig for solving linear programming problems of all sizes.

Most real-world linear programming problems have more than two variables and thus are too complex for graphical solution. A procedure called the **simplex method** may be used to find the optimal solution to such problems. The simplex method is actually an algorithm (or a set of instructions) with which we examine corner points in a methodical fashion until we arrive at the best solution—highest profit or lowest cost. Computer programs (such as POM for Windows) and Excel spreadsheets are available to solve linear programming problems via the simplex method.

For details regarding the algebraic steps of the simplex algorithm, see Tutorial 3 on the CD-ROM that accompanies this book or refer to a management science textbook.[2]

SUMMARY

This module introduces a special kind of model, linear programming. LP has proven to be especially useful when trying to make the most effective use of an organization's resources.

The first step in dealing with LP models is problem formulation, which involves identifying and creating an objective function and constraints. The second step is to solve the problem. If there are only two decision variables, the problem can be solved graphically, using the corner-point method or the iso-profit/iso-cost line method. With either approach, we first identify the feasible region, then find the corner point yielding the greatest profit or least cost.

All LP problems can also be solved with the simplex method, using software such as POM for Windows or Excel. This approach produces valuable economic information such as the shadow price, or dual, and provides complete sensitivity analysis on other inputs to the problem. LP is used in a wide variety of business applications, as the examples and homework problems in this module reveal.

KEY TERMS

Linear programming (LP) *(p. 518)*
Objective function *(p. 519)*
Constraints *(p. 519)*
Graphical solution approach *(p. 520)*
Decision variables *(p. 520)*
Feasible region *(p. 521)*
Iso-profit line method *(p. 522)*

Corner-point method *(p. 524)*
Parameter *(p. 525)*
Sensitivity analysis *(p. 525)*
Shadow price (or dual) *(p. 526)*
Iso-cost *(p. 527)*
Simplex method *(p. 534)*

USING EXCEL SPREADSHEETS TO SOLVE LP PROBLEMS

Excel and other spreadsheets offer the ability to analyze linear programming problems using built-in problem-solving tools. Excel OM is not illustrated because Excel uses a *built-in* tool named Solver to find LP solutions. Solver is limited to 200 changing cells (variables), each with 2 boundary constraints and up to 100 additional constraints. These capabilities make Solver suitable for the solution of complex, real-world problems.

We use Excel to solve the Shader Electronics problem in Program B.2. The objective and constraints are repeated here:

Objective function: Maximize profit =

$7(No. of Walkmans) + $5(No. of Watch-TVs)

Subject to: 4(Walkmans) + 3(Watch-TVs) ≤ 240

2(Walkmans) + 1(Watch-TV) ≤ 100

The Excel screen in Program B.3 shows Solver's solution to the Shader Electronics Company problem. Note that the optimal solution is now shown in the *changing cells* (cells B8 and C8, which served as the variables). The Reports selection performs more extensive analysis of the solution and its environment. Excel's sensitivity analysis capability was illustrated earlier in Program B.1.

[2]See, for example, Barry Render, Ralph M. Stair, and Michael Hanna, *Quantitative Analysis for Management*, 8th ed. (Upper Saddle River, NJ: Prentice Hall, 2003): Chapters 7–9 or Barry Render, Ralph M. Stair, and Raju Balakrishnan, *Managerial Decision Modeling with Spreadsheets* (Upper Saddle River, NJ: Prentice Hall, 2003): Chapters 2–4.

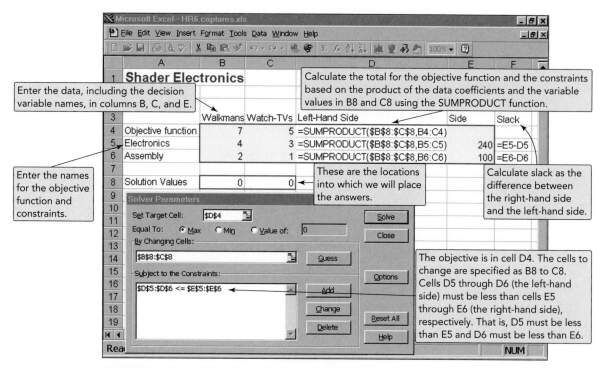

PROGRAM B.2 ■ Using Excel to Formulate the Shader Electronics Problem

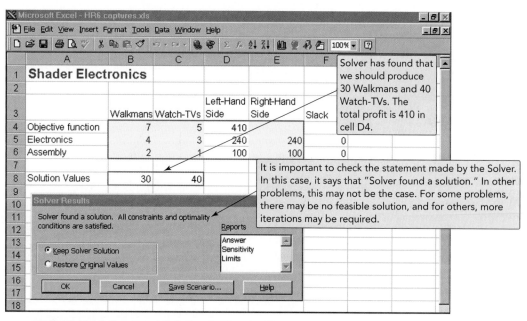

PROGRAM B.3 ■ Excel Solution to Shader Electronics LP Problem

 # USING POM FOR WINDOWS TO SOLVE LP PROBLEMS

POM for Windows can handle LP problems with up to 22 constraints and 99 variables. As output, the software provides optimal values for the variables, optimal profit or cost, and sensitivity analysis. In addition, POM for Windows provides graphical output for problems with only two variables. Please see Appendix V for further details.

SOLVED PROBLEMS

Solved Problem B.1

Smitty's, a clothing manufacturer that produces men's shirts and pajamas, has two primary resources available: sewing-machine time (in the sewing department) and cutting-machine time (in the cutting department). Over the next month, Smitty can schedule up to 280 hours of work on sewing machines and up to 450 hours of work on cutting machines. Each shirt produced requires 1.00 hour of sewing time and 1.50 hours of cutting time. Producing each pair of pajamas requires .75 hour of sewing time and 2 hours of cutting time.

To express the LP constraints for this problem mathematically, we let

$$X_1 = \text{number of shirts produced}$$
$$X_2 = \text{number of pajamas produced}$$

SOLUTION

First constraint: $1X_1 + .75X_2 \le 280$ hours of sewing-machine time available—our first scarce resource

Second constraint: $1.5X_1 + ② X_2 \le 450$ hours of cutting-machine time available—our second scarce resource

Note: This means that each pair of pajamas takes 2 hours of the cutting resource.

Smitty's accounting department analyzes cost and sales figures and states that each shirt produced will yield a $4 contribution to profit and that each pair of pajamas will yield a $3 contribution to profit. This information can be used to create the LP *objective function* for this problem:

Objective function: maximize total contribution to profit $= \$4X_1 + \$3X_2$

Solved Problem B.2

We want to solve the following LP problem for Failsafe Computers using the corner-point method:

Maximize profit $= \$9X_1 + \$7X_2$

$$2X_1 + 1X_2 \le 40$$

$$X_1 + 3X_2 \le 30$$

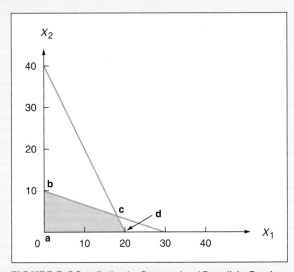

FIGURE B.10 ■ Failsafe Computers' Feasible Region

SOLUTION

Figure B.10 illustrates these constraints:

Corner-point **a**: $(X_1 = 0, X_2 = 0)$ Profit $= 0$
Corner-point **b**: $(X_1 = 0, X_2 = 10)$ Profit $= 9(0) + 7(10) = \$70$
Corner-point **d**: $(X_1 = 20, X_2 = 0)$ Profit $= 9(20) + 7(0) = \$180$

Corner-point **c** is obtained by solving equations $2X_1 + 1X_2 = 40$ and $X_1 + 3X_2 = 30$ simultaneously. Multiply the second equation by -2 and add it to the first.

$$2X_1 + 1X_2 = 40$$
$$\underline{-2X_1 - 6X_2 = -60}$$
$$-5X_2 = -20$$

Thus $X_2 = 4$.

$$X_1 + 3(4) = 30 \quad \text{or} \quad X_1 + 12 = 30 \quad \text{or} \quad X_1 = 18$$
$$\text{Corner-point } \textbf{c}: (X_1 = 18, X_2 = 4) \quad \text{Profit} = 9(18) + 7(4) = \$190$$

Hence the optimal solution is

$$(x_1 = 18, x_2 = 4) \quad \text{Profit} = \$190$$

Solved Problem B.3

Holiday Meal Turkey Ranch is considering buying two different types of turkey feed. Each feed contains, in varying proportions, some or all of the three nutritional ingredients essential for fattening turkeys. Brand Y feed costs the ranch $.02 per pound. Brand Z costs $.03 per pound. The rancher would like to determine the lowest-cost diet that meets the minimum monthly intake requirement for each nutritional ingredient.

The following table contains relevant information about the composition of brand Y and brand Z feeds, as well as the minimum monthly requirement for each nutritional ingredient per turkey.

COMPOSITION OF EACH POUND OF FEED

INGREDIENT	BRAND Y FEED	BRAND Z FEED	MINIMUM MONTHLY REQUIREMENT
A	5 oz.	10 oz.	90 oz.
B	4 oz.	3 oz.	48 oz.
C	.5 oz.	0	1.5 oz.
Cost/lb	$.02	$.03	

SOLUTION

If we let

X_1 = number of pounds of brand Y feed purchased

X_2 = number of pounds of brand Z feed purchased

then we may proceed to formulate this linear programming problem as follows:

$$\text{Minimize cost (in cents)} = 2X_1 + 3X_2$$

subject to these constraints:

$$5X_1 + 10X_2 \geq 90 \text{ oz.} \quad (\textit{ingredient A constraint})$$
$$4X_1 + 3X_2 \geq 48 \text{ oz.} \quad (\textit{ingredient B constraint})$$
$$\tfrac{1}{2}X_1 \geq 1\tfrac{1}{2} \text{ oz.} \quad (\textit{ingredient C constraint})$$

Figure B.11 illustrates these constraints.

The iso-cost line approach may be used to solve LP minimization problems such as that of the Holiday Meal Turkey Ranch. As with iso-profit lines, we need not compute the cost at each corner point, but instead draw a series of parallel cost lines. The lowest cost line (that is, the one closest in toward the origin) to touch the feasible region provides us with the optimal solution corner.

For example, we start in Figure B.12 by drawing a 54¢ cost line, namely, $54 = 2X_1 + 3X_2$. Obviously, there are many points in the feasible region that would yield a lower total cost. We proceed to

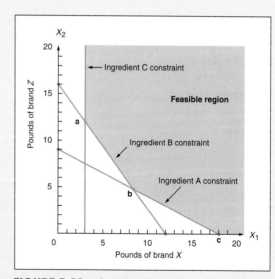

FIGURE B.11 ■ Feasible Region for the Holiday Meal Turkey Ranch Problem

move our iso-cost line toward the lower left, in a plane parallel to the 54¢ solution line. The last point we touch while still in contact with the feasible region is the same as corner point **b** of Figure B.11. It has the coordinates $(X_1 = 8.4, X_2 = 4.8)$ and an associated cost of 31.2 cents.

FIGURE B.12 ■

Graphical Solution to the Holiday
Meal Turkey Ranch Problem Using
the Iso-Cost Line

*Note that the last line parallel to the 54¢
iso-cost line that touches the feasible
region indicates the optimal corner point.*

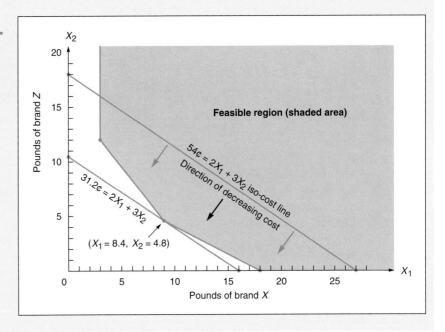

INTERNET AND STUDENT CD-ROM EXERCISES

Visit our home page or use your student CD-ROM to help with material in this module.

 On Our Home Page, www.prenhall.com/heizer

- Self-Tests
- Practice Problems
- Internet Exercises
- Current Articles and Research
- Internet Homework Problems
- Internet Cases

 On Your Student CD-ROM

- PowerPoint Lecture
- Practice Problems
- Active Model Exercise

ADDITIONAL CASE STUDIES

See our Internet homepage at www.prenhall.com/heizer **for these three additional case studies:**

- **Mexicana Wire Works**: This capacity case at a Mexican electrical manufacturer addresses backordering versus using temporary workers.

- **Coastal States Chemical**: This production planning case deals with handling cuts in natural gas availability.

- **Chase Manhattan Bank**: This scheduling case involves finding the optimal number of full-time versus part-time employees at a bank.

Quantitative Module

Transportation Models

Module Outline

LEARNING OBJECTIVES

When you complete this module you should be able to

IDENTIFY OR DEFINE:

   Transportation modeling

   Facility location analysis

EXPLAIN OR BE ABLE TO USE:

   Northwest-corner rule

   Stepping-stone method

The problem facing rental companies like Avis, Hertz, and National is cross-country travel. Lots of it. Cars rented in New York end up in Chicago, cars from L.A. come to Philadelphia, and cars from Boston come to Miami. The scene is repeated in over 100 cities around the U.S. As a result, there are too many cars in some cities and too few in others. Operations managers have to decide how many of these rentals should be trucked (by costly auto carriers) from each city with excess capacity to each city that needs more rentals. The process requires quick action for the most economical routing, so rental car companies turn to transportation modeling.

Because location of a new factory, warehouse, or distribution center is a strategic issue with substantial cost implications, most companies consider and evaluate several locations. With a wide variety of objective and subjective factors to be considered, rational decisions are aided by a number of techniques. One of those techniques is transportation modeling.

The transportation models described in this module prove useful when considering alternative facility locations *within the framework of an existing distribution system.* Each new potential plant, warehouse, or distribution center will require a different allocation of shipments, depending on its own production and shipping costs and the costs of each existing facility. The choice of a new location depends on which will yield the minimum cost *for the entire system.*

TRANSPORTATION MODELING

Transportation modeling

An iterative procedure for solving problems that involve minimizing the cost of shipping products from a series of sources to a series of destinations.

Transportation modeling finds the least-cost means of shipping supplies from several origins to several destinations. *Origin points* (or *sources*) can be factories, warehouses, car rental agencies like Avis, or any other points from which goods are shipped. *Destinations* are any points that receive goods. To use the transportation model, we need to know the following:

1.   The origin points and the capacity or supply per period at each.
2.   The destination points and the demand per period at each.
3.   The cost of shipping one unit from each origin to each destination.

The transportation model is actually a class of the linear programming models discussed in Quantitative Module B. As it is for linear programming, software is available to solve transportation problems. To fully use such programs, though, you need to understand the assumptions that underlie the model. To illustrate one transportation problem, in this module we will look at a company called Arizona Plumbing, which makes, among other products, a full line of bathtubs. In our example, the firm must decide which of its factories should supply which of its warehouses. Relevant data for Arizona Plumbing are presented in Table C.1 and Figure C.1. Table C.1 shows, for example, that it costs Arizona Plumbing $5 to ship one bathtub from its Des Moines factory to its Albuquerque warehouse, $4 to Boston, and $3 to Cleveland. Likewise, we see in

TABLE C.1 ■

Transportation Costs per Bathtub for Arizona Plumbing

FROM \ TO	ALBUQUERQUE	BOSTON	CLEVELAND
DES MOINES	$5	$4	$3
EVANSVILLE	$8	$4	$3
FORT LAUDERDALE	$9	$7	$5

FIGURE C.1 ■

Transportation Problem

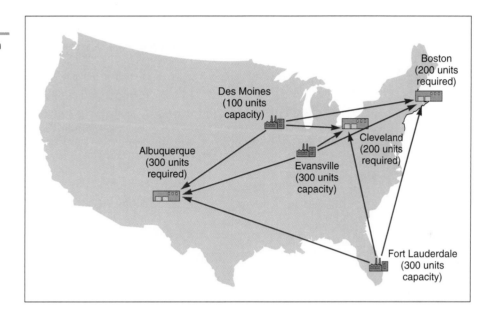

Figure C.1 that the 300 units required by Arizona Plumbing's Albuquerque warehouse might be shipped in various combinations from its Des Moines, Evansville, and Fort Lauderdale factories.

The first step in the modeling process is to set up a *transportation matrix*. Its purpose is to summarize all relevant data and to keep track of algorithm computations. Using the information displayed in Figure C.1 and Table C.1, we can construct a transportation matrix as shown in Figure C.2.

FIGURE C.2 ■

Transportation Matrix for Arizona Plumbing

From \ To	Albuquerque	Boston	Cleveland	Factory capacity	
Des Moines	$5	$4	$3	100	← Des Moines capacity constraint
Evansville	$8	$4	$3	300	← Cell representing a possible source-to-destination shipping assignment (Evansville to Cleveland)
Fort Lauderdale	$9	$7	$5	300	
Warehouse requirement	300	200	200	700	

Cost of shipping 1 unit from Fort Lauderdale factory to Boston warehouse

Cleveland warehouse demand

Total demand and total supply

DEVELOPING AN INITIAL SOLUTION

Once the data are arranged in tabular form, we must establish an initial feasible solution to the problem. A number of different methods have been developed for this step. We will discuss two of them, the northwest-corner rule and the intuitive lowest-cost method.

The Northwest-Corner Rule

Northwest-corner rule
A procedure in the transportation model where one starts at the upper left-hand cell of a table (the northwest corner) and systematically allocates units to shipping routes.

The **northwest-corner rule** requires that we start in the upper left-hand cell (or northwest corner) of the table and allocate units to shipping routes as follows:

1. Exhaust the supply (factory capacity) of each row (e.g., Des Moines: 100) before moving down to the next row.
2. Exhaust the (warehouse) requirements of each column (e.g., Albuquerque: 300) before moving to the next column on the right.
3. Check to ensure that all supplies and demands are met.

Example C1 applies the northwest-corner rule to our Arizona Plumbing problem.

Example C1

In Figure C.3 we use the northwest-corner rule to find an initial feasible solution to the Arizona Plumbing problem. To make our initial shipping assignments, we need five steps:

1. Assign 100 tubs from Des Moines to Albuquerque (exhausting Des Moines's supply).
2. Assign 200 tubs from Evansville to Albuquerque (exhausting Albuquerque's demand).
3. Assign 100 tubs from Evansville to Boston (exhausting Evansville's supply).
4. Assign 100 tubs from Fort Lauderdale to Boston (exhausting Boston's demand).
5. Assign 200 tubs from Fort Lauderdale to Cleveland (exhausting Cleveland's demand and Fort Lauderdale's supply).

The total cost of this shipping assignment is $4,200 (see Table C.2).

FIGURE C.3 ■ Northwest-Corner Solution to Arizona Plumbing Problem

From \ To	(A) Albuquerque	(B) Boston	(C) Cleveland	Factory capacity
(D) Des Moines	$5 100	$4	$3	100
(E) Evansville	$8 200	$4 100	$3	300
(F) Fort Lauderdale	$9	$7 (100)	$5 200	300
Warehouse requirement	300	200	200	700

Means that the firm is shipping 100 bathtubs from Fort Lauderdale to Boston

The northwest-corner rule is easy to use, but totally ignores costs.

TABLE C.2 ■ Computed Shipping Cost

ROUTE				
FROM	TO	TUBS SHIPPED	COST PER UNIT	TOTAL COST
D	A	100	$5	$ 500
E	A	200	8	1,600
E	B	100	4	400
F	B	100	7	700
F	C	200	5	$1,000
				Total: $4,200

The solution given is feasible because it satisfies all demand and supply constraints.

The Intuitive Lowest-Cost Method

Intuitive method
A cost-based approach to finding an initial solution to a transportation problem.

The **intuitive method** makes initial allocations based on lowest cost. This straightforward approach uses the following steps:

1. Identify the cell with the lowest cost. Break any ties for the lowest cost arbitrarily.
2. Allocate as many units as possible to that cell without exceeding the supply or demand. Then cross out that row or column (or both) that is exhausted by this assignment.
3. Find the cell with the lowest cost from the remaining (not crossed out) cells.
4. Repeat steps 2 and 3 until all units have been allocated.

Example C2

When we use the intuitive approach on the data in Figure C.2 (rather than the northwest-corner rule) for our starting position we obtain the solution seen in Figure C.4.

The total cost of this approach is = $3(100) + $3(100) + $4(200) + $9(300) = $4,100.
(D to C)   (E to C)   (E to B)   (F to A)

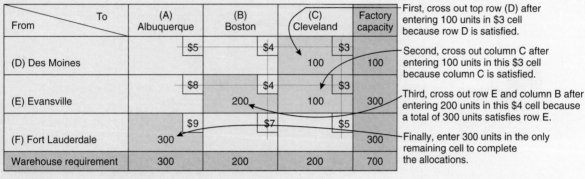

FIGURE C.4 ■ Intuitive Lowest-Cost Solution to Arizona Plumbing Problem

While the likelihood of a minimum cost solution *does* improve with the intuitive method, we would have been fortunate if the intuitive solution yielded the minimum cost. In this case, as in the northwest-corner solution, it did not. Because the northwest-corner and the intuitive lowest-cost approaches are meant only to provide us with a starting point, we often will have to employ an additional procedure to reach an *optimal* solution.

THE STEPPING-STONE METHOD

Stepping-stone method
An iterative technique for moving from an initial feasible solution to an optimal solution in the transportation method.

The **stepping-stone method** will help us move from an initial feasible solution to an optimal solution. It is used to evaluate the cost effectiveness of shipping goods via transportation routes not currently in the solution. When applying it, we test each unused cell, or square, in the transportation table by asking: "What would happen to total shipping costs if one unit of the product (for example, one bathtub) were tentatively shipped on an unused route?" We conduct the test as follows:

1. Select any unused square to evaluate.
2. Beginning at this square, trace a closed path back to the original square via squares that are currently being used (only horizontal and vertical moves are permissible). You may, however, step over either an empty or an occupied square.
3. Beginning with a plus (+) sign at the unused square, place alternate minus signs and plus signs on each corner square of the closed path just traced.
4. Calculate an improvement index by first adding the unit-cost figures found in each square containing a plus sign, and then by subtracting the unit costs in each square containing a minus sign.
5. Repeat steps 1 through 4 until you have calculated an improvement index for all unused squares. If all indices computed are *greater than or equal to zero*, you have reached an optimal solution. If not, the current solution can be improved further to decrease total shipping costs.

Example C3 illustrates how to use the stepping-stone method to move toward an optimal solution. We begin with the northwest-corner initial solution developed in Example 1.

Example C3

We can apply the stepping-stone method to the Arizona Plumbing data in Figure C.3 (see Example 1) to evaluate unused shipping routes. As you can see, the four currently unassigned routes are Des Moines to Boston, Des Moines to Cleveland, Evansville to Cleveland, and Fort Lauderdale to Albuquerque.

Steps 1 and 2. Beginning with the Des Moines–Boston route, first trace a closed path *using only currently occupied squares* (see Figure C.5). Place alternate plus signs and minus signs in the corners of this path. In the

Excel OM Data File
ModCExC3.xla

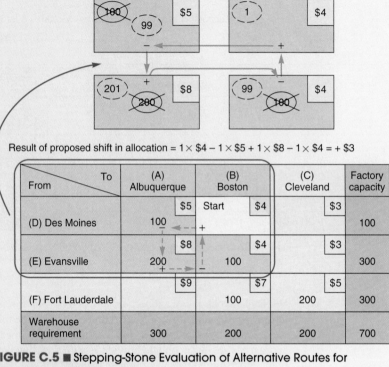

FIGURE C.5 ■ Stepping-Stone Evaluation of Alternative Routes for Arizona Plumbing

There is *only* one closed path that can be traced for each unused cell.

upper left square, for example, we place a minus sign because we have *subtracted* 1 unit from the original 100. Note that we can use only squares currently used for shipping to turn the corners of the route we are tracing. Hence, the path Des Moines–Boston to Des Moines–Albuquerque to Fort Lauderdale–Albuquerque to Fort Lauderdale–Boston to Des Moines–Boston would not be acceptable because the Fort Lauderdale–Albuquerque square is empty. It turns out that *only one closed route exists for each empty square.* Once this one closed path is identified, we can begin assigning plus and minus signs to these squares in the path.

Step 3. How do we decide which squares get plus signs and which squares get minus signs? The answer is simple. Because we are testing the cost effectiveness of the Des Moines–Boston shipping route, we try shipping 1 bathtub from Des Moines to Boston. This is 1 *more* unit than we *were* sending between the two cities, so place a plus sign in the box. However, if we ship 1 more unit than before from Des Moines to Boston, we end up sending 101 bathtubs out of the Des Moines factory. Because the Des Moines factory's capacity is only 100 units, we must ship 1 bathtub less from Des Moines to Albuquerque. This change prevents us from violating the capacity constraint.

To indicate that we have reduced the Des Moines–Albuquerque shipment, place a minus sign in its box. As you continue along the closed path, notice that we are no longer meeting our Albuquerque warehouse requirement for 300 units. In fact, if we reduce the Des Moines–Albuquerque shipment to 99 units, we must increase the Evansville–Albuquerque load by 1 unit, to 201 bathtubs. Therefore, place a plus sign in that box to indicate the increase. You may also observe that those squares in which we turn a corner (and only those squares) will have plus or minus signs.

Finally, note that if we assign 201 bathtubs to the Evansville–Albuquerque route, then we must reduce the Evansville–Boston route by 1 unit, to 99 bathtubs, in order to maintain the Evansville factory's capacity constraint of 300 units. To account for this reduction, we thus insert a minus sign in the Evansville–Boston box. By so doing we have balanced supply limitations among all four routes on the closed path.

Step 4. Compute an improvement index for the Des Moines–Boston route by adding unit costs in squares with plus signs and subtracting costs in squares with minus signs.

$$\text{Des Moines–Boston index} = \$4 - \$5 + \$8 - \$4 = +\$3$$

This means that for every bathtub shipped via the Des Moines–Boston route, total transportation costs will increase by \$3 over their current level.

From \ To	(A) Albuquerque	(B) Boston	(C) Cleveland	Factory capacity
(D) Des Moines	$5 — 100	$4	Start $3	100
(E) Evansville	$8 200+	$4 — 100	$3	300
(F) Fort Lauderdale	$9	$7 + 100	$5 — 200	300
Warehouse requirement	300	200	200	700

FIGURE C.6 ■ Testing Des Moines to Cleveland

Let us now examine the unused Des Moines–Cleveland route, which is slightly more difficult to trace with a closed path (see Figure C.6). Again, notice that we turn each corner along the path only at squares on the existing route. Our path, for example, can go through the Evansville–Cleveland box but cannot turn a corner; thus we cannot place a plus or minus sign there. We may use occupied squares only as stepping-stones:

$$\text{Des Moines–Cleveland index} = \$3 - \$5 + \$8 - \$4 + \$7 - \$5 = +\$4$$

Again, opening this route fails to lower our total shipping costs.

Two other routes can be evaluated in a similar fashion:

$$\text{Evansville–Cleveland index} = \$3 - \$4 + \$7 - \$5 = +\$1$$
$$(\text{Closed path} = \text{EC} - \text{EB} + \text{FB} - \text{FC})$$
$$\text{Fort Lauderdale–Albuquerque index} = \$9 - \$7 + \$4 - \$8 = -\$2$$
$$(\text{Closed path} = \text{FA} - \text{FB} + \text{EB} - \text{EA})$$

Because this last index is negative, we can realize cost savings by using the (currently unused) Fort Lauderdale–Albuquerque route.

Because the cities in the tables are in random order, crossing an unoccupied cell is fine.

In Example C3, we see that a better solution is indeed possible because we can calculate a negative improvement index on one of our unused routes. *Each negative index represents the amount by which total transportation costs could be decreased if one unit were shipped by the source-destination combination.* The next step, then, is to choose that route (unused square) with the *largest* negative improvement index. We can then ship the maximum allowable number of units on that route and reduce the total cost accordingly.

What is the maximum quantity that can be shipped on our new money-saving route? That quantity is found by referring to the closed path of plus signs and minus signs drawn for the route and then selecting the *smallest number found in the squares containing minus signs*. To obtain a new solution, we add this number to all squares on the closed path with plus signs and subtract it from all squares on the path to which we have assigned minus signs.

One iteration of the stepping-stone method is now complete. Again, of course, we must test to see if the solution is optimal or whether we can make any further improvements. We do this by evaluating each unused square, as previously described. Example C4 continues our effort to help Arizona Plumbing arrive at a final solution.

Example C4

To improve our Arizona Plumbing solution, we can use the improvement indices calculated in Example C3. We found in Example C3 that the largest (and only) negative index is on the Fort Lauderdale–Albuquerque route (which is the route depicted in Figure C.7).

The maximum quantity that may be shipped on the newly opened route, Fort Lauderdale–Albuquerque (FA), is the smallest number found in squares containing minus signs—in this case, 100 units. Why 100 units? Because the total cost decreases by $2 per unit shipped, we know we would like to ship the maximum

To From	(A) Albuquerque	(B) Boston	(C) Cleveland	Factory capacity
(D) Des Moines	100 $5	$4	$3	100
(E) Evansville	200 $8	100 $4	$3	300
(F) Fort Lauderdale	$9	100 $7	200 $5	300
Warehouse demand	300	200	200	700

FIGURE C.7 ■ Transportation Table: Route FA

possible number of units. Previous stepping-stone calculations indicate that each unit shipped over the FA route results in an increase of 1 unit shipped from Evansville (E) to Boston (B) and a decrease of 1 unit in amounts shipped both from F to B (now 100 units) and from E to A (now 200 units). Hence, the maximum we can ship over the FA route is 100 units. This solution results in zero units being shipped from F to B. Now we take the following four steps:

1. Add 100 units (to the zero currently being shipped) on route FA.
2. Subtract 100 from route FB, leaving zero in that square (though still balancing the row total for F).
3. Add 100 to route EB, yielding 200.
4. Finally, subtract 100 from route EA, leaving 100 units shipped.

Note that the new numbers still produce the correct row and column totals as required. The new solution is shown in Figure C.8.

To From	(A) Albuquerque	(B) Boston	(C) Cleveland	Factory capacity
(D) Des Moines	100 $5	$4	$3	100
(E) Evansville	100 $8	200 $4	$3	300
(F) Fort Lauderdale	100 $9	$7	200 $5	300
Warehouse demand	300	200	200	700

FIGURE C.8 ■ Solution at Next Iteration (Still Not Optimal)

Dummy sources
Artificial shipping source points created in the transportation method when total demand is greater than total supply in order to effect a supply equal to the excess of demand over supply.

Total shipping cost has been reduced by (100 units) × ($2 saved per unit) = $200 and is now $4,000. This cost figure, of course, can also be derived by multiplying the cost of shipping each unit by the number of units transported on its respective route, namely: 100($5) + 100($8) + 200($4) + 100($9) + 200($5) = $4,000.

Looking carefully at Figure C.8, however, you can see that it, too, is not yet optimal. Route EC (Evansville–Cleveland) has a negative cost improvement index. See if you can find the final solution for this route on your own. (Programs C.1 and C.2, at the end of this module, provide an Excel OM solution.)

Dummy destinations
Artificial destination points created in the transportation method when the total supply is greater than the total demand; they serve to equalize the total demand and supply.

SPECIAL ISSUES IN MODELING

Demand Not Equal to Supply

A common situation in real-world problems is the case in which total demand is not equal to total supply. We can easily handle these so-called *unbalanced* problems with the solution procedures that we have just discussed by introducing **dummy sources** or **dummy destinations.** If total supply is greater than total demand, we make demand exactly equal the surplus by creating a dummy destination. Conversely, if total demand is greater than total supply, we introduce a dummy source (factory) with a supply equal to the excess of demand. Because these units will not in fact be shipped, we assign cost coefficients of zero to each square on the dummy location. In each case, then, the cost is zero. Example C5 demonstrates the use of a dummy destination.

Example C5

Let's assume that Arizona Plumbing increases the production in its Des Moines factory to 250 bathtubs, thereby increasing supply over demand. To reformulate this unbalanced problem, we refer back to the data presented in Example C1 and present the new matrix in Figure C.9. First, we use the northwest-corner rule to find the initial feasible solution. Then, once the problem is balanced, we can proceed to the solution in the normal way.

Total cost = 250($5) + 50($8) + 200($4) + 50($3) + 150($5) + 150(0) = $3,350

**Excel OM Data File
ModCExC5.xla**

From \ To	(A) Albuquerque	(B) Boston	(C) Cleveland	Dummy	Factory capacity
(D) Des Moines	$5 250	$4	$3	0	250
(E) Evansville	$8 50	$4 200	$3 50	0	300
(F) Fort Lauderdale	$9	$7	$5 150	0 150	300
Warehouse requirement	300	200	200	150	850

New Des Moines capacity

FIGURE C.9 ■ Northwest-Corner Rule with Dummy

Degeneracy

Degeneracy
An occurrence in transportation models when too few squares or shipping routes are being used so that tracing a closed path for each unused square becomes impossible.

To apply the stepping-stone method to a transportation problem, we must observe a rule about the number of shipping routes being used: *The number of occupied squares in any solution (initial or later) must be equal to the number of rows in the table plus the number of columns minus 1.* Solutions that do not satisfy this rule are called degenerate.

Degeneracy occurs when too few squares or shipping routes are being used. As a result, it becomes impossible to trace a closed path for one or more unused squares. The Arizona Plumbing problem we just examined was not degenerate as it had 5 assigned routes (3 rows or factories + 3 columns or warehouses − 1).

When the navy in Thailand drafts a young man, he first reports to the induction center closest to his home. From one of 36 centers, he is transported by truck to one of four naval bases. The problem of deciding how many men should be assigned and transported from each center to each base is solved using the transportation model. Each base gets the number of recruits it needs, and costly extra trips are avoided.

To handle degenerate problems, we must artificially create an occupied cell: That is, we place a zero or a *very* small amount (representing a fake shipment) in one of the unused squares and *then treat that square as if it were occupied.* Remember that the chosen square must be in such a position as to allow all stepping-stone paths to be closed. We illustrate this procedure in Example C6.

Example C6

Check the unused squares to be sure that: *no. of rows + no. of columns* − 1 equals the number of filled squares.

Martin Shipping Company has three warehouses from which it supplies its three major retail customers in San Jose. Martin's shipping costs, warehouse supplies, and customer demands are presented in the transportation table in Figure C.10. To make the initial shipping assignments in that table, we apply the northwest-corner rule.

The initial solution is degenerate because it violates the rule that the number of used squares must equal the number of rows plus the number of columns minus 1. To correct the problem, we may place a zero in the unused square that permits evaluation of all empty cells. Some experimenting may be needed because not every cell will allow tracing a closed path for the remaining cells. Also, we want to avoid placing the 0 in a cell that has the negative sign in a closed path. No reallocation will be possible if we do this.

For this example, we try the empty square that represents the shipping route from Warehouse 2 to Customer 1. Now we can close all stepping-stone paths and compute improvement indices.

**Excel OM Data File
ModCExC6.xla**

From \ To	Customer 1	Customer 2	Customer 3	Warehouse supply
Warehouse 1	$8 100	$2	$6	100
Warehouse 2	$10 0	$9 100	$9 20	120
Warehouse 3	$7	$10	$7 80	80
Customer demand	100	100	100	300

FIGURE C.10 ■ Martin's Northwest-Corner Rule

SUMMARY

The transportation model, a form of linear programming, is used to help find the least-cost solutions to systemwide shipping problems. The northwest-corner method (which begins in the upper-left corner of the transportation table), or the intuitive lowest-cost method may be used for finding an initial feasible solution. The stepping-stone algorithm is then used for finding optimal solutions. Unbalanced problems are those in which the total demand and total supply are not equal. Degeneracy refers to the case in which the number of rows + the number of columns − 1 is not equal to the number of occupied squares. The transportation model approach is one of the four location models described earlier in Chapter 8. Additional solution techniques are presented on your CD in Tutorial 4.

KEY TERMS

Transportation modeling *(p. 540)*
Northwest-corner rule *(p. 542)*
Intuitive method *(p. 542)*
Stepping-stone method *(p. 543)*

Dummy sources *(p. 546)*
Dummy destinations *(p. 546)*
Degeneracy *(p. 547)*

USING EXCEL OM TO SOLVE TRANSPORTATION PROBLEMS

Excel OM's Transportation module uses Excel's built-in Solver routine to find optimal solutions to transportation problems. Program C.1 illustrates the input data (from Arizona Plumbing) and total cost formulas. To reach an optimal solution, we must go to Excel's *Tools* bar, request *Solver*, then select *Solve*. The output appears in Program C.2.

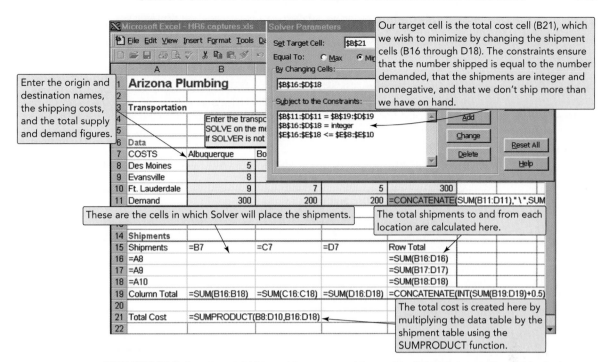

Enter the origin and destination names, the shipping costs, and the total supply and demand figures.

Our target cell is the total cost cell (B21), which we wish to minimize by changing the shipment cells (B16 through D18). The constraints ensure that the number shipped is equal to the number demanded, that the shipments are integer and nonnegative, and that we don't ship more than we have on hand.

These are the cells in which Solver will place the shipments.

The total shipments to and from each location are calculated here.

The total cost is created here by multiplying the data table by the shipment table using the SUMPRODUCT function.

PROGRAM C.1 ■ Excel OM Input Screen and Formulas, Using Arizona Plumbing Data

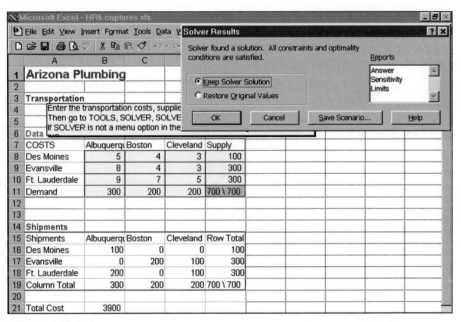

PROGRAM C.2 ■ Output from Excel OM with Optimal Solution to Arizona Plumbing Problem

USING POM FOR WINDOWS TO SOLVE TRANSPORTATION PROBLEMS

POM for Windows' Transportation module can solve both maximization and minimization problems by a variety of methods. Input data are the demand data, supply data, and unit shipping costs. See Appendix V for further details.

SOLVED PROBLEMS

Solved Problem C.1

Williams Auto Top Carriers currently maintains plants in Atlanta and Tulsa to supply auto top carriers to distribution centers in Los Angeles and New York. Because of expanding demand, Williams has decided to open a third plant and has narrowed the choice to one of two cities— New Orleans and Houston. Table C.3 provides pertinent production and distribution costs as well as plant capacities and distribution demands.

Which of the new locations, in combination with the existing plants and distribution centers, yields a lower cost for the firm?

TABLE C.3 ■ Production Costs, Distribution Costs, Plant Capabilities, and Market Demands for Williams Auto Top Carriers

| | TO DISTRIBUTION CENTERS | | | |
FROM PLANTS	LOS ANGELES	NEW YORK	NORMAL PRODUCTION	UNIT PRODUCTION COST
Existing plants				
Atlanta	$8	$5	600	$6
Tulsa	$4	$7	900	$5
Proposed locations				
New Orleans	$5	$6	500	$4 (anticipated)
Houston	$4	$6[a]	500	$3 (anticipated)
Forecast demand	800	1,200	2,000	

[a]Indicates distribution cost (shipping, handling, storage) will be $6 per carrier between Houston and New York.

SOLUTION

To answer this question, we must solve two transportation problems, one for each combination. We will recommend the location that yields a lower total cost of distribution and production in combination with the existing system.

We begin by setting up a transportation table that represents the opening of a third plant in New Orleans (see Figure C.11). Then we use the northwest-corner method to find an initial solution. The total cost of this first solution is $23,600. Note that the cost of each individual "plant-to-distribution-center" route is found by adding the distribution costs (in the body of Table C.3) to the respective unit production costs (in the right-hand column of Table C.3). Thus, the total production-plus-shipping cost of one auto top carrier from Atlanta to Los Angeles is $14 ($8 for shipping plus $6 for production).

Total cost = (600 units × $14) + (200 units × $9)
+ (700 units × $12) + (500 units × $10)
= $8,400 + $1,800 + $8,400 + $5,000
= $23,600

Is this initial solution (in Figure C.11) optimal? We can use the stepping-stone method to test it and compute improvement indices for unused routes:

Improvement index for Atlanta–New York route

= + $11 (Atlanta–New York) − $14 (Atlanta–L.A.)
+ $9 (Tulsa–L.A.) − $12 (Tulsa–New York)
= −$6

Improvement index for New Orleans–Los Angeles route

= + $9 (New Orleans–L.A.)
− $10 (New Orleans–New York)
+ $12 (Tulsa–New York)
− $9 (Tulsa–L.A.)
= $2

Because the firm can save $6 for every unit shipped from Atlanta to New York, it will want to improve the initial solution and send as many units as possible (600, in this case) on this currently unused

FIGURE C.11 ■

Williams Transportation Table for New Orleans

From \ To	Los Angeles	New York	Production capacity
Atlanta	$14 / 600	$11	600
Tulsa	$9 / 200	$12 / 700	900
New Orleans	$9	$10 / 500	500
Demand	800	1,200	2,000

FIGURE C.12 ■

Improved Transportation Table for Williams

From \ To	Los Angeles	New York	Production capacity
Atlanta	$14	$11 600	600
Tulsa	$9 800	$12 100	900
New Orleans	$9	$10 500	500
Demand	800	1,200	2,000

route (see Figure C.12). You may also want to confirm that the total cost is now $20,000, a savings of $3,600 over the initial solution.

Next, we must test the two unused routes to see if their improvement indices are also negative numbers:

Index for Atlanta–Los Angeles

$$= \$14 - \$11 + \$12 - \$9 = \$6$$

Index for New Orleans–Los Angeles

$$= \$9 - \$10 + \$12 - \$9 = \$2$$

Because both indices are greater than zero, we have already reached our optimal solution using the New Orleans plant. If Williams elects to open the New Orleans plant, the firm's total production and distribution cost will be $20,000.

This analysis, however, provides only half the answer to Williams's problem. The same procedure must still be followed to determine the minimum cost if the new plant is built in Houston. Determining this cost is left as a homework problem. You can help provide complete information and recommend a solution by solving Problem C.8 in your Student Lecture Guide.

Solved Problem C.2

In Solved Problem C.1, we examined the Williams Auto Top Carriers problem by using a transportation table. An alternative approach is to structure the same decision analysis using linear programming (LP), which we explained in detail in Quantitative Module B.

SOLUTION

Using the data in Figure C.11 (p. 550), we write the objective function and constraints as follows:

$$\text{Minimize total cost} = \$14X_{\text{Atl,LA}} + \$11X_{\text{Atl,NY}} + \$9X_{\text{Tul,LA}} + \$12X_{\text{Tul,NY}} + \$9X_{\text{NO,LA}} + \$10X_{\text{NO,NY}}$$

Subject to:

$X_{\text{Atl,LA}} + X_{\text{Atl,NY}}$	≤ 600	(production capacity at Atlanta)
$X_{\text{Tul,LA}} + X_{\text{Tul,NY}}$	≤ 900	(production capacity at Tulsa)
$X_{\text{NO,LA}} + X_{\text{NO,NY}}$	≤ 500	(production capacity at New Orleans)
$X_{\text{Atl,LA}} + X_{\text{Tul,LA}} + X_{\text{NO,LA}}$	≥ 800	(Los Angeles demand constraint)
$X_{\text{Atl,NY}} + X_{\text{Tul,NY}} + X_{\text{NO,NY}}$	≥ 1200	(New York demand constraint)

INTERNET AND STUDENT CD-ROM EXERCISES

Visit our home page or use your student CD-ROM to help with material in this module.

On Our Home Page, www.prenhall.com/heizer

- Self-Tests
- Practice Problems
- Internet Exercises
- Current Articles and Research
- Internet Homework Problems
- Internet Cases

On Your Student CD-ROM

- PowerPoint Lecture
- Practice Problems
- ExcelOM
- Excel OM Example Data Files

ADDITIONAL CASE STUDIES

Internet Case Studies: Visit our Web site at www.prenhall.com/heizer **for these free case studies:**

- **Consolidated Bottling (B)**: This case involves determining where to add bottling capacity.

- **Northwest General Hospital**: This case involves minimizing the time to distribute hot food in a hospital.

Waiting-Line Models

LEARNING OBJECTIVES

When you complete this module you should be able to

IDENTIFY OR DEFINE:

The assumptions of the four basic waiting-line models

DESCRIBE OR EXPLAIN:

How to apply waiting-line models

How to conduct an economic analysis of queues

Paris's EuroDisney, Tokyo's Disney Japan, and the U.S.'s Disney World and Disneyland all have one feature in common—long lines and seemingly endless waits. However, Disney is one of the world's leading companies in the scientific analysis of queuing theory. It analyzes queuing behaviors and can predict which rides will draw what length crowds. To keep visitors happy, Disney makes lines appear to be constantly moving forward, entertains people while they wait, and posts signs telling visitors how many minutes until they reach each ride.

Queuing theory
A body of knowledge about waiting lines.

Waiting line (queue)
Items or people in a line awaiting service.

The body of knowledge about waiting lines, often called **queuing theory**, is an important part of operations and a valuable tool for the operations manager. **Waiting lines** are a common situation—they may, for example, take the form of cars waiting for repair at a Midas Muffler Shop, copying jobs waiting to be completed at a Kinko's print shop, or vacationers waiting to enter Mr. Frogg's Wild Ride at Disney. Table D.1 lists just a few OM uses of waiting-line models.

Waiting-line models are useful in both manufacturing and service areas. Analysis of queues in terms of waiting-line length, average waiting time, and other factors helps us to understand service systems (such as bank teller stations), maintenance activities (that might repair broken machinery), and shop-floor control activities. As a matter of fact, patients waiting in a doctor's office and broken drill presses waiting in a repair facility have a lot in common from an OM perspective. Both use human and equipment resources to restore valuable production assets (people and machines) to good condition.

TABLE D.1 ■

Common Queuing Situations

SITUATION	ARRIVALS IN QUEUE	SERVICE PROCESS
Supermarket	Grocery shoppers	Checkout clerks at cash register
Highway toll booth	Automobiles	Collection of tolls at booth
Doctor's office	Patients	Treatment by doctors and nurses
Computer system	Programs to be run	Computer processes jobs
Telephone company	Callers	Switching equipment to forward calls
Bank	Customers	Transactions handled by teller
Machine maintenance	Broken machines	Repair people fix machines
Harbor	Ships and barges	Dock workers load and unload

CHARACTERISTICS OF A WAITING-LINE SYSTEM

In this section, we take a look at the three parts of a waiting-line, or queuing, system (as shown in Figure D.1):

1. Arrivals or inputs to the system. These have characteristics such as population size, behavior, and a statistical distribution.
2. Queue discipline, or the waiting line itself. Characteristics of the queue include whether it is limited or unlimited in length and the discipline of people or items in it.
3. The service facility. Its characteristics include its design and the statistical distribution of service times.

We now examine each of these three parts.

Arrival Characteristics

The input source that generates arrivals or customers for a service system has three major characteristics:

1. *Size* of the arrival population.
2. *Behavior* of arrivals.
3. *Pattern* of arrivals (statistical distribution).

Size of the Arrival (Source) Population   Population sizes are considered either unlimited (essentially infinite) or limited (finite). When the number of customers or arrivals on hand at any given moment is just a small portion of all potential arrivals, the arrival population is considered **unlimited**, or **infinite**. Examples of unlimited populations include cars arriving at a big city carwash, shoppers arriving at a supermarket, and students arriving to register for classes at a large university. Most queuing models assume such an infinite arrival population. An example of a **limited**, or **finite**, population is found in a copying shop that has, say, eight copying machines. Each of the copiers is a potential "customer" that might break down and require service.

Pattern of Arrivals at the System   Customers arrive at a service facility either according to some known schedule (for example, 1 patient every 15 minutes or 1 student every half hour) or else they arrive *randomly*. Arrivals are considered random when they are independent of one another and their occurrence cannot be predicted exactly. Frequently in queuing problems, the number of

Unlimited, or infinite, population
A queue in which a virtually unlimited number of people or items could request the services, or in which the number of customers or arrivals on hand at any given moment is a very small portion of potential arrivals.

Limited, or finite, population
A queue in which there are only a limited number of potential users of the service.

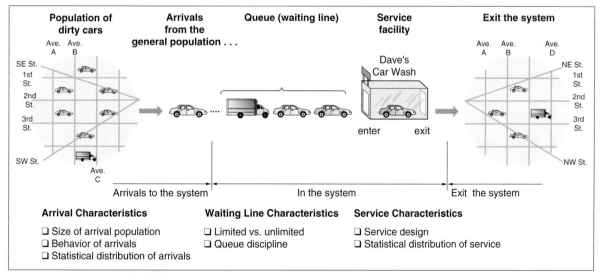

FIGURE D.1 ■ Three Parts of a Waiting Line, or Queuing System, at Dave's Car Wash

Poisson distribution

A discrete probability distribution that often describes the arrival rate in queuing theory.

arrivals per unit of time can be estimated by a probability distribution known as the **Poisson distribution**.[1] For any given arrival time (such as 2 customers per hour or 4 trucks per minute), a discrete Poisson distribution can be established by using the formula

$$P(x) = \frac{e^{-\lambda}\lambda^x}{x!} \quad \text{for } x = 0, 1, 2, 3, 4, \ldots \qquad (D\text{-}1)$$

where $P(x)$ = probability of x arrivals
           x = number of arrivals per unit of time
           λ = average arrival rate
           e = 2.7183 (which is the base of the natural logarithms)

With the help of the table in Appendix III, which gives the value of $e^{-\lambda}$ for use in the Poisson distribution, these values are easy to compute. Figure D.2 illustrates the Poisson distribution for $\lambda = 2$ and $\lambda = 4$. This means that if the average arrival rate is $\lambda = 2$ customers per hour, the probability of 0 customers arriving in any random hour is about 13%, probability of 1 customer is about 27%, 2 customers about 27%, 3 customers about 18%, 4 customers about 9%, and so on. The chances that 9 or more will arrive are virtually nil. Arrivals, of course, are not always Poisson distributed (they may follow some other distribution). Patterns, therefore, should be examined to make certain that they are well approximated by Poisson before that distribution is applied.

FIGURE D.2 ■

Two Examples of the Poisson Distribution for Arrival Times

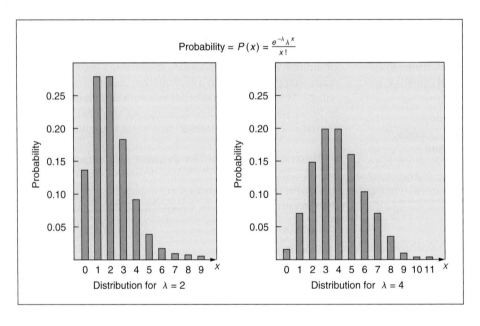

Probability = $P(x) = \frac{e^{-\lambda}\lambda^x}{x!}$

Distribution for $\lambda = 2$

Distribution for $\lambda = 4$

"The other line always moves faster."
          Etorre's Observation

"If you change lines, the one you just left will start to move faster than the one you are now in."
          O'Brien's Variation

Behavior of Arrivals   Most queuing models assume that an arriving customer is a patient customer. Patient customers are people or machines that wait in the queue until they are served and do not switch between lines. Unfortunately, life is complicated by the fact that people have been known to balk or to renege. Customers who *balk* refuse to join the waiting line because it is too long to suit their needs or interests. *Reneging* customers are those who enter the queue but then become impatient and leave without completing their transaction. Actually, both of these situations just serve to highlight the need for queuing theory and waiting-line analysis.

Waiting-Line Characteristics

The waiting line itself is the second component of a queuing system. The length of a line can be either limited or unlimited. A queue is *limited* when it cannot, either by law or because of physical restrictions, increase to an infinite length. A small barbershop, for example, will have only a limited

---

[1]When the arrival rates follow a Poisson process with mean arrival rate λ, the time between arrivals follows a negative exponential distribution with mean time between arrivals of $1/\lambda$. The negative exponential distribution, then, is also representative of a Poisson process, but describes the time between arrivals and specifies that these time intervals are completely random.

number of waiting chairs. Queuing models are treated in this module under an assumption of *unlimited* queue length. A queue is *unlimited* when its size is unrestricted, as in the case of the toll booth serving arriving automobiles.

A second waiting-line characteristic deals with *queue discipline*. This refers to the rule by which customers in the line are to receive service. Most systems use a queue discipline known as the **first-in, first-out (FIFO) rule**. In a hospital emergency room or an express checkout line at a supermarket, however, various assigned priorities may preempt FIFO. Patients who are critically injured will move ahead in treatment priority over patients with broken fingers or noses. Shoppers with fewer than 10 items may be allowed to enter the express checkout queue (but are *then* treated as first-come, first-served). Computer-programming runs also operate under priority scheduling. In most large companies, when computer-produced paychecks are due on a specific date, the payroll program gets highest priority.[2]

Service Characteristics

The third part of any queuing system are the service characteristics. Two basic properties are important: (1) design of the service system and (2) the distribution of service times.

Basic Queuing System Designs   Service systems are usually classified in terms of their number of channels (for example, number of servers) and number of phases (for example, number of service stops that must be made). A **single-channel queuing system**, with one server, is typified by the drive-in bank with only one open teller. If, on the other hand, the bank has several tellers on duty, with each customer waiting in one common line for the first available teller, then we would have a **multiple-channel queuing system**. Most banks today are multichannel service systems, as are most large barbershops, airline ticket counters, and post offices.

In a **single-phase system**, the customer receives service from only one station and then exits the system. A fast-food restaurant in which the person who takes your order also brings your food and takes your money is a single-phase system. So is a driver's license agency in which the person taking your application also grades your test and collects your license fee. However, say the restaurant requires you to place your order at one station, pay at a second, and pick up your food at a third. In this case, it is a **multiphase system**. Likewise, if the driver's license agency is large or busy, you will probably have to wait in one line to complete your application (the first service stop), queue again to have your test graded, and finally go to a third counter to pay your fee. To help you relate the concepts of channels and phases, Figure D.3 presents four possible channel configurations.

Service Time Distribution   Service patterns are like arrival patterns in that they may be either constant or random. If service time is constant, it takes the same amount of time to take care of each customer. This is the case in a machine-performed service operation such as an automatic car wash. More often, service times are randomly distributed. In many cases, we can assume that random service times are described by the **negative exponential probability distribution**.

Figure D.4 shows that if *service times* follow a negative exponential distribution, the probability of any very long service time is low. For example, when an average service time is 20 minutes (or three customers per hour), seldom if ever will a customer require more than 1.5 hours in the service facility. If the mean service time is 1 hour, the probability of spending more than 3 hours in service is quite low.

Measuring the Queue's Performance

Queuing models help managers make decisions that balance service costs with waiting-line costs. Queuing analysis can obtain many measures of a waiting-line system's performance, including the following:

1. Average time that each customer or object spends in the queue.
2. Average queue length.
3. Average time that each customer spends in the system (waiting time plus service time).
4. Average number of customers in the system.
5. Probability that the service facility will be idle.
6. Utilization factor for the system.
7. Probability of a specific number of customers in the system.

First-in, first-out (FIFO) rule
A queuing discipline in which the first customers in line receive the first service.

Single-channel queuing system
A service system with one line and one server.

Multiple-channel queuing system
A service system with one waiting line but with several servers.

Single-phase system
A system in which the customer receives service from only one station and then exits the system.

Multiphase system
A system in which the customer receives services from several stations before exiting the system.

Negative exponential probability distribution
A continuous probability distribution often used to describe the service time in a queuing system.

[2]The term *FIFS* (first-in, first-served) is often used in place of FIFO. Another discipline, LIFS (last-in, first-served) also called last-in, first-out (LIFO), is common when material is stacked or piled so that the items on top are used first.

FIGURE D.3 ■ Basic Queuing System Designs

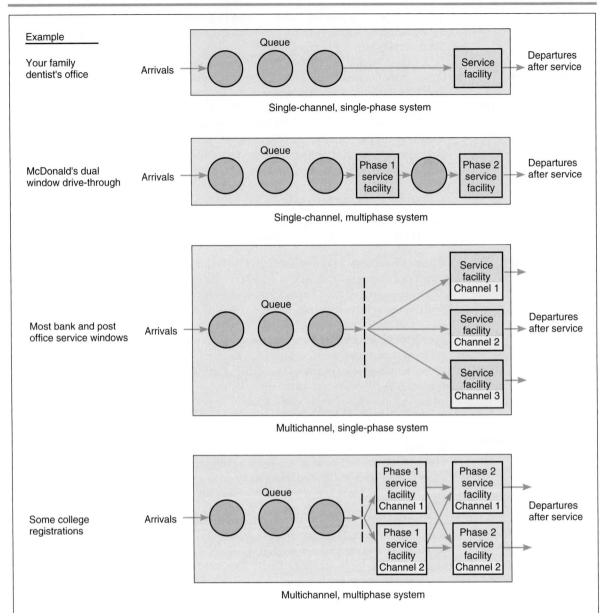

Example

Your family dentist's office — Single-channel, single-phase system

McDonald's dual window drive-through — Single-channel, multiphase system

Most bank and post office service windows — Multichannel, single-phase system

Some college registrations — Multichannel, multiphase system

FIGURE D.4 ■

Two Examples of the Negative Exponential Distribution for Service Times

Although Poisson and exponential distributions are commonly used to describe arrival rates and service times, you should not take them for granted; Normal and Erlang distributions, or others, may be more valid.

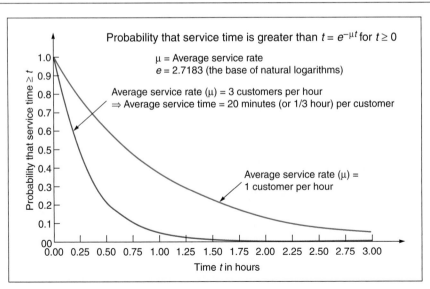

Probability that service time is greater than $t = e^{-\mu t}$ for $t \geq 0$

μ = Average service rate
e = 2.7183 (the base of natural logarithms)

Average service rate (μ) = 3 customers per hour
⇒ Average service time = 20 minutes (or 1/3 hour) per customer

Average service rate (μ) = 1 customer per hour

Probability that service time ≥ t

Time t in hours

OM IN ACTION

L.L. Bean Turns to Queuing Theory

L.L. Bean faced severe problems. It was the peak selling season, and the service level to incoming calls was simply unacceptable. Widely known as a high-quality outdoor goods retailer, about 65% of L.L. Bean's sales volume is generated through telephone orders via its toll-free service centers located in Maine.

Here is how bad the situation was: During certain periods, 80% of the calls received a busy signal, and those who did not often had to wait up to 10 minutes before speaking with a sales agent. L.L. Bean estimated it lost $10 million in profit because of the way it allocated telemarketing resources. Keeping customers waiting "in line" (on the phone) was costing $25,000 per day. On exceptionally busy days, the total orders lost because of queuing problems approached $500,000 in gross revenues.

Developing queuing models similar to those presented here, L.L. Bean was able to set the number of phone lines and the number of agents to have on duty for each half hour of every day of the season. Within a year, use of the model resulted in 24% more calls answered, 17% more orders taken, and 16% more revenues. The new system also meant 81% fewer abandoned callers and an 84% faster answering time. The percent of callers spending less than 20 seconds in the queue increased from 25% to 77%. Needless to say, queuing theory changed the way L.L. Bean thought about telecommunications.

Sources: Modern Material Handling (December 1997): S12–S14; and *Interfaces* (January/February 1991): 75–91 and (March/April 1993): 14–20.

QUEUING COSTS

As described in the *OM in Action* box, "L.L. Bean Turns to Queuing Theory," operations managers must recognize the trade-off that takes place between two costs: the cost of providing good service and the cost of customer or machine waiting time. Managers want queues that are short enough so that customers do not become unhappy and either leave without buying or buy but never return. However, managers may be willing to allow some waiting if it is balanced by a significant savings in service costs.

One means of evaluating a service facility is to look at total expected cost. Total cost is the sum of expected service costs plus expected waiting costs.

What does the long wait in the typical doctor's office tell you about the doctor's perception of your cost of waiting?

As you can see in Figure D.5, service costs increase as a firm attempts to raise its level of service. Managers in *some* service centers can vary capacity by having standby personnel and machines that they can assign to specific service stations in order to prevent or shorten excessively long lines. In grocery stores, for example, managers and stock clerks can open extra checkout counters. In banks and airport check-in points, part-time workers may be called in to help. As the level of service improves (that is, speeds up), however, the cost of time spent waiting in lines decreases. (Refer again to Figure D.5.) Waiting cost may reflect lost productivity of workers while tools or machines await repairs or may simply be an estimate of the cost of customers lost because of poor service and long queues. In some service systems (for example, an emergency ambulance service), the cost of long waiting lines may be intolerably high.

FIGURE D.5 ■

The Trade-Off between Waiting Costs and Service Costs

As we look at Figure D.5, it becomes clear that different organizations place different values on their customers' time (with many colleges and motor vehicle offices placing minimal cost on waiting time).

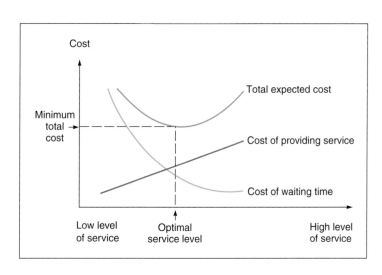

TABLE D.2 ■ Queuing Models Described in This Chapter

MODEL	NAME (TECHNICAL NAME IN PARENTHESES)	EXAMPLE	NUMBER OF CHANNELS	NUMBER OF PHASES	ARRIVAL RATE PATTERN	SERVICE TIME PATTERN	POPULATION SIZE	QUEUE DISCIPLINE
A	Simple system (M/M/1)	Information counter at department store	Single	Single	Poisson	Exponential	Unlimited	FIFO
B	Multichannel (M/M/S)	Airline ticket counter	Multi-channel	Single	Poisson	Exponential	Unlimited	FIFO
C	Constant service (M/D/1)	Automated car wash	Single	Single	Poisson	Constant	Unlimited	FIFO
D	Limited population (finite population)	Shop with only a dozen machines that might break	Single	Single	Poisson	Exponential	Limited	FIFO

THE VARIETY OF QUEUING MODELS

Visit a bank or a drive-through restaurant and time arrivals to see what kind of distribution (Poisson or other) they might reflect.

A wide variety of queuing models may be applied in operations management. We will introduce you to four of the most widely used models. These are outlined in Table D.2, and examples of each follow in the next few sections. More complex models are described in queuing theory textbooks[3] or can be developed through the use of simulation (which is the topic of module F). Note that all four queuing models listed in Table D.2 have three characteristics in common. They all assume:

1. Poisson distribution arrivals.
2. FIFO discipline.
3. A single-service phase.

In addition, they all describe service systems that operate under steady, ongoing conditions. This means that arrival and service rates remain stable during the analysis.

Model A: Single-Channel Queuing Model with Poisson Arrivals and Exponential Service Times

The most common case of queuing problems involves the *single-channel*, or single-server, waiting line. In this situation, arrivals form a single line to be serviced by a single station (see Figure D.3 on p. 558). We assume that the following conditions exist in this type of system:

1. Arrivals are served on a first-in, first-out (FIFO) basis, and every arrival waits to be served, regardless of the length of the line or queue.
2. Arrivals are independent of preceding arrivals, but the average number of arrivals (*arrival rate*) does not change over time.
3. Arrivals are described by a Poisson probability distribution and come from an infinite (or very, very large) population.
4. Service times vary from one customer to the next and are independent of one another, but their average rate is known.

What is the impact of equal service and arrival rates?

5. Service times occur according to the negative exponential probability distribution.
6. The service rate is faster than the arrival rate.

When these conditions are met, the series of equations shown in Table D.3 can be developed. Examples D1 and D2 illustrate how Model A (which in technical journals is known as the M/M/1 model) may be used.

---

[3]See, for example, N. U. Prabhu, *Foundations of Queuing Theory*, Klewer Academic Publishers (1997).

TABLE D.3 ■

Queuing Formulas for
Model A: Simple System,
Also Called M/M/1

λ = mean number of arrivals per time period

μ = mean number of people or items served per time period

L_s = average number of units (customers) in the system (waiting and being served)

$$= \frac{\lambda}{\mu - \lambda}$$

W_s = Average time a unit spends in the system (waiting time plus service time)

$$= \frac{1}{\mu - \lambda}$$

L_q = Average number of units waiting in the queue

$$= \frac{\lambda^2}{\mu(\mu - \lambda)}$$

W_q = Average time a unit spends waiting in the queue

$$= \frac{\lambda}{\mu(\mu - \lambda)}$$

ρ = Utilization factor for the system

$$= \frac{\lambda}{\mu}$$

P_0 = Probability of 0 units in the system (that is, the service unit is idle)

$$= 1 - \frac{\lambda}{\mu}$$

$P_{n>k}$ = Probability of more than k units in the system, where n is the number of units in the system

$$= \left(\frac{\lambda}{\mu}\right)^{k+1}$$

Example D1

**Excel OM Data File
ModDExD1.xla**

Active Model D.1

Example D1 is further
illustrated in Active
Model D.1 on your
CD-ROM.

Tom Jones, the mechanic at Golden Muffler Shop, is able to install new mufflers at an average rate of 3 per hour (or about 1 every 20 minutes), according to a negative exponential distribution. Customers seeking this service arrive at the shop on the average of 2 per hour, following a Poisson distribution. They are served on a first-in, first-out basis and come from a very large (almost infinite) population of possible buyers.

From this description, we are able to obtain the operating characteristics of Golden Muffler's queuing system:

$$\lambda = \text{2 cars arriving per hour}$$

$$\mu = \text{3 cars serviced per hour}$$

$$L_s = \frac{\lambda}{\mu - \lambda} = \frac{2}{3 - 2} = \frac{2}{1}$$

$$= \text{2 cars in the system, on average}$$

$$W_s = \frac{1}{\mu - \lambda} = \frac{1}{3 - 2} = 1$$

$$= \text{1-hour average waiting time in the system}$$

$$L_q = \frac{\lambda^2}{\mu(\mu - \lambda)} = \frac{2^2}{3(3 - 2)} = \frac{4}{3(1)} = \frac{4}{3}$$

$$= \text{1.33 cars waiting in line, on average}$$

$$W_q = \frac{\lambda}{\mu(\mu - \lambda)} = \frac{2}{3(3 - 2)} = \frac{2}{3} \text{ hour}$$

$$= \text{40-minute average waiting time per car}$$

$$\rho = \frac{\lambda}{\mu} = \frac{2}{3}$$

$$= \text{66.6\% of time mechanic is busy}$$

$$P_0 = 1 - \frac{\lambda}{\mu} = 1 - \frac{2}{3}$$

$$= \text{.33 probability there are 0 cars in the system}$$

Probability of More Than *k* Cars in the System

k	$P_{n>k} = (2/3)^{k+1}$
0	.667 ← Note that this is equal to $1 - P_0 = 1 - .33 = .667$.
1	.444
2	.296
3	.198 ← Implies that there is a 19.8% chance that more than 3 cars are in the system.
4	.132
5	.088
6	.058
7	.039

Once we have computed the operating characteristics of a queuing system, it is often important to do an economic analysis of their impact. Although the waiting-line model described above is valuable in predicting potential waiting times, queue lengths, idle times, and so on, it does not identify optimal decisions or consider cost factors. As we saw earlier, the solution to a queuing problem may require management to make a trade-off between the increased cost of providing better service and the decreased waiting costs derived from providing that service.

Example D2 examines the costs involved in Example D1.

Example D2

The owner of the Golden Muffler Shop estimates that the cost of customer waiting time, in terms of customer dissatisfaction and lost goodwill, is $10 per hour of time spent *waiting* in line. Because the average car has a 2/3-hour wait (W_q) and because there are approximately 16 cars serviced per day (2 arrivals per hour times 8 working hours per day), the total number of hours that customers spend waiting each day for mufflers to be installed is

$$\frac{2}{3}(16) = \frac{32}{3} = 10\frac{2}{3} \text{ hour}$$

Hence, in this case,

Although many parameters are computed for a queuing study, L_q and W_q are the two most important when it comes to actual cost analysis.

$$\text{Customer waiting-time cost} = \$10\left(10\frac{2}{3}\right) = \$107 \text{ per day}$$

The only other major cost that Golden's owner can identify in the queuing situation is the salary of Jones, the mechanic, who earns $7 per hour, or $56 per day. Thus:

$$\text{Total expected costs} = \$107 + \$56$$
$$= \$163 \text{ per day}$$

This approach will be useful in Solved Problem D.2 on page 570.

A $P_{n>3}$ of .0625 means that the chance of having more than 3 customers in an airport check-in line at a certain time of day is 1 in 16. If this British Airways office can live with 4 or more passengers in line about 6% of the time, one service agent will suffice. If not, more check-in positions and staff will have to be added.

Model B: Multiple-Channel Queuing Model

Now let's turn to a multiple-channel queuing system in which two or more servers or channels are available to handle arriving customers. We still assume that customers awaiting service form one single line and then proceed to the first available server. Multichannel, single-phase waiting lines are found in many banks today: A common line is formed, and the customer at the head of the line proceeds to the first free teller. (Refer to Figure D.3 on p. 558 for a typical multichannel configuration.)

The multiple-channel system presented in Example D3 again assumes that arrivals follow a Poisson probability distribution and that service times are exponentially distributed. Service is first-come, first-served, and all servers are assumed to perform at the same rate. Other assumptions listed earlier for the single-channel model also apply.

The queuing equations for Model B (which also has the technical name of M/M/S) are shown in Table D.4. These equations are obviously more complex than those used in the single-channel model; yet they are used in exactly the same fashion and provide the same type of information as the simpler model. (*Note:* The POM for Windows and Excel OM software described later in this chapter can prove very useful in solving multiple-channel, as well as other, queuing problems.)

TABLE D.4 ■

Queuing Formulas for Model B: Multichannel System, Also Called M/M/S

M = number of channels open
λ = average arrival rate
μ = average service rate at each channel

The probability that there are zero people or units in the system is

$$P_0 = \frac{1}{\left[\sum_{n=0}^{M-1} \frac{1}{n!}\left(\frac{\lambda}{\mu}\right)^n\right] + \frac{1}{M!}\left(\frac{\lambda}{\mu}\right)^M \frac{M\mu}{M\mu - \lambda}} \quad \text{for } M\mu > \lambda$$

The average number of people or units in the system is

$$L_s = \frac{\lambda\mu(\lambda/\mu)^M}{(M-1)!(M\mu - \lambda)^2} P_0 + \frac{\lambda}{\mu}$$

The average time a unit spends in the waiting line or being serviced (namely, in the system) is

$$W_s = \frac{\mu(\lambda/\mu)^M}{(M-1)!(M\mu - \lambda)^2} P_0 + \frac{1}{\mu} = \frac{L_s}{\lambda}$$

The average number of people or units in line waiting for service is

$$L_q = L_s - \frac{\lambda}{\mu}$$

The average time a person or unit spends in the queue waiting for service is

$$W_q = W_s - \frac{1}{\mu} = \frac{L_q}{\lambda}$$

Example D3

**Excel OM Data File
ModDExD3.xla**

The Golden Muffler Shop has decided to open a second garage bay and hire a second mechanic to handle installations. Customers, who arrive at the rate of about $\lambda = 2$ per hour, will wait in a single line until 1 of the 2 mechanics is free. Each mechanic installs mufflers at the rate of about $\mu = 3$ per hour.

To find out how this system compares to the old single-channel waiting-line system, we will compute several operating characteristics for the $M = 2$ channel system and compare the results with those found in Example D1:

$$P_0 = \frac{1}{\left[\sum_{n=0}^{1} \frac{1}{n!}\left(\frac{2}{3}\right)^n\right] + \frac{1}{2!}\left(\frac{2}{3}\right)^2 \frac{2(3)}{2(3) - 2}}$$

$$= \frac{1}{1 + \frac{2}{3} + \frac{1}{2}\left(\frac{4}{9}\right)\left(\frac{6}{6-2}\right)} = \frac{1}{1 + \frac{2}{3} + \frac{1}{3}} = \frac{1}{2}$$

$$= .5 \text{ probability of zero cars in the system}$$

Active Model D.2

Examples D2 and D3 are further illustrated in Active Model D.2 on the CD-ROM and in the Exercise located in your Student Lecture Guide.

Then,

$$L_s = \frac{(2)(3)(2/3)^2}{1![2(3)-2]^2}\left(\frac{1}{2}\right) + \frac{2}{3} = \frac{8/3}{16}\left(\frac{1}{2}\right) + \frac{2}{3} = \frac{3}{4}$$

$\quad\quad = .75$ average number of cars in the system

$$W_s = \frac{L_s}{\lambda} = \frac{3/4}{2} = \frac{3}{8} \text{ hour}$$

$\quad\quad = 22.5$ minutes average time a car spends in the system

$$L_q = L_s - \frac{\lambda}{\mu} = \frac{3}{4} - \frac{2}{3} = \frac{1}{12}$$

$\quad\quad = 0.83$ average number of cars in the queue

$$W_q = \frac{L_q}{\lambda} = \frac{.083}{2} = .0415 \text{ hour}$$

$\quad\quad = 2.5$ minutes average time a car spends in the queue

We can summarize the characteristics of this 2-channel model and compare them to those of the single-channel model as follows:

	SINGLE CHANNEL	TWO CHANNELS
P_0	.33	.5
L_s	2 cars	.75 car
W_s	60 minutes	22.5 minutes
L_q	1.33 cars	.083 car
W_q	40 minutes	2.5 minutes

The increased service has a dramatic effect on almost all characteristics. In particular, time spent waiting in line drops from 40 minutes to only 2.5 minutes.

Queues exist not only in every industry but also around the world. Here, the Moscow McDonald's on Pushkin Square, four blocks from the Kremlin, boasts 700 indoor and 200 outdoor seats, employs 800 Russian citizens, and generates annual revenues of $80 million. In spite of its size and volume, it still has queues and has had to develop a strategy for dealing with them.

Model C: Constant Service Time Model

Some service systems have constant, instead of exponentially distributed, service times. When customers or equipment are processed according to a fixed cycle, as in the case of an automatic car wash or an amusement park ride, constant service times are appropriate. Because constant rates are certain, the values for L_q, W_q, L_s, and W_s are always less than they would be in Model A, which has variable service rates. As a matter of fact, both the average queue length and the average waiting time in the queue are halved with Model C. Constant service model formulas are given in Table D.5. Model C also has the technical name of M/D/1 in the literature of queuing theory.

TABLE D.5 ■

Queuing Formulas for Model C: Constant Service, Also Called M/D/1

Average length of queue: $L_q = \dfrac{\lambda^2}{2\mu(\mu - \lambda)}$

Average waiting time in queue: $W_q = \dfrac{\lambda}{2\mu(\mu - \lambda)}$

Average number of customers in system: $L_s = L_q + \dfrac{\lambda}{\mu}$

Average waiting time in system: $W_s = W_q + \dfrac{1}{\mu}$

Example D4 gives a constant-service time analysis.

Example D4

**Excel OM Data File
ModDExD4.xla**

Active Model D.3

Example D4 is further illustrated in Active Model D.3 on your CD-ROM.

Garcia-Golding Recycling, Inc., collects and compacts aluminum cans and glass bottles in New York City. Its truck drivers currently wait an average of 15 minutes before emptying their loads for recycling. The cost of driver and truck time while they are in queues is valued at $60 per hour. A new automated compactor can be purchased to process truckloads at a constant rate of 12 trucks per hour (that is, 5 minutes per truck). Trucks arrive according to a Poisson distribution at an average rate of 8 per hour. If the new compactor is put in use, the cost will be amortized at a rate of $3 per truck unloaded. The firm hires a summer college intern, who conducts the following analysis to evaluate the costs versus benefits of the purchase:

Current waiting cost/trip = (1/4 hr. waiting now) ($60/hr.cost) = $15/trip

New system: λ = 8 trucks/hr. arriving μ = 12 trucks/hr. served

Average waiting time in queue = $W_q = \dfrac{\lambda}{2\mu(\mu - \lambda)} = \dfrac{8}{2(12)(12 - 8)} = \dfrac{1}{12}$ hr.

Waiting cost/trip with new compactor = (1/12 hr. wait) ($60/hr. cost) = $ 5/trip

Savings with new equipment = $15 (current system) – $5 (new system) = $10/trip

Cost of new equipment amortized: = $ 3/trip

Net savings: $ 7/trip

Model D: Limited Population Model

When there is a limited population of potential customers for a service facility, we must consider a different queuing model. This model would be used, for example, if we were considering equipment repairs in a factory that has 5 machines, if we were in charge of maintenance for a fleet of 10 commuter airplanes, or if we ran a hospital ward that has 20 beds. The limited population model allows any number of repair people (servers) to be considered.

This model differs from the three earlier queuing models because there is now a *dependent* relationship between the length of the queue and the arrival rate. Let's illustrate the extreme situation: If your factory had five machines and all were broken and awaiting repair, the arrival rate would drop to zero. In general, then, as the *waiting line* becomes longer in the limited population model, the *arrival rate* of customers or machines drops.

Table D.6 displays the queuing formulas for the limited population model. Note that they employ a different notation than Models A, B, and C. To simplify what can become time-consuming calculations, finite queuing tables have been developed that determine *D* and *F*. *D* represents the probability that a machine needing repair will have to wait in line. *F* is a waiting-time efficiency factor. *D* and *F* are needed to compute most of the other finite model formulas.

TABLE D.6 ■

Queuing Formulas and
Notation for Model D:
Limited Population
Formulas

Service factor: $X = \dfrac{T}{T + U}$

Average number waiting: $L = N(1 - F)$

Average waiting time: $W = \dfrac{L(T + U)}{N - L} = \dfrac{T(1 - F)}{XF}$

Average number running: $J = NF(1 - X)$

Average number being serviced: $H = FNX$

Number of population: $N = J + L + H$

NOTATION

D = probability that a unit will have to wait in queue
F = efficiency factor
H = average number of units being served
J = average number of units not in queue or in
  service bay
L = average number of units waiting for service
M = number of service channels

N = number of potential customers
T = average service time
U = average time between unit service requirements
W = average time a unit waits in line
X = service factor

Source: L. G. Peck and R. N. Hazelwood, *Finite Queuing Tables* (New York: John Wiley, 1958).

A small part of the published finite queuing tables is illustrated in this section. Table D.7 provides data for a population of $N = 5$.[4]

To use Table D.7, we follow four steps:

1. Compute X (the service factor, where $X = T/(T + U)$).
2. Find the value of X in the table and then find the line for M (where M is the number of service channels).
3. Note the corresponding values for D and F.
4. Compute L, W, J, H, or whichever are needed to measure the service system's performance.

Example D5 illustrates these steps.

Example D5

**Excel OM Data File
ModDExD5.xla**

Past records indicate that each of the 5 laser computer printers at the U.S. Department of Energy, in Washington, DC, needs repair after about 20 hours of use. Breakdowns have been determined to be Poisson-distributed. The one technician on duty can service a printer in an average of 2 hours, following an exponential distribution. Printer downtime costs $120 per hour. Technicians are paid $25 per hour. Should the DOE hire a second technician?

Assuming the second technician can repair a printer in an average of 2 hours, we can use Table D.7 (because there are $N = 5$ machines in this limited population) to compare the costs of 1 versus 2 technicians.

1. First, we note that $T = 2$ hours and $U = 20$ hours.
2. Then, $X = \dfrac{T}{T + U} = \dfrac{2}{2 + 20} = \dfrac{2}{22} = .091$ (close to .090).
3. For $M = 1$ server, $D = .350$ and $F = .960$.
4. For $M = 2$ servers, $D = .044$ and $F = .998$.
5. The average number of printers *working* is $J = NF(1 - X)$.
   For $M = 1$, this is $J = (5)(.960)(1 - .091) = 4.36$.
   For $M = 2$, it is $J = (5)(.998)(1 - .091) = 4.54$.
6. The cost analysis follows:

NUMBER OF TECHNICIANS	AVERAGE NUMBER PRINTERS DOWN ($N - J$)	AVERAGE COST/HR. FOR DOWNTIME ($N - J$)($120/HR.)	COST/HR. FOR TECHNICIANS (AT $25/HR.)	TOTAL COST/HR.
1	.64	$76.80	$25.00	$101.80
2	.46	$55.20	$50.00	$105.20

This analysis suggests that having only one technician on duty will save a few dollars per hour ($105.20 − $101.80 = $3.40).

[4]Limited, or finite, queuing tables are available to handle arrival populations of up to 250. Although there is no definite number that we can use as a dividing point between limited and unlimited populations, the general rule of thumb is this: If the number in the queue is a significant proportion of the arrival population, use a limited population queuing model. For a complete set of N values, see L. G. Peck and R. N. Hazelwood, *Finite Queuing Tables* (New York: John Wiley, 1958).

TABLE D.7 ■ Finite Queuing Tables for a Population of N = 5

X	M	D	F	X	M	D	F	X	M	D	F	X	M	D	F	X	M	D	F	X	M	D	F
.012	1	.048	.999		1	.404	.945		1	.689	.801	.330	4	.012	.999		3	.359	.927	.650	4	.179	.972
.019	1	.076	.998	.110	2	.065	.996	.210	3	.032	.998		3	.112	.986	.520	2	.779	.728		3	.588	.850
.025	1	.100	.997		1	.421	.939		2	.211	.973		2	.442	.904		1	.988	.384		2	.918	.608
.030	1	.120	.996	.115	2	.071	.995		1	.713	.783		1	.902	.583	.540	4	.085	.989		1	.998	.308
.034	1	.135	.995		1	.439	.933	.220	3	.036	.997	.340	4	.013	.999		3	.392	.917	.700	4	.240	.960
.036	1	.143	.994	.120	2	.076	.995		2	.229	.969		3	.121	.985		2	.806	.708		3	.678	.815
.040	1	.159	.993		1	.456	.927		1	.735	.765		2	.462	.896		1	.991	.370		2	.950	.568
.042	1	.167	.992	.125	2	.082	.994	.230	3	.041	.997		1	.911	.569	.560	4	.098	.986		1	.999	.286
.044	1	.175	.991		1	.473	.920		2	.247	.965	.360	4	.017	.998		3	.426	.906	.750	4	.316	.944
.046	1	.183	.990	.130	2	.089	.933		1	.756	.747		3	.141	.981		2	.831	.689		3	.763	.777
.050	1	.198	.989		1	.489	.914	.240	3	.046	.996		2	.501	.880		1	.993	.357		2	.972	.532
.052	1	.206	.988	.135	2	.095	.993		2	.265	.960		1	.927	.542	.580	4	.113	.984	.800	4	.410	.924
.054	1	.214	.987		1	.505	.907		1	.775	.730	.380	4	.021	.998		3	.461	.895		3	.841	.739
.056	2	.018	.999	.140	2	.102	.992	.250	3	.052	.995		3	.163	.976		2	.854	.670		2	.987	.500
	1	.222	.985		1	.521	.900		2	.284	.955		2	.540	.863		1	.994	.345	.850	4	.522	.900
.058	2	.019	.999	.145	3	.011	.999		1	.794	.712		1	.941	.516	.600	4	.130	.981		3	.907	.702
	1	.229	.984		2	.109	.991	.260	3	.058	.994	.400	4	.026	.977		3	.497	.883		2	.995	.470
.060	2	.020	.999		1	.537	.892		2	.303	.950		3	.186	.972		2	.875	.652	.900	4	.656	.871
	1	.237	.983	.150	3	.012	.999		1	.811	.695		2	.579	.845		1	.996	.333		3	.957	.666
.062	2	.022	.999		2	.115	.990	.270	3	.064	.994		1	.952	.493						2	.998	.444
	1	.245	.982		1	.553	.885		2	.323	.944	.420	4	.031	.997					.950	4	.815	.838
.064	2	.023	.999	.155	3	.013	.999		1	.827	.677		3	.211	.966						3	.989	.631
	1	.253	.981		2	.123	.989	.280	3	.071	.993		2	.616	.826								
.066	2	.024	.999		1	.568	.877		2	.342	.938		1	.961	.471								
	1	.260	.979	.160	3	.015	.999		1	.842	.661	.440	4	.037	.996								
.068	2	.026	.999		2	.130	.988	.290	4	.007	.999		3	.238	.960								
	1	.268	.978		1	.582	.869		3	.079	.992		2	.652	.807								
.070	2	.027	.999	.165	3	.016	.999		2	.362	.932		1	.969	.451								
	1	.275	.977		2	.137	.987		1	.856	.644	.460	4	.045	.995								
.075	2	.031	.999		1	.597	.861	.300	4	.008	.999		3	.266	.953								
	1	.294	.973	.170	3	.017	.999		3	.086	.990		2	.686	.787								
.080	2	.035	.998		2	.145	.985		2	.382	.926		1	.975	.432								
	1	.313	.969		1	.611	.853		1	.869	.628	.480	4	.053	.994								
.085	2	.040	.998	.180	3	.021	.999	.310	4	.009	.999		3	.296	.945								
	1	.332	.965		2	.161	.983		3	.094	.989		2	.719	.767								
.090	2	.044	.998		1	.638	.836		2	.402	.919		1	.980	.415								
	1	.350	.960	.190	3	.024	.998		1	.881	.613	.500	4	.063	.992								
.095	2	.049	.997		2	.117	.980	.320	4	.010	.999		3	.327	.936								
	1	.368	.955		1	.665	.819		3	.103	.988		2	.750	.748								
.100	2	.054	.997	.200	3	.028	.998		2	.422	.912		1	.985	.399								
.100	1	.386	.950	.200	2	.194	.976		1	.892	.597	.520	4	.073	.991								
.105	2	.059	.997																				

Source: From L. G. Peck and R. N. Hazelwood, *Finite Queuing Tables* (New York: John Wiley, 1958):4. © 1985, John Wiley & Sons, Inc.

OTHER QUEUING APPROACHES

Many practical waiting-line problems that occur in service systems have characteristics like those of the four mathematical models described above. Often, however, *variations* of these specific cases are present in an analysis. Service times in an automobile repair shop, for example, tend to follow the normal probability distribution instead of the exponential. A college registration system in which seniors have first choice of courses and hours over other students is an example of a first-come, first-served model with a preemptive priority queue discipline. A physical examination for military recruits is an example of a multiphase system, one that differs from the single-phase models discussed earlier in this module. A recruit first lines up to have blood drawn at one station, then waits for an eye exam at the next station, talks to a psychiatrist at the third, and is examined by a doctor for medical problems at the fourth. At each phase, the recruit must enter another queue and wait his or her turn. Many models, some very complex, have been developed to deal with situations such as these. One of these is described in the *OM in Action* box, "Shortening Arraignment Times in New York's Police Department."

OM IN ACTION

Shortening Arraignment Times in New York's Police Department

At one time, people arrested in New York City averaged a 40-hour wait (some more than 70 hours) prior to arraignment. They were kept in crowded, noisy, stressful, unhealthy, and often dangerous holding facilities and, in effect, denied speedy court appearances. The New York Supreme Court has since ruled that the city must attempt to arraign within 24 hours or release the prisoner.

The arrest-to-arraignment (ATA) process, which has the general characteristics of a large queuing system, involves these steps: arrest of suspected criminal, transport to a police precinct, search/fingerprinting, paperwork for arrest, transport to a central booking facility, additional paperwork, processing of fingerprints, a bail interview, transport to either the courthouse or an outlying precinct, checks for a criminal record, and finally, an assistant district attorney drawing up a complaint document.

To solve the complex problem of improving this system, the city hired Queues Enforth Development, Inc., a Massachusetts consulting firm. The firm's Monte Carlo simulation of the ATA process included single- and multiple-server queuing models. The modeling approach successfully reduced the average ATA time to 24 hours and resulted in an annual cost savings of $9.5 million for the city and state.

Source: R. C. Larson, M. F. Cahn, and M. C. Shell, "Improving the New York Arrest-to-Arraignment System," *Interfaces* 23, no. 1 (January–February 1993):76–96.

SUMMARY

Queues are an important part of the world of operations management. In this module, we describe several common queuing systems and present mathematical models for analyzing them.

The most widely used queuing models include Model A, the basic single-channel, single-phase system with Poisson arrivals and exponential service times; Model B, the multichannel equivalent of Model A; Model C, a constant service rate model; and Model D, a limited population system. All four models allow for Poisson arrivals, first-in, first-out service, and a single-service phase. Typical operating characteristics we examine include average time spent waiting in the queue and system, average number of customers in the queue and system, idle time, and utilization rate.

A variety of queuing models exists for which all of the assumptions of the traditional models need not be met. In these cases, we use more complex mathematical models or turn to a technique called simulation. The application of simulation to problems of queuing systems is addressed in Quantitative Module F.

KEY TERMS

Queuing theory *(p. 554)*
Waiting line (queue) *(p. 554)*
Unlimited, or infinite, population *(p. 555)*
Limited, or finite, population *(p. 555)*
Poisson distribution *(p. 556)*
First-in, first-out (FIFO) rule *(p. 557)*

Single-channel queuing system *(p. 557)*
Multiple-channel queuing system *(p. 557)*
Single-phase system *(p. 557)*
Multiphase system *(p. 557)*
Negative exponential probability distribution *(p. 557)*

USING EXCEL OM FOR QUEUING

Excel OM's Waiting-Line program handles all four of the models developed in this module. Program D.1 illustrates our first model, the M/M/1 system, using the data from Example D1.

USING POM FOR WINDOWS FOR QUEUING

There are several POM for Windows queuing models from which to select in that program's Waiting-Line module. The program can include an economic analysis of cost data, and, as an option, you may display probabilities of various numbers of people/items in the system. See Appendix V for further details.

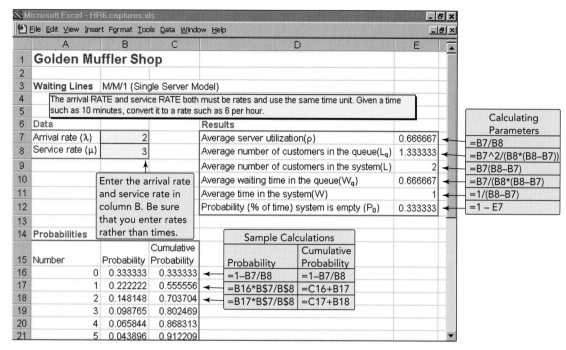

PROGRAM D.1 ■ Using Excel OM for Queuing

Example D1's (Golden Muffler Shop) data are illustrated in the M/M/1 model.

SOLVED PROBLEMS

Solved Problem D.1

Sid Das Brick Distributors currently employs 1 worker whose job is to load bricks on outgoing company trucks. An average of 24 trucks per day, or 3 per hour, arrive at the loading platform, according to a Poisson distribution. The worker loads them at a rate of 4 trucks per hour, following approximately the exponential distribution in his service times.

Das believes that adding a second brick loader will substantially improve the firm's productivity. He estimates that a two-person crew at the loading gate will double the loading rate from 4 trucks per hour to 8 trucks per hour. Analyze the effect on the queue of such a change and compare the results to those achieved with one worker. What is the probability that there will be more than 3 trucks either being loaded or waiting?

SOLUTION

	NUMBER OF BRICK LOADERS	
	1	2
Truck arrival rate (λ)	3/hr.	3/hr.
Loading rate (μ)	4/hr.	8/hr.
Average number in system (L_s)	3 trucks	.6 truck
Average time in system (W_s)	1 hr.	.2 hr.
Average number in queue (L_q)	2.25 trucks	.225 truck
Average time in queue (W_q)	.75 hr.	.075 hr.
Utilization rate (ρ)	.75	.375
Probability system empty (P_0)	.25	.625

Probability of More Than k Trucks in System		
	PROBABILITY $n > k$	
k	ONE LOADER	TWO LOADERS
0	.75	.375
1	.56	.141
2	.42	.053
3	.32	.020

These results indicate that when only one loader is employed, the average truck must wait three-quarters of an hour before it is loaded. Furthermore, there is an average of 2.25 trucks waiting in line to be loaded. This situation may be unacceptable to management. Note also the decline in queue size after the addition of a second loader.

Solved Problem D.2

Truck drivers working for Sid Das (see Solved Problem D.1) earn an average of $10 per hour. Brick loaders receive about $6 per hour. Truck drivers waiting *in the queue or at the loading platform* are drawing a salary but are productively idle and unable to generate revenue during that time. What would be the *hourly* cost savings to the firm if it employed 2 loaders instead of 1?

Referring to the data in Solved Problem D.1, we note that the average number of trucks *in the system* is 3 when there is only 1 loader and .6 when there are 2 loaders.

SOLUTION

	NUMBER OF LOADERS	
	1	**2**
Truck driver idle time costs [(average number of trucks) × (hourly rate)] = (3)($10) = $30		$ 6 = (.6)($10)
Loading costs	6	12 = (2)($6)
Total expected cost per hour	$36	$18

The firm will save $18 per hour by adding a second loader.

Solved Problem D.3

Sid Das is considering building a second platform or gate to speed the process of loading trucks. This system, he thinks, will be even more efficient than simply hiring another loader to help out on the first platform (as in Solved Problem D.1).

Assume that workers at each platform will be able to load 4 trucks per hour each and that trucks will continue to arrive at the rate of 3 per hour. Then apply the appropriate equations to find the waiting line's new operating conditions. Is this new approach indeed speedier than the other two that Das has considered?

SOLUTION

$$P_0 = \frac{1}{\left[\displaystyle\sum_{n=0}^{1}\frac{1}{n!}\left(\frac{3}{4}\right)^n\right] + \frac{1}{2!}\left(\frac{3}{4}\right)^2 \frac{2(4)}{2(4)-3}}$$

$$= \frac{1}{1 + \dfrac{3}{4} + \dfrac{1}{2}\left(\dfrac{3}{4}\right)^2\left(\dfrac{8}{8-3}\right)} = .454$$

$$L_s = \frac{3(4)(3/4)^2}{(1)!(8-3)^2}(.4545) + \frac{3}{4} = .873$$

$$W_s = \frac{.873}{3} = .291 \text{ hr.}$$

$$L_q = .873 - 3/4 = .123$$

$$W_q = \frac{.123}{3} = 0.41 \text{ hr.}$$

Looking back at Solved Problem D.1, we see that although length of the *queue* and average time in the queue are lowest when a second platform is open, the average number of trucks in the *system* and average time spent waiting in the system are smallest when two workers are employed at a *single* platform. Thus, we would probably recommend not building a second platform.

Solved Problem D.4

St. Elsewhere Hospital's Cardiac Care Unit (CCU) has 5 beds, which are virtually always occupied by patients who have just undergone major heart surgery. Two registered nurses are on duty in the CCU in each of the three 8-hour shifts. About every 2 hours (following a Poisson distribution), one of the patients requires a nurse's attention. The nurse will then spend an average of 30 minutes (exponentially distributed) assisting the patient and updating medical records regarding the problem and care provided.

Because immediate service is critical to the 5 patients, two important questions are: What is the average number of patients being attended by the nurses? What is the average time that a patient spends waiting for one of the nurses to arrive?

SOLUTION

$N = 5$ patients

$M = 2$ nurses

$T = 30$ minutes

$U = 120$ minutes

$$X = \frac{T}{T + U} = \frac{30}{30 + 120} = .20$$

From Table D.7 (p. 567), with $X = .20$ and $M = 2$, we see that

$F = .976$

$H =$ average number being attended to $= FNX$

$\quad = (.976)(5)(.20) = .98 \approx 1$ patient at any given time

$W =$ average waiting time for a nurse $= \dfrac{T(1 - F)}{XF}$

$$\quad = \frac{30(1 - .976)}{(.20)(.976)} = 3.69 \text{ minutes}$$

INTERNET AND STUDENT CD-ROM EXERCISES

Visit our home page or use your student CD-ROM to help with material in this module.

 On Our Home Page, www.prenhall.com/heizer

- Self-Tests
- Practice Problems
- Internet Exercises
- Current Articles and Research
- Internet Homework Problems
- Internet Case

 On Your Student CD-ROM

- PowerPoint Lecture
- Practice Problems
- Active Model Exercises
- ExcelOM
- Excel OM Example Data Files

ADDITIONAL CASE STUDY

See our Internet homepage at www.prenhall.com/heizer **for this additional free case study:**

- **Pantry Shopper**: The case requires the redesign of a checkout system for a supermarket.

Learning Curves

Module Outline

LEARNING OBJECTIVES

When you complete this module you should be able to

IDENTIFY OR DEFINE:

What a learning curve is

Example of learning curves

The doubling concept

DESCRIBE OR EXPLAIN:

How to compute learning curve effects

Why learning curves are important

The strategic implications of learning curves

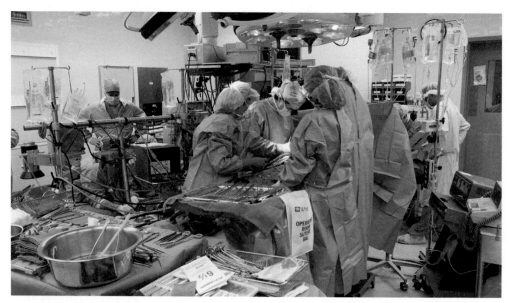

Medical procedures such as heart surgery follow a learning curve. Research indicates that the death rate from heart transplants drops at a 79% learning curve, a learning rate not unlike that in many industrial settings. It appears that doctors and medical teams improve, as do your odds as a patient, with experience. If the death rate is halved every three operations, practice may indeed make perfect.

Learning curves
The premise that people and organizations get better at their tasks as the tasks are repeated; sometimes called experience curves.

Most organizations learn and improve over time. As firms and employees perform a task over and over, they learn how to perform more efficiently. This means that task times and costs decrease.

Learning curves are based on the premise that people and organizations become better at their tasks as the tasks are repeated. A learning curve graph (illustrated in Figure E.1) displays labor-hours per unit versus the number of units produced. From it we see that the time needed to produce a unit decreases, usually following a negative exponential curve, as the person or company produces more units. In other words, *it takes less time to complete each additional unit a firm produces.* However, we also see in Figure E.1 that the time *savings* in completing each subsequent unit *decreases.* These are the major attributes of the learning curve.

Learning curves were first applied to industry in a report by T. P. Wright of Curtis-Wright Corp. in 1936.[1] Wright described how direct labor costs of making a particular airplane decreased with learning, a theory since confirmed by other aircraft manufacturers. Regardless of the time needed to produce the first plane, learning curves are found to apply to various categories of air frames (e.g.,

FIGURE E.1 ■

The Learning-Curve Effect States That Time per Repetition Decreases as the Number of Repetitions Increases

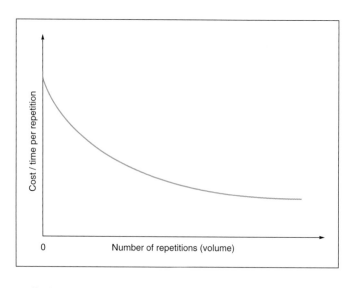

[1]T. P. Wright, "Factors Affecting the Cost of Airplanes," *Journal of the Aeronautical Sciences* (February 1936).

jet fighters versus passenger planes versus bombers). Learning curves have since been applied not only to labor but also to a wide variety of other costs, including material and purchased components. The power of the learning curve is so significant that it plays a major role in many strategic decisions related to employment levels, costs, capacity, and pricing.

The learning curve is based on a *doubling* of production: That is, when production doubles, the decrease in time per unit affects the rate of the learning curve. So, if the learning curve is an 80% rate, the second unit takes 80% of the time of the first unit, the fourth unit takes 80% of the time of the second unit, the eighth unit takes 80% of the time of the fourth unit, and so forth. This principle is shown as

$$T \times L^n = \text{Time required for the } n\text{th unit} \qquad \text{(E-1)}$$

where T = unit cost or unit time of the first unit
L = learning curve rate
n = number of times T is doubled

If the first unit of a particular product took 10 labor-hours, and if a 70% learning curve is present, the hours the fourth unit will take require doubling twice—from 1 to 2 to 4. Therefore, the formula is

$$\text{Hours required for unit } 4 = 10 \times (.7)^2 = 4.9 \text{ hours}$$

LEARNING CURVES IN SERVICES AND MANUFACTURING

Try testing the learning-curve effect on some activity you may be performing. For example, if you need to assemble four bookshelves, time your work on each and note the rate of improvement.

Different organizations—indeed, different products—have different learning curves. The rate of learning varies depending on the quality of management and the potential of the process and product. *Any change in process, product, or personnel disrupts the learning curve.* Therefore, caution should be exercised in assuming that a learning curve is continuing and permanent.

As you can see in Table E.1, industry learning curves vary widely. The lower the number (say 70% compared to 90%), the steeper the slope and the faster the drop in costs. By tradition, learning curves are defined in terms of the *complements* of their improvement rates. For example, a 70% learning curve implies a 30% decrease in time each time the number of repetitions is doubled. A 90% curve means there is a corresponding 10% rate of improvement.

Stable, standardized products and processes tend to have costs that decline more steeply than others. Between 1920 and 1955, for instance, the steel industry was able to reduce labor-hours per unit to 79% each time cumulative production doubled.

Learning curves have application in services as well as industry. As was noted in the caption for the opening photograph, 1-year death rates of heart transplant patients at Temple University Hospital follow a 79% learning curve. The results of that hospital's 3-year study of 62 patients

TABLE E.1 ■

Examples of Learning-Curve Effects

EXAMPLE	IMPROVING PARAMETER	CUMULATIVE PARAMETER	LEARNING-CURVE SLOPE (%)
1. Model-T Ford production	Price	Units produced	86
2. Aircraft assembly	Direct labor-hours per unit	Units produced	80
3. Equipment maintenance at GE	Average time to replace a group of parts	Number of replacements	76
4. Steel production	Production worker labor-hours per unit produced	Units produced	79
5. Integrated circuits	Average price per unit	Units produced	72[a]
6. Hand-held calculator	Average factory selling price	Units produced	74
7. Disk memory drives	Average price per bit	Number of bits	76
8. Heart transplants	1-year death rates	Transplants completed	79

[a]Constant dollars.

Sources: James A. Cunningham, "Using the Learning Curve as a Management Tool," *IEEE Spectrum* (June 1980): 45. © 1980 IEEE; and David B. Smith and Jan L. Larsson, "The Impact of Learning on Cost: The Case of Heart Transplantation," *Hospital and Health Services Administration* (spring 1989): 85–97.

receiving transplants found that every three operations resulted in a halving of the 1-year death rate. As more hospitals face pressure from both insurance companies and the government to enter fixed-price negotiations for their services, their ability to learn from experience becomes increasingly critical. In addition to having applications in both services and industry, learning curves are useful for a variety of purposes. These include:

1. Internal: labor forecasting, scheduling, establishing costs and budgets.
2. External: supply chain negotiations (see the SMT case study at the end of this module).
3. Strategic: evaluation of company and industry performance, including costs and pricing.

APPLYING THE LEARNING CURVE

A mathematical relationship enables us to express the time required to produce a certain unit. This relationship is a function of how many units have been produced before the unit in question and how long it took to produce them. Although this procedure determines how long it takes to produce a given unit, the consequences of this analysis are more far-reaching. Costs drop and efficiency goes up for individual firms and the industry. Therefore, severe problems in scheduling occur if operations are not adjusted for implications of the learning curve. For instance, if learning-curve improvement is not considered when scheduling, the result may be labor and productive facilities being idle a portion of the time. Furthermore, firms may refuse additional work because they do not consider the improvement in their own efficiency that results from learning. From a supply-chain perspective, our interest is in negotiating what our suppliers' costs should be for further production of units based on the size of our order. The foregoing are only a few of the ramifications of the effect of learning curves.

With this in mind, let us look at three ways to approach the mathematics of learning curves: arithmetic analysis, logarithmic analysis, and learning-curve coefficients.

Arithmetic Approach

The arithmetic approach is the simplest approach to learning-curve problems. As we noted at the beginning of this module, each time that production doubles, labor per unit declines by a constant factor, known as the learning rate. So, if we know that the learning rate is 80% and that the first unit produced took 100 hours, the hours required to produce the second, fourth, eighth, and sixteenth units are as follows:

NTH UNIT PRODUCED	HOURS FOR NTH UNIT
1	100.0
2	$80.0 = (.8 \times 100)$
4	$64.0 = (.8 \times 80)$
8	$51.2 = (.8 \times 64)$
16	$41.0 = (.8 \times 51.2)$

As long as we wish to find the hours required to produce N units and N is one of the doubled values, then this approach works. Arithmetic analysis does not tell us how many hours will be needed to produce other units. For this flexibility, we must turn to the logarithmic approach.

TABLE E.2 ■

Learning Curve
Values of b

LEARNING RATE (%)	b
70	−.515
75	−.415
80	−.322
85	−.234
90	−.152

Logarithmic Approach

The logarithmic approach allows us to determine labor for *any* unit, T_N, by the formula

$$T_N = T_1(N^b) \qquad \text{(E-2)}$$

where T_N = time for the Nth unit
T_1 = hours to produce the first unit
b = (log of the learning rate)/(log 2) = slope of the learning curve

Some of the values for b are presented in Table E.2. Example E1 shows how this formula works.

Example E1

**Excel OM Data File
ModEExE1.xla**

The learning rate for a particular operation is 80%, and the first unit of production took 100 hours. The hours required to produce the third unit may be computed as follows:

$$T_N = T_1(N^b)$$
$$T_3 = (100 \text{ hours})(3^b)$$
$$= (100)(3^{\log .8/\log 2})$$
$$= (100)(3^{-.322}) = 70.2 \text{ labor-hours}$$

The logarithmic approach allows us to determine the hours required for *any* unit produced, but there *is* a simpler method.

Learning-Curve Coefficient Approach

The learning-curve coefficient technique is embodied in Table E.3 and the following equation:

$$T_N = T_1 C \qquad \text{(E-3)}$$

where T_N = number of labor-hours required to produce the Nth unit
T_1 = number of labor-hours required to produce the first unit
C = learning-curve coefficient found in Table E.3

The learning-curve coefficient, C, depends on both the learning rate (70%, 75%, 80%, and so on) and the unit of interest.

TABLE E.3 ■ Learning-Curve Coefficients, where Coefficient = $N^{(\log \text{ of learning rate}/\log 2)}$

UNIT NUMBER (N) TIME	70%		75%		80%		85%		90%	
	UNIT TIME	TOTAL TIME	UNIT TIME	TOTAL TIME	UNIT TIME	TOTAL TIME	UNIT TIME	TOTAL TIME	UNIT TIME	TOTAL TIME
1	1.000	1.000	1.000	1.000	1.000	1.000	1.000	1.000	1.000	1.000
2	.700	1.700	.750	1.750	.800	1.800	.850	1.850	.900	1.900
3	.568	2.268	.634	2.384	.702	2.502	.773	2.623	.846	2.746
4	.490	2.758	.562	2.946	.640	3.142	.723	3.345	.810	3.556
5	.437	3.195	.513	3.459	.596	3.738	.686	4.031	.783	4.339
6	.398	3.593	.475	3.934	.562	4.299	.657	4.688	.762	5.101
7	.367	3.960	.446	4.380	.534	4.834	.634	5.322	.744	5.845
8	.343	4.303	.422	4.802	.512	5.346	.614	5.936	.729	6.574
9	.323	4.626	.402	5.204	.493	5.839	.597	6.533	.716	7.290
10	.306	4.932	.385	5.589	.477	6.315	.583	7.116	.705	7.994
11	.291	5.223	.370	5.958	.462	6.777	.570	7.686	.695	8.689
12	.278	5.501	.357	6.315	.449	7.227	.558	8.244	.685	9.374
13	.267	5.769	.345	6.660	.438	7.665	.548	8.792	.677	10.052
14	.257	6.026	.334	6.994	.428	8.092	.539	9.331	.670	10.721
15	.248	6.274	.325	7.319	.418	8.511	.530	9.861	.663	11.384
16	.240	6.514	.316	7.635	.410	8.920	.522	10.383	.656	12.040
17	.233	6.747	.309	7.944	.402	9.322	.515	10.898	.650	12.690
18	.226	6.973	.301	8.245	.394	9.716	.508	11.405	.644	13.334
19	.220	7.192	.295	8.540	.388	10.104	.501	11.907	.639	13.974
20	.214	7.407	.288	8.828	.381	10.485	.495	12.402	.634	14.608
25	.191	8.404	.263	10.191	.355	12.309	.470	14.801	.613	17.713
30	.174	9.305	.244	11.446	.335	14.020	.450	17.091	.596	20.727
35	.160	10.133	.229	12.618	.318	15.643	.434	19.294	.583	23.666
40	.150	10.902	.216	13.723	.305	17.193	.421	21.425	.571	26.543
45	.141	11.625	.206	14.773	.294	18.684	.410	23.500	.561	29.366
50	.134	12.307	.197	15.776	.284	20.122	.400	25.513	.552	32.142

Example E2 uses the preceding equation and Table E.3 to calculate learning-curve effects.

Example E2

Excel OM Data File
ModEExE2.xla

Active Model E.1

Examples E2 and E3 are further illustrated in Active Model E.1 on the CD-ROM and in the Exercise located in your Student Lecture Guide.

It took a Korean shipyard 125,000 labor-hours to produce the first of several tugboats that you expect to purchase for your shipping company, Great Lakes, Inc. Boats 2 and 3 have been produced by the Koreans with a learning factor of 85%. At $40 per hour, what should you, as purchasing agent, expect to pay for the fourth unit?

First, search Table E.3 for the fourth unit and a learning factor of 85%. The learning-curve coefficient, C, is .723. To produce the fourth unit, then, takes

$$T_N = T_1 C$$
$$T_4 = (125,000 \text{ hours})(.723)$$
$$= 90,375 \text{ hours}$$

To find the cost, multiply by $40:

$$90,375 \text{ hours} \times \$40 \text{ per hour} = \$3,615,000$$

Table E.3 also shows *cumulative values*. These allow us to compute the total number of hours needed to complete a specified number of units. Again, the computation is straightforward. Just multiply the table value times the time required for the first unit. Example E3 illustrates this concept.

Example E3

Example E2 computed the time to complete the fourth tugboat that Great Lakes plans to buy. How long will *all four* boats require?

Looking this time at the "total time" column in Table E.3, we find that the cumulative coefficient is 3.345. Thus, the time required is

$$T_N = T_1 C$$
$$T_4 = (125,000)(3.345) = 418,125 \text{ hours in total for all 4 boats}$$

For an illustration of how Excel OM can be used to solve Examples E2 and E3, see Program E.1 at the end of this module.

As later times become available, it can be useful to revise the basic unit; this is especially so when the first unit is estimated prior to production.

Using Table E.3 requires that we know how long it takes to complete the first unit. Yet, what happens if our most recent or most reliable information available pertains to some other unit? The answer is that we must use these data to find a revised estimate for the first unit and then apply the table to that number. Example E4 illustrates this concept.

Example E4

Great Lakes, Inc., believes that unusual circumstances in producing the first boat (see Example E2) imply that the time estimate of 125,000 hours is not as valid a base as the time required to produce the third boat. Boat number 3 was completed in 100,000 hours.

To solve for the revised estimate for boat number 1, we return to Table E.3, with a unit value of $N = 3$ and a learning-curve coefficient of $C = .773$ in the 85% column. To find the revised estimate, we divide the actual time for boat number 3, 100,000 hours, by $C = .773$

$$\frac{100,000}{.773} = 129,366 \text{ hours}$$

Applications of the learning curve:
1. Internal → determine labor standards and rates of material supply required.
2. External → determine purchase costs.
3. Strategic → determine volume-cost changes.

STRATEGIC IMPLICATIONS OF LEARNING CURVES

So far, we have shown how operations managers can forecast labor-hour requirements for a product. We have also shown how purchasing agents can determine a supplier's cost, knowledge that can help in price negotiations. Another important application of learning curves concerns strategic planning.

An example of a company cost line and industry price line are so labeled in Figure E.2. These learning curves are straight because both scales are log scales. When the *rate* of change is constant, a log-log graph yields a straight line. If an organization believes its cost line to be the "company cost" line and the industry price is indicated by the dashed horizontal line, then the company must have costs at the points below the dotted line (for example, point *a* or *b*) or else operate at a loss (point *c*).

FIGURE E.2 ■

Industry Learning Curve for Price Compared with Company Learning Curve for Cost

Note: Both the vertical and horizontal axes of this figure are log scales. This is known as a log-log graph.

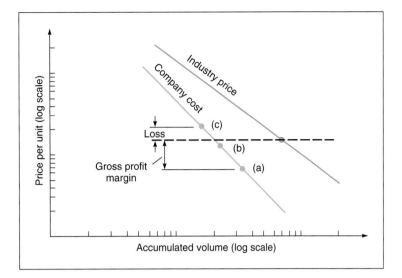

Lower costs are not automatic; they must be managed down. When a firm's strategy is to pursue a curve steeper than the industry average (the company cost line in Figure E.2), it does this by:

1. Following an aggressive pricing policy.
2. Focusing on continuing cost reduction and productivity improvement.
3. Building on shared experience.
4. Keeping capacity growing ahead of demand.

Costs may drop as a firm pursues the learning curve, but volume must increase for the learning curve to exist. Moreover, managers must understand competitors before embarking on a learning-curve strategy. Weak competitors are undercapitalized, stuck with high costs, or do not understand the logic of learning curves. However, strong and dangerous competitors control their costs, have solid financial positions for the large investments needed, and have a track record of using an aggressive learning-curve strategy. Taking on such a competitor in a price war may help only the consumer.

LIMITATIONS OF LEARNING CURVES

Before using learning curves, some cautions are in order:

- Because learning curves differ from company to company, as well as industry to industry, estimates for each organization should be developed rather than applying someone else's.
- Learning curves are often based on the time necessary to complete the early units; therefore, those times must be accurate. As current information becomes available, reevaluation is appropriate.
- Any changes in personnel, design, or procedure can be expected to alter the learning curve. And the curve may spike up for a short time even if it is going to drop in the long run.
- While workers and process may improve, the same learning curves do not always apply to indirect labor and material.
- The culture of the workplace, as well as resource availability and changes in the process, may alter the learning curve. For instance, as a project nears its end, worker interest and effort may drop, curtailing progress down the curve.

SUMMARY

The learning curve is a powerful tool for the operations manager. This tool can assist operations managers in determining future cost standards for items produced as well as purchased. In addition, the learning curve can provide understanding about company and industry performance. We saw three approaches to learning curves: arithmetic analysis, logarithmic analysis, and learning-curve coefficients found in tables. Software can also help analyze learning curves.

KEY TERM   Learning curves (p. 574)

 # USING EXCEL OM FOR LEARNING CURVES

Program E.1 shows how Excel OM develops a spreadsheet for learning-curve calculations. The input data come from Examples E2 and E3. In cell B7, we enter the unit number for the base unit (which does not have to be 1), and in B8 we enter the time for this unit.

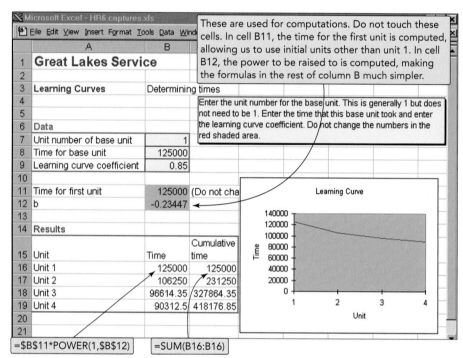

These are used for computations. Do not touch these cells. In cell B11, the time for the first unit is computed, allowing us to use initial units other than unit 1. In cell B12, the power to be raised to is computed, making the formulas in the rest of column B much simpler.

Enter the unit number for the base unit. This is generally 1 but does not need to be 1. Enter the time that this base unit took and enter the learning curve coefficient. Do not change the numbers in the red shaded area.

PROGRAM E.1 ■ Excel OM's Learning-Curve Module, Using Data from Examples E2 and E3

USING POM FOR WINDOWS FOR LEARNING CURVES

POM for Windows' Learning Curve module computes the length of time that future units will take, given the time required for the base unit and the learning rate (expressed as a number between 0 and 1). As an option, if the times required for the first and Nth units are already known, the learning *rate* can be computed. See Appendix V for further details.

SOLVED PROBLEMS

Solved Problem E.1

Digicomp produces a new telephone system with built-in TV screens. Its learning rate is 80%.

(a) If the first one took 56 hours, how long will it take Digicomp to make the eleventh system?

(b) How long will the first 11 systems take in total?

(c) As a purchasing agent, you expect to buy units 12 through 15 of the new phone system. What would be your expected cost for the units if Digicomp charges $30 for each labor-hour?

SOLUTION

(a) $T_N = T_1 C$ from Table E.3—80% unit time

T_{11} = (56 hours)(.462) = 25.9 hours

(b) Total time for the first 11 units = (56 hours)(6.777) = 379.5 hours

from Table E.3—80% total time

(c) To find the time for units 12 through 15, we take the total cumulative time for units 1 to 15 and subtract the total time for units 1 to 11, which was computed in part (b). Total time for the first 15 units = (56 hours)(8.511) = 476.6 hours. So, the time for units 12 through 15 is 476.6 − 379.5 = 97.1 hours. (This figure could also be confirmed by computing the times for units 12, 13, 14, and 15 separately using the unit-time column and then adding them.) Expected cost for units 12 through 15 = (97.1 hours) ($30 per hour) = $2,913.

Solved Problem E.2

If the first time you performed a job took 60 minutes, how long will the eighth job take if you are on an 80% learning curve?

SOLUTION

Three doublings from 1 to 2 to 4 to 8 implies $.8^3$. Therefore, we have

$$60 \times (.8)^3 = 60 \times .512 = 30.72 \text{ minutes}$$

or, using Table E.3, we have $C = .512$. Therefore:

$$60 \times .512 = 30.72 \text{ minutes}$$

INTERNET AND STUDENT CD-ROM EXERCISES

Visit our home page or use your student CD-ROM to help with material in this module.

 On Our Home Page, www.prenhall.com/heizer

- Self-Tests
- Practice Problems
- Internet Exercises
- Current Articles and Research
- Internet Homework Problems

 On Your Student CD-ROM

- PowerPoint Lecture
- Practice Problems
- Active Mode Exercise
- ExcelOM
- Excel OM Example Data Files

F

Simulation

Module Outline

LEARNING OBJECTIVES

When you complete this module you should be able to

IDENTIFY OR DEFINE:

Monte Carlo simulation

Random numbers

Random number interval

Simulation software

DESCRIBE OR EXPLAIN:

The advantages and disadvantages of modeling with simulation

The use of Excel spreadsheets in simulation

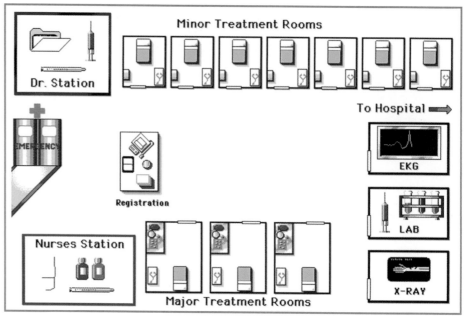

When Bay Medical Center faced severe overcrowding at its outpatient clinic, it turned to computer simulation to try to reduce bottlenecks and improve patient flow. A simulation language called Micro Saint analyzed current data relating to patient service times between clinic rooms. By simulating different numbers of doctors and staff, simulating the use of another clinic for overflow, and simulating a redesign of the existing clinic, Bay Medical Center was able to make decisions based on an understanding of both costs and benefits. This resulted in better patient service at lower cost.

Source: Micro Analysis and Design Simulation Software, Inc., Boulder, CO.

There are many kinds of simulations, and although this module stresses Monte Carlo simulations, you should be aware of "physical" simulations (such as a wind tunnel model) as well.

Simulation models abound in our world. The city of Atlanta, for example, uses them to control traffic. Europe's Airbus Industries uses them to test the aerodynamics of proposed jets. The U.S. Army simulates war games on computers. Business students use management gaming to simulate realistic business competition. And thousands of organizations like Bay Medical Center develop simulation models to help make operations decisions.

Most of the large companies in the world use simulation models. Table F.1 lists just a few areas in which simulation is now being applied.

WHAT IS SIMULATION?

Simulation

The attempt to duplicate the features, appearance, and characteristics of a real system, usually via a computerized model.

Simulation is the attempt to duplicate the features, appearance, and characteristics of a real system. In this module, we will show how to simulate part of an operations management system by building a mathematical model that comes as close as possible to representing the reality of the system. The model will then be used to estimate the effects of various actions. The idea behind simulation is threefold:

1. To imitate a real-world situation mathematically,
2. Then to study its properties and operating characteristics, and
3. Finally to draw conclusions and make action decisions based on the results of the simulation.

TABLE F.1 ■

Some Applications of Simulation

Ambulance location and dispatching	Bus scheduling
Assembly-line balancing	Design of library operations
Parking lot and harbor design	Taxi, truck, and railroad dispatching
Distribution system design	Production facility scheduling
Scheduling aircraft	Plant layout
Labor-hiring decisions	Capital investments
Personnel scheduling	Production scheduling
Traffic-light timing	Sales forecasting
Voting pattern prediction	Inventory planning and control

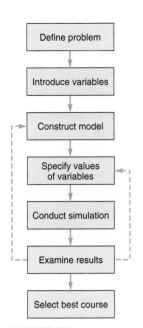

Define problem

Introduce variables

Construct model

Specify values
of variables

Conduct simulation

Examine results

Select best course

FIGURE F.1 ■

The Process of
Simulation

In this way, a real-life system need not be touched until the advantages and disadvantages of a major policy decision are first measured on the model.

To use simulation, an OM manager should

1. Define the problem.
2. Introduce the important variables associated with the problem.
3. Construct a numerical model.
4. Set up possible courses of action for testing.
5. Run the experiment.
6. Consider the results (possibly modifying the model or changing data inputs).
7. Decide what course of action to take.

These steps are illustrated in Figure F.1.

The problems tackled by simulation may range from very simple to extremely complex, from bank-teller lines to an analysis of the U.S. economy. Although small simulations can be conducted by hand, effective use of the technique requires a computer. Large-scale models, simulating perhaps years of business decisions, are virtually all handled by computer.

In this module, we examine the basic principles of simulation and then tackle some problems in the areas of waiting-line analysis and inventory control. Why do we use simulation in these areas when mathematical models described in other chapters can solve similar problems? The answer is that simulation provides an alternative approach for problems that are very complex mathematically. It can handle, for example, inventory problems in which demand or lead time is not constant.

ADVANTAGES AND DISADVANTAGES OF SIMULATION

Simulation is a tool that has become widely accepted by managers for several reasons. The main *advantages* of simulation are as follows:

1. Simulation is relatively straightforward and flexible.
2. It can be used to analyze large and complex real-world situations that cannot be solved by conventional operations management models.
3. Real-world complications can be included that most OM models cannot permit. For example, simulation can use *any* probability distribution the user defines; it does not require standard distributions.
4. "Time compression" is possible. The effects of OM policies over many months or years can be obtained by computer simulation in a short time.
5. Simulation allows "what-if" types of questions. Managers like to know in advance what options will be most attractive. With a computerized model, a manager can try out several policy decisions within a matter of minutes.
6. Simulations do not interfere with real-world systems. It may be too disruptive, for example, to experiment physically with new policies or ideas in a hospital or manufacturing plant.
7. Simulation can study the interactive effects of individual components or variables in order to determine which ones are important.

The cost of simulating a frontal car crash at Ford was $60,000 in 1985. By 1998, the event could be simulated for $200. It now costs under $10. Using Ford's new supercomputer, the simulation takes just minutes.

Can you think of a real-world business application in which a math model would be much better than playing with the actual operation of the firm?

The main *disadvantages* of simulation are as follows:

1. Good simulation models can be very expensive; they may take many months to develop.
2. It is a trial-and-error approach that may produce different solutions in repeated runs. It does not generate optimal solutions to problems (as does linear programming).
3. Managers must generate all of the conditions and constraints for solutions that they want to examine. The simulation model does not produce answers without adequate, realistic input.
4. Each simulation model is unique. Its solutions and inferences are not usually transferable to other problems.

Monte Carlo method

A simulation technique that uses random elements when chance exists in their behavior.

MONTE CARLO SIMULATION

When a system contains elements that exhibit *chance* in their behavior, the **Monte Carlo method** of simulation may be applied. The basis of Monte Carlo simulation is experimentation on chance (or *probabilistic*) elements by means of random sampling.

Computer simulation models have been developed to address a variety of productivity issues at fast-food restaurants such as Burger King. In one, the ideal distance between the drive-through order station and the pickup window was simulated. For example, because a longer distance reduced waiting time, 12 to 13 additional customers could be served per hour—a benefit of over $10,000 in extra sales per restaurant per year. In another simulation, a second drive-through window was considered. This model predicted a sales increase of 15%, $13,000 per year per restaurant.

The technique breaks down into five simple steps:

1. Setting up a probability distribution for important variables.
2. Building a cumulative probability distribution for each variable.
3. Establishing an interval of random numbers for each variable.
4. Generating random numbers.
5. Actually simulating a series of trials.

Let's examine these steps in turn.

Step 1. Establishing Probability Distributions. The basic idea in the Monte Carlo simulation is to generate values for the variables making up the model under study. In real-world systems, a lot of variables are probabilistic in nature. To name just a few: inventory demand; lead time for orders to arrive; times between machine breakdowns; times between customer arrivals at a service facility; service times; times required to complete project activities; and number of employees absent from work each day.

Cumulative probability distribution

The accumulation of individual probabilities of a distribution.

One common way to establish a *probability distribution* for a given variable is to examine historical outcomes. We can find the probability, or relative frequency, for each possible outcome of a variable by dividing the frequency of observation by the total number of observations. Here's an example.

The daily demand for radial tires at Barry's Auto Tire over the past 200 days is shown in columns 1 and 2 of Table F.2. Assuming that past arrival rates will hold in the future, we can convert this demand to a probability distribution by dividing each demand frequency by the total demand, 200. The results are shown in column 3.

Random-number intervals

A set of numbers to represent each possible value or outcome in a computer simulation.

Step 2. Building a Cumulative Probability Distribution for Each Variable. The conversion from a regular probability distribution, such as in column 3 of Table F.2, to a **cumulative probability distribution** is an easy job. In column 4, we see that the cumulative probability for each level of demand is the sum of the number in the probability column (column 3) added to the previous cumulative probability.

Random number

A series of digits that have been selected by a totally random process.

Step 3. Setting Random-Number Intervals. Once we have established a cumulative probability distribution for each variable in the simulation, we must assign a set of numbers to represent each possible value or outcome. These are referred to as **random-number intervals**. Basically, a **random number** is a series of digits (say, two digits from 01, 02, . . . , 98, 99, 00) that have been selected by a totally random process—a process in which each random number has an equal chance of being selected.

TABLE F.2 ■

Demand for
Barry's Auto Tire

To establish a probability distribution for tires, we assume that historical demand is a good indicator of future outcomes.

(1) DEMAND FOR TIRES	(2) FREQUENCY	(3) PROBABILITY OF OCCURRENCE	(4) CUMULATIVE PROBABILITY
0	10	10/200 = .05	.05
1	20	20/200 = .10	.15
2	40	40/200 = .20	.35
3	60	60/200 = .30	.65
4	40	40/200 = .20	.85
5	30	30/200 = .15	1.00
	200 days	200/200 = 1.00	

TABLE F.3 ■

The Assignment of Random-Number Intervals for Barry's Auto Tire

DAILY DEMAND	CUMULATIVE PROBABILITY	INTERVAL OF PROBABILITY	RANDOM NUMBERS
0	.05	.05	01 through 05
1	.10	.15	06 through 15
2	.20	.35	16 through 35
3	.30	.65	36 through 65
4	.20	.85	66 through 85
5	.15	1.00	86 through 00

You may start random number intervals at either 01 or 00, but the text starts at 01 so that the top of each range is the cumulative probability.

If, for example, there is a 5% chance that demand for Barry's radial tires will be 0 units per day, then we will want 5% of the random numbers available to correspond to a demand of 0 units. If a total of 100 two-digit numbers is used in the simulation, we could assign a demand of 0 units to the first 5 random numbers: 01, 02, 03, 04, and 05.[1] Then a simulated demand for 0 units would be created every time one of the numbers 01 to 05 was drawn. If there is also a 10% chance that demand for the same product will be 1 unit per day, we could let the next 10 random numbers (06, 07, 08, 09, 10, 11, 12, 13, 14, and 15) represent that demand—and so on for other demand levels.

Similarly, we can see in Table F.3 that the length of each interval on the right corresponds to the probability of 1 of each of the possible daily demands. Thus, in assigning random numbers to the daily demand for 3 radial tires, the range of the random-number interval (36 through 65) corresponds *exactly* to the probability (or proportion) of that outcome. A daily demand for 3 radial tires occurs 30% of the time. All of the 30 random numbers greater than 35 up to and including 65 are assigned to that event.

Step 4. Generating Random Numbers. Random numbers may be generated for simulation problems in two ways. If the problem is large and the process under study involves many simulation trials, computer programs are available to generate the needed random numbers. If the simulation is being done by hand, the numbers may be selected from a table of random digits.

Step 5. Simulating the Experiment. We may simulate outcomes of an experiment by simply selecting random numbers from Table F.4. Beginning anywhere in the table, we note the interval in

TABLE F.4 ■ Table of Random Numbers

52	06	50	88	53	30	10	47	99	37	66	91	35	32	00	84	57	07
37	63	28	02	74	35	24	03	29	60	74	85	90	73	59	55	17	60
82	57	68	28	05	94	03	11	27	79	90	87	92	41	09	25	36	77
69	02	36	49	71	99	32	10	75	21	95	90	94	38	97	71	72	49
98	94	90	36	06	78	23	67	89	85	29	21	25	73	69	34	85	76
96	52	62	87	49	56	59	23	78	71	72	90	57	01	98	57	31	95
33	69	27	21	11	60	95	89	68	48	17	89	34	09	93	50	44	51
50	33	50	95	13	44	34	62	64	39	55	29	30	64	49	44	30	16
88	32	18	50	62	57	34	56	62	31	15	40	90	34	51	95	26	14
90	30	36	24	69	82	51	74	30	35	36	85	01	55	92	64	09	85
50	48	61	18	85	23	08	54	17	12	80	69	24	84	92	16	49	59
27	88	21	62	69	64	48	31	12	73	02	68	00	16	16	46	13	85
45	14	46	32	13	49	66	62	74	41	86	98	92	98	84	54	33	40
81	02	01	78	82	74	97	37	45	31	94	99	42	49	27	64	89	42
66	83	14	74	27	76	03	33	11	97	59	81	72	00	64	61	13	52
74	05	81	82	93	09	96	33	52	78	13	06	28	30	94	23	37	39
30	34	87	01	74	11	46	82	59	94	25	34	32	23	17	01	58	73
59	55	72	33	62	13	74	68	22	44	42	09	32	46	71	79	45	89
67	09	80	98	99	25	77	50	03	32	36	63	65	75	94	19	95	88
60	77	46	63	71	69	44	22	03	85	14	48	69	13	30	50	33	24
60	08	19	29	36	72	30	27	50	64	85	72	75	29	87	05	75	01
80	45	86	99	02	34	87	08	86	84	49	76	24	08	01	86	29	11
53	84	49	63	26	65	72	84	85	63	26	02	75	26	92	62	40	67
69	84	12	94	51	36	17	02	15	29	16	52	56	43	26	22	08	62
37	77	13	10	02	18	31	19	32	85	31	94	81	43	31	58	33	51

Source: Reprinted from *A Million Random Digits with 100,000 Normal Deviates,* Rand (New York: The Free Press, 1995). Used by permission.

[1]Alternatively, we could have assigned the random numbers 00, 01, 02, 03, and 04 to represent a demand of 0 units. The two digits 00 can be thought of as either 0 or 100. As long as 5 numbers out of 100 are assigned to the 0 demand, it does not make any difference which 5 they are.

Table F.3 into which each number falls. For example, if the random number chosen is 81 and the interval 66 through 85 represents a daily demand for 4 tires, then we select a demand of 4 tires. Example F1 carries the simulation further.

Example F1

Let's illustrate the concept of random numbers by simulating 10 days of demand for radial tires at Barry's Auto Tire (see Table F.3). We select the random numbers needed from Table F.4, starting in the upper left-hand corner and continuing down the first column:

DAY NUMBER	RANDOM NUMBER	SIMULATED DAILY DEMAND
1	52	3
2	37	3
3	82	4
4	69	4
5	98	5
6	96	5
7	33	2
8	50	3
9	88	5
10	90	5

39   Total 10-day demand

39/10 = 3.9 = Tires average daily demand

It is interesting to note that the average demand of 3.9 tires in this 10-day simulation differs substantially from the *expected* daily demand, which we may calculate from the data in Table F.3:

$$\text{Expected demand} = \sum_{i=1}^{5} (\text{probability of } i \text{ units}) \times (\text{demand of } i \text{ units})$$

$$= (.05)(0) + (.10)(1) + (.20)(2) + (.30)(3) + (.20)(4) + (.15)(5)$$

$$= 0 + .1 + .4 + .9 + .8 + .75$$

$$= 2.95 \text{ tires}$$

However, if this simulation were repeated hundreds or thousands of times, the average *simulated* demand would be nearly the same as the *expected* demand.

Naturally, it would be risky to draw any hard and fast conclusions about the operation of a firm from only a short simulation like Example F1. Seldom would anyone actually want to go to the effort of simulating such a simple model containing only one variable. Simulating by hand does, however, demonstrate the important principles involved and may be useful in small-scale studies.

OM IN ACTION

Simulating Taco Bell's Restaurant Operation

Determining how many employees to schedule each 15 minutes to perform each function in a Taco Bell restaurant is a complex and vexing problem. So Taco Bell, the $5-billion giant with 6,500 U.S. and foreign locations, decided to build a simulation model. It selected MODSIM as its software to develop a new labor-management system called LMS.

To develop and use a simulation model, Taco Bell had to collect a substantial amount of data. Almost everything that takes place in a restaurant, from customer arrival patterns to the time it takes to wrap a taco, had to be translated into reliable, accurate data. Just as an example, analysts had to conduct time studies and data analysis for every task that is part of preparing every item on the menu. To the researcher's surprise, the hours devoted to collecting data greatly exceeded those needed to actually build the LMS model.

Inputs to LMS include staffing, such as number of people and positions. Outputs are performance measures, such as mean time in the system, mean time at the counter, people utilization, and equipment utilization. The model paid off. More than $53 million in labor costs were saved in LMS's first 4 years of use.

Sources: OR/MS Today (June 2000): 30, and (October 1997): 20–21; and Interfaces 28, 1 (January–February 1998): 75–91.

SIMULATION OF A QUEUING PROBLEM

Barge arrivals and unloading rates are both probabilistic variables. Unless they follow the queuing probability distributions of Module D, we must turn to a simulation approach.

An important use of simulation is in the analysis of waiting-line problems. As we saw in Module D, the assumptions required for solving queuing problems are quite restrictive. For most realistic queuing systems, simulation may be the only approach available.

Example F2 illustrates the use of simulation for a large unloading dock and its associated queue. Arrivals of barges at the dock are not Poisson-distributed, and unloading rates (service times) are not exponential or constant. As such, the mathematical waiting-line models of Quantitative Module D cannot be used.

Example F2

Following long trips down the Mississippi River from industrial midwestern cities, fully loaded barges arrive at night in New Orleans. The number of barges docking on any given night ranges from 0 to 5. The probability of 0, 1, 2, 3, 4, and 5 arrivals is displayed in Table F.5. In the same table, we establish cumulative probabilities and corresponding random-number intervals for each possible value.

TABLE F.5 ■ Overnight Barge Arrival Rates and Random-Number Intervals

NUMBER OF ARRIVALS	PROBABILITY	CUMULATIVE PROBABILITY	RANDOM-NUMBER INTERVAL
0	.13	.13	01 through 13
1	.17	.30	14 through 30
2	.15	.45	31 through 45
3	.25	.70	46 through 70
4	.20	.90	71 through 90
5	.10	1.00	91 through 00
	1.00		

A study by the dock superintendent reveals that the number of barges unloaded also tends to vary from day to day. In Table F.6, the superintendent provides information from which we can create a probability distribution for the variable *daily unloading rate*. As we just did for the arrival variable, we can set up an interval of random numbers for the unloading rates.

TABLE F.6 ■ Unloading Rates and Random-Number Intervals

DAILY UNLOADING RATES	PROBABILITY	CUMULATIVE PROBABILITY	RANDOM-NUMBER INTERVAL
1	.05	.05	01 through 05
2	.15	.20	06 through 20
3	.50	.70	21 through 70
4	.20	.90	71 through 90
5	.10	1.00	91 through 00
	1.00		

The relation between random number intervals and cumulative probability is that the top end of each interval is equal to the cumulative probability percentage.

Barges are unloaded on a first-in, first-out basis. Any barges not unloaded on the day of arrival must wait until the following day. However, tying up barges in dock is an expensive proposition, and the superintendent cannot ignore the angry phone calls from barge owners reminding him that "time is money!" He decides that, before going to the Port of New Orleans controller to request additional unloading crews, he should conduct a simulation study of arrivals, unloadings, and delays. A 100-day simulation would be ideal, but for purposes of illustration, the superintendent begins with a shorter 15-day analysis. Random numbers are drawn from the top row of Table F.4 to generate daily arrival rates. To create daily unloading rates, they are drawn from the second row of Table F.4. Table F.7 shows the day-to-day port simulation.

TABLE F.7 ■ Queuing Simulation of Port of New Orleans Barge Unloadings

(1) Day	(2) NUMBER DELAYED FROM PREVIOUS DAY	(3) RANDOM NUMBER	(4) NUMBER OF NIGHTLY ARRIVALS	(5) TOTAL TO BE UNLOADED	(6) RANDOM NUMBER	(7) NUMBER UNLOADED
1	—[a]	52	3	3	37	3
2	0	06	0	0	63	0 [b]
3	0	50	3	3	28	3
4	0	88	4	4	02	1
5	3	53	3	6	74	4
6	2	30	1	3	35	3
7	0	10	0	0	24	0 [c]
8	0	47	3	3	03	1
9	2	99	5	7	29	3
10	4	37	2	6	60	3
11	3	66	3	6	74	4
12	2	91	5	7	85	4
13	3	35	2	5	90	4
14	1	32	2	3	73	3 [d]
15	0	00	5	5	59	3
	20		41			39
	Total delays		Total arrivals			Total unloadings

[a]We can begin with no delays from the previous day. In a long simulation, even if we started with five overnight delays, that initial condition would be averaged out.

[b]Three barges could have been unloaded on day 2. Yet because there were no arrivals and no backlog existed, zero unloadings took place.

[c]The same situation as noted in footnote b takes place.

[d]This time, 4 barges could have been unloaded, but because only 3 were in queue, the number unloaded is recorded as 3.

The superintendent will likely be interested in at least three useful and important pieces of information:

$$\begin{pmatrix} \text{Average number of barges} \\ \text{delayed to the next day} \end{pmatrix} = \frac{20 \text{ delays}}{15 \text{ days}}$$

$$= 1.33 \text{ barges delayed per day}$$

$$\text{Average number of nightly arrivals} = \frac{41 \text{ arrivals}}{15 \text{ days}}$$

$$= 2.73 \text{ arrivals per night}$$

$$\text{Average number of barges unloaded each day} = \frac{39 \text{ unloadings}}{15 \text{ days}}$$

$$= 2.60 \text{ unloadings per day}$$

When the data from Example F2 are analyzed in terms of delay costs, idle labor costs, and the cost of hiring extra unloading crew, the dock superintendent and port controller can make a better staffing decision. They may even choose to resimulate the process assuming different unloading rates that would correspond to increased crew sizes. Although simulation cannot guarantee an optimal solution to problems such as this, it can be helpful in recreating a process and identifying good decision alternatives.

SIMULATION AND INVENTORY ANALYSIS

In Chapter 12, we introduced inventory models. The commonly used EOQ models are based on the assumption that both product demand and reorder lead time are known, constant values. In most real-world inventory situations, though, demand and lead time are variables, so accurate analysis becomes extremely difficult to handle by any means other than simulation.

In this section, we present an inventory problem with two decision variables and two probabilistic components. The owner of the hardware store in Example F3 would like to establish *order quantity* and *reorder point* decisions for a particular product that has probabilistic (uncertain) daily demand and reorder lead time. He wants to make a series of simulation runs, trying out various order quantities and reorder points, in order to minimize his total inventory cost for the item. Inventory costs in this case will include ordering, holding, and stockout costs.

Example F3

Simkin's Hardware sells the Ace model electric drill. Daily demand for the drill is relatively low but subject to some variability. Over the past 300 days, Simkin has observed the sales shown in column 2 of Table F.8. He converts this historical frequency into a probability distribution for the variable daily demand (column 3). A cumulative probability distribution is formed in column 4 of Table F.8. Finally, Simkin establishes an interval of random numbers to represent each possible daily demand (column 5).

TABLE F.8 ■ Probabilities and Random-Number Intervals for Daily Ace Drill Demand

(1) DEMAND FOR ACE DRILL	(2) FREQUENCY	(3) PROBABILITY	(4) CUMULATIVE PROBABILITY	(5) INTERVAL OF RANDOM NUMBERS
0	15	.05	.05	01 through 05
1	30	.10	.15	06 through 15
2	60	.20	.35	16 through 35
3	120	.40	.75	36 through 75
4	45	.15	.90	76 through 90
5	30	.10	1.00	91 through 00
	300 days	1.00		

When Simkin places an order to replenish his inventory of drills, there is a delivery lag of from 1 to 3 days. This means that lead time may also be considered a probabilistic variable. The number of days that it took to receive the past 50 orders is presented in Table F.9. In a fashion similar to the creation of the demand variable, Simkin establishes a probability distribution for the lead time variable (column 3 of Table F.9), computes the cumulative distribution (column 4), and assigns random-number intervals for each possible time (column 5).

TABLE F.9 ■ Probabilities and Random-Number Intervals for Reorder Lead Time

(1) LEAD TIME (DAYS)	(2) FREQUENCY	(3) PROBABILITY	(4) CUMULATIVE PROBABILITY	(5) RANDOM-NUMBER INTERVAL
1	10	.20	.20	01 through 20
2	25	.50	.70	21 through 70
3	15	.30	1.00	71 through 00
	50 orders	1.00		

The first inventory policy that Simkin wants to simulate is an order quantity of 10 with a reorder point of 5. That is, every time the on-hand inventory level at the end of the day is 5 or less, Simkin will call his supplier that evening and place an order for 10 more drills. Note that if the lead time is 1 day, the order will not arrive the next morning, but rather at the beginning of the following workday.

The entire process is simulated in Table F.10 for a 10-day period. We assume that beginning inventory (column 3) is 10 units on day 1. We took the random numbers (column 4) from column 2 of Table F.4.

Table F.10 was filled in by proceeding 1 day (or line) at a time, working from left to right. It is a four-step process:

1. Begin each simulated day by checking to see whether any ordered inventory has just arrived. If it has, increase current inventory by the quantity ordered (10 units, in this case).
2. Generate a daily demand from the demand probability distribution by selecting a random number.
3. Compute: ending inventory = beginning inventory minus demand. If on-hand inventory is insufficient to meet the day's demand, satisfy as much demand as possible and note the number of lost sales.
4. Determine whether the day's ending inventory has reached the reorder point (5 units). If it has, and if there are no outstanding orders, place an order. Lead time for a new order is simulated by choosing a random number and using the distribution in Table F.9.

TABLE F.10 ■ Simkin Hardware's First Inventory Simulation. Order Quantity = 10 Units; Reorder Point = 5 units

(1) DAY	(2) UNITS RECEIVED	(3) BEGINNING INVENTORY	(4) RANDOM NUMBER	(5) DEMAND	(6) ENDING INVENTORY	(7) LOST SALES	(8) ORDER?	(9) RANDOM NUMBER	(10) LEAD TIME
1		10	06	1	9	0	No		
2	0	9	63	3	6	0	No		
3	0	6	57	3	(3)ᵃ	0	Yes	(02)ᵇ	1
4	0	3	(94)ᶜ	5	0	2	(No)ᵈ		
5	(10)ᵉ	10	52	3	7	0	No		
6	0	7	69	3	4	0	Yes	33	2
7	0	4	32	2	2	0	No		
8	0	2	30	2	0	0	No		
9	(10)ᶠ	10	48	3	7	0	No		
10	0	7	88	4	3	0	Yes	14	
					Totals: 41	2			

ᵃThis is the first time inventory dropped to the reorder point of 5 drills. Because no prior order was outstanding, an order is placed.

ᵇThe random number 02 is generated to represent the first lead time. It was drawn from column 2 of Table F.4 as the next number in the list being used. A separate column could have been used from which to draw lead-time random numbers if we had wanted to do so, but in this example, we did not do so.

ᶜAgain, notice that the random digits 02 were used for lead time (see footnote b). So the next number in the column is 94.

ᵈNo order is placed on day 4 because there is an order outstanding from the previous day that has not yet arrived.

ᵉThe lead time for the first order placed is 1 day, but as noted in the text, an order does not arrive the next morning, but rather the beginning of the following day. Thus, the first order arrives at the start of day 5.

ᶠThis is the arrival of the order placed at the close of business on day 6. Fortunately for Simkin, no lost sales occurred during the 2-day lead time before the order arrived.

Simkin's first inventory simulation yields some interesting results. The average daily ending inventory is

$$\text{Average ending inventory} = \frac{41 \text{ total units}}{10 \text{ days}} = 4.1 \text{ units/day}$$

We also note the average lost sales and number of orders placed per day:

$$\text{Average lost sales} = \frac{2 \text{ sales lost}}{10 \text{ days}} = .2 \text{ unit/day}$$

$$\text{Average number of orders placed} = \frac{3 \text{ orders}}{10 \text{ days}} = .3 \text{ order/day}$$

Example F4 shows how these data can be useful in studying the inventory costs of the policy being simulated.

Example F4

Simkin estimates that the cost of placing each order for Ace drills is $10, the holding cost per drill held at the end of each day is $.50, and the cost of each lost sale is $8. This information enables us to compute the total daily inventory cost for the simulated policy in Example F3. Let's examine the three cost components:

$$\begin{aligned}\text{Daily order cost} &= (\text{cost of placing 1 order}) \\ &\quad \times (\text{number of orders placed per day}) \\ &= \$10 \text{ per order} \times .3 \text{ order per day} = \$3\end{aligned}$$

$$\begin{aligned}\text{Daily holding cost} &= (\text{cost of holding 1 unit for 1 day}) \\ &\quad \times (\text{average ending inventory}) \\ &= 50¢ \text{ per unit per day} \times 4.1 \text{ units per day} = \$2.05\end{aligned}$$

Daily stockout cost = (cost per lost sale)
$$\times \text{(average number of lost sales per day)}$$
$$= \$8 \text{ per lost sale} \times .2 \text{ lost sales per day} = \$1.60$$

Total daily inventory cost = Daily order cost + Daily holding cost
$$+ \text{ Daily stockout cost} = \$6.65$$

Now that we have worked through Example F3, we want to emphasize something very important: This simulation should be extended many more days before we draw any conclusions as to the cost of the order policy being tested. If a hand simulation is being conducted, 100 days would provide a better representation. If a computer is doing the calculations, 1,000 days would be helpful in reaching accurate cost estimates. (Moreover, remember that even with a 1,000-day simulation, the generated distribution should be compared with the desired distribution to ensure valid results.)

Let us say that Simkin *does* complete a 1,000-day simulation of the policy from Example F3 (order quantity = 10 drills, reorder point = 5 drills). Does this complete his analysis? The answer is no—this is just the beginning! Simkin must now compare *this* potential strategy to other possibilities. For example, what about order quantity = 10, reorder point = 4? Or order quantity = 12, reorder point = 6? Or order quantity = 14, reorder point = 5? Perhaps every combination of values—of order quantity from 6 to 20 drills and reorder points from 3 to 10—should be simulated. After simulating all reasonable combinations of order quantities and reorder points, Simkin would likely select the pair yielding the lowest total inventory cost. Problem F.12 later in this module gives you a chance to help Simkin begin this series of comparisons.

THE ROLE OF COMPUTERS IN SIMULATION

The explosion of personal computers has created a wealth of computer simulation languages and broadened the use of simulation. Now, even spreadsheet software can be used to conduct fairly complex simulations.

Computers are critical in simulating complex tasks. They can generate random numbers, simulate thousands of time periods in a matter of seconds or minutes, and provide management with reports that make decision making easier. A computer approach is almost a necessity in order to draw valid conclusions from a simulation.

Computer programming languages can help the simulation process. *General-purpose languages*, such as BASIC or C++, constitute one approach. *Special-purpose simulation languages*, such as GPSS and SIMSCRIPT, have a few advantages: (1) they require less programming time for large simulations, (2) they are usually more efficient and easier to check for errors, and (3) random-number generators are already built in as subroutines.

OM IN ACTION

Simulating Jackson Memorial Hospital's Operating Rooms

Miami's Jackson Memorial Hospital, Florida's largest with 1,576 inpatient beds, is also one of the U.S.'s finest. In 1996, it received the highest accreditation score of any public-sector hospital in the country. Jackson's operations management team is constantly seeking ways of increasing hospital efficiency, and the construction of new operating rooms (ORs) prompted the development of a simulation of the existing 31 ORs.

The OR section of the hospital includes a patient holding area and a patient recovery area, both of which were experiencing problems due to ineffective scheduling of OR services. A simulation study, modeled using the ARENA software package, sought to maximize use of OR rooms and staff. Inputs to the model included (1) the amount of time a patient waits in the holding area, (2) the

specific process the patient undergoes, (3) the staff schedule, (4) room availability, and (5) time of day.

The first hurdle that the management team had to deal with at Jackson was the vast amount of records to review to extract the information necessary for the simulation model. The second hurdle was the *quality* of the data. A thorough analysis of the records determined which were good and which had to be discarded. In the end, Jackson's carefully screened databases led to a good set of data inputs for the model. The simulation model then successfully developed five measures of performance: (1) number of procedures a day, (2) average case time, (3) staff utilization, (4) room utilization, and (5) average waiting time in the holding area.

Sources: M. A. Centeno et al., "Challenges of Simulating Hospital Facilities," *Proceedings of the 12th Annual Conference of the Production and Operations Management Society,* Orlando, FL, March 2001; and *Knight Ridder Tribune Business Service* (January 29, 2002): 1.

Commercial, easy-to-use prewritten simulation programs are also available. Some are generalized to handle a wide variety of situations ranging from queuing to inventory. These include programs such as Extend, Modsim, Witness, MAP/1, Slam II, Simfactory, Taylor II, Micro Saint, and ARENA. The *OM in Action* box, "Simulating Jackson Memorial Hospital's Operating Rooms," describes one application of ARENA software.

Spreadsheet software such as Excel can also be used to develop simulations quickly and easily. Such packages have built-in random-number generators and develop outputs through "data-fill" table commands.

SUMMARY

Simulation involves building mathematical models that attempt to act like real operating systems. In this way, a real-world situation can be studied without imposing on the actual system. Although simulation models can be developed manually, simulation by computer is generally more desirable. The Monte Carlo approach uses random numbers to represent variables, such as inventory demand or people waiting in line, which are then simulated in a series of trials. Simulation is widely used as an operations tool because its advantages usually outweigh its disadvantages.

KEY TERMS

Simulation *(p. 584)*
Monte Carlo method *(p. 585)*
Cumulative probability distribution *(p. 586)*

Random-number intervals *(p. 586)*
Random number *(p. 586)*

 ## SIMULATION WITH EXCEL SPREADSHEETS

The ability to generate random numbers and then "look up" these numbers in a table in order to associate them with a specific event makes spreadsheets excellent tools for conducting simulations. Excel OM does not have a simulation module because we are able to model all simulation problems directly in Excel. Program F.1 illustrates an Excel simulation for Example F1.

Notice that the cumulative probabilities are calculated in column D of Program F.1. This procedure reduces the chance of error and is useful in larger simulations involving more levels of demand.

The = VLOOKUP function in column I looks up the random number (generated in column H) in the leftmost column of the defined lookup table ($C$3:$E$8). The = VLOOKUP function moves downward through this column until it finds a cell that is bigger than the random number. It then goes to the previous row and gets the value from column E of the table.

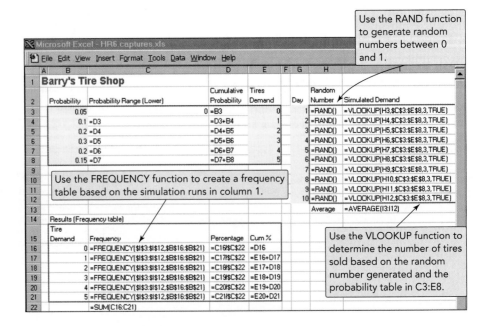

PROGRAM F.1 ■

Using Excel to Simulate Tire Demand for Barry's Auto Tire Shop

PROGRAM F.2 ■

Excel Simulation Results for Barry's Auto Tire Shop

The spreadsheet output in Program F.2 shows a simulated average of 3 tires per day.

	A	B	C	D	E	F	G	H	I
1	**Barry's Tire Shop**								
2		Probability	Probability Range (Lower)	Cumulative Probability	Tires Demand		Day	Random Number	Simulated Demand
3		0.05	0	0.05	0		1	0.963575	5
4		0.1	0.05	0.15	1		2	0.48135	3
5		0.2	0.15	0.35	2		3	0.549832	3
6		0.3	0.35	0.65	3		4	0.112751	1
7		0.2	0.65	0.85	4		5	0.539579	3
8		0.15	0.85	1	5		6	0.42122	3
9							7	0.012112	0
10							8	0.165232	2
11							9	0.956794	5
12							10	0.882812	5
13								Average	3
14		Results (Frequency table)							
15		Tires Demanded	Frequency	Percentage	Cum %				
16		0	1	10%	10%				
17		1	1	10%	20%				
18		2	1	10%	30%				
19		3	4	40%	70%				
20		4	0	0%	70%				
21		5	3	30%	100%				
22			10						

In the output screen of Program F.2, for example, the second random number shown is .481. Excel looked down the left-hand column of the lookup table ($C$3:$E$8) of Program F.2 until it found .65. From the previous row it retrieved the value in column E which is 3. Pressing the F9 function key recalculates the random numbers and the simulation.

USING POM FOR WINDOWS FOR SIMULATION

POM for Windows is capable of handling any simulation that contains only one random variable, such as Example F1. For further details, please refer to Appendix V.

SOLVED PROBLEMS

Solved Problem F.1

Higgins Plumbing and Heating maintains a stock of 30-gallon hot-water heaters that it sells to homeowners and installs for them. Owner Jerry Higgins likes the idea of having a large supply on hand to meet any customer demand. However, he also recognizes that it is expensive to do so. He examines hot-water heater sales over the past 50 weeks and notes the following:

HOT-WATER HEATER SALES PER WEEK	NUMBER OF WEEKS THIS NUMBER WAS SOLD
4	6
5	5
6	9
7	12
8	8
9	7
10	3
	50 weeks total data

(a) If Higgins maintains a constant supply of 8 hot-water heaters in any given week, how many times will he stock out during a 20-week simulation? We use random numbers from the seventh column of Table F.4 (on p. 587), beginning with the random digit 10.

(b) What is the average number of sales per week over the 20-week period?

(c) Using an analytic nonsimulation technique, what is the expected number of sales per week? How does this compare to the answer in part (b)?

SOLUTION

HEATER SALES	PROBABILITY	RANDOM-NUMBER INTERVALS
4	.12	01 through 12
5	.10	13 through 22
6	.18	23 through 40
7	.24	41 through 64
8	.16	65 through 80
9	.14	81 through 94
10	.06	95 through 00
	1.00	

(a)

WEEK	RANDOM NUMBER	SIMULATED SALES	WEEK	RANDOM NUMBER	SIMULATED SALES
1	10	4	11	08	4
2	24	6	12	48	7
3	03	4	13	66	8
4	32	6	14	97	10
5	23	6	15	03	4
6	59	7	16	96	10
7	95	10	17	46	7
8	34	6	18	74	8
9	34	6	19	77	8
10	51	7	20	44	7

With a supply of 8 heaters, Higgins will stock out three times during the 20-week period (in weeks 7, 14, and 16).

(b) Average sales by simulation = total sales/20 weeks = 135/20 = 6.75 per week

(c) Using expected values,

$$E \text{ (sales)} = .12(4 \text{ heaters}) + .10(5)$$
$$+ .18(6) + .24(7) + .16(8)$$
$$+ .14(9) + .06(10) = 6.88 \text{ heaters}$$

With a longer simulation, these two approaches will lead to even closer values.

Solved Problem F.2

Random numbers may be used to simulate continuous distributions. As a simple example, assume that fixed cost equals $300, profit contribution equals $10 per item sold, and you expect an equally likely chance of 0 to 99 units to be sold. That is, profit equals $-\$300 + \$10X$, where X is the number sold. The mean amount you expect to sell is 49.5 units.

(a) Calculate the expected value.

(b) Simulate the sale of 5 items, using the following double-digit random numbers:
     37   77   13   10   85

(c) Calculate the expected value of part (b) and compare with the results of part (a).

SOLUTION

(a) Expected value = $-300 + 10(49.5) = \$195$

(b) $-300 + \$10(37) = \$70$
     $-300 + \$10(77) = \$470$
     $-300 + \$10(13) = -\$170$
     $-300 + \$10(10) = -\$200$
     $-300 + \$10(85) = \$550$

(c) The mean of these simulated sales is $144. If the sample size were larger, we would expect the two values to be closer.

INTERNET AND STUDENT CD-ROM EXERCISES

Visit our home page or use your student CD-ROM to help with material in this module.

 On Our Home Page, www.prenhall.com/heizer

- Self-Tests
- Practice Problems
- Internet Exercises
- Current Articles and Research
- Internet Homework Problems
- Internet Cases

On Your Student CD-ROM

- PowerPoint Lectures
- Practice Problems

ADDITIONAL CASE STUDIES

See our Internet homepage at www.prenhall.com/heizer **for these additional free case studies:**

- **Saigon Transport**: This Vietnamese shipping company is trying to determine the ideal truck fleet size
- **Bialis Waste Disposal**: This case involves a German firm that is debating whether to continue operating an office in Italy

Appendices

APPENDIX I   NORMAL CURVE AREAS

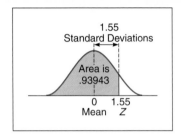

To find the area under the normal curve, you can apply either Table I.1 or Table I.2. In Table I.1, you must know how many standard deviations that point is to the right of the mean. Then, the area under the normal curve can be read directly from the normal table. For example, the total area under the normal curve for a point that is 1.55 standard deviations to the right of the mean is .93943.

TABLE I.1

	.00	.01	.02	.03	.04	.05	.06	.07	.08	.09
.0	.50000	.50399	.50798	.51197	.51595	.51994	.52392	.52790	.53188	.53586
.1	.53983	.54380	.54776	.55172	.55567	.55962	.56356	.56749	.57142	.57535
.2	.57926	.58317	.58706	.59095	.59483	.59871	.60257	.60642	.61026	.61409
.3	.61791	.62172	.62552	.62930	.63307	.63683	.64058	.64431	.64803	.65173
.4	.65542	.65910	.66276	.66640	.67003	.67364	.67724	.68082	.68439	.68793
.5	.69146	.69497	.69847	.70194	.70540	.70884	.71226	.71566	.71904	.72240
.6	.72575	.72907	.73237	.73536	.73891	.74215	.74537	.74857	.75175	.75490
.7	.75804	.76115	.76424	.76730	.77035	.77337	.77637	.77935	.78230	.78524
.8	.78814	.79103	.79389	.79673	.79955	.80234	.80511	.80785	.81057	.81327
.9	.81594	.81859	.82121	.82381	.82639	.82894	.83147	.83398	.83646	.83891
1.0	.84134	.84375	.84614	.84849	.85083	.85314	.85543	.85769	.85993	.86214
1.1	.86433	.86650	.86864	.87076	.87286	.87493	.87698	.87900	.88100	.88298
1.2	.88493	.88686	.88877	.89065	.89251	.89435	.89617	.89796	.89973	.90147
1.3	.90320	.90490	.90658	.90824	.90988	.91149	.91309	.91466	.91621	.91774
1.4	.91924	.92073	.92220	.92364	.92507	.92647	.92785	.92922	.93056	.93189
1.5	.93319	.93448	.93574	.93699	.93822	.93943	.94062	.94179	.94295	.94408
1.6	.94520	.94630	.94738	.94845	.94950	.95053	.95154	.95254	.95352	.95449
1.7	.95543	.95637	.95728	.95818	.95907	.95994	.96080	.96164	.96246	.96327
1.8	.96407	.96485	.96562	.96638	.96712	.96784	.96856	.96926	.96995	.97062
1.9	.97128	.97193	.97257	.97320	.97381	.97441	.97500	.97558	.97615	.97670
2.0	.97725	.97784	.97831	.97882	.97932	.97982	.98030	.98077	.98124	.98169
2.1	.98214	.98257	.98300	.98341	.98382	.98422	.98461	.98500	.98537	.98574
2.2	.98610	.98645	.98679	.98713	.98745	.98778	.98809	.98840	.98870	.98899
2.3	.98928	.98956	.98983	.99010	.99036	.99061	.99086	.99111	.99134	.99158
2.4	.99180	.99202	.99224	.99245	.99266	.99286	.99305	.99324	.99343	.99361
2.5	.99379	.99396	.99413	.99430	.99446	.99461	.99477	.99492	.99506	.99520
2.6	.99534	.99547	.99560	.99573	.99585	.99598	.99609	.99621	.99632	.99643
2.7	.99653	.99664	.99674	.99683	.99693	.99702	.99711	.99720	.99728	.99736
2.8	.99744	.99752	.99760	.99767	.99774	.99781	.99788	.99795	.99801	.99807
2.9	.99813	.99819	.99825	.99831	.99836	.99841	.99846	.99851	.99856	.99861
3.0	.99865	.99869	.99874	.99878	.99882	.99886	.99899	.99893	.99896	.99900
3.1	.99903	.99906	.99910	.99913	.99916	.99918	.99921	.99924	.99926	.99929
3.2	.99931	.99934	.99936	.99938	.99940	.99942	.99944	.99946	.99948	.99950
3.3	.99952	.99953	.99955	.99957	.99958	.99960	.99961	.99962	.99964	.99965
3.4	.99966	.99968	.99969	.99970	.99971	.99972	.99973	.99974	.99975	.99976
3.5	.99977	.99978	.99978	.99979	.99980	.99981	.99981	.99982	.99983	.99983
3.6	.99984	.99985	.99985	.99986	.99986	.99987	.99987	.99988	.99988	.99989
3.7	.99989	.99990	.99990	.99990	.99991	.99991	.99992	.99992	.99992	.99992
3.8	.99993	.99993	.99993	.99994	.99994	.99994	.99994	.99995	.99995	.99995
3.9	.99995	.99995	.99996	.99996	.99996	.99996	.99996	.99996	.99997	.99997

Source: From Richard I. Levin and Charles A. Kirkpatrick, *Quantitative Approaches to Management*, 4th ed. Copyright © 1978, 1975, 1971, 1965 by McGraw-Hill, Inc. Used with permission of McGraw-Hill Book Company.

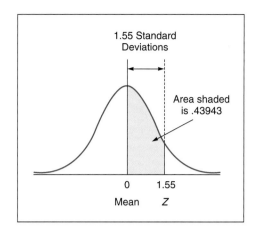

As an alternative to Table I.1, the numbers in Table I.2 represent the proportion of the total area away from the mean, μ, to one side. For example, the area between the mean and a point that is 1.55 standard deviations to its right is .43943.

TABLE I.2

z	.00	.01	.02	.03	.04	.05	.06	.07	.08	.09
0.0	.00000	.00399	.00798	.01197	.01595	.01994	.02392	.02790	.03188	.03586
0.1	.03983	.04380	.04776	.05172	.05567	.05962	.06356	.06749	.07142	.07535
0.2	.07926	.08317	.08706	.09095	.09483	.09871	.10257	.10642	.11026	.11409
0.3	.11791	.12172	.12552	.12930	.13307	.13683	.14058	.14431	.14803	.15173
0.4	.15542	.15910	.16276	.16640	.17003	.17364	.17724	.18082	.18439	.18793
0.5	.19146	.19497	.19847	.20194	.20540	.20884	.21226	.21566	.21904	.22240
0.6	.22575	.22907	.23237	.23565	.23891	.24215	.24537	.24857	.25175	.25490
0.7	.25804	.26115	.26424	.26730	.27035	.27337	.27637	.27935	.28230	.28524
0.8	.28814	.29103	.29389	.29673	.29955	.30234	.30511	.30785	.31057	.31327
0.9	.31594	.31859	.32121	.32381	.32639	.32894	.33147	.33398	.33646	.33891
1.0	.34134	.34375	.34614	.34850	.35083	.35314	.35543	.35769	.35993	.36214
1.1	.36433	.36650	.36864	.37076	.37286	.37493	.37698	.37900	.38100	.38298
1.2	.38493	.38686	.38877	.39065	.39251	.39435	.39617	.39796	.39973	.40147
1.3	.40320	.40490	.40658	.40824	.40988	.41149	.41309	.41466	.41621	.41174
1.4	.41924	.42073	.42220	.42364	.42507	.42647	.42786	.42922	.43056	.43189
1.5	.43319	.43448	.43574	.43699	.43822	.43943	.44062	.44179	.44295	.44408
1.6	.44520	.44630	.44738	.44845	.44950	.45053	.45154	.45254	.45352	.45449
1.7	.45543	.45637	.45728	.45818	.45907	.45994	.46080	.46164	.46246	.46327
1.8	.46407	.46485	.46562	.46638	.46712	.46784	.46856	.46926	.46995	.47062
1.9	.47128	.47193	.47257	.47320	.47381	.47441	.47500	.47558	.47615	.47670
2.0	.47725	.47778	.47831	.47882	.47932	.47982	.48030	.48077	.48124	.48169
2.1	.48214	.48257	.48300	.48341	.48382	.48422	.48461	.48500	.48537	.48574
2.2	.48610	.48645	.48679	.48713	.48745	.48778	.48809	.48840	.48870	.48899
2.3	.48928	.48956	.48983	.49010	.49036	.49061	.49086	.49111	.49134	.49158
2.4	.49180	.49202	.49224	.49245	.49266	.49286	.49305	.49324	.49343	.49361
2.5	.49379	.49396	.49413	.49430	.49446	.49461	.49477	.49492	.49506	.49520
2.6	.49534	.49547	.49560	.49573	.49585	.49598	.49609	.49621	.49632	.49643
2.7	.49653	.49664	.49674	.49683	.49693	.49702	.49711	.49720	.49728	.49736
2.8	.49744	.49752	.49760	.49767	.49774	.49781	.49788	.49795	.49801	.49807
2.9	.49813	.49819	.49825	.49831	.49836	.49841	.49846	.49851	.49856	.49861
3.0	.49865	.49869	.49874	.49878	.49882	.49886	.49889	.49893	.49897	.49900
3.1	.49903	.49906	.49910	.49913	.49916	.49918	.49921	.49924	.49926	.49929

APPENDIX II   POISSON DISTRIBUTION VALUES

$$P(X \le c; \lambda) = \sum_{0}^{c} \frac{\lambda^x e^{-\lambda}}{x!}$$

The following table shows 1,000 times the probability of c or fewer occurrences of an event that has an average number of occurrences of λ.

VALUES OF C

λ	0	1	2	3	4	5	6	7	8	9	10
.02	980	1000									
.04	961	999	1000								
.06	942	998	1000								
.08	923	997	1000								
.10	905	995	1000								
.15	861	990	999	1000							
.20	819	982	999	1000							
.25	779	974	998	1000							
.30	741	963	996	1000							
.35	705	951	994	1000							
.40	670	938	992	999	1000						
.45	638	925	989	999	1000						
.50	607	910	986	998	1000						
.55	577	894	982	998	1000						
.60	549	878	977	997	1000						
.65	522	861	972	996	999	1000					
.70	497	844	966	994	999	1000					
.75	472	827	959	993	999	1000					
.80	449	809	953	991	999	1000					
.85	427	791	945	989	998	1000					
.90	407	772	937	987	998	1000					
.95	387	754	929	984	997	1000					
1.00	368	736	920	981	996	999	1000				
1.1	333	699	900	974	995	999	1000				
1.2	301	663	879	966	992	998	1000				
1.3	273	627	857	957	989	998	1000				
1.4	247	592	833	946	986	997	999	1000			
1.5	223	558	809	934	981	996	999	1000			
1.6	202	525	783	921	976	994	999	1000			
1.7	183	493	757	907	970	992	998	1000			
1.8	165	463	731	891	964	990	997	999	1000		
1.9	150	434	704	875	956	987	997	999	1000		
2.0	135	406	677	857	947	983	995	999	1000		

Source: Adapted from E. L. Grant, *Statistical Quality Control*, McGraw-Hill Book Company, New York (1964). Reproduced by permission of the publisher.

APPENDIX II   POISSON DISTRIBUTION VALUES

VALUES OF C

λ	0	1	2	3	4	5	6	7	8	9	10	11	12	13	14	15	16	17	18	19	20	21	22
2.2	111	359	623	819	928	975	993	998	1000														
2.4	091	308	570	779	904	964	988	997	999	1000													
2.6	074	267	518	736	877	951	983	995	999	1000													
2.8	061	231	469	692	848	935	976	992	998	999	1000												
3.0	050	199	423	647	815	916	966	988	996	999	1000												
3.2	041	171	380	603	781	895	955	983	994	998	1000												
3.4	033	147	340	558	744	871	942	977	992	997	999	1000											
3.6	027	126	303	515	706	844	927	969	988	996	999	1000											
3.8	022	107	269	473	668	816	909	960	984	994	998	999	1000										
4.0	018	092	238	433	629	785	889	949	979	992	997	999	1000										
4.2	015	078	210	395	590	753	867	936	972	989	996	999	1000										
4.4	012	066	185	359	551	720	844	921	964	985	994	998	999	1000									
4.6	010	056	163	326	513	686	818	905	955	980	992	997	999	1000									
4.8	008	048	143	294	476	651	791	887	944	975	990	996	999	1000									
5.0	007	040	125	265	440	616	762	867	932	968	986	995	998	999	1000								
5.2	006	034	109	238	406	581	732	845	918	960	982	993	997	999	1000								
5.4	005	029	095	213	373	546	702	822	903	951	977	990	996	999	1000								
5.6	004	024	082	191	342	512	670	797	886	941	972	988	995	998	999	1000							
5.8	003	021	072	170	313	478	638	771	867	929	965	984	993	997	999	1000							
6.0	002	017	062	151	285	446	606	744	847	916	957	980	991	996	999	999	1000						
6.2	002	015	054	134	259	414	574	716	826	902	949	975	989	995	998	999	1000						
6.4	002	012	046	119	235	384	542	687	803	886	939	969	986	994	997	999	1000						
6.6	001	010	040	105	213	355	511	658	780	869	927	963	982	992	997	999	999	1000					
6.8	001	009	034	093	192	327	480	628	755	850	915	955	978	990	996	998	999	1000					
7.0	001	007	030	082	173	301	450	599	729	830	901	947	973	987	994	998	999	1000					
7.2	001	006	025	072	156	276	420	569	703	810	887	937	967	984	993	997	999	999	1000				
7.4	001	005	022	063	140	253	392	539	676	788	871	926	961	980	991	996	998	999	1000				
7.6	001	004	019	055	125	231	365	510	648	765	854	915	954	976	989	995	998	999	1000				
7.8	000	004	016	048	112	210	338	481	620	741	835	902	945	971	986	993	997	999	1000				
8.0	000	003	014	042	100	191	313	453	593	717	816	888	936	966	983	992	996	998	999	1000			
8.5	000	002	009	030	074	150	256	386	523	653	763	849	909	949	973	986	993	997	999	999	1000		
9.0	000	001	006	021	055	116	207	324	456	587	706	803	876	926	959	978	989	995	998	999	1000		
9.5	000	001	004	015	040	089	165	269	392	522	645	752	836	898	940	967	982	991	996	998	999	1000	
10.0	000	000	003	010	029	067	130	220	333	458	583	697	792	864	917	951	973	986	993	997	998	999	1000

APPENDIX III   VALUES OF $e^{-\lambda}$ FOR USE IN THE POISSON DISTRIBUTION

VALUES OF $e^{-\lambda}$

λ	$e^{-\lambda}$	λ	$e^{-\lambda}$	λ	$e^{-\lambda}$	λ	$e^{-\lambda}$
.0	1.0000	1.6	.2019	3.1	.0450	4.6	.0101
.1	.9048	1.7	.1827	3.2	.0408	4.7	.0091
.2	.8187	1.8	.1653	3.3	.0369	4.8	.0082
.3	.7408	1.9	.1496	3.4	.0334	4.9	.0074
.4	.6703	2.0	.1353	3.5	.0302	5.0	.0067
.5	.6065	2.1	.1225	3.6	.0273	5.1	.0061
.6	.5488	2.2	.1108	3.7	.0247	5.2	.0055
.7	.4966	2.3	.1003	3.8	.0224	5.3	.0050
.8	.4493	2.4	.0907	3.9	.0202	5.4	.0045
.9	.4066	2.5	.0821	4.0	.0183	5.5	.0041
1.0	.3679	2.6	.0743	4.1	.0166	5.6	.0037
1.1	.3329	2.7	.0672	4.2	.0150	5.7	.0033
1.2	.3012	2.8	.0608	4.3	.0136	5.8	.0030
1.3	.2725	2.9	.0550	4.4	.0123	5.9	.0027
1.4	.2466	3.0	.0498	4.5	.0111	6.0	.0025
1.5	.2231						

APPENDIX IV   TABLE OF RANDOM NUMBERS

52	06	50	88	53	30	10	47	99	37	66	91	35	32	00	84	57	07
37	63	28	02	74	35	24	03	29	60	74	85	90	73	59	55	17	60
82	57	68	28	05	94	03	11	27	79	90	87	92	41	09	25	36	77
69	02	36	49	71	99	32	10	75	21	95	90	94	38	97	71	72	49
98	94	90	36	06	78	23	67	89	85	29	21	25	73	69	34	85	76
96	52	62	87	49	56	59	23	78	71	72	90	57	01	98	57	31	95
33	69	27	21	11	60	95	89	68	48	17	89	34	09	93	50	44	51
50	33	50	95	13	44	34	62	64	39	55	29	30	64	49	44	30	16
88	32	18	50	62	57	34	56	62	31	15	40	90	34	51	95	26	14
90	30	36	24	69	82	51	74	30	35	36	85	01	55	92	64	09	85
50	48	61	18	85	23	08	54	17	12	80	69	24	84	92	16	49	59
27	88	21	62	69	64	48	31	12	73	02	68	00	16	16	46	13	85
45	14	46	32	13	49	66	62	74	41	86	98	92	98	84	54	33	40
81	02	01	78	82	74	97	37	45	31	94	99	42	49	27	64	89	42
66	83	14	74	27	76	03	33	11	97	59	81	72	00	64	61	13	52
74	05	81	82	93	09	96	33	52	78	13	06	28	30	94	23	37	39
30	34	87	01	74	11	46	82	59	94	25	34	32	23	17	01	58	73
59	55	72	33	62	13	74	68	22	44	42	09	32	46	71	79	45	89
67	09	80	98	99	25	77	50	03	32	36	63	65	75	94	19	95	88
60	77	46	63	71	69	44	22	03	85	14	48	69	13	30	50	33	24
60	08	19	29	36	72	30	27	50	64	85	72	75	29	87	05	75	01
80	45	86	99	02	34	87	08	86	84	49	76	24	08	01	86	29	11
53	84	49	63	26	65	72	84	85	63	26	02	75	26	92	62	40	67
69	84	12	94	51	36	17	02	15	29	16	52	56	43	26	22	08	62
37	77	13	10	02	18	31	19	32	85	31	94	81	43	31	58	33	51

Source: Excerpted from *A Million Random Digits with 100,000 Normal Deviates*, The Free Press (1955): 7, with permission of the Rand Corporation.

APPENDIX V USING EXCEL OM AND POM FOR WINDOWS

Two approaches to computer-aided decision making are available with this text: **Excel OM** and **POM** (Production and Operations Management) **for Windows**. These are the two most user-friendly software packages available to help you learn and understand operations management. Both programs can be used either to solve homework problems identified with a computer logo or to check answers you have developed by hand. Both software packages use the standard Windows interface and run on any IBM-PC compatible 486 or higher with at least 4-MB RAM and operating Windows 95 or better.

Excel OM

Excel OM also has been designed to help you to better learn and understand both OM and Excel. Even though the software contains 17 modules and over 35 submodules, the screens for every module are consistent and easy to use. The modules are illustrated in Program V.1. This software is provided by means of the CD-ROM that is included in the back of this text at no cost to purchasers of this textbook. Excel 97 or better must be on your PC.

To install *Excel OM*:

1. Insert the CD-ROM.
2. Open My Computer from the desktop and double-click on the CD drive.
3. Open the ExcelOM2 folder.
4. Open the ExcelOM2.Heizer program.
5. Follow the setup instructions on the screen.

Default values have been assigned in the setup program, but you may change them if you like. The default folder into which the program will be installed is named C:\ExcelOM2 and the default name for the program group placed in the START menu is Excel OM 2. Generally speaking it is simply necessary to click NEXT each time the installation asks a question.

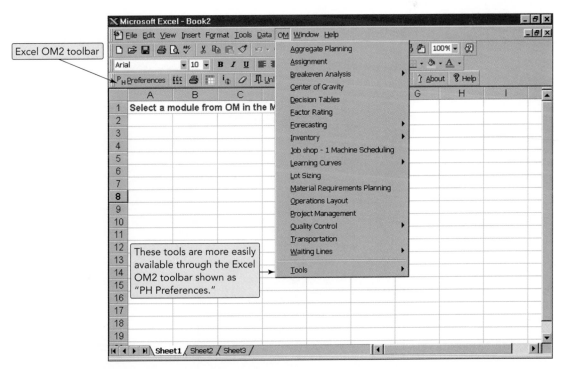

PROGRAM V.1 ■ Excel OM Modules

Starting the Program If you do not already have Excel open, then to start Excel OM, double-click on the Excel OM 2 shortcut placed on the desktop at installation. Alternatively, you may click on START, PROGRAMS, EXCEL OM 2. If you already have Excel open, then simply load the file ExcelOM2.xla, which is in the directory C:\ExcelOM2, if you did not change this directory at the time of installation.

It is also possible to install Excel OM as an Excel add-in that is loaded each time you start Excel. To do this, simply go to TOOLS, ADDINS, BROWSE and select Excel OM2.xla from the C:\ExcelOM2 folder. Uninstalling adds-ins in Excel is more difficult than installing add-ins, so this method is not suggested unless every time you open Excel it is for your OM homework.

Excel OM serves two purposes in the learning process. First, it can simply help you solve homework problems. You enter the appropriate data, and the program provides numerical solutions. POM for Windows operates on the same principle. However, Excel OM allows for a second approach; that is, noting the Excel *formulas* used to develop solutions and modifying them to deal with a wider variety of problems. This "open" approach enables you to observe, understand, and even change the formulas underlying the Excel calculations, hopefully conveying Excel's power as an OM analysis tool.

POM for Windows

POM for Windows is decision support software that may be ordered as an option with this textbook. Program V.2 shows a list of 24 OM programs on the CD that will be installed on your hard drive. Once you follow the standard setup instructions, a POM for Windows program icon will be added to your start menu and desktop. The program may be accessed by double-clicking on the icon. Upgrades to POM for Windows are available on the Internet through the Prentice Hall download library, found at http://www.prenhall.com/weiss.

To illustrate the ease-of-use of POM for windows, we include Programs V.3 to V.6. Program V.3 shows one aspect of the forecasting module, exponential smoothing, as applied to the Port of New Orleans data in Chapter 4.

Programs V.4 and V.5 illustrate the process of Assembly-Line Balancing, using data from Chapter 9. The first screen, V.4, provides input data, while V.5 shows the results of the line balance.

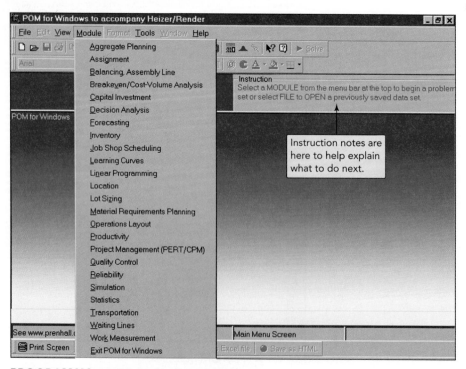

PROGRAM V.2 ■ POM for Windows Module List

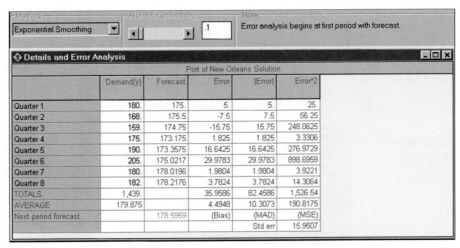

Method		Alpha for smoothing			Note	
Exponential Smoothing ▼		◄	►	.1	Error analysis begins at first period with forecast.	

◇ Details and Error Analysis   _ □ ✕

Port of New Orleans Solution

	Demand(y)	Forecast	Error	|Error|	Error*2
Quarter 1	180.	175.	5.	5.	25.
Quarter 2	168.	175.5	-7.5	7.5	56.25
Quarter 3	159.	174.75	-15.75	15.75	248.0625
Quarter 4	175.	173.175	1.825	1.825	3.3306
Quarter 5	190.	173.3575	16.6425	16.6425	276.9729
Quarter 6	205.	175.0217	29.9783	29.9783	898.6959
Quarter 7	180.	178.0196	1.9804	1.9804	3.9221
Quarter 8	182.	178.2176	3.7824	3.7824	14.3064
TOTALS	1,439.		35.9586	82.4586	1,526.54
AVERAGE	179.875		4.4948	10.3073	190.8175
Next period forecast		178.5959	(Bias)	(MAD)	(MSE)
				Std err	15.9507

PROGRAM V.3 ■ POM for Windows Forecasting Example Using Chapter 4 Data

---

Five different heuristics are available.

Method		Cycle time computation					Task time unit
Longest operation time ▼		○ Given 40 units		8	○ seconds	minutes ▼	
		● Computed per			○ minutes		
					● hours		

Example

TASK	Minutes	Predecessor 1	Predecessor 2	Predecessor 3	Predecessor 4
A	10				
B	11	a			
C	5	b			
D	4	b			
E	12	a			
F	3	c	d		
G	7	f			
H	11	e			
I	3	g	h		

Only enter the immediate predecessor(s).

PROGRAM V.4 ■ POM for Windows Assembly-Line Balancing Module, Using Input Data from Chapter 9

Balancing, Assembly Line Results   _ □ ✕

Example Solution

Station	Task	Time (minutes)	Time left (minutes)	Ready tasks
				A
1	A	10.	2.	B,E
2	E	12.	0.	B,H
3	B	11.	1.	H,C,D
4	H	11.	1.	C,D
5	C	5.	7.	D
	D	4.	3.	F
	F	3.	0.	G
6	G	7.	5.	I
	I	3.	2.	
Summary Statistics				
Cycle time	12	minutes		
Time allocated (cycle time * # stations)	72	minutes/cycle		
Time needed (sum of task times)	66	minutes/unit		
Idle time (allocated-needed)	6	minutes/cycle		
Efficiency (needed/allocated)	91.66666%			
Balance Delay (1-efficiency)	8.333333%			
Min (theoretical) # of stations	6			

PROGRAM V.5 ■ Output Screen to Accompany Program V.4's POM for Windows Line Balancing Example

Finally, Program V.6 is an example of POM for Windows' Job Shop Scheduling module. It uses data from Chapter 15. You will find that all of this powerful program's modules are easy to run. Just follow the prompts that appear on the top of each screen.

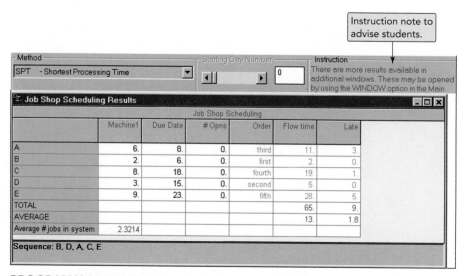

Instruction note to advise students.

Method			Starting Day Number		Instruction
SPT - Shortest Processing Time			◄| ►	0	There are more results available in additional windows. These may be opened by using the WINDOW option in the Main

Job Shop Scheduling Results   _ □ ✕

Job Shop Scheduling

	Machine1	Due Date	# Opns	Order	Flow time	Late
A	6.	8.	0.	third	11.	3.
B	2.	6.	0.	first	2.	0.
C	8.	18.	0.	fourth	19.	1.
D	3.	15.	0.	second	5.	0.
E	9.	23.	0.	fifth	28.	5.
TOTAL					65.	9.
AVERAGE					13.	1.8
Average # jobs in system	2.3214					

Sequence: B, D, A, C, E

PROGRAM V.6 ■ POM for Window's Job Shop Scheduling Module, Using Chapter 15 Data

Name Index

Note: Any page number preceded by a T indicates that the topic is included on the Student CD-ROM.

General Index

NOTE: Any page number preceded by a T indicates that the topic is included on the Student CD-ROM.

Get Better Business Results

with Microsoft Project 2002

Take control of your projects with Microsoft® Project 2002, the world's leading project management program. Microsoft Project 2002 makes it easier than ever to manage schedules and resources, communicate project status effectively, and analyze project information quickly. And now the whole team can actively participate throughout the project life cycle — new, Web-based tools* make collaboration easy. The bottom line is that Microsoft Project 2002 keeps everyone informed so that projects stay on track and within budget. No wonder project managers have made Microsoft Project their tool of choice for years.

To learn more about Microsoft Project 2002, visit
http://www.microsoft.com/project.